Design, Operation, and Control of Insect-Rearing Systems

Design, Operation, and Control of Insect-Rearing Systems

Design, Operation, and Control of Insect-Rearing Systems

Science, Technology, and Infrastructure

Allen Carson Cohen

CRC Press
Taylor & Francis Group
Boca Raton London New York

CRC Press is an imprint of the
Taylor & Francis Group, an Informa business

First edition published 2021

by CRC Press
6000 Broken Sound Parkway NW, Suite 300, Boca Raton, FL 33487-2742

and by CRC Press
2 Park Square, Milton Park, Abingdon, Oxon, OX14 4RN

ISBN: 978-1-138-57125-9 (hbk)
ISBN: 978-0-367-77581-0 (pbk)
ISBN: 978-0-203-70295-6 (ebk)

Typeset in Times
by KnowledgeWorks Global Ltd.

Contents

Preface

PREFACE TO DESIGN, OPERATION, AND CONTROL OF INSECT-REARING SYSTEMS

I found a dimpled spider, fat and white,
On a white heal-all, holding up a moth
Like a white piece of rigid satin cloth–
Assorted characters of death and blight
Mixed ready to begin the morning right,
Like the ingredients of a witches' broth–
A snow-drop spider, a flower like a froth,
And dead wings carried like a paper kite.
What had that flower to do with being white,
The wayside blue and innocent heal-all?
What brought the kindred spider to that height,
Then steered the white moth thither in the
night?
What but design of darkness to appall?–
If design govern in a thing so small.

Design, by Robert Frost

I thought it is appropriate to begin this book with a wonderful poem about design in the natural world. I discuss this poem in a paper in the *American Entomologist* (Cohen 1999), but suffice it to say that Frost's question about design in nature should not be a matter of question or doubt in the insect-rearing system!

P.1 THE CONCEPTS OF DESIGN, OPERATION, AND CONTROL IN A GENERALIZED REARING SYSTEM CONTEXT

Most of the title, *Design, Operation, and Control of Insect-Rearing Systems*, was suggested by Dr. Greg Simmons (USDA, APHIS). In my proposal to write this book, I wanted to call it "Infrastructure of Insect-Rearing Systems," but in reviewing that proposal, Dr. Simmons suggested the terms "design" and "operation" to better reflect the intended contents of this book. My immediate response to Dr. Simmons' suggestion was, "Wow, this really focuses me on the core of what I was intending to treat in writing this book." As I worked on the book for the past couple of years, the concept of "control" became a prominent part of what I was learning about what we strive for when we develop rearing systems for any given insect. This being said, I want to put forward at the front of this book the concept that insect-rearing systems **must** be treated with a science-based perspective. I insist throughout my teaching and writing that rearing systems are best—most effective—when they are based on sound scientific principles. Therefore, *design* and the incumbent *operation* and *control* must come from science.

Simply speaking, "**design**" reflects the idea that rearing systems are engineered treatments of organisms that we culture for various reasons (to be discussed throughout this book). "**Operation**" includes the processes that evolved from efforts to make the system work to allow us to produce suitable insects—with "*process*" being a key, operant term. Processes are developed through the **algorithm** approach. Algorithms, in the context of rearing systems, are sets of rules applied to solving problems involved in the steps of a rearing process. The "**control**" concept is a direct outgrowth of a 1910 paper by Delcourt and Guyenot who argued that the way to proceed with effective research into the genetics of *Drosophila* was to develop fine control over the rearing process (Figure FM.1). The title of this paper, "The Possibility of Studying Certain Diptera in a Defined Environment," clearly indicates that a "defined environment" is a desirable, virtually essential feature of the rearing process the authors describe. It is essential to note that "**defined**" equals "**controlled**."

Figure FM.1 illustrates the relationships that guide rearing-system algorithms. All rearing operations are products of a set of concepts that envisioned how the systems would work. The envisioning is the design or the intention of the system's designer. However, often the operation does not meet the exact vision that was originally intended. This deviation from intention is what requires control to bring the system into a capable mode. The designer, for example, may have intended to keep the rearing temperature at 25°C, and the designer called for a heating and cooling arrangement that was aimed at the set-point temperature. A well-designed system requires not only the actuators (heating and cooling units) to meet the set-point standards, but the system also needs a comparator (in this case a thermostat) that answers the question, "what is the current temperature?" If the temperature

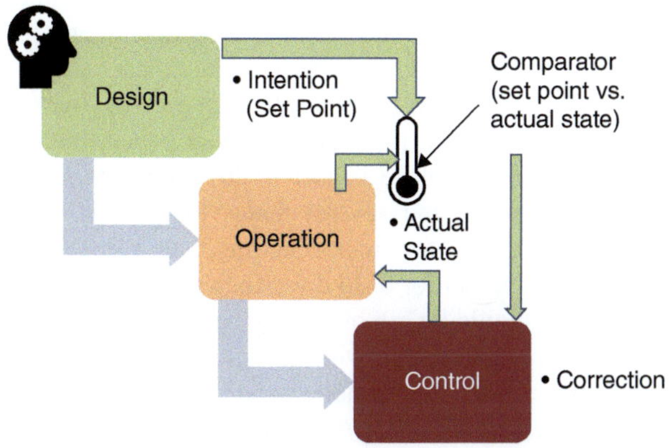

Figure FM.1 The relationship between design, operation, and control in systems.

is below the set-point of 25°C, the message goes to the control system, which activates a heater that raises the temperature to the set-point. However, if the heating overshoots the set-point, then the control unit signals a cooling system component to add cool air that lowers the temperature, and the process proceeds continuously. Finer-tuned systems have several features that make them more precise than coarsely tuned systems. The sensor and comparator units are more sensitive to fluctuations (deviations) from the set-point. The actuators have the power to make rapid changes (vs. sluggish responses) to departures from the set-point (design ideal). Also, stable and precise systems have features that add inertia to the desired conditions; in temperature-regulation systems, the kinds of inertia are the insulation of the room, the circulation of the air, and the conductivity of the rearing containers (as a few examples). It is important to recognize that the temperature regulation example is only one of many design components in the rearing systems' operations. The insects' diet components, other environmental features (besides temperature), microbial profiles, insects' genetic diversity, etc. are also parts of the design: operation: control model that compose the subject of this book.

P.2 THE REARING-SYSTEM ALGORITHM CONCEPT

The following is a simple rearing algorithm that is abstracted from several of the early publications (Morgan et al. 1915; Bridges 1932; Bridges and Darby 1933) on rearing *Drosophila melanogaster*:

(1) *Containers:* use clean half-pint or pint milk bottles with cotton stoppers; → (2) *environment:* keep culture at 25°C (or as close thereto as possible); → (3) obtain *fruit flies:* catch flies with rotting fruit, such as over-ripe bananas, which attract *D. melanogaster* from the environment, or use previously cultured flies; → (4) *diet:* make diet by cutting pieces of ripened bananas, steeping them in aqueous extracts of ripened, fermented bananas (Morgan et al. 1915, "old juice" containing yeast); keep food stored in wide-mouthed glass jars; → (5) *feeding:* place a teaspoon of the macerated, partially fermented food on a piece of absorbent paper (such as a paper towel), and place in jar; → (6) *inoculate* jar with flies (one female and several males for genetic experiments, 10s of flies for stock breeding, and allow them to develop on the diet for 8–14 days; then remove adult flies). Adult flies should be removed to prevent previous generation adults from mating with newly emerged adults (this precaution pertains to genetic experiments). All the earlier researchers in *Drosophila* genetics used diethyl ether as an anesthetic; however, more recently, other means of anesthetizing these and other insects have been utilized, including cold, CO_2, and triethylamine (FlyNap) (Chen and Hillyer 2013). For successive generations, this process (Steps 1 through 6) is repeated. →→→

Figure FM.2 Original bottles used by T. H. Morgan and his group at Columbia University. These are the ½ pint and 1 pint bottles discussed by Morgan et al. 1915 and Bridges and Darby 1933.

Researchers and educators using *Drosophila* as subjects use what is fundamentally this procedure with a few modifications, such as gelled, agar-based diets (derived from Baumberger 1917b and discussed further in Appendix 1) and nutrients derived from processed bananas or other fruit, often with corn meal as an added nutrient and texture-enhancer. The old-fashioned glass milk bottles are no longer readily available and have been replaced with plastic vials (Figures FM.2–FM.4).

Figure FM.3 *Drosophila melanogaster* larvae feeding on a diet mentioned in the rearing algorithm described in section "The Rearing System Algorithm Concept." The fly larvae are lined up at the interface of the banana agar and yeast diet. The algorithms were developed by early researchers in *D. melanogaster* genetics, including Bridges, Morgan, Guyenot, Delcourt, and Baumberger. The account of the algorithmic thinking behind early diet development and the meticulous rationale for vetting each step of the diet algorithm is explained further in Appendix 1.

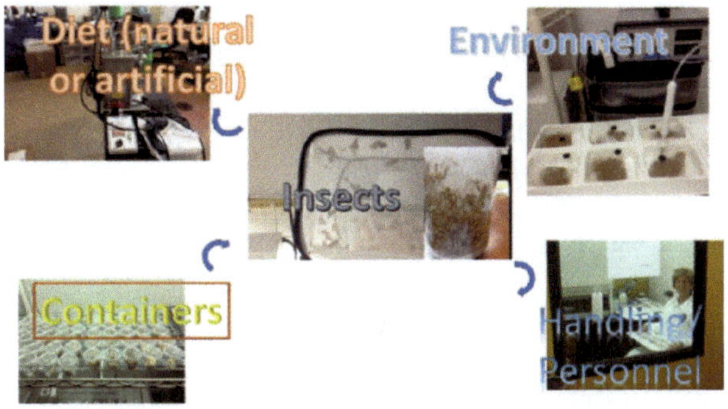

Figure FM.4 A generalized diagram of the major rearing-system components (insects, diets, containers, and personnel).

P.3 SOME PERSPECTIVES ON INSECT-REARING SYSTEMS: MY POINT OF VIEW

As I write the final touches of this book in 2020, I realize that I have been focused on insect rearing for the past 40+ years, and have done hundreds of experiments, read thousands of papers, and have done countless hours of observations to various insectaries or other rearing systems. During this period, I have learned a great deal about how rearing systems had been developed, although authors of rearing papers are most often discouraged from giving very much detail on how they came to incorporate or test various features of their rearing system (often very important features). Also during this long period of observation and study, I have developed opinions about how things should be done. In writing this book, I have tried to draw heavily and honestly from the various papers and books that I cite, and in this way I have tried to remain objective and present insect-rearing systems from the perspective of reality, truth, and scientific basis. However, as a human with all the flaws of human frailty, I know that my opinions are slanted. It would have been helpful to the reader to be able to distinguish between the objective truth (or reality if the word "truth" is uncomfortable) and the "Cohen perspective." I tried to signal the reader as much as possible when the narrative was my own first-person point of view vs. what authors or systems were telling us.

In defense of my personal perspective, I have tried consciously to be honest about my suggestions and conclusions; and furthermore, I have tried to keep my opinions to what I feel is to the benefit of the scientific community and the community of rearing professionals.

P.4 INSECT REARING IS A SCIENTIFIC DISCIPLINE

One of the most important foundational concepts underlying this book is that insect rearing best serves the entomological community if it is treated as a scientific discipline. As a science, insect rearing must be expected to be understood on a mechanistic level with knowable causes and effects. As a science, insect rearing is subject to what a guiding document from the National Institutes of Health (NIH) describes as Scientific Rigor and Reproducibility (NIH 2019, p. 13).

P.5 SCIENTIFIC RIGOR IN INSECT-REARING RESEARCH AND EDUCATION

The NIH document presents the following discussion under **Scientific Rigor and Reproducibility:**

> Scientific Rigor is the robust and unbiased application of the scientific method to well-defined research questions. To facilitate robust results, valid data are obtained by utilizing independent approaches to address research questions and ensuring that experiments are sufficiently controlled and documented to be reproducible. Reproducibility or replication of experiments, observations, and results is an integral component of the scientific method and allows science to be "self-correcting." Combined, rigor and reproducibility are essential to the generation of knowledge and the elaboration of new principles. Rigor involves four areas of focus:
>
> 1. Careful assessment of prevailing knowledge in field of study to identify strengths and weaknesses in prior research and any gaps in knowledge.
> 2. How applications of the scientific methodology will ensure robust and unbiased experimental design, methodology, analysis, interpretation, and dissemination of information.
> 3. The careful consideration of biologic variables that can influence experimental design and scientific methodology; i.e., species, sex, age, weight, animal models, and environmental considerations.
> 4. Validation of reagents, chemicals, biologics, and tests used in the research.

While the NIH elucidation of Scientific Rigor and Reproducibility is discussed in considerable detail in Chapter 14 ("Research Integrity in Insect-Rearing Studies"), it bears some explanation at the beginning of this book. I have found that from re-reading the discussion of scientific rigor from NIH 2019, I had become increasingly convinced that the greatest strengths in our existing rearing systems came as results of adhering to the principles outlined in the discussion; and the greatest failings in the discipline came from departures from these tenets. In fact, ideas that I had struggled with about rearing (such as not being sufficiently respected, difficulties in publishing rearing papers, inadequate funding for rearing research, etc.) would be improved and corrected by the adoption and adherence to the NIH principles of Scientific Rigor. However, what is even more important to the scientific community who are stakeholders in the qualities of the insects that are products of rearing systems is the fact that the "biologic variables" mentioned in Focus Areas 3 and 4 are too often not appropriately **considered** and **validated**. A notable exception to the previous statement is to be found in the rearing systems where subjects for sterile insect techniques (SIT) are produced. The SIT community has been a pioneering entity. This is discussed in detail in Chapter 10 ("Relationships between Fitness, Quality Control, Health, Homeostasis, and Stress"). However, I feel that programs where biologic variables of insects are well-considered and validated are islands of quality control in a sea of lax standards (Cohen 2001, 2019).

An observation that I have made over my 40+ year career in insect rearing is that researchers who try to investigate questions of how rearing-system protocols affect various qualities of the insect subjects are discouraged by lack of funding, difficulties in publishing papers that deal with cause-and-effect relationships in rearing systems and their outcomes, and by prevailing attitudes of supervisors and peers of rearing specialists who are indifferent to efforts to ask the question: what is the biological consequence of a rearing process? The very community of scientists and other stakeholders who should most welcome rigorous, scientifically-based insect-rearing inquiry is not only indifferent but also sometimes hostile to such inquiries as I have tried to document in many of my teachings and publications such as Cohen 2001 and 2019. My hope is that the present book will help call attention to the need for broadening and deepening the efforts to enhance science-based principles into insect-rearing systems for improvement of rearing research and production programs.

P.6 TEN REARING PRINCIPLES TO MAXIMIZE AND OPTIMIZE DESIGN, OPERATION, AND CONTROL

1. Rearing is an amalgam of applied and basic science.
2. Rearing is (or should be) a science based on discovery of cause-and-effect relationships.
3. Rearing is a system (behaves systematically, adhering to the engineering precepts of systems).
4. Knowledge of target insects' biology leads to control.
5. Error (variability or uncertainty) is counter to quality.
6. Environmental components of rearing systems are interdependent and interactive.
7. Reared insects (cultured insects) are domesticated.
8. Automation is a pathway to improved, more feasible mass-rearing success.
9. Rearing knowledge (advancements) should be public (shared).
10. Rearing research (and education) needs to be supported by public funding.

P.6.1 Rearing Is an Amalgam of Applied and Basic Science

While rearing has the practical application of producing insects for various programs such as research, biological control, SIT, etc., the process of rearing lends itself to many levels of discovery about the basic biology of the target insects. There is a reciprocal relationship between what knowledge is needed to rear insects (thermal tolerance, diet requirements, gas exchange dynamics, etc.) and what can be learned by rearing insects under controlled experimental conditions.

P.6.2 Rearing Is (or Should Be) a Science That Is Based on Discovery of Cause-and-Effect Relationships

Though many people (including some rearing experts) regard insect rearing as an art, it is best served when it is treated as an empirical science. All aspects of rearing system practice are most productive when they are based on experiments or carefully vetted practices. Although many of the practices were determined by scientific research previously established, it is mainly science-driven practices which succeed in supporting reliable rearing of high-quality insects.

P.6.3 Rearing Is a System (i.e., It Conforms to the Engineering Precepts of Systems Theory)

Included below is a quotation, which describes the dynamic features of systems theory, including the fact that systems are conglomerations of interrelated and interdependent parts. Systems may be natural (such as the organ systems of higher animals) or they may be human-made (such as an automobile). The quotation points out that systems exist in and are bounded by space and time, and especially important: they are influenced by their environment and defined by their purpose. It is also notable that systems may be greater than the sum of their parts, and they may express synergy or emergent behavior:

> **Systems theory** is the interdisciplinary study of systems. A system is a cohesive conglomeration of interrelated and interdependent parts which can be natural or human-made. Every system is bounded by space and time, influenced by its environment, defined by its structure and purpose, and expressed through its functioning. A system may be more than the sum of its parts if it expresses synergy or emergent behavior. (Systems Theory. 16 December 2017, at 13:21 (UTC) (https://en.wikipedia.org/wiki/Systems_theory)).

P.6.4 Knowledge of Target Insects' Biology, Including Homeostasis and Stress, Leads to Control

This is the "know your insect" feature of rearing systems, the concept that the more you know about your insect's total biology, the better your rearing efforts will be toward developing and maintaining a rearing process that meets the needs of your insect. **This is also to be understood as the fact that rearing personnel's primary duty is to reduce stress factors in their rearing operations**. If you understand the feeding biology of the insect, you are more prone to provide a diet that meshes with the insect's feeding adaptations (the right texture, the right concentration, the right nutritional content, the right diet presentation method, etc.).

P.6.5 Error (Variability or Uncertainty) Is Counter to Quality

One of the chief duties of rearing personnel is to identify, quantify, and reduce error. Inherently, the reduction of error means a more reliable system will emerge. The various types of error (gross error, systematic error, and random error; discussed in Chapter 7) must be understood as major causes of loss of fitness of the insects, in a biological sense, or failure of the system in a statistical sense.

P.6.6 Environmental Components of Rearing Systems Are Interdependent and Interactive

The rearing room or rearing facility can be thought of as the artificial environment or habitat of our target insects. This is in accord with the concept advanced in Chapter 6 that rearing systems can be considered as parallel or analogous to the insect's ecological niche. The better we understand the natural (ecological) niche and the artificial (rearing system) niche, the more we can accept the responsibility of playing the role of nature in our insect colonization. The interdependence and interactions are so far-reaching and extensive that it is impossible to exaggerate their importance and subtlety. A change in temperature of the rearing room affects the humidity of the insects' air, the metabolic rate of the insect, the inclination toward microbial growth, the quality of temperature-sensitive diet components, and even the rate of gas exchange between the inside and outside air in the rearing container. This simple but far-reaching example is a mere suggestion of the complex "downstream" consequences of seemingly simple, benign changes.

P.6.7 Reared Insects (Cultivated Insects) Are Domesticated

This very important point is a good concept to keep in mind constantly. Rearing personnel and the customers of the rearing products are better served if they realize that the moment that insects are taken into captivity, they are undergoing intense selection that makes them different from their wild counterparts. Some of the selection is deliberate, but the vast majority of it is unintended. Furthermore, we are becoming growingly aware not only about the genetic selection and truncation that our rearing processes engender, but there are also epigenetic changes that we impose in our rearing efforts. **The bottom line of the domestication process is that we must recognize that reared insects, as products of unnatural rearing conditions, are different from wild counterparts**.

P.6.8 Automation Is a Pathway to Improved, More Feasible Mass-Rearing Success

Since the 1950s and 1960s, automation of rearing processes has been a goal of large-scale rearing programs. In fact, the scale of rearing where automation has not been developed is anemic compared to systems that have been thoroughly automated. It is also axiomatic that automation reduces error that comes from human labor. The reduction in error includes microbial control (sanitation), measurement errors, and other processing steps where human inattention or distraction causes deviations from standards.

P.6.9 Rearing Knowledge Should Be Made Public

The core knowledge of insect-rearing systems exists in the published literature, standard operating procedures (SOPs) that exist in rearing facilities or in internal reports of those facilities, and in the minds of rearing specialists. The only part of this infrastructure that makes rearing information available to the scientific community as a whole is the body of publications. In the oldest rearing system, silkworm rearing in ancient China, the knowledge of rearing was a deeply guarded secret. Even when some of the secrets leaked out, they were held secret in the countries that adopted silkworm rearing. Similarly, in modern times, many rearing discoveries are held as secrets by companies that hold their rearing information as proprietary. I have found that many of the students who have taken my rearing courses and workshops were not allowed to divulge even what species of insects they were working with. Furthermore, in doing consultation with companies where I am asked to help them with their rearing systems, I am obliged to sign a confidentiality agreement about not divulging information about the companies' rearing processes. Yet all of these companies and the rest of the rearing community profit from the information that is openly published and available to everyone in the rearing community on an equal basis. While it is understandable that for-profit organizations wish to retain as proprietary information that which would give them a financial advantage over their competitors, it is also a reality that everyone involved in insect-rearing programs would be able to advance their enterprises by having a reliable and rich source of information about all aspects of insect rearing.

P.6.10 Rearing Research (and Rearing Education) Should Be *Directly* Supported by Public Funding

In line with making public rearing information (communication infrastructure), rearing research and rearing education need much more funding than is currently spent on this important aspect of rearing progress, and the funding should be mainly provided by the public sector, especially federal or national governments. The word "directly," in the opening statement, is very important. It implies that rearing research and education are part of a primary program, rather than an outgrowth of some larger and implicitly more important program. The implication of this proposal is that rearing is an independent discipline (a scientific and technological entity), rather than a subsystem under the umbrella of biological control, SIT, host plant resistance, research, or other programs that depend on reliable insect-rearing systems. This will be discussed in more detail in Chapter 13 on "Future of Insect Rearing," but for now, I emphasize that rearing research and education will never get the whole-hearted attention and quality practices that they need for sweeping accomplishments and improvements unless the infrastructure provides all the benefits of an independent discipline, including professional standing for the practitioners, a reliable and well-respected outlet for rearing research communications, dedicated and specialized education in insect-rearing science, and high-quality facilities. There is a need on a worldwide basis for centers of excellence, institutes, and adequately funded programs in insect rearing. **When we consider the accomplishments in science and technology that were made possible by insect rearing (even without adequate support or respect), it should be obvious that insect rearing deserves the kind of respect and attention that I am describing**:

A Little Learning (from *An Essay on Criticism*)
A little learning is a dangerous thing;
Drink deep, or taste not the Pierian spring:
There shallow draughts intoxicate the brain,
And drinking largely sobers us again.
Fired at first sight with what the Muse imparts,
In fearless youth we tempt the heights of Arts;
While from the bounded level of our mind
Short views we take, nor see the lengths behind,
But, more advanced, behold with strange surprise
New distant scenes of endless science rise!
So pleased at first the towering Alps we try,
Mount o'er the vales, and seem to tread the sky;
The eternal snows appear already past,
And the first clouds and mountains seem the last;
But those attained, we tremble to survey
The growing labours of the lengthened way;
The increasing prospect tires our wandering eyes,
Hills peep o'er hills, and Alps on Alps arise!

Alexander Pope (1688–1744)

Despite Pope's warning about the dangers of just "a little learning," I have tried, throughout this book, to use data-driven approaches to design, operation, and control of rearing systems. I came to value statistics too early in my life to disregard it but too late to master it. Yet the more that I worked with insect-rearing systems and the problems that these complex systems present to those of us who have devoted our lives to producing high quality, healthy, useful insects, the need for scientific treatment of our rearing processes (both in developing them and in maintaining them) is ever-apparent. The treatment of process control in this chapter and throughout this book was done through use of SAS JMP programs. However, there are several other statistical programs that treat the components of statistical process control such as R, QI Macros, Analyse It, Infinity QS, and several other commercially available packages. As a word of caution, this chapter and the others where I used statistical analysis to help us understand control in real-world rearing systems are not intended to be in-depth statistical engineering tutorials. In all the uses of statistical programs to help deal with variability (error, uncertainty) and to help us discover and understand the basis of any and all departures from standards of fitness and quality, the treatment of data with established engineering statistics is indispensable. Therefore, even at the risk of the dangers of "a little learning," I tried to guide readers to take the "statistical plunge" into the "Pierian spring."

P.7 ERROR (VARIANCE OR UNCERTAINTY) IN PROCESS CONTROL

The topic of statistical error is covered in greater depth in Chapter 7 on "Designing Error (Variability) Reduction in Rearing Systems: Reducing Uncertainty," but it is such an integral part of process control that it bears mention here in the context of SPC. I find it helpful to think of continuous awareness of error in developing rearing systems and in establishing process control strategies as analogous to the awareness of form in an athletic performance. For skilled athletes, there is a perpetual adherence to following the guidelines of proper form, be it a baseball batting swing, a quarterback's throwing motion of a football, or a basketball player's positioning of head, hands, and knees in a free throw. Like a trained athlete, a rearing professional is tuned to the sources of error or departures from the expected and desired outcome. The three types of error are relentlessly present and at work to undermine the regularity and quality of rearing processes. These types of error include:

1. Gross error (also called "operator error" or "human error," such as when an instrument is misread or a pipette is misused by such actions as sucking up extra liquid with the discard function of the pipette)
2. Systematic error (error that is repeatedly in the same direction and about the same scope, such as a balance that is not calibrated and reads 3% less than the "true" weight or a pipette that consistently delivers 97 μL, rather than the 100 μL where that it is set to deliver)
3. Random error (where we get unpredictable variation from sources other than systematic error or gross error—in random error the variations are not in one direction but vary around a mean). Causes of random error can be non-directional instrument drift, placement of the object being weighed on different parts of the balance pan, air currents, vibrations at the bench where the balance is stationed, or gains and losses in the moisture content of the sample

The correction recommended for gross error is training of personnel to use more careful practices. A worker can be trained as to proper pipetting techniques such as learning to pipette at a certain speed (not a sudden, jerky uptake or discarding of liquid), using the proper stops on the pipette, learning to avoid bubbles or how to treat bubbles or other surface/liquid interactions that distort transfer. Proper leveling of balances can reduce instances of gross error, and appropriate use of the tare function to make sure that the weighing starts with the appropriate zero can reduce gross errors. A very common (and very serious) gross error takes place when a worker forgets to re-tare the balance to account for a different weigh boat (which is one of several weigh boat/tare errors).

Finally, throughout this book, I have used a combination of sources from the literature on insect rearing and rearing support materials. In several cases, I have used examples from my own research and teaching experience where I felt that the reader would be given clear guidance from these case studies. Also, throughout this book, the topics of error (variability or uncertainty) are tied to process control and quality control, and these topics are related to them and the title of the book: *Design, Operation, and Control of Insect Rearing Systems*. In synthesizing this work, I believe that if rearing specialist (and students learning to become rearing specialists) use the premises of this book, that rearing should be treated as science and that the rigor in the attempted teaching in this text are followed, insect rearing will be elevated to a higher level discipline that will better serve the practitioners and their audience.

Acknowledgments

I dedicate this book to my wife, Jacqueline Louise Cohen, without whom, I could never have completed this work. Her support of my pursuits over the past 48 years has taken her into the field and the lab, being research partner, business associate, IT specialist, sounding board, artistic designer, mother to our children, and loving partner. Jackie has given me limitless encouragement and countless patient hours of listening and discussing virtually every idea in the entire body of my work, including those in this book. My son, Benjamin Aaron Cohen, has been a supporting and inspiring source in my work over the past 40 years.

I have approached insect-rearing systems here with an eclectic approach, trying to integrate various disciplines, such as nutrition, physiology, ecology, biochemistry, morphology, food science, systems theory, and other disciplines to provide a coordinated sense of the conditions for the insects we domesticate. My sense of the need to understand the disparate factors that make up the organisms and their relationship to their environment was developed by several people who influenced this kind of thinking. The earliest influences on my sense of the importance of seeing organisms from a broad point of view were some of my teachers and colleagues, including Lon McClanahan, Bob Vandergrift, Brian Myres, and Florence Tamas. They saw structure and function, ecology and biochemistry, natural history and physical chemistry as virtually inseparable dimensions in living systems. John Pinto took me into his laboratory and was a model of a kind of Renaissance scientist who integrated every possible feature of insects into his systematics perspectives. He and his wife, Diana (and their sons Derrin and Trevin), were also kind friends to me and my family. In my early work in insect rearing at the USDA, ARS Biological Control Laboratory in Tucson and in later work at Phoenix and Mississippi State, I had gotten to work with several technicians and technical assistants who were tremendous supporters, including Don Brummett, Lisa Smith, Delphine Harris, Diana Munoz Yang, and Nancy Sego. In my early days in the Tucson lab, I had the opportunity to work with world-class rearing experts, including Ray Patana, C. Glenn Jackson, Harry Graham, and Jack Debolt.

There are others who were mentors, teachers, and reviewers who helped me develop and improve my knowledge and ideas. But I am especially indebted to some outstanding workers, without whose diligence and capability I would not have had the successes in developing insect diets (or I would not have known that the diets were successful without their competent bioassays and culture handling). Those whose contributions most profoundly affected my work are Nina M. Urias, Lisa Smith, Gay McCain, Brenda Woods, and Patrick Crittenden. Nina, Lisa, Gay, and Brenda helped me build a grasp of how the bioassay fits inextricably into the diet-development/diet-assessment paradigm. Patrick has tirelessly worked at the analytical aspects of my work in taking apart diets so that we could understand what the components do individually and collectively. Amanda Lawrence and Bill Monroe contributed to my microscopic studies of diet and insect feeding systems and the nature of diet/insect interactions. Recently, Mr. Chuck Mooney guided my research with electron microscopy. Discussions with Gerald Baker changed my approach to looking at diets visually (via microscopy) which led to looking at diets as dispersions, a concept that so permeates much of my work. Many fruitful and enjoyable discussions with G. Doug Inglis expanded my understanding of diet contaminants, symbionts, and were supportive of my progress in the field. In fact, Doug Inglis was a major contributor to my vision of the complex interactions that I am calling the eclectic approach to "knowing your insect." Jack Debolt's work was a model for diet development techniques. Nelson Thompson was a mentor and early influence on my thinking about insect nutrition. The late Ken Hagen was an intellectual stimulant, model of top notch research, and the kindest of encouragers and supporters for my work. And Margaret Connor, as the patent advisor for my patents on diets and diet delivery systems, forced me to think in the most analytical and mechanistic terms how and why diets worked or did not work; her rigorous approach forced me to develop a deeper understanding of what we do when we feed insects artificial diets. The pioneers of insect-rearing system accomplishments laid down the foundation of all the successes that are taken for granted the entomology community. Robert T. Staten encouraged my efforts over nearly two decades, and his own work has been an impetus to my grasp of technological approaches to scaling-up diet-based mass-rearing systems. I can say with confidence the Bob Staten set into motion my thinking about development of mass-rearing systems.

Between the period when I wrote the 1st edition of *Insect Diets: Science and Technology* and now (2020), I have had numerous experiences with insect rearing in projects for government and private business organizations (consulting, contracts, and grants). As I mentioned in the 2nd edition of *Insect Diets: Science and Technology*, these projects gave me experience with dozens of species of insects that were new to me and that gave me whole, new perspectives about insect diets and the rearing context of these diets. My assignments included baits for yellow jackets, diets and rearing systems for house crickets, house flies, mealworms, hemlock woolly adelgids, balsam woolly adelgids, predators of the adelgids (coccinellids and derodontids), painted lady and monarch butterflies, western bean cutworms, cactus moths, various weed-eating insects, especially beetles such as *Rhinoncomimus latipes*, a predator of mile-a-minute weeds, brown lacewings, hymenopteran gall-forming parasites of *Arundo*, silkworms, drosophilid flies, honeybees, buprestid and long-horned beetles, stink bugs, and–through – interactions with workshop participants, students (many of whom are represented in Chapter 12), and professional colleagues–many other insects (and a few mites). Some of my richest learning experiences were with my work with honeybees (from the earliest to the most recent part of my career). The honeybee research gave me the chance to work with the great Lonnie Standifer and others from the Carl Hayden Bee Research Laboratory, including Eric Erickson and Gerald Loper. More recently, I had the opportunity to work with Samuel Ramsey on feeding dynamics of Varroa mites.

In my more recent experiences at North Carolina State University, I have had the enriching and growth-promoting experiences of working with people from the Entomology Department and from other departments across the campus, including the Department of Food, Bioprocessing, and Nutritional Science (where Professor Chris Daubert and Chris Pernell brought me into the world of food rheology and Mr. Michael Bumgardner and Professor K. P. Sandeep guided me through operation and application of a twin-screw extruder), the Department of Industrial and Systems Engineering (where Thom Hodgson helped me learn rearing-related facets of systems engineering), Plant Sciences, Biomanufacturing Training and Education Center (BTEC where Mr. Rick Lawless and Dr. Lucas Vann helped me learn applications of bioreactor technology to improvement of insect-diet components), Forestry Biomaterials, and the Department of Biological Sciences.

Also, as I mentioned in *Insect Diets: Science and Technology*, CRC Press, Boca Raton, FL, some of my richest learning experiences came from teaching insect rearing courses in on-site and online contexts. Preparing to teach rearing courses clarified my perspective and understanding to meet the scrutiny of an intelligent and critical-thinking audience. Furthermore, the students in these courses (and in the rearing workshops that I have taught over the two decades) were stimulating in the questions asked from a fresh viewpoint, and in the answers to test questions and especially in their term projects. Students caused me to examine my perspectives of diets and rearing from a 360° perspective. Many of the students that I've taught over the years are represented in Chapter 12 on "Rearing Education." Included in my teaching experiences were contacts with rearing professionals from private industry and public sector rearing labs. Often, these people had perspectives and techniques that I had never experienced, so interaction with these people was incredibly enriching. Also, my opportunities to do on-site visits to industrial and government rearing laboratories were beyond an eye-opening, mind-expanding experience. I got to see first-hand how FMC, Monsanto, BASF, Pioneer-DuPont, Beneficial Insectary, S.C. Johnson, and other industrial labs performed their rearing operations, often with great competence, good equipment, and remarkable personnel. Likewise, my visits to state laboratories such as the PABIL facility in New Jersey, the State of Connecticut Experimental Station, the Florida DPI, and the North Carolina Laboratory of Biological Control conduct their excellent and highly professional operations (this includes the excellent work of Dr. Kathy Kidd). And of course, the leadership provided by the USDA-sponsored laboratories is under-appreciated in my opinion in terms of the remarkable advancements that they made in rearing and diet technology, including the APHIS laboratories at Otis Air Base (Massachusetts), the Mission, Texas Mexican Fruit Fly Laboratory (where Dr. Hugh Conway has been an excellent cooperator), and of course the Pink Bollworm Rearing Facility, in Phoenix, AZ where they kindly hosted visits from my workshop participants and showed all aspects of their rearing accomplishments (this includes Dr. Fred Stewart and Ernie Miller, and more recently, Greg Simmons). The USDA, ARS (where I got my start for the first 23 years of my rearing career) has excellent facilities and professional personnel whose mission is to advance insect rearing in ways that are of service to the people of the United States and the world.

A very important part of my recent learning experiences and being introduced into a new realm was the considerable time that I spent with forest entomology, where Dr. Fred Hain included me in his group of students and colleagues. Thanks to Fred's inclusiveness, I got to serve on several graduate students' committees, and I found the programs in forest entomology a kind of second home. Through my association with the forest entomology group, I worked with a wonderful group of students who contributed to my learning about the basics of insect feeding systems, including Dr. Kelly F. Oten, Dr. David Bednar, Micah Gardner, Dr. Chen Zha, and Mary Talley. During my interactions with the forest entomology group at NCSU, I was most privileged to work with Mr. John Strider whose constant curiosity and energy fueled many efforts to study one of the most challenging insects that I have ever encountered, the adelgids. At the same time (starting in 2004), I became a research partner of Dr. Carole Cheah (from the Connecticut Experimental Station), and with my interactions with Fred and Carole, we were supported (especially by the USDA, Forest Service and the USDA, APHIS) to do research on hemlock woolly adelgids, balsam woolly adelgids, and the predators of these two species. Over the course of my career, I have had the opportunity to work with several post-doctoral research associates who added depth and perspective to my career: James Hagler, Fanrong Zeng, Jason Hanson, and Molly Robinett. The USDA, APHIS had supported my work with other forest insects, including the emerald ash borer in a project and the spotted lanternfly project. Through this project sponsored by the USDA, APHIS, I have been able to work with Dr. Hannah Nadel, David Lance, and Ben Slager (who worked with me on a contract to advance rearing of the emerald ash borer and the spotted lanternfly, which included work with Ms. Mauri Hickin and Liam Sullivan). During this period, I also had the support and productive learning experiences from Dr. Kenneth Bloem (USDA, APHIS) who brought me into the cactus moth project and the emerald ash borer project. Ken's rearing background was a tremendous asset to my learning, and his further support for the USDA, APHIS webinar series on insect rearing was a further opportunity to work with many of the top notch rearing experts in the world: the staffs of the rearing labs run by APHIS. The statistical analysis underlying the design of experiments portion of this research was made possible by Drs. Marcia Gumpertz and David Dickey as part of the service of the NCSU Department of Statistics. Recently, working with Dr. Gary McGuire on a contract with Rivis Inc., I have had the chance to deepen my understanding of the rearing conditions required for SIT.

Over the time frame when I started and ended my business venture (Insect Diet and Rearing Research, LLC), I had been funded by several companies (Clearwater Butterflies, The Bug Company, MGK, Beneficial Insectary, FMC, DuPont Pioneer, and Monsanto) and several government organizations (USDA, APHIS; USDA, ARS; USDA, Forest Service; USDA, NIFA – National Institute of Food and Agriculture grants, or SBIR grants). I have also gotten support in terms of facility use, provision of insects, and scientific advice from the State of Connecticut Experimental Station, the State of New Jersey Beneficial Insect Rearing Lab, the State of Florida Department of Plant Industry, and the State of North Carolina Department of Agriculture and Consumer Services (Plant Industry—Plant Protection Section Biological Control Laboratory).

Also, I offer my special thanks to Dr. Sergei F. Krasnyanski of the Plant Transformation Laboratory in the Department of Horticulture at NCSU, José M. Bruno-Bárcena, James W. Brown (Department of Biological Sciences and formerly Department of Microbiology), Tom Dorsey, Amy Dierks, Kathleen Kidd and Rebecca Norris (quarantine at the Biological Control Program of the North Carolina Department of Agriculture and Consumer Services), Thom Hodgson (Department of Industrial and Systems Engineering), Roger Thompson (retired, USDA, ARS), Den Truong (USDA, ARS, NC State), Richard Lawless and Lucas Vann (BTEC, NCSU), James Carpenter and Susan Dawdry (USDA, ARS in Tifton, Georgia), Erica Simbro and Leigh Christensen (Pioneer-DuPont), Gordon Vadis and Hong Ji (The Bug Company), Sinthya Penn (Beneficial Insectary), Robert Suryani (MGK), Chris Daubert, Chris Pernell, and Sharon Ramsey (NCSU, Department of Food Science), and Jenny DeSio, Tom Dorsey, Amy Dierks, and Cynthia Detweiler (of the State of New Jersey PABIL facility), Peter Ebling (Great Lakes Forestry Centre, Sault Ste. Marie, Ontario), Kenneth Johnson and Mary Wallin (FMC Corporation), and Hugh Conway and Donald Vacek (from the USDA, APHIS). Two dear friends, Dr. Ann Black and Brad Heidinger from ImmunoReagents, Inc. (our former neighbors on the NCSU Centennial Campus), added greatly to my learning about potential immunological and biochemical tests that could support rearing and diet quality measurements; plus, they were always there to help us with students and workshops.

Several of my colleagues from the Entomology Department have been instrumental and supportive of my role at NCSU, and in serving on graduate student committees, having research partnerships, and other interactions with these people, I have expanded my thinking about many ideas in this book: Mike Waldvogel, George Kennedy, Ed Vargo, Wes Watson, Fred Gould, David Orr, Marce Lorenzen, Clyde Sorenson, Yasmin Cardoza, Dominic Reisig, Arun Babu, and Coby Schal. The support personnel have been of such support that it is hard to properly credit them, including Ms. Michele Sabateli, Pat Batchelor, Katie Youngs, and Deanine Johnson.

Many of my richest learning experiences have come from interactions with students, especially the graduate students at the University of Arizona (where I began my insect diet and rearing career), Mississippi State University (where I worked for several of my last years with the USDA, ARS and where I had the chance to work with my friend Dr. Frank Davis and with many other outstanding scientists and rearing staff), and most recently and currently with North Carolina State University (where I came to pursue efforts to make valuable rearing accomplishments through the private business sector from 2005 until now). Both at Mississippi State University and here at NCSU, I have co-taught workshops in insect rearing, and I have taught several courses in insect rearing at NCSU (as in-person courses and as online courses). One of the leaders in the field of insect rearing for about five decades is Dr. Norm Leppla. Because of sheer numbers and logistic constraints, I cannot mention all the students who influenced me and helped me be a better diet and rearing specialist, but there is no substitute for the experience that teaching these people has provided to me and most especially learning from them. One of the long-standing supporters of my research and someone from whom I learned a great deal about insect physiology, insect/plant interactions, and some wonderful insights into rearing is W. S. Bowers. He has not only been a mentor to me but also a great advocate of insect rearing and artificial diets. He helped advance the field in ways that even he may not appreciate, especially with his pioneering work on tobacco hornworms. My good fortune in having John Law as a guide through some of my journeys into digestive enzyme biochemistry was of great value.

Special mention is in order for two people who had enriched the field of insect rearing: the late Dr. Alan C. Bartlett and the late Professor Henry Hagedorn. I was privileged to work with Dr. Bartlett for much of my career, partly when we were stationed in Phoenix and later in insect-rearing workshops where he taught the genetics of insect rearing. Dr. Bartlett was a pioneer in the discipline of genetics of reared insects. He brought to this field the understanding of population genetics and its application to mass-rearing systems. His research and teaching shed light on many of the "black box" features of what happens to insects under conditions of domestication.

Henry Hagedorn was not only a great insect biochemist and teacher (and leader of the Center for Insect Science for many years), he also founded the open access *Journal of Insect Science*. His vision was to establish a journal that would give free access to all aspects of entomology to scientists and students from around the world, including emerging nations where funds for publishing and access to insect research were limited. He was inclusive to a greater degree than anyone can imagine—helping authors with their writing and science to make *JIS* a teaching system, as well as an outlet for new research. He invited me to become editor of the section in *JIS* dealing with insect diets, rearing systems, and feeding biology, and in that function, he encouraged researchers to produce papers on all aspects of insect rearing. Henry truly manifested the spirit of what I speak of in this book about establishing a home for insect rearing in a reputable journal with worldwide circulation. Everyone in entomology will miss him, especially the people in the rearing community who need someone of his stature and vision to further advance the field.

About the Author

Allen Carson Cohen completed his high school education in Chicago, IL and went to a university in California for BA and MA in English and a teaching credential. He later returned to college for an MA in biology, specializing in the physiological ecology of rattlesnakes. He completed his PhD in entomology at the University of California, Riverside (1978). He has served for seven years as editor for insect diets, insect rearing, and insect-feeding biology for the *Journal of Insect Science* and serves as a mentor for students, researchers, and professionals in diet and rearing science and technology. He currently serves as an editor for insect rearing and insect-feeding biology for the journal *Insects*. He holds six US patents, has published more than 125 refereed papers, and has received several technology transfer awards for his work on insect diets. He is currently a research professor and coordinator of the Insect Rearing Education and Research Program at North Carolina State University, where he teaches online and in-person courses on insect rearing and conducts research on developing and improving insect-rearing systems. Professor Cohen currently teaches live courses in insect diets and the design, operation, and control of rearing systems to a worldwide audience.

CHAPTER 1

Approaches to Developing Rearing Systems, Creating an Algorithm for Developing or Problem Solving Rearing Systems, Heuristics, and Reliable, Reproducible Insect Production

1.1 BACKGROUND ON INSECT-REARING INFRASTRUCTURE

Insect rearing is a discipline that incorporates science and technology for practical purposes: production of insects for various reasons (discussed in Chapter 2 "Who's Who in Insect Rearing?"). The discipline of insect rearing is composed of an infrastructure described by the diagram in Figure 1.1.

1.1.1 The Infrastructure of Insect Rearing

The diagram in Figure 1.1 shows six components with **funding** being central because all the other components are driven by or made possible by funding. It is also important to realize that the nature of the funding has a huge impact on the nature of the other five components. This will be discussed further in the following paragraphs, and it is also covered in detail by Cohen 2019. Once funding is established, the other components are so interdependent that separating them is like a chicken/egg phenomenon. **Personnel** are essential to rearing programs. It is the staff that does the rearing setups, ordering, cleaning, harvesting, etc. The personnel must have a place to work on their rearing system, which means that however small or large, there must be a **facility** for the rearing process to take place. This leads to the centerpiece of insect rearing, the **rearing system**. The rearing system consists of the insects themselves and all the satellite operations that must be in place for the system to be perpetuated. We think of rearing systems as species-specific features of rearing infrastructure. While there is confluence of rearing procedures from one system to another, a rearing system for a given insect species will be different from all other systems in various ways. Because the system is the central focus of this book, it will be covered in far more detail than can be afforded here. Back to the personnel: the people who are doing the rearing were not born with knowledge about their rearing process, meaning that they had to be **educated** in rearing practices. The education may have been (and usually is) on-the-job training, or it may have included formal education such as coursework, workshops, webinars, and reading. The reading in insect-rearing domains comes mainly from published literature—books, technical/scientific papers, pamphlets, etc., all of which make up **communication** of rearing knowledge.

1.1.2 The Scale of Insect-Rearing Systems

To suggest two simple examples of the potential scale of rearing infrastructure, (1) a high-school student may choose or be assigned to rear some *Drosophila melanogaster* for her/his interest or for a class assignment. The student and her/his teacher may purchase a kit of *D. melanogaster* from a supplier such as Carolina Biologicals to start a small colony in a small corner of the biology classroom laboratory. The kit contains materials including the insects, diets (pre-made by the supplier), extra rearing containers (plastic vials with spongy plastic stoppers), paint brushes for gently transferring the flies, and possibly a proprietary anesthetic (called "Fly-Nap") so that flies could be examined for phenotypic determinations and sexing. The kit contains instructions for rearing the flies and depending on the type of kit, possibly instructions for carrying out genetic, nutritional, or ecological experiments. The rearing system in the example was developed and optimized prior to the student's access to it. The student and her/his teacher (and possibly fellow students) constitute the personnel, and the facility is the high-school lab with a bench or chamber for fly-rearing. The communication is from the instruction sheet that comes with the kit and whose information had been drawn from the

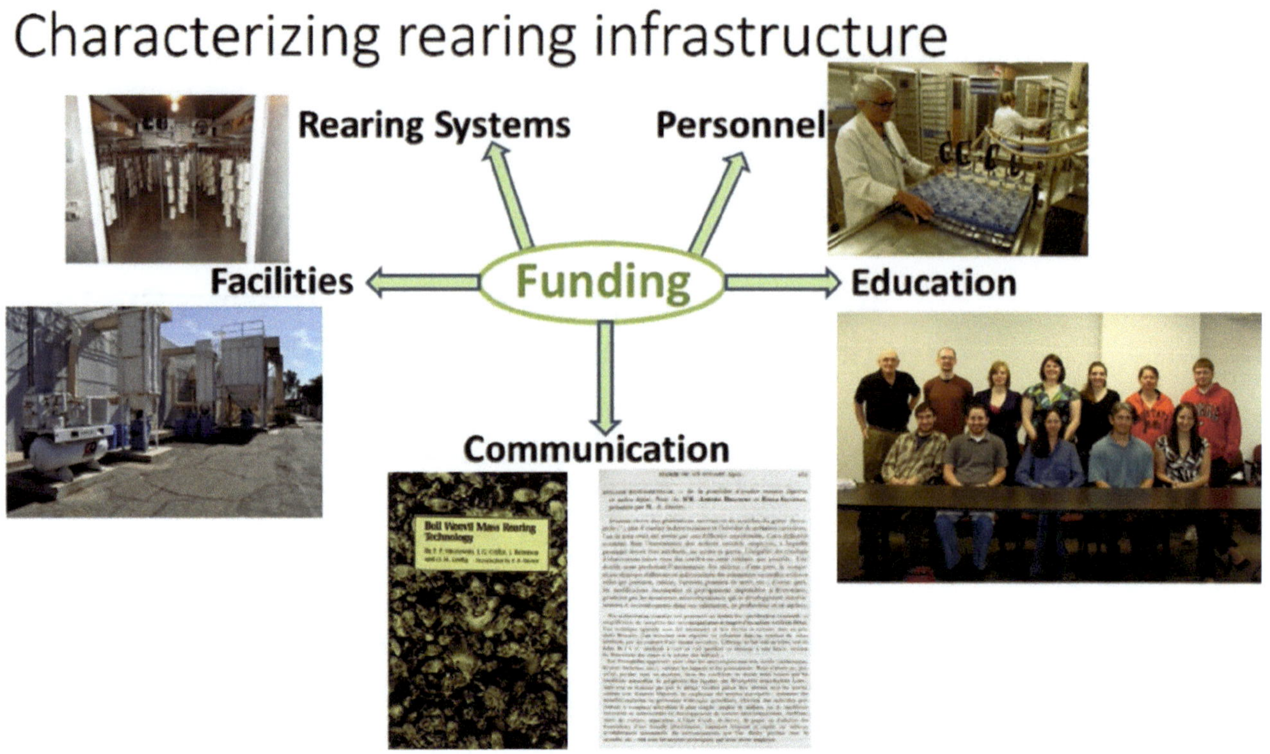

Figure 1.1 Components of insect-rearing infrastructure.

literature on rearing *Drosophila*-type flies dating back to the 1910 paper by Delcourt and Guyenot. The education in this case is a continuing process that begins possibly with the teacher starting the student(s) with the basics of rearing *Drosophila* and/or the student reading the instructions and possibly delving further into the literature or doing an internet search of "*Drosophila* rearing." The funding came from the school (or the taxpayers who fund the school's science program) (Figures 1.2 and 1.3).

Example 2 might include a new worker who was hired by a large rearing laboratory such as one of the United States Department of Agriculture (USDA), Animal and Plant Health Inspection Service (APHIS), or industry rearing programs for sterile insect technique (SIT) or pesticide screening operations. The new hire perhaps has no rearing background and is starting as an entry-level worker. His/her laboratory manager (insectary manager) schedules on-the-job training for the new person who will be preparing diets and handling the insects (e.g., gypsy moths at the Otis Airbase Facility, see Figure 1.1). In typical cases, the new employee is educated with on-the-job training. Often entry-level new hires have little or no prior education or background with insects or insect rearing. Sometimes, such employees are sent to workshops on insect rearing, and/or they are encouraged to take courses related to rearing (discussed further in Chapter 12 "Rearing Education: The Pathway to Improving Rearing Science and Technology"). However, despite efforts from various institutions such as Mississippi State University (Schneider et al. 2018) or North Carolina State University (Cohen

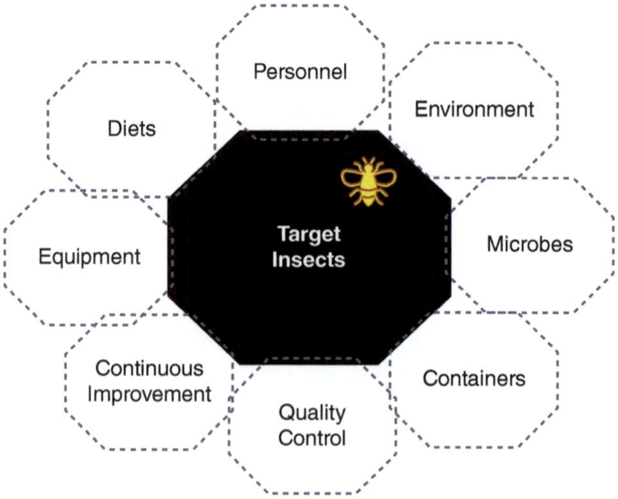

Figure 1.2 Diagram of a rearing system's components. This highly simplified diagram shows the target insects at the center of the system's components. In this model, there are nine components that interact in ways designed to produce insects that fit well into the outcome expectations to meet the purpose of the system.

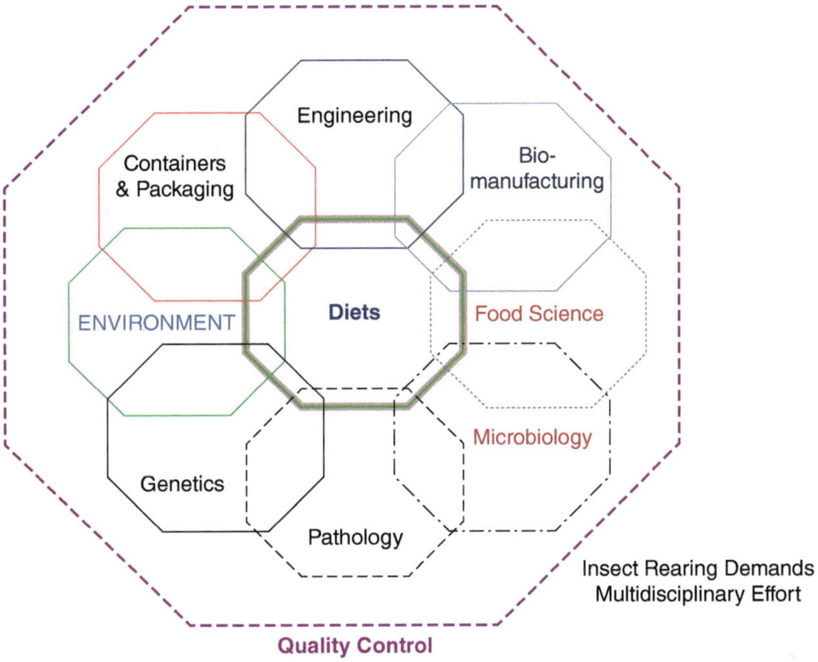

Figure 1.3 The interdisciplinary nature of insect diets.

2019), there remains an inadequate opportunity for formal rearing education; therefore, the rearing novice or incipient rearing professional is burdened with a haphazard base of information on insect rearing. Often the facilities for insect rearing in larger programs (such as those in industry and the USDA, APHIS, or Agricultural Research Service or Forest Service) are modern, very well-equipped laboratories or insect-rearing factories. Often, too, the insect-rearing systems in the large, well-equipped facilities are well-established and are guided by detailed work plans and standard operating procedures (SOPs). However, despite having well-designed facilities and well-defined insect-rearing system parameters, the lack of education can be a great hindrance to peak operation and efficient trouble-shooting/problem solving. This points to the recourse to communication through consultation with the literature.

1.1.3 Further Examination of Rearing Communication

Over the years of my involvement in insect rearing, I have been impressed with the number of publications that describe the technical details of hundreds of rearing systems. However, finding the great number and diversity of papers is not easy, especially for a novice who is not adept at literature searches. The fact that rearing information that does manage to get published is often hidden in a publication whose title, abstract, or keywords do not reveal the information pertinent to the new learner's interest. My own experience with publishing papers on insect rearing is that editors and reviewers are generally impatient with detailed explanations of rearing systems' algorithmic details. Therefore, in recent and contemporary publications, rearing details are bare-boned and skimpy. In contrast, older papers on insect rearing were often very rich with details, including speculation, interpretation, and extensive attention to details and mechanisms. Three excellent examples of papers that contain rich discussion that followed the twists and turns in the researchers' path of discovery are Baumberger 1919, Michelbacher et al. 1932, and Bridges and Darby 1933. Though these papers are discussed further in several places in this book, suffice it to say that Baumberger and Michelbacher et al. discussed in granular detail the nutritional requirements for several insects, especially flies. These authors teased apart the evidence for deeming live microbes (especially yeasts) as essential vs. strictly chemical factors that were essential or valuable to insects. These authors were working in a time when microbial interactions were in their early days of being appreciated. The thinking and experiments that these authors used or cited from other authors were elegant examples of critical thought and using cleverly devised experiments to help decide the role(s) of microbes in the lives of insects. The other paper (by Bridges and Darby) reviewed all the existing studies of *Drosophila* nutrition and feeding biology, especially in the context of pH and physical (textural) aspects of diets. The type of thinking in the papers by these authors, and of many other pioneers in insect-rearing system research, is heuristic as explained in the section that follows.

1.2 HEURISTIC APPROACH TO REARING SYSTEM DEVELOPMENT AND IMPROVEMENT

1.2.1 Heuristic Definitions

The Online Dictionary defines heuristic as follows: "Derived from a Greek word that **means** 'to discover,' **heuristic** describes a rule or a **method** that comes from experience and helps you think through things, like the **process** of elimination, or the **process** of trial and error. You can think of a **heuristic** as a shortcut." The concept of heuristic in psychology is further defined by Wikipedia as: "In psychology, **heuristics** are simple, efficient rules which people often use to form judgments and make **decisions**. They are mental shortcuts that usually involve focusing on one aspect of a complex problem and ignoring others." And for computer technology, heuristic analysis is described as follows: "**Heuristic** analysis is a method employed by many computer antivirus programs designed to detect previously unknown computer **viruses**, as well as new variants of **viruses** already in the 'wild.'" In all these descriptions and definitions of the concept of heuristics, there is a trial and error factor and a set of "rules" that go to the heart of a process or organized phenomenon.

1.2.2 Pure Trial Error vs. Modified, Feedback-Based Discovery

By way of explanation of how algorithms are developed (especially with a heuristic approach), the natural recourse to algorithmic thinking is explained. It is valuable to add that if an algorithm is to be useful, low error, with highly predictable results, it must be arrived at with critical thinking and by scientific means. This is to say that the most useful algorithms are arrived at by critical thinking and meticulous vetting.

The following example should illustrate the point: if you were playing a game of hide and seek where a person hid an item (say a coin), and the seekers randomly walked around the game arena (classroom, playground, etc.) and turned up objects, looked in drawers, felt the tops of door jambs, etc., the search could be considered completely random and fully based on trial and error. This method may lead to the discovery of the hidden object, but it is not based on critical thinking and not very effective, especially in a complex environment. However, if the person who hid the coin gave clues such as "warmer" and "colder" as a seeker came nearer or farther from the hiding site, the feedback would help the seeker build an ever-increasingly resolved concept of where the coin was hidden. Eventually such feedback (which could be considered a negative feedback loop under the constraints of systems theory) would give the seeker an almost certain sense of the exact location of the coin. Hunters and trackers use the modified feedback system to locate (pinpoint) their quarry. They find tracks, droppings, patches of fur, and broken twigs to refine their search. Animals with a keen sense of smell can use olfactory cues to track their quarry, where molecules of various degrees of volatility can be detected from a distance or from close-by. The strength of the odor signal gives the tracking animal feedback about how close the quarry might be. An excellent example of such tracking by olfaction can be found in male moths such as pink bollworms seeking the pheromone scent ((*Z*,*Z*)-7,11- and (*Z*,*E*)-7,11-hexadecadienyl acetates) of the calling female. The tracking of the pheromone plume may be either orthokinesis or klinokinesis (orthotaxis or klinotaxis) or both. The tracker (the male moth) flies through the pheromone plume either comparing the concentration of scent molecules as it zigzags to the higher concentration or speeds or slows its flying in accord with the concentration gradient. The bottom line of this discussion is that the surest way of obtaining quality, high-resolution outcomes is to use high-information search systems. In the discussion of rearing system algorithms that is to follow, we will see that the most high-information, high-resolution systems are science based.

1.3 WHAT PROBLEM SOLVING A REARING SYSTEM?

1.3.1 Deeper Perspective on Rearing Systems

The phrase "*rearing system*" consists of the qualifier "*rearing*" that specifies that we are dealing with a type of system. This phrase guides us to certain rules that guide us in dealing with the central concept of a system, which is a collection of parts assembled for a given purpose. The parts may be components such as insects, diets, containers, heating/cooling equipment, lighting sources, timers, racks, controls, etc. Another essential part of the concept of the system is that it involves processes such as selecting insects that we wish to rear, mixing diets, selecting or making containers (e.g., putting holes in condiment cups to provide suitable gas exchange), setting and checking temperatures, setting and checking lighting features (photoperiod, light intensity, and light quality), and all other activities/processes that rearing personnel perform.

In a definition of "*system*," online dictionary suggests the phrases "interconnecting network" and "an organized scheme or procedure." These phrases are very applicable to what I am suggesting about rearing systems, that there are *interconnections* between the components and that the procedures are *organized*. In Chapters 7 (Designing Error (Variability) Reduction in Rearing Systems: Reducing Uncertainty), 9 (Tutorials on Using Design of Experiments with SAS, JMP Protocols), and 10 (Relationships between Fitness, Quality Control, Health, Homeostasis, and Stress), approaches to understanding and analyzing the interconnections and the organization of rearing systems are discussed in detail. However, for now, it is helpful to understand that the many of the characteristics of our insects depend on the nature of the diets, environment, and other features of the rearing process. In this vein, we are well-served to remember that because of the interconnectedness of rearing system components, when we modify one component, we are likely to affect at least one other component. For example, even a small temperature change (e.g., 1°C) can change the drying power of the air significantly. Between 20°C and 30°C, the maximum water content of air goes from 17.3 to 30.4 g/m^3 which translates to about 1.3 g/m^3 for every degree centigrade of temperature change. This further translates to greater (or lesser, depending on the direction of the temperature change) drying power of the air with every temperature change, and this drying power affects the insects, diet, frass, and microbes in our rearing system. In association with the change in temperature affecting the drying power of the air, there is the issue of volume of air in the rearing container and the rate of air exchange as governed by factors such as the number and size of openings in the rearing container. Some rearing containers have small pin-holes for gas exchange while others have larger holes or mesh, which allows very high rates of gas exchange. The positioning of the container in the insectary can influence gas exchange where, for example, the container is bathed in circulating air. This is further complicated by the question of rate of air movement over the container. This is discussed further in Chapters 6 (Rearing Systems as Ecological Niches and N-Dimensional Hypervolumes) and 8 (Building a Process Control System). All of these factors constitute the concept of the algorithms that underlie rearing system designs and that contribute to the decisions about interactions and outcomes of rearing systems.

1.3.2 Steps in Rearing Systems' Algorithms

In Figure 1.4, the steps in the rearing process (discussed in detail in Chapter 7) for painted lady butterflies are presented. The steps are the chronological sequence of processes that make up the rearing system, shown on a single generation basis. Starting with Step 1 where the newly enclosed adults are feeding on artificial nectar, for example, in the F_1 generation, the steps are executed by the rearing staff in the specified sequence. Once Step 15 occurs, the process begins again with the next (F_2) generation and proceeds through the F_n generation. Looking ahead at a principal goal of this book and of rearing personnel, the factors that contribute to maintaining or degrading quality are to be understood, so that the ideal rearing procedures involve, as much as possible, the detection of problems in the earlier steps that lead to reduction of quality in the later steps. To use a simple example, if there is a problem with the nectar eaten by the adults in Step 1, the consequences of that problem may not become apparent until Step 15 is reached. The problem may be an insufficient amount of artificial nectar or mold in the nectar that leads to the adults in Step 1 not getting an adequate amount of certain nutrients (e.g., amino acids or minerals). This could cause a malnutrition disorder in the adults that may reduce the amount of the key nutrients, preventing the adult females from being able to allocate

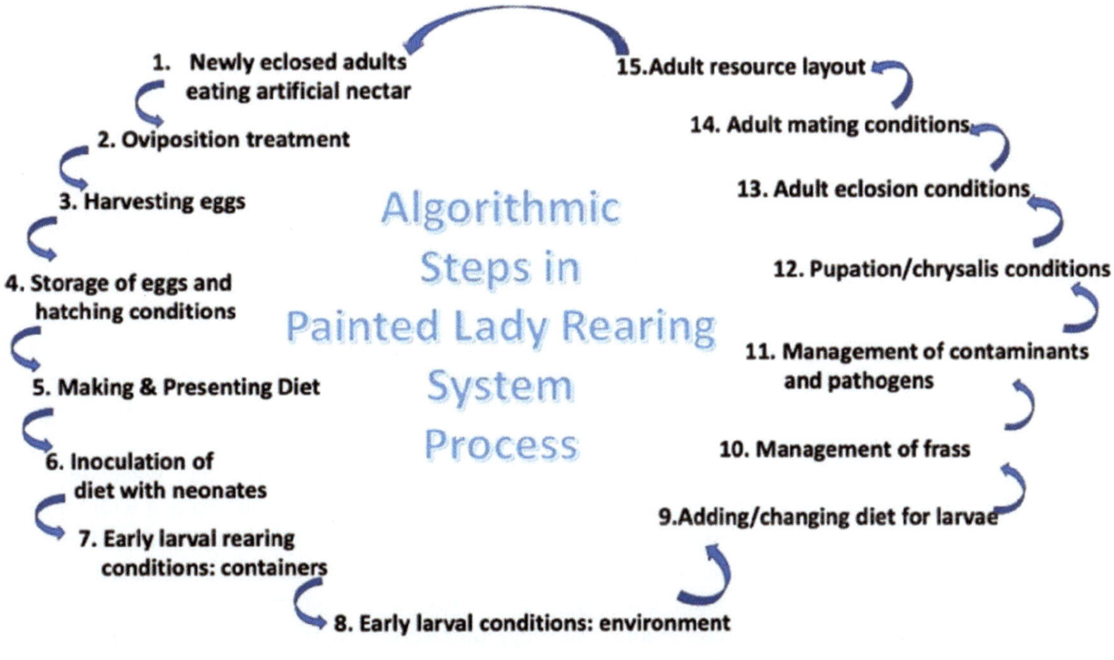

Figure 1.4 An example of the steps in a rearing system algorithm.

the full maternal contribution to the nutrition of the eggs. This nutritional deficiency may be expressed as a reduction in the reproductive potential (where a female that has a biotic potential of 300 eggs only produces 200 eggs), or it may be manifested as a reduction in the nutritional load that the butterfly allots to each egg (e.g., depositing 70 µg of yolk nutrients per egg, rather than 80 µg). This "short-changing" of the eggs may not be apparent in the initial steps of the process, and the resulting defect may not be expressed for more than 30 days after the causational event took place. Potential failures of the system can take many forms, and in the complexities of each insect species' biology, developing a system of symptomology is one of the most challenging factors in insect rearing. The idea of developing an extensive symptomology that allows rearing personnel (the trouble shooters or problem solvers) to diagnose problems with nutrition, pathology/contaminants, environment, or genetics is a kind of Holy Grail of science-based insect rearing. While the idea that a complete symptomology will ever be developed is overly optimistic, the goal of having a fairly extensive set of diagnostic tools serves as a basis for continuous improvement in rearing systems, and the idea of having step-by-step diagnosis and assessment of rearing processes is realistic. This discussion should motivate the reader to seek understanding of the algorithms that drive rearing systems.

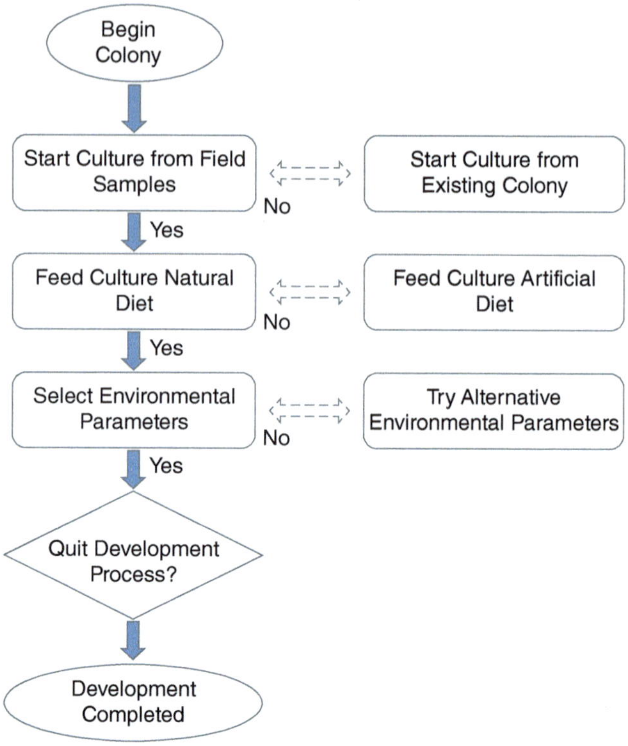

Figure 1.5 A schematic diagram of an algorithm for rearing an insect.

1.3.3 What Is a Rearing Algorithm?

It's a set of instructions, a pathway of sorts, for any process. It's a set of steps that are involved in establishing or running a process. Another way of looking at it is that an algorithm is a set of instructions to solve a problem. The term is derived from computer systems, but it has far wider applications, including use in insect rearing (see Figure 1.5).

The diagram in Figures 1.4–1.6 are designed to provide a simple starting point to acquaint the reader with the "algorithm concept" as it applies to insect rearing, but before applying specifics about rearing paradigms, I offer an everyday example of how we use algorithmic thinking almost constantly. I also suggest that **while algorithms for computer technology are generally digital** and therefore binary ("yes/no" decisions), **real-life algorithms are often analog** (i.e., in a continuous gradation distribution). A digital framework is yes/no, on/off, up/down, left/right, etc.; while an analog framework is like the volume knob (connected to a rheostat) on a radio where the volume (actually the voltage inducing volume changes) is a continuum from very quiet → louder → still louder → until the loudest volume is reached. The older radios used a variable selector that tuned into stations, but the tuning, even with respect to a given station, was continuous. As we tuned our old radios, we would search (by turning the dial) not only for the station we desired but also for the best reception of that station. In the following examples (leading to our decision-making processes in developing and improving rearing systems), I encourage the reader to think in terms of which improvements or changes are strictly digital and which are analog. This leads to the example of using algorithms in everyday life (Figure 1.6).

1.3.4 How Are Algorithms Used in Everyday Life?

In day-to-day living, we use algorithms to do our everyday tasks. For example, if we are making breakfast, we decide what foods will we be using, and how will we prepare the foods? If you have hard-boiled eggs prepared already, then you may decide to get the cooked egg out of its storage container (I like pickled eggs that sit in a vinegar and beet juice and seem to stay palatable for months!) If I'm in a hurry, I can pop the egg in my mouth, chew, and swallow, and *voila*! I've eaten breakfast! The algorithm is simple, go to the kitchen shelf where the pickled eggs are stored; open the jar; jab and egg with a fork; place the egg in my mouth; chew, swallow—end of process. Oops, I dribbled vinegar and beet juice on the kitchen counter and floor. Next time, I may plan to use a napkin or to scoop the egg into a bowl where I can cut it, and use a fork to pick up the cut pieces and neatly and patiently put the pieces into my mouth—far less messy.

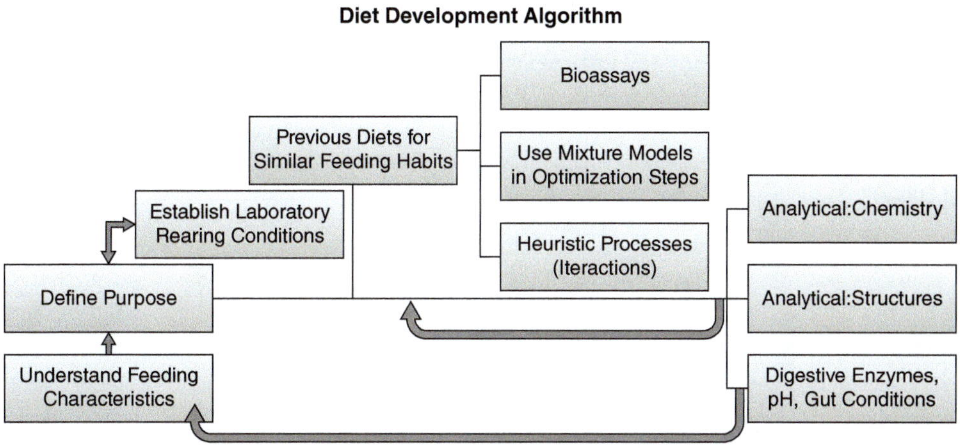

Figure 1.6 An algorithmic model of diet development.

All the steps in this simple example are parts of an algorithm that humans do regularly without adorning the process with the name, "algorithm." Let's make the process of breakfast preparation (and eating) a little more complex. Let's say that we want fried eggs, toast, and coffee. This breakfast algorithm has the major components of (1) frying eggs, (2) making toast, and (3) making coffee. Uh-oh, this is getting complicated. If I fry the eggs first, then make the toast, then brew the coffee, the eggs will be cold by the time the toast and the coffee are ready. So, I have a decision about the order of the major sub-processes. By experience, I have learned that it takes me about 5 minutes to brew coffee, and with the coffee maker that I have, I can keep the coffee warm and fresh-tasting for nearly an hour. My other experience taught me that the toast takes about 2 minutes from bread to toast, another minute to spread cream cheese on the toast. During this 3-minute toast-making component, I can be cooking the eggs the way I like them—over medium well (which takes about 3 minutes with another minute after I turn over the eggs in the frying pan). So if I plan my algorithm properly, I can (1) make my coffee (and even drink some while I'm working on the other parts of my process), (2) start my eggs, (3) start my toast, then have everything converge into a completed eggs, toast and coffee breakfast, which I can eat in about 2–4 minutes while watching the news on TV.

But this simple story is really **much** more complex than I let on. I had to decide how long to toast the bread to get it to my optimum (how I like it best). The dial on the toaster is helpful, but not infallible, and I had to do some trial and error (incremental testing) over several days to determine the ideal time. I also learned (using either my superior powers of reasoning or by unfortunate experiences) that it is better to apply cream cheese after the bread was toasted, rather than before. I also learned that when I had to take a slice of bread from the freezer, rather than from my room-temperature pantry, it took much longer to get the toast where I wanted it in terms of brownness. I learned something similar about the eggs and that taking the eggs out of the refrigerator the night before to let them warm to room temperature made the cooking time for the eggs less than it was for cold eggs (i.e., eggs starting at 4°C take longer to cook than eggs starting at 22°C.) I also learned from reading that eggs can have the pathogenic bacteria *Salmonella* on their shells, so I routinely wash the eggs with tap water (though I'm not confident that I should not use some soap in the egg-washing process). I also learned somewhere that placing a cover on my frying pan helped the eggs cook faster and more evenly. All of these factors are part of an algorithm of making breakfast of three simple components: eggs, toast with cream cheese, and coffee. I won't belabor the situation to discuss the decisions about making coffee, the brand, the grind, the coffee-maker brand, or the ratio of coffee grounds to water.

If we think about what we will prepare for breakfast the night before, we might include removing eggs or butter from the refrigerator to allow them to reach room temperature overnight, thus making the eggs more prone to cook evenly and quickly and helping the butter to spread more easily. If we are really thinking ahead and critically, we may consider the effects of warming and cooling butter in terms of its stability.

This example, though simplistic, should help us realize how much our day-to-day (minute-by-minute) activities involve constant development of "life-algorithms" and how adept we are with algorithmic thinking. The example can also be considered in terms of the digital vs. analog practices we execute. We may have asked "do I eat breakfast/or do I not eat breakfast?" (digital or analog?): if "yes, eat breakfast," then the next question arises: eat at home/not eat at home? (digital or analog?). If as part

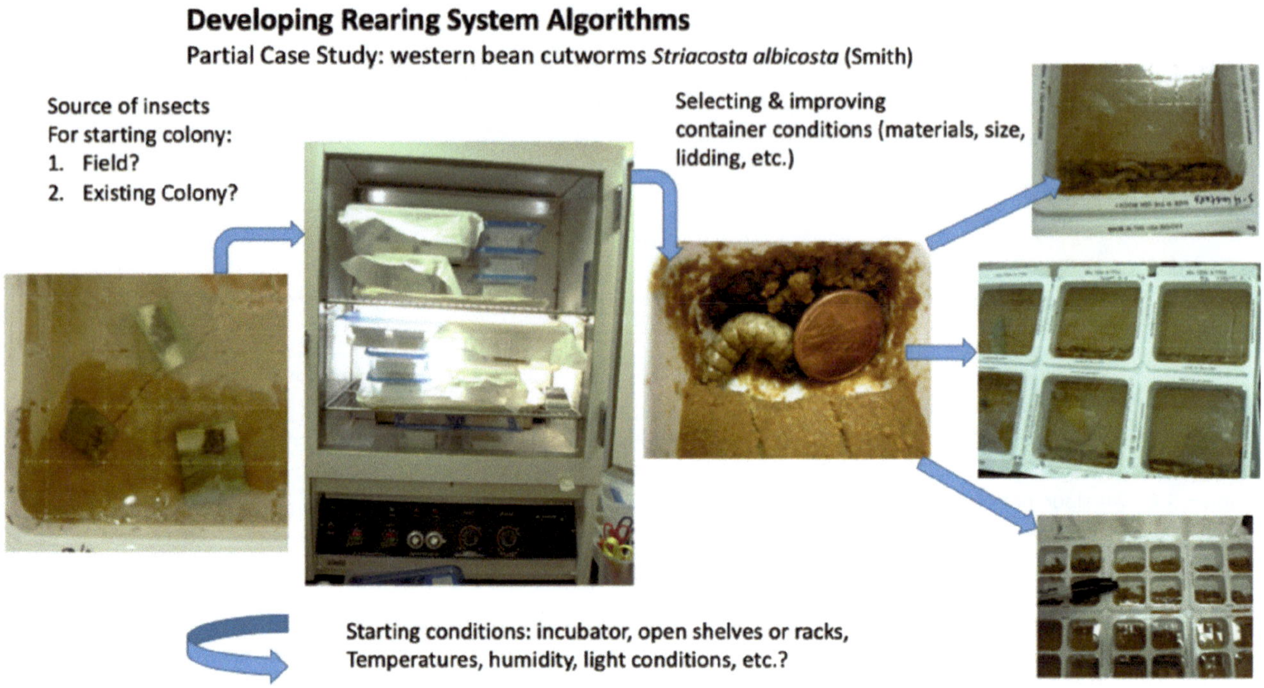

Figure 1.7 Photographic depiction of an incipient (developmental) rearing process for western bean cutworms.

of the breakfast-making at home question had arisen (and the decision was made to cook eggs by frying them), then we had to decide how high to turn the heat (assume the use of a gas stove with a continuously adjustable flame adjusting knob). There are many different levels of flame that we control, and the flame intensity is incrementally controlled (clearly an analog operation). To translate these everyday life concepts to rearing systems, Figures 1.7–1.9

Figure 1.8 Examples of major components of rearing systems diagrammed in Figures 1.4 and 1.5.

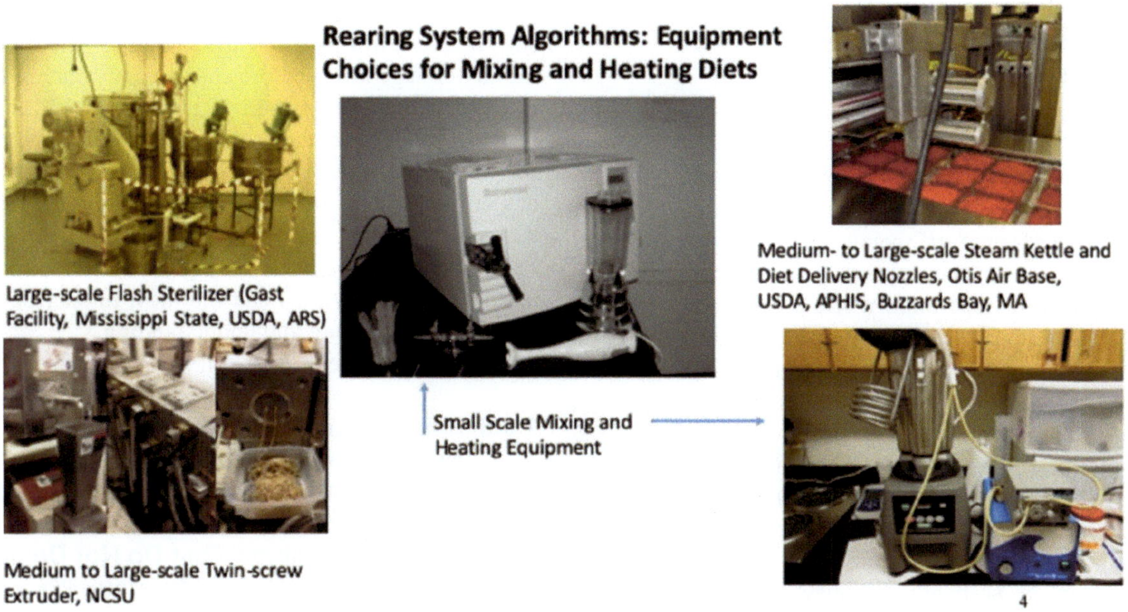

Figure 1.9 Examples of diet mixing and heating equipment for small-, medium-, and large-scale rearing systems.

provide examples of rearing system processes and components that become part of the algorithm framework.

1.3.5 Getting Back to Rearing and Diet Algorithms

Figures 1.8 and 1.9 illustrate several aspects of the algorithmic choices that rearing system developers must make. Figure 1.1 is a diagram of a highly simplified selection process where the questions of (1) where will the insects originate (field-collected or existing colonies)? (2) Which diets will we start with (natural food or artificial diets)? (3) Which environmental parameters will we use (which temperature, humidity, light conditions, etc.)? The diagram shows that the environmental parameters may be changed according to experimental results. If, for example, we started with 27°C and the development was too slow, we may change the temperature to 29°C; this change could be made while keeping all the other parameters the same, or multiple factors could be changed at once giving us a factorial design experiment (discussed in detail in Chapter 9).

1.3.6 Algorithms for Insect-Rearing Systems Are Best If They Are Science-Based

This raises an important question that haunts rearing professionals: is insect rearing an art or a science? Is it best for the advancement of insect rearing to treat it as an art or as a science?

1.3.7 Definitions of Art, Science, and Technology

1.3.7.1 Art vs. Science and Technology

1.3.7.1.1 Art

1. The quality, production, expression, or realm, according to aesthetic principles of what is beautiful, appealing, or of more than ordinary significance
2. The class of objects subject to **esthetic** criteria
3. A field, genre, or category of art
4. The fine arts collectively
5. Any field using the skills or techniques of art

1.3.7.1.2 Science

1. A branch of knowledge or study dealing with a body of facts or truths systematically arranged and showing the operation of general laws
2. **Systematic knowledge** of the physical or material world gained through **observation and experimentation**
3. Any of the branches of natural or physical science
4. Systemized knowledge in general
5. Knowledge as of facts or principles; **knowledge gained by systematic study**

1.3.7.1.3 Technology

1. a. The **practical application of knowledge**, especially in a particular area
   b. A capacity given by application of knowledge

2. **A manner of accomplishing a task, especially using technical processes, methods of knowledge**
3. The specialized aspects of a particular field

From Online Merriam Webster Dictionary

These definitions help make the case that insect rearing belongs in the realms of science and technology rather than art. The term "art" being applied to rearing probably comes from the concepts that rearing practices are a kind of black box operation where unexplained practices result in outcomes that we desire, but we do not care to understand. If you use a given mixture of ingredients to serve as a diet, and provide conditions which worked, in the past, to produce a species of insect, then no more need be understood. It seems that the intention of people who insist that rearing is an art is to credit rearing specialists with a special skill that was developed (learned) or was possibly innate (a "green thumb" of insect husbandry). The concept of husbandry is relevant to our discussion of rearing as a science and technology vs. art. Merriam Webster defines husbandry as follows:

1. **a:** The cultivation or production of plants or animals: agriculture
   b: The scientific control and management of a branch of farming and especially of domestic animals
2. The control or judicious use of resources: conservation

Note the definitions under (1) "cultivation or production" and "scientific control and management … especially of domestic animals." I am especially impressed with the features of these definitions that include the science of cultivation and production. This leads to a dissection of the terms, "science" and "art." These concepts are more fully explored in the following section.

Table 1.1 is compiled from statements that I have gleaned from extensive reading about the philosophical thinking about the two domains. In the comparison and contrast, I have tried to capture the most fundamental and pervasive concepts that help to clearly and cleanly differentiate the domains of science and art. I also hasten to add that I am not judging the intellectual value or importance of science, technology, and art in any kind of hierarchy. In fact, I began my academic life in the humanities, having a bachelor's and master's degree in English with two years of graduate work toward a PhD in that subject and about six years of teaching English at the high school and community college level (before changing my career direction into vertebrate biology). In fact, I have found that some of the most rigorous critical thinking was to be found in the humanities where philosophers, art critics, and literary critics display an equally clear and deep thinking as the bulk of science and technology scholars. I would also be hard put to argue convincingly which domain (art or science/technology) has had the most impact on shaping the world of thinking and understanding.

How do science and art compare?

Table 1.1 Comparing Science and Art

Science	Art
Mechanistic	Non-mechanistic
Must be logically consistent	Logic not necessary
Rational	Emotional
Repeatability essential	Repeatability NOT essential
Predictive value	No predictions
Seeks correction	Does not seek correction

1.3.8 More Discussion of Why Rearing Is Not (or Should Not Be Art). Science Seeks Meaning Through Understanding Mechanisms or Causes-and-Effects. Art Has Meanings That Do Not Demand Cause-and-Effect Understanding

It can be argued that the most salient feature of science is the quest for causality. In contrast, "meaning" in art (be it visual art, music, or literature) comes from factors that succeed in communicating a feeling or thought without recourse to knowing why the communication "works." Without wanting to be overly philosophical, I can argue that when artists (say poets, for example) use their medium to give an emotional or thought-provoking impression (i.e., to convey an artistic meaning to their world), they may succeed by using devices of which they are not even conscious, or that they use deliberately because they have a developed sense that words, rhythms, sounds, etc. evoke what the artist (poet) strives to evoke. When the artists use their craft to convey some kind of meaning (a thought about a subject or an emotion), they are not concerned with inquiry or understanding of how meaning is communicated. With science and technology, the concern is for how the meaning is related to the real world.

1.3.9 Science and Technology Demand Formally Logical Systems

Art does not. The logical system of science is data-driven empiricism. The logic of science includes building on previous knowledge and correcting the knowledge-base as more understanding and information become available. Art may build on previously existing systems such as movements, but it may be completely independent of logical systems.

1.3.10 Science Is by Its Very Core Rational

Merriam Webster's definition of rational is something that is in agreement with reason or logic. Art does not require reason or internal consistency. Often, art departs from reason or rationality to make its point. Art may have self-contradictions, paradoxes, or inconsistencies; in fact, much of what emerges as great art involves such disparities to make its point. In contrast, science struggles with inconsistencies and tries to resolve them with the understanding that inconsistencies probably mean that the science has not gone far enough to fill in gaps or to resolve incongruities.

1.3.11 Repeatability Is Essential to Science and Technology

It can be argued that a large part of the truth of science lies in the ability of other scientists to repeat what scientific inquiry has taught us. With artistic creations, uniqueness is a driving force and central factor. While artists may borrow from past works, using allusions to previous literature, musical themes or treatments of sound, or copying styles of previous visual artists, the greatness of a work such as *Hamlet* comes, in part, from its uniqueness as all components of this play interact in character development, the sounds of the language (cadence of the verse, rhymes, and other spoken poetic devices), and the interplay of themes that evoke the sense of tragedy that is special in this creation. Even translation of the Shakespearean (Elizabethan) English to modern English or some other language causes the work to lose some of its artistic uniqueness. In contrast, with science and technology, the ability to replicate a work that illustrates a principle fortifies the "truth" or universality of the principle. In fact, one of the great strengths of science is that the repeatability engenders the ability to **predict**.

1.3.12 Predictive Value Is Another Hallmark of Science

Once a scientific principle is established, it has predictive value, meaning that by properly using the knowledge of the principle scientists (or properly educated lay-people) can make predictions, including forming and using models that guide inquiries to predicted outcomes. Knowing that an antimicrobial agent (such as potassium sorbate) has properties that kill or inhibit, let's say, fungal (mold) growth, it becomes a predictable outcome that using potassium sorbate in established concentrations will inhibit fungal growth. If the range of antifungal activity is established as part of the inquiry, then we can expect that the outcome (fungal growth suppression) will occur. Other facets of the predictions would come to play if, for example, the humidity was exceptionally high or the fungal load of spores was too high. These factors might constitute an exception or unexpected outcome of mold growth despite inclusion of an antifungal agent. Other intervening variables could include improper pH or a defective batch of potassium sorbate. Therefore, the power of the predictive ability is enhanced or depressed according to how well core or satellite factors are obeyed. In accord with this generalization about prediction is an all-important aspect of science-driven rearing procedures, that is, the ability to form a **correcting** feature of the rearing algorithm.

1.3.13 Self-Correction Is the Feature of Science That Pursues the Concept of Continuous Improvement

It is axiomatic that all levels of science are subject to (in fact **demand**) self-correction. Such self-correction is ever-pursued by competent rearing personnel. Throughout rearing processes, outcomes crop up that were contrary to expectations or were otherwise in need of change. Skillful rearing personnel notice departures from standards or desired outcomes, and they seek the sources (causes) of the deviations and then seek to correct them. The example above where mold cropped up in a rearing system that had not previously suffered from mold problems should prompt the rearing personnel to determine the likely causes (excess moisture, high microbial load, pH problems, or defective antifungal agents, for example). Then, the personnel must correct the problem. The process of self-correction is so universal in rearing systems that it becomes a routine factor where there is a continual evaluation of the data or outcomes in various steps in the rearing process. Self-correction is fueled by the feedback system that is part of day-to-day living. One possible problem with self-correction, however, is that the correction process must be **documented** and **communicated**. Responsible rearing personnel should never change things that were part of the established SOPs (protocols) without properly consulting with supervisors or without establishing clear records of the change. It is also a cannon of responsible self-correction of established rearing processes that any changes being considered are **tested** in a controlled comparison, **not** simply made in an across the board draconian modification. The extent, complexity, and duration of the testing are matters of decision-making that should involve all responsible personnel in the rearing system's chain of command.

1.4 CASE STUDIES: DEVELOPMENT OF MASS-REARING SYSTEMS FOR FOUR SPECIES OF PIERCING AND SUCKING INSECTS

1.4.1 Introduction to Case Studies of Developing Mass-Rearing Systems for Two Species of Predators and Two Species of Plant Bugs

In organizing an explanation of the process of developing rearing system algorithms, I felt that the most effective and authoritative narrative would be drawn from my own experiences. The case studies here are drawn from my first two decades of research in diets and rearing systems for insects with piercing and sucking mouthparts. These case studies chronicle development of four very successful rearing programs in mass-rearing for the predators *Geocoris punctipes* and *Chrysoperla rufilabris* and for the plant bugs *Lygus hesperus* and *L. lineolaris*. The rearing systems treated here have following points in common: (1) all four species utilize extra-oral digestion (EOD) as their mode of feeding; (2) successful mass-rearing systems were developed for each of the four species; and (3) each case involved an insect for which the development of an artificial diet and a diet-centered rearing system had proved difficult for previous researchers. All four successful diets had in common the characteristics that they were suitable to accommodate the feeding mechanisms (EOD) in following three major ways: (a) the diets were semisolid with a consistency that mimics the internal structures of prey (in the case of the predators) or plant materials (in the case of the plant bugs); (b) the nutrient composition was very concentrated rather than dilute as would characterize simple liquids such as insect hemolymph or plant saps (xylem sap, phloem sap, or nectar); and (c) the semisolid diets could be presented or packaged in ways that lent themselves to leakage-free arrangements and were subject to automation in diet-packet filling. One more preliminary point deserves explanation: in all four insects' rearing systems, once the diets were developed, the non-diet based components needed to be worked out. Simply having a working diet was not sufficient to make the transition from small-scale rearing to mass-rearing. In all four cases, many of the rearing system requirements had already been worked out and put in place when the right diet was invented. Although, thanks to the excellent and insightful work of Dr. Jack Debolt, the diet and diet presentation system for *L. hesperus* (the western tarnished plant bug) was already well on its way to being incorporated into a mass-rearing system when I arrived at the Tucson, USDA, ARS Biological Control of Insects Laboratory, I will begin this narrative with my journey of discovery with the big-eyed bug, *G. punctipes*.

1.4.2 Background for the Case Study on *Geocoris punctipes* Artificial Diet-Based Rearing System

The narrative begins with a story of initial failure. For the first four years that I tried to develop a liquid artificial diet for predatory Hemiptera, my efforts failed. The failure came largely from my and the scientific community's lack of understanding of the feeding biology of big-eyed bugs, lacewings, and many other insects with piercing/sucking mouthparts. In my first job with the USDA, ARS, I was hired to develop an artificial diet for native predators of the native pests, *Lygus hesperus* Knight (western tarnished plant bugs; Hemiptera: Miridae). As is often the case, my choices were somewhat narrow thanks to the demands of the job, to use an artificial diet to rear native predators of the plant bugs—predators that occurred in the western United States and that were documented to consume *L. hesperus*. My choices had to come from native lygaeids, reduviids, nabids, and pentatomids (and later lacewings = chrysopids). My colleagues at the Biological Control of Insects Laboratory in Tucson, AZ, had already experienced rearing of all the common species that were Lygus predators, and they suggested that the big-eyed bug, *Geocoris punctipes* (Hemiptera: Lygaeidae), might be the easiest and most convenient.

The state of knowledge when I started my research in 1979 was that because big-eyed bugs and species with similar feeding habits had "liquid feeding" mouthparts (=piercing and sucking mouthparts), they must be hemolymph feeders (see figure *Geocoris* in DeBach). Subscribing to this concept for the first three years of my research (1979–1982), I kept trying to model diets after hemolymph of prey—a process that I approached with biochemical analysis of prey hemolymph. I documented the results of my biochemical analyses of prey hemolymph in several publications (Cohen 1981, 1982, 1983a; Cohen and Patana 1982). I had succeeded in developing a liquid diet that supported development of a single generation of *G. punctipes* with no reproduction (Cohen 1981). The diet trials were frustrating, with high mortality and high rates of deformities such as malformed wings and other abnormal cuticular structures. The series of liquid diets described in Cohen 1981 and 1983a were clearly not adequate to support the required goal of producing successive generations of fit predators, but because this represented more progress than had ever been made with hemipteran predators, I went forward with publication efforts. The work with liquid diets also included novel feeding systems to deliver liquid diets for predatory arthropods, and it also included improvements of liquid presentation systems derived from liquid feeding systems developed by the late Professor Ken Hagen. The far-reaching implications of the relationships that I discuss here are to be found in many subsequent research studies, including a dramatic paradigm shift of the true target of Varroa mite feeding being fat body, rather than hemolymph of honeybees (Ramsey et al. 2019).

1.5 HOW DO WE DEVELOP SCIENTIFICALLY-BASED REARING SYSTEM ALGORITHMS?

1.5.1 Case Study: Developing Scientifically-Based Algorithms of *Geocoris punctipes* Rearing System

Figures 1.10–1.17 and the "*Geocoris punctipes* Algorithm" provide simple starting places for formal development of rearing system algorithms. Like the above

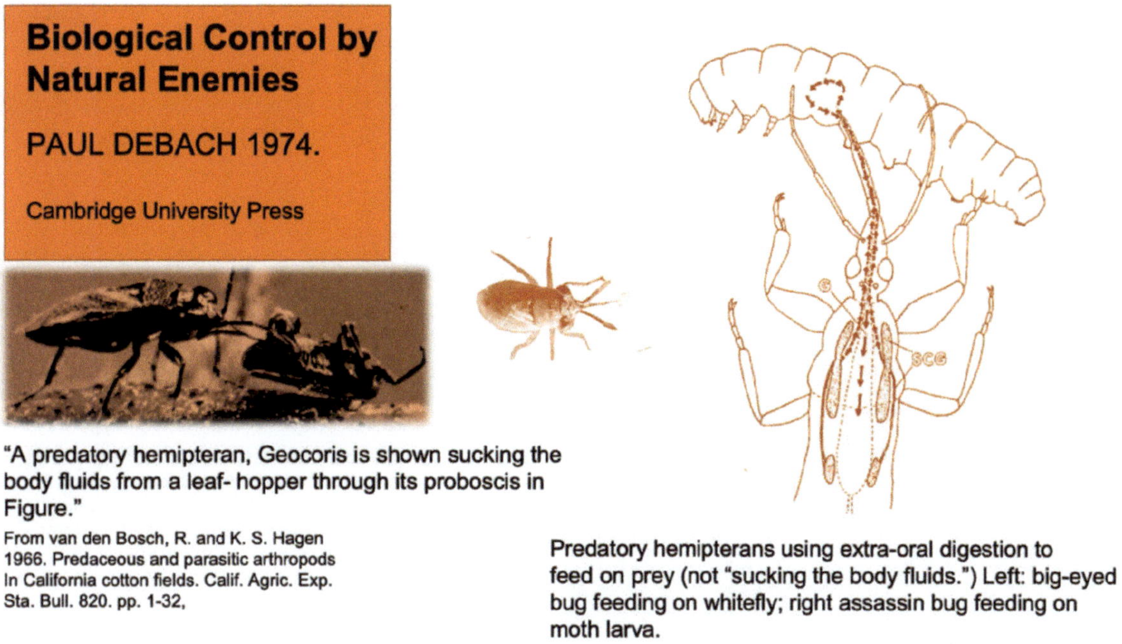

Figure 1.10 Geocoris in DeBach. The classic text in biological control of insects by Paul DeBach reflects the state of knowledge of feeding mechanisms by big-eyed bugs and other predators in 1964. DeBach repeats the standing wisdom of the time, writing that *Geocoris* was shown sucking body fluids from its prey (a leafhopper). The thinking was typical of the early era of biological control where the standing paradigm about other predaceous Hemiptera, Neuroptera, and Coleoptera was that they feed on prey hemolymph. On the right side of this figure, the true nature of feeding by predatory Hemiptera (and many other kinds of predatory arthropods) is shown to be extra-oral digestion (Cohen 1995, 1998).

Figure 1.11 *Agricultural Research* (June 1997) feature story showing the development of predator diets through the "eclectic approach" of understanding feeding dynamics and mechanisms, knowing nutritional contents of prey, and using components of the feeding niche to develop and optimize artificial diets and other features of predators' biology to develop mass-rearing systems.

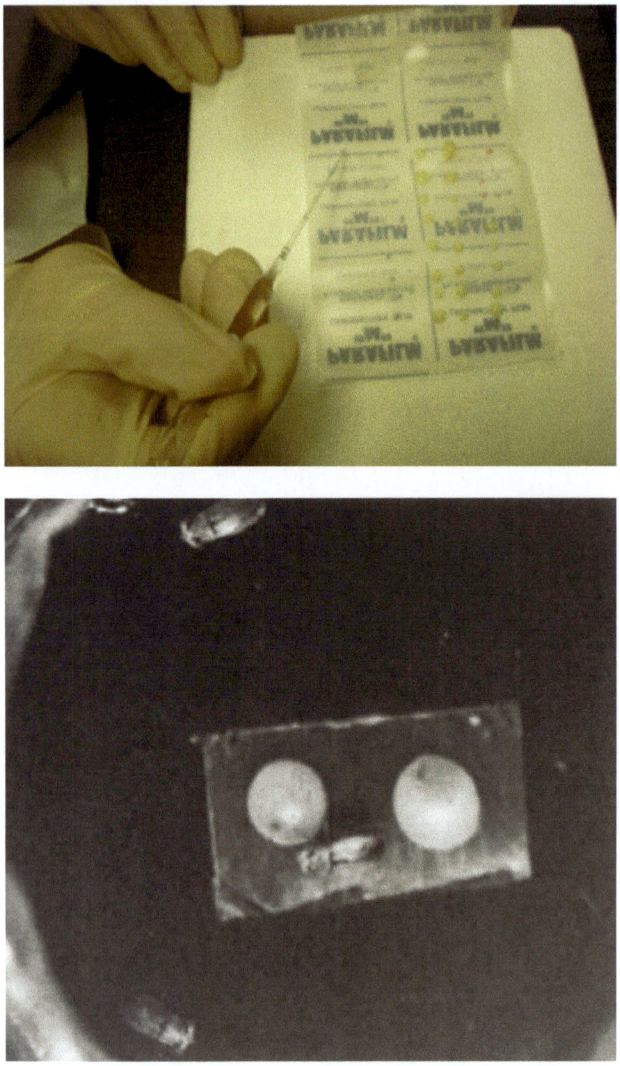

Figure 1.12 **(a) and (b)** Diet delivery system for the liquid diet paradigm. The encapsulation system shown in the figure involves a procedure developed by Cohen 1983a (derived from a previously described encapsulation system developed by Hagen and Tassan 1965). Figure (a) shows wax-coated liquid diet droplets being formed on Parafilm®; and figure (b) shows the Reacti-Therm controlled heating block where diet and wax mixture is prepared for delivering wax-coated liquid diet.

Figure 1.13 *Geocoris punctipes* consuming a cotton aphid on a cotton leaf (picture taken by Jack Dykinga at the USDA, ARS Western Cotton Laboratory, Phoenix, AZ). This big-eyed bug, which is a generalist predator, was reared for hundreds of generations on the artificial diet and in the rearing system developed by the author (A. C. Cohen). The rearing system for *G. punctipes* was adopted by the USDA, APHIS by Dr. Robert T. Staten and mass-reared for about ten years for use as a biological control agent for cotton whiteflies (*Bemisia tabaci*; Hemiptera: Aaleurodidae). An important feature of this image is the manner of feeding seen here where stylets (mouthparts consisting of mandibular stylets and maxillary stylets forming a salivary canal and a food canal) are inserted into the prey (or other food source such as plant tissues). The discovery that this feeding with piercing/sucking mouthparts is not a true liquid feeding, but rather an extra-oral digestion (EOD) led to the development of a semisolid diet consisting of complex foods (beef liver, ground beef, cooked chicken eggs, etc.), which are very concentrated with nutrient-rich components. This discovery became a paradigm shift from hemolymph feeding to semisolid tissue feeding, and that shift became the transition point from nutrient-dilute materials limited to components that could be dissolved in water to lipids and proteins that were closer to the natural food of big-eyed bugs (and later, lacewings and other predators) than the liquid diets. The discovery of EOD in these predators also enabled the research group to depart from the problems and drawbacks of trying to present liquid diets in some kind of encapsulated or coated form of liquid (Figure 1.14). The presentation of the diet in stretched Parafilm® packets (Figures 1.15–1.17) allowed the diet to be presented as a greater biomass and in a semisolid form that lent itself to large-scale rearing. In fact, to this day, it seems that the only commercial production of predators on an artificial diet was accomplished with derivations of the diet (what Cohen and Urias 1986 called a meat-based diet) and the presentation system discussed here.

example about breakfast algorithms (do I eat or do I not eat?), the question of whether or not to rear an insect may have many possible purposes: "I'm a student who wants to do a science fair project for 4H;" "I'm a researcher who wants to study a metabolic pathway in some kind of conveniently available animal;" "I'm an entrepreneur who wants to colonize live food for feeding pets, etc." This raises the question (once it was decided to rear a population of insects), which insect species will I rear? If I am trying to rear a biological control agent to manage a given pest, the choices of rearing subjects are narrowed by the question of which predators or parasitoids attack the pest in question.

1.5.1.1 *Rearing Components in a* Geocoris punctipes *Rearing Algorithm*

1. Containers
   a. Cardboard ice cream cartons
   b. Plastic storage boxes
   c. Air exchange (organdy/organza cloth)

Figure 1.14 Mass-rearing system for *Geocoris punctipes* Say at USDA, APHIS Methods Development, Phoenix, AZ. A.C. Cohen (left) and R. T. Staten (right) inspect the rearing containers to determine age strata of the insects.

2. Environmental components
   a. Temperature
   b. Humidity
   c. Light (photoperiod, intensity, quality)
3. Diet
   a. Artificial (meat-based paste)
   b. Watering
4. Diet presentation
   a. Parafilm® packets
   b. Wax coating
   c. Frequency of food and water changes/additions
5. *Geocoris punctipes* population

Figure 1.15 The Vac-U-Form Machine. The thermoforming machine was used with wooden molds as templates for the containers seen in Figure 1.11. Note the cylindrical top on both template blocks that was designed to receive cardboard ice cream carton lids. The plastic cage top was cut away to allow access to the inside of the containers for technicians to change diets, access insects, etc.

6. Oviposition
   a. Site (flannel cloth)
   b. Harvesting
   c. Inoculation
7. Successive generations
   a. Selection, population size, and other Hardy-Weinberg factors
   b. Domestication

In this example, the rearing system development for the predatory insect known as the big-eyed bug (*Geocoris punctipes* Say Geocoridae: Hemiptera) is discussed. Later in the quest for diet-development for predators, the Geocoris feeding system, with a few modifications, was expanded to include green lacewings (*Chrysoperla rufilabris* Burmeister Chrysopidae: Neuroptera) showing how one rearing system's protocols can lead to another system (in what Cohen 2019 calls the "confluence of established rearing information.")

1.5.1.2 Making Choices: Which Insect to Rear

My career with the USDA, ARS began in 1979 when as a postdoctoral research associate in a program (United States Department of Agriculture, Agricultural Research Service USDA, ARS Biological Control of Insect Laboratory) where efforts were being made to develop augmentative mass-rearing of predators and parasites of cotton and alfalfa pests. The major predators in these systems were known to be hemipterans including big-eyed bugs, minute pirate bugs, assassin bugs, nabid bugs, and stink bugs. The most commonly found species among these were the big-eyed bugs (family Geocoridae); and this factor (large populations that were associated with the target pests) led to our choice of targeting the big-eyed bug, *Geocoris punctipes*. A further advantage in selecting G. punctipes as our primary target for diet-development is that our laboratory staff had already developed rearing conditions for this predators using beet armyworms as prey with green bean supplements; so the transition to artificial diet-based rearing was seemingly a simple step.

1.5.1.3 Geocoris Rearing Algorithm

Once the species of predator was decided, the rearing system had to be developed and optimized. To my advantage, the laboratory that I had joined when I entered the world of insect rearing was an established rearing facility with four scientists (Ray Patana, Jack Debolt, Glenn Jackson, and Harry Graham) who had collectively more than 100 years of rearing experience, and with their four technicians, there was 50 more years of rearing knowledge). The team was already rearing big-eyed bugs on "natural" diets (green beans and heat-killed [coddled] beet armyworms); so, the basic rearing algorithm was

Figure 1.16 (a) and (b) Lacewing oviposition units from Ridgway et al. 1970. Figure (a) shows the adult cage made of fiberglass (though various insectaries use cardboard ice cream cartons for adult holding and oviposition units). The cotton wick shown in (a) provides water at the bottom of the cage, and the walls of the cage are streaked as a paste made with a blend of Food Wheast (Knudsen Dairy, Los Angeles, CA) and sugar. Currently, Wheast is not available as the original fermentation product described by Hagen and Tassan 1965 as sweet dairy whey cultured with *Saccharomyces fragilis*, a dairy yeast that is known to be able to hydrolyze lactose sugar (Cohen 2015). Therefore, in more current rearing of lacewings, the adult diets usually consist of mixtures of sugar or honey and brewer's yeast (Cohen and Smith 1998). Eggs, such as those seen in (b), are harvested from either brown paper or organdy placed on top of the cage. Various systems of egg harvest have been devised, and range from scraping the eggs with an abrasive pad or chemical removal with a solution of about 1% bleach. The harvested lacewing eggs are distributed into the Verticel (or Hexcel) compartments whose top and bottom openings are covered with organdy (Figure 1.18) to begin the next generation. The cells, especially those made of Verticel, are excellent sites for pupation where the late 3rd instar larvae spins a silk cocoon and remains in the cell until it is ready to undergo adult eclosion. In anticipation of the adult eclosion, the organdy layers are removed from the Verticel, and the rearing unit is placed in an adult emergence container where it can find food, water, and go through the mating process that starts the next generation.

already in place. We had a suitable environmental regimen, which was derived from previous studies in our laboratory. The rearing algorithm was developed by (1) **adopting containers** already being used in the laboratory (cardboard ice cream cartons with organza cloth replacing under the lid which was partially cut out to allow air exchange); (2) **environmental factors** (27°C, ± 1.5; RH 60% ± 10% 14:10 light: dark provided with overhead fluorescent lighting); (3) **a diet** of green beans and coddled (heat-killed beet armyworms), later to be replaced by artificial diet; (4) **diet was presented** first by dropping larvae and washed green beans into rearing containers, then placing diet in wax-coated capsule (initially), later Parafilm-wrapped artificial, meat-based diet; (5) **the big-eyed bugs** were collected in local alfalfa fields with sweep nets and vacuum sweeping (DeVac), and once the colony was started and conditions worked out to provide growth, development, and successive generations, the **laboratory-adapted** colony was used for most of the research with occasional recourse to collecting field-fresh individuals; (6) **oviposition** was initially handled by allowing the insects to oviposit on green beans and inner surfaces of the rearing container (most eggs being attached to the green beans); however, we discovered that female *G. punctipes* would select, preferentially, the fuzzy side of flannel cloth squares for oviposition sites; and (7) **successive generations** were produced following these protocols, thus allowing a moderately stable colony to be established. Each generation took about 4–5 weeks from egg to egg-laying adults, allowing us to produce about 10 generations per year. As part of the initial phases of these studies, we kept records of the development time and weights (biomass) of individuals from each instar. This

Figure 1.17 Lacewing feeding on the Cohen and Smith 1998 diet. Diet is packaged in Parafilm® that is stretched to enhance penetration capability by lacewing larvae. See Figure 1.1 for description of the presentation method.

information was used as a baseline for comparison of artificial diet-rearing outcomes with the predators reared on "natural" foods.

1.5.1.4 Domestication and Fitness

We realized early on that of about 500 adult or late nymphal *G. punctipes* that we would collect from the field and start in our cultures, only about 100–300 would survive long enough to mate and produce progeny. The losses were attributed to cannibalism, starvation, and damage accrued during collecting and transport (the field collection site was about 30 km from the laboratory, and even with air conditioning, the Tucson, AZ, heat took its toll on the insects). The high mortality of freshly collected insects and the fairly high death rate of 1st and 2nd generation captive big-eyed bugs served as a kind of bottleneck or "winnowing effect" described by Bartlett (1984). Once collection damage (including transport) has been eliminated, it seems likely that the differential survival of insects that have had a chance to acclimate to the laboratory conditions is a type of selection. This selection is evident in general with insects that are subjected to laboratory constraints, and because it probably involves some genetic selection (and possibly epigenetic selection), inadvertent selection in early phases of insect colonization should be considered as early stages of domestication. Alan Bartlett discussed this in his writing and especially in rearing workshops that he taught with me. He made a very important point that most of the selection, genetic truncation, and domestication that occur in a laboratory population of insects are phenomena that happen early in the colonization process. In the case of the big-eyed bug colonies, the insects that we worked with to develop diets and other rearing conditions had been colonized for about 3.5 years (about 30+ generations) before we were able to develop an artificial diet that would support their complete development. Therefore, the diets that we developed were accomplished with fairly well-domesticated insects.

1.5.1.5 Understanding Feeding Dynamics of Geocoris punctipes and Other Insects: The Scientific Basis of the Rearing Advancements

During the first 15 years of my work with artificial diets for predators, I had become growingly aware of the complexities of diets and their rearing system context. These complexities led me to study such diverse topics as EOD, digestive enzyme dynamics, predator feeding habits, nutritional ecology, physiological ecology in rearing systems, and many other topics, not the least of which was the complex dynamics in diets. It was my discovery that *G. punctipes* and many other predator species depended largely on EOD that led to my success with a diet for big-eyed bugs, lacewings, and several other predators (e.g., Cohen 1985a, 1995; Cohen and Smith 1998). Throughout the early part of my career (the first 15–20 years), I continued to study the basic science and the applications toward diets, and it continued that the basic science papers were readily published (e.g., Cohen 1984, 1989, 1993, 1995). As it turns out, the discovery that big-eyed bugs and many other predators rely on EOD to handle their food reaches far beyond predators and includes insects and other arthropods from a wide variety of feeding guilds. Too often, interpretations of feeding by insects have been a simplistic model of either biting and swallowing food that is digested within the digestive system, or that insects insert their mouthparts like drinking straws and merely suck out the liquids that are in the food sources. This was the model that we held for plant bugs such as *Lygus hesperus* and *L. lineolaris*. I will discuss how this overly simplistic model failed us initially in developing artificial diet for these mirids and for other insects such as pentatomids.

1.5.1.6 Algorithm for Upscaling the Scale of the Geocoris Rearing System

The rearing system developed by Cohen 1985a was adopted by USDA, APHIS Methods Development under the direction of Dr. R. T. Staten to mass-rear native generalist predators (*G. punctipes*) for use as biological control agents to suppress exotic pests such as the tobacco whitefly (*Bemisa tabaci*). This rearing algorithm was the modification of the small/medium-scale rearing system for *G. punctipes* to a larger scale system. A part of the upscaled mass-rearing algorithm was to optimize the containers to facilitate labor involvement. The container development algorithm was (1) to make larger containers than those used in the original system (Cohen 1985a) by using a plastic thermoforming machine and 2) to include a lidding system that allowed quick, simplified access to the container interiors (for adding water, changing diet packets, and removing oviposition cloths). The original containers were 0.5 and 4.0 L cardboard cartons and 33 × 20 × 13 cm plastic storage boxes (6.0 L). The thermoformed boxes were designed to approximate the dimensions of the storage boxes, but with a modification of the top to fit the lid of a 4 L cardboard ice cream carton to fit snugly (Figure 1.10). The design of the containers was a crucial factor in the upscaling of the rearing system. The process of upscaling is an excellent example of a rearing system step-by-step algorithm. Here are the steps: (1) Because the plastic storage box used in the original small-scale system in the USDA, ARS laboratory in Tucson was found to be adequate for various aspects of *Geocoris* rearing, the approximate configuration was used, but it had to be modified, especially for easy-opening lidding. This gave rise to the use of cage-making with a plastic-shaping thermoforming system. (2) The lidding needed to be easy to open and close because the *Geocoris* needed a change of diet every 2–3 days (Cohen 1985a). Having to lift off a large plastic lid allowed too much opening leading to escape of individuals. (3) Also, the removable top of the original storage containers allows escape, so a better-sealed system was needed—with the cardboard bottom glued to the thermoformed top proved to be a reusable container that could be made cheaply and replaced every year or so as the container deteriorated with time. The need to have adequate air exchange was met by cutting an opening of the cardboard lids, and using organza cloth, which gave a completely escape-proof closure, yet allowed technicians to reach into the cage to change the diet packets and water supply. (4) The food was the identical diet with that described by Cohen 1985a, but instead of using a watering system with wetted cotton in a Petri dish, the Phoenix-APHIS modification reverted to using green beans as a source of water. Though it was demonstrated (Cohen and Debolt 1983) that *G. punctipes* did not require plant materials to supplement the Cohen diet, the APHIS Methods Development personnel determined that it was more economically feasible to use green beans, rather than the labor-intense watering system in the original Cohen 1985a paper. Under the conditions at the Phoenix-APHIS laboratory, the green beans were found to retain acceptable freshness for the entire 10-day holding period for adults. We had begun to develop a drip-irrigation watering system to automate the rearing process that was a weak point, but before we could do this on a large scale, the rearing program was terminated. (5) The oviposition system developed by Cohen 1985a was continued in the large-scale system, where pieces of flannel cloth (5 × 5 cm flannel squares) were placed in the adult cages and harvested every two days along with the food changes.

1.5.1.7 *Extension of the* Geocoris punctipes *Rearing System to* Chrysoperla rufilabris *Mass Rearing*

Ridgway et al. 1970 and Morrison et al. 1975 developed methods of rearing green lacewings on the eggs of factitious prey (insects such as *Ephestia kuehniella* [the Mediterranean flour moth] or *Sitotroga cerealella* [Angoumois grain moth] that are not the natural hosts of the predators. The adult handling methods included placing pre-emergent pupae in 4–12 L cardboard ice cream cartons with a mixture of Food Wheast ® or brewer's yeast and sucrose with a cotton wick with water at the bottom of the container. Morrison et al. 1975 used a resinous material called Hexcel, and later Ridgway et al. 1970 used a cardboard-based modification of Hexcel called Verticel as rearing containers whose cellular character served as compartments to separate the highly cannibalistic lacewing larvae. The ceilings and floors of the larval containers were covered with hot-glued organza, which allowed

Table 1.2 Diets for Big-Eyed Bugs (Cohen 1985a) and Lacewings (Cohen and Smith 1998)

Component	Big-Eyed Bug Diet (g)	Lacewing Diet (g)
Beef liver	100	100
Ground beef (minimum 17% fat)	100	100
Hen's eggs (whole)	0	100
Water	10	45
Sucrose	2	15
Honey (25% in 75% water)	0	25
Brewer's yeast	0	10
Acetic acid (10%: 90% water)	0	5
Sodium propionate	0	0.6
Potassium sorbate	0	06
Streptomycin sulfate	0	0.1
Chlortetracycline	0	0.1

larvae to feed through openings in the organza, where they could reach *Ephestia* or *Sitotroga* eggs (Figures 1.16 and 1.17). The novel diet developed for lacewing larvae by Cohen and Smith 1998 and Cohen 1998 (Table 1.2) was placed in Parafilm® containers where the film had been stretched to about 1/4 to 1/8 of its original thickness, allowing the lacewings to reach their sickle-shaped mouthparts through the organza and then through the Parafilm® to access the diet.

The discovery that big-eyed bugs and many other predators made extensive use of extra-oral digestion (EOD) led to a new paradigm (Cohen and Smith 1998) which suggested that if the natural feeding biology of the predatory insects was a kind of "solid-to-liquid feeding," and if the portions of the prey that were the major targets of pre-ingestion digestion and extraction of a slurry consistency from the semisolid internal structures, and if the feeding targets were very concentrated with macromolecules (proteins, complex lipids, complex carbohydrates such as glycogen, and nucleic acids), then making a diet that simulated the internal structures of prey in terms of nutrient profiles, concentrations, and consistency (texture) would be a far more suitable diet than strictly liquid diets that were designed to imitate the dilute hemolymph of prey. This EOD-driven model became the rationale for the new diets and the hypothesis emerging from these observations and this reasoning provided the experimental basis for testing various complex, high-nutrient solid and semisolid foods. These foods included various meat products (internal organs such as liver, ground muscle/fat tissue such as ground beef and chicken eggs).

The simple diet described by Cohen 1985a represented a major breakthrough in artificial diet development for predators. This raises an important issue regarding the use of chicken eggs. Cohen and Smith 1998 and Cohen 2015 discuss at length the efficacy of chicken eggs for predator nutrition, especially for lacewing larvae. Briefly, the high nutrient concentration of egg yolks make them especially appropriate food for predators

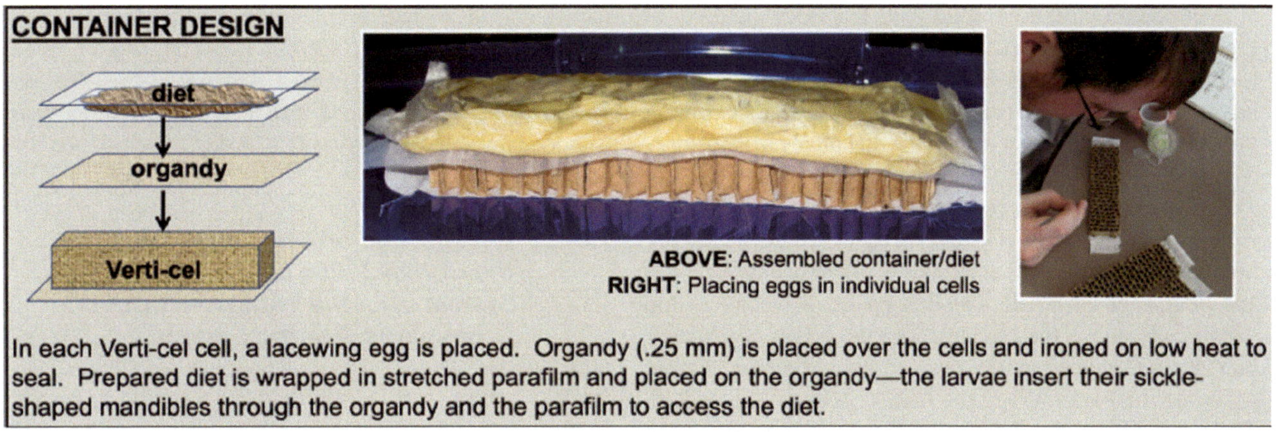

Figure 1.18 Modified Cohen and Smith 1998 diet being presented to lacewing larvae in Verticel cardboard rearing unit. This photo and diagrams are from a class poster (see Figure 1.12) by K. L. Oten, M. J. Gardner, J. Hanley, and A. C. Cohen.

whose requirements for high protein concentrations, abundant sterols, fatty acids, and complete vitamin and mineral concentrations are well documented. Chicken egg yolks have been tried by several researchers as diet components for several entomophagous insects, including Hagen and Tassan 1965 and Cohen 1981. In most of these earlier cases of incorporation of chicken eggs, the standing paradigm was that these insects were liquid feeders that required all their nutrients in a liquid form to accommodate the insects' piercing and sucking mouthparts. In presenting the yolk components as rather dilute liquids, though they showed some degree of efficacy, they never proved to be the "breakthrough" food component the researchers hoped to find. Also, Cohen and Smith 1998 noted that in many earlier efforts by Cohen to incorporate raw eggs and egg yolks into predator diets, there were extensive problems with microbial contamination associated with the rich nutritional composition of the egg yolks. However, a key process that Cohen and Smith introduced into the diet production for lacewings was to cook the eggs to nearly 90°C to get them to form a stringy semisolid material that when mixed hot with the ground beef and ground liver, induced a pasteurization of the meat/egg mixture, killing most of the vegetative bacterial and fungal contaminants. Thus, using properly cooked egg provided: (1) the nutritional base, (2) the texture, and (3) the microbial destruction that made the diet meet the requirements of nutritional value, palatability, and stability. As assurance against the early onset of microbial growth, the Cohen and Smith 1998 diet was further protected with acetic acid (which has antimicrobial characteristics and lowers the pH to levels incompatible with common microbial contaminants). However, the fact remains that no matter how nutritional, palatable, bioavailable, and stable a diet may be, it is of little use unless the diet presentation or delivery system is suitable. Therefore, diet packaging systems became a central focus of the research on predator mass-rearing as illustrated by the Parafilm-stretching device designed in the predator mass-rearing laboratory (Figures 1.17–1.19)

1.5.1.8 Scaling Up the Lacewing Diet for Mass Production

The change in paradigm regarding the feeding strategy (or feeding mechanism) employing EOD by a huge number of predatory arthropods (discussed extensively and reviewed by Cohen 1995 and 1998) made possible the development of a semisolid diet that met the consistency and concentration issues that had hindered successful development of practical diets. The change in concept from a liquid diet that simulated hemolymph to one that simulated the soft internal tissues of insects, including fat body, reproductive structures, muscle, and egg nutrients posed a new problem: delivery

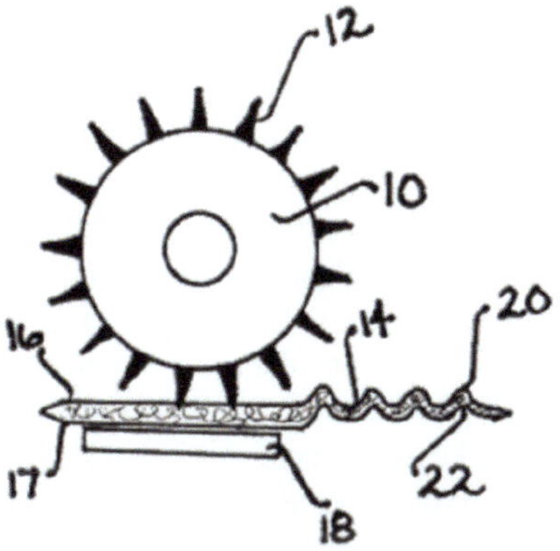

Figure 1.19 Arthropod diet delivery system. Figure 1.18 is an illustration of a diet packet dimpling system for making diet-filled protrusions in a flat, diet-filled Parafilm® sandwich (Cohen et al. 2003).

or presentation of the semisolid diet. The lacewing larva in Figure 1.16 is seen using its sickle-shaped paired mouth parts to penetrate the Parafilm® package to access the diet. The larvae pump digestive secretions into the packet, beginning the digestive process that makes the diet an ingestible slurry of partially digested (and soluble) proteins combined with the other nutrients. The rearing system seen in Figure 1.17 includes the cardboard cells (Verticel) where the larvae are housed individually and enclosed by organdy (organza) top and bottom of each cell. The diet packet is placed on top of the cells above the organdy, and the larvae reach their paired sickle-shaped mouth parts through the organdy and penetrate the Parafilm® in the manner seen in Figure 1.16. The reach of the small 1st instar larvae is limited, so the Parafilm® must be stretched to about 1/4 the original thickness (which is 125 µm), and the packet must be pressed tightly against the feeding unit to prevent gaps that would prevent larvae from reaching an adequate amount of food. The device pictured in Figure 1.18 was devised to automate and standardize the stretching of the Parafilm®, which otherwise must be stretched by hand, a process that reduced the efficiency of the diet delivery.

1.5.2 Case Study II: Development of a Mass-Rearing System for Tarnished and Western Tarnished Plant Bugs

The saltatory progress came from my detailed studies of feeding biology of various predatory Hemiptera and work with Dr. Jack Debolt who had already succeeded in developing an artificial diet for *Lygus hesperus* Knight (Hemiptera: Miridae), which was a remarkable accomplishment. Although all prior work on Lygus species was conducted under the assumption that plant bugs (Miridae) fed on liquids from the host plants' leaves, stems, and fruits, Dr. Debolt prepared semisolid diets that were composed of components such as lima bean meal, whole (raw) poultry eggs, and other largely non-water-soluble materials. Cohen 1985a, 1989, 1990, 1995, and 1998 showed that the original standing model of Hemiptera feeding was incorrect. The predatory Hemiptera and many (most) of the plant-eating Hemiptera could not derive adequate nutrition entirely on liquid (hemolymph or plant "juice") ingestion, but they needed more concentrated (solid or semisolid) food. In fact, all the hemipteran predators clearly were using EOD. They were using potent digestive enzymes derived from their salivary glands and were delivered to the interior of the prey via the predators' salivary duct that extended through their interlocking stylets. The saliva was mixed with the muscles, fat body, reproductive structures, and other internal solids, and the action of the saliva and the mandibular stylets (armed with impressive dentition) chemically and mechanically macerated the internal structures of the prey and was rendered into a fluid slurry that was then ingested as a concentrated food—through the food canal in the stylet bundle. All these facts of biochemistry, morphology, and behavior had to be grasped for me to finally see that these predators needed a semisolid or solid food. Once I understood this, the ideas flowed about sources of food such as ground beef, liver, beef, or pork internal organs as potential diet materials for these and other predators.

1.5.2.1 Background on the Derivation of the Cohen 2000c Lygus Rearing System from the Patana and Debolt L. hesperus Rearing System

When I began working on diets for predators in the early years of my career, Dr. Jack Debolt had already succeeded in developing an artificial diet (Debolt 1982) for the western tarnished plant bug, *Lygus hesperus* Knight (Hemiptera: Miridae). At the same time, Mr. Ray Patana, who was also with our research group, had developed a diet presentation and oviposition system (Patana 1982; Figure 1.20) for enhancing the scale of

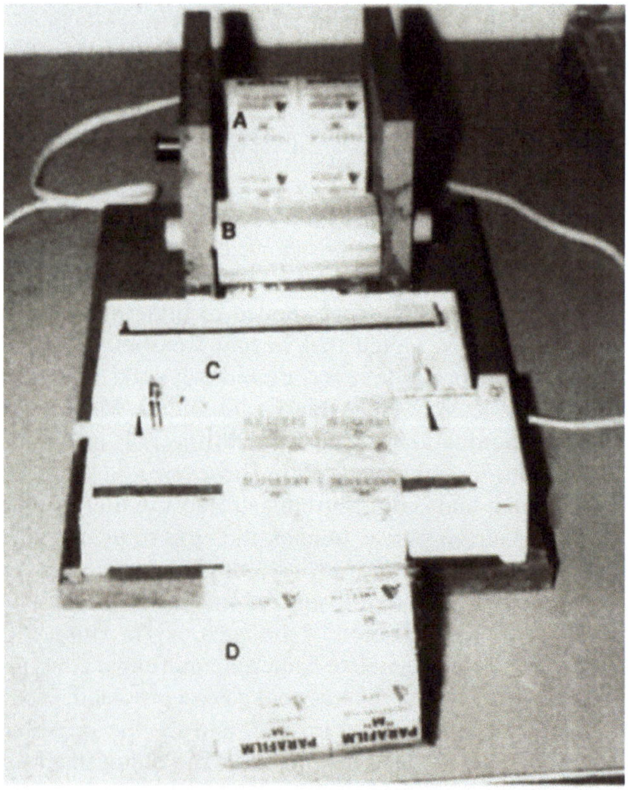

Figure 1.20 Original packet-making apparatus (from Patana 1982) used for making continuous strip of diet and oviposition packets. Parafilm® roll (a); roll of plastic film (b); heat sealer (c); and roll of Parafilm® and plastic film sealed on two sides (d). The automation of diet and oviposition packet making became further developed at the Gast Facility (at Mississippi State, USDA, ARS, BCMRRU).

rearing *L. hesperus*. Debolt and Patana had solved many of the problems that had been barriers to mass-rearing *L. hesperus* and presented the summary of the rearing protocols in a chapter (Patana and Debolt 1985) of the Singh and Moore's 1985 *Handbook of Insect Rearing*. Several earlier researchers had attempted with little success to develop artificial diets for *Lygus* species. Debolt reviewed many of these efforts (e.g., Strong and Kruitwagen 1969; 1970; Vanderzant 1967) as potential starting points for his assignment to develop an artificial diet and a diet-based rearing system for *L. hesperus*. However, in agreement with several of the earlier authors, the diets that they developed did not succeed in producing successive generations of *Lygus* bugs, a result that was partially masked by the communal rearing of this mirid, which displays cannibalism during periods of low nutrition. Many of the "sages" advised Debolt to realize that developing a diet for an insect with "picky" feeding habits would be impossible, and even if a diet were successfully developed, mass-rearing would be impossible because of the dependence of mirids on host plants for oviposition. An essential part of the mass-rearing system for Lygus species was development of a system that simulated plant materials where plant bugs normally oviposit. The system developed by Patana 1982 and Patana and Debolt 1985 was based on previous work by Vanderzant 1967, who reported Lygus ovipositing through Parafilm into rolled, diet-soaked filter paper and newsprint paper. Both agar and carrageenan gels were tested, with carrageenan yielding superior results in terms of amount of oviposition and survival of the eggs.

1.5.2.2 Rationale for the Cohen 2000c Diet System

After extensive discussions with Jack Debolt over the decade of 1980–1990, it started becoming apparent that we needed a better grasp of the feeding dynamics and mechanisms of *Lygus* and other hemipteran insects. Both Debolt and I had arrived at the idea that Lygus bugs were not simply inserting straw-like mouthparts and sucking liquids from their food. Jack Debolt and I performed numerous experiments where we tried to isolate the hypothesized liquids that the plant bugs were feeding on—both in natural foods and artificial diets. For example, we knew that Lygus bugs could develop from neonate nymphs to adults on green beans alone (though they thrived better on green beans supplemented with heat-killed beet armyworms or other insects such as aphids). We reasoned that there was interstitial juice in the green beans that was supporting the complete development of the plant bugs. Based on the hypothesis that Lygus nymphs feed and develop on bean juices, we tried a variety of extraction techniques to prepare "bean juice," which we fed to the Lygus bugs in the same kinds of membrane-based feeding chamber described by Debolt 1982. However, we were never able to get development beyond one or two instars from such feeding. With the Debolt diet in the same kinds of chambers, we could elicit complete development. We tried extensive maceration and filtration techniques to make "ultra-pure" bean juice, nearly wearing out vacuum pumps to extract the bean juices with 0.22 μm filters, and we found that the more we filtered, the poorer the results—thus disproving the hypothesis that *Lygus* was using the juices from the beans. The scanning electron micrograph inset in Figure 1.20 holds much of the answer to our dilemma. The two openings—the food canal and the salivary canal—are about 17 and 8 μm in diameter, far larger than the opening needed to extract simple liquids such as plant saps. Plus, the prominent salivary canal was a strong indication that much of the Lygus bug's feeding dynamics depended on a considerable application of saliva into the food. The 17 μm food canal indicates that the food being ingested either had large particles in it or was of a viscous consistency that required a rather large opening to allow adequate flow of ingested material. Indeed, the morphology of the Lygus (and other hemipteran feeding) structures was suited to EOD (what Cohen 1998 called "solid-to-liquid" feeding).

This in-depth consideration of feeding morphology and physiology helps us understand why the original Debolt diet was more successful than were diets of his predecessors. The Debolt diet (and, in turn, the Cohen diet) for Lygus bugs contained substantial amounts of solids that were nutritionally quite complete and complex. Components such as lima bean meal and wheat germ provided a rich array of complex carbohydrates, proteins, lipids, nucleic acids, as well as minerals and smaller organic molecules. The feeding apparatus of Lygus bugs (and many other hemipterans) involves the insertion of stylets (Figure 1.21), flushing the food source's biomass with digestive enzymes, macerating the tissues with mandibular stylet tips, and turning the solid foods and their interstitial liquids into a highly nutritious slurry that could be ingested and further digested, then absorbed. The feeding mechanism allows pinpoint selection of required nutrients from pockets or compartments of the food source where the nutrients reside (Figures 1.20, 1.22, and 1.23).

The modified diet was not only simpler to produce and less expensive than the Debolt diet, but its more nutrient-dense character made it completely suitable for other *Lygus* species, including *L. lineolaris* (for which the Debolt diet could not sustain complete development and successive generations). It must be added, however, that without the Debolt diet as an important step in hemipteran dietetics, the Cohen 2000c diet would never have been developed.

Figure 1.21 (a) and (b) Elements of the Gast Lygus rearing system+ derived from Debolt 1982.

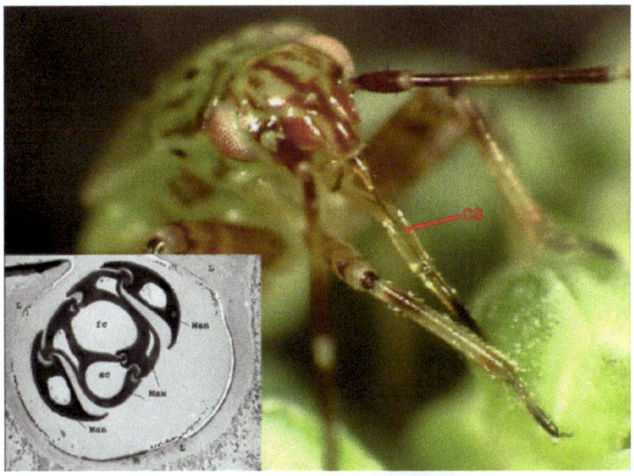

Figure 1.22 Dr. Jack Warner Debolt's system of maintaining colonies of *Lygus hesperus* and *L. lineolaris* modified to meet the needs of the mass-rearing facility at USDA, ARS Biological Control and Mass Rearing Research Unit (Gast Facility) at Mississippi State, MS. Dr. Debolt was the originator of the first diet and rearing system to produce successive generations of *L. hesperus*, allowing mass-rearing of this mirid pest for production of parasitoids (*Anaphes iole* [Hymenoptera: Mimaridae], *Peristenus relictus* [*stygicus*] [Hymenoptera: Braconidae], and *Leiophron uniformis* [Gahan] [Hymenoptera: Braconidae]). The oviposition packets were used by the Tucson (USDA, ARS) Biological Control of Insects Laboratory and by the BCMRRU (Mississippi State, USDA, ARS) laboratory for mass-rearing *Anaphes iole* and the nymphs of the *L. hesperus* and *L. lineolaris* were used as hosts for the mass-rearing of the braconid parasitoids that were involved in several pilot tests and other field tests for the management of Lygus species.

Table 1.3 Lygus Diets Developed by Debolt 1982 and by Cohen 2000c

Component	Debolt Diet 1982	Cohen 2000c Diet (g)
Chicken egg yolks	0	120
Whole chicken eggs	144	55
Wheat germ (toasted)	36	80
Lima bean meal (coarsely ground)	36	120
Soy flour (re-fatted, toasted)	0	20
Sucrose	22	10
Lecithin (soy in soy oil)	0.04	4
Vitamins (Vanderzant)	7.2 (special mixture, see Debolt 1982)	3.2
Brewer's yeast	0	1.8
Honey solution (50%)	0	9
Salt mixture	2.3 (special mixture, see Debolt 1982)	0
Casein hydrolyzate	14.3	0
Gelcarin 812	2.2	0
Water	730	573
Acetic acid (10% glacial acetic: 90% water)	0	3
Chlortetracycline	0.4	0.01
Streptomycin sulfate	0.01	0.01
Formalin (Deleted in later batches of Cohen Diet)	0.4	0.4
Niacin	0.8	0
p-Aminobenzoic acid	0.8	0
Cholesterol	0.4	0
RNA	0.4	0
Tween 80	0.7	0
Linoleic acid	0.14	0
Propionic acid	0	0.3

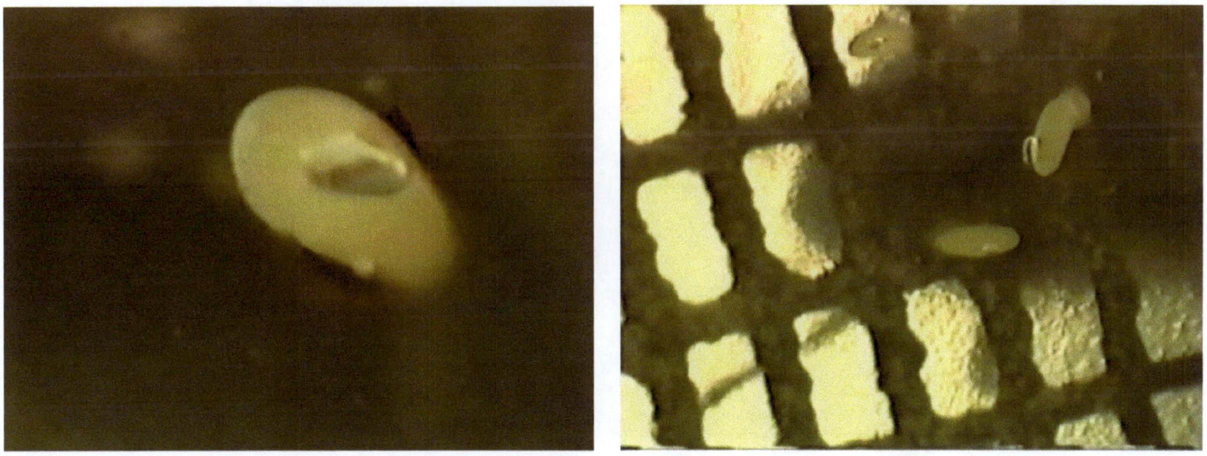

Figure 1.23 **(a) and (b)** *Lygus hesperus* eggs being inserted into carrageenan gel in oviposition packet. (a) Shows an individual egg being released by the ovipositor. Figure (b) shows the shadow of a female Lygus underneath the wire screen on which the carrageenan packet had been set. Three *L. hesperus* eggs can be seen in the transparent gel. In the Patana gel pack, a 2% Gelcarin 812 (Kappa carrageenan from FMC Biopolymer, Princeton, NJ). The use of carrageenan as an oviposition substrate was developed by Patana and Debolt 1985 after they tested agar and carrageenan and found that the carrageenan stimulated far more oviposition than did agar. In the modifications of the *L. hesperus* rearing system developed by Debolt and Patana, Cohen 2000c modified the Debolt diet (see Table 1.3) substantially based on Cohen's studies of extra-oral digestion (Cohen 1985a, 1989, 1990, 1995, 1998; Cohen and Smith 1998). The modifications included using highly nutrient-dense components such as soy flour, larger proportions of wheat germ, chicken egg yolks, etc.

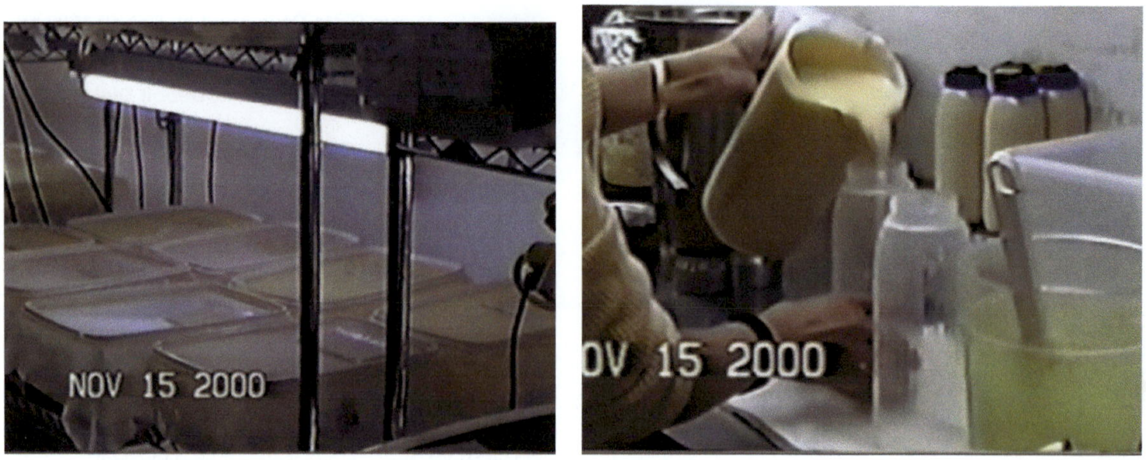

Figure 1.24 **(a) and (b)** Arrangement for mass-rearing *Lygus hesperus* and *L. lineolaris* at the USDA, ARS R. T. Gast Laboratory. (a) Shows eight cages with screen tops and diet and oviposition packets on top of cages. (b) Technician is pouring the Lygus diet (Cohen 2003a), which she will be pouring into Parafilm® and plastic packets. Each cage holds about 300–400 *Lygus* bugs, and the lidding of the cage does not have to be removed between inoculation of cage with infested egg packet and the harvesting of eggs from adult cages. This system is devised by Debolt 1982 and Patana 1982 and allows a single technician to handle several hundred thousand Lygus bugs per day. Handling includes changing diet packets every other day, and placing and harvesting oviposition packets daily. Figure (b) shows the final stages of making the Lygus diet with the semiliquid diet being poured into squeeze bottles for delivery (by hand) into Parafilm and polyethylene plastic packets.

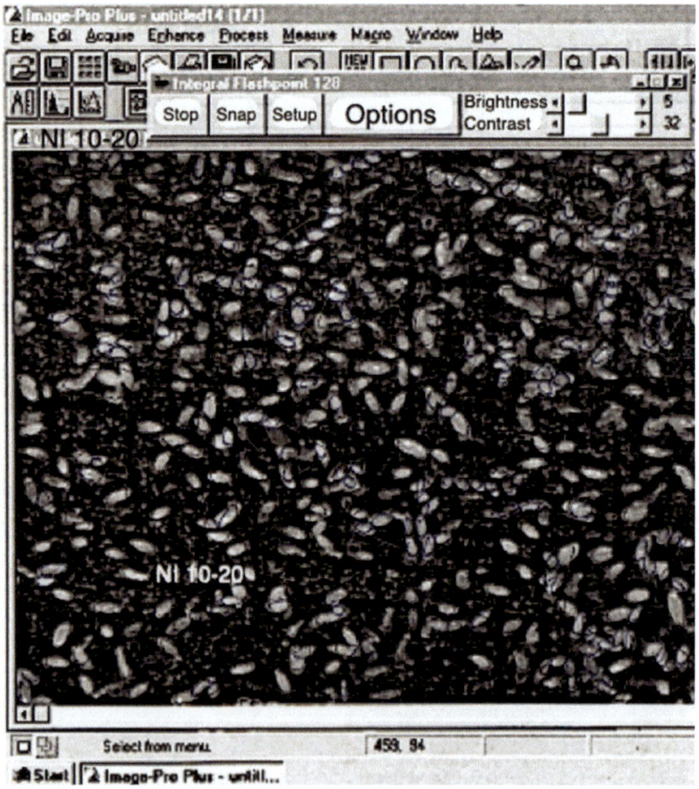

Figure 1.25 Lygus eggs in a packet of Gelcarin (carrageenan) enclosed in Parafilm®. This image (from Image Pro software) represents the automatic counting system that was used at the USDA, ARS, BCMRRU laboratory to evaluate the numbers of eggs oviposited into gel packs. The egg counts per cage were used as a quality assessment tool to determine whether the colony was performing at established baseline levels. The other parameters for quality control evaluation were the biomass per cage, and the mortality in smaller, sentinel cages. Freshly emerged adults were placed in cages, and their mortality over a two-week period was determined. These quality standards were used to evaluate any planned changes in the colonization conditions as reported in Alverson and Cohen (2002).

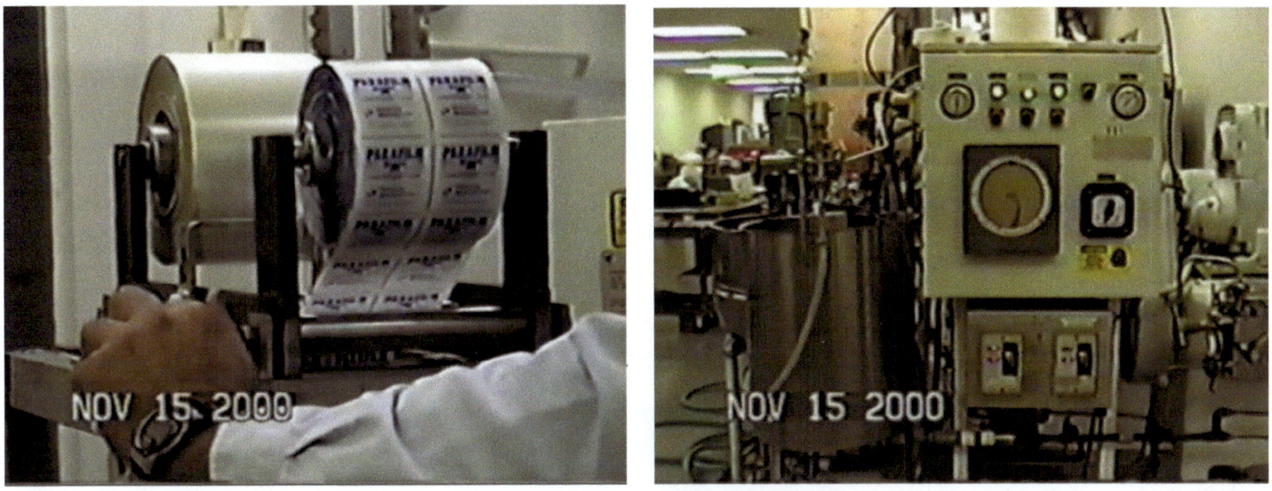

Figure 1.26 **(a) and (b)** Making diet and oviposition packets in a semi-automated packet-making and packet-filling machine (which was custom-made at the USDA, ARS Gast Facility by Donald Nordlund and Dana Harsh). The packet material composed of Parafilm® and polyethylene plastic film is seen in (a). Diet or carrageenan oviposition material is pumped through the heating/pumping apparatus (modified from the flash sterilizer described for boll weevil rearing and Lepidoptera rearing in Chapter 5). This packet-filling system was adapted from the diet-making apparatus, which was a pasteurization system.

Besides the proof of principle that *Lygus* bugs could be reared on artificial diet and in turn they could be induced to oviposit readily in gel packets (Figures 1.20, 1.22, and 1.23), the Debolt diet (and later the Cohen diet [Figures 1.22 and 1.24]) for *Lygus* bugs demonstrated a reality about the need for a highly concentrated solid or semisolid food for *Lygus* (and for phytophagous stink bugs, which was demonstrated later). The rearing systems for Lygus bugs, G. punctipes, and C. rufilabris demonstrated that the coordination was essential between diets, diet packaging, and diet/oviposition systems (Figures 1.24–1.28).

The theme and central point of this book is that rearing systems are developed in as engineering processes where the rearing personnel make a constant effort to design, operate, and control their rearing systems. The four systems considered here were subject to incremental improvements toward achieving the design improvements and increase of control, but the

Figure 1.27 **(a) and (b)** Pump for loading the diet/carrageenan (Gelcarin 812 solution). The molten carrageenan solution is pumped into the Parafilm packets. The molten, viscous diet or oviposition gel material is transferred through tubing and into awaiting diet packet envelopes seen in figure on the right. Once each packet is filled, it is cut from the string of other packets and sealed with a hot wire sealer.

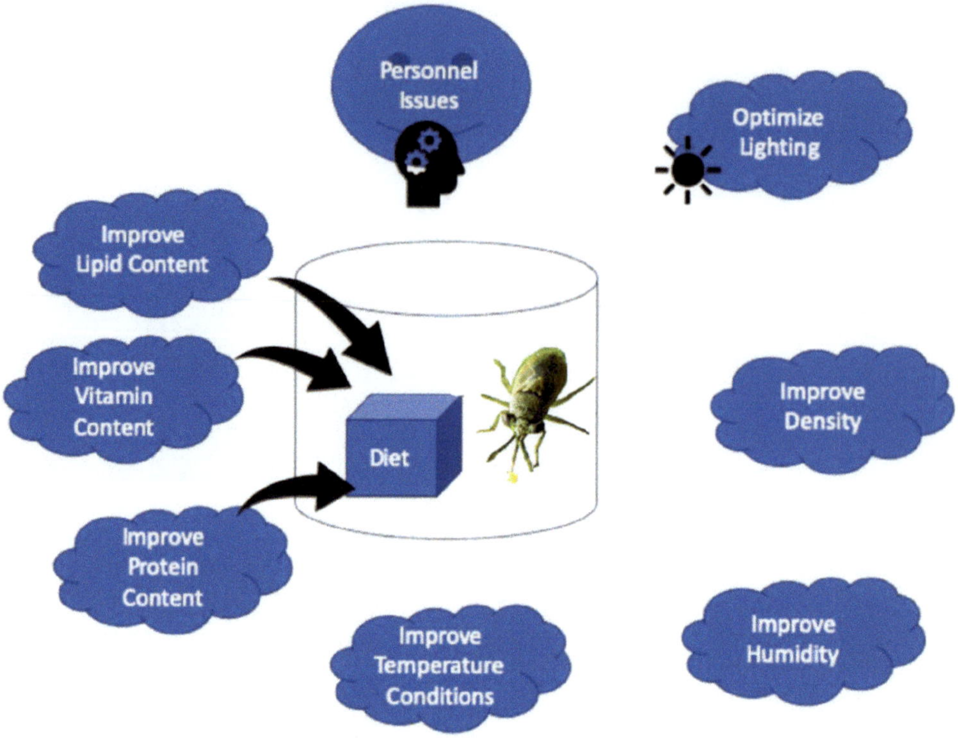

Figure 1.28 Working on algorithm sub-components toward continuous improvement. The figure is a mind map that shows some of the rearing system components and subsets of the components as part of a plan to improve the quality (fitness) of the insects and/or to improve the efficiency and economic feasibility of the system. The diagram is intended to suggest the seemingly disparate factors that may be improved to improve the overall rearing system. In the rearing systems discussed here for the predators, *Geocoris punctipes* and *Chrysoperla rufilabris* as well as for the phytozoophagous plant bugs, *Lygus hesperus* and *L. lineolaris*, all of these factors and many more were taken into consideration for programs of continuous improvement in all four rearing systems.

researchers (especially myself) were not conversant with the methods of design of experiments (DOE) when we developed these systems 20–40 years ago. If I were returning to the development and improvement of these systems today, I would put first and foremost the charge of using DOE for all initial research on rearing system development and for all efforts in continuous improvement in the process control of each system. Hopefully, the remainder of this book will serve as a roadmap to using DOE, process control, and other scientifically based approaches achieving higher levels of quality in the insects we produce.

CHAPTER 2

Who Is Who in Insect Rearing?

2.1 WHY ARE INSECTS REARED?

Rearing operations by type include the following: (1) research, (2) sterile insect technique (SIT), (3) biological control, (4) feeder insects, (5) insect products (silk, honey, shellac, wax, biochemical products such as toxins and antimicrobial substances), (6) waste treatment (e.g., soldier flies), (7) celebrations/ornamentation, (8) insects as pets, (9) educational purposes, (10) conservation, (11) forensic analysis, and (12) medical purposes. Briefly, here is a summary of these rearing operations. **It is crucial to recognize the fact that there are tremendous successes in rearing science and technology that are to be found in the organizations that have performed these rearing functions. These successes did not just happen; but rather, they exemplify decades of development (research and development = R & D), which are characterized throughout this book.**

2.1.1 Rearing Insects for Research

Rearing insects for research is the broadest application, considering the many kinds of research that utilize insects such as genetics, testing pesticides or toxins, ecological explorations, using insects as model systems for discovery and evaluation of microbial characteristics (such as virility), biochemical and metabolic studies, substitutes for vertebrates in studies of nutrition, and many other research functions. Research purposes of rearing can be classified as using reared insects for (a) research on management of insects, especially insect pests, and (b) as tools or model systems for research problems other than insect management. An example of the first function is rearing pests such as wax worms to determine the most efficacious pesticide to suppress wax worm attacks on beehives. An example of the second function is rearing wax worms to study the basis of diet-induced thermogenesis as part of a basic research inquiry into the nature of metabolism and heat production.

2.1.2 Sterile Insect Technique (SIT) and Biological Control

Both SIT (Figure 2.1) and biological control (Figure 2.2) demand large rearing systems to operate efficiently. These approaches are applications of managing pest species that require well-vetted rearing system technology (Figure 2.3a and b).

The *S. tsugae* program (Figure 2.2 a) has been in operation for more than two decades and is responsible for releasing thousands of the beetles, which are specialist predators of the invasive and most destructive pest, hemlock woolly adelgids. The PABIL (Philip Alampi Beneficial Insects Laboratory, in Trenton, NJ) shown in Figure 2.2 b is another noteworthy laboratory that conducts programs to protect natural and agricultural systems in New Jersey. In these images, a biological control program is aimed at protecting New Jersey (and other northeastern ecosystems) from the spread of the invasive weed, mile-a-minute, using the imported beetle *Rhinoncomimus latipes.* Other state-run rearing systems for various programs from several US states have small to medium rearing facilities for biological control and other programs for crop and consumer protection. Many private companies are present worldwide, including many companies that make up the **ANBP (Association of Natural Biocontrol Producers [nbp.org/index.php/members-products]**). The ANBP includes at least 50 North American companies and organizations that produce and distribute insects and other biological control products used to manage insect pests. Many of the companies are members of the ANBP mass-rear thousands of insects per day for sale and release in biological control programs in greenhouses and in the field. Predators such as lacewings, minute pirate bugs, and lady beetles and parasitoids such as *Trichogramma* spp. and whitefly parasites are sold for control of several pest species.

2.1.3 Feeder Insects

Feeder insects have been reared by several companies in North America and worldwide for use with insect-eating pets and zoo animals (Figures 2.3–2.5). The variety of insects used as food for other animals is increasing as the market calls for increasing variety in the diets of pets such as birds, lizards, frogs, and tarantulas. The standard market item had been crickets (Indian house crickets, mainly), mealworms, wax worms, and several other species. Recently, the market has opened to hornworms

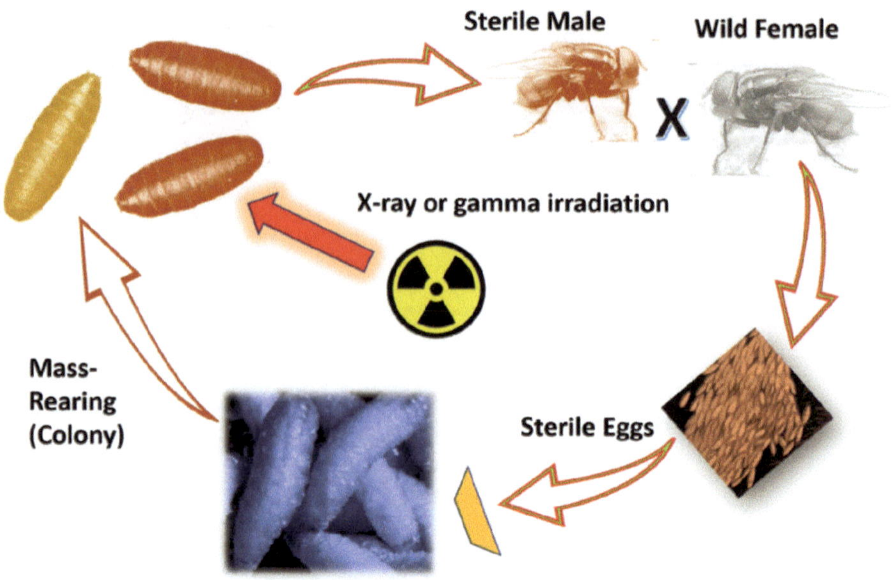

Figure 2.1 The sterile insect technique (SIT) for screwworms, as devised by E. F. Knipling, was so successful that it became a "proof of principle", as illustrated in the figure. The technique was then applied in many other programs such as tephritid fruit flies (discussed in Chapter 3) and coddling moth SIT as described in excellent detail by Dyck et al. 2005. This book contains a comprehensive perspective on the various facets of SIT, including history, rearing dynamics, extensive quality control guidelines, specific and highly useful information about the technology of sterilization (equipment, dosage, safety precautions, etc.), and also a perspective of the four major strategic options in SIT: suppression, containment, prevention, and eradication. The book also provides several case studies that chronicle efforts to use SIT with specific insects. The perspectives of area-wide integrated pest management (AW-IPM) are useful to other IPM practitioners and planners. From the rearing perspective, the chapter by Andrew Parker is useful; plus the other references to rearing throughout the book provide useful information that helps supplement other works on rearing for SIT, such as Dyck 2010.

Figure 2.2 (a) Biological control and rearing program at the Connecticut Experimental Station (Valley Laboratory, Windsor, CT). Dr. Carole Cheah is shown in her rearing lab where she developed a mid-sized rearing system for the hemlock woolly adelgid predator, *Sasajiscymnus tsugae*. Figure 2.2 (b) The PABIL (Philip Alampi Beneficial Insects Laboratory, in Trenton, NJ).

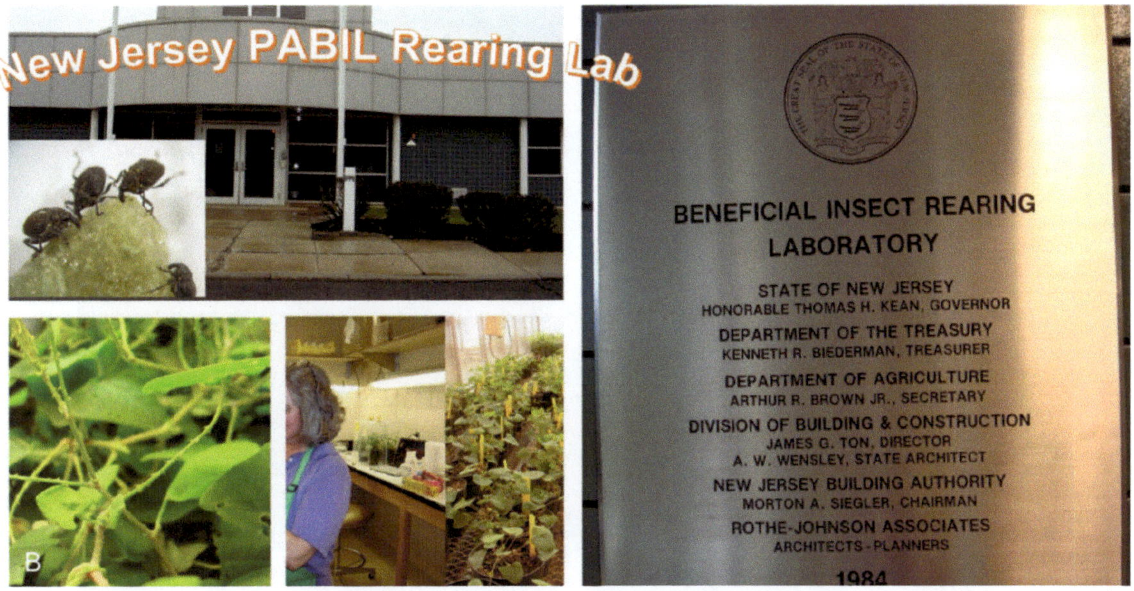

Figure 2.2 (*Continued*)

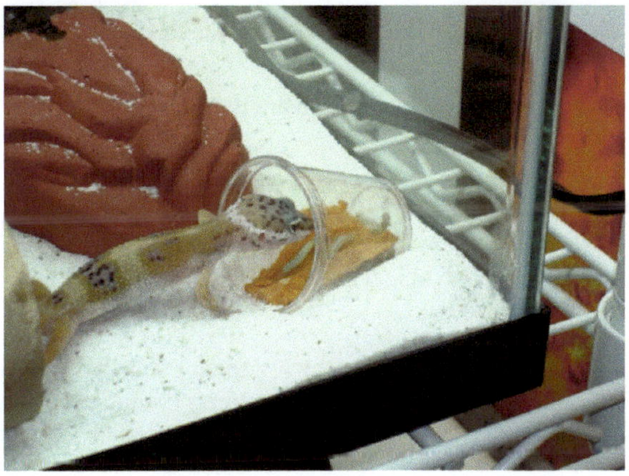

Figure 2.3 A leopard gecko is feeding on cabbage loopers enriched with plant pigments and calcium.

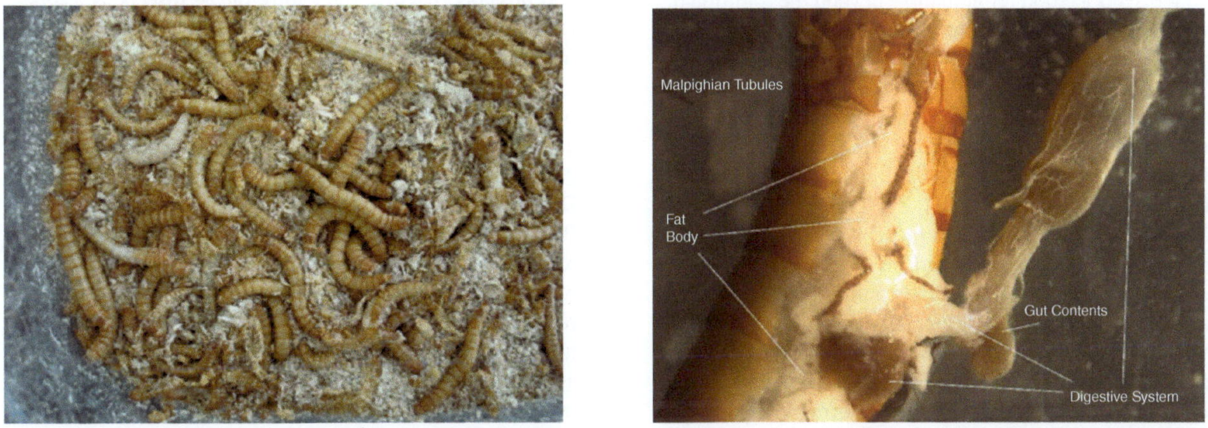

Figure 2.4 (a) and (b) Mealworm larvae, Tenebrio molitor. (a) A colony of mealworms feeding flour and wheat germ. (b) A dissected mealworm, showing the fat body, Malpighian tubules, and the digestive system, including some gut contents shown leaking out of the ruptured GI tract.

Figure 2.5 Live crickets being sold in a pet store. The containers are Bug Boxes (products of The Bug Company), containing crickets conveniently packaged.

and silkworms for larger pets. Several dozen companies have been producers of feeder insects, but now that the market is further expanded into production of insects to feed poultry and now even humans.

There is a flurry of efforts and publications regarding direct use of insects and nutrients derived from insects to meet the growing need for human food security (Figure 2.6). This is discussed by Cohen 2015, where the following summary explains the background:

> Recently, there has been a resurgence of thinking about using insects as food for humans—beyond the traditional insect cuisine as a novelty or seasonal supplement in tribal or cultures in rural settings (Klunder et al. 2012, Carrington 2011, and van Huis et al. 2013). There are several recent papers that present information on the nutritional value of insects including four papers by Finke 2002, 2005, 2007 and 2013, and papers by Mishra et al. 2003, Longvah et al. 2011, Zhou and Han 2006, and Xia et al. 2012. These papers present information about the nutritional value of selected insects in the context of use as vertebrate foods, especially humans. The paper by Longvah et al. 2011 is an especially good model of how information about insects as human food should be presented because these authors included experiments on net protein utilization (NPU) of the silkworm prepupae and pupae. Although the work was done with rats as model vertebrates, the evaluation of protein utilization is much more meaningful than a simple presentation of protein composition.

This increased attention to entomophagy by humans follows some modern pioneers as advocates of the potential benefits of using insects as food for cultural, economic and environmental benefits, including Bodenheimer 1951, Taylor 1973, and Defoliart 1975 and 1999. The monograph by van Huis et al. 2013 and published by FAO treats the subject extensively, including documentation of the nutritional value of a wide range of insects and strategies for encouraging programs that help facilitate the use of insects either as staples or supplements, including ideas about converting insect biomass into a flour, which would at once enhance the palatability of the insects and possibly disguise the nature of the insects as a food for people whose culture has not traditionally espoused insects as food.

Although there is considerable enthusiasm for the possibility of using insects as a possible way to fill a void that is anticipated with a burgeoning human population, estimated by the UN (United Nations Department of Economic and Social Affairs. World Population to 2300. 2004. Executive Summary, Page 2) to grow from seven billion in 2014 to more than nine billion in 2050 (an increase of almost 30% of the current level in 36 years), there is a desperate need to find new ways to feed people. Many advocates of entomophagy by humans express optimism about the potential of insects to increase their biomass quickly, and they cite analyses of the nutritional composition of insects, especially their relatively high content of protein of quality comparable to some of the best foods currently in wide use (dairy products, chicken eggs, and plants such as

Figure 2.6 Insects (crickets) used as human food. This stir-fried dish was made by a local Raleigh, North Carolina restaurant for sale at the annual Raleigh Bug Fest. Long lines of people await opportunities to sample various insect-containing dishes from salads, to entrees, to deserts such as cricket-containing ice cream.

soy with high levels of protein). These entomophagy enthusiasts point to the long-standing practices of various peoples throughout the world who eat a large variety of insects (van Huis et al. 2013).

This topic is further discussed in Chapter 13 (Future of Insect Rearing) in this book. There are entire issues of ESA journals that are dedicated to insects as food, and there are a few publications that add a cautionary note to the enthusiasm about the degree to which farmed or reared insects will be able to fill a major need for human food security (Lundy and Parella 2015; Cohen 2015 and Cohen's website: www.werearinsects.com).

2.1.4 Rearing Insects for Products

The oldest insect-rearing systems are those connected to production of insect products such as silk, honey, shellac, and wax. As explained in great detail in Chapter 4 (Silkworm Rearing Systems), the mulberry silkworm has been domesticated and reared for 5,000 years, mainly for the production of silk, but also as food—where pupae that had been heat-killed during the silk processing were collected and eaten after various cooking methods were applied. Honeybee colonization shares with silkworms the distinction of being used by humans for millennia where domesticated bees were exploited not only for honey production, but also for use of beeswax (dating back to Ancient Egypt), and propolis which has also been valued as a glue or resin. Gambichler 2004 mentions that Antonio Stradivari was said to have used propolis as a component in the coating for his violins. Another very useful product from cultured insects is shellac made from the lac insect, *Kerria lacca* (Hemiptera: Coccoidea: Kerriidae) as described by Raman 2014.

2.1.5 Waste Treatment

A creative new enterprise is emerging in the technology of using soldier flies (and other insects now entering the picture) for treatment of waste. Generally, the insect biomass derived from waste treatment is intended for other uses, especially as food for other organisms. For example, black soldier flies (*Hermetia illucens*, Stratiomyidae: Diptera) are well known to use waste products such as enormous amounts of poultry waste produced in farming chickens, turkeys, and other fowl. The black soldier fly industries are increasingly being used as providers for feeders and even as human food. Several sites are available that publicize and further explain benefits and methods of mass rearing these insects (e.g., https://www.forbes.com/sites/ariellasimke/2019/12/01/black-soldier-flies-are-the-new-superstars-of-sustainable-aquaculture/#3978889e3ac1). Increasing numbers of suppliers are selling black soldier flies as feeders (e.g., Flukers at flukers.com, Symton at https://symtonbsf.com/collections/all/products/popworms-premium). Across the world, sophisticated rearing research is being conducted in ways to increase the efficiency of rearing black soldier flies and other Diptera that are well known as feeders of a broad range of dead and decaying matter. The attractive features of these projects and developing systems are that at once they are a way of reducing waste from other processes (such as food processing and farming), and producing a higher nutrient product (= insects) that can be used to feed other organisms.

2.1.6 Insects Used in Celebrations/ Ornamentation

The North American Butterfly Association website (https://www.naba.org) contains extensive photos and descriptions of their activities, including companies that sell butterflies for celebrations and for ornamental purposes. Often, the purpose of the organizations under the umbrella of the North American Butterfly Association include conservation purposes to help preserve the insects themselves and their habitats and plants that are associated with the butterflies. Therefore, it is worthwhile to recognize that there are conservation efforts as well as the simple esthetic appreciation of butterflies and other insects. This purpose is related to the next topic: insects as pets.

2.1.7 Insects as Pets

In several cultures in the world, various insects are valued for esthetic reasons, including the custom in various Asian countries to keep crickets in small cages, possibly for good luck. Several sites found on Google advertise cricket cages for sale, and in China there are thriving businesses that sell cages for various insects that people keep as pets.

2.1.8 Insects Used for Educational Purposes

Several companies provide insects for schools or other organizations to use for educating students about insects in particular and science, in general. Some of these companies include Carolina Biological Supply (https://www.carolina.com), Educational Science (https://www.educationalscience.com/monarch-artificial-diet-kit-prepared-30-cups-x-10ml-md300), and Mulberry Farms (www.mulberryfarms.com). A few companies such as Carolina Biologicals not only sell insects such as *Drosophila*, silkworms, painted lady butterflies, wax worms, and hornworms, but they also provide a rich curriculum with instructions about how to rear the insects and how to include them in scientific demonstrations and experiments. Other companies such as the biological control suppliers and butterfly suppliers provide the insects that the educators can apply to experiments or demonstration of scientific principles. Several highly effective and far-reaching educational and conservation efforts have been made through various organizations such as The North Carolina Museum of Natural Sciences, The Sonoran Desert Museum (near Tucson, AZ), and many others, worldwide. Displays and educational treatment of insects and other arthropods are incredibly popular, especially where people can enter enclosures where various local and exotic butterflies and sometimes other insects are on display and people can come in contact with these charismatic creatures. Having a butterfly land on your hand can turn an entomophobe into an avid entomophile instantly! The Cincinnati Zoo and Botanical Garden was a pioneer in dedicated insect and other arthropod display. The website discusses their collection and display as described below (http://cincinnatizoo.org/plan-your-visit/exhibits/world-of-the-insect/):

> In 1978, the first exhibit building devoted to insects in any U.S. zoo was built at the Cincinnati Zoo & Botanical Garden and was awarded the American Zoo and Aquarium Association's (AZA) Exhibit Award. Many significant achievements in husbandry, breeding and display of particular species, including the bullet ant, Peruvian fire stick and giant water bug, have received special awards over the years.

2.1.9 Rearing for Insect Conservation

In most of the organizations (zoos, aquaria, botanical gardens, and museums) where insects are reared for display, the purpose is twofold education and conservation. Much of what was said in the previous section applies equally to conservation. In fact, the idea of conservation goes hand-in-hand with the understanding that comes from a good educational program. I saw the changes that took place in people's thinking about insects (and other arthropods) in my classroom teaching, but nowhere was the gulf between indifference or even dislike for insects and support for them bridged so effectively as in the volunteer work I did in various butterfly gardens. As soon as people got close to the butterflies and got to see some of their diversity, their beauty, and their grace, their squeamishness was overcome. Some of my most rewarding experiences in my life as an entomologist came from the Raleigh Bug Fest where people got to be "up close and personal" with various insects, even the "creepy-crawly" ones, and the inherently interesting nature of the insects overrode the prejudices, and both children and adults became fascinated. Always to my amazement, even the Madagascar hissing cockroaches received rapt attention and even some affectionate handling! Once a person has watched silkworms spinning, eating, and, yes, even defecating, they find it hard not to accept the benefits of conservation and preservation of diversity.

2.1.10 Rearing Insects for Forensic Analysis

Cammack et al. 2016 summarized the basis of forensic entomology as follows: "The body of literature on decomposition ecology is large and useful in homicide investigations" (Reed 1958; Payne 1965; Anderson and VanLaerhoven 1996; Tomberlin and Adler 1998; Watson and Carlton 2003; Tabor et al. 2004 and 2005; Hobischak et al. 2006; Benbow et al. 2013; Malainey and Anderson. 2020).

2.1.11 Rearing Insects for Medical Purposes

Medicinal maggot (*Lucilia sericata*, Calliphoridae, Green Bottles): how they work and who sells them? In their review of maggot therapy, Sherman et al. 2000 wrote:

> Certain fly larvae can infest corpses or the wounds of live hosts. Those which are least invasive on live hosts have been used therapeutically, to remove dead tissue from wounds, and promote healing. This medicinal use of maggots is increasing around the world, due to its efficacy, safety and simplicity. Given our low cultural esteem for maggots, the increasing use and popularity of maggot therapy is evidence of its utility. Maggot therapy has successfully treated many types of chronic wounds, but much clinical and basic research is needed still. In this review, the biology of myiasis and the history of maggot therapy are presented, the current status of our understanding and clinical use of medicinal maggots is discussed, and opportunities for future research and applications are proposed.

The rationale for using maggot therapy (efficacy, safety, and simplicity) offers compelling reasons for pursing the production of flies that can be used for therapeutic purposes. Rearing maggots for maggot therapy: Reviewing "maggot therapy" in the Web of Science yields 450 publications from 1934 to 2019 (August 31). Maggot therapy (the application of larvae of flesh-eating flies to help promote wound healing) has been used for millennia having come in and out of vogue. Recently (over the past 40 years or so), the scientific basis of using maggots has helped usher in a renaissance of maggot-based wound treatment. The mechanisms have been better elucidated recently where we now understand that the appropriate species of fly larvae not only eat necrotic tissue without damaging the healthy tissue, but the maggots also apply antimicrobial substances that further clean the wounds, and they provide exudates that induce host defenses and wound-healing mechanisms.

More recently, Stadler 2020 reviewed the complete literature on maggot therapy, including updating of the review by Sherman et al. 2000 and dealing with questions of transport and handling of the maggots at the sites where they are to be applied. Stadler points out that:

> the literature to date has focused on isolated discussions of echelon-specific issues such as diet improvement and sterilization protocols in the production echelon, and the relative effectiveness of medicinal maggot application methods in the treatment echelon. There is little knowledge in the public domain regarding the transport and distribution of medicinal maggots, but existing supply chains for vaccines, blood and pathology specimens may provide learning and supply chain integration opportunities. Maggot therapy knowledge across the treatment echelon is generally substantive but there is still insufficient knowledge regarding patients' and health care providers' attitudes toward the therapy, and their experiences of receiving and administering MT. Moreover, there is no research concerned with the humane disposal of medicinal flies during production and after treatment.

Some of the major companies that produce therapy maggots include:

- International Biotherapy Society (http://biotherapysociety.org/)
- BioMonde (https://biomonde.com/en/) in Great Britain
- Monarch Labs (https://www.monarchlabs.com/)

2.1.12 Who Rears Insects (i.e., What Are the Organizations That Practice or Support Insect Rearing)?

In breaking down the major participants in rearing operations, the summary from Cohen 2019, explained in the following sections, provides a general overview and scope.

2.1.12.1 Government (Federal and State)

Many countries have **federal laboratories** devoted to insect rearing, and virtually every country has a presence in insect management most of which have at least some insect rearing as part of their research systems. Some of the major national organizations are:

- United States Department of Agriculture (USDA)—Animal and Plant Health Inspection Service (APHIS), Forest Service, Agricultural Research Service (ARS)
- Canadian Forest Service, Ag Canada
- Brazilian Agricultural Research Corporation—EMBRAPA
- Australia—Centre for International Agricultural Research
- France—National Institute for Agricultural Research (INRA)
- Great Britain—Department of Agriculture and Rural Development
- Spain—Institute for Research in Food and Technology (IRTA) in Catalonia
- Africa, Asia, Worldwide—Consultative Group for International Agricultural Research (CGIAR)
- Iran Agricultural Research (IAR), Agricultural Research Organization
- Israel—Volcani Center
- Egypt—Agricultural Research Center (ARC)
- China and eastern Asia—International Food Policy Research Institute
- Japan—National Agriculture and Food Research Organization (NARO)
- Indian Council of Agricultural Research (ICAR)

State or province: In the United States, every state has its own department of agriculture, each with insect-rearing programs.

State (or province) programs: In the United States, for example, every state has its own department of agriculture or plant industry, and many of these have significant rearing facilities. As examples of state programs in the United States, the PABIL (Phillip Alampi Beneficial Insect Lab, in the New Jersey Department of Agriculture) is a model of excellent rearing technology for the biological control of pests in New Jersey. A second example of a state rearing program is the Biological Control Program of the North Carolina Department of Agriculture and Consumer Services (NCDA&CS). A third example of medium-scale rearing is the Florida Department of Agriculture and Consumer Services Biocontrol Rearing Facilities. The California Department of Food and Agriculture (CDFA) is larger, in terms of insect rearing, than national facilities in smaller nations, and the State of New York has considerable resources in rearing. These are just a few examples of programs that include insect rearing in the various US states. Similar organizations are rearing-based efforts that exist in various provinces of several countries and that serve as supplements to national systems where insect rearing is a prominent part of pest management programs.

2.1.12.2 Non-Government Organizations

Non-government organizations (non-profit entities) that support rearing efforts include the United Nations, Food and Agriculture Organization (FAO), World Health Organization (WHO), and the International Atomic Energy Agency (IAEA). An example of direct involvement in rearing is the work done by the Joint FAO/IAEA Programme on Nuclear Techniques in Food and Agriculture (https://www.iaea.org/topics/sterile-insect-technique), which oversees programs in SIT and genetic pest management (GPM) for management of biting flies and agricultural pests. The philosophy of FAO/IAEA is captured here: "Jointly with the FAO, the IAEA assists its Member States in developing and adopting nuclear-based technologies for optimizing agricultural insect pest management practices that support the intensification of crop production and the preservation of natural resources." Other non-government, non-profit organizations such as the Bill and Melinda Gates Foundation can be influential in insect rearing by providing funding for certain rearing-based projects. The Gates Foundation recently put out a call for proposals for researchers to develop and artificial vertebrate blood to help advance mass-rearing of mosquitoes for programs in malaria control. Other such foundations will occasionally offer funding opportunities in projects that involve rearing as a component of the overall program. However, it must be emphasized that such opportunities are rare, and very often the rearing component becomes buried in the project as a whole.

2.1.12.3 Private Industry

Many companies rear insects for internal research and as food for other organisms, e.g., for biological control, education. Companies such as BASF, Bayer, Corteva, FMC, Monsanto, and S. C. Johnson rear insects for internal research. Companies that rear insects as food for other organisms include butterfly suppliers and companies with programs in integrated pest management (IPM) that produce entomophages and weed-eaters, including the ANBP. Some companies produce millions of pollinators such as bumble bees, and others rear billions of crickets, mealworms, and other "feeder" insects for pet stores and zoos. Others rear insects for sale as human food. These topics are covered in preceding paragraphs.

Many companies produce insects that are the end-products rather than research subjects. Those companies include suppliers of butterflies for release at festivities; in the United States alone, more than 50 companies are listed by the International Butterfly Breeders Association (IBBA), whose website lists the companies that sell monarchs, painted ladies, swallowtails, and other butterflies intended for release at weddings, anniversaries, birthdays, and other occasions. The Association of Natural Biocontrol Producers (ANBP) lists more than 50 members in North America. This is a partial representation of producers of predatory and parasitic insects used for various types of IPM programs, including biological control. Probably, the largest biomass of reared insects include products of the companies that sell crickets, meal worms, and other "feeder" insects used as foods for a huge variety of pets, zoo animals, and even human food. In fact, the topic of using insects as food for fish and poultry and for direct use as human food is emerging as a major industry. Several organizations in the private and public sector are supporting projects in using insects as intermediaries for processing low-nutrient quality materials (often waste substances) to produce higher quality biomass.

A special case of insect rearing that supersedes all other insect-rearing efforts in historical standing, economic value, and cultural importance is the ancient practice of silk production, mainly through use of mulberry silkworms. For example, Baruah and Gogoi 2013 state that in India alone, 35 million people participate in a 55 billion dollar industry! The industry known as sericulture is a historical first for insect rearing. Mulberry silkworms (*Bombyx mori*) were domesticated in China starting at least 4,000 to 5,000 years ago (Hamamura 2001). The domestication process involved what must be the first rearing system for an insect, and has taken myriad forms over the millennia, including more recent efforts to develop artificial diets to replace mulberry leaves. The deliberate genetic selection of silkworms has been aimed at improvement of the quality and quantity of silk, the efficiency of food utilization, and other features of selected silkworms that made them more convenient for silk farming and more economically sound. Other work on silkworms includes the management of silkworm diseases, including the famous work of Louis Pasteur, who is considered the "father of modern insect pathology" (Tanada and Kaya 1993). Pasteur was consulted in the emerging silk industry in Europe to help solve the problem of pebrine disease, a microsporidian infection caused by *Nosema bombycis* (Tanada and Kaya 1993). Many other aspects of the base of knowledge in entomology came from silkworms and the research that went into rearing them, but probably the most fruitful and influential outcomes came from the nutritional knowledge and feeding biology that was produced by researchers trying to solve problems with replacement of mulberry leaves with artificial diet (reviewed by Hamamura 1959, 2001).

2.1.12.4 Educational Institutions

Educational institutions have been prominent in rearing advancements. Many universities have insectaries in which insects are reared for research and teaching purposes. For example, the University of California, Riverside, where I was a graduate student (1975–78), had a main insectary and at least 20 sub-colonies of various insects, including a rearing room where Ralph B. March and George P. Georghiou kept their colony of DDT-susceptible houseflies, a colony originated in the 1940s for pesticide resistance studies.

At North Carolina State University, we have a central insectary where researchers in the Department of Entomology and Plant Pathology rear 30–50 species, and several laboratories have colonies of aphids, many Lepidoptera, forest insects, bed bugs, mosquitoes, and dozens of other species. In addition to the research function of these colonies, they are important for education. Although I have presented examples from just a few educational institutions, there are hundreds of other major rearing efforts at the various colleges and universities around the world.

Many educators from pre-school to graduate school rear insects for demonstrations. At NCSU, members of our Entomology Graduate Student Association (EGSA) and faculty mentors invest considerable amounts of time taking insects to local classrooms in an outreach effort that is rewarded with hundreds of students per year getting special exposure to the wonderful world of insects and the enthusiastic presentations by entomology's future leaders.

2.1.13 Rearing Operations for Sterile Insect Technique (SIT)

Because SIT is such a prominent force in insect rearing, it deserves a special attention to its infrastructure. Without question, the largest efforts in mass-rearing are the various programs that are dedicated to SIT. This includes the "insect factories" that produce screwworms (*Cochliomyia hominivorax* [Coquerel]), various species of tephritid fruit flies, including Mediterranean fruit fly (*Ceratitis capitata* [Wiedemann]), Mexican fruit fly (*Anastrepha ludens* [Loew]), melon fly (*Bactrocera cucurbitae* [Coquillett]), codling moth (*Cydia pomonella* [L.]), pink bollworm (*Pectinophora gossypiella* [Saunders]), which are no longer in mass-production, thanks to the success in the eradication program, diamondback moth (*Plutella xylostella* [L.]), and various biting flies such as mosquitoes and tsetse flies.

Dowell et al. 2005, in expressing the relationship between mass-rearing systems and SIT, summarized these points: "Insect mass-rearing for a sterile insect technique (SIT) programme is designed to move beyond the largescale rearing of insects in a laboratory to the industrial production of consistently high-quality insects for sterilization and release. Each facility reflects the unique biology of the insect reared within it, but there are some generalities for all rearing facilities." In the same book, Parker 2005 provides a comprehensive and most useful discussion of mass-rearing systems in SIT programs. He introduces the topic with this statement, "As the sterile insect technique (SIT) relies upon released sterile male insects efficiently competing with wild males to mate with wild females, it follows that mass-rearing of insects is one of the principal steps in the process." This statement implies that the quality or fitness of the insect-products of the SIT-related rearing systems is of paramount importance, and in this chapter, Parker explains all aspects of the rearing system, and how and why it must function within constraints that assure products comparable to the attributes of the wild counterparts.

The topics covered by Parker in relationship to rearing for SIT programs include:

- Introduction
- Economics
- Facility Design and Location
- Escapes and Environmental Concerns
- Strain Management
- Production, Process, and Product Control
- Automation of Rearing
- Diet
- Sex Separation
- Marking
- Storage

2.1.13.1 Introduction

Parker gives an overview of the requirements in a mass-rearing system whose purpose is SIT. He writes:

> As the sterile insect technique (SIT) relies upon released sterile male insects efficiently competing with wild males

to mate with wild females, it follows that mass-rearing of insects is one of the principal steps in the process. Mass-rearing for the SIT presents both problems and opportunities due to the increased scale involved compared with rearing insects for most other purposes.

Parker points out that of the 1300+ species of species rearing on artificial diet (as reported by Singh 1985), only a few species have been reared for SIT and have had the detailed and demanding criteria applied to their colonization and mass-rearing. Parker provides details of the standards and stringent criteria that is required for mass-rearing for SIT.

2.1.13.2 Economics

Parker explains that economics of rearing for SIT is of greater importance than rearing for most other purposes. He points out that large rearing facilities, such as the El Pino factory in Guatemala, produce more than two billion sterile male Mediterranean fruit flies *Ceratitis capitata* (Wiedemann) per week, leading to many economic issues related to labor, automation, diet, etc. Parker, citing Caceres et al. 2004 as an exception, points out that for SIT-based rearing, there is very little published information on the economics of rearing. This is unfortunate because throughout the world of mass-rearing, there is a tremendous amount of in house knowledge that was developed in response to the need to economize. For example, under M. Chaudhury's influence (e.g., Chaudhury et al. 1998, 2000, 2011; Chaudhury and Alvarez 1999) there have been several cost-saving measures developed in mass-production of screwworm flies (*Cochliomyia hominivorax*, Diptera: Calliphoridae), but finding such papers requires special search strategies. Parker 2005 also points out that there is perpetually the important issue of balancing economic improvements while maintaining quality. These points about economics and other components of rearing systems discussed here and elsewhere in this book should lead us to application of science-based rearing system optimization (Cohen 2001, 2015, 2018a, 2018b, 2019).

2.1.13.3 Facility Design and Location

Parker lists the exhaustive criteria for planning SIT-based mass-rearing facilities, including:

- Logistical access ease of delivery of rearing supplies and equipment
- Geological stability earthquake risk, hurricanes, flooding, etc.
- Political stability
- Acceptance of the facility by the local community
- Local government requirements or restrictions
- Access to water and utilities, reliability of supplies
- Construction and maintenance costs
- Availability of support and repair services
- Labor costs and availability
- Waste disposal
- Access to a suitable airport for rapid delivery of insects
- Distance from release area
- Quarantine considerations—can the insect survive in the surroundings?
- Proposed progression in case eradication is the objective—will the facility become stranded behind the eradication front?

2.1.13.4 Escapes and Environmental Concerns

An important consideration that rearing facilities must face is the prevention of escapes of destructive and sometimes invasive pests from the mass-rearing facilities. Efforts are a constant challenge to design containers that are at once hospitable to the target insects but resistant to escape. The handling procedures such as harvesting the insects to be transferred to the next step in the rearing procedures can lead to escapes, and these procedures must be carefully designed. In the event of inevitable escapes from various procedural steps, there must be safeguards such as black light traps, sticky traps, and in some insects: pheromone traps. The facility's infrastructure should be designed to be "escape proof" including the use of weather stripping that seals all openings such as doors or windows (Figure 2.7).

Well-designed containment facilities utilize structural features such as screens in all drains and extensive caulking to close off all openings. This includes regular inspections that help detect new violations of the system's integrity such as rodent gnawing or bird activities that create new opportunities for escape. The hallmark of this kind of security is well-trained and well-motivated personnel who will remain ever-vigilant in their efforts to maintain facility security.

2.1.13.5 Strain Management

Parker 2005 points out that the features insured by proper strain management are: (1) fidelity to "normal" behavior for the species in question, (2) fecundity, and (3) genetic diversity—all of which can be lost due to inbreeding or other violations of Hardy-Weinberg Equilibrium conditions. Parker reviews the various discussions (from Mackauer 1972; Bartlett 1985; Nunney 2002) regarding the numbers of colonized insects are appropriate for starting and maintaining a diverse and healthy colony. These points are discussed elsewhere in the current book, especially Chapter 10 on Relationships between Fitness, Quality Control, Health, Homeostasis, and Stress.

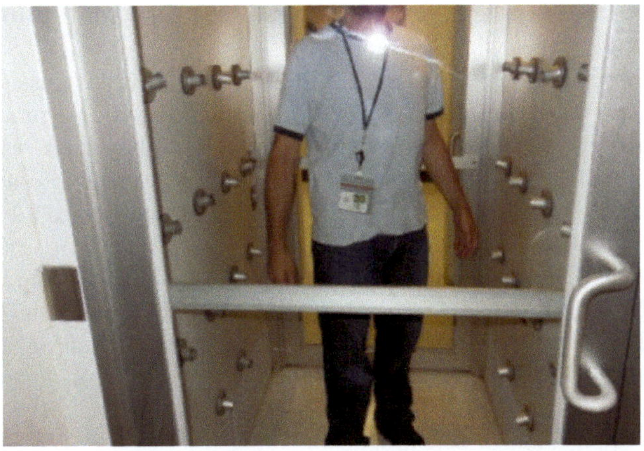

Figure 2.7 Researcher in a transition chamber at the USDA, APHIS Otis Air Base Quarantine Facility. The chamber scrubs entering and leaving personnel to help assure removal of potential insect "passengers" that could gain entry or exit from the facility.

2.1.13.6 Production, Process, and Product Control

Parker 2005 divides the overall concept of control into these three areas: (1) **production control** ("monitoring all rearing operations in terms of personnel, materials, equipment, schedules, and environment," (2) **process control** ("sampling immature insect stages to predict quality and determine sources of variability"), and (3) **product control** ("both at the output from the facility and in the field"). I like to think of the control concept as consisting of two areas: (1) **process control** and **product** (or **quality**) **control**. In my concept, all components of the processes in the rearing system are under the heading of process control. This includes the personnel and their background, training, attitude, etc.; the equipment for all phases of the rearing operations, including environmental systems (heating, cooling, and air circulating equipment); diet-making equipment (flash sterilizers, blenders, mixers, etc.); and raw materials (diet components, containers, oviposition materials, etc.). The process control factors even include the insects that go into the systems as part of the colonization procedure. The final outcome is under the auspices of product control. The final outcome (the nature of the insects, their qualities in terms of biological and physical characteristics) is the sum total of all the insects' characteristics. The final outcome is the overall quality of the insects that our processes have produced.

A very interesting and potentially useful concept that Parker presents in his "production, process and product control" is the filter rearing system (FRS) where a small sub-colony is maintained under less rearing pressure (and stress) than the large colony. The small, low density colony allows various treatments that could improve the stock (filtering out adverse factors) and allow supplementation with colony-improving factors such as host plants, improvement of pheromone responses, etc. The small size of the colony allows more controlled and individualized conditions than would be possible with the large colony, and the small "filter" colony could be used for selection applications that could improve the stock. In my observation, the Pasteur approach to screening out diseased individuals (such as silkworms infected with Nosema microsporidia) can be thought of as an example of a filter system.

2.1.13.7 Automation of Rearing

Throughout the history of mass-rearing, especially for programs such as SIT-based systems, automation and robotics have been a driving force in rearing system improvements (reviewed by Parker 2005; Cohen 2015). Parker points out that the motivation to develop automated rearing procedures are: "(1) reducing the expenses of rearing, (2) reduction in human error, with increased product performance and consistency," (3) "reduction in microbial contamination from personnel," and (4) "increased space utilization efficiency, resulting in lower costs—for building and for energy for environmental control." These points agree with the conclusions of Cohen 2015, but there are further issues that automation addresses. As we use the philosophy of automation and robotics, we approach questions of implications of making jobs done by humans replaceable by machines. The highly monotonous, yet indispensable aspects of rearing make for nearly inhumane jobs in terms of repetitiveness. For example, weighing out (or otherwise measuring) ingredients to make diets and the packaging of those diets is so mind-numbingly repetitive that they are often considered by insectary workers as unfulfilling jobs, leading to high rates of personnel turnover. This high turnover problem becomes the bane of the insectary manager's life where she/he is not only facing the errors that inevitably stem from repetition, but then they are also constantly interviewing new applicants and having to train those applicants.

For the reasons described above, it is desirable to automate or add robotics to all aspects of the rearing system including egg collection, diet-making/diet delivery, handling (such as sexing and quality assessment) of pupae, adult components such as sex ratios, and regulation of all environmental factors. This topic is discussed in detail by Smith 1999 and Smith and Nordlund 1999. In fact, for a period of time between 1995 and 2005, the USDA, ARS laboratory Biological Control of Mass Rearing Research Unit (BCMRRU) at the USDA facility at Mississippi State University was dedicated to application of scientifically based development of mass-rearing protocols. Smith and Nordlund 1999 discuss the types of approaches, including intensive engineering protocols that were developed to further automation. A more recent treatment of this topic is presented by Dowell et al. 2005. In conclusion on this topic, it is important to add that the accomplishments in automation are unfortunately under-represented in the literature that is freely accessible to the scientific community. The difficulty in published rearing system papers is discussed by Cohen 2019.

2.1.13.8 Diet

Pointing out that diets are arguably the most important component of rearing systems, Parker 2005 goes on to make the following statements about diets:

> It is important to ensure the quality of diet ingredients before purchase or use, e.g. by monitoring the physical and chemical variability of the ingredients, bacterial load, and any insecticide residues. Poor quality control of diet ingredients is a major problem in mass-rearing, including chemical and biological assays that the ingredients must pass. In the end, a cheaper ingredient may in fact be more expensive due to frequent "crashes" in production caused by low-quality ingredients.

These statements by Parker cannot be overemphasized, and while they are the focus of much of the book on insect diets by Cohen 2015, it is worthwhile to express here that the procedures for checking the diet's chemistry and biological efficacy on and ingredient-by-ingredient basis can be greatly facilitated by application of the design of experiments procedures emphasized in the current book (e.g., Chapters 8 [on Process Control] and 9 [on DOE procedures]).

2.1.13.9 Sex Separation

Parker 2005 discusses the importance of separating the sexes for SIT, a practice that is rarely required in any of the other mass-rearing systems. He explains that while males are the gender that actually performs the control function in SIT, simultaneous release of females can have a minor positive effect as what he calls "sperm sinks" that divert some of the wild fertile males from mating with wild females. Parker also discusses the strategies of separation such as by external morphology and by other means such as automated sorting by size or weight.

2.1.13.10 Marking

It is important to be able to distinguish the sterile insects from the wild ones to determine the ratios for proper proportioning. Although some SIT programs do not use marking, for those that do, various internal dyes such as Calco Red (Griffin 1984) and rubidium; while external dyes, especially fluorescent ones, have received considerable use. Internal dye markers include rare elements or isotopes (both radioactive and stable), and more recently immuno-marking has been used (Hagler and Cohen 1990; Hagler et al. 1992). Even more recently, genetic markers have been used, including fluorescent proteins that can be assessed with the aid of UV light. It is essential that whatever the marking strategy, the substance must not damage the insect, and it must be easy to administer and not interfere with the rearing process.

2.1.13.11 Storage

The advantage is tremendous of being able to produce and store an excess of insects to be released when the seasonal demand is greatest. Taking advantage of natural storage features such as diapause is very appealing as this possibility plays into the biology of the insects in question and causes very low amounts of stress. The use of cryopreservation (Leopold 1998, 2000) has been used with considerable success, especially when this storage strategy has been worked out in concert with the insect's physiology. Parker also reviews the influences of such physiological responses (stress responses) such as heat shock proteins, cold hardening, and dealing with osmotic stress (Lee et al. 1987; Denlinger et al. 1992; Parsell and Lindquist 1993; Denlinger and Lee 1998; Rinehart et al. 2000; Denlinger 2002).

CHAPTER 3

Design and Operation of Insect-Rearing Systems

Science, Technology, and Infrastructure

Upon entering one of the mass-rearing facilities, such as the ones discussed in this section, we notice the scope of the rearing processes and the intricacies of the operations. What we are seeing in rearing operations is what I have been calling throughout this book, ***infrastructure***. The infrastructure includes the facilities, equipment, materials, target insects, and the base of knowledge that exists in personnel. The infrastructure of a given rearing system generally has implicit literature on what led to the protocols that are being utilized.

It is to be greatly respected how many scientific and technological details have been integrated into these systems where every step in the process has been worked out in terms of efficiency optimization and cost effectiveness. The interface of the different engineering components with the living systems that are at the heart of mass-rearing is nothing short of amazing. Insects that originated in nature have been brought into captivity where they are supplied with artificial foods that are alien to the characteristics of the actual, native foods but serve to offer palatability, nutritional value, bioavailability to the insects' feeding systems, and stability in the face of various forces of entropy and deterioration. Besides diet considerations, mass-rearing systems provide accommodations for all the requirements of the target insects, including environmental features, gas exchange, waste removal or management, reproductive factors, and various specialized needs such as proper densities, provision of healthy microbial relationships, and for some insects, availability of social conditions. In short, the rearing systems must be microcosms of the insects' natural world. And the further remarkable feature of rearing systems is that they are carefully engineered to meet the requirements of rearing personnel to allow them to do their jobs efficiently, economically, and within the scope of often very meagre training. The systems are designed to reduce one of the great enemies of quality rearing outcomes: error or unacceptable variation in the process and product! Of course, it is vital that the product of rearing systems (insects to be used for specific purposes) is capable of doing the job for which the rearing system is designed, such as usefulness in SIT systems, biological control programs, use of the insects as food for other species, as ornaments for human esthetics, as producers of products such as bio-manufactured materials, or as research tools.

3.1 THE REARING SYSTEM AS A CENTRAL CONCEPT

To organize the features common to rearing systems in general, we must include the "starter" insects (their population characteristics, including their genetics, their interactions, and their biology), the food (natural, factitious, or artificial diets), the environment (covered in detail below), and containment (all reviewed by Cohen 2018a and 2019). It is important to recognize that rearing systems have evolved under the goal of developing processes that are organized for the purpose of producing abundant numbers of insects that are healthy, capable of producing successive generations of offspring. The underlying need to meet these desired outcomes is for rearing personnel to have **control over all the rearing processes**. The concept of control was originally expressed by Delcourt and Guyenot 1910 and is discussed and developed below; however, it can be expressed simply that *control* is what defines engineered systems. For rearing systems, control specifies an understanding of all the needs of the target insect and the requirements for the rearing system to function. It implies the need for rearing specialists to know how to meet the demands of all parts of the rearing process.

3.2 THE CENTERPIECE OF MODERN INSECT-REARING SYSTEMS: *DROSOPHILA*

Besides the thousands of papers written about Drosophila genetics since the early 20th century, thousand more papers have been written about all aspects of Drosophila biology on topics ranging from ecology of these flies (Starmer and Fogelman 1986; Starmer et al. 1986; Anagnostou et al. 2011), and further studies on gerontology, nutrition, and countless other aspects of Drosophila biology such as feeding choice studies (Ja et al. 2009; Deshpande et al. 2015). The papers by Deshpande et al. and Ja et al. cover the important topics of food selection in relationship to fitness of *Drosophila*, and they also contribute to the growing body of information about relationships between diet and fitness in the context of aging (geriatrics). The papers of Bridges and Darby 1933 and Bridges 1932, respectively, treat the evolution of diets for *Drosophila* and other aspects of rearing systems such as containers, home-made incubators, and basics of the ecology of rearing *Drosophila* as a laboratory organism (Figure 3.1). These papers are remarkably in-depth treatments of the progress that took place in the laboratory of Thomas Hunt Morgan with special emphasis on the evolution of rearing conditions, with the most comprehensive and in-depth explanations of the rearing system developed over two decades from the early 1910s through the 1930s. The comprehensive work by Morgan et al. 1913 treats the genetics discoveries with *Drosophila* and other species used as models of Mendelian genetics. The work done in Morgan's lab at Columbia University in the early 20th century shaped the field of genetics and other areas of biology, including insect rearing, to such an extent that many authors have regarded the "Fly Room" where the genetic research of Morgan's group (including, Bridges, Muller, and Sturtevant) became the subject of popular books and a movie called "The Fly Room," discussed in Anonymous 2015.

The Morgan team collectively generated two Nobel Prizes (for Morgan and another for Muller) and was responsible for establishing chromosomal mapping, X-ray induced mutations, sex linkage, and many other genetic discoveries. A less-recognized and less-heralded set of accomplishments were the many rearing advances that set the stage for much of today's insect rearing, not only for *Drosophila*, but also for countless other insects. This is especially true Calvin Bridges' detailed and meticulous scientific studies of several aspects of the *D. melanogaster* rearing system.

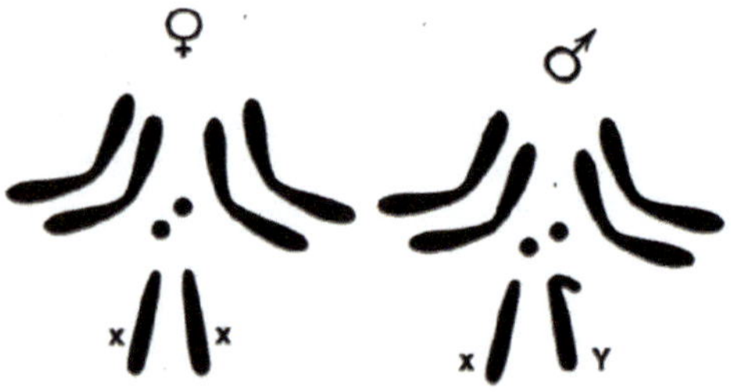

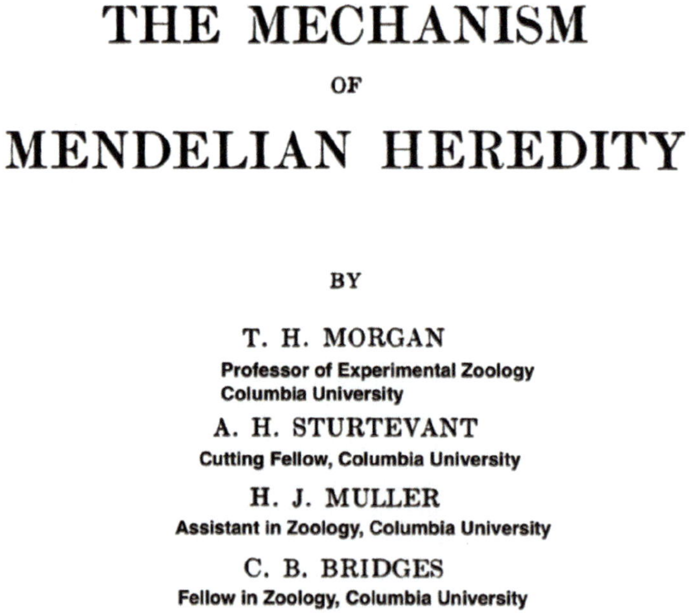
THE MECHANISM

OF

MENDELIAN HEREDITY

BY

T. H. MORGAN
Professor of Experimental Zoology
Columbia University

A. H. STURTEVANT
Cutting Fellow, Columbia University

H. J. MULLER
Assistant in Zoology, Columbia University

C. B. BRIDGES
Fellow in Zoology, Columbia University

Figure 3.1 Excerpts from the publications from the productive laboratory known as "The Fly Room" at Columbia University. The team referred to as "The Fly Boys, consisting of Alfred Sturtevant, Herman Muller, and Calvin Bridges—led by T. H. Morgan in their discoveries of much of the foundation of modern genetics. The figure includes the four color varieties of *Drosophila melanogaster* drawn by Ms. E. M. Wallace. The color phases of *D. melanogaster* from Morgan 1912 are: normal or gray female (upper left), a black female (lower left), a brown female (lower right), and a yellow female (upper right). The contrast between black, yellow, and brown flies is well brought out in the figures. The figure (lower left) includes the chromosomes (karyotypes) of *D. melanogaster*, showing the eight chromosomes that characterize the karyotype of this species, and on the right is the title page of the 1915 book by Morgan, Sturtevant, Muller, and Bridges. The book contains an early synopsis of the rearing system that was used for two decades in the Morgan lab and that was the continued topic of research in rearing system improvements by Calvin Bridges (along with his other excellent contributions to rearing system development).

3.2.1 *Drosophila* Diet and Rearing-System Development as a Historically First Modern Rearing System

While the two oldest papers that advanced the concept of artificial diet for flesh flies were by Bogdanow 1906 and 1908, the first paper that lays the foundation of controlled conditions in rearing and diet presentation was the Delcourt and Guyenot 1910 study that explains the importance of controlling all aspects of rearing conditions. This paper and several other pioneering diet studies by Guyenot 1913a, 1913b, 1913c, 1914a, 1914b made significant advances in *Drosophila* dietetics. For example, in Guyenot 1913a, the subject of the role of live microbes vs. aseptic diets was explored—a topic that had gained extensive attention in insect rearing from the earliest inquiries to the present day when insect gut microbiome evaluations abound. It is valuable to note that in the early days of diet development, the scientific community was in its "Golden Age" of microbiology and genetics, and some of the early studies on insects reflected the contemporaneous interest in the role of microbes in insect nutrition, leading to numerous studies of the essentiality of various species of microbes (e.g., Loeb 1915; Baumberger 1917a, 1917b, 1919; Baumberger and Glaser 1917; Loeb and Northrop 1917; Michelbacher et al. 1932).

3.2.2 Development of *Drosophila* Diets and Diet-Based Rearing Systems

Delcourt and Guyenot 1910 explained that in their efforts to study the heredity of Drosophila over successive generations, a requisite for accurate determination of virtually all genetic questions was the development of controlled conditions in the rearing process. They explained that the inconsistencies of the foods used to rear the flies, where the qualities and differences in fermentation progress of raisins, bananas, and apples caused variability that was too great to allow interpretation of the genetic results that they were testing. These inconsistencies and difficulties demanded a control of the rearing system—mainly through control of diets. It was also obvious that there was a microbial involvement in *Drosophila* rearing as was addressed by Delcourt and Guyenot 1910, Guyenot 1913a, 1913b, 1913c, 1914a, Baumberger 1917a, 1917b, Baumberger and Glaser 1917, Loeb and Northrup 1917, and Morgan et al. 1913 (as explained further by Bridges and Darby 1933). The sequence of development of artificial diets for *Drosophila* was as follows: fermented fruit (generally bananas) → mashed over-ripe bananas with yeast added → banana puree extract with yeast and agar → various fruit (raisins, apples, etc.) purees with yeast and agar. Two of the major advancements that were profoundly influential to not only *Drosophila* rearing but also rearing countless other insect species were the addition of yeast (live or dead) and agar to provide a gelled diet.

3.2.3 The Evolution of Controlled Diets for *Drosophila*: Microbial Relationships

For the period of 1910 through 1933, one of the liveliest subjects of inquiry was the relationship between insects and microbes in terms of dietary issues. There was a preoccupation with establishing the role and extent of microbial involvement in the feeding biology of Drosophila and other insects, especially other Diptera. A major portion of this progression was explained in elegant detail by Baumberger 1919 who reviewed his own and other researchers' experiments with yeasts vs. bacteria as food sources (as external or internal symbionts vs. nutrients) and live vs. heat-killed microbes. An important and illuminating part of these studies was the series of experiments that sought to tease apart the nutritional requirements that were part of the foods prior to microbial action vs. the foods post-fermentation (meant here as microbial action on media or raw materials) vs. the microbes themselves (Baumberger 1919; Michelbacher et al. 1931). After extensive review of the literature on the roles of microbes in feeding success on natural and artificial diets, it remains difficult to conclude how much of a diet's success in meeting target insects' needs is from direct effects of microbes (changing the nutritional quality of raw materials or media components), the contribution of the original raw materials, or the nutritional value of the microbes themselves (Cohen 2015). Cohen 2015 concluded that these complex questions about microbe/media/microbe-media interactions must be determined on a per species basis across all insects being reared. In summation, the many diets for *Drosophila* that are currently being used were evolved through a series of refinements of a fruit base with yeast added along with other nutritional and functional components such as corn meal, agar, and antifungal agents. Despite considerable effort to develop completely defined (holidic) diets for *Drosophila* to help establish the basic nutritional requirements for this insect, and in this context, all practical diets for this insect are based on complete foods in conjunction with yeast—mainly baker's or brewer's yeast *Saccharomyces cerevisiae*. The insistence of the rearing community to use *S. cerevisiae* is unfortunate in light of the extensive literature on the nutritional ecology of *Drosophila* (and tephritid fruit flies) where other species of yeast have been shown to be superior to *S. cerevisiae* as diet ingredients (Heed et al. 1976; Vacek et al. 1979; Starmer and Fogleman 1986; Starmer et al. 1986; and more recently Anagnostou et al. 2011; Meshrif et al. 2016, and the group working out of the Ja laboratory, Ja et al. 2009; Deshpande et al. 2015).

3.2.4 Other Aspects of *Drosophila*

Other aspects of *Drosophila* rearing systems are treated in this review in separate sections on temperature and humidity, containers, and miscellaneous environmental factors (such as light, radiation, etc.).The container aspect of rearing may seem to be trivial or simple, but because the container has to be an imitation of the insect's natural environment and because containers must be compatible with all the environmental needs of the insect, the container becomes a key component of rearing system development. Using the key words, "Insect Rearing" and "containers," 76 papers found in a search of Web of Science. What may seemingly be a minor part of insect-rearing systems—containers—must be recognized as one of the most consequential parts of the rearing system/ecology concept. A good container must serve the target insect's needs, including thermal features, humidity accommodation, gas exchange, being hospitable to developmental and reproductive needs, and any other microhabitat factors inherent in the insect's biology. The container must also serve the needs of the rearing personnel, including availability of the containers at a reasonable cost, storage accommodation, ease in access to the insects, and cleaning requirements. The comments of one of the earliest papers on rearing (Burgess 1908) these comments were made:

> One of the serious problems which it was necessary to solve in order to successfully rear the parasitic and predaceous insects which were being shipped from Europe to prey on the gypsy and brown-tail moths was to secure apparatus by means of which these insects, as well as their hosts, could be successfully reared in large numbers. All of the old style equipment in general use by entomologists for rearing work was tested, but in many cases it was found that radical improvements were necessary in order to accomplish the results desired. It was of primary importance to place the insects under as nearly as possible natural conditions and at the same time to keep them in confinement where they could be studied and observed and not allowed to escape from captivity.

In the last paragraph of the paper, Burgess elegantly stated the concept that the rearing conditions must simulate nature—but with conditions that meet the convenience of the investigator (or the rearing system personnel):

> It is perfectly true that breeding devices must be used which are especially adapted to the habits and character of the insects to be studied, and it may be desirable to modify these cages in such a way as to suit the special needs of the investigator.

From these earliest treatments of containers (or rearing devices) throughout the history of insect rearing, a variety of container-types is remarkable. As is the case for many rearing innovations, the rearing system for *Drosophila* spp. was often the earliest model. For example, Bridges 1932 wrote about the early rearing system in the laboratory of T. H. Morgan (a famous facility that came to be known as "The Fly Room," which was located at Columbia University and was used for most of Morgan's career [discussed above]).

In the early work on the heredity of *Drosophila* the cultures were raised in quart fruit jars, in quart milk bottles, and in a very miscellaneous collection of museum jars and other laboratory glassware. The great difficulty was in getting the flies out of these cultures for etherization and examination. By 1913 pint and half-pint milk bottles had become standard, largely on account of their uniformity of mouths, their heavy, strong glass and the ease with which they could be procured.

Bridges 1932 went on to explain in elaborate detail the other advantages of half-pint milk bottles and the rationale behind the refinements, including the importance of being able to clean and even sterilize the milk bottles to prevent carryover of insects from a previous generation as well as for sanitation. Bridges 1932 went into further detail about the optimization of container design, including his developing an agreement with a bottle manufacturing company to produce a special, custom-made square version of the half-pint bottle. Many laboratories continued to use the glass milk bottles until the later 20th century when the convenience and low expense of plastic containers overtook glass bottles in popularity. For example, he describes the modification of the standard round milk bottle to a more convenient and efficient square shape:

> In 1916 I made experiments to determine the optimum area of culture surface, and concluded that the pint milk bottle was best for the flies, but the half-pint, which was not far inferior in that respect, was greatly superior in ease of handling, etc. Accordingly in 1917 I drew designs and made a wooden model of a special bottle, having the floor area of the pint bottle and the height (135 mm) and other good features of the half-pint bottle.

Bridges 1933 went into further detail about the optimization of container design, including his developing an agreement with a bottle manufacturing company to produce a special, custom-made square version of the half-pint bottle. Many laboratories continued to use the glass milk bottles until the later 20th century when the convenience and low expense of plastic containers overtook glass bottles in popularity.

After Bridges' 1933 paper, there are very few papers that treat containers as a specific, central topic. However, as is the case with many aspects of rearing systems publications, there is a wealth of information about rearing containers that can be found in publications if they are scrutinized. Several papers that treated container specifications and that were influential or that presented novel containment systems among the larger body of rearing papers are Shorey 1963, Shorey and Hale 1965, Cothran and Gyrisco 1966, Ignoffo and Boening 1970, and Singh and Surrey 1980. The Ignoffo and Boening 1970 paper introduced the multiple cell rearing tray which became a tremendous improvement in mass-rearing containment systems, including the innovation of using the food industry's form-fill-seal technology to make various plastic trays such as the 32-cell tray and the 8-cell tray, which greatly reduce handling of individual cups. Several laboratories invested in the purchase of tray-making form-fill apparatus to ultimately reduce costs of tray purchases (Tillman et al. 1997) (discussed in greater detail in Chapter 5 "Case Study of the USDA, ARS R. T. Gast BCMRRU: Boll Weevil and Lepidoptera Mass-Rearing") (Cohen and Staten 1994). As Tillman et al. 1997 describe, the multiple-cell trays required specially adapted lidding materials that consisted of a plastic material with a glue adhesive that was either a self-sticking glue or a heat-activated glue. Such lidding was designed to include pre-perforated zones fitted to each rearing unit, allowing adequate gas exchange while reducing the loss of water from the diet and insects. Of course, these gas exchange demands must be met and balanced with protection from excessive water loss through evaporation from the diets and the insects. Also, appropriate containers must be designed to allow for gas exchange and protection from excess water loss such as those described by Tillman et al. 1997.

3.3 EVOLUTION OF SCREWWORM REARING SYSTEMS

The most celebrated and oldest system of sterile insect technique (SIT) is the one associated with the mass production of sterile screwworm flies, *Cochliomyia hominivorax* (Coquerel) (Babilonia and Maki Website; Brown 1984; Parker 2005) (Figure 3.2). In fact, the demonstration that SIT is feasible is unquestionabe, thanks to the various successes of the screwworm program. This success starts with the concepts advanced by Knipling 1955 and became one of the most successful and well-recognized insect control strategies, largely due to the remarkable success in mass-rearing technology developed between 1936 and the 1960s. Babilonia and Maki discuss the eradication of this pest in the United States, Mexico, and now throughout most of Central America, and these authors also describe the responses of the screwworm program to outbreaks of this pest in Africa. Parker's chapter on the development of screwworm rearing in Dyck et al. 2005 updates many of the changes adopted in screwworm rearing programs, updating the treatment of the evolution of the mass-rearing system described by Brown 1984. The Parker chapter, along with the other contents of the Dyck et al. 2005 book, offers a valuable resource in summarizing the dynamics of changes in the mass-rearing techniques, including the process control and quality control applications to the highly successful mass-rearing systems for screwworms, and other insect species that are produced for SIT programs. The Parker chapter presents valuable economic data in the context of total program effectiveness and in terms of the great leverage of seemingly incremental changes in rearing procedures that effect significant cost savings that help make the SIT programs financially feasible.

Parker 2005 includes wide-ranging information as well as granular details in matters of insectary design, management of genetic diversity, process and quality control, as well as details of automation of sexing, diet-production (including diet development issues), insectary environmental issues, treatment of risks of escape of fertile individuals, and well-being of personnel who spend nearly a third of their working lives in the mass-rearing facility.

Figure 3.2 Original screwworm mass-rearing facility at Mission, TX.

3.4 EVOLUTION OF SCREWWORM DIETS AND DIET PRESENTATION

The earliest diets for screwworms (Melvin and Bushland 1936, 1940) consisted of water, ground beef, citrated beef blood, and formalin which supported 88 consecutive generations over four years with a yield of about 4,000 pupae per week. Melvin and Bushland 1940 also reported smaller scale rearing of screwworms with a sterile diet of egg yolk and whole milk in a soft agar kept sterile with formalin. The early diet's yield of 4,000 individuals per week was amplified to more than a half billion per week, thanks to the many improvements in diet, automation, and other features of rearing advancements (Brown 1984). Although the modifications in diet and the evolutionary progress, which was often seemingly incremental, are too many to cover adequately here, many of the changes in rearing media include substitution of other ingredients for the meat, changes in the semisolid diet to a liquid form, using whole egg and calf milk supplements, and media for stabilizing the liquid diet, such as cellulose, cellulose acetate, shredded paper, cotton, bagasse, corn cob fractions (grits), or other locally available materials. Several improvements in the liquid-stabilizing domain of diet components were discussed by Chaudhury and Alvarez 1999 and Chaudhury et al. 2011. These works document the importance of texture and physical matching of the insects' feeding requirements, beyond simply meeting nutritional needs. Besides the mass-rearing science and technology in the screwworm SIT system, scores of research studies on the physiology and genetics of mass-reared insects have been stimulated by the impetus of the huge, economically important rearing programs. There are reciprocal benefits of such studies as Bush et al. 1976 and the technological advancements of the large-scale systems where insects have become highly, though inadvertently, domesticated (Figure 3.3).

It should be noted that the screwworm mass-rearing system faced many problems that needed to be addressed as the scope, economic and biological feasibility of the program could be satisfied. Some of these improvements or breakthroughs were described in seemingly incremental reports in the peer-reviewed literature (most often in the *Journal of Economic Entomology*).

3.5 THE EVOLUTION OF REARING SYSTEMS FOR PINK BOLLWORM

Currently, the largest program in rearing pink bollworms (*Pectinophora gossypiella* Gelechiidae: Lepidoptera) is the USDA, APHIS facility in Phoenix, AZ (USA). The Pink Bollworm Facility has been in operation since the early 1960s dedicated to production of pink bollworms for several research projects and largely for programs dedicated to the management of these serious pests. As well as being a model for innovation and progress in insect-rearing systems, the Phoenix facility has been a model of coordinated support from the private sector, several state governments, and national governments

Figure 3.3 **(a–m)** Pink bollworm rearing at the USDA, APHIS in Phoenix, AZ. Figure (a) shows the larval pink bollworm rearing room; figure (b) shows a worker handling the Hexcel pupation collection sheets; figure (c) shows the pupal weighing setup for quality control; figure (d) shows the adult collection configuration; figure (e) shows adult collection and chilling; figure (f) shows adult layer cages with vacuum system for scale removal; figure (g) shows diet leaving extruder and being conveyed to the chiller belt; figure (h) shows cut/chopped diet being added to the form-fill rearing units; figure (i) shows infested trays (with Calco Red marked diet) being hanged on rearing racks; figure (j) shows outside system for scale collection and pre-disposal; figure (k) shows eggs being dried on paper towels after oviposition; figure (l) shows pupae being loaded onto trays for adult collection. Figure (m) was kindly provided by Dr. Greg Simmons and Dr. Neil Morrison and represents genetically modified pink bollworms used in field trials as reported by Simmons et al. (2011). The photos for figures a through l were kindly provided by Dr. Hannah Nadel, USDA, APHIS, Otis Laboratory.

(Continued)

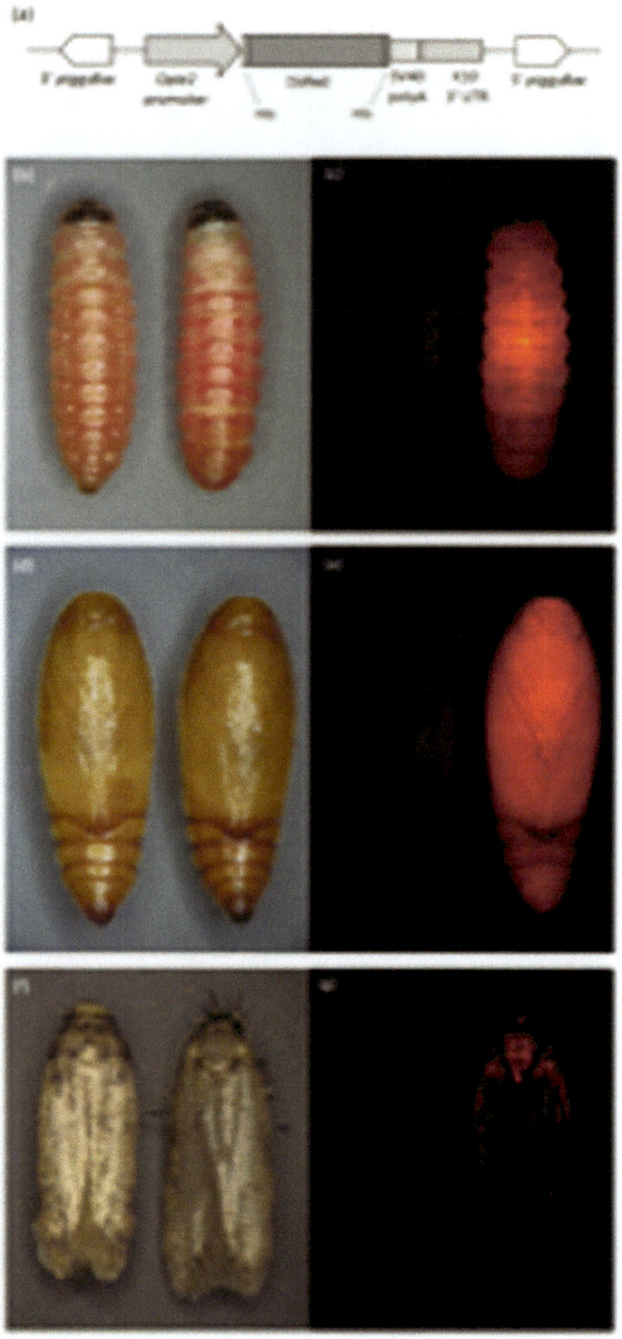

Figure 3.3 *(Continued)*

in the United States and Mexico. The management programs include very successful eradication efforts in Arizona, California, New Mexico, Texas, and several states in Mexico. The management strategies include SITs, pheromones, cultural practices, biological control, genetic selection/transgenic technology, and small volume applications of pesticides in an overall program of integrated pest management (IPM). Tens of millions of dollars per year have been dedicated to the pink bollworm program, and hundreds of workers at all levels from entry level personnel to senior leadership positions. A recent article in Western Farm Press (http://www.westernfarmpress.com/cotton/pink-bollworm-eradication-cotton-2017) explains the history of pink bollworm management and ultimately eradication in the southwestern cotton belt. From my perspective as a student of insect-rearing history, the single most indispensable feature of this program's success is the mass-rearing system for pink bollworms. The characterization of the evolution of this rearing system is a prime example of the issues that need to be understood.

3.5.1 The Early Days of Pink Bollworm Rearing: Pioneering Studies in Lepidoptera Rearing Systems

During the 1950s, several papers were published on the basics of pink bollworm diets and rearing, starting with a nutrition study by Beckman et al. 1953. Erma Vanderzant was a major contributor to nutritional research on cotton boll weevils and pink bollworms much of her attention was dedicated to development of our understanding of the basic nutrition of these insects, with a stated goal of applying nutrition to practical mass-rearing systems (Vanderzant and Reiser 1956a, 1956b; Vanderzant 1957, 1958; Vanderzant et al. 1957, 1959). One of the most successful advancements in insect rearing, the introduction of wheat germ to pink bollworm diets, is a product of the Vanderzant-led research efforts. It is an important fact that the introduction of wheat germ to insects diets was in a paper on boll weevil diets (Vanderzant et al. 1959), though in this report the wheat germ did not lead to the remarkable success that it did in the debut of this material in the pink bollworm diet (Adkisson et al. 1960b). In fact, the only positive effect in the boll weevil diet report was that wheat germ had unexpectedly stimulated oviposition, which previously could only be induced with extracts from cotton bolls. Though the rationale for using wheat germ in the boll weevil diet and the pink bollworm diet was not stated by Vanderzant et al. 1959 nor by Adkisson et al. 1960b, the thinking that led the Adkisson team (which included Erma Vanderzant) to use wheat germ in the pink bollworm diet turned out to be in retrospect revolutionary in terms of the hundreds of species reared on diets based on wheat germ and the billions (possibly trillions) of insects whose production is successful thanks to this single component. It is of further interest that wheat germ later turned out to be a key component in boll weevil diets, once these wheat germ diets were supplemented with cholesterol (Vanderzant 1967).

3.5.2 Significant Later Accomplishments in Pink Bollworm Diets

Other important steps in the evolution of pink bollworm rearing include the addition of soy flour to the diet as mentioned by Stewart 1984 and the adoption of several antimicrobial agents (Ouye 1962; Stewart et al. 1976). Tracing the origins of soy components now used in pink bollworm diet presents a complex forensic story. Stewart 1984 did not discuss the reasons or sources that influenced adding soy flour to the diet he described as stemming from Ouye 1962 (who used a wheat germ diet based on Adkisson et al. 1960a) It is also noteworthy that the Ouye 1962 paper was the first to introduce various antimicrobial agents—both antifungals and antibiotics—into conventional artificial diets. Stewart 1984 cites Shaver and Raulston 1971, so it seems that the later paper influenced that adoption of soy flour into the pink bollworm diet, making production of the insects more economical and reliable than if the diet was limited to casein as a protein source. Soy flour is a major component of the current pink bollworm diet as it is an important part of dozens of other insect diets. The nutritional and functional basis of soy flour is discussed thoroughly by Cohen 2015; but suffice it to say that it is an exceptionally nutritious material, which supplies all the amino acids, many vitamins, minerals, and offers textural features such as fiber and feeding stimulants. It is evident that use of soy components in insect diets began in Japan in an effort to develop artificial diets for silkworms (*Bombyx mori*) as described by Hamamura 2001 (a translation of a book on silkworm diets written in the 1970s) and Ito and Horie 1962. Redfern 1963 reported using soy protein isolate in a diet for red leaf roller, and the first report of soy flour in what I am calling Western literature was by Shaver and Raulston 1971, who reported excellent success in using soy flour (raw, toasted, and soy meal) to replace dairy casein. In several reports from the Japanese literature or from Japanese research groups, a material called soy casein was used as a protein source for silkworm diets. These comments raise a point that has become apparent to me that there is a wealth of information on excellent insect diet and nutrition science that resides in Japanese, Chinese, Indian, and possibly other Asian literature but remains unavailable to those of us who have not accessed this body of information. The fortunate experience that I had in getting access to the Hamamura translation (2001) of a remarkable book on *Silkworm Rearing on Artificial Diet*.

3.5.3 Scaling up Pink Bollworm Rearing

The diet for pink bollworms was initially produced in kg or sub-kg quantities, then progressed to 50–100 kg batches made in steam kettles (Stewart 1984). The development of rearing technology in the early stages (1960s) is reviewed by Martin 1966. Between the 1970s and the mid-1990s, the production of pink bollworms was limited to about a million adult insects per day, largely due to scope of diet production. The transition to diet production through adoption of the twin screw extruder derived from the human food industry allows production of diet adequate to support 25,000,000 pink bollworms per day (Edwards et al. 1996). However without the automation (e.g., Miller et al. 1996; Richmond and Martin 1966) the large-scale production of the pink bollworm would not be possible. The pink bollworm rearing system includes scores of innovations such as automated egg surface sterilization, scale collection systems that protect workers and insects from the hazards and pathogens in scales, Honeycomb (=Hexcel, Stamford, CT 06901) for pupation sites and pupal collection, form-fill seal machine-produced containers for larvae, refrigerated adult collection sites, disposal of tons of scales per year, oviposition technology using sterile paper toweling fitted to vacuum-connected adult mating and oviposition containers, and many other features. Most of these features are illustrated in Figure 3.3(a–m). Stewart 1984 summarized the progress toward automation for a period from 1970 to 1979 when production went from about 0.5 million per day to about 1.5 million per day. But when the extruder technology was adopted (Edwards et al. 1996), production was able to leap to 10–20 times the number produced by the steam-kettle method in 1979.

3.5.4 The Twin-Screw Extruder Is Introduced to Insect Rearing

Edwards et al. 1996 provided a revolutionary framework in mass-rearing, not only for the pink bollworm but also for many other species (Table 3.1). The statements of objectives for adopting twin-screw extrusion for insect diets were as follows:

1. Develop a screw profile and extrusion conditions for heat treating pink bollworm diet (PBW) diet containing 63 to 67% moisture to temperatures of 150°C
2. Develop forming techniques producing particles of extruded diet with greater surface roughness than the PBW facility process
3. Measure the extent of bacterial contamination in the extruded diet to determine heat treatments which would eliminate bacterial growth
4. Measure the extent of thiamine losses during processing to determine whether additional vitamins may be needed in the diet to make up for losses during processing

3.5.5 Details of the Extruder Engineering Procedures

Some explanation of the configuration of the extruder and screw configuration is in order. Edwards et al. 1996 explain that the extruder used for these tests included two co-rotational screws of a 37.4 mm diameter and a length of 1480 mm, with nine zones or barrels whose temperature could be controlled independently of one another (Figure 3.4, normal screw configuration). While the details of the extruder and screws are discussed at length in the Edwards et al. 1996 paper, it is informative to note that there is a potentially great diversity of portions of the screws that can lead to features such as improved kneading, shearing, and mixing. Within the range of simultaneous or contiguous actions that the screw configuration can lend to extruder processing is a degree of friction that adds heating and mechanical force to modifying the product on a micro- and molecular level.

It must also be noted that the complexity of the various factors and forces in the extrusion process leads to dynamics and results that are difficult to predict. The simple heating of a material without the shearing, kneading, cavitation, etc. would lead to a more static, predictable outcome than the combined heating along with all the other forces and dynamics. An added difficulty in predicting the outcomes of extrusion of insect diets is that each component in the diet (Table 3.2, PBW Diet reported by Edwards et al. 1996) includes 21 different substances plus water. It is impossible to predict in advance how the ingredients will interact with one another during the extrusion process. For example, taking just four components (wheat germ, cellulose, agar, and corn oil), it is likely (or possible) that corn oil will behave far differently in mingling with wheat germ, agar, and cellulose at higher temperatures and pressures than it would at lower levels of these factors. The outcomes from these rather stochastic and chaotic interactions can be a product that is far more desirable than what is made from the same materials but processed under more conventional conditions for diet-making such as with a steam kettle and paddle-type mixer. It is also possible that the complex actions of the extruder can lead

Table 3.1 Profiles of Autoclave, Steam Kettle, Flash Sterilizer, and Twin-Screw Extruder (From Cohen 2015)

Equipment Type	Throughput (kg/h)	Microbial Kill Capacity	Cooking Homogeneity	Destruction of Ascorbic Acid	Relative Cost	Versatility
Autoclave	1–5	High	Low	High	Low to moderate	Low
Steam kettle	5–25	Not spore-formers	Low	Moderate	Low to moderate	Low
Flash sterilizer	50–100	High	High[a]	Low	Moderate to high	Moderate
Twin-screw extruder	50–200	High	High	Low	High	Very high

[a] Mixing homogeneity is moderate.

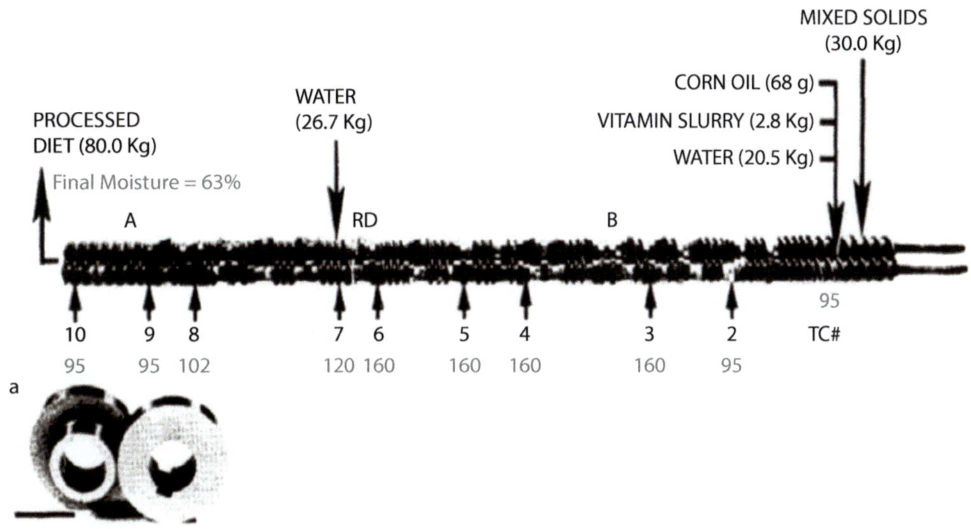

Figure 3.4 Normal screw configuration noting screw elements (A), kneading blocks (B), and restriction discs (RD, and inset A), with a typical mass balance at the extruder. TC marks the position of the product temperature thermocouple in the zone of the same number. TC number 10 is the diet temperature. The screws from the Continua 37 (Werner and Pfleiderer Corp., Ramsey, NJ) twin screw extruder used by Edwards et al. 1996 to produce artificial diet for pink bollworms (redrawn by Cohen 2015). The screw conformation, the contents being processed, and the temperatures of the twin-screw extruder used by Edwards et al. 1996 to produce the initial tests of extruded diet for pink bollworms are shown here.

to an inferior product to the conventionally prepared ones. This discussion leads us to the conclusion that even with expert engineering, the development of diets and diet modifications and optimization, must be done empirically (Figure 3.5). This topic is covered in more detail in Chapter 7 (Designing Error (Variability) Reduction in Rearing Systems: Reducing Uncertainty).

Other features of optimization of the extruded diets are the interplay between cooking temperatures along the extruder's barrels, the moisture content, pressure, diet residence time within the extruder, and exit temperatures and pressures. It is important to note that the extruded diet's lower moisture content (63% vs. >75% for steam

Table 3.2 Pink Bollworm Diet Reported by Edwards et al. 1996

Ingredient	Amount (kg/100 kg)	Ingredient	Amount (g/100 kg)
Toasted soy flour	44.90	Tetrachlorocycline	270
Wheat germ	19.55	Corn oil	200
Agar	13.00	Fumidil B	112
Sugar	9.85	Calcium pantothenate	51
Wesson salt mixture	6.5	Nicotinic acid	26.1
Powdered cellulose	1.69	Riboflavin	13.0
Potassium sorbate	1.34	Folic acid	13.0
Methyl paraben	1.05	Thiamine HCl	6.5
Locust bean gum	0.69	Pyridoxine HCl	6.5
Choline chloride	0.65	Biotin	0.52
		Vitamin B12	0.027

This diet is slightly modified from the formulation reported by Stewart 1984. However, unlike the diet prepared in steam kettles, having a final moisture content of about 75%, the extruded diet has a moisture content of 65%.

Figure 3.5 A small twin-screw extruder producing a diet "noodle." The target insect is the PBW, which was the first insect species to be reared on extruded diet. (a–c) Cryo-SEM micrographs of PBW diet with different exit die temperatures, progressively higher from (a–c). The fragmentation of the diet was greater with successively higher exit die temperatures, thus creating a greater surface area of available edges that allow neonate PBW to feed more efficiently. Edwards et al. 1996 point out that neonates have a difficult time initiating feeding when diet is too smooth and too few surfaces are narrow enough to be spanned by the narrow gape of the mandibles.

kettle diets) was not only acceptable to the PBW larvae, but it also proved to be superior to the higher water content, conventionally cooked diets. In fact the same extruded diet used for PBW has also been used successfully to rear several other species of insects, including *Amyelois transitella*, the navel orangeworm (Lepidoptera: Pyralidae) (Dr. Matt Ciomperlik, USDA, APHIS; aphis.usda.gov).

3.6 CODLING MOTH MASS REARING

The SIT program and the mass-rearing system for codling moth (*Carpocapsa pomonella*) (Lepidoptera: Olethreutidae) is one of the crown jewels of insect mass-rearing coupled with highly successful control programs (Hamilton and Hathaway 1966; Rogers and Winks 1993; Bloem and Bloem 2000). The codling moth rearing system that is currently in operation and described by Dyck 2010 demonstrates a confluence of rearing techniques that evolved over two decades. This topic is thoroughly reviewed by Dyck 2010, but a few comments about the evolution involved in this system are worth mentioning. As Dyck points out, diets for codling moths were derived from diets previously developed for other species, including the earliest diets described by Redfern 1964, Hamilton and Hathaway 1966, and Rock 1967, and several diets reported by Howell 1970, 1971, 1972. Most of these diets were not unique to codling moths, except for the use of soy protein by Redfern, which seems to have led to the use of soy products in insect diets among Western entomology researchers. This point is made in the context that soy products in silkworm diets had been used for several decades by Asian researchers as discussed by Ito 1961a, 1961b, Ito et al. 1974, 1975, and Hamamura 2001. Also, the development of mass-rearing systems for codling moth introduced several other important breakthroughs that influenced not only codling moth rearing technology but also other systems. For example, the use of sawdust by Brinton et al. 1969 was an important foundation for work with wood-eating insects and is a valuable model for managing texture in insect diets with materials other than hydrocolloids (Cohen 2015) (Figure 3.6). Codling moth rearing was also influential in the overall scope and approaches to the advancement of mass-rearing technology with respect to quality control and automation (Rogers and Winks 1993). Another important aspect of the mass-rearing and SIT program for codling moths is the development and advancements in quality control associated with this rearing program. Several of the papers that are excellent examples of applications of quality control principles and that are based on codling moth mass-rearing include Bloem et al. 1999a, 1999b, 2001 and Calkins et al. 2000.

A special commentary is in order regarding the Dyck monograph. The publication is a scholarly treatment of the methods and rationale behind the methods

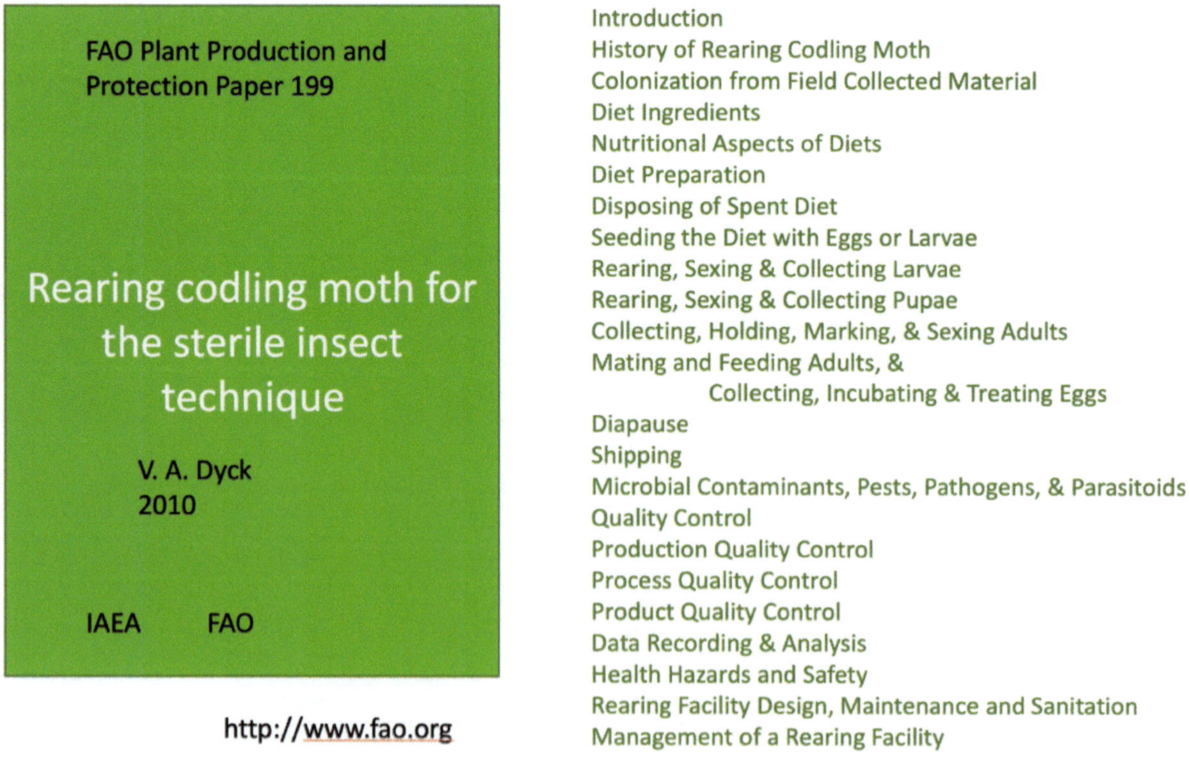

Figure 3.6 Monograph by Dyck 2010 on mass-rearing codling moths.

of rearing codling moths with nearly 400 cited references that deal directly or indirectly with the science and technology behind rearing these important insects. Dyck provides the details needed for a meticulous set of standard operating procedures (SOPs) and a rich treatment of the reasoning behind the procedures. The treatment of codling moth that I have included in this chapter is of very limited scope because the procedures and processes are so well expressed and documented in the Dyck monograph.

3.7 GYPSY MOTH

3.7.1 Gypsy Moth Introductory Comments

ODell et al. 1984 provide one of the most complete accounts of gypsy moth (*Lymantria dispar* L. Lepidoptera: Erebidae {formerly Lymantriidae or Noctuidae} mass-rearing. They begin their account with the history of gypsy moths in the United States, explaining that the insect had been brought to the United States (Massachusetts) from France by Leopold Trouvelot who tried to rear the notorious pest in his basement from which some moths escaped in 1868 into the Medford, MA neighborhood and established the US infestation that remains a serious pest today. Trouvelot, an astronomer and natural historian, was seeking a replacement for silkworms (*Bombyx mori*) to establish a silk-production system for commercial purposes. ODell et al. continued with the history of gypsy moth rearing in 1957–1960 at the Northeastern Forest Insect and Disease Laboratory (US Forest Service) and at the USDA, APHIS Gypsy Moth Methods Development Laboratory at the Otis Air National Guard Base, MS, starting in 1965. Both laboratories were involved in various programs involving research on gypsy moth control by SIT, biological control (including pathogens), and pesticides. ODell et al. describe the more recent efforts to cultivate the gypsy moth as follows:

> In 1975, with the support from the US Department of agriculture's expanded gypsy moth program a USDA interagency team began investigating how to improve mass rearing technology for the gypsy moth. The main stimulus for developing a large-scale rearing capability at that time was the potential of a virus specific to the gypsy moth this virus is now registered under the name GYPCHEK. The goal of the team was to develop the best possible system for mass rearing, specifically: a low-cost diet and containerization, methods for handling insects, procedures for control of diseases, and the best possible rearing environment, quality control procedures, procedures for controlling the health and safety of workers, and pilot scale facilities for insect and virus production. The accomplishments of this program through 1977 have been reviewed by Bell et al. 1981 the rapid improvement of the mass-rearing capability by 1977 was directly responsible for re-evaluation of the sterile male technique it has also been largely responsible for the success of several other studies that require large numbers of uniform healthy insects.

3.7.2 Gypsy Moth Rearing Facilities

ODell et al. 1984 describe the rearing facilities as having begun with some mobile homes (a common recourse to facilities for insect rearing) and advancing to more permanent buildings. In 1975, the prototype facility was capable of producing 25,000 insects per day. ODell et al. explain that the colony selection began in the early 1870s, and they point out that what they call the North American strain, derived from the insects that escaped from the Trouvelot colony, were a highly in-bred laboratory stock. The authors site Ernst Mayr 1942 in considering that a strong founder effect was operant in this colony and all colonies derived from it.

The strain of gypsy moth widely reared and the stock in the Connecticut Forest Service Laboratory is called the NJSS (New Jersey Standard Strain), which was colonized from batches of eggs collected near Blairstown, NJ, in the winter of 1967. This strain had become laboratory-adapted (domesticated), and the eggs and larvae from the original collection in New Jersey had several instances of pathogen attack; therefore, ODell et al. conclude, the selected strain had some degree of presumed resistance to disease. The NJSS strain was further truncated genetically by researchers who saved only 5% of each generation for colony production. The researchers selected the pupae that were placed in the stream of colonization succession by observing vigor which led to improved viability of the colony.

> Factors affecting virus quality—human error, which enhances microbial and fungal contamination during production, virus challenge, and harvest, is the main cause of poor virus quality. Airborne contamination caused by poor sanitary conditions is the result of human error, as are poor harvesting conditions, which increased microbial and fungal contamination in dead insects. Deviation from production procedures is likely to damage virus quality. For example, increasing insect density per cup or changing environmental conditions would likely increase variations in insect development and decrease polyhedral inclusion bodies program; a change in diet composition for example a reduction in vitamin C concentration might increase insect susceptibility to bacterial contamination, producing more contamination in the final product. Quality of the virus may be measured indirectly by gypsy moth susceptibility. The degree of susceptibility to NJSS produced virus depends on the geographic distribution, age, and physiological conditions of the hosts piece of the host, species of the hosts food plant, and climatic conditions.

3.7.3 Quality Control of Gypsy Moths at the Northeastern Forest Experiment Station (NFES)

The environmental conditions in the NFES laboratory are exemplary and deserve special mention and attention here. "Airflow in the rearing room is distributed downward in a vertical laminar-flow pattern with a complete clean-air exchange—HEPA (high-efficiency particulate air) filtered, 95% DOP (dioctyl phthalate aerosol penetration) rating—every 3-5 minutes." The rearing conditions include air temperature maintained at 25°C, 55–60% relative humidity with a photoperiod of 16:8 (photophase:scotophase). The conditions are recorded continually with a wet-dry bulb recorder and hygrothermographs, which are checked daily. Larval development, under these conditions, takes 29–30 days for males and 30–31 days for females, and the rearing team checks the synchrony of development every seven days post implant of eggs. The records are recorded and graphed.

Thanks to the exceptionally meticulous record-keeping, the research teams at NFES and at USDA, APHIS (Otis Airbase) discovered a departure from what had been very stable colonies that were producing high quality offspring for about 27 successive generations over the period of 1967–1980 (ODell et al. 1984). However, in 1980, irregularities in the development of larvae were noticed by Mr. John Tanner and B. P. Weeks in the Otis (APHIS) laboratory. ODell 1992 notes that Tanner and Weeks were the first to use the expression "straggling," describing it as "the failure of many larvae to develop past the first or second instar." They reported that between November of 1979 and February of 1980, the number of larvae that failed to pupate increased from less than 5% to between 40% and 50%. Personnel in the APHIS laboratory began a process of selection of larvae to include in the mass-rearing system, using only larvae that had reached 2nd instar within seven days, post-hatching to start the successive generations of the colony.

This selection measure temporarily seemed to solve the straggling problem, but it soon re-emerged, and it was also noticed that the Forest Service (Hamden, CT) colony also began showing signs of straggling or what was coming to be known as "abnormal performance syndrome." This syndrome is discussed in more detail in Chapter 8 ("Building a Process Control System"), but the approach to attacking this problem was enhanced by the very fastidious quality assessment procedures that were applied by meticulous and conscientious scientists and technicians in the two gypsy moth laboratories. This story of careful adherence to SOPs that were designed via application of rationale and science-driven protocols is a model for effective insect-rearing systems. Included in the SOPs and quality control efforts were an impressive array of tests that included behavioral assessments such as flights in wind tunnels, mating choices, life history parameters, analytical features—such as egg yolk nutritional content, microbial assessments of intracellular bacteria, screening for viruses, and many other types of analyzing the reared insects and the effects of virtually all features of the rearing system. For example, before a container configuration was adopted, it was put through rigorous comparisons to help optimize the ultimate outcome.

3.7.4 Further Discussion of the Gypsy Moth Diet: Chronicling the Abnormal Performance Syndrome

The diet in Table 3.3 is an example of the application of the meticulous comparisons to arrive at all the ingredients from wheat germ to water. Yet, despite the high level of care given to selecting the types and amounts of ingredients, factors that were beyond the control or anticipation of the rearing specialists caused departures from quality. The most notable of these departures was in the Wesson salt (mineral) mixture. As described by Keena 1993, ODell et al. 1997, and Keena et al. 1998 the change of the iron compound from amorphous to crystalline iron phosphate helped cause an iron deficiency that was exacerbated by other features of the gypsy moth rearing system. ODell 1992 also pointed out several other possible contributors to unstable populations of lab-reared gypsy moths in the Forest Service and APHIS rearing systems.

Among the valuable suggestions in the chapter by ODell et al. 1984 are the following which are applicable to many other species besides gypsy moths:

1. Determine and maintain optimal storage conditions and shelf life for perishable dietary ingredients, particularly wheat germ, casein, vitamins, and gelling agents.
2. Standardize through automation the processing and dispensing of diet (mainly to eliminate day-to-day variation).

Table 3.3 Composition of Diet Used for Rearing Gypsy Moths

Ingredient	Source	Grams per Liter
Wheat germ, raw	Mennel Milling Co., Fostoria, OH	120
Casein	New Zealand Milk Products, Rosemont, IL	25
Salt mix, Wesson	ICN Biochemicals, Cleveland, OH	8
Methyl paraben	Tenneco Chemical, Piscataway, NJ	1
Sorbic acid	Dirigo Corp., South Boston, MA	2
Vitamin premix (26862)	Roche Chemicals, Division of Hoffman-LaRoche	10
Agar	Moorehead and Co.	15
Water		800

Figure 3.7 Current gypsy moth rearing at USDA, APHIS, Otis Airbase, MA. Upper left, technicians Hannah Landers and Christine McCallum adding diet and eggs to containers; upper center, rearing containers with diet labeled with Calco Red; to center: gypsy moth adults in plastic containers, diet production room; right: quality control operations performed by Sue Lane and supervisor, Dr. Hannah Nadel; and lower left: automated cage-handling in diapause room. (Photos courtesy of Dr. Hannah Nadel, USDA, APHIS).

3. Automate implantation of eggs and larvae on diet, and automate harvesting of pupae.
4. Simplify monitoring of development and reproductive performance.
5. Determine how many insects can be successfully reared per unit of space if the rearing system is further refined.
6. Determine what environmental conditions will foster the best rate and synchrony of development (particularly for egg hatch, pupation, and adult eclosion).
7. Develop a method for rearing mostly males for sterile-male production and mostly females for virus production. (The female produces more virus because it is larger than the male.)
8. Develop and explore the use of nondiapause or short-diapause genotypes to augment or replace the existing diapause strain (to reduce or eliminate the requirement for chilling and to shorten generation time).
9. Develop bioassays, and establish standards for evaluating insects for consistency of performance in tests for standard toxicity and activity.
10. Develop systematic methods for investigating deviations in product quality.
11. Develop a communication system for informing users of production and product change, and encourage the establishment of internal standards for periodic evaluation of the product received.
12. Develop an antiserum for desensitizing personnel acutely allergic to gypsy moth setae and scales.
13. Investigate relative chilling effect of air movement on eggs being held at various temperatures (0–7°C) and at 70–100% relative humidity.

The suggestions by ODell et al. of these factors (which I have italicized) are especially applicable to many other rearing situations: (1) diet component freshness, (2) and (3) automation of diet processing and egg-implantation, (4) simplification (and standardization) of development assessment, (5) optimization of numbers per rearing unit, (6) determination of environmental factors that lead to synchronous development (this should include other factors such as strategies to harvest cohorts of similar developmental stages and ages), and suggestions (9–11) are concerned with improvements in process and quality control. All of these points reinforce the title and theme of this book, "design, operation, and control" of all rearing systems (Figure 3.7).

3.8 TEPHRITID FRUIT FLIES

3.8.1 Introduction to Tephritid Mass Rearing

Some of the largest facilities for insect mass-rearing are centers for production of sterile tephritid fruit flies (Cáceres and Rendón 2008; Gersabeck et al. 2010;

Cáceres et al. 2012). An important feature of the various tephritid fruit fly mass-rearing systems is that they serve as examples and models for other mass-rearing technology in terms of: (a) quality control/process control engineering statistics (Boller and Chambers 1977; Boller et al. 1981; Brazzel et al. 1986; Calkins et al. 1994; Anonymous 1999; FAO/IAEA/USDA 2014) and (b) modeling of automation and other features that are engineering strategies for economical and biologically feasible mass production (Leppla et al. 1983; Schwarz et al. 1985; Vargas 1989; Enkerlin 2005; Cáceres and Rendón 2008; Gersabeck et al. 2010; Ekesi and Mohamed 2011; Cáceres et al. 2012; FAO/IAEA/USDA 2014) (Figure 3.8).

One of the most useful and complete characterizations of rearing systems for tephritid fruit flies used for SIT is FAO/IAEA/USDA 2014, especially for *Ceratitis capitata* (Wied.), *Anastrepha* spp., and *Bactrocera* spp. To explain the overarching rationale for the detailed quality control system, the authors write:

> Problems in rearing, irradiation, packing, shipping, handling and releasing procedures, and strain quality can be often identified by closely monitoring QC results thus allowing for corrections to be made before field programmes are adversely affected. Although routine and periodic QC testing can reveal measurable changes (both positive and negative), it is less definitive in pinpointing the cause or causes of reduced performance or quality. For example, insect weight, flight ability, and mating propensity can be affected by one or multiple causes. Nevertheless, any sudden decrease or observable changes over time that result in lower than expected results should stimulate a search to identify possible causes. QC tests also play a role in monitoring and measuring recovery to normal levels.

The rationale is further explained as follows:

> It became evident in several SIT fruit fly projects in the 1960s and 1970s that mass-reared fruit flies were not performing in the field as expected. In an effort to define and measure the performance of mass-reared flies and wild flies, tests were developed by various workers (Chambers 1975). Scientists gave particular attention to the effects of gamma radiation on the insect and its ability to mate successfully. Correlations could be made in some instances, but not all. In some instances non-irradiated mass-reared insects also performed poorly shifting attention to seek other causes for poor performance especially in the production processes. For almost 35 years (see Appendix A: Chronology of Product Quality Control of Tephritid Flies for Use in SIT Programmes), systematic efforts by many groups and individual scientists successfully developed a series of simple and useful tests to compare the quality of sterile tephritid flies produced at different rearing facilities. Those efforts led to the production of several fruit fly quality control manuals (e.g., Orozco et al. 1983, Boller and Chambers 1984, Brazzel et al. 1986).

3.8.2 Details of Quality-Control Procedures from Several Tephritid Rearing Facilities

The authors of FAO/IAEA/USDA 2014 and several other authors (e.g., Vargas 1989; Enkerlin 2005; Cáceres and Rendón 2008; Gersabeck et al. 2010; Ekesi and Mohamed 2011; Cáceres et al. 2012) suggest adoption of several simple, but robust tests of fitness of insects that are to be evaluated after mass-rearing and also after

Mexfly and Medfly adult fly emergence and release activities at eight different ERFs: 2007

ERFs	Annual cost	Released per week (millions)	APHIS Share	Cooperator Share
Medfly (*C. capitata*)				
Los Alamitos	$13.3 m	283	40%	60%
Retalhuleu	$4.0 m	700	100%	
Tapachula	$2.7 m	374		100%
Sarasota	$2.2 m	85	100%	
Medfly (*A. ludens*)				
Edinburg	$0.7 m	95	95%	5%
Harlingen	$0.6 m	36	95%	5%
Tijuana	$0.5 m	14	100%	
Reynosa	$0.3 m	18	100%	
Total	$24.3 m	1,682	100%	

Figure 3.8 Mexfly and Medfly raring at several USDA, APHIS facilities in North America.

irradiation treatments. These tests include **pupal weight**, **flight capability**, and **tolerance to stress** from water and food depravation. It is important to recognize that the tests are applicable to detect deficiencies (losses of fitness or quality) relating to the rearing procedures and the irradiation treatments, including handling of the insects in preparation of SIT release. It is also valuable to recognize that these types of tests can be applied to assessment of the quality or fitness of insects to be used for other than SIT purposes such as in biological control and as research subjects.

3.8.3 Pupal Weight

Pupal weight is one of the most robust predictors of fitness or quality in tephritids and many other insects. The document FAO/IAEA/USDA 2014 explains how the values such as those in Table 3.4 are used in the various rearing programs where downward trends in mean pupal weight are taken as indicators of problems in: (1) nutrition, (2) overcrowding during larval stage, (3) high temperatures in the diet, (4) inappropriate holding conditions during pupal maturation, or (5) other factors if the specified ones are not detected. The authors also specify that small size of pupae is strongly associated with poor performance in the field. The correlation between weight and performance of other biological characteristics is high, and generally weight of given stages is a good predictor of general quality characteristics such as flight capacity, reproductive performance, tolerance to stressful field conditions, and other features. To further reinforce the adherence to quality standards, several of the tephritid programs perform emergence, stress, and flight ability tests using standardized materials such as Petri dishes, Plexiglass tubing, temperature cabinets, and aspirators.

Table 3.4 Specifications for Mean Pupal Weight of Tephritid Flies Produced for SIT Programs (From FAO/IAEA/USDA 2014)

Species	Pupal Weight (mg)	Minimum Acceptable Mean
Ceratitis capitata		
Bisexual strain	7.00	7.50
Genetic sexing strain	7.50	7.80
Anastrepha ludens	17.08	17.71
Anastrepha obliqua	13.52	14.22
Anastrepha suspensa	10.00	14.00
Bactrocera curcurbitae	13.00	13.50
Bactrocera dorsalis	12.30	12.90
Bactrocera oleae	6.00	6.54
Bactrocera tryoni	8.50	10.00

3.8.4 The Emergence and Flight Test

The emergence and flight test is performed by placing pre-emergent pupae (ones about to eclose as adults) in petri dishes in Plexiglas cylinders lined with talc (to prevent the flies from crawling out). The containers are inspected after a few days to determine how many flies have escaped by flying out, and this flight ability is taken as an indicator of fitness. This is a simple, yet very definitive tests from FAO/IAEA/USDA 2014.

3.8.4.1 Objective

To obtain an accurate estimate of the percentage of adults that will emerge and produce flies that are able to fly.

3.8.4.2 Discussion

Sterile flies must be able to disperse from the point of release to locate shelter, food, water and, most importantly, mates. If they are unable to fly, they will therefore be useless to the SIT programme.

An accurate estimate of the number of fliers in a batch of sterile pupae delivered to a fly emergence and release facility assists in planning releases to achieve a desired overflooding ratio. The test is performed routinely on pupae before and after irradiation, and again after shipment for programmes with remote fly emergence and release facilities.

The test involves placing pupae in a tube and calculating the number of fliers from what is left in the tube at a set time after the adults have emerged. Examination of what remains in the tubes gives information on how many flies have emerged, have only partially emerged, are deformed, or are fully emerged but are non-fliers. This information is useful to evaluate and correct problems in rearing, age of irradiation, handling, and shipping.

The test has been refined several times since it was first designed and now requires the flies to escape from the tubes unaided, rather than being lured.

A different procedure is used to assess the flight ability of sterile flies after chilling and after release. For those stages the test is called absolute fliers since uses adults as a starting point.

3.8.4.3 Equipment

- Plexiglas tubes: outside diameter 8.9 cm with 3 mm thick walls; painted black (or of opaque black Plexiglas) so that light enters only at the top; 10 cm high (Figure 2.2). Plexiglas has been chosen over cardboard or glass because it is unbreakable and can be washed and reused indefinitely. Care must be taken to prevent scratching the interior of the tubes whilst cleaning them. Avoid using an abrasive cleaner

- Petri dish lids, 90–100 mm in diameter, should be painted black or the bottom overlaid with black paper (such as black filter paper)
- Strip of porous paper (such as construction or blotter paper), 1 cm wide, and formed into a ring 6 cm in diameter
- Miscellaneous equipment: unscented talcum powder, hand counter, soft forceps, and micro-spoon spatula
- Room/chamber with a controlled environment

3.8.4.4 Test Conditions

- Temperature: 25 ± 1°C
- Humidity: 65 ± 15% RH
- Light intensity: 1,500 lux (top of tubes)
- Photoperiod 14 hours light: 10 hours dark
- Test frequency: Emergence and flight ability are assessed for every irradiation batch or shipment.

3.8.4.5 Procedure

"Two days before emergence, 100 pupae are placed within the ring of paper, which is centered in the bottom of the Petri dish. Before each use, the inside of the tube is lightly coated with unscented talcum powder to prevent the flies walking out. Tubes are tapped on a firm surface to remove excess talc, and the talc should be wiped off of the bottom 1 cm (*Ceratitis* and *Bactrocera*) or 3 cm (*Anastrepha*) of the inside of the tube to provide resting places for newly emerged flies to set their wings. The wiping should be done with a gloved finger or a cloth to avoid introducing moisture or oil to the tube. The Plexiglas tube with talc is placed in the darkened Petri dish lid. Five replicates (five tubes with 100 pupae each) are set up for each lot to be tested."

Results of the emergence and flight test are presented in Table 3.5.

Table 3.5 Summary of Quality Control Standards

	Minimum			Acceptable Mean		
Species/Strain	**Pre-irradiation**	**Post-irradiation**	**Post-shipment**	**Pre-irradiation**	**Post-irradiation**	**Post-shipment**
Emergence (%)						
Ceratitis capitata bisexual strain	85	80	75	93	90	88
Ceratitis capitata genetic sexing strain	75	70	70	80	75	70
Anastrepha ludens	91	90	80	93	92	86
Anastrepha obliqua	92	90	86	94	92	89
Anastrepha suspensa	85	80	75	93	90	88
Bactrocera cucurbitae	90	80	70	92	90	85
Bactrocera dorsalis	82	74	70	90	79	75
Bactrocera oleae	80	75	55	85	80	60
Bactrocera tryoni	80	70	65	85	80	75
Fliers (%)						
Ceratitis capitata bisexual strain	75	70	65	85	82	78
Ceratitis capitata genetic sexing strain	70	65	65	75	70	65
Anastrepha ludens	88	86	71	90	89	79
Anastrepha obliqua	89	85	77	90	88	80
Anastrepha suspensa	75	70	65	85	80	75
Bactrocera cucurbitae	80	75	–	85	83	–
Bactrocera dorsalis	75	69	62	83	77	72
Bactrocera oleae	70	65	50	77	72	55
Bactrocera tryoni	70	65	60	75	70	65

Specifications for percentages of pupae producing adult flies (emergence) and flies capable of flight for tephritid flies produced for SIT programs (from FAO/IAEA/USDA 2014). These tables provide a model of a fitness/QC system for flies (and other insects being subjected to rearing and irradiation or other types of potential stress). The establishment of minima and acceptable means is performed with empirical tests and on standards that users of the insects help establish by field-testing. For example, the Mexican fruit fly (*Anastrepha ludens*) was demonstrated under optimized conditions to have a minimal emergence of 91%, 90%, and 80% under pre-irradiation, post-irradiation, and post-shipment conditions, respectively. However, the mean observed emergences under optimal conditions were 93%, 92%, and 86%, respectively for pre-irradiated, post-irradiated, and post-shipped flies. These standards should be considered a flexible guideline to continuous improvement where optimization of rearing, handling in irradiation, and shipment are improved. For example, an improvement in diet may translate to increased minima and means, as may improvements in handling insects during the irradiation process—for example, reducing the oxygen in the atmosphere of the flies during pre-irradiation periods, which can translate into reduction of free-radical damage to flies.

It is evident from Table 3.5 ("Specifications ...") that some species of tephritid fruit flies demonstrate lower performance parameters than others. For example, *Bactrocera tryoni* (the Queensland fruit fly) and *B. oleae* (the olive fruit fly) perform more poorly than several of the other species in terms of emergence and flying capability percentages under rearing and irradiation/shipment conditions. Many of the percentages are in the 70s and 60s percentages, and for *B. oleae*, their post-shipment percentages are 50% and 55%, respectively, for minimum and acceptable means for fliers, and 55% and 60% for emergence rates. There seems to be considerable room for improvement of these parameters for these two species, and for that matter, for the other species of tephritids. The research that would be needed to make such improvements would undoubtedly be able to more than pay for itself over a few years by raising the quality/fitness levels of the insects in their mass-rearing, irradiation, and post-shipment performance.

Regarding the use of anoxia to reduce the damages ensued by irradiation due to radiation-caused generation of free radicals, FAO/IAEA/USDA 2014 authors write:

> Use of reduced-oxygen atmospheres during irradiation of tephritid pupae is mandatory to reduce the generation of free radicals by gamma or X rays that induce collateral intracellular damage during irradiation, thereby allowing higher levels of sterility to be attained without unduly reducing the quality and competitiveness of the resulting flies.

The authors also point out that there is a trade-off between duration or extent of the imposed anoxia and mortality of flies probably due to oxygen deprivation. Therefore, the use of anoxia is not without risks to the quality/fitness of the insects in SIT programs. Clearly, any improvements toward reduction of fitness under the circumstances of rearing, irradiation, and shipment would be welcome and economically beneficial. Therefore, experiments in rearing system improvements that would make insects in SIT programs more fit or of higher quality would be beneficial. For example, to reduce damage from free radicals generated during radiation and to reduce the extent of anoxia imposition, a higher level of anti-free radical compounds and precursors for free radical scavenging enzymes such as catalase and superoxide dismutases could have extensive benefit.

3.8.5 Stress Test

The stress test of survival under stress is described as follows:

> Objective of Survival under Stress Test: This test is a relative measure of reserves available to the adult fly at different critical periods: at emergence, post-handling/chilling, and post-release.

3.8.5.1 Discussion

The ability of flies to survive under stress, without food or water, is important in an SIT programme because released sterile insects must survive long enough to become sexually mature and seek mates. Some facilities test survival of flies individually in small cells and have reported better survivorship than when the test is conducted with many flies in a Petri dish as described in the procedure. However, the ability to survive the stress caused by crowding in the pre-release containers is important; hence the standard test requires that the flies be tested in Petri dishes.

This test is conducted at the rearing facility on non-irradiated pupae. For comparative purposes pupae are also tested post-irradiation, and a further three times at the fly emergence and release facility: on arrival, post-handling/chilling and post-release.

3.8.5.2 Procedure

The **Stress Test** procedure is as follows:

> A sample of several thousand pupae is placed in an emergence cage without food or water. Within two hours of emergence, 50 males and 50 females (or 100 flies from a production lot if a genetic sexing strain is used) are transferred using an aspirator, or preferably a suction pump, to each of five Petri dishes without food or water. The dishes are held in darkness until the end of the test, the duration of which varies with species (see Table 3.4). After the pre-set time, dead flies are removed by tilting the Petri dish, removing the stopper and shaking out the dead flies, taking care not to allow live flies to escape. Both dead and live flies are recorded by sex. Results are expressed as percentage survival. A sample recording sheet is at 9.3 Survival Under Stress Assessment Form.
>
> A similar procedure is followed when the pupae arrive at the emergence and release facility. When the test is performed post-handling/chilling and post-release, samples of flies are collected after the chilling process and from the release bags or boxes.

The flies are held at 25 ± 1°C, 60–70% RH in total darkness for 4–5 days. The authors of FAO/IAEA/USDA 2014 describe the survival of three tephritid species (one from two different strains) in Table 3.6. This test

Table 3.6 Specifications for Survival Under Stress for Tephritid Flies Produced for SIT Programs

Species	Test Interval (h)	Minimum Survival (%)
Ceratitis capitata		
Bisexual strain	48	50
Genetic sexing strain (*tsl*)	48	65
Anastrepha ludens	72	55
Anastrepha obliqua	48	40

The table is based on information from FAO/IAEA/USDA 2014.

Table 3.7 Diets for the Mexican Fruit Fly, *Anastrepha ludens* (Loew)

APHIS Diet Components	Methods Development (Mission, TX)	Guatemala Facility	Mass-Rearing Facility (Mission, TX)
Corncob fractions	157.6	180.3	157.6
Sugar	82.7	87.7	82.7
Torula yeast	63	67.4	63
Wheat germ	35.5	8.9	59.1
Soy flour	35.5	0	23.6
Corn meal (Maseca)	11.8	47.6	0
Na benzoate	1 (~1,000 ppm)	3.6 (~3,600 ppm)	1 (~1,000 ppm)
Methyl paraben	2.3 (~2,300 ppm)	1.8 (~1,800 ppm)	2.3 (~2,300 ppm)
HCl	9.8	8.9	9.8
Water	600.8	594.4	600.2
Total	1000	1000.6	999.3

is a stringent and robust test of fitness, yet one that is simple to perform.

3.8.6 Diets for Mexican Fruit Flies

The diets for Mexican fruit flies (presented in Table 3.7) are fairly simple meridic formulations. The corncob fractions are largely a bulking agent which adds texture and suitable consistency to the diet. The torula yeast, wheat germ, soy flour, and corn meal are all complex nutrients that contain proteins, carbohydrates, lipids, vitamins, and minerals. The sodium benzoate and methyl paraben are preservatives that inhibit microbial growth, especially fungal growth, and the hydrochloric acid lowers the pH, thus enhancing the effects of the antimicrobial compounds. Rather than a designated gelling-hydrocolloid such as agar, carrageenan or alginate, the texture of the diet is greatly influenced by matric potential where the wheat germ, soy flour, corn meal, and especially the corncob fractions bind free water by surface tension effects (see Passioura 1980; Tyree and Karamanos 1980; Tyree and Jarvis 1982). Taking advantage of the binding power of the matric potential of many substances, especially diet components that are insoluble and macromolecular entanglements or otherwise complex matrices, can be a highly useful way of replacing the dependence on gels and the expensive components used as gelling agents such as agar, carrageenan, alginates, and modified starches. The pH of the diets is about 3.5 (personal communication from Dr. Hugh E. Conway PPQ S&T Mission Lab, Mission, TX). The antimicrobial (antifungal) agents used in the Mexfly diets in Table 3.7 are from 1,000 to 3,600 ppm for benzoate and 1,800 to 2,300 ppm for methyl paraben, and Table 3.8 indicates that the effective range of antifungal activity for these two compounds is 400 (pH 3.0) to 2,000 (at pH 5.0) for benzoic acid and 800 to 1,000 ppm for methyl paraben (at 3.0 and 5.0, respectively). Therefore, it is evident that the concentrations of these two compounds are appropriately anti-fungal, but within ranges that are safe to the fly larvae. This is an important point because the balance between effective suppression of microbial growth and the lack of toxicity to the target insects can tip either to under-dosage, which leads to fungal outbreaks vs. over-dosage, which leads to toxicological effects in the insects (Alverson and Cohen 2002) (Table 3.8).

Table 3.9 shows the actual composition of each of three diets tested, and Tables 3.10 and 3.11 show the costs per ingredient and the biological responses (quality or fitness factors) per unit of diet cost for a large-scale production of Mexican fruit flies. Table 3.11 shows that the number of flying insects produced per kg of diet and the highest absolute yield (though not statistically significant) is from the agar diet. The lowest yield of flying insects per kg is from the carrageenan diet. The only difference in the costs of the three diets is attributable to the relative cost of agar vs. carrageenan vs. the standard mass-rearing diet, which contains the conventionally used bulking agent, corn cob fractions. The value of corn cob fractions in terms of fitness of the insects produced, cost of the diet, and convenience in handling the diet makes this material a remarkable contribution to tephritid dietetics. However, obtaining clean and properly textured corn cob fractions has become difficult since the major supplier has gone out of business (personal communication from Dr. Hugh Conway, USDA, APHIS, Mission, TX Methods Development).

In further efforts to reduce costs of rearing *Anastrepha ludens* and to avoid using corncob fractions, Pascacio-Villafán et al. 2018 tested a liquid diet that consisted of the components in Table 3.12.

3.8.7 Liquid Diets for Tephritids

The liquid diet was tested to eliminate corncob fractions, which were replaced by various diet carriers or support materials such as sponge, cloth rags, carpet felt, and other materials that could be cleaned and recycled. The concept of eliminating bulking materials from the tephritid diets was advanced by Chang et al. 2004 and 2006 and Chang 2009 for various species of *Bactrocera*. The rationale for eliminating solid bulking agents from tephritid diets is that bulking agents: (1) add to the expense of the diets, (2) may increase the waste that must be disposed of after pupae have been harvested from diets, and (3) bulking agents may carry contaminants such as toxins (such as mycotoxins reported by Aceituno-Medina et al. 2016) and microbes. While the authors report promising finding regarding use of recyclable replacements for bulking agents used with liquid diets, Pascacio-Villafán et al. 2018 also mention that there are further improvements

Table 3.8 Effective and "Safe" Levels of Antimicrobial Components of Insect Diets (From Cohen 2015)

Antifungal Agent	Effective Concentration (PPM) at pH 3.0	Effective Concentration (PPM) at pH 5.0 {"Safe" Levels: "Toxic" Levels for *Agria affinis*}	Effective Concentration (PPM) at pH 7.0
Methyl paraben	800	1000 {100:1000}	1500
Propyl paraben	200	300	500
Benzoic acid	400	2000 {1000: 10,000}	Not effective
Sorbic acid	400	800 {2000: 15,000}	Not effective
Propionic acid	800	800 {4000: 15,000}	Not effective

Table 3.9 Diet Formulations for *Anastrepha ludens* Tested by Pascacio-Villafán et al. 2020 (Amounts Are Presented in % on a w/w Basis)

Diet Type	Yeast	Sugar	Corn Flour	Sodium Benzoate	Methyl Paraben	Citric Acid	Water	Agar	Carrageenan	Corncob Powder	Guar Gum
Gel-agar	5.3	8.2	6	0.4	0.1	0.44	79.33	0.234			
Gel-carrageenan	5.3	8.2	6	0.4	0.1	0.44	79.14		0.424		
Mass-rearing	6.1	8.3	5.4	0.41	0.1	0.45	62.84			16.3	0.1

Table 3.10 Cost of Ingredients in Diets for *Anastrepha ludens* (From Pascacio-Villafán et al. 2020)

Ingredient	Cost Per kg (US $)
Torula yeast	6.03
Sugar	1.00
Corn flour	0.52
Corncob powder	1.4
Citric acid	1.65
Sodium benzoate	3.1
Methyl paraben	10.8
Guar gum	5.3
Pregelatinized corn starches A and B	1.005
Agar	40.67
Carrageenan	23.76
Gelatin	9.52

Table 3.12 Components of the Experimental Liquid Diet and the Control Solid Diet Used by Pascacio-Villafán et al. 2018

Ingredient	Solid Diet (% by Weight)	Liquid Diet (% by Weight)[a]
Torula yeast[b]	6.0	6.0
Sugar[c]	8.2	8.2
Corn flour[d]	5.3	5.3
Corncob fractions[e]	19.0	0
Guar gum[f]	0.1	0.1
Citric acid[g]	0.44	0.44
Sodium benzoate[h]	0.4	0.4
Nipagin (methyl paraben)[i]	0.1	0.1
Water	60.46	79.46

Note: Exchange rate at the time of writing was US$1 = $18.8 Mexican pesos.

[a] Liquid diet used either synthetic sponge Scotch Brite (a package with three 20- × 18-cm sponges cost US$1.7), or carpet felt (1 m² cost US$3.5) as support material.
[b] Lallemand Mexico SA de CV (1 kg cost US$5.1).
[c] Ingenio de Huixtla SA de CV Mexico (1 kg cost US$0.78).
[d] Maíz Industrializado del Sureste, Arriaga, Chiapas, Mexico (1 kg cost US$0.46).
[e] Mt. Corncob fractions 100 Pulaski, Chicago, (1 kg cost US$0.82).
[f] Tic Gums, Belcamp, MD, USA (1 kg cost US$5.9).
[g] Cava Nutrielimentos SA de CV, Mexico (1 kg cost US$1.4).
[h] Joaquín Lambaren Valencia, Mexico (1 kg cost US$3.2).
[i] Joaquín Lambaren Valencia, Mexico (1 kg cost US$9.4).

Table 3.11 Optimization Criteria Used to Find an Optimal Diet Formulation for Rearing of *Anastrepha ludens* (From Pascacio-Villafán et al.)*

Diet	Estimated Number of Pupae/kg Diet	Estimated Number of Flying Insects/kg Diet	Cost of 1 kg of Diet (US $)	Cost of the Diet Required to Produce 1 Million of Flying Insects
Mass-rearing	3593.3 (±259.7)[a]	2982.6 (±321.6)[a]	0.74	248.0
Gel-agar	3544.6 (±177.8)[a]	3085.8 (±259.7)[a]	0.559	181.1
Gel-carrageenan	3065.8 (±119.9)[b]	2741.1(±259.7)[a]	0.564	205.9

* Means labeled with identical letters within each column indicate no significant differences for one-way ANOVA (t-test contrasts, $p > 0.05$).

and optimization efforts required to make the liquid diet-based rearing process practical on a large scale. Pascacio-Villafán et al. 2018 also report the importance of further optimization of larval density to reconcile the needs for minimizing space required for rearing millions of larvae while reducing adverse effects of competition waste product build-up from larvae that are too densely distributed.

3.8.8 Diagnosis of Rearing Problems

However robust and pertinent quality assurance measurements may be, they cannot translate automatically to process control improvements that lead to correcting fitness deviations. That is to say that detection of a reduction in pupal weight, tolerance to stress or in flight capability does not pinpoint the source or cause of these problems. It does not tell us that the wheat germ in the diet had a reduced lipid content or that the yeast had a compromised vitamin content. It does not tell us (directly) if insects were subjected to stress from waste products that were overly accumulated due to excessive crowding of larvae or thermal stress related to crowding or periodic surges in temperature of rearing rooms. However, with the right approach, it should be possible to develop a scheme of symptomology, or the correlation of symptoms and their most like likely causes.

Taking human health/medical practice as an example of effective, systematic diagnosis, we know that there are suites of symptoms and diagnostic tools that help medical practitioners ascertain or identify diseases and other medical problems. With disorders such as diabetes, symptoms include extreme thirst, frequent urination, fatigue, slow healing of sores, severe itching, and various other general indications. These symptoms, however, are so general that many of them can be associated with several other disorders rather than diabetes. As a physician considers these symptoms, she/he may administer tests of fasting blood glucose and/or urinary glucose concentrations; these tests can help diagnose the presence of diabetes. Further testing of factors such as insulin levels can further pinpoint diabetes as a cause of these symptoms. Other disorders have sets of stereotypic symptoms, which can pin down a diagnosis, often without ambiguity. Suites of symptoms such as counts of various white blood cells, metabolic panels (levels of blood sodium, potassium, calcium, uric acid, bilirubin, creatinine, urea, and various enzymes) help medical personnel diagnose a broad variety of diseases or disorders.

Similarly, rearing problems (= disorders) may have suites of symptoms such as lower pupal weight, longer than normal development times, diminished tolerance to environmental stresses, poor flight ability, and many other. These problems, like many medical disorders, may be ambiguous in terms of being caused by potentially several factors (thermal stress during rearing, diet deficiencies, microbial contamination, genetic truncation, etc.). For useful diagnostic applications, the disorders need to be analyzed by scientifically based, cause-and-effect connection of a symptom or suite of symptoms and a specific unambiguous cause. For example, once a broad symptom such as poor flight ability, reduced tolerance to stress, or excessive wing deformities is recognized in the colony, more pin-pointing investigations would be triggered, such as measurement of hemolymph chemistry (sugar concentration, lipid profiles, blood cell counts, antioxidant capacity of hemolymph, etc.), and those specific features could then be used as diagnostic tools to further isolate the causative agent for the disorder. It is similar to a patient going to a doctor and saying, "I don't feel well; I feel dizzy when I stand, I feel numbness in my extremities, I have swelling in my extremities, I don't sleep well, I have nausea and vomiting," etc. A physician who encounters some of these symptoms might order a metabolic panel for blood work, and the results of the panel include levels of calcium, sodium, potassium, etc., which can help narrow the range of possible causes. For example, if the sodium level of the blood was below a normal concentration, this may be an indication of hyponatremia, which may be an indication of kidney or liver failure, which is further pinpointed with other tests, including other components in the metabolic panel.

The fact that human medicine has been intensively studied for centuries, and diagnostic tests have been refined to a high degree, makes diagnosis of human diseases and disorders more clearly and precisely accomplished. However, with insects, far less attention has been paid to diagnosis and symptomology. In Chapter 10 ("Relationships between Fitness, Quality Control, Health, Homeostasis, and Stress"), I discuss many examples of relationships between various components or aspects of rearing systems and symptoms of the stresses. For example, visual impairment in boll weevils was shown to be related to carotenoid (retinol) deficiency. Chronic oxygen deprivation manifested itself in excessive production of tracheoles and mitochondria. Diets deficient in lipids are known to cause reduction in amounts of fat body and can further lead to reduction of hemolymph vitellin titer. In several lepidopterans, water deprivation, thermal stress, and food deprivation led to pronounced changes in hemolymph proteins, carbohydrates, and free amino acid titers, as well as specific changes in calcium and potassium levels in the hemolymph (Cohen and Patana 1982). There are numerous publications that report responses of internal milieu chemistry and morphology to various types of stress stimuli; however, there remains a need for further review and compilation of this causal/symptomatic relationship. There is also a further need to draw such relationships directly to tie symptoms with causes on a per species basis. Further work with a given species in a well-prescribed rearing system would help rearing personnel to know what is the normal range of hemolymph chemistry profiles, morphological features, and even behavioral phenomena. The knowledge of what is normal, along with studies of deviations from the norms could become helpful bases of diagnosis of rearing problems. One such approach to profiling the small molecules in biological systems has come to be known as metabolomics. With tools such as HPLC, GC, GC-Mass Spec, H NMR analysis, capillary electrophoresis, and other analytical systems, it is possible to analyze hundreds of small molecules in insects metabolic systems (Tomita and Nishioka 2005; Weckwerth 2006). Coupling these analytical methods with huge data bases and data mining capabilities, there is a wealth of potential systematic information that can be used to help pinpoint what is "normal" metabolically and what deviations in the rearing system lead to specific abnormalities. A recent example of the application of metabolomics to a better understanding of rearing systems is found in Melis et al. 2019. Melis et al. 2019 showed using metabolomic analysis with H NMR that using dried brewers' spent grain could increase desirable nutritional contents of yellow meal worms (Tenebrio molitor) over conventionally used wheat bran. In a study of the metabolic effects of stress from low oxygen (hypoxia) and high carbon dioxide atmospheres (hypercapnia), Cui et al. 2019 found that the two facets of gas exchange stress caused strong shifts in several key metabolites. These studies are excellent examples of application of metabolomics to help us better understand the implications of stress factors that derive from various aspects of insect-rearing systems, including diets, gas exchange, thermal stress, and virtually every other deviation from normal homeostasis induced in our rearing systems.

A further note must be made for microbial relationships with normal, healthy insects. Pathology texts such as Tanada and Kaya 1993 list many relationships between insects and their symbionts and pathogens. These lists include compelling and explicit symptoms or outcomes of violations of the normal relationships. For example, when a mutualistic microbe has been lost from its host due to some kind of rearing problem, such as over-administration of antibiotics or failure to allow normal transmission of the symbiont to its host, the symptoms are clear and relatively unambiguous. Likewise, insects that are undergoing a malaise due to infection of a pathogen display responses that can be further diagnosed with direct observation (such as gut dissections that reveal packets of polyhedral viruses), microbial culture methods, or as of late, molecular tests of DNA or RNA that pinpoint the causative agent. With the growing availability of molecular methods and extensive access to sequencing data, molecular testing is an ever-increasing tool for diagnosis.

Ekesi and Mohamed 2011 provide an extensive, detailed description of rearing techniques and quality control parameters for the major species of tephritid fruit flies including *Bactrocera*, *Ceratitis*, *Dacus*, and *Tirrhithrum*. Many of the conditions and standards described by Ekesi and Mohamed are addressed to facilities that are less sophisticated than those used in mass-rearing tephritid species from Asian and western mass-rearing facilities, but the practices described by Ekesi and Mohamed are of practical value, especially in smaller facilities worldwide (Table 3.13).

Table 3.13 Area-Wide IPM Programs, Integrating SIT (From Enkerlin 2005)

Strategic Option	Program
Prevention	Mediterranean Fruit Fly Preventive Release Programme (Los Angeles Basin, California, USA)
	Mediterranean Fruit Fly Preventive Programme (Tampa-Miami, Florida, USA)
	Mexican Fruit Fly Preventive Programme (Rio Grande Valley, Texas, Mexico—US border)
	Mexican Fruit Fly Preventive Programme (Tijuana, Baja California, Mexico—US border)
	National Fruit Fly Control Programme (Chile)
Containment	Mediterranean Fruit Fly Programme "Programa Moscamed" (Guatemala, Mexico, USA, 1983–present)
	Binational Chile-Peru Programme for Mediterranean Fruit Fly Eradication (1996–present)
	Queensland Fruit Fly SIT Control Programme (South-Eastern Australia)
Eradication	Mediterranean Fruit Fly Programme "Programa Moscamed" (Guatemala, Mexico, USA, 1978–1982)
	Melon Fly Eradication Programme (Okinawa, Japan, 1982–1994)
	Oriental Fruit Fly Eradication Programme (Guam)
	National Mediterranean Fruit Fly Control Programme (Chile 1992–1995)
	Mediterranean Fruit Fly SIT Control Programme (Southern Australia)
	Mexican Fruit Fly and West Indian Fruit Fly Eradication Programme (North- West Mexico)
	Mediterranean Fruit Fly Eradication Programme "PROCEM" (Mendoza—Patagonia, Argentina, 1992–present)
Suppression	Mediterranean Fruit Fly Suppression Programme (South Africa) Oriental Fruit Fly Suppression in Pilot Areas (Thailand)
	Melon Fly, Oriental Fruit Fly, and Mediterranean Fruit Fly Area-Wide Integrated Pest Management Program (Hawaii)
	Mediterranean Fruit Fly Suppression Programme "Programa Madeira-Med" (Portugal)
	Bactrocera spp. Pilot Suppression Programme (Guimaras Island, Philippines) Mediterranean Fruit Fly Suppression Programme (Israel/Jordan) Mediterranean Fruit Fly SIT Control Programme (Peru)
	National Fruit Fly Campaign (Mexico)
	Mediterranean Fruit Fly Suppression Programme (Spain)

Table 3.2 shows the extensive applications of mass-rearing based SIT on a worldwide basis as of 2005. More recent publications indicate that the application of mass-rearing for SIT and for other programs in genetic pest management are even more prominent today.

CHAPTER 4

Silkworm Rearing Systems

4.1 OVERVIEW AND BACKGROUND

4.1.1 Relating Silkworm Husbandry to Insect Rearing as a Scientific Process

Throughout this book, insect rearing is treated as a science or scientifically derived process. The advantage of using technology that is based on sound rationale and scientific inquiry is that such technology can be understood as a system that is reliable, predictable, and fixable when things go wrong. This is implicit in the title and theme of this book which includes the concepts of ***design***, ***operation***, and ***control*** in rearing systems. I originally came to the topic of insect rearing (when my career in the USDA, Agricultural Research Service began in 1979) with a prejudice that most of the science aspects of insect rearing were recent advances. Even recently (Cohen 2018a, 2019), I had considered the "birth announcement" of scientifically-based insect rearing as the paper by Delcourt and Guyenot 1910, but the more I studied discussions of silkworm rearing, the more impressed I became as to the antiquity of scientifically and rationale-based rearing technology that existed in the ancient practice of sericulture.

I suspect that the algorithm of rearing systems processes and process sequences was developed or was evolving as the ancient Chinese sericulturists practiced rearing of silkworms and got rather natural feedback from their practices and noticing how the silkworms responded to the specific features of those practices. If the ancient Chinese rearing personnel gave the silkworms too much food, they would have noticed that the excess leaves began to rot, or if they provided too few leaves, they would have observed behaviors that indicated that the larvae were restless or foraging for more food. If the conditions were too humid because of poor ventilation, they might have noticed the tendency for mold to form (rotting), but leaves that were too dry led to lower growth rates. Assuming that many of the diseases known today were present 4,000 or 5,000 years ago, there was probably some kind of recognition that factors such as poor sanitation, poor ventilation, no opportunity to reduce contact between larvae and their waste products, and untimely feeding resulted in sickly larvae. However, we also know that the early practices of sericulture in ancient China were kept secret, so we have no opportunity to reconstruct a reliable history of the evolution of silkworm rearing. Yet we also must acknowledge the fact that the processes of sericulture were so successful that China had been and remains today the world's leading and most successful producer of silk. Therefore, however and whenever the silkworm rearing practices were developed and perfected, they could stand as models of efficient mass-rearing systems.

4.1.2 Mulberry Silkworms: The Most Highly Domesticated Insects

In investigating silkworm rearing, it becomes evident that *design*, *operation*, and *control* have long been a prominent part of sericulture and specifically silkworm rearing systems. As we review the history of silkworm rearing, we see implications of these concepts as they must have been introduced into the evolution of processes started in China and carried to the other countries that copied and modified the Chinese systems. I regret that due to lack of my access to the vast information that was passed on in the various silkworm rearing programs, I must neglect the specific systematic developments that began in China (and later in India, Japan, the Persian Empire and other regions in Asia and Europe), were lively and essential features of the silkworm production systems throughout much of the world. Part of my failing to completely or fairly represent the accomplishments in silkworm rearing are due to several factors including: (1) lack of formal communication of standard operating procedures (SOPs) in silk production, (2) the cryptic nature of rearing procedures due to the necessity to keep secret innovations that proved successful as rearing components, (3) my own lack of language ability to find writings that may exist, and (4) the diversity of the cultural nature of silkworm rearing as it appeared in the various countries where silkworm rearing became economically and socially important.

Therefore, in this chapter, I have drawn heavily from the monographs by Anonymous 1828 and Kelly 1903 and from the book by Hamamura (translated from the Japanese book published in 1974 and published in English in 2001).

Unfortunately, all of these sources are out of print and not readily available, despite their remarkable value in explaining the scientific and empirically well-vetted basis of silkworm rearing. Because of the limited availability of these sources, I have included rather extensive portions of these sources as quotations from the Anonymous 1828 letter (which is actually a monograph) and the Hamamura 2001 book as well as a complete copy of the Kelly 1903 monograph as Appendix (Kelly 1903).

4.2 HISTORICAL BACKGROUND

4.2.1 Report from Anonymous 1828

There are many accounts of the origin and history of sericulture with respect to the mulberry silkworm. Most accounts include stories of the discovery of silk attributed to Empress Si Ling Shi, some stories having her seated beneath a mulberry tree drinking tea. When a cocoon of *B. mori* happened to fall into her hot tea, she saw the silk entwine in her cup, and the idea of using silk as a basis for making cloth was the beginning of sericulture. A slightly less romantic account is given by Roberts 1945:

> The first use of silk as a fiber has been traced to China during the reign of Huang·Ti, a famous emperor. About the year 2640 B. C., his wife, the empress Si Ling Shi, is credited with being the first person to encourage the cultivation of the mulberry tree, the rearing of the worms, and the reeling of silk. It is claimed that she devoted herself personally to the care of silkworms, and the Chinese have attributed to her the invention of the silk cloth weaving loom, which to a modified extent, is still used today. Her name "Si" is the Chinese word for "silk," the English word probably being a derivative. Because of her discovery, she was deified and has ever since been known as "The Goddess of Silkworms." The silk industry is believed to have originated in the valley of the Yellow River, and from there spread to different parts *of* China, the provinces of Chekiang, Kiagsu, Szechwen, Shatung, and Kwantug, and Manchuria being ·the best breeding and rearing sections.

Silkworms (*Bombyx mori* L.) have been cultivated for about five millennia (~5,000 years), making them the most thoroughly domesticated insect species (Figure SRS 1). Though honeybees (*Apis mellifera* L.) are known to be pictured in cave paintings and other records of early humans, cultivation and subsequent domestication of honeybees is most likely a later event in human history (More 1976; Crane 1999; Weber 2012; and see Figure 4.1). One conspicuous form of evidence of the degree of domestication of *B. mori* is the relative helplessness of the silkworms or total dependence upon their human keepers. For example, though adult

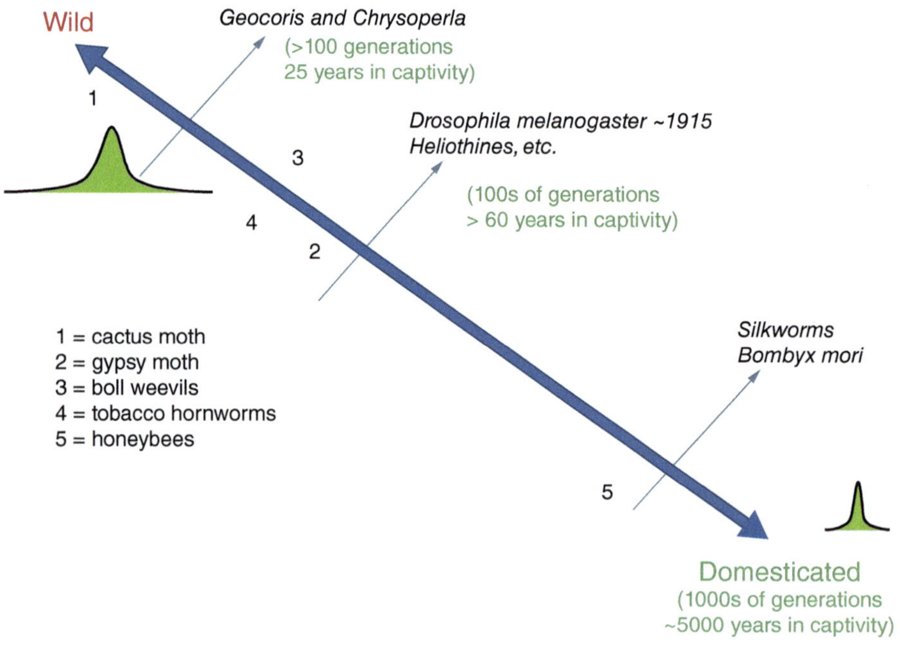

Figure 4.1 Diagram of domestication of five species of insects. The disproportional timeline represents a rough approximation of the onset of domestication for populations from each species, and the Gaussian curves signify the truncation of genetic diversity that takes place as a consequence of the inbreeding that accompanies domestication.

silkworms of both sexes have wings, neither females nor males can fly. In fact, the adults are nearly immobile, except for weak crawling. In fact, there are no longer either wild or feral *B. mori* in existence. It is hard to fathom that an organism has become so extremely domesticated as to be unable to sustain itself in nature or even in an agroecosystem. In contrast with the helplessness of *B. mori* throughout their lifecycle, honeybees are often present in nature as feral colonies. *Apis mellifera* populations are able to swarm to establish new colonies, find food independent of beekeepers, and carry on all their biological needs to produce successive generations without recourse to humans. In this vein, the subject of domestication will be discussed further in this chapter and in several other places in this book.

For a succinct but thorough perspective on the history of silkworm rearing, I provide background information taken from Yasuji Hamamura 2001 who gave the following history of silkworm rearing:

> In 1927, an excavation of the ruins of Yanshiao era was carried out in a village close to the capital city Changnan in Chansi province of ancient China. Yanshiao is a very old prehistoric era, about 4000-4500 years ago which is equivalent to 2500 B.C.
>
> Partially cut open cocoon shells were found in the excavated articles. Along with them, objects resembling the spinning wheels made of rocks were also found. These cocoon shells are at present preserved in the Kogu Museum in Taiwan.
>
> Nunome of Kyoto University of Textile Technology had procured the photographs of the cocoon shells and after studying them carefully he reported the following observations:
>
> (a) The cocoons were smaller than those of silkworms. They were 15 mm long and 7 mm in diameter and appeared to be those of *Rondotia menciana*. '[**note by Cohen: this species is in the family Bombycidae and includes wild populations extant today**]'. This insect is said to be living even now in the coastal areas of Ageshie. (b) Since the objects which resemble the spinning wheels were also excavated simultaneously it gives the impression that the spinning of threads was being carried out. (c) From the way the cocoons are sharply cut it seems that the pupae were used as food and at the same time the silk floss was Io being produced.
>
> From the observations of Nunome the relationship between man and silkworms can be archaeologically traced back to about 4500 years from now.

Hamamura went on to write:

> According to the biography of Hidari Haruaki the rearing of silkworms was popularized in the era of Hoangli Dynasty around the same period. Although this record was based on the communication by word of mouth it seems to be well supported by archaeological evidence.
>
> "There is no clue to suggest whether during this period the silkworms were reared to collect the cocoons or the cocoons were naturally produced. However 1000 years later the engravings on the tortoise shells discovered from the ruins of the era of In Dynasty had the impressions of a hand plucking the mulberry leaf with 4 cocoons adhering to it. This indicated that in this era also the naturally produced cocoons were collected." These statements still leave the mystery of the point where total domestication occurred, leaving all extant *B. mori* to be totally dependent on human custodians of this species.

Hamamura continued a detailed account of the evidence regarding the history of silk and the transfer of silk technology to Japan whose records first appeared in 239 AD. He continues, "Thus the relation between man and silkworms as per currently available evidence dates back to about 4500 years. Perhaps in certain cases it may even be a few thousand years older. If the ancestors of silkworms were associated with mulberry the history will descend down to several million years. This in other words will lead to a never ending journey of the eternal past. Mulberry and silkworms appear to have existed ever since their origin as inseparable organisms." The concluding paragraphs almost poetically express the history of the relationship, "During the past 4500 years human beings have greatly benefited from the silk produced jointly by mulberry and silkworms and this has contributed to the grand cultural advancements. Subsequently efforts were made to add the possible artificial devices both in the mulberry cultivation and silkworm rearing. ***However, nothing could be done to disturb the relationship between mulberry and silkworm.***" [The italics and bold are Cohen's.]

Finally, regarding the history of silkworms and the relationship to humans, Hamamura turns our attention to the science behind artificial diet development with the following statements, "Nevertheless, science has to be applied to find out the secret truth behind such a wonderful blessing of nature. In 1935, the question 'why silkworms feed exclusively on mulberry' was put forward, and this stretched the hands to catch the mystery of mulberry-silkworm relation."

Though Hamamura's accounting of the history of silkworm domestication is thorough and fascinating, it does not account for *Bombyx mori*'s departure from wild, natural populations to their current state of being entirely dependent on human oversight. Throughout this book, I have tried to emphasize the importance of both deliberate and inadvertent selection to encourage vigilance in recognizing how many ways our cultivated/domesticated insects depart from their original wild or ancestral counterparts.

4.2.2 The Consequences of Excessive Domestication

In discussing domestication, I use the perspective of the late Dr. Alan C. Bartlett (1984 and personal discussions) to include all aspects of the capture of wild insects and bringing them into captivity for human uses. The enigma of the super-domestication syndrome is a subject of several academic treatments such as Yoshitake 1968, Yukuhiro et al. 2002, and Arunkumar et al. 2006 who point out that a close relative of *B. mori*, *B. mandarina*, is found in nature and is thought to be the original species from which *B. mori* was derived. There is considerable biological evidence that *B. mori* populations were derived from common ancestors with *B. mandarina*, including evidence from genetic studies and hybridization capabilities between these two congeneric species. However, despite this taxonomic closeness, there is little question that *B. mori* are a truly separate species and that their dependence on humans is remarkable, even amongst the cadre of domesticated species. To this point, we know that dogs and cats can become feral and independent of human care as can cattle, goats, and sheep. The same concept applies to plants and microbes.

The extreme selection that was done with *B. mori* results in several consequences that are both advantageous and disadvantageous in our efforts to cultivate these insects for silk production (sericulture). The very restricted diet is not merely specialization to mulberry (*Morus* spp.: Moraceae), but specifically, white mulberry, *M. alba* is the most suitable host species for silkworms. As Kelly 1903 points out, the subspecies or varieties of *M. alba* and their phenological and physiological conditions also play important roles in the suitability of the mulberry as hosts for silkworms. Other features and possible outcomes of extreme domestication of silkworms are the high efficiency of food utilization displayed by larvae as discussed by Waldbauer 1968. Another aspect of domestication is the deliberate effort to select for high silk production efficiency (Figure 4.2). These efforts include selection for strains of silkworms that produce higher quality of silk, such as strength, softness, color, etc.

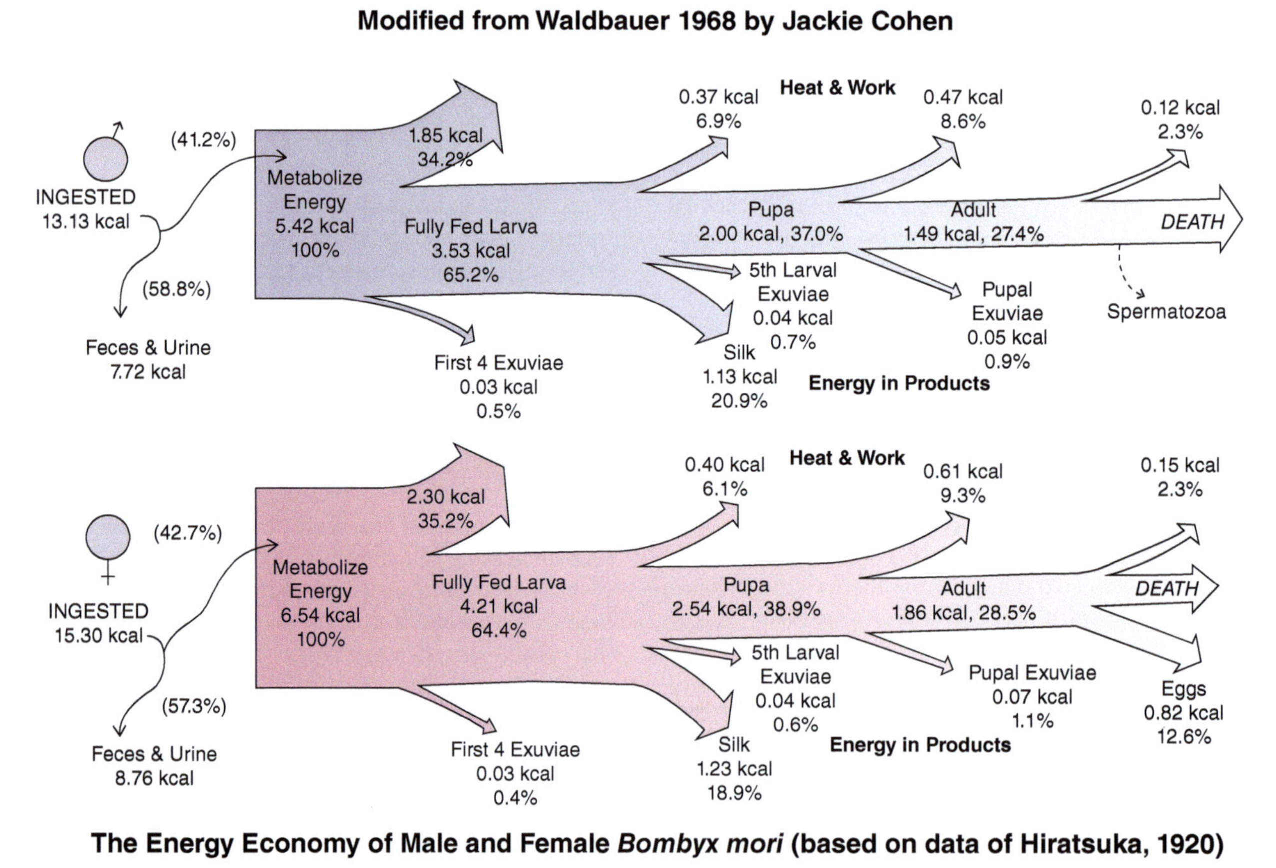

Figure 4.2 Silkworm energy budgets. Representation of energy budget of silkworms feeding on mulberry (*Morus alba*) leaves from Hiratsuka 1920 and reported by Waldbauer 1968. Note the silk cocoon represents about 20% of the total energy budget. (Redrawn by J. L. Cohen.)

4.3 BACKGROUND AND THE VISION OF SERICULTURE

4.3.1 Silkworm Rearing Systems and Sericulture in the United States

The topic of silkworm rearing is too large to cover adequately in this book on design, operation, and control of insect-rearing systems. However, taking one small portion of this topic will allow a useful overview of the oldest (by far) and most vetted rearing system in existence. Because of the extensive literature and other information bases on rearing silkworms in the United States, I am confining most of the discussion to efforts, largely by the US government, to encourage the American public to undertake sericulture as a promising and potentially lucrative enterprise. An exception to my focus on US publications on silkworm rearing is my treatment of development of artificial diets for silkworms, which is a subject dominated by Japanese researchers and admirably described by Professor Hamamura 2001. The body of information on sericulture and the rearing aspects of this enterprise seem to be most readily available in western literature, especially English, Italian, and French. For practical reasons, I rely upon the English documents, including translations of other writings into English. This is not to say that the major contributions to the current state of silkworm rearing were developed by workers writing in English. To explain my respect for the remarkable developments in silkworm rearing, I refer the readers to the historical treatments of sericulture explained in this chapter to express the state of knowledge developed by the Chinese, Indians, Japanese, Persians, Turks, and European countries where sericulture was a major success by the middle ages or thereabouts.

The document by an anonymous author (Anonymous 1828) and Kelly's 1903 monograph are the two major sources of information that I consider highly systematic treatments of silkworm rearing. Before I present details from Kelly's vision and her practical approach to attaining that vision of a commercially successful sericulture program, I will offer some history of silkworm rearing in the United States from pre-Civil War years, including the early US Colonial times. Most of the information here is derived from a remarkable monograph that is part of a report to the US Congress (Anonymous 1828). In the section titled "History of Silk in the United States," the report states:

> The culture of silk first commenced in Virginia. Upon the settlement of that colony, it was deemed an object of the first importance; and the attention of the settlers was strongly directed to it by the British government, by which silkworm eggs, white mulberry trees, and printed instructions, were sent over and distributed. King James the First, in the 20th year of his reign, having, doubtless, seen the defeat of his plan to encourage the silk culture at home, was induced to attempt it in Virginia; and, "having understood that the soil naturally yieldeth store of excellent mulberries," gave instructions to the Earl of Southampton, to urge the cultivation of silk in the colony, in preference to tobacco, "which brings with it many disorders and inconveniences."

This made the beginning of silkworm cultivation in the United States an early 17th century event! The Report continues:

> In obedience to the command, the Earl wrote an express letter on the subject, to the Governor and Council, in which he desired them to compel the colonists to plant mulberry trees, and also vines. Accordingly, as early as the year 1623, the colonial assembly directed the planting of mulberry trees; and in 1656, another act was passed, in which the culture of silk is described as the most profitable commodity for the country; and a penalty of ten pounds of tobacco is imposed upon every planter who should fail to plant at least ten mulberry trees for every hundred acres of land in his possession.

It is interesting that we can think of it as an amazing fact that an insect-rearing project was actually mandated by law in US Colonial days, and it is even more amazing that the success of the silkworm rearing effort obviated the law as further described in Anonymous 1828:

> In the same year a premium of 4,000 pounds of tobacco was given to a person as an inducement to remain in the country, and prosecute the trade in silk; and, in the next year, a premium of 10,000 pounds of tobacco was offered to anyone who should export £200 worth of the raw material of silk. About the same time, 5,000 pounds of the same article was promised "to anyone who should produce 1,000 pounds of wound silk in one year." The act of 1656, coercing the planting of the mulberry trees, was repealed, in the year 1658, but was revived two years after; and the system of rewards and penalties was steadily pursued until the year 1666, when it was determined that all statutory provisions were thereafter unnecessary, as the success of divers persons in the growth of silk, and other manufactures, "evidently demonstrated how beneficial the same would prove." Three years after, legislative encouragements were revived, but subsequently to the year 1669, the interference of government seems entirely to have ceased. The renewal of the premiums after the act of the year 1658, was, doubtless, owing to the recommendation of Charles II: for, in the year 1661, among the instructions given to Sir Wm. Berkeley, upon his reappointment as governor, and while in England on a visit, the king recommended the cultivation of silk, and mentioned, as an inducement to the colonists to attend to his advice, "that he had formerly worn some of the silk of Virginia, which he found not inferior to that raised in other countries."

The reader is reminded that the Pilgrims are said to have landed in Plymouth Rock in 1620, so the direction of the colonial assembly to plant mulberry is clearly an early event in US history! The Report goes on to account for how the various colonies came to explore sericulture as an economic foundational practice and that there was considerable success whereby nearly every household was said to be engaged in cultivating mulberry and raising silkworms. There were various laws and official measures to encourage or enforce the pursuit of silkworm rearing, including rewards and penalties of tobacco, which was a major part of the colonial currency (Anonymous 1828). Among many other indications of the extent of the practice of sericulture in early America, the report mentions that King Charles II wore a robe made of Virginia silk at his coronation in April of 1661! The history of sericulture in the United States contains many equally interesting facts about the key role of silk in the early history of the United States as a colony of England and then as part of the history of the new nation. The account in the Report includes a state-by-state record of efforts to adopt sericulture as a major part of US agriculture and textile programs. This report and others such as Skinner's 2018 discussion of sericulture in Connecticut explain reasons why sericulture had failed to persist as a staple of US agriculture (in contrast with such successes as corn/maize, wheat, soy beans, and dairy) where the United States emerged as and remains today a world leader. Throughout this chapter and elsewhere in this text, I suggest that failures in rearing programs (silkworms or other insect-based systems) stem from lack of educated practitioners and a lack of infrastructure. In fact, both Bloom's 2018 and Skinner's 2018 accounts mention the existence of numerous pamphlets, mainly in Italian and French but some in English on various aspects of sericulture. The intention of the pamphlets was to convey to lay people vetted and systematic techniques to reduce chances of failure in all aspects of silk culture. While the Kelly monograph is a thorough and systematic treatment of silkworm rearing and a remarkable contribution to early efforts to systematize insect (silkworm) rearing, it is not alone in the domain of older literature on silkworm rearing technology. Thanks to my learning about the site, www.wormspit.com, I was able to download a paper even older than Kelly's monograph—a monograph (report by Anonymous 1828) published by the US Congress and Department of the Treasury. It is also noteworthy that some very modern, up-to-date, and authoritative information about silkworm rearing and other aspects of silk culture exist. One especially helpful and beautiful site is run by Michael Cook (www.wormspit.com). The Cook site (Cook 2019) contains a wealth of information that amounts to highly detailed and well-vetted SOPs about silkworm rearing along with priceless other information about silk processing. The Cook site also contains a "library" of PDFs on silk culture and "Links" pages, which provide site visitors with valuable connections with all aspects of sericulture. In an excellent article on how and why silk production in Connecticut (and other northeastern states) failed, Taryn Skinner 2018 wrote:

> In October 1789, during a trip to Connecticut, U.S. President George Washington described some "exceeding good" silk lustring and "very fine" silk thread that were part of a growing domestic industry. In fact, by the time Washington wrote those words in his journal, the area that became the state of Connecticut in 1788 had been practicing raw silk production, known as sericulture, for over half a century—and silk was on the rise.

Despite this auspicious start, silk production failed to reach anywhere near the expected outcomes. Skinner points out that part of the failure was largely attributable to: (1) lack of expertise in the silkworm rearing system, where huge amounts of mulberry leaves must be provided several times a day, (2) failure of the amateur sericulturists to match the correct species and variety of mulberry to: (a) the climate of the Northeast and (b) the means of harvesting the leaves in an efficient manner, and (3) failure of the amateur sericulturists to recognize the technically challenging nature of reeling silk (and other post-harvest processing steps). Skinner points out that:

> By 1826, three out of every four households in Mansfield, Connecticut, were raising silkworms, and by 1826, Congress commissioned a report on the potential for a U.S. silk industry. By 1840, Connecticut outpaced other states in raw silk production by a factor of three. Within the next two decades, however, the industry would collapse, leaving the country to wonder what went wrong.

Skinner notes that the silk industry in Connecticut was instituted by Ezra Stiles, the 7th president of Yale University. Stiles raised silkworms as a hobby and provided mulberry seeds and silkworm eggs to many people across the state. He encouraged farmers to plant mulberry as a staple for the intended silkworm industry. Stiles and others in the early 19th century tried to solve the problem of feeding silkworms in cold weather—an especially daunting challenge in light of the silkworms' immense appetite. Skinner wrote:

> Inspired by Connecticut's raw silk production, local entrepreneurs invested in machinery to manufacture silk thread and fabric from reeled silk filaments. In 1810, the Hanks brothers opened the United States' first silk-mill in Mansfield, Connecticut, and in 1838, the Cheney brothers opened a mill which would eventually expand to 38,000 spindles, and become the largest silk manufacturer in the U.S. The future looked bright for silk.

However, problems emerged in this seemingly bright future. Skinner explained:

> "The problems began with a new variety of mulberry and ended with lumpy thread. Beginning with Stiles, Connecticut sericulturists had always used an Italian variety of white mulberry, Morus alba...." "However, in the 1830s ..., farmers and investors latched onto a Chinese variety, *Morus multicaulis*, a subspecies of black mulberry which produced larger leaves and more of them per tree ...." "It could also be harvested more often. The price of *M. multicaulis* skyrocketed as speculators sought to profit from selling cuttings from these fast-growing trees."

Skinner explained the downfall of the Connecticut (and northeastern United States) silk industry as having begun, in part, with the "charismatic and unreliable" Samuel Whitmarsh a sericulturist had helped lead silkworm growers to rely too heavily on an inferior host plant (*M. multicaulis*). The overstated potential of *M. multicaulis* brought about a frenzy of speculation and investment in the untested mulberry species, leading to the situation "the popping of the bubble," described by Skinner as follows:

> In the bubble's aftermath in the early 1840s, companies along the East Coast went bankrupt, as did Whitmarsh, and angry farmers tore out their orchards. Joshua Grant, a silk producer in Baltimore, called the collapse a "dire disaster that has overspread the land like a funeral pall." Then a series of harsh winters, followed by a blight in 1843-44, killed many of the remaining mulberry trees.
>
> By 1881, sericulture in Connecticut had been entirely abandoned. The now much older Elder Sharp, who had valued his mulberry tree so highly, said, "Our silk was good, bright and strong, needing only patience to better understand the reeling ... let us do what we can at this late day to repair our error." Instead, silk-mills continued to import from Asia and continued to manufacture silk fabric through the mid-20th century. Today, the legacy of Connecticut's silk industry can be seen in the white mulberry trees which have spread everywhere and are now considered an invasive species.

Furthermore, there were several aspects of sericulture infrastructure in place and showing promise to make the rather new nation (the United States) a strong player in the world silk market. Skinner's 2018 excellent account of the rise and demise of sericulture in 19th century Connecticut mentions several people who were instrumental in the "silk movement," including Ezra Stiles, the seventh president of Yale University whose enthusiasm for silk included his advocacy of supplying mulberry trees and healthy silkworm eggs to people in the state. As part of his extension function (an unusual status for a university president), he even wore academic robes made of silk that he had cultivated. Skinner also discussed the actions of the Hanks brothers who opened a major silk mill in Mansfield, CT, leading to a larger mill by the Cheney brothers, which Skinner described as having 38,000 spindles, making it the largest silk manufacturer in the United States.

In Skinner's treatment of the failure of sericulture to reach its full potential and in the perspectives of Henrietta Aiken Kelly (discussed below), some crucial factors and general principals emerge about the potential of insect-rearing systems to meet the expectations behind mass-rearing programs. First, there must be a driving force that compels or fuels the rearing system. In the case of silkworm rearing, the driving force is the production of silk as an economically feasible product. This means that there must be a demand for silk, and it is implied that the silk must be of such suitable quality that it is desirable. This requirement, Skinner pointed out, was not met. One cause of this failure was the lack of industrial-scale silk reeling. Reeling a strong, smooth, well-wrought silk was not an easy task for amateurs. Skinner wrote, "Despite everything, in 1847, Stebbins remained hopeful about the 'sequel of the silk industry.' But the region's sericulture had one insurmountable flaw that prevented this revival: Stiles' gowns aside, Connecticut's silk was not industrial grade, so silk-mills could not use it to manufacture fabric. According to cultural anthropologist Dr. Janice Stockard, who has interviewed silk reelers in South China, reeling—the practice of unwinding the filaments of silk from their cocoons—requires observation, training, and practice. In 19th-century Connecticut households, women were expected to learn the skill from pamphlets." Skinner went on to conclude, "In pamphlets, the term 'spinning' described the critical technique of reeling silk from cocoons," Stockard 2002 says. "Women in farming households improvised, based on their experience spinning wool and using technology found in the home, including the wool wheel."

"The product they ended up with was adequate for sewing thread, but not strong enough for the industrial-silk-manufacturing infrastructure that Connecticut had begun to build." According to one scathing assessment, "Connecticut women in 70 years have not improved their knowledge of reeling." Another issue, Stockard 2002 says, was the expectation that women could reel silk "whenever leisure from other duties permitted." In other words, women were supposed to wedge a high-skill, time-intensive task into their already full workloads, and the result was sub-par silk.

"Simultaneously unwinding several cocoons from a basin of near-boiling water while twisting these filaments into one even thread and reeling it onto a wheel was hard," Stockard 2002 says, "If reeling was interrupted to tend to a child or chore, the silk would gum up and lump." Faced with this weak, lumpy thread, Connecticut manufacturers began to import raw silk from China, Japan, and Italy.

Therefore, the driving engine of silkworm rearing—demand for the product—was not in place. However, the failure of the sericulture system went beyond the problems with reeling. The quality and ease of mass-production of silk were hindered by the availability of an abundant supply of suitable quality of mulberry leaves (discussed above). Beyond this, the rearing process per se presented several unresolved problems. These problems are discussed by Anonymous 1828, Kelly 1903, and Kelly in letters to L. O. Howard and others as presented in Bloom 2019. The rearing problems included the rearing system factors that include: (a) provision of optimum diet (ages of mulberry leaves, storage, changing techniques, and contamination, (b) environment, (c) diseases of silkworms, and (d) genetics).

4.4 SILKWORM REARING SYSTEMS FROM KELLY'S PERSPECTIVE

4.4.1 The Kelly Monograph

This section includes a critical discussion of the 1903 monograph, *The Culture of the Mulberry Silkworm* by H. A. Kelly. Because of its value as an example of a highly practical manual of SOPs enhanced by a treatment of rationale for most of the procedures, I have provided the monograph here in Appendix 2 Kelly. Henrietta Aiken Kelly, was appointed as a "special investigator," whose job was to develop and communicate a knowledge base on silkworm rearing for potential commercial applications in the United States. Kelly covered a complete range of information about silkworm rearing as evidenced by the Table of Contents, which included these section titles: The life of the silkworm, The food of the silkworm, Implements necessary for silkworm culture, Silkworm eggs: how to winter and hatch them, The rearing of silkworms (General Directions, Ventilation, Disinfection, Space Required, Temperature), The ages of the silkworm, Preparations for spinning, Preparing cocoons for the market, and Diseases of silkworms (Pebrine; Flacherie, or flaccidity; Gattine; Calcino or muscaradine; and Grasserie). The Contents image reveals the overview of this treatise. It must be noted that this reference does not contain references used by Kelly, and though she often credits sources of some of the materials drawn from other silkworm specialists, it is not always clear which information is from translations of Italian and French writings that Kelly studied in her stay in Europe.

4.4.2 Kelly and Her Purpose-Driven Rationale for Insect-Rearing (Systems)

Besides being the first domesticated insect and the leading-edge of the rearing system development, silkworms (*Bombyx mori*) and the rearing systems built around them serve as an important example of how the purpose of rearing drives the nature of the systems in question. The narrative written by Bloom 2019 provides the most eloquently stated rationale and plan for developing an insect-rearing system. Later in this chapter, I will focus on the treatment of silkworm rearing by Henrietta Aiken Kelly as an excellent illustration of how insect rearing can and should be treated scientifically. I also present an updating of some of the obviously outdated aspects of the system that Kelly presents. However, the vision of Henrietta Kelly for using sericulture centered on silkworm rearing is a model for tying rationale for perfecting the rearing process to economic gain and social well-being.

4.4.3 Overview of Henrietta Aiken Kelly's Work

In her remarkable monograph, there is an explanation of the purpose of the bulletin (**KFN 1**) by L. O. Howard (The Entomologist for the USDA from 1878 to 1931 and Kelly's supervisor); the bulletin was intended for the audience being, "persons who wish to begin the culture of the mulberry silkworm in different parts of the United States" toward the goal of providing a new (in the United States) crop that could create opportunities lost during the upheavals related to the Civil War and its consequences. L. O. Howard's role in the efforts to bring sericulture to the United States is documented by Bloom 2019 and Russell 1978. The background of the Kelly 1903 monograph is a forceful reminder of the context essential to complete success in insect-rearing systems. Though Kelly had modest support (from the USDA) for her vision of expansive and economically significant silkworm rearing (and the overall sericulture) program, there were several factors that worked against the success of silkworm rearing and sericulture as a major industry. The comprehensive scope and depth of this monograph are evident from the "Contents" section (under **KFN 1**).

4.4.4 Discussion of the Kelly Monograph (Also see Appendix 2: Kelly Silkworm Treatise)

In 1903, Henrietta Aiken Kelly produced a manuscript explaining a complete rearing system for silkworms to the chief entomologist of the USDA, L. O. Howard who was the Entomologist for the USDA from 1878 to 1931 and Kelly's supervisor. Much of Kelly's vision for reestablishing sericulture in the United States was made clear in her letters to Howard presented in Bloom's 2019 book about *Southern Women in the Progressive Era*. The letters in Bloom's book tell the exact details of the purpose of Kelly's changing the course of her life to learn and later teach the cultivation of silkworms toward improving the social and economic status of people impacted adversely by the aftermath of the US Civil War. Kelly's involvement and her background provide a fascinating and pertinent explanation of the relevance of her treatment of silkworm rearing systems. Briefly, she was an educator of high standing in the post-Civil War Charleston, South Carolina community (Bloom 2019). The letter indicated that Howard supported the efforts steered by Kelly to introduce silk production to the US agriculture community. The brief Preface further explains the intentions of Kelly to provide "instruction in silk culture on a *modern scientific basis*." She credits the many Italian and French authorities from whom she gained her background after spending five to six years in Europe learning first-hand the scope and details of silkworm rearing. The italics and bold font are mine to provide emphasis to the point that Kelly makes throughout her writing and speaking about silkworm culture—that it must be scientifically based. As a longtime advocate of treating rearing as a science, rather than an art, I rejoiced in reading Kelly's imprecations to potential rearing personnel to use vetted scientific practices.

4.4.5 Cotton and Silk: Purpose-Based Determinants of a Silkworm Mass-Rearing Technology

The rationale for developing a silkworm rearing program in the United States was based on economic and social benefits that were especially promising in the post-Civil War era (prior to World War I). Among the causes of the Civil War, was the need to produce cotton for export to Europe and for domestic purposes. This was discussed by Dattel 2019 in his essay "Cotton and the Civil War":

> "If slavery was the corner stone of the Confederacy, cotton was its foundation. At home its social and economic institutions rested upon cotton; abroad its diplomacy centered around the well-known dependence of Europe … upon an uninterrupted supply of cotton from the southern states."

In relation to the efforts to replace or supplement cotton in the Reconstruction South, Debra Bloom introduced Kelly and provided background as follows:

> Henrietta Aiken Kelly devoted her career to the education of South Carolina's most privileged daughters, first as a teacher and vice-principal of the Charleston Normal School and later as founder of the Charleston Female Seminary. In 1896 a new adventure beckoned. Closing her school, fifty-two-year-old Kelly boarded the SS Majestic and sailed for Europe. The marriage of her friend Janie Johnston Perry Scaife to the Duc de Litta-Visconti-Arese was one reason for the trip, but it was silk culture that Kelly observed at the Duc's medieval Italian manor in Casale Litta that enticed her to stay. As the Duc's special guest, local farmers invited Kelly to watch the eggs of the *Bombyx mori* moth hatch and cycle through their molting stages. Tiny caterpillars grew exponentially, and Kelly carefully fed them ever-increasing amounts of the *Morus alba's* white fruited mulberry leaves. As the final stage began, Kelly prepared the spinning racks with layered tree limbs for the sightless young insects to mount as they settled down to spin their cocoons. When the season came to an end, she walked with the peasants as they brought 'great golden heaps' of dried, harvested cocoons to the marketplace to be sold. Kelly noted the similarities between Lombardy's agricultural conditions and her native South Carolina and became convinced that sericulture was the key to reviving the rural communities of her homeland. Buoyed by her discovery, Kelly studied botany at Europe's finest institutions. Over the next decade, she became a one-woman crusader for the establishment of a raw silk industry in the South.

Bloom goes on to explain that when the 52-year old Kelly closed down the Charleston Female Seminary in 1896, she felt that the major social and economic change that was needed in the South, after the disruption of cotton economy by the Civil War, was to replace or supplement "King Cotton" with an alternative industry—silk. In several of her letters, Kelly reveals her concern that with the resources going into educating and employing freed slaves and their offspring, impoverished white people were falling behind the African-Americans and needed an industry that could rescue the disenfranchised whites from poverty. Kelly saw silk production as the panacea, and she addressed this demographic as the target of her writing and her education efforts in silkworm rearing and post-harvest silk processing. Though the letters provided by Bloom 2019 are more explicit about the motivation than the 1903 monograph, much of Kelly's vision is presented in the monograph as explained in Kelly's Preface.

4.4.6 Narrative on Kelly's Treatise

In footnote **KFN 1 (in Appendix 2)**, the basis of the work of Henrietta Aiken Kelly is stated as a vehicle to help introduce silkworm culture into the United States. It represents a convergence of opportunity for the intention of certain persons such as the Secretary of Agriculture, James Wilson, Howard, and others who saw the possibility that silk production could replace (at least, in part) the economic and social reliance on cotton. During the post-Civil War period from 1865 through the first few decades of the 20th century, the upheaval of economic and social institutions was a burden to the southern states. This was discussed by Dattel and indirectly by Kelly in several of her letters and in the "Preface" to her monograph. What is implied in the various sources that are mentioned is that the high cash value of cotton, both in domestic use and in export, became a vast hole in the economy of the southern states in the United States. The visionaries such as Secretary Wilson, L. O. Howard, and especially H. A. Kelly saw the high cash value of silk as a means to filling the economic gap left by the abolition of slavery. Kelly further saw silk production as a panacea for the poor southern whites to gain employment and economic improvement over the meagre incomes of sharecroppers. It was, in a very intriguing way, a potential to allow the South to regain the economic and social losses that had accompanied the end of slavery. Kelly spoke often, in her letters, of the people—older adult women and young children—to gain a lucrative means of supplementing their families' income with silk, much as the Italian peasants that she met in the Duc de Litta-Visconti-Arese's manor in northern Italy. She saw that the production of silk could be fairly lucrative, and she realized that elsewhere, such as in Italy, France, in the Middle-East, and in Asian countries, silk was produced without recourse to slavery. Apparently, Kelly had planned to learn whatever she could of sericulture as she closed her school in Charleston, South Carolina and set out on her voyage to Europe. She spent some time in France and Switzerland, but most of her learning took place in Italy, where she participated in the practices to learn all aspects of silkworm rearing as well as the agricultural technology applied to cultivation and harvesting of mulberry.

In her preface (in **KFN 1**), Kelly explained that the purpose of the monograph was "To inaugurate the public instruction in silk culture on a modern scientific basis ..."(**KFN 1**). Howard, in turn, recommended, in a "letter of transmittal," that the Secretary of Agriculture (James Wilson) authorize the publication of the monograph (**KFN 2**). Kelly had written the monograph on silkworm rearing to help establish silk production (sericulture) as a part of the post-Civil War efforts to gain a foothold for poor southern people and an economically feasible alternative to cotton. Despite the fact that silk had been viewed since pre-Revolutionary times as an economically and culturally feasible enterprise in the Colonies, the production of silk had been introduced and reintroduced many times, this time to fill in the gap that was in place after the elimination of slavery and the hitherto practices of cotton production was disrupted by the abolition of slavery in the United States. The bulletin was intended for the audience of "persons who wish to begin the culture of the mulberry silkworm in different parts of the United States" toward the goal of providing a new (in the United States) crop that could create opportunities lost during the upheavals related to the Civil War and its consequences. L. O. Howard's role in the efforts to bring sericulture to the United States is documented by Bloom 2019 and Russell 1978 (Figures 4.3 and 4.4).

Figure 4.3 "Races of silkworm" from Pasteur. This illustration shows three of the races of silkworms delineated by color phases of adults.

4.4.7 Further Background of Kelly's Motivation

In the section on "the culture of the mulberry silkworm," ← (**KFN 2**) Kelly gives a little history of silkworm domestication, a topic which I cover later in this chapter. She mentions the antiquity of silkworm rearing from China and the European cultivation since the sixth century, AD. It is interesting that Kelly recognized that races of silkworms and the cultivation and domestication behind these races was crucial to silkworm biology. Kelly's recognition of silkworm races in 1903 preceded William Bateson's first use of the word, "genetics" (Ananya Mandal, https://www.news-medical.net/life-sciences/Gene-History.aspx). Kelly points out that while the origin of the mulberry silkworm was the foot of the Himalayas, the recent races were results of domestication. Kelly goes on to characterize the anatomical characteristics of silkworms along

Figure 4.4 Two races of silkworm with highly visible color markings ("Plain" and "Zebra"). Besides varietal or racial differences in adults and larvae, there are differences in colors of cocoons that have been bred over the centuries, and many other varietal differences pertaining to efficiency of conversion of food to cocoon biomass, degree of resistance to disease, and tolerance to different varieties or species of mulberry.

with explanations of basic physiology and developmental features.

4.4.8 The Scientific Basis of Kelly's Monograph

An extremely important facet of the Kelly monograph is that it is entirely dedicated to a scientific treatment (**KFN1**) of rearing an insect—something that is to be appreciated for its forward thinking, especially in such an early (1903) document. It will be seen elsewhere in this chapter that an even earlier treatment of silkworm rearing (Anonymous 1828) was even more detailed than the Kelly document; however, though the older document (although it is empirically based) does not reflect the meticulous insistence on keeping the rearing process scientific as does the Kelly 1903 monograph.

The comprehensiveness of the Kelly rearing system monograph is evident in the Contents page (**KFN 3**) where she points to the many aspects of silkworm rearing that are treated, including life history, food, implements used in silkworm culture, treatment of eggs, and the details of rearing conditions, including ventilation, disinfection (which should NOT be applied in modern silkworm rearing facilities), spatial requirements, temperature (treated in exquisite and useful detail), specifics about the various larval stages (which Kelly calls "ages" rather than the modern term, "instars"), preparation for accommodating the pre-pupa's spinning, and diseases of silkworms, which are amazingly well described in terms of present-day understanding of silkworm pathologies. Also in **KFN 3**, Kelly presents a table that details the duration of each "age" or instar of larval silkworms under standardized conditions. One of the most important features of the Kelly monograph is the explicitness of these standard conditions and the importance or rationale for adhering to them, making this document a superb example of SOPs.

In the section marked **KFN 4**, Kelly provides details of the developmental stage that she calls the "chrysalis" or "chrysalid state." She had described in the previous section the spinning activities, and she discusses the fact that though the chrysalis is a sort of "sleeping stage," it is also a period of "great activity, in which the entire being is transformed." Herein she also provides drawings derived from Maillot.

4.4.9 The Timeliness of Kelly's Comments Today

The details and accuracy of Kelly's accounts of the silkworm's biology remain valid today. For example, in **KFN 5**, she describes the silk that is extruded from the silk glands, mentioning that the silk has a varnish that composes 1/5 the mass of the thread. She goes on to describe the remarkable rate of growth of silkworm larvae, pointing out that in 30 days, larvae increase their biomass to 14,000 times their original weight as neonates. The point about the rapid and massive growth rate is an essential point for the rearing technicians to grasp to reinforce the importance of providing adequate amounts of food. This issue is not simply a matter of providing huge amounts of mulberry leaf biomass; as she goes on to explain in several places in the monograph, the food must be the correct type of mulberry provided at the proper phenological period. Furthermore, there are periods when it is advisable to have breaks in the supply of food. Kelly further explains that the larvae can become diseased if the highly regimented feeding protocol is not obeyed. This point will be covered below in the discussion of silkworm diseases.

In the section noted as **KFN 6**, Kelly describes accurately a complete accounting of the details of silkworm (and larval Lepidoptera, in general) molting, even including the "feeding dormancy" of the worms and the changes in cuticle. Also in this section, Kelly describes the five instars of the silkworm, including the time spent in each stage. She calls the instars "ages" of the larvae, and the importance of understanding the duration of each instar (age) becomes abundantly clear throughout the monograph, where Kelly warns the rearing technician about how mismatching the feeding protocol with the larval stage can lead to diseases and other maladies that result from deviation from the SOPs. A most crucial point here is that Kelly presents a solid understanding that stands as a rationale for all the seemingly "picky" precautions. What Kelly has done, master teacher that she was, is to provide for the incipient rearing specialists the

Figure 4.5 Picking mulberry leaves (from Pasteur 1870).

scientifically- based rationale for each meticulous step (Figure 4.5).

4.4.10 The Food of Silkworms

In section **KFN 7**, Kelly describes the "food of the silkworm," including essential details about species and strains of mulberry appropriate to silkworms. She also explains the massive food intake in the 5th age (5th instar), and the prodigious amount of frass (fecal matter) that the worm produces. During this period, the worms typically lose 12% of their body weight as frass deposition, and they prepare for spinning. She describes the movements of the silkworm, especially the head movements, involved in spinning its silk cocoon. She points out the importance of having an appropriately dry and roomy place where it can spin, explaining that if, for example, conditions are too crowded, the worms can spin either smaller cocoons or produce double cocoons whose worth is diminished. In the section marked **KFN 8**, she explains the characteristics of the pupa (chrysalis) in the silk cocoon, mentioning that all the organs of the adult are developed in this stage, and the time-frame of the development is 8–20 days, depending on the temperature. Elsewhere in the monograph, Kelly goes into intricate detail about how and why the temperature is to be regulated in the rearing situation. The leaf of the white mulberry is the natural and the best food for the silkworm. There are many varieties of the white mulberry—some much better adapted than others to commercial silk culture, and some better suited to certain localities.

Kelly explains, "The race, the size of the worms, the variety and age of the mulberry, the nutritive quality of the leaf, the year, the season, and the climate make the requisite quantity of leaf very variable. The following is given as a basis of calculation, all circumstances being considered, and the leaves not being cleaned: For the larvae hatched from 1ounce of eggs, during the first age [instar], 11pounds of leaves; during the second age[instar], 26 to 33 pounds; during the third age [instar], 88 to 145 pounds; during the fourth age [instar], 264 to 352 pounds; during the fifth age [instar], 1,540 to 1,760 pounds; or about 2,200 pounds in all, of which one-half is consumed in the last six or seven days of the fifth age [instar]."

Kelly's writing is an impressive combination of scientific accuracy, fine detail, and conversational style. She points out that certain varieties of white mulberry (*Morus alba* as compared with *M. nigra* and *M. rubra*, black and red mulberry, respectively) are the best food for silkworms. This point about the importance of mulberry variety in rearing/sericulture success is discussed further in this chapter in the section on failures of sericulture efforts in the northeastern United States. Kelly also points out that the race of silkworms and the varieties and age of mulberry must be matched. She provides readers with the impressive appetites of the various instars ("ages") of silkworm larvae, which increase 14,000-fold from neonate to late 5th instar larvae, and she mentions that one ounce (~28.4 g) of *B. mori* eggs produce larvae that consume about 2,200 pounds (~1,000 kg) of fresh leaves. Kelly goes on to note that matching the age of the leaves and that of the silkworms is essential for nutritional matching and to reduce the chances of disease, which is a menacing problem in silkworm production. The practical details Kelly provides include advice to rearing personnel not to change the leaves abruptly but, rather, to introduce them gradually and with caution to use only fresh leaves. She further advises the equilibration of leaves as to temperature of the rearing room and that the leaves be kept moist but never so wet as to promote fermentation, which she explained leads to the disease called flacherie (French for flaccidness). The management of diseases in silkworm rearing, as described by Kelly, is discussed below.

Kelly explains the importance of such details as matching the age of the leaf and the age of the larva writing, "young worms fed on old leaf, or old worms fed on young leaf, are apt to become diseased, and even though they may not die, will scarcely mount or will spin indifferently." She explains further that, "a change of leaf, too, should, if possible, be avoided, or made gradually. Fresh leaf only should be used, and never when wet with dew or heated, or before it has in a measure acquired the temperature of the room in which the worms are being reared. It should be gathered early in the morning or in the evening and should not be bruised or torn, nor should the baskets or aprons used to carry it be the same which are used to remove litter. To avoid fermentation the leaves must be spread out in a cellar or cool, darkened room. Cut up only a limited quantity of leaf at a time and cover with a damp cloth to keep fresh, but never submerge the leaf in water, as this is apt to occasion flacherie, a very destructive disease. From the fourth age on there should always be a day's supply ahead, so that in case of rain the worms will not have to fast." Furthermore, the meticulous detail and step-by-step protocol offer a practical set of SOPs. Besides the specifics on potential for disease that may stem from mismatch of age of leaves and age of larvae, Kelly also provides all the non-intuitive factors such as allowing the leaves' temperatures to come to equilibrium with rearing room temperatures, and that wet leaves should never be provided to larvae due to their propensity to decay (ferment, in Kelly's terms). She even specifies the time of day when it is best to collect the leaves, a point that relates to potential surface moisture and the nutritional state of the leaves. Finally, as discussed in the section labeled "**KFN 8**," Kelly specifies the tree cultivation parameters that affect optimum rearing conditions. For example, she specifies that the height of the trunk of the mulberry should be restricted to 5–6 feet for the convenience of the workers who must harvest leaves of appropriate age and condition. She includes a time-tested observation that the best trees are those that bear little or no fruit, in agreement with the rationale that a tree that is spreading its energy and material resources to both fruit and leaves will compromise the nutrients that are diverted to leaves. This would result in diminishing nutritional value of the leaves.

4.4.11 Economic Perspective of Kelly's Monograph

Kelly provides an economic perspective and lays out a model of the economic constraints that would allow success in the rearing process (**KFN 9**). In the section of the monograph where she discusses the economics is where she suggests the equipment or implement needed for rearing silkworms, including shelves, wrapping paper (or newspaper), a small ladder, knives, baskets (for the leaves), small rearing trays, tulle (a fine-weave of cloth that is similar to organza or organdy), knives (to cut leaves), baskets for transporting leaves, brush (or straw or shaving) for a place for worms to spin, and finally a thermometer. Kelly points out that these implements can be used repeatedly, so in time they will more than pay for themselves and the original investment. In this section, she specifies the height of the shelves (14 inches or about 36 cm), and she is explicit about the need to keep shelves far enough from the walls to allow passage of personnel. Throughout the monograph, Kelly reminds readers of the labor-intensive nature of proper rearing of silkworms. In Figure 4.6 (KFN 9), the rearing room configuration is illustrated with three racks with six shelves each shown. The racks are about 7 feet tall (about 2.2 m), and the need for the ladder becomes apparent from this picture where the women carrying out

Fig.1–adult silkworm: 1, head; 2, 3, 4, 5, 6, 7, 8, 9, 10, 12, rings; 11, horn; 13, 3 pairs of articulated legs; 14, 4 pairs of abdominal or false legs; 15, a pair of false legs on the last ring.

Figure 4.6 (a) and (b) Silkworm larvae on mulberry leaves and external structure of larva.

rearing responsibilities help illustrate the relative proportions of the system. Besides illustrating the proportions and placement of the rearing shelves, Figure 4.6 (KFN 9) also shows the trays and the arrangement of the larvae and their pupation sites (spinning sites) on the bottom two shelves of the center rack (Figures 4.7–4.9).

Figure 4.6 (a) shows healthy larval silkworms feeding on freshly picked mulberry leaves, and Figure SRS 4b is Kelly's drawing of a 5th (final) instar larva with its external parts labeled. Interestingly Kelly incorrectly labels the larva (shown in figure (b)) as "Adult silkworm." Figure 4.7(a) shows the cocoons and an emerged adult.

Figures 4.7 (a) Photograph of cocoons (and a recently emerged adults) reared in the Cohen lab and (b) drawings of an adult male and female (from Kelly). Note the differences between the male and female, with the male's abdominal terminus being truncated and the female's abdomen being distended and rounded at the tip.

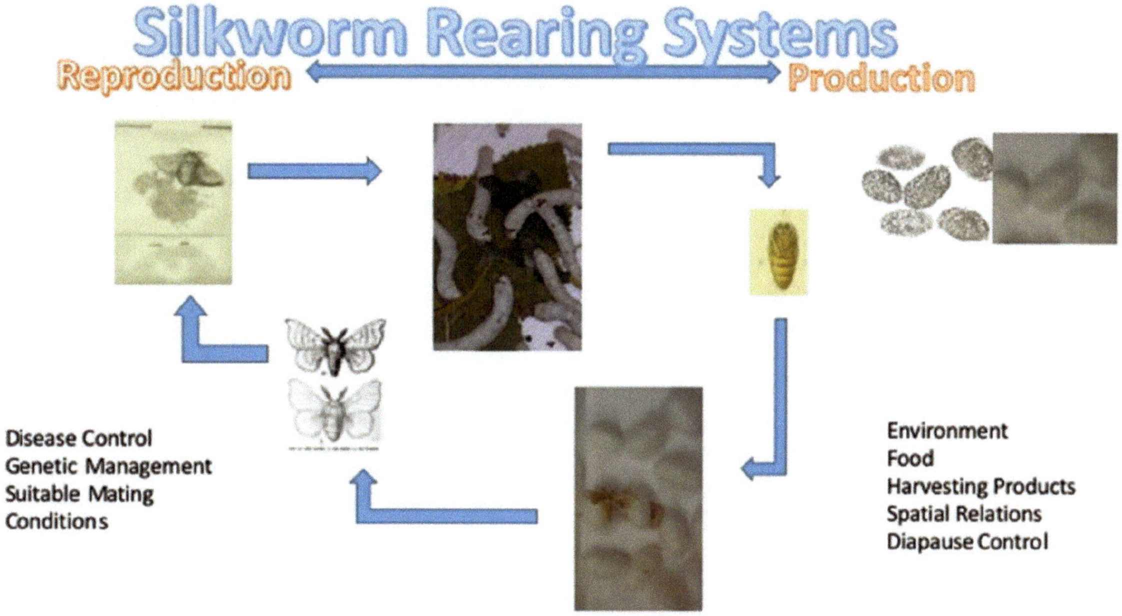

Figure 4.8 "Silkworm rearing systems" (reproduction and production). This figure illustrates the stages of the silkworm life cycle and how the rearing system SOPs provided by Kelly (and elsewhere by Anonymous 1828). On the lower left, the factors specific to continuous production of healthy eggs or "seed" are presented, and on the lower right the factors involved in the production of the product of silkworm rearing for production of the main product: cocoons.

In Kelly's treatment of the various life stages of silkworms in the rearing context, the treatment of eggs (**KFN 10**) is of special importance. She discusses the two kinds of rearing: production and reproduction, and she makes a sharp contrast between the difficulties in providing eggs that are healthy, laying great emphasis on the special conditions and special training involved in "seed generation." She uses the term "seed" to mean eggs, and this is in keeping with earlier works such as those of Anonymous 1828 and Dandolo. In the section marked **KFN 11**, Kelly explains that for univoltine races (or strains) of silkworms, there are three phases undertaken by eggs: an active phase, an inactive phase (from October to February), and a phase of intense developmental activity from Kelly

Figure 4.9 (a) and (b) Hatching silkworm eggs. (a) Shows several eggs in various stages of hatching, and (b) shows three silkworm eggs, one hatched (lower left), one at a pre-hatching stage (upper), and one in the process of chewing its way out of the egg. The hatching process takes about an hour and involves the application of digestive enzymes and intense mandibular actions to cut through the chorion.

makes several important points about silkworm eggs, one of them pertaining to the phenomenon that we now know of as diapause. The term "diapause" was not used until about 14 years after the Kelly monograph was published, and that first use pertained to earthworms (Schmidt 1917). Also, Kelly explains that there are two kinds of silkworm cultures: production and reproduction. Production cultures are used to yield silk fiber. Reproduction cultures have special requirements that involve a considerable expertise because control of diseases and grasping issues pertaining to the races of silkworms demand skills beyond the layman's level. In a broader sense, production is a purpose that underlies cultures of most species insects that are maintained for various purposes such as research, feeders, SIT, and biological control. In general, as in the specific case of silkworms, facilities where insects are reared operate under the intention that the cultured insects have some kind of purpose in accord with the overarching program, and there is a need for maintaining a continuous production of successive generations of the insects in question. However, with silkworms (*B. mori*), there are such dominating and potentially destructive issues with disease and other departures from the basic quality that is desired, special measures must be taken to produce the seed for all other operations.

In the section marked **KFN 12**, Kelly discusses the treatment of eggs required to break the "inactive period" (= breaking diapause) and also the measures that must be taken to assure uniformity of hatch period. Here, in **KFN 12**, she gives especially helpful SOPs that guide the rearing personnel to matching the egg hatch with the period of availability of mulberry leaves that are suitable for providing adequate nutrition for optimal larval growth. She is especially emphatic about the importance of not imposing periods of famine on neonate worms that are especially susceptible to bouts of starvation. Kelly provides exquisite details on the construction of incubators (how "ordinary tinsmiths" can construct suitable units) as well as proper use of such incubators, including hanging a thermometer behind the glass door so that the door does not need to be opened to check the temperature. It now seems quaint to those of us accustomed to having access to "high tech" incubators with electrical heating, cooling, and temperature regulation around a set point to read Kelly's instructions about using a gas or oil lamp as a heating source, but in the late 19th century the incubator design that Kelly described was state of the art.

In Figure 4.6 of "Appendix Kelly" a silkworm rearing room is illustrated (the figure from a source that Kelly identifies as Gobin). Kelly, as noted in the Appendix marked **KFN 13**, provided a complete and adequate narrative that would enable personnel who are not rearing experts to become competent at rearing silkworms as a commercial enterprise. I have quoted a paragraph on ventilation to illustrate the sense of the detail and practical wisdom that H. A. Kelly provides throughout. This paragraph also illustrates the kind of rationale-driven SOPs that were afforded to rearing silkworms with a technology driven by scientific principles. Most remarkably, this paper was written in 1903! Kelly writes:

> Ventilation.
>
> The domesticated worm should be surrounded continually by pure air. The amount of carbonic-acid gas given out by the worms and their attendants is very considerable; besides this, a quantity of deleterious gas is generated by the litter of the beds, and the lights and fires consume a great deal of oxygen. Myriads of spores and germs of organic-life float in the air of the rearing room, and their influence paralyzes the vital energy of the skin and of the organs of respiration, on whose normal functions the robustness of the worm so much depends. Hence, it is evident that the quantity of vitiated air which should be expelled from the room requires the introduction of a large quantity of fresh air. For this, a double system of ventilation is necessary, which may be obtained by double openings in the windows, to allow the heated bad air to pass out above and the cool fresh air to come in below. To renew the air in every part of the room, and to avoid a single and often violent current, there should be more than one window.

The opening statement about assuring that the "domesticated worms" be continually surrounded by pure air is further explained in terms of removal of carbon dioxide (which identifies as carbonic-acid gas given out by the worms and the attendants). I have measured carbon dioxide in containers in insect-rearing containers as high as 15,000 to 20,000 ppm, where fresh air is normally 350–400 ppm (Cohen unpublished data). Although I have not been able to find a reliable source on the levels of CO_2 that constitute stress for silkworms, it is reasonable to suspect that levels of this gas very much above concentrations in fresh air would have deleterious effects on target organisms. Kelly also correctly points out that the leaf litter in the beds of food provided for the silkworms would deplete oxygen (as well as being a further source of CO_2). Kelly's prescription for circulation of fresh air is reasonable for her time, though in currently available air handling systems, many of Kelly's points are obviated. However, her cautioning us about the "Myriads of spores and germs of organic life" that are present in the rearing setting certainly remains relevant where we deal constantly with microbial contaminants in our rearing systems.

Kelly also provided detailed instructions on sanitation and various measures for preventing the contamination and disease problems that are associated with all rearing situations, especially silkworm production. I hasten to caution readers that the system of disinfection should NOT be applied in today's insect-rearing setting; however, the assiduous efforts at disinfecting accumulating microbes should be taken seriously:

> Disinfection.
>
> Eight or ten days before introducing the worms into their quarters all the shelves and implements should be washed in a solution of chloride of lime (11 pounds of chloride of lime to 88 quarts of water), or in a solution of sulphate of copper (1 to 100 by weight). When everything is in order—tools, perforated paper, material for the worms to spin their cocoons on, etc., each in its own room—close the doors and windows as tightly as possible and fumigate the rooms with sulphur (11 pounds of sulphur to every 100 cubic yards of space). To fumigate properly, powder the sulphur and place it in an earthen or metallic vessel over a slow fire. The sulphur will gradually melt and take fire. Place it immediately in the rearing room and leave it there, with the doors and windows completely shut, for twenty-four hours. Nets should not be exposed to sulphur fumes, for this would soon rot them, but should be washed in a solution of sulphate of copper, and immediately afterwards in plain water. Twenty -four hours after the fumigation the floors should be washed with a solution of chloride of lime or sulphate of copper, and the walls should be whitewashed with lime. When dead worms are seen on the shelves, change the beds and create in the rearing room sulphurous gas by burning a pound of sulphur during six hours, or make a strong wood smoke, which is a good disinfectant and will not harm the worms. [Please note that I used Kelly's spelling of sulfur.]

Her suggestion of chloride of lime [$Ca(ClO)_3$] or calcium hypochloride (known today as bleaching powder or chlorine powder) remains an effective and fairly safe disinfectant. It should be noted that the most established disinfectant used today is a solution of sodium hypochlorite ($NaOCl$). **The fumigation with sulfurous (sulphurous) gas is not a safe practice and is NOT recommended for modern insectaries!** In modern insectaries, various safe disinfectants are available, and sanitation practices such as those derived from hospitals are both effective in terms of antimicrobial capabilities as well as safety for humans and the insects that they are cultivating. The need for fumigation can be largely obviated by providing surfaces that lend themselves to effective cleaning of floors, walls, ceilings, carts, and other possible sources of chronic contamination by microbes. These surfaces and the conditions for disinfection in a modern rearing facility are discussed in detail in the chapter on the USDA, ARS, Robert T. Gast Facility.

Details of the silkworm rearing system are especially well communicated in the section labeled **KFN 14** and **KFN 15** where Kelly expresses the details of "**Space Required**" and "**Temperature**." The details (SOPs) regarding space and temperature required are described on the basis of instar ("age") of the larvae. In discussing the space required per 1 ounce of eggs, Kelly relates the SOPs in terms of square yards required per instar, and she reinforces the rationale by asserting that by providing the geometrical progression of space per instar, the yield of cocoons can be raised from 60 kg of silk to as much as 90, and she further adds that the quality of the silk will be enhanced by simply providing more of the resource: space. As for temperature, Kelly's specificity is highly precise, explaining that silkworms (being non-tropical organisms) thrive between 68°F (20°C) and 77°F (25°C), with a middle temperature of about 72–73°F (22.2–22.8°C) as optimal.

In footnotes **KFN 16**, **KFN 17**, **KFN 18**, **KFN 19**, and **KFN 20**, Kelly provides step-by-step instructions (SOPs) for rearing personnel to handle the various larval stages from 1st through 5th instar larvae (Ages, 1–5, respectively). I note here that an elegant summary of the treatment of larval stages in terms of amounts of food and other factors is summarized by Anonymous 1828 and is presented below as Figure "Table of Specific Treatment by Larval Stages."

THE FIRST AGE: In **KFN 16**, Kelly explains the characteristics of the neonates (the 1st instar larvae) that hatch in the early morning and then crawl through openings in the tulle (fine cloth) or paper to start feeding. She notes that the neonates should be held at 68–70°F to prevent them from accelerated feeding and growth, which translates to reduction of the number of feedings per day to a more manageable level. Kelly further instructs rearing personnel to mince the leaves for better access by the neonates, and she further cautions personnel to keep same aged cohorts to be kept together, with special caution not to mix worms of different ages. The extremely high labor required of insectary workers is underscored by Kelly's instructions to provide 4–8 meals a day for the first age worms! She explains that if batches of 1st instar larvae are found to be of slightly lesser size than their cohorts, the technicians should move the shelves to a location closer to the heat source and increase their ration of food by 1 or 2 feedings per day. It was illuminating to me that one of the most common sources of insectary problems, developmental straggling, was discussed by Kelly in her early paper. In fact, there are similar observations about what we now call straggling syndrome in the Anonymous 1828 publication.

THE SECOND AGE: For both 2nd and 3rd instars (ages), Kelly is emphatic (**KFN 17**) about the need to change the tulle cloth (Figure 4.8) or the perforated paper (Kelly Appendix, Figure 4.9), moving the worms to their new location. The movement to the new location is facilitated by placing a small amount of food on the fresh paper or tulle so that the worms are attracted to move on their own seeking the fresh food. Then, once they are established on the paper or tulle cloth, they can simply be lifted to their new feeding site. She explains the features that are "announced" by the worms as to their readiness to mold (triangular spots on the head and grayish cuticle). She adds that it is important to allow all the worms to molt before providing new food.

THE THIRD AGE: Kelly explains in **KFN 18** that the silkworms double their size between the 2nd and 3rd instars (ages), and that to accommodate this dramatic change, their space should be doubled from 8 to 16 square yards (6.7–13.4). She also points out that the new, larger rearing room should be pre-heated to 80°F (26.7°C), then dropped down to the intended rearing room temperature. She does not explain the rationale for the "over-heating," but it seems likely that she has experienced potentially negative effects of moving the larvae at this vulnerable age to a room that is too cool. Throughout her narrative, Kelly is very attuned to the importance of temperature in the silkworm rearing system. Kelly is also attuned to the amount and nature of the leaves provided to the various ages of larvae, instructing workers to provide six feedings per day for 3rd instar larvae. She also guides the rearing workers to use lighter amounts of leaf material at the beginning and end of each instar and cautions personnel to "read" the needs of the worms by paying attention to their appetite. She advises workers to realize that the 3rd instar is the most vulnerable to diseases and other disruptions of the rearing process. She writes that if the larvae pass without problems through the 3rd instar, the remainder of the rearing process is bound to be successful. Kelly goes on to emphasize that because of the susceptibility of 3rd instar larvae, workers should strive to keep the worms "sparse" and the leaves free of excess moisture but fresh enough to assure complete nutrition and hydration. Also, in **KFN 18**, Kelly offers a special treatment of developmental stratification, which she calls "three divisions." She advises using the netting or hand selection of the newly molted, larger larvae when they reach about 1/3 of the population. Then, the other 2/3 are given fresh provisions and allowed to complete their molts. Kelly further suggests fine-tuning of the separation of the larvae, where she points out that the delays in molting, if they affect a large fraction of the larvae, translate to problems with disease and ultimately with lower yield and quality of cocoons. I am impressed with the sensitivity or fine-tuning of the practices that Kelly tries to impose on the rearing personnel, especially in terms of using the homogeneity of developmental rates as key indicators of the quality of rearing operations. This point of uniformity is emphasized throughout this book in terms of attuning rearing personnel to the importance of reduction of variability (= error or uncertainty).

THE FOURTH AGE: In the section marked **KFN 19**, Kelly points out that upon becoming 4th instars, the larvae cease to be gray and display the color specific to their race. Here, she also adds further admonition to pay attention to temperature, mentioning that if the temperature is too low, below 68°F (20°C), the development will be retarded, and the efficiency of the rearing process will be greatly reduced. Also, Kelly points out that during the early 4th instar, the worms require a period of recovery from the molt and that adding excessive food will hinder their growth. Lastly, she further reinforces the importance of good ventilation, which allows the worms ability to reach their full potential due to good gas exchange and reduction of excess moisture. If the space is adequate, the feedings done at correct intervals, the air is clean and readily accessible, and the temperature is within the proper narrow range, the worms will be far more prone to reach the last larval instar (the fifth age).

THE FIFTH AGE: In **KFN 20**, Kelly is especially emphatic that workers not be too hasty in changing the beds, which involves potentially excessive handling. She points out that the "new organs" need time to become consistent and fully formed. She further exhorts workers to hold back on food as the newly molted worms cannot digest food until their digestive systems have fully matured, so she suggests keeping the worms in the beds of the 4th instars and giving a meal of lightly sprinkled leaves during the first part of the 5th instar. Kelly instructs insectary workers to space the worms adequately with a prescription that worms be provided with a space of the width of one worm between two others. She adds that with such attention to adequate spacing, along with four to six feedings per day, will assure the highest possible number of high quality cocoons. In this section (**KFN 20**), she points out that during the first five days of the 5th instar, the larvae feed voraciously and grow enormously, but in the last three days of this age, the feeding rate diminishes. She explains further that during the last three days of the 5th age, the circumference of the worms begins to diminish along with an increase in fecal (frass) production—all terminating with a movement of the worms away from the food source. Kelly explains further that the consistency of the fecal material becomes moist and soft; and, importantly, during this period of extra release of moisture, the worms become more subject to disease. She also notes other tell-tale signs of the worms' being in a pre-pupal state where the head becomes prolonged and a transparent gold or white. An important comment that Kelly makes here is that the synchronous timing of the population is especially crucial because stragglers that are not ready to pupate become especially susceptible to disease or other problems that significantly reduce yield of cocoons.

Figure 4.10 shows the recommended configuration of the feeding beds in relation to the twigs that are provided for pupation as described by Kelly in **KFN 20** and **KFN 21** (Preparing Cocoons for the Market). In this figure, the ingenious system of using arched mulberry twigs as pupation (spinning) sites. This configuration allows rearing personnel to re-use the twigs that previously held leaves that were fed to the worms. Once the twigs had been denuded of the foliage, they serve as ideal spinning sites

Figure 4.10 Pupation sites (from Pasteur 1870).

where the mature worms can abandon the leaf bed and go through a natural "wandering" process characteristic of many species of pre-pupating Lepidoptera. The twigs, if assembled properly in the arch conformation, provide a great deal of well-ventilated surface area with an optimal surface for commencement of spinning.

The large amount of surface area further allows the worms to separate from one another, which Kelly points out is important to prevent over-crowding of cocoons which elevates the chances of worms forming double cocoons—a factor that greatly reduces the value of the silk. In **KFN 21**, Kelly provides what was then state of the art methods of killing the pupae in ways that are safe to the silk product being harvested. She suggests the most efficient method as an application of steam from a boiling cauldron of water with steam being circulated through a screened tube.

The last part of the Kelly 1903 monograph is her treatment of the diseases of silkworms (noted as **KFN 22** in the Kelly Appendix). The diseases that she discusses in some detail include Pebrine; Flacherie, or flaccidity; Gattine; Calcino or Muscaradine; and Grasserie. Although Kelly does not always use conventional citations as is found in later literature, she makes sufficient reference to Pasteur's works, such as Pasteur 1870, that it seems apparent that her sources mainly boil down to Pasteur's works, especially the 1870 book. Because of Kelly's strong reliance on Pasteur, and because it will be useful for me to update the works on silkworm pathology, I will defer discussion of Kelly's treatment of disease in concert with other citations in "Section 4.6.."

While this remarkable monograph is an excellent introduction to silkworm rearing, it barely scratches the surface of the extent of knowledge that is associated with the ancient practices of rearing *Bombyx mori*. Throughout Asia and many parts of Europe silkworm rearing has been a highly fruitful and economically important practice that is as sophisticated as any aspect of sustainable agriculture, and in this important status there is a tremendous body of scientific background on genetics, nutrition, and ecology of *B. mori* in a rearing setting. The extent of this body of information, though it pertains to the subject of this book—insect rearing, renders the topic of silkworm rearing beyond the scope of this book and the brief treatment that I have given it here.

4.5 SECTION BY ANONYMOUS 1828

One of the most remarkable documents that I have encountered in my 40+ years of study of insect rearing is the US Congressional Report, Anonymous 1828. The report contains an amazing history of efforts to initiate a full-blown program in sericulture in the United States, as well as SOPs for mulberry production, silkworm rearing, silk processing (including reeling, dying, and other aspects of sericulture). In the context of the present book on insect rearing, I have focused most of our attention to the rearing portion of sericultural infrastructure. The following list of questions were posed in 1826 by the agricultural committee of the 20th US Congress and were the basis for the document referred to as "The Report," or Anonymous 1828:

1. What efforts have been made in the State of ___, if any, and at what periods of time, to raise silk?
2. Whence was the silkworm obtained? Is this insect a native of ___, and, if so, in what respects does it differ, if any, from the oriental or foreign silkworm? Are there any varieties of this insect known in ___? Please to describe whatever kinds there may be.
3. Does the mulberry flourish in ___, in its different varieties; what soil suits it best; is the white mulberry an indigenous tree? If not, has it been found to thrive as well in the United States as the red and black mulberry? Which of the latter is indigenous, or, are both?
4. Does the silk-worm feed as well upon the red and black mulberry as upon the white? Is there any other leaf or plant known in this country upon which it does feed; and, above all, will it yield silk of equal quality when fed upon any other leaf or food, as upon that of the white mulberry?
5. Be pleased to state the best methods of raising and multiplying the several species of the mulberry tree; how it is propagated; how old it should be before transplanting; what particular modes of treatment are required; the age at which it should begin to furnish food for the worm; whence trees or cuttings can be

obtained for transplantation; the price per hundred; how many trees should be planted per acre; and would it, or would it not be profitable to cultivate the mulberry, for the purpose of feeding the worm, in the form of a bush or a shrub?

6. What quantity of raw silk ought to be produced from an acre of full-grown mulberry trees, planted and reared in the best manner? How many silkworms are required to produce a pound of silk? What quantity of raw silk can be produced from an acre of mulberry trees by other modes of cultivation than the full-grown tree?
7. Be pleased to state, as much in detail as may be in your power, the methods of treating the silkworm, with a view of producing the best silk, and in the greatest quantity; embracing, particularly, the habits of the insect, as to appetite, cleanliness, position, or accommodations, whilst the cocoon is forming; the species of twig or branch on which it is best that it should be spun; temperature, and degree of light most favorable to the insect, and all other circumstances which may affect its health and strength, and its capacity to yield a perfect cocoon; also the process of forming the silk by the insect until the cocoon is completed; the best modes of treating the cocoon and of obtaining the silk; and how far the labors of women, children, and old men, may be usefully employed in the culture of silk.
8. What difference, if any, exists between the Chinese and Italian or Spanish worm; which country is reputed to produce the best insect, and what particular part of the country?
9. Does the climate of _______ admit of raising silk with full advantage in the open air; or is it best that the insect should always be housed? If the latter, what size and form of building or apartment is best fitted to its operations? If the silk be raised in the open air, is any shelter, and of what kind, required, as well from the rays of the sun as from wet, and to protect the insects from birds and other animals that prey upon them?
10. Is there reason to suppose that, in some parts of the United States, the climate may admit of more than one crop of silk being raised from the silkworm, in the course of the year, as in Asia? How will this inquiry apply to________?
11. Have lightning and thunder been found to affect the silkworm in our climate, and how far, if at all? Has any insect of our climate, or animal of any kind, been found to prey upon, or injure it, beyond those that are known to do so in other climates?
12. What is the greatest quantity of silk that has been raised in the State of ________, in any one year; into what articles or forms has it been manufactured—and have the manufactured articles been consumed at home or sent abroad and where; where has the raw silk been sent when not wrought into manufactures?
13. What prices have the raw or manufactured silks commanded in the market of the United States, as compared with similar articles imported from Europe, China, or the East Indies? What progress has been made, in the drawing, spinning, and twisting of sewing silk in any part of ______? Have its strength and evenness been equal to the French, English, or Italian sewing silk? How have we succeeded in the coloring or dying? All information on this head will be acceptable.
14. Will you be pleased to describe, with as much particularity as may be, the machinery most approved for winding off the silk from the cocoon; for forming it into organzine or thrown silk; and for manufacturing silk into all its different forms (including velvets) whether for clothing, furniture, or any other purpose? If within your power and convenience to add a sketch or drawing of any or all of the above machinery, it would be acceptable.
15. Describe the difference in the quality of the silk produced from all the different varieties of the silkworm, and annex, if in your power, a drawing (colored) of all the varieties of this insect, in its moth and other states. Is the insect brought from abroad supposed to degenerate in the United States, and if so, to what causes is it imputed, and how are they best to be obviated? What are the proper rules to be observed in the choice of eggs, in the choice of cocoons for seed, and in the choice of the insect for laying eggs?
16. If silk has been raised at any periods heretofore in _______, but not at present, what are supposed to be the causes that have led to its discontinuance?
17. Please to state any other facts or circumstances that may bear upon the objects or spirit of the resolution above referred to of the House of Representatives of the United States, though not especially comprehended under any of the foregoing inquiries.

Richard Rush
Treasury Department
July 29, 1826

Richard Rush (1780–1859) served as Secretary of the Treasury, as well as US Minister to France and England, Secretary of State, and US Attorney General. It was during his tenure as Treasury Secretary (1825–1829) that he commissioned the 232 page report on sericulture in response to a mandate from the US House of Representatives. I included the questions put forth by Secretary Rush to illustrate the sophistication of scientific thinking about rearing silkworms. I found an amazing degree of insight into questions of environmental factors (population density, optimal food—including understanding nuances of differences between the species, varieties, and field conditions of the host plant, white mulberry—temperature, light issues, etc.) as well as attention to issues of degree of education/training that would be required of rearing personnel, and highly sophisticated questions about system of housing the silkworms. There was in the questions and the Report (Anonymous 1828) clear vision of the quality control of the insects themselves, the food, and other conditions that went into the rearing process and the quality of the silk, especially within the context of the world market. I include here, by way of explanation, some comments quoted from the Preface to the Report:

> The directions and plans of this experienced author, refer to a large establishment of five ounces of eggs, but the intelligent cultivator will readily apply them to a scale of any size; in all cases, the practical results will be similar.
>
> To the directions and observations of Dandolo, many others are added, which were derived from an extensive correspondence instituted by the Secretary of the Treasury, in all parts of the Union, on the subject matter of the resolution of the House of Representatives, as well as from information taken from the most valuable and recent publications obtained by the Secretary, from different parts of Europe. To these abundant sources of knowledge, were added others, which the writer himself was able to command; among these was a free access, during the past year, to a large establishment, conducted upon the plan pointed out by Count Dandolo. To the practical part are prefixed a short history of silk in ancient times; an account of the former and present state of its culture in North America; and the natural history of the silkworm.

I have included some further selections of the text in the Report (Anonymous 1828) to illustrate what I consider a highly sophisticated explanation of the rearing system. I hasten to add that the treatment of information about rearing conditions was intended to inform prospective rearing personnel, how to perform all the rearing functions. Therefore, this information must be considered an SOP of the silkworm rearing system. I ask the reader to observe that the methods are not only very specific, but also they are based on some kind of systematic observations that are what Kelly 1903 describes as "scientific" rearing procedures—ones that had been arrived at with empirical means such as the algorithms that I describe in later chapters (e.g., "The Algorithmic Basis of Rearing Systems").

> When the cultivator has observed the state of the vegetation in the mulberry leaf, and imagined it fitting to have his silkworms hatched in ten days, he will put the eggs in the boxes in the proper quantities. He must weigh them, and keep a register, in which he must note his observations upon the course and progress of the insects, and thus at once secure theory and practice: beginning by marking the day and hour upon which he sets the box in the stove room, and also the number of the box: and, in short, everything that may be worthy of notice. The wicker hurdles should have paper laid inside of them and the distance before prescribed to be left between the boxes, is to prevent the silkworm from going from one to the other. If the temperature of the stove room should not reach 64°, on the day fixed upon to put in the eggs, it is necessary to light a little fire that it may rise to that degree which ought to be continued during two days. If the thermometer indicate that the exterior air is above 64°, the shutters should be closed, and the door and ventilator opened, to create a draught and cool the stove room. The third day, the temperature should be raised to 66°, the fourth day to 68°, the fifth day to 71°, the sixth day to 73°, the seventh day to 75°, the eight day to 77°, the ninth day to 80°, the tenth, eleventh, and twelfth days to 81°.
>
> The following are the signs of the speedy vivification of the silkworm:
>
> The ash-gray color of the eggs grows bluish; then purplish; it then again grows gray, with cast of yellow, and finally, of a dingy white.
>
> If the eggs of silkworms belonging to different persons, are put into the same stove-room, great differences will be observable, not only in changes of color in the eggs, but also as the period of hatching the worms. The insects of the eggs that have been preserved, through the winter, in an even and gentle temperature, and those of the eggs which have undergone maceration, come forth in four or five days sooner, namely, at the 71st, 73rd, or 75th degree of temperature; whilst those that have been kept in a very cold atmosphere, appear some days later.
>
> (The propriety of attending to the gradual increase of heat in hatching the eggs, is so reasonable, and so conformable to all experience, that it cannot be too strongly insisted on. Great heat, suddenly applied and continued, never fails to push on the worms too fast, and to render them red when they first come out. The nearer the American cultivator is able to keep to the degrees of the thermometer, which Dandolo has pointed out, the more certain will be success.
>
> By maceration is commonly understood eggs preserved in bugs under cushions or mattresses or blankets, and similar things, until the moment of putting them in the stove-house. This uncertain method must be injurious to the regular and secure development of the worms. It happens that great quantities are spoiled by maceration, the worms coming forth, and shortly dying.
>
> It appears rational that when there is a certain and regular method, we should not adopt another, the result of which is uncertain and irregular, particularly when this secure method is not attended by any great expense.

Elsewhere in the treatise, the Report's author explains the start of the rearing process with this description of handling the eggs that are being removed from their diapause to begin the rearing process.

It should be noted that much of the 1828 Report includes information from a treatise by Count Vincenzo Dandolo (1758–1819) whose work on silkworms proved to be extremely authoritative and the basis of much of the body of procedures used in Europe and the New World. Presented here is a brief section from Chapter VI. Of the 1828 Report, Document 158:

CHAPTER VI.

THE ART OF REARING SILK WORMS

Of the care necessary previously to the hatching of the silk-worm five ounces.

The first thing necessary to commence the operations of the year, is to detach the eggs of the silk-worm from the cloths upon which they were deposited, and to prepare them for hatching.

These directions to scrape off the eggs from the cloth or paper upon which they were laid by the moth, are given in the European books on silk-worms, and yet would seem to be unnecessary. Careful observation of the insects when hatching, has satisfied the writer that they even find less difficulty in leaving their shells when they are remained fixed, than when they were detached. In this opinion he is sustained by two persons familiar with the rearing of silk-worms.

The report continues:

In past times, the cultivators imagined that the silkworm might be hatched at random, and spontaneously; and that, if it were necessary to make an artificial climate, it was enough to use the heat of beds, or the natural heat of the body, or the kitchen fire, etc. and similar means. It is now allowed that these methods, at best uncertain, are often pernicious to the insect. However, since luxury has invented the hot-house, to enable us, by an artificial atmosphere, to raise exotics, it was surely natural to apply this invention to the improvement of the cultivation of silkworms; and yet it is but very lately, that this application of the invention has been thought of, which enables us, in a few days to hatch, with ease and certainty, any given quantity of silkworms, and rear them favorably.

It is proposed to state, in this chapter, the care which the eggs require, to prepare them for the favorable development of the worm, and the care necessary to fix and continue the requisite degree of temperature. We shall treat of

First. The preliminary preparation of the eggs.
Second. Of the necessity of fixing, by the thermometer, the temperature calculated to favor the hatching of the eggs, and the rearing of the silkworm.
Third. Of the hot-house or stove-room, in which they must be hatched.
Fourth. Of the hatching.

I point out the explicitness of the instructions, including hints of rationale-based procedures, and I especially emphasize the concept of CONTROL of the rearing—in this case in relation to the temperature. Figure 4.11 is a table included in the Report, Document 158. Note that the table provides information on the feeding demands laid out in a day-by-day basis for all five larval instars (which are called "ages" in this treatise and others written in the early works on insects' developmental stages). Note further that the instars are further sub-divided by day for each instar (1st instar with 5 days, 2nd instar with 4 days, and so on as specified in the table). Also, the logistics of rearing space (Column 2) is worked out for the rearing personnel where the number of feet required for each instar is specified. It's interesting to note that the standard one ounce of eggs is about 30,000 eggs, so if nearly 100% of the eggs hatch and reach the end of the 5th instar, there are nearly 30,000 larvae weighing nearly 5 g each (Cohen, unpublished observation) or 150 g of larvae to feed in the end of the larval period! The 3rd column provides the prescribed temperature for each instar, and the 4th column provides the weight (in pounds) of leaves required for the daily ration. The last column (Column 6) provides notes on the specific features of the feeding materials, including the ages of the leaves to be used, how they should be cut, and how they should be provided with branches (which gives the worms a place to climb, separate themselves from other worms (= spacing), and helps keep the leaves from drying out. It is especially noteworthy to me that elsewhere in this report, the rationale for the spacing of the silkworms is explained by the mentioning of the fact that the worms breathe through spiracles that would be blocked by close contact with other worms under crowded conditions. The author of the Report, Document 158, was aware that the circulation of air was associated with prevention of diseases. In my 40+ years of study of insect rearing in the 20th and 21st centuries, including visits to scores of rearing facilities, I have never seen such explicit and scientifically -based SOPs as those in this document!

4.6 DISEASES OF SILKWORMS STEMMING FROM REARING CONDITIONS

It is a reasonable assumption that the diseases that currently affect *B. mori* were present in ancient times when the Chinese first began to culture silkworms. Interestingly, the diseases described by Kelly 1903 were the center of Pasteur's 1870 investigations, and many of them were discussed by Anonymous 1828. Once Pasteur provided the sericulture community with an understanding of the microbial basis of these diseases, it became a matter of proper diagnosis and both prophylaxis and treatment that allowed sericulturists to gain control over what was previously a matter of fate.

In covering the information about pathogens and sanitation issues with silkworms as presented by Kelly 1903, it is worthwhile to update the excellent commentary with up-to-date information. I have therefore included several summaries of references, including the following abstract from Li et al. 2017:

Silkworm pathogens have been heavily impeding the development of sericultural industry and play important roles in lepidopteran ecology, and some of which are

Doc 158-page 172

Table of rearing of Silk-worms to the Sixth Age, from one ounce of Eggs.

Ages		Space occupied by the worms on the feeding frames or hurdles.	Temperature Fahrenheit's scale.	Quantity of leaves.	Total of leaves for each age.	OBSERVATIONS
		Feet. In.		lb. oz.	lbs.	
FIRST AGE	1st day			0 14		Tender young leaves, chopped fine; four meals, progressively increased,
	2d do			1 6		" " " the first the largest, the last the smallest.
	3d do	9 6	75°	8 0	7	" " " four meals.
	4th do			1 6		The first meal of 9 oz. the other less, if the leaves have not been eaten.
	5th do			0 6		Tender leavers, chopped fine. Worms casting their first skins.
SECOND AGE	1st day			4 8		Half tender branches and half leaves, cut fine; the first meal 12 ounces, the other leaves in two meals.
	2d do	19	73° to 75°	6 12	21	Tender leaves, four meals, the two first less than the two last. Enlarge the spaces.
	3d do			7 8		" " the two first the largest.
	4th do			2 4		" " distributed as wanted. Worms casting their second skins.
THIRD AGE	1st day			6 12		Half tender branches, half leaves a little chopped; second meal 1 lb. 14 oz.
	2d do			21 8		Chopped leaves, four meals; the two first less than the two last. Enlarge the spaces.
	3d do	46	71° to 73°	22 8	69	" " the two first the largest.
	4th do			12 8		" " " " the last, least.
	5th do			6 8		" " distributed as required. Third casting of their skins.
	6th do			" "		" " " " Worms are roused.
FOURTH AGE	1st day			23 4		9 lbs. of branches, 14 lbs. of leaves, cut coarsely; give the branches first.
	2d do			39 0		Leaves coarsely cut, four meals; the two first the smallest. Enlarge the spaces.
	3d do			52 8		" " three meals; the three first the least, the last 17 lbs. 4 oz.
	4th do	109	68° to 71°	59 4	210	Whole leaves, four meals; the three first 16 lbs. 4 oz. the last 104 lbs.
	5th do			29 4		Picked leaves distributed as wanted; the first meal the larges.
	6th do			6 12		" " " " Worms prepare to cast their skins.
	7th do			" "		Leaves as wanted. Worms are roused.
FIFTH AGE	1st day			42 0		Half branches and half leaves. Enlarge the spaces.
	2d do			65 10		Picked leaves, four meals; the first the smallest, 12 lbs. the last 22 lbs.
	3d do			93 0		" the first meal 22 lbs. the last 27 lbs. 12 ounces.
	4th do			130 0		" " 27 lbs. 12 oz. the last 37½ lbs.
	5th do			185 8		" " 37½ lbs. the second 46 lbs. 14 ounces.
	6th do	239	68° to 69°	223 0	1,281	" four meals, the last the most abundant.
	7th do			214 0		" " the first meal the largest, the rest to lessen gradully.
	8th do			150 0		" four or five meals, the first the largest, 46 lbs. 14 ounces.
	9th do			120 14		" distributed as wanted. Worms approach maturity.
	10th do			56 4		" " " if not enough, give more.

Figure 4.11 "Table of specific treatment by larval stages" (from Anonymous 1828).

used as biological insecticides. Rapid advances in studies on the omics of silkworm pathogens have produced a large amount of data, which need to be brought together centrally in a coherent and systematic manner. This will facilitate the reuse of these data for further analysis. We have collected genomic data for 86 silkworm pathogens from 4 taxa (fungi, microsporidia, bacteria and viruses) and from 4 lepidopteran hosts, and developed the open-access Silkworm Pathogen Database (SilkPathDB) to make this information readily available. The implementation of SilkPathDB involves integrating Drupal and GBrowse as a graphic interface for a Chado relational database which houses all of the datasets involved. The genomes have been assembled and annotated for comparative purposes and allow the search and analysis of homologous sequences, transposable elements, protein subcellular locations, including secreted proteins, and gene ontology. We believe that the SilkPathDB will aid researchers in the identification of silkworm parasites, understanding the mechanisms of silkworm infections, and the developmental ecology of silkworm parasites (gene expression) and their hosts.

It is also noteworthy that the paper by Li et al. 2017 describes important information on the topic of the genetics of silkworms themselves. An even more comprehensive treatment of genetics of silkworms and the evolutionary history of domesticated *B. mori* is found in Xiang et al. 2018. The diseases discussed by Kelly 1903 are virtually the same ones listed by Li et al. 2017 as well as by Balavenkatasubbaiah and Narasimha Nayaka 2008, Bebitha et al. 2016, and Tayal and Chauhan 2017, including the following diseases (Figures 4.12–4.18):

- **Flacherie:** Causative agent. The main causative pathogens are Infectious flacherie virus (BmIFV), Densonucleosis virus (BmDNV) and bacteria such as Streptococcus sp., *Staphylococcus* sp., *Bacillus thuringiensis* and *Serratia marcescens.* Flacherie is also caused by the combined infection of bacteria and viruses.
- **Grasserie** disease is also known as nuclear polyhedrosis, milky disease and hanging disease

Figure 4.12 (a) Pasteur's pupae and (b) microscopic closeup of illustrating various disease conditions. The microscopic view shows microbial corpuscles or bodies.

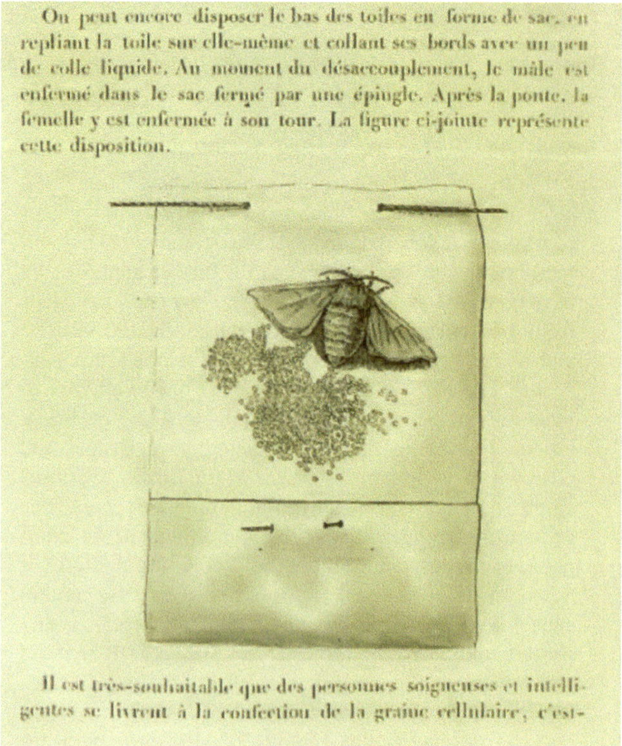

On peut encore disposer le bas des toiles en forme de sac, en repliant la toile sur elle-même et collant ses bords avec un peu de colle liquide. Au moment du désaccouplement, le mâle est enfermé dans le sac fermé par une épingle. Après la ponte, la femelle y est enfermée à son tour. La figure ci-jointe représente cette disposition.

Il est très-souhaitable que des personnes soigneuses et intelligentes se livrent à la confection de la graine cellulaire, c'est-

Figure 4.13 Silkworm female with eggs (from Pasteur 1870). Pasteur's instructions for performing the examination of eggs from pairs of silkworm moths. Note that the pair of mating moths are cut from the laying (oviposition paper, and an envelope is made of the paper where the male is confined (note silhouette of the male is seen near the pin). Once the female has completed ovipositing her clutch of eggs, she is also captured. Finally, the internal structures of the two adults are examined, and if they are shown to be free of the pebrine bodies, the eggs can be considered free of the pebrine-causing microsporidian. In Pasteur's words, "We can still arrange the bottom of the canvases in the form of a bag, by folding the sour canvas itself and sticking its edges with a pen of liquid glue. At the time of uncoupling, the male is locked in the bag closed by a pin. After laying, the female is locked in turn. The attached figure represents this arrangement. It is very desirable that careful and intelligent people engage in the preparation of the cell seed."

- **White muscardine** (or **calcino**) is caused by the fungus, *Beauveria bassiana.*
- **Pebrine:** Causative agent. *Nosema bombycis* and other strains of microsporidia such as *Vairimorpha, Pleistophora,* or *Thelohania.* (Also see Anonymous 1922, He et al. 2019, Hatakeyama and Hayasaka 2003, Kajiwara and Murakami 2019)

The insidious condition of flacherie was first investigated by Pasteur (in 1865) and later tied to several causative microbes, including several viruses and at least two species of bacteria as described by Tanada and Kaya 1993.

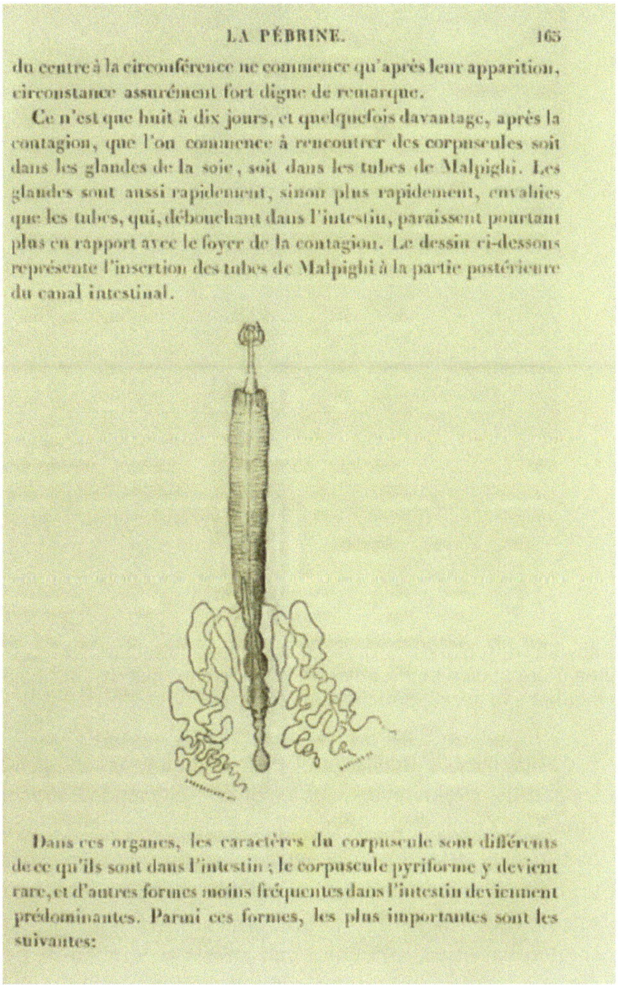
LA PÉBRINE. 165

du centre à la circonférence ne commence qu'après leur apparition, circonstance assurément fort digne de remarque.

Ce n'est que huit à dix jours, et quelquefois davantage, après la contagion, que l'on commence à rencontrer des corpuscules soit dans les glandes de la soie, soit dans les tubes de Malpighi. Les glandes sont aussi rapidement, sinon plus rapidement, envahies que les tubes, qui, débouchant dans l'intestin, paraissent pourtant plus en rapport avec le foyer de la contagion. Le dessin ci-dessous représente l'insertion des tubes de Malpighi à la partie postérieure du canal intestinal.

Dans ces organes, les caractères du corpuscule sont différents de ce qu'ils sont dans l'intestin ; le corpuscule pyriforme y devient rare, et d'autres formes moins fréquentes dans l'intestin deviennent prédominantes. Parmi ces formes, les plus importantes sont les suivantes:

Figure 4.14 "Pasteur GI tract of pebrine-infested silkworm." In this drawing, Pasteur shows (with the accompanying description) the progress of Pebrine infection ten days after inoculation. The disease organisms (now known to be microsporidia, mainly *Nosema bombycis*), though other genera may be involved with silkworms such as *Vairimorpha, Pleistophora,* or *Thelohania.*

Figure 4.15 "Pasteur silkworms feeding on mulberry leaves." Among many careful observations of silkworm diseases, Pasteur considered the possibility that contaminated food could lead to some of the silkworms' diseases. This is known today to be an important route for entry of viruses, bacteria, microsporidia (various protozoans), and fungi per os.

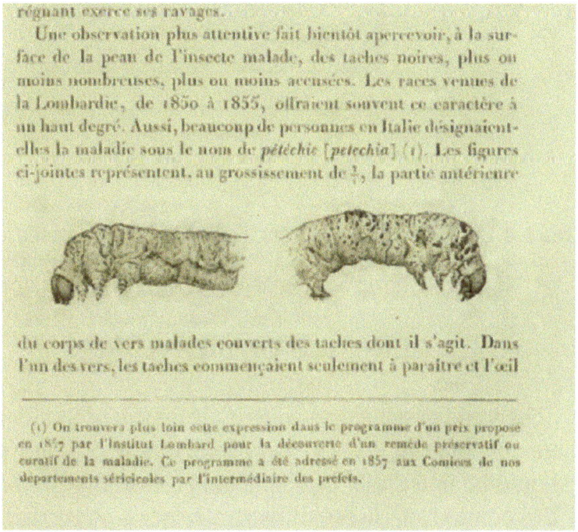
régnant exerce ses ravages.

Une observation plus attentive fait bientôt apercevoir, à la surface de la peau de l'insecte malade, des taches noires, plus ou moins nombreuses, plus ou moins accusées. Les races venues de la Lombardie, de 1850 à 1855, offraient souvent ce caractère à un haut degré. Aussi, beaucoup de personnes en Italie désignaient-elles la maladie sous le nom de *pétéchie* [*petechia*] (1). Les figures ci-jointes représentent, au grossissement de $\frac{3}{1}$, la partie antérieure du corps de vers malades couverts des taches dont il s'agit. Dans l'un des vers, les taches commençaient seulement à paraître et l'œil

(1) On trouvera plus loin cette expression dans le programme d'un prix proposé en 1857 par l'Institut Lombard pour la découverte d'un remède préservatif ou curatif de la maladie. Ce programme a été adressé en 1857 aux Comices de nos départements séricicoles par l'intermédiaire des préfets.

Figure 4.16 "Pasteur's anterior of pebrine-infected silkworm larvae." A closer observation soon reveals more or less numerous, more or less marked black spots on the surface of the skin of the sick insect. The Venetian races of Lombardy, from 1850 to 1855, often exhibited this character to a high degree. Also, many people in Italy called the disease petechie. The attached figures represent at an enlargement of 3/1 the anterior part of the body of sick worms covered with the spots in question.

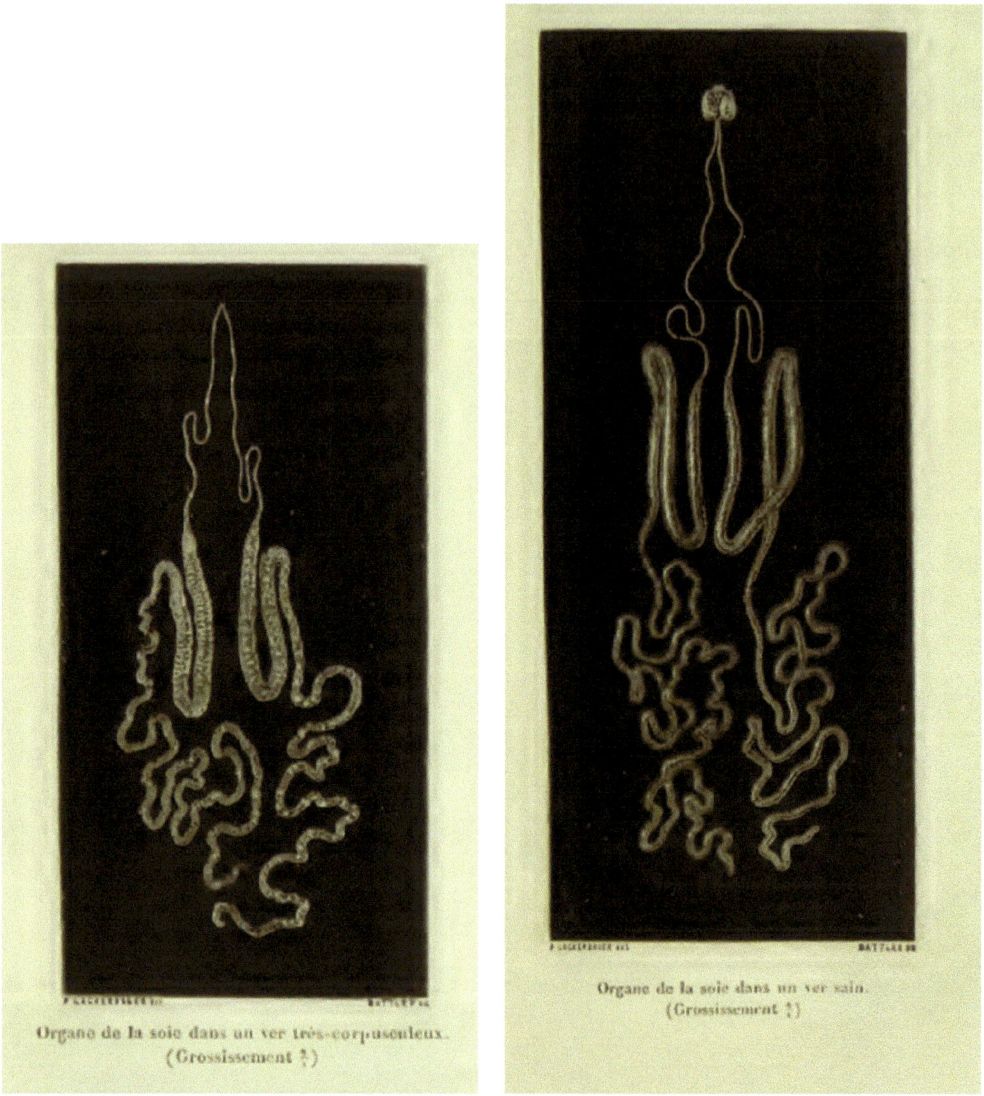

Figure 4.17 (a) and (b) Pasteur's depiction of diseased and healthy salivary glands. Figure (a) shows pebrine-infected worm's salivary glands with microsporidia-containing structures (note bodies in the anterior and posterior glands) and figure (b) shows healthy salivary gland complex lacking the mottled and misshapen tubules.

4.7 GENETICS OF SILKWORMS

The topic of silkworm genetics is far too broad for adequate coverage in this chapter (or in this book); however, it should be noted that this subject is especially important in the context of insect/silkworm rearing. As treated in other chapters in this book, we must recognize that the population biology of insects in rearing systems includes the genetic characteristics and overall, the gene pool of the insects being reared. In the case of silkworms, the fact that these insects have been domesticated to an extreme point with deliberate and inadvertent selection has shaped the gene pool over the 5,000 year domestication process. Much of this is treated by dozens of authors including Yukuhiro et al. 2002 and Arunkumar et al. 2006. A comprehensive treatment of genetics of silkworms' genetics and the evolutionary history of domesticated *B. mori* is found in Xiang et al. 2018. Among the most conspicuous features of silkworms' genetic selection and domestication are the races or populations whose colors have been selected such as the zebra race, pink silkworm larvae, and spotted varieties. Silk color can have a genetic basis, and various silk colors have been developed in certain strains as has silk fiber characteristics (physical features such as tensile strength, diameter, and biochemical composition = amino acid sequences in fibroin and sericin, amount of fiber in the cocoon, metabolic efficiency of food utilization, immune system features, and countless other traits that have been affected through the process of selection and domestication). Among these, the selection of

Figure 4.18 "Pasteur's infected silkworm larvae" (Pasteur 1870). This figure illustrates several phases of silkworms infected by flacherie (or flaccidity). In Kelly's 1903 treatment of flacherie (**KFN 22** in "Kelly Appendix 2"), she described this disease as "the most dreaded disease among European silkworms." She notes that seemingly healthy worms are struck with the disease later in their larval lives, sometimes manifesting itself during or immediately prior to spinning.

"docile" larvae and adults was probably an early step in the domestication process, making these insects easier to handle for commercial sericulture. However, in the selection process, pleiotropic factors crept into the genome of *B. mori*, that is instances where genes that had multiple phenotypic expressions—some undesirable—became an indelible part of the silkworm's domestication. This is likely to be at least part of the character of silkworms' being extremely susceptible to the ubiquitous diseases described above and plaguing sericulture.

4.8 SILKWORM ARTIFICIAL DIETS

The topic of artificial diets for silkworms has occupied the major attention of several groups of researchers. The most notable and remarkable account is Hamamura's description of his and his research group's involvement in silkworm diet research over a 40-year period, and his integration of the studies of silkworms' specialization to mulberry provides us with one of the most complete and productive chronicles of artificial diet development and feeding dynamics. His overview and explanation of how he came to be involved in silkworm diet development should be studied by every student of insect rearing. He begins the section on "Feeding Mechanisms and Artificial Diet of Silkworms" as follows, starting with the question "Why Do Silkworms Feed Exclusively on Mulberry?"

> In 1935 when I was teaching biochemistry at the Agriculture Faculty of Kyoto University, I was transferred to the School of Sericulture in Kyoto as Professor of Sericulture. In this school all the students had to undergo practical training in sericulture twice a year in May and September. In May when the training started, I saw for the first time the rearing of silkworms on a mass scale. Daily the students had to go dragging the carts to the mulberry field which was 4 kilometres away from the school to collect mulberry. When the larvae attained the 5th instar stage followed by cocoon harvesting and cocoon drying.
>
> It was during this period when one day it suddenly occurred to me why the silkworms feed exclusively on mulberry leaves. Unable to contain my curiosity, I asked an expert on the subject and was disappointed to receive the following reply: "what can one say excepting that it is called KAIKO (meaning silkworm in Japanese) because it feeds on mulberry." I compromised by accepting the reply as in 1935 one cannot expect anything more on this subject.
>
> As I had joined the School of Sericulture I thought I should study the physiological chemistry and nutritional chemistry of the silkworms. However, at that time no details were available on the chemical contents of mulberry on which the silkworms exclusively feed. Under such condition, it was not possible to directly take up the physiological chemistry or nutritional chemistry for serious studies. If only the clue to why the silkworms feed only on mulberry is worked out, then it will be possible to prepare the diet with the components artificially judged to be ideal and this would pave the way for a totally new field of research study on silkworms. I decided to work on this research theme.

I am amazed at the elegance of Hamamura's experimental design, and I am no less impressed with his humility and his nearly poetic writing (with kudos to the translator). He goes on to describe his journey into silkworm feeding dynamics:

> Till then I had no experience of handling the insects and hence felt like a complete stranger in this field. For a while, I wondered about the starting point of the study. As a chemist, I thought that the collection of silkworms on the mulberry leaves is apparently associated with the odour of mulberry. I was too impatient to refer to any literature and instead straight away started the experiment. I brought 2 or 3 fresh mulberry leaves and placed them in a petri-dish. I poured cold ethanol on the leaves in an effort to bring out the odour of the mulberry. The leaves turned white after 3–4 hours and the alcohol displayed refreshing green colour. When this solution was absorbed in filter paper and the alcohol evaporated, the green filter paper emitted the fresh odour of mulberry.
>
> When the alcohol evaporated, the mulberry leaves from which the odour had been removed turned in the form of brownish wilted residue. The filter paper with the odour and the leaf residue were kept at the two opposite ends of a necktie box and 10 larvae of molted 5th instar were placed at the centre of the box. The box was closed and keep for 30 minutes. When the lid was opened, all the 10 larvae were found to have collected on the filter paper. In all the repeated experiments, none of the larvae moved towards the end with mulberry residue. I reacted with excitement by saying, "what a simple thing to happen!"
>
> However, the excitement lasted only for a short while. On close and sustained observation, it was found that the larvae which had got attracted towards the filter paper, merely collected on the filter paper but did not show any sigh of biting as they would have done in the case of mulberry leaves. All the larvae raised their heads and shook them right and left. In the hope of finding the larvae displaying the biting behaviour at least once, I continued to watch them throughout the night. The larvae refused to oblige. Though I consoled myself with the idea that the larvae should not be expected to eat the filter paper, I was desperate to see the larvae bite the filter paper at least once. As the morning dawned, I gave up and cursed the silkworms for their adamant Puritanism.
>
> At that time it was not possible to know that the odour of mulberry can only attract the silkworms. The reason for not biting the filter paper was considered to be the difference in the tactile feeling between the filter paper and the mulberry leaves. In order to confirm this view, experiments with poplar leaves and cherry leaves were carried out. After removing the specific odour of these leaves, the odour of mulberry was induced into them, but the larvae did not show any reaction. In spite of all my efforts to find out the tactile sense of the silkworms to the leaves of mulberry, I could not get a satisfactory solution to the problem.
>
> As nothing whatsoever is known about this aspect of silkworms, I decided to deal with it in my own way. At the research laboratory, I was busy handling certain problems on sugar chemistry, which I had started in Kyoto University. Others in the laboratory were busy studying the copper content of the gastric juice of silkworms, enzyme chemistry of ecdysone of silkworms, and Vitamin C synthesis in silkworms. As the problems relating to silkworms could be studied only during the silkworm seasons, there were gap periods when the work could not be carried out. As a result, the time passed rapidly without much progress in the study. Unfortunately, that was also the period when the war was entering a serious phase, and finally in 1941, it erupted in the Pacific War. Under the disturbing circumstances, the work could not be carried out in a consolidated manner. This was followed by the radical change in the contents of the research studies. The research solely concentrated on the method of producing dry edible paste and fat from the silkworm pupae and the method of producing sugar from chrysanthemum bulbs. All the findings were patented. Thus, the research study on silkworms was suspended temporarily.

For the remainder of the chapter, Hamamura provides remarkable details and insights into his "dissection" of silkworm feeding mechanisms. He launched his discussion with a disarmingly candid and admirably humble admission of his and his students' misgivings about commencing the onerous task of developing a diet that had eluded Hamamura and his colleagues for ten years (1935–1945) prior to the new efforts:

> However, there was much apprehension among the students regarding my resuming the search for an answer to this question "why silkworms fed exclusively on mulberry leaves" They wondered how long I will be involved in this problem and that too without proper planning. At times, I also felt that I could not do anything worthy for nearly 10 years and that research study with Nature as the opponent may turn out to be meaningless as in such case one is compelled to merely arrange the data learnt from Nature. However, this may also reveal the great tricks of Nature. I was aware that by asserting my ideas as rational and well planned, I might have to face disappointment at the end and give up by calling "sour grapes." Keeping all these in mind, I decided not to give up my ordeal.

After years of having spirits elevated with remarkable discoveries and a series of blind alleys, Hamamura and his team dissected the overall feeding biology into three steps: (1) attraction, (2) biting, and (3) swallowing.

Hamamura 2001 discuses at length the functional components that both elicit normal feeding responses and nutritional value, and he integrates these observations into the narrative that chronicles the development of the diets (M_0 though M_{50}+). In a most diligent fashion, Hamamura mentions the sources of and rationale for each diet component. One of the most successful diets, M_{15} is presented in Table 4.2. Another diet that was devised to address the feeding stimulation factors is presented here in Table 4.1, which was developed via meticulous analysis of fresh mulberry leaves and individual addition of

Table 4.1 Hamamura's Three Phases of the Feeding Response

Attracting factors	Citral, linalool, linalyl acetate, and terpinyl acetate (all terpenes)
Biting factors	β-Sitosterol, isoquercitrin, and morin
Swallowing factors	Cellulose, sugars/inositol, and potassium phosphate

each component. In fact, the narrative in Chapters 1 and 2 of Part 1 of Hamamura 2001 is one of the most comprehensive documentations of extensive integration of analysis and bioassay techniques for developing artificial diets. The protocol described by Hamamura involves: (1) analysis of the "signature" compounds in mulberry (such compounds as isoquercitrin which is widely distributed and morin which is fairly specific to mulberry), (2) presenting the compounds in gels (jellies described by Hamamura), and (3) observing the responses of the insects to the gels with or without the suspected chemostimuli. Once a suspected chemo-stimulus is identified, it is then incorporated into the complex artificial diet.

Ultimately, for practical reasons, Hamamura and his cooperators (and fellow diet researchers from other laboratories) resorted to using mulberry powder to include the token stimuli and nutritional stimulants (such as sugars, cellulose, and β-sitosterol). Since Hamamura published a series of works on the factor(s) that determine the silkworms' feeding specificity, many papers have appeared attempting to solve the problems of explaining these insects feeding specificity (Zhang et al. 2019).

The diet in Table 4.2 was used for small-scale rearing of successive generations of silkworms, and it or one similar to it was used in attempted efforts in larger-scale rearing to explore the feasibility of replacing fresh mulberry leaves with artificial diet for commercial silk production. However, the diet developed over nearly 30 years of research by several teams of researchers imposed major problems that have never been surmounted (Table 4.3).

Table 4.2 Hamamura's Composite (M_{15}) Best-Case Artificial Diet

Ingredient	Weight (g)
Mulberry leaf powder	15
Soyabean protein	30
Starch	15
Agar	2
Cellulose powder	35
B-Sitosterol	0.5
Vitamin mixture	0.5
Salt mixture	0.5
Gallic acid	0.5
Lecithin	0.5
Choline chloride	0.05
Water	230

Table 4.3 The Cappellozza et al. 2005 Diet: This Diet Is Widely Used in Small-Scale *Bombyx mori* Rearing

Ingredient	Weight (g)
Dried mulberry leaf powder	25
Defatted soybean meal	36
Wheat meal	15
Corn starch	4.0
Soybean fiber	5.0
Citric acid	4.0
Ascorbic acid	2.0
Salt mixture	3.0
Agar	4.2
Vitamin mixture	0.399
Sorbic acid	0.200
Propionic acid	0.691
Chloramphenicol	0.010
β-Sitosterol	0.500
Water	260

4.9 FUTURE PERSPECTIVES (DIETS, DISEASE MANAGEMENT, GENETIC MODIFICATION)

Although systems of silkworm rearing have been developed over the span of about 5,000 years and in nearly every nation in the world, there remain some obstacles to even more economical and practical utilization of the silkworm, *Bombyx mori*. One of the major obstacles is a completely satisfactory artificial diet. Despite the efforts described by Hamamura and various reports that appear to have solved the diet problem, there is still no artificial diet reported to be equal to fresh mulberry leaves for sustaining consecutive generations of mulberry silkworms.

Hamamura summarized the major factors that needed to be addressed to make feasible and practical silkworm mass-rearing on artificial diet as:

1. Synchronization of growth
2. Synchronization of molting
3. Improvement of cocoon shell weight
4. Improvement of egg production
5. Improvement of fiber quality
6. Countermeasures against disease
7. Reduction in cost of diet
8. Issues for the future (including nutrition, natural defenses against disease such as the fluorescent red substance in silkworm guts, hormonal control of growth and molting

The International Sericultural Commission (ISC) (Anonymous 2019a) provides information such as that in Table 4.4 as well as a wealth of information on all issues pertaining to sericulture. The ISC provides an important organizational structure that is aimed at improving economic and social well-being of nations throughout the world by advocating and supporting sericulture. The ISC website (Anonymous 2019a) provides an excellent history

Table 4.4 List of Silk-Producing Countries and Their Silk Production in Metric Tons

#	Countries	2014	2015	2016	2017	2018
1	Bangladesh	44.5	44	44	41	41
2	Brazil	560	600	650	600	650
3	Bulgaria	8	8	9	10	10
4	China	1,46,000	1,70,000	1,58,400	1,42,000	1,20,000
5	Colombia	0.5	0.5	-	-	-
6	Egypt	0.8	0.8	1.2	1.1	1.25
7	India	28,708	28,523	30,348	31,906	35,261
8	Indonesia	10	8	4	2.5	2.5
9	Iran	110	120	125	120	110
10	Japan	30	30	32	20	20
11	North Korea	320	350	365	365	350
12	South Korea	1.2	1	1	1	1
13	Philippines	1.1	1.2	1.82	1.5	2
14	Syria	0.5	0.3	0.25	0.25	0.25
15	Thailand	692	698	712	680	680
16	Tunisia	4	3	2	2	2
17	Turkey	32	30	32	30	30
18	Uzbekistan	1,100	1,200	1,256	1,200	1,800
19	Vietnam	420	450	523	520	680
20	Madagascar	15	5	6	7	7
	Total	**178057.62**	**202072.83**	**192512.27**	**177507.35**	**159648.00**

of silk and various other resources such as publications, education, and other infrastructural services. An important contribution to understanding why sericulture succeeds or why it often fails is advanced in a publication (Anonymous 2019b) on the development of sericulture in Latin American regions. This publication points out several reasons why efforts to develop a significant presence of the silk industry have failed in Latin American countries, except for Brazil, which has emerged as a major silk producer. Among the reasons for failures in establishing and maintaining silk culture on an economically feasible basis are lack of government coordination, backup and encouragement of all aspects of sericulture, including establishing the proper type and amount of mulberry, the development and perpetuation of skills in silkworm rearing, and the post-harvest treatment of cocoons—including all aspects of fiber production (such as reeling, luster, and dye-processing).

CHAPTER 5

Case Study of the USDA, ARS R. T. Gast BCMRRU: Boll Weevil and Lepidoptera Mass Rearing

5.1 BACKGROUND OF THE GAST FACILITY'S BOLL WEEVIL AND LEPIDOPTERA REARING SYSTEMS

One of the most successful, useful, and extensively automated rearing operations was the system developed at the USDA, ARS mass rearing facility that was in operation for more than three decades, from the 1960s to the early 2000s, at the Robert T. Gast Facility at Mississippi State University. What I am calling the Gast Facility System, developed by several scientists cited within this chapter, is actually two systems: a boll weevil mass-rearing system, and a spin-off of the boll weevil technology, a lepidopteran rearing system. Both systems ultimately came to use the same type of flash sterilizer and the same type of form-fill seal technology. The systems are discussed in the following pages, and all photos and descriptions are explained as portions of the system diagrammed in Figure 5.1. Also, the standard operating procedures (SOPs) are presented in detail in Appendix 3: Gast Mass-Rearing System

As shown in Figure 5.1, both the boll weevil system and the lepidopteran (moth) system were kept in adjacent rooms (two rooms for the boll weevil system and two different rooms for the moth system). Figure 5.1 also shows that there were mixing kettles, which fed unheated diet into steam jacketed cylinders, constituting a flash sterilizer. Also in both systems, heated and cooled diet was carried through a stainless steel tube to the form-fill seal machine where appropriate diet containers/cages were produced with a heating and pressing system where molds were used with pressure to form the rearing trays. The form-fill seal machine was also outfitted with automated accoutrements that performed such functions as diet cooling (to set the gel), addition of eggs, particulate absorbent materials (such as corn cob grits or vermiculite), applying lidding materials (perforated plastic sheets with hot glue), and cutting the strips of plastic cage materials into individual cages or multiple cell trays. Once all these functions were carried out, the inoculated, diet-containing cages could be placed in the larval rearing room under appropriate environmental conditions (temperature, humidity, light periods, and air flow), and after incubation, pupation could take place and pupae could be moved to the adult handling room. Details of this process are further described in the following pages (Figure 5.2).

5.2 BACKGROUND ON THE BIOLOGY OF THE BOLL WEEVIL: THE FOUNDATION OF THE MASS-REARING SYSTEM

It is imperative that a rearing system is designed to mesh with the fundamental biology and life history of the target species. Because boll weevils are among the most fastidiously specialized insects tuned exquisitely to exploit cotton, it is a remarkable feat to design and operate a rearing system that meets the requirements of these insects. Before exploring the papers on development of mass-rearing technology, it is valuable to discuss major features of the specialized biology of *Anthonomus grandis*.

The complex biology and life history of the boll weevil are at the core of the successful mass-rearing program and eradication efforts that depended upon the success of the mass-rearing process (Figure 5.3). Briefly, the complexities of the life cycle begin with the egg that is successfully inserted into cotton squares (Figures 5.4 and 5.5). Upon hatching, neonate weevil larvae eat their way into the developing cotton bolls where they select the most nutritious portions of the cotton bolls, especially developing seeds and their surrounding tissues. Pupation takes place in the bolls, and after completion of the pupal period, adults emerge ready to mate, with some going into a state of diapause (Brazzel and Newsom 1959). Adults, after emerging from pupation or diapause, must feed on an adequate nutritional source to provide biomass that is to be converted into reproductive materials, especially eggs.

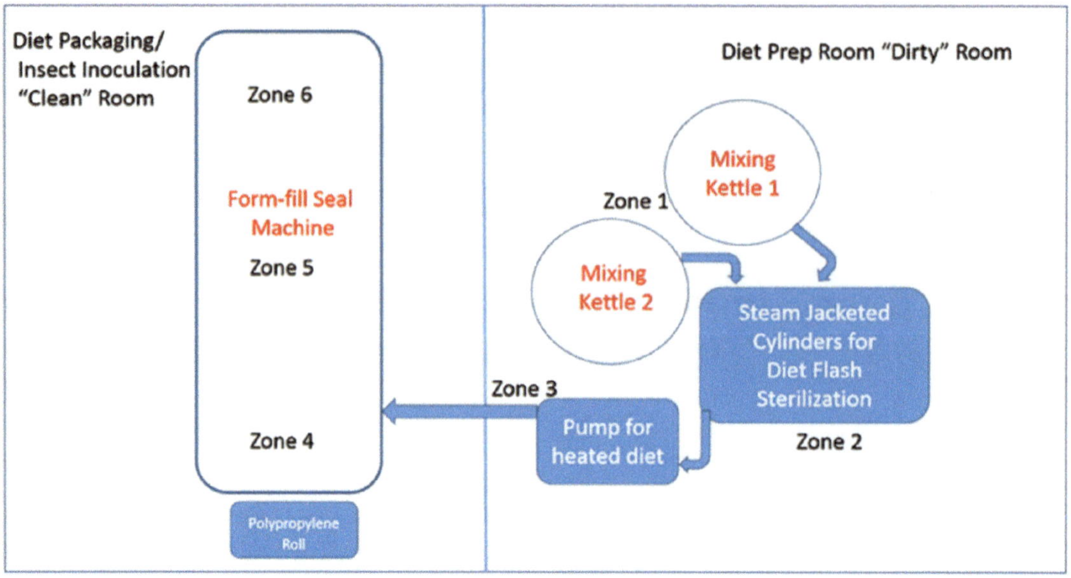

Figure 5.1 A schematic diagram of the flash sterilizer system in the "dirty" room on the right and the tandem form-fill seal machine on the left. The flow of the system is in a clockwise series of five major zones. Zones 1, 2, and 3 are shown in Figures 5.2 and 5.3, which show the various parts of the flash sterilizer. It should be noted that this system is nearly identical to the separate boll weevil rearing system and the lepidopteran rearing systems, both discussed in this chapter.

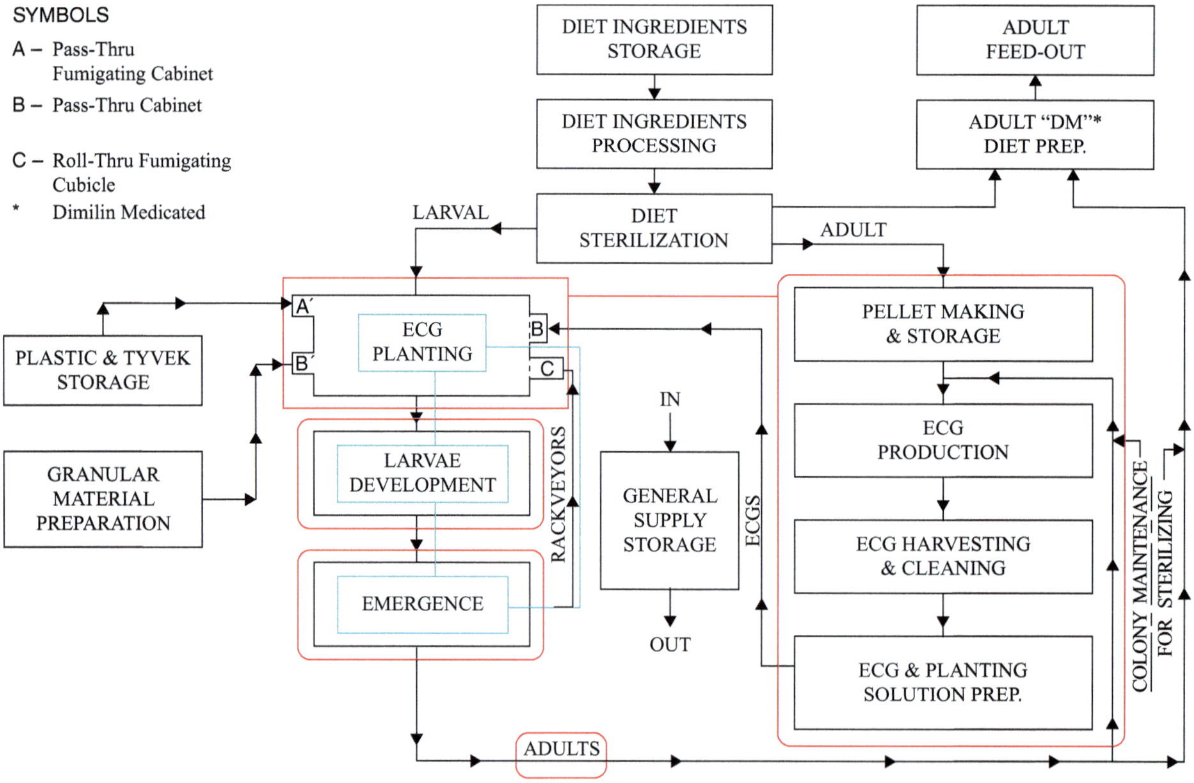

Figure 5.2 Diagram of the layout of the Gast Boll Weevil Facility (from Griffin 1984). The diagram in the figure represents the stepwise production scheme that accommodates all the features of the boll weevil life cycle, from egg → adult → egg, etc. The sequential (successive) generation of adults from the previous generation's eggs is represented in blue. The steps taken to process the eggs and the successive "core" operations are represented in red, and the peripheral operations are shown as satellite boxes and arrows. For example, the storage, processing, and sterilization of diets for adults and larvae are represented in the middle, top of the diagram. The upper right boxes represent the treatment of adults with the chemical sterilant D.

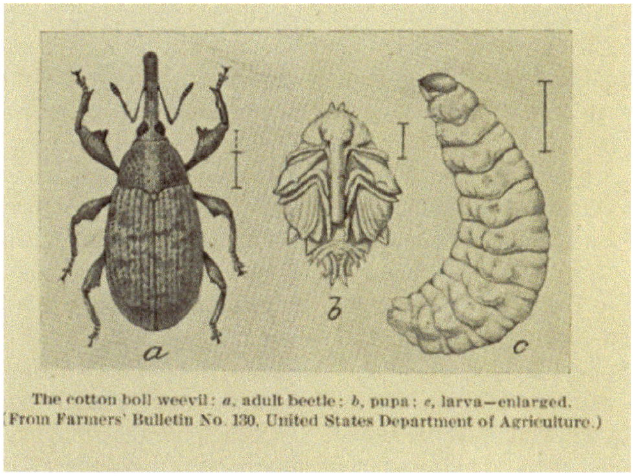

Figure 5.3 Boll weevil life cycle image. (National Agricultural Library) showing adult, pupal, and larval boll weevil ((a–c), respectively).

5.2.1 Aspects of Boll Weevil Mass Rearing: Interdigitation of Biology and Technology

I. Oviposition and eggs
   A. Chemical stimuli/responses
   B. Texture stimuli/responses
   C. Shape stimuli/responses
II. Larval development and larval diets
   A. Implanting/scarification
   B. Diet chemical and physical stimuli
   C. Larval frass management in diet (tunneling, microbial control, etc.)
III. Handling pupae
IV. Adult Emergence → Adult Handling (diapause, mating, feeding, oviposition)

Brazzell and Newsom 1959 discovered that boll weevils undergo diapause, and since that observation, several other authors (e.g., Gast, etc.) researched the factors that induce diapause as well as the existence of different diapause biology in different strains of insects. Walker 1967 showed that boll weevils can move from diapause to reproductive capability either gradually or all at once, where some females oviposit a single egg or a few eggs as they are moving physiologically into diapause. It was important in the mass-rearing program for boll weevils to have a thorough understanding of the nature of diapause as a prerequisite to maximization of boll weevil mass-rearing. Clearly, an uncontrolled and unexpected entry into diapause of a substantial portion of the colony could derail daily operations.

5.2.1.1 *Oviposition in Boll Weevil Rearing: Matching Engineering to Life History*

Davich explained the life history of the boll weevil in its association with cotton phenology as follows: "Eggs are oviposited inside the floral bud (square) and hatch in 3-5 days (Figure 5.4). The larval stage lasts 5-7 days. The square abscises from the plant and drops to the ground 4–8 days after the egg is laid and insect development is completed inside the fallen square. The pupal stage lasts 4-5 days followed by emergence of the adult from the shell of the fallen square. Four-five days elapse

Figure 5.4 Boll Weevil Development Linked to Cotton Phenology (BWDLCP). (Cotton figures kindly provided by Dr. Arun Babu from Professor Dominic Reisig's Laboratory); Boll weevil images redrawn from USDA: Farmers Bulletin No. 130: National Agricultural Library.

prior to the beginning of egg-laying by the adult. Thus, under favorable conditions, a generation may be produced in 2½-3 weeks." However, the life cycle may be more complex than what Davich explained, for example where the eggs are laid late in the square (floral bud) stage, and the neonate develops in an intact flower that has not abscised. Or a further complexity occurs when a gravid female oviposits into an immature boll, and the larva completes its entire development within the cotton boll. The presence of a developing weevil in the boll may or may not cause the boll to undergo abscission. Finally, as pointed out by several authors including Bazzell and Newsom 1959 and Lloyd et al. 1967, under certain conditions (such as low nightly temperatures, short—11 hours—photoperiod, and availability of only bolls as food), adult boll weevils may undergo diapause. During diapause, the reproductive system is atrophied, and the biomass inside the weevil becomes predominated by fat body for energy and materials storage

The key point of this discussion of life history and phenology of the boll weevil is that all of the stimuli and resources that the insect requires must be provided in the rearing system, and because of the complexities of the life cycle, the demand of the rearing system designers is to provide all the nuances of the boll weevil's ecological niche specifications. It is axiomatic that the closer that the rearing system's components fit the ecological needs (food, environment, social requirements, etc.), the more apt the insects are to display fitness or what we are calling quality. It is also important to recognize that as we place captive insects into the context that we provide in rearing—inherently a very artificial process compared to nature where the insect evolved over hundreds of millions of years, the more must expect that the effects of selection and domestication will have changed the insect from its natural biological character.

This figure (derived from Everett and Ray 1964) shows a hypothesized stimulus-response matrix where various "factors" (chemical or physical components) govern the behavior of female weevils that (1) search, (2) probe, (3) feed, (4) seek feeding hole, (5) insert ovipositor, and (6) lay egg. The researchers who succeeded in developing the complete rearing system at the Gast boll weevil rearing facility had to provide all the components depicted in this diagram to meet the oviposition needs (as well as all other needs) of the boll weevils that they were charged with mass-rearing. Importantly, the various components had to be made available to the weevils in a system that was capable of being assembled in an automated series of processes

The diagram in Figure 5.5 illustrates the complexities of boll weevil oviposition behavior which was based on previous research and the current (Everett and Ray 1964) study led to the hypothesis that five factors had to be present to stimulate the full process of oviposition ("Factors A–E"). The factors were thought to be chemical and/or physical/textural stimuli. It is noteworthy that the correct combination of all stimuli has to be present to induce oviposition. The scheme is a classical stimulus-response "releaser mechanism" of animal behavior, where each stimulus induces ("releases") the appropriate behavioral step leading to successful oviposition. The recognition that emanated from the various studies cited in Figure 5.5, along with other background papers, was a complex behavioral construct that had to be in place with the right stimuli and the right characteristics of those stimuli all falling within the boundaries genetically programmed into boll weevils. For example, something(s) in the water-soluble extract of squares helped stimulate both feeding and oviposition, and the shape of the oviposition substrate was further involved in successful induction of oviposition (Figure 5.5).

The large, powerful mandibles are at the forefront of the image (Figure 5.6), and the sensory structures (paired maxillary and labial palps) are evident in a position ventral (below) the mandibles. The numerous chemoreceptor sensilla on the maxillary and labial palps are evident in this photograph, and the function of these chemoreceptors is to discriminate chemical stimuli that are responsible for feeding and oviposition choices.

5.3 THE USDA, ARS R. T. GAST BCMRRU: BOLL WEEVIL AND LEPIDOPTERA MASS REARING

Mass-rearing of boll weevils and several species of Lepidoptera were among the most well-automated, well-developed, and well-documented rearing systems. The details of the Gast rearing systems have been documented in several publications culminating in the book, *Boll Weevil Mass Rearing Technology* by Sikorowski et al. 1984, with dozens of shorter publications (spanning the 1960s through the early 2000s, several of which are referenced here) that explain steps or subsets of the entire rearing process that was developed between the 1960s and the early 2000s when the boll weevil was fully eradicated from the US cotton belt. The story of boll weevil eradication began with a paper by Brazzel and Newsom 1959 and concludes with an explanation by Smith and Swink 2003.

Jones 2001 discussed the origins of boll weevils and congeneric species in Mexico where this genus (*Anthonomus*) originated in association of host plants from the plant genus *Hampea* (members of the mallow family, Malvaceae), but boll weevils have shown a predisposition to thrive in various members of the genus *Gossypium*. The relationships between *A. grandis* and

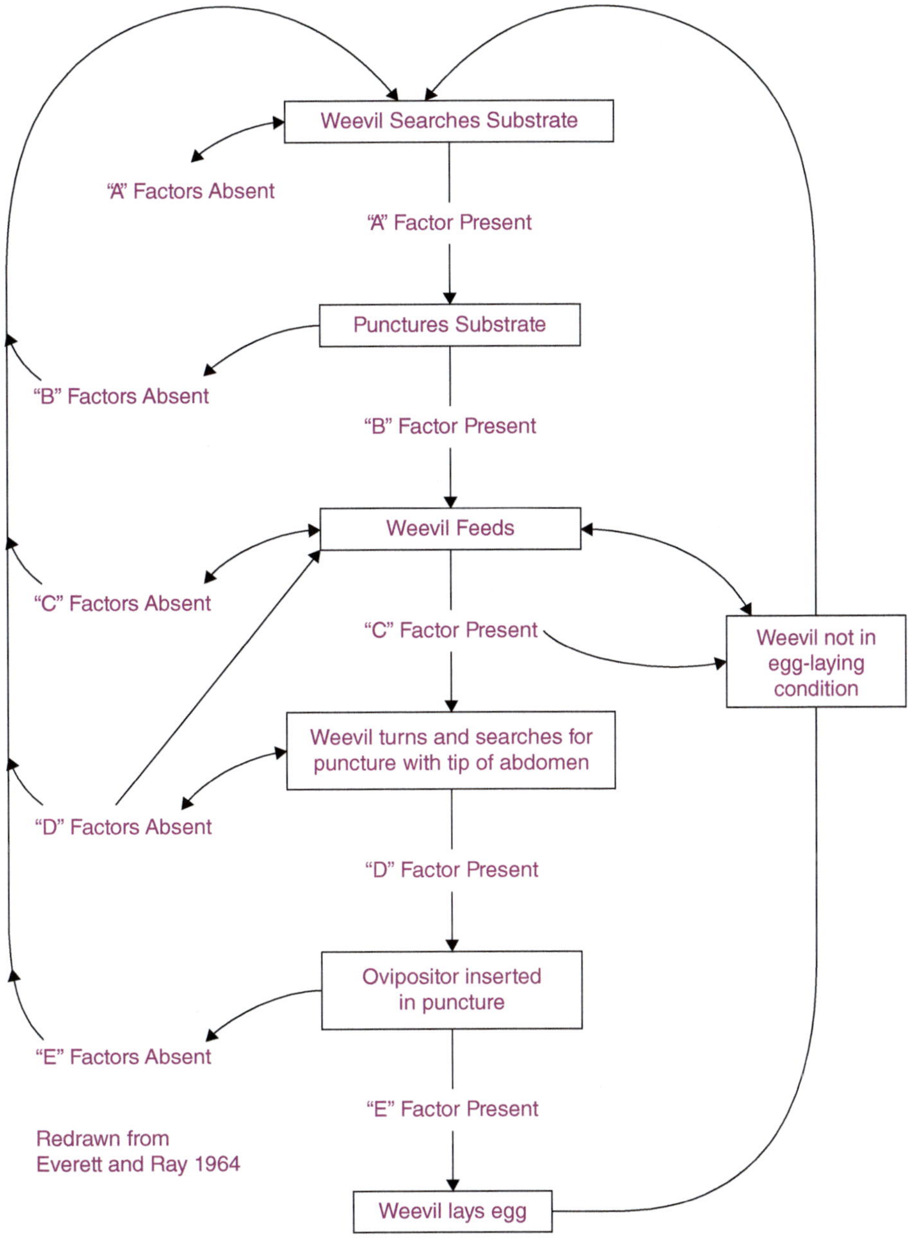

Figure 5.5 "Weevil Oviposition Cycle.".

various mallow plants are the basis for the ultimate insect/host interactions with cotton plants (*Gossypium*) and are the underpinnings of the successful spread of cotton boll weevils throughout much of Central and South America, as well as the invasion of US and Mexican cotton fields. It seems likely that the features (chemical and physical cues) of mallow family members that allow *A. grandis* to succeed in infestation of cotton plants are general enough to permit simulation of feeding and ovipositional stimuli to be achieved with artificial diets and wax coatings—all lacking in actual cotton squares or cotton bolls.

5.3.1 Cotton Squares and Boll Structures (in Ritchie et al. 2007)

Ritchie et al. 2007 explain that the sequence of reproductive structures proceeds from squares of various ages and sizes → blooms of various ages → bolls of various stages, the bolls being the fruiting structures of the cotton plant. They write, "Examples of square and boll shedding. Shedding can be related to several environmental factors such as population density, water, nitrogen, insect pressure, and disease. Newly formed bolls and young squares tend to be the most susceptible to shedding." Davich 1984

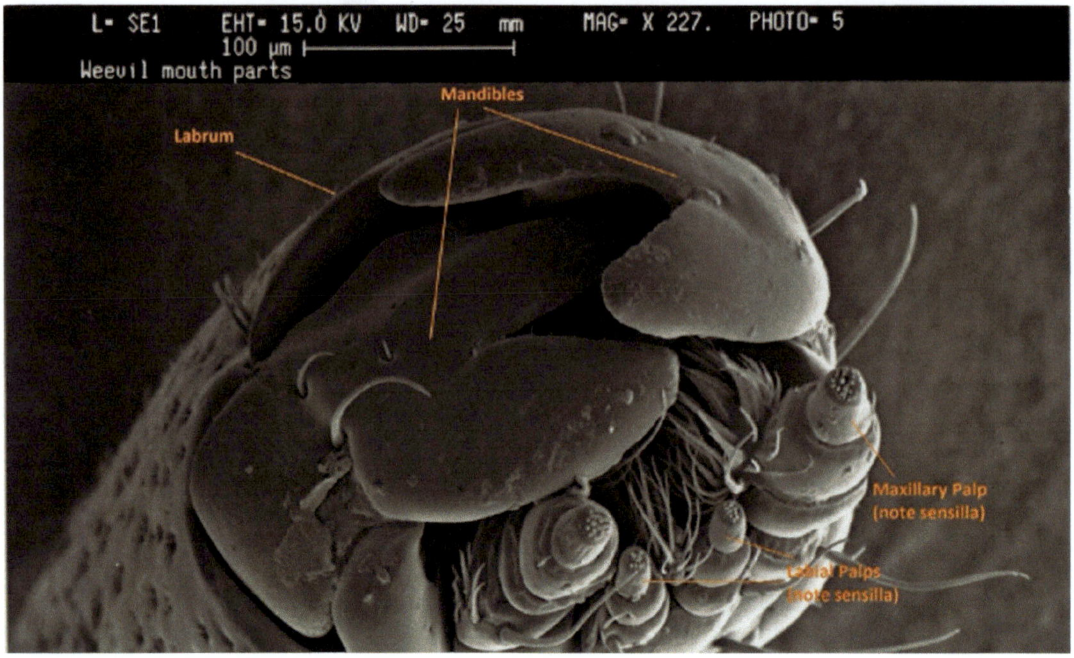

Figure 5.6 Scanning electron micrograph of boll weevil (*Anthonomus grandis*) mouthparts.

gives an extensive narrative of the boll weevil's life history in conjunction with the cotton plant's development. He points out that the major food and oviposition target of boll weevil adults is the square stage. Boll weevil-infested squares—ones that have been used as oviposition sites—may remain on the plant for a period of time that allows weevils to complete their larval development. Some squares drop, for a variety of causes described by Ritchie et al. 2007 such as density of plants, nutritional stress, water stress, and pressures from insect feeding. The stress imposed by the boll weevils was shown to be another factor inducing square drop.

5.4 FEATURES COMMON TO THE BOLL WEEVIL SYSTEM AND THE LEPIDOPTERA REARING SYSTEMS

Both systems employed **flash sterilizers** for diet-preparation and a **form-fill seal machine** for packaging the diet and introducing the insects in formed plastic rearing trays. Trays are covered with hot-glue-coated Tyvek lidding that has been perforated with small air exchange holes. The boll weevil system was dedicated only to production of *Anthonomus grandis* while the lepidopteran system was used for beet armyworms, *Spodoptera exigua*, corn earworms *Helicoverpa zea*, and tobacco budworms, *Heliothis virescens*. From time to time, attempts were made to rear other insects such as other species of Lepidoptera such as southwestern corn borers *Diatraea grandiosella* and even such diverse insects as coffee berry borer *Hypothenemus hampei*. Clean areas were maintained by following standards of clean rooms which included establishment of an environment that included automated processes to greatly reduce operator input. The automation reduced human error and introduction of contaminants by personnel. It is important to note that the design of the system allowed the diet to be produced in one room (a "dirty room") and to be flash-sterilized, then delivered through sterile tubing to the next room ("clean room") where the molten diet was delivered into trays, cooled, implanted with eggs via the egg-planting liquid, covered with sterile granular material (corn cob grits), covered with lidding, sealed into the 32-cell trays with a heat sealer, and then cut into free section—all with no handling by personnel. All of this is shown in Figures 5.7–5.13.

5.4.1 Boll Weevil Mass rearing

5.4.1.1 Example of a Highly Successful Rearing System: Boll Weevil Mass Rearing at the Robert T. Gast Facility (A Case Study)

This bulletin (Figure 5.7) is an excellent example of how SOPs can be presented with explicit and exhaustive details, and especially with rationale for many of the steps. It is my observation over 40+ years of rearing experience that insectary workers (as is the case with all laboratory technical personnel) will do a better job of conforming to the explicit requirements in SOPs if they understand a rational, scientific basis for the steps that they are asked to undertake.

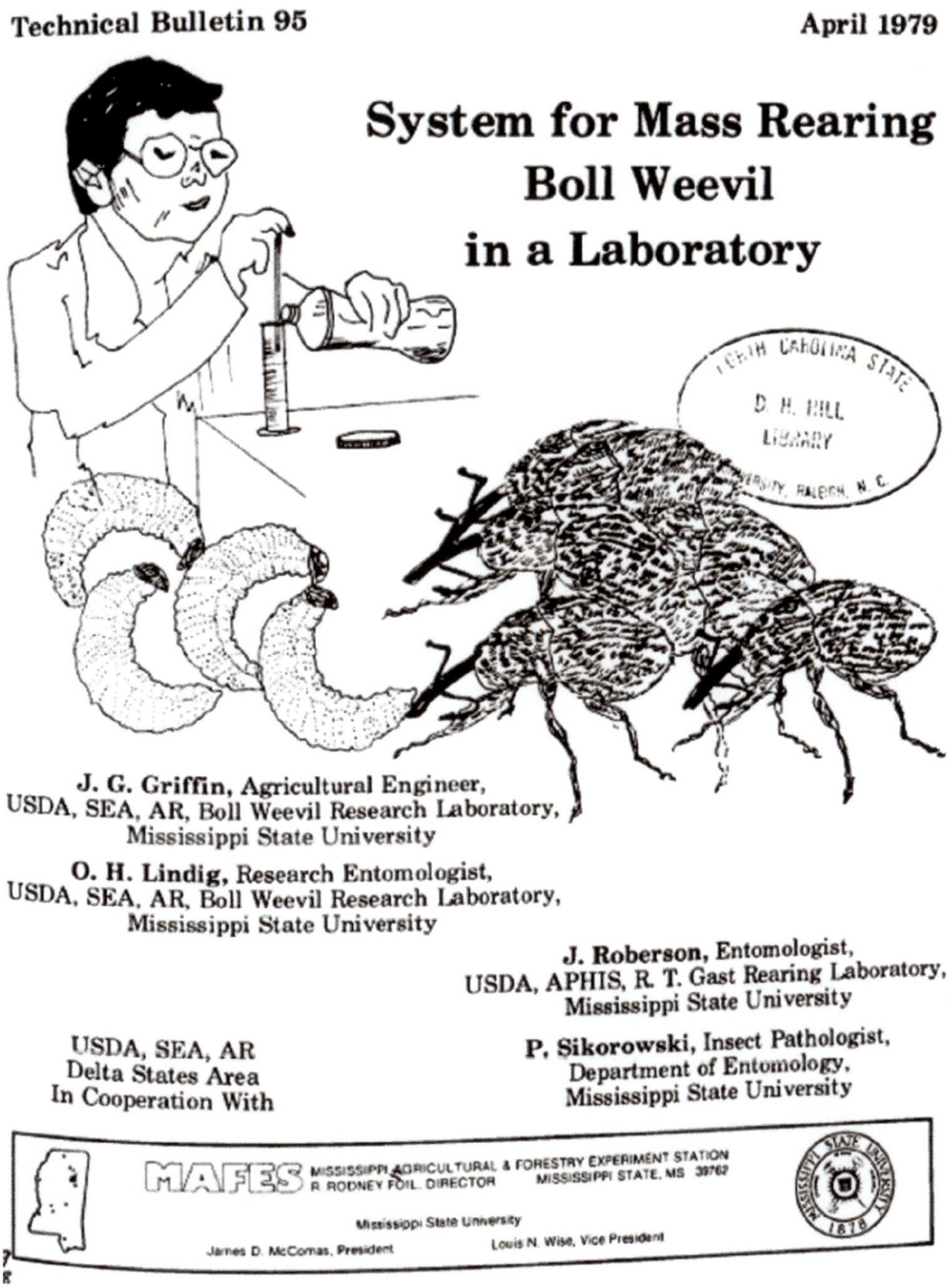

Technical Bulletin 95 April 1979

System for Mass Rearing Boll Weevil in a Laboratory

J. G. Griffin, Agricultural Engineer,
USDA, SEA, AR, Boll Weevil Research Laboratory,
Mississippi State University

O. H. Lindig, Research Entomologist,
USDA, SEA, AR, Boll Weevil Research Laboratory,
Mississippi State University

J. Roberson, Entomologist,
USDA, APHIS, R. T. Gast Rearing Laboratory,
Mississippi State University

USDA, SEA, AR
Delta States Area
In Cooperation With

P. Sikorowski, Insect Pathologist,
Department of Entomology,
Mississippi State University

MAFES MISSISSIPPI AGRICULTURAL & FORESTRY EXPERIMENT STATION
R. RODNEY FOIL, DIRECTOR MISSISSIPPI STATE, MS 39762

Mississippi State University
James D. McComas, President Louis N. Wise, Vice President

Figure 5.7 Technical Bulletin 95 describing the system for boll weevil mass-rearing (by Griffin et al. 1979).

5.4.1.2 The Boll Weevil Rearing System at the R. T. Gast Facility

One of the all-time most successful modern rearing systems was developed over a couple decades (1960s through 1980s) at the USDA, ARS Robert T. Gast Rearing Laboratory (GRL) at Mississippi State University. Although dozens of background papers led to this system's completion, the details of this system, in the form of SOPs, are presented in a single document written by Griffin et al. 1979 and whose cover is pictured in Figure 5.3. For the readers' convenience, I have provided this document in its entirety as Appendix "Gast Boll Weevil Mass-Rearing System." Throughout this section of the current chapter, I use notations "GFN" to indicate the portion of the Appendix to which I refer. I have also used shaded text to call readers' attention to key sections, but I suggest that for readers who wish to develop a rearing system and SOPs for that system, a detailed

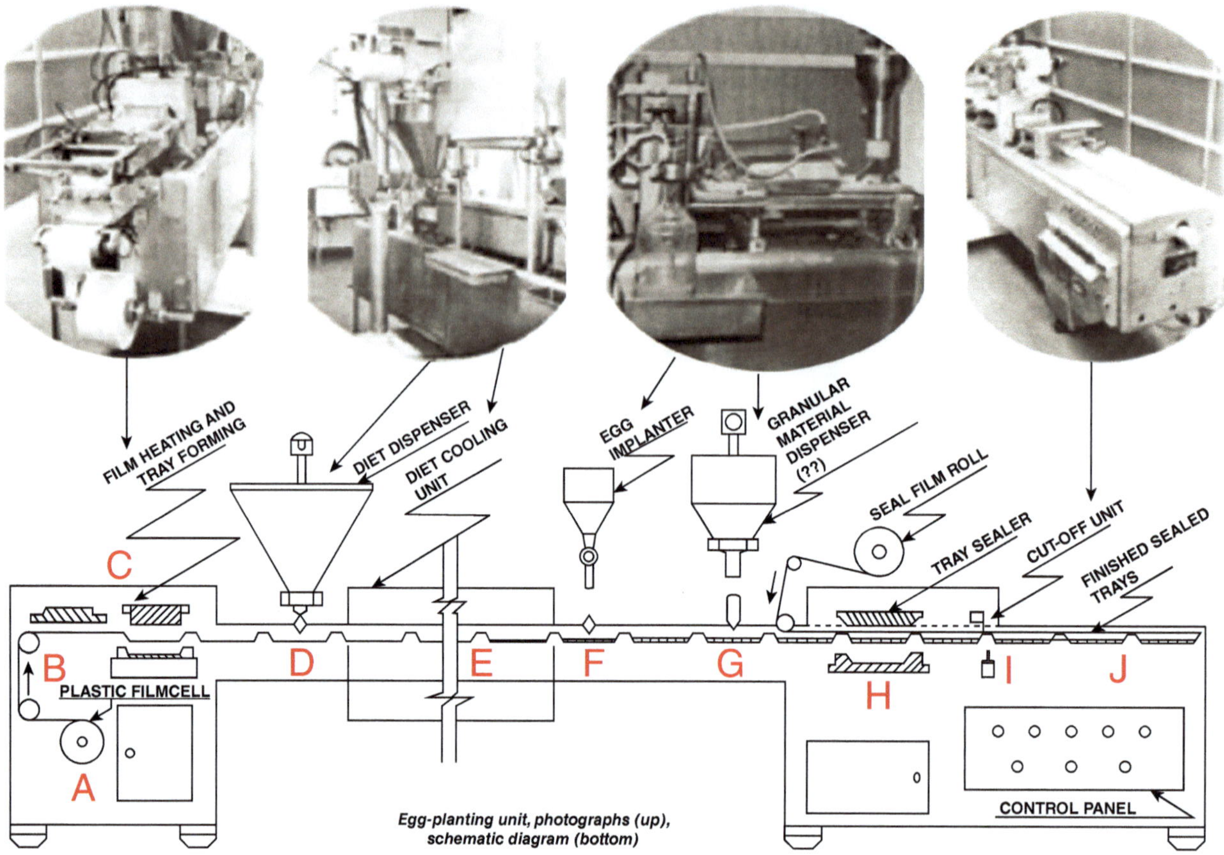

Figure 5.8 Egg planting unit for boll weevil egg → diet inoculation.

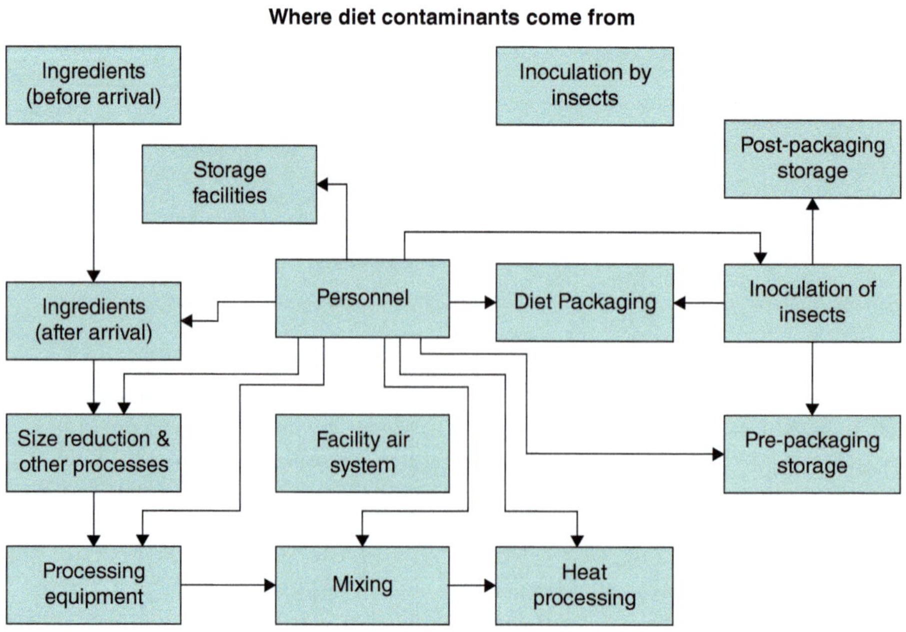

Figure 5.9 Sources of diet contaminants (from Cohen 2015).

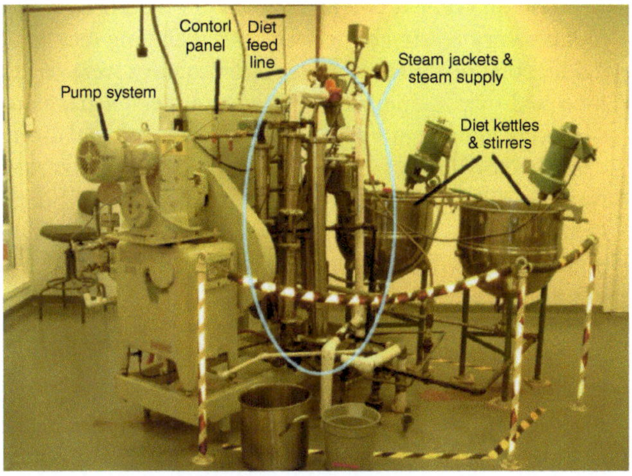

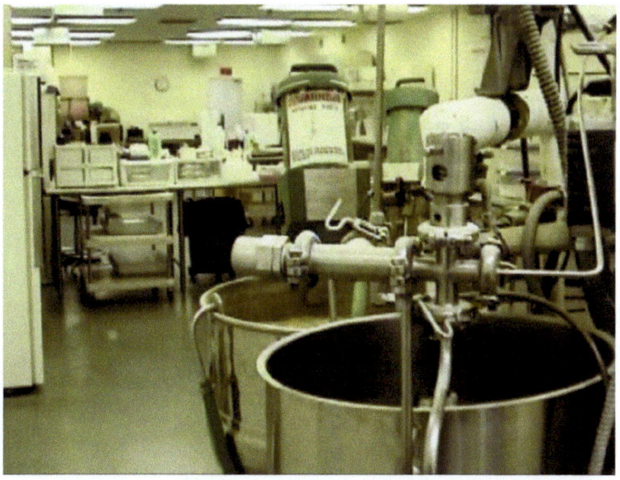

Figure 5.10 (a) and (b) Flash sterilizer. Figures include the diet-making apparatus used in the Gast Facility for making boll weevil diet and lepidopteran diets.

reading of the Griffin et al. paper would be rewarding and an excellent guide to developing and maintaining a well-run rearing system. I also suggest that another treatment of the Gast's boll weevil mass-rearing system was treated in more detail than the Griffin et al. monograph, a book by Sikorowski et al. 1984. In tandem, these works are some of the best-documented models of the rationale and scientific basis of very successful insect-rearing systems. Between the two documents Griffin et al. 1979 and Sikorowski et al. 1984, the reader has the most granular, detailed, and authoritative accounts of a mass-rearing system for a given insect (Figure 5.7).

The history of diet development for boll weevils is treated in useful detail by Lindig 1984. Lindig summarizes the pioneering works of Gast, Vanderzant, and several others who applied analytical techniques as well as trial and error (or heuristic methods) to find replacements for fresh cotton squares and bolls. Among the analytical techniques, boll weevil diet researchers used extracts of cotton squares and bolls as well as many defined and semi-defined ingredients. One tragic outcome of the intensive research to find cotton plants' extracts for artificial diets was the death of the dedicated researcher Robert T. Gast. Briefly, Gast was making extracts of squares using organic solvents, and early one Sunday evening (please see Figure Gast Obituary), Gast was working alone in the laboratory, and upon removal of autoclaved, solvent-extracted squares, the materials exploded and Gast was severely burned as he stood in front of the door of the autoclave.

In the Introduction **(GFN 1)**, Griffin et al. point out the purpose of the mass-rearing system—to provide millions of

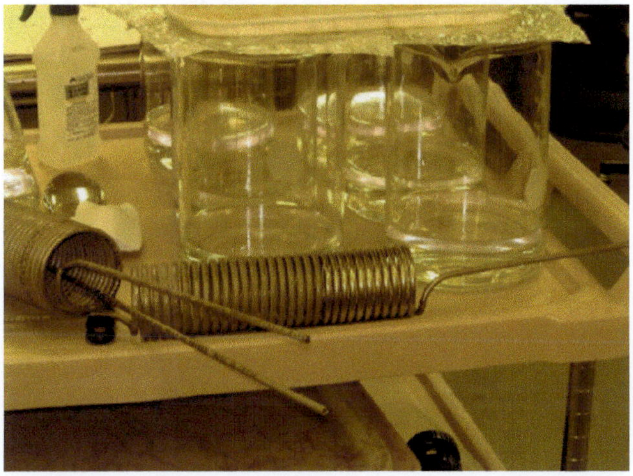

Figure 5.11 (a) Shows the stainless-steel coils (ca. 0.5 cm diameter openings) that carry diet through the steam jackets pictured in (b).

Figure 5.12 Diet feed line carrying molten diet from the flash sterilizer to an entry port in the form-fill seal machine.

insects per week to be used in a boll weevil eradication program. It should be noted that while sterile insect technique (SIT) was one of the primary targets for the Gast operations, the high quality and numerous weevils that were produced by this system were also useful for other programs such as biological control and research on insect biology and pest management. In **GFN 1**, the authors state a fundamental character of successful mass-rearing systems: "Producing millions of boll weevils for research and control programs means caring for successive generations throughout the life cycle, from egg to adult collection. The operation must be efficient if mass rearing … is to be economically feasible. ***The important factors are equipment to mechanize the various steps, environment and diet appropriate to growth and reproduction and sanitation to protect weevils from microbial contamination***." [Cohen's bold and italicized font]

Also, in **GFN 1**, Griffin et al. 1979 give a brief history of laboratory rearing of boll weevils, including mention of the fact that in earlier studies, cotton squares were used as food and oviposition sites, and eggs dissected from such sites were individually placed on incipient artificial diets as described by Vanderzant and Davich 1958 and Brazzel and Newsom 1959. Clearly, works such as those by Griffin and Lindig 1977 were essential components that allowed the high levels of efficiency needed to perform true mass-rearing of boll weevils. Interestingly, the journal where these two seminal papers were published, *The Transactions of the American Society of Agricultural Engineers* (now *The American Society of Agricultural and Biological Engineers*), is not an entomological journal, nor one that would be readily available to entomologists. In a January 5, 2020 search of Web of Science, I found that these papers were cited only two and six times, respectively, most being self-citations. The point I am making here and elsewhere in this book is that too much of insect-rearing science and technology is lost to the scientific community because of negligent or even hostile treatment by journal editors and reviewers of rearing accomplishments.

The development of boll weevil diets reflects an exceptional stepwise (incremental) series of discoveries. The section on diets (**GFN 2**) mentions that two diets are used in the Gast mass-rearing system, one for egg-hatching and larval growth, the other for adult feeding and oviposition. The formulations for the diets are presented in Appendix Gast Boll Weevil Mass-Rearing System in Tables 5.1 and 5.2. It should be noted that the diets in more recent iterations contained a commercial protein from cotton known as Pharmamedia® as described by Lindig 1979. The specifications for pre-mixing the diet and for each step in synthesizing the diet are explicitly stated in this section, including the specifications for the flash sterilizer. This includes the proper suspension of dry materials in water that is to be

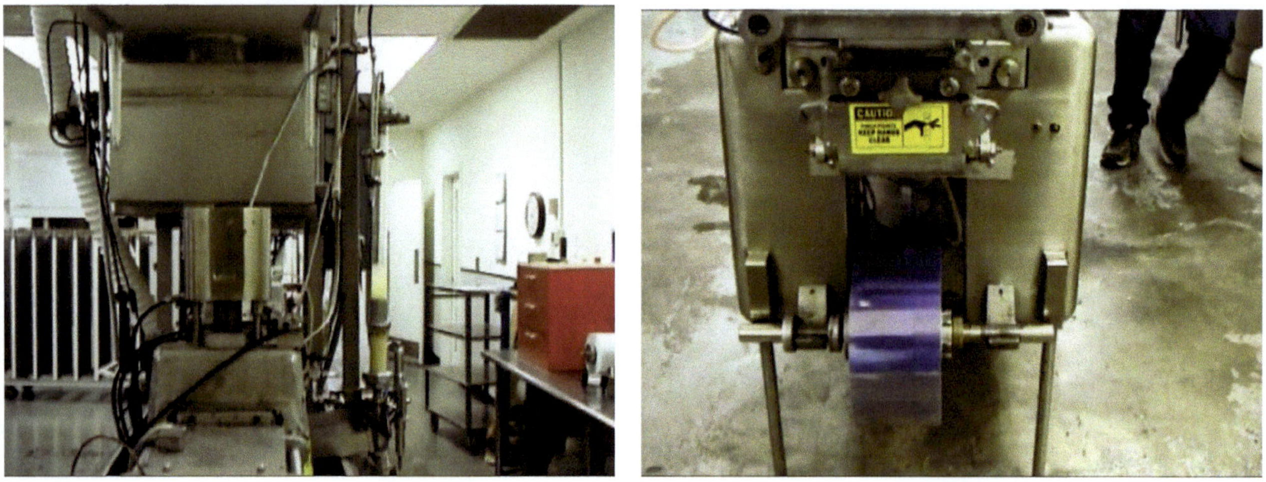

Figure 5.13 (a) and (b) Gast Lepidoptera diet/tray-forming setup. The tray-forming equipment is a modified form-fill seal machine.

Table 5.1 Background Information Characterizing the Components of Boll Weevil Oviposition (From Everett and Ray 1964)

Factor and Reaction Responsible for	Reference	Observation
A. Puncturing substrate	Keller et al. 1962	Water-soluble extract of cotton arrested searching activity
	Everett and Earle 1964	Height and shape of paraffin shell influenced number of punctures made in shell
B. Feeding	Vanderzant et al. 1959	Casein and wheat germ more attractive for feeding than protein hydrolysate
	Vanderzant and Davich 1961	Some sugars stimulated feeding more than others
	Keller et al. 1962	Water-soluble extract caused increase in feeding on treated materials
C. Turning and searching for punctures with tip of abdomen	Vanderzant et al. 1959	More eggs laid in diets containing squares than in those germinated cottonseed or mixed pollen
	Vanderzant and Davich 1961	More eggs laid in cylinders with hemispherical tops than in flat dishes
	Everett and Earl 1963	Water extract of squares stimulated oviposition
	Present Paper (Everett and Ray 1964)	Cotton squares, water extracts of squares, and saline solutions stimulated oviposition
D. Inserting ovipositor in puncture	Present Paper (Everett and Ray 1964)	Cotton squares and extracts of squares aided weevil in detecting puncture with ovipositor
E. Laying eggs	Present Paper (Everett and Ray 1964)	Probing with ovipositor indicated the weevil is receptive to stimulation

Table 5.2 Ingredients and Quantity of Each Used to Make a Batch of the Boll Weevil Diets (From Griffin et al. 1979)

Ingredient	Adult Diet (g)	Larval Diet (g)
Cottonseed meal (sifted)	28,500	31,800
Agar	8,400	8,280
Sugar	6,850	7,800
Promine D	10,800	12,280
Corn cob grits (60 mesh)	6,000	
Cholesterol	300	360
Wesson salt	1,125	1,260
Vitamin mix 1	1,170	1,260
K sorbate	585	630
Tegosept (methyl paraben)	375	400
Ascorbic acid	270	290
Methenamine	60	
Cottonseed meats	10,800	10,724

mixed to the point where no lumps are present, and then diet is passed through the heating coils, which are set at 141°C for sterilization before diet is pumped into the cooling coils at 49°C and 41°C, respectively for adult and larval diets. The ingredients listed in Tables 5.1 and 5.2 are generally conventional components used in several insect diets, except for Proamine D and methenamine (an antibiotic). After extensive searching, I could not determine the nature of Proamine D, though from its proportions in the diets and the amount used in the diets in Tables 5.1 and 5.2, I surmise that it was a nutritional supplement, possibly a protein or protein hydrolysate. The cottonseed "meats" are de-hulled cottonseed kernels, which can be further processed by flaking, expansion, and extractions (oil removal) to form cottonseed cake, meal, and pellets (Glasi and Drouillard 2002). It is noteworthy that the cottonseed meats include oil which contains triglycerides and sterols, which may be of great nutritional value to insects, which can tolerate the secondary compounds such as gossypol. The cottonseed meal, which is also used in boll weevil diets, contains a higher concentration of protein than does cottonseed meat.

5.4.1.2.1 Critique of Diet-Making Methods Section

Despite the very explicit instructions for parts of the diet-making section of this monograph, the authors fail to express that rates of flow of diets as they are passing through the sterilizing and cooling coils. The narrative by Roberson 1984 gives several of the missing details that would be necessary for insectary personnel to duplicate or repeat the SOPs provided in Griffin et al. 1979. For example, here is a paragraph from Roberson that conveys the essential details about heating time for the boll weevil diet in the flash sterilizer:

> The operator supervised equipment operation, and it was his responsibility to assure that adequate temperature and steam pressures were present for the diet sterilization. It took about one minute for the diet mixture to enter the pump and go through all processing lines, for sterilization.

The significance of the term "flash sterilizer" is apparent when we read that the total residence time for diet from the pump to the output of the cooling coils is about one minute! However, despite the seemingly detailed explanation provided by Roberson, it is still not clear what the steam pressures were. The equipment that constitutes the major portion of the custom-designed flash sterilizers in the Gast facility were the Cherry-Burrell No-Bae Uniterm IV® sterilizer unit and the Barber Coleman 520 series automatic, solid-state temperature control and auxiliary modulating valve for controlling heating and cooling temperatures. The other components for volume and

Table 5.3 Updated Version of Gast Diets for Adult and Larval Boll Weevils (From Lindig 1984)

Gast Adult and Larval Diet Components	Adult Diet	Larval Diet
Agar	80	78.6
Sugar	65	74
Wesson salts	10.5	12
Cholesterol	2.8	3.4
Vitamin mix	10.8	12
Ascorbic acid	2.5	2.8
Potassium sorbate	5.4	6
Methyl paraben	3.5	3.8
Cottonseed meal[a]	265	311
Cottonseed meats	100	102
Proamine D	100	162
Corn Cob Grits	56	
Methenamine	0.8	
Biotin solution	0.6 mL	0.6 mL
B12 solution	0.6 mL	0.6 mL
HCl (6 N)	4.0 mL	4.0 mL
Water	3,800	3,800

[a] Pharmamedia (Archer Daniels Midland).

weight measurements are also presented. The *Procedure* is fully described in a 9-step list of explicit directions (Tables 5.3 and 5.4).

5.4.1.2.2 Preparing Wax-Coated Adult Diet Pellets

The efficient handling of adults to garner optimum oviposition and egg recovery involves the preparation of wax-coated cylinders of diet. This is described in **GFN3** and **GFN4**. The wax mixture is a combination of beeswax and paraffin (40:60 by weight). The beeswax though expensive is important to give the wax coating a softer (less brittle) character and a lower melting point than the paraffin alone. The melting point of the prescribed mixture is about 60°C. The pellet-making procedure includes steriliation (for 10–15 minutes) of the equipment that is to handle the molten diet (Griffin et al. 1979, Figure 5.1). The diet is pumped into forming tubes, which are 5/16 of an inch in diameter (12.7/40.64 cm = 0.31 cm). The cylinder of diet that is forced through the tube is cooled (with 7.5°C water in a heat exchanger) to the set point of the agar (45–50°C) so that the diet cylinders can be cut to 1.27 cm in length, dropped into the molten (66°C) wax, and then removed from the wax bath with a screen upon which the pellets are allowed to cool to solidify the wax coating. The pellets can then be placed in the adult oviposition container where the adult weevils can utilize the pellets as food and as oviposition substrate.

Table 5.4 Updated Version of Gast Diets for Adult and Larval Boll Weevils (From Lindig 1984)

Vitamin Mixture[a]	
Niacinamide	20
Ca Pantothenate	20
Riboflavin	10
Thiamine HCl	5
Pyridoxine HCl	5
Folic acid	5
Inositol	40
Sucrose	2,800

[a] Pulverized/triturated in a ball mill for 1 hour.

5.5 HOW THE PELLET DESIGN MESHES WITH THE BOLL WEEVIL OVIPOSITION BIOLOGY

5.5.1 Oviposition

With boll weevils, as well as other insects, the biological and physical factors that invite successful oviposition must be understood and applied. With boll weevils in nature (or an agro-ecosystems), adult weevils, after mating, use cotton squares (flower buds that are between the blossom and fruit stage of development of cotton reproductive structures) to insert their fertilized eggs into the plant tissue. It is established that boll weevils show preference for squares, which induce insertion of the female weevil's ovipositor in strong preference over other plant parts, including maturing bolls. The basic biology of oviposition by boll weevils was explored by Vanderzant and Davich 1958 and 1961, as well as Moore 1967, and more recently by Esquivel 2007 (with several other investigations of boll weevil oviposition). Briefly, the female weevil, which is about to oviposit, feeds on cotton squares or bolls, mainly squares, then bites a hole into the square and changes her position to allow her to probe the opening with her abdominal tip terminating with her ovipositor. Upon sensing the proper sequence of stimuli (described in detail by Everett and Ray 1964 and Everett and Earle 1964), the female weevil ultimately lays an egg onto the diet and within the opening in the diet (both for natural plant tissue and artificial diet). Once the egg is laid, the female can resume the sequence to oviposit the next egg, and if she is not physiologically prepared to lay another egg, she can resume feeding, tunneling into the square, and resuming the sequence described in Figure "oviposition cycle."

In efforts to meet the need to match with artificial media and devices the biological factors that lead to successful oviposition, Gast 1965, Moore et al. 1967, and Ludvik 1972 developed forms of diet/oviposition substrate-presentation (Gast's dome-shaped pellets, Ludvik's

"Disposo-Trays"), which ultimately led to the pellets that Griffin and his colleagues developed at the Gast rearing facility. The final, and most successful form of oviposition medium, was the adult, gelled diet in the form of a cylinder coated with wax. The automation procedures for boll weevil mass-rearing, including the aspects of oviposition, were developed by several researchers whose reports on the development of the mass-rearing process were chronicled in excellent detail in tens of papers on this topic in various journals and USDA reports, as well as reports by state universities. For example, Griffin and Lindig 1975 wrote about the portion of the process where they tested and adopted use of granular materials as part of the rearing:

> Equipment and methods used to implant boll weevil, *Anthonomus grandis* Boheman, eggs in or on artificial diets for laboratory rearing of the weevil were described by Gast and Vardell (1963), Gast (1966), and Gast and Davich (1966). These equipment and methods, while adapted for laboratory or small-size rearing operations, require enormous amounts of time and labor, limit capacity, and increase the possibility for microbial contamination in the mass rearing of weevils. The research reported here was conducted at the Robert T. Gast Rearing Laboratory to determine the possible use of a granular material to cover the eggs and diet in the development dishes, which, if successful, would eliminate the jobs of diet surface scarifying and drying in the egg planting operation and make the operation more adaptable for mechanization.

The pellet configuration or design is of far-reaching importance because it meets the needs of the boll weevils to give continuity in their life history and also because the technology has potential for use with countless other insects that have similar modes of ovipostion—especially insertion of eggs into plant materials. The design of the wax pellet was a product of a sequence of incremental observations decribed in Figures 5.4 and 5.5 where the various steps of boll weevil behavior had to be satisfied with appropriate designs that met the stimulus-response "releaser" behavior. For example, crucial contributions to boll weevil mass rearing were the discoveries by Vanderzant et al. 1959 that boll weevils were stimulated to oviposit by "factors" from wheat germ and the further discovery that presentation of curved, rather than flat, shapes were stimulating to oviposition. The pellet design included wax coating, which stimulated feeding "interest" by the adult weevils as well as protecting the pellets' diet from desiccating and further protected the diet from oxidative and microbial degradation. Furthermore, the pellets could be easily harvested and processed for automated recovery of the weevil eggs without recourse to labor-intensive handling described by Vanderzant and Davich 1958.

Continuing to **GFN5**, once the pellets are made, the diet gelled, and the wax solidified, the pellets are presented to adults in an oviposition cage where thousands of adult weevils are allowed to feed, mate, and oviposit. A feature of the weevils' natural biology is that they can tolerate communal, crowded conditions. This character may have been enhanced by domestication of the weevils whose populations are far more dense than they would have been in the field. When I looked into adult weevil cages, I was always amazed that the normal mating sequences were apparently not disrupted by the teeming masses of about 9,000 weevils, many of which were paired in copulo. The directions in **GFN5** are to place 2,250 pellets measured by volume of about 3.2 L into adult cages with 108 g of weevils. The environmental conditions for the adult oviposition room are 30 +/– 1°C, 50 +/– 5% RH, and 24 hours of total darkness. The pellets are removed and replaced every 24 hours, and egg infested pellets are processed for extraction as described in **GFN6a/GNF6b**.

The egg extraction process **GFN6a** is nothing less than an engineering masterpiece. The pellets are placed in a hopper where they are sprayed with cool water to harden the wax (making it brittle), and the spray also begins the wax-cracking process. The partially cracked, wet pellets are flushed through a wax cracker which further dislodges the wax from the diet (the eggs being held by the diet gel); and this mixture is agitated with alternate skimming of the wax, which because of its lesser density than diet and water floats to the top where it is skimmed in repeated steps of agitation, rinsing, and skimming. Once all the wax is removed, the mixer turns the mixture into a thick slurry of eggs, diet, and rinse water, which then goes into a disintegrator which chops the pellets into smaller size, thus releasing the eggs. The eggs are denser than the diet and wash water, and they sink to the bottom of the disintegrator where they are captured after having been separated from the coarser (non-egg) particles, which is discarded. The bottom mixture of eggs and fine particles from the diet are then subjected to a saturated sodium chloride (brine), which, after mixing, allows the eggs to float to the surface (being less dense than saturated brine), and they can then be skimmed off the top of the brine, and washed. The next wash of the eggs is with fresh water, which causes the eggs to settle to the bottom (they are more dense than fresh water), and the remaining particles of wax and diet debris float to the surface and can be skimmed and discarded. Eggs are further separated from the surrounding medium with separatory funnels and further use of brine. Once all the cleaning is done, the drained but moist eggs can be placed into graduated cylinders where they are estimated by volume where 1 mL = about 4,100 eggs.

In **GFN6b**, the egg surface sterilization process is explained, where cleaned eggs are subjected to agitation in an approximately 7.4% formaldehyde solution (500 mL of 37% formaldehyde plus 2,000 mL fresh water). Because of the toxic properties of formaldehyde, the work is carried out under a fume hood with the sash as low as possible. The liquid is decanted off the eggs, and the rinsing is repeated two more times; then eggs are stored in graduated cylinders (4,100 eggs/mL) awaiting suspension in planting solution.

The sections **GFN7a**, **GFN7b**, and **GFN7c**, provide description of the post-sterilization/planting solution treatment of the eggs. The planting solution consists of a 2.2% corn starch solution/supension and a 22.2% solution of sucrose, a very viscous carrier for the eggs, whose density is so close to that of the planting solution, that the eggs remain suspended. The planting solution is heat treated both by boiling and by autoclaving, both treatments accomplishing both the dissolution of the solutes (starch and sugar), the activation of the hydrocolloid properties of the starch (Cohen 2015), and the sterilization of the egg planting liquid. In **GFN7b**, the process of suspending eggs in planting solution is described. The eggs, 400 mL are placed in a sterile jar where 2,200 mL of planting solution is added, after which 8.67 mL of 37% formaldehyde solution is added. The suspension/mixture is shaken on a laboratory shaker until the eggs are homogeneously suspended. It is to be assumed that the osmotic potential of the sugar/starch solution is close to being at equilibrium with the osmotic potential of the eggs. Otherwise, eggs would either swell and possibly burst if the solution were at a higher water activity than the eggs, or the eggs would shrink and shrivel if the solution were at substantially lower water activity than the eggs. The balance in terms of osmotic potential and density is evidently a near-perfect match.

Because of the large amount of time and labor involved in scarifying (making groves or incisions) in the diet to prepare it for neonate feeding, Griffin and Lindig 1975 tested a variety of materials (vermiculite, corn cob grits, and sand) to determine if any granular material placed over the egg-infested diet would preclude the need for scarification. They found that several of these materials improved the rate of neonate acceptance of the diet, and they determined that sand was most economical and of completely suitable character. Therefore, the Boll Weevil Research Laboratory staff adopted the use of what ultimately became a sterilized and medicated sand covering for the eggs after they were sprayed onto the diet in the planting medium. In **GFN7c**, the treatment of sand is provided and includes drying the sand in the sun or in a drying oven, screening the sand to assure appropriate particle size, autoclaving it (for 6 hours!), and mixing it with antimicrobial agents, including streptomycin sulfate, neomycin sulfate, and an antifungal agent called Folpet from Ortho, Chevron Chemical Co.

All the procedures described in **GFN5-GFN7c** staged the preparations for planting eggs onto the diet. The many precautions that were developed over the two decades of research were designed, incrementally, to maximize efficiency and egg/neonate vitality in light of potential problems with rate of hatch, acceptance of diet by neonates, prevention of microbial contamination, and other biological and practical issues that met economic demands. The next phase, planting eggs, is described in **GFN8**. The planting phase is a highly automated process that lends itself to adoption by programs that are also aimed at high efficiency, biological accommodation, and high throughput. The centerpiece of the implanting system was the Formseal® in-line form-fill-seal machine that forms trays from Tyvek® plastic. The tray-former is coupled with the flash sterilizer that allows diet made in a separate room to be fed into the trays, after which the molten diet is cooled to below the set point of the agar in the diet. The egg suspension and the sand are staged in separate jars in the Filmatic Model FKS filler unit (described in **GFN8**), and the egg suspension is sprayed onto the diet surface then covered with the sterile, medicated sand. Each tray (11.4 × 25.9 × 1.9 cm) is filled with 185 mL of molten diet, which is cooled to set the gel, and then has 4 mL of egg planting fluid (= 2,550 eggs) evenly spread over the surface of the diet; then 48 cc of sand is placed on the egg/diet surface, resulting in a 1–2-mm layer of sand over the eggs. The trays are automatically fed to the next station where the Tyvek lidding coated with heat activated glue, and tray lid is heat-sealed to safely contain developing larvae. The equipment for all of these processes are illustrated in Figure 5.8.

The egg planting (Figure 5.8) unit contains the core equipment where many of the most important processes take place. Starting at the lower left part of the diagram is the plastic (Tyvek) roll (A) that serves as the tray-forming material. The lead edge of the Tyvek is fed up (arrow, B) and then to the right (C) where it is heated to soften it, then is pressed with the tray-forming die (also C), making an indentation of 11.4 × 25.9 × 1.9 cm, which is then fed to the next position (D) where it is filled with molten diet from the diet dispenser (above D) and then fed to the diet cooling unit (E) where the diet is chilled to under 45–50°C. The newly formed tray containing cooled diet is fed to the next position (F) under the egg planter, which delivers the planting fluid with eggs suspended. The tray, with eggs implanted, moves to the next position (G) where it is layered with granular material (sand or sand plus corn cob grits). The tray is then conveyed to the tray sealer area (H) where a strip of Tyvek, coated with a heat-activated glue, is fed over the diet tray and sealed with heat (all in area H). Finally, the tray is cut from the Tyvek roll (I) and sent to the end of the conveyor where it can be removed and stored in a rearing tray holding cart where the eggs and developing larvae are incubated for the remainder of the pre-adult developmental process.

Once the trays are completed with finished trays sealed and cut into pairs of trays (containing about 5,000 eggs per pair of trays), the trays are placed in the rearing tray holding cart (Appendix SMRBW, Figure 5.9). At this point, the main rearing activity is begun with **GFN9** "Hatching Eggs and Developing Weevils." In this part of the process, the planted trays are held for 13 days at 30 ± 1°C at 45–55% relative humidity and in continuous light. Evidently the development in continuous light helps assure that the boll weevils will not be triggered to enter diapause (Lloyd et al. 1967; Walker 1967). Under the environmental conditions specified here, about 20% of the boll weevils will have started eclosion (emergence) of adults after 13 days. This was determined to be the optimal conditions for beginning adult collection (**GFN10**). The Tyvek covers of the trays are removed, and the trays are placed into emergence boxes (Appendix SMRBW, Figure 5.10). The details of the emergence boxes are described in **GFN10** and the rationale for these specific procedures is explained by Griffin 1982a, 1982b and 1984. In Appendix SMRBW, Figure 5.10, the emergence box is shown to be partially opened to allow visualization of the shelves of half the unit. There were problems with escapes from the emergence box (described by Roberson 1984). Roberson 1984 reported that between 5,000 and 15,000 adults were escaping per day, and the measure of adding weather stripping to seal the box improved the containment, but did not fully solve the problem. In years after the Roberson chapter was published, Gast personnel changed to an emergence box that was similar to a chest freezer, and this better-sealed container greatly reduced escapes. Unfortunately, this modification was not documented in a publication and my report is anecdotal, stemming from my observations. Roberson's very thoughtful and useful summary of the updated rearing system is worth thorough reading, but for this discussion, I present a few of Roberson's comments and critique.

The evaluation and critique of the rearing efforts between 1976 and 1979 are provided by Roberson 1984, based on Table 5.5.

Table 5.5 Mean Production of Boll Weevils 1976–1979 (From Roberson 1984)

	1976	1977	1979
Trays planted per week	13,749	17,880	7,027
Eggs/tray	2,595	2,415	2,100
Adults/tray	346	349	643
Eggs/female/day	3.6	3.4	5.6
Percent adult/egg production unit	13.3	14.5	30.6
Percent production for maintenance	32	25	14
Percent microbial tray contamination	14	0.9	5.8

Roberson 1984 described the 1978 period as a "relaxed production period" to allow adjustments and corrections of the problems discovered in the two previous years. The reduction in total trays implanted (from nearly 18,000 to about 7,000) along with several other changes helped increase the efficiency and production values seen in 1979. For example, the number of adults per tray and the number of eggs/female nearly doubled from 1976/1977 to 1979. The rearing personnel were also able to greatly reduce the number of contaminated trays and the percent of total trays needed for maintenance of production. Most notably, the percent of adult yield per production unit was more than doubled by the improvements made between 1976 and 1979. Interestingly, Roberson noted that the incorporation of corn cob grits with the sand improved the uniformity of development of neonates and early instar larvae. He also noted that treatment of racks that increased air flow around trays, including better spacing of trays, improved the uniformity of development. This suggests that the larvae were not receiving adequate gas exchange (please see Chapter 10 ("Relationships between Fitness, Quality Control, Health, Homeostasis, and Stress").

Roberson 1984 wrote:

> "Too often the technology employed to implement an action program such as the Boll Weevil Eradication Trial (BWET) is accepted matter-of-factly, with little consideration of the time and money investments involved. Similarly, the organizational role of management in relation to personnel, materials, and deadlines is vital to program objectives. By documenting the data and experience gathered while working with this program, the rearing process can be evaluated and its positive aspects appropriately applied to rearing programs." This statement suggests the importance of data-driven analysis and retrospective in a process of what TQM leaders call "continuous improvement."
>
> The first major boll weevil rearing program in which scientists attempted scheduled, massive field releases was the pilot Boll Weevil Eradication Experiment in 1971 to 1972. At its end, a critique of the laboratory rearing technology identify technical deficiencies in many areas. Most serious were difficulties with new equipment, many unskilled workers assigned to critical tasks, and difficulties in maintaining effective sanitation. After completion of the experiment, scientists sought improvement of boll Weevil laboratory rearing procedures. They studied automation to reduce the need for human labor in areas critically susceptible to microbial-contamination, and worked toward higher quality of laboratory reared insects through sanitation techniques and use of antibiotics.

The Gast system is a model of continuous improvement practices applied to rearing systems. Among the dozens of changes made in response to the problems that were detected, Roberson reported the following major reorganizational steps.

5.5.2 Corrective Actions of Rearing Procedures 1977

1. Installation of backup heating and chilling units
2. Reorganization of workrooms for greater separation of clean rooms from contaminated areas
3. Installation of more cooling in tray forming area; installation of positive pressure and absolute filters in exhaust duct of larval holding areas. Improvement of air circulation in larval holding areas
4. Transfer or enclosing milling area for bulk diet mixing. If contained in the same building, a dust collection system is needed
5. Stocking of essential equipment for standby purposes
6. Installation of direct communications systems from diet preparation to tray-forming areas. Provide signal system for direct contact to supervisor
7. Designing a production quality control system that will enable supervisory evaluation at all times. Records must interrelate with all work areas for rapid assessment and initiation of immediate or long range modifications or adjustments of processes that are out of control

An important part of the Griffin et al. 1979 document (Appendix SMRBW) is the detailed SOPs is for "Sanitation." (**GFN11**). This section advises procedures for: (1) personnel hygiene, (2) equipment, and (3) facilities. The personnel hygiene includes standard practices (covered by Good Laboratory Practices or GLP), including proper hand cleaning, use of laboratory clothes, including shoes that are used only in the clean lab spaces, freshly laundered lab coats, proper face masks, and protective eye wear. In general, the sanitation procedures, especially those pertaining to personnel hygiene, are mutually protective to the insects and to the personnel: it's a two-way phenomenon. Not specified in the Griffin et al. 1979 document are prohibitions about smoking, eating, applying makeup, or other personal actions that might contaminate the insects and the personnel. It is interesting and informative to consider that in the design of the Boll Weevil Research Laboratory (later renamed the Robert T. Gast Boll Weevil Research Laboratory, or the Gast Facility) there were showers and dressing rooms in several portions of the building. When the Gast building was completed in 1972 (about five years after the death of R. T. Gast), there were four dressing rooms equipped with showers, and several wash areas for personnel hygiene.

J. G. Griffin 1984 wrote:

> Of the many problems encountered, probably the most important ones were diet contamination; inexperience, quality and efficiency of Labor; and poor weevil and equipment performance. The greatest visible cause contributing to the poor results was the microbial contamination of diets and its detrimental effects on the production and performance of the weevils. It was difficult at the time to determine any main source of contamination, since many possible contributing sources were present, but some sources were obviously e.g., contaminants reaching the critical clean production areas from the dry diet ingredient storage and processing areas, equipment used for handling the sterile diet was manually handled and was not easily cleaned or sterilized, workers in close contact with the sterile diets, and a high concentration of workers in the rooms where the sterile diet was handled.

For decades, Gast Boll Weevil Laboratory, typical of all rearing systems, was plagued with microbial contamination problems. The most significant source of the incidents of contamination and the ensuing loses of quality and productivity were attributed to personnel (Griffin 1984; Roberson 1984; Cohen 2015). As depicted in Figure 5.9, personnel contact almost every step of the rearing process. It is especially important that the boll weevil rearing system at the Gast Laboratory was one of the most fully automated rearing systems in existence. The researchers at the Gast facility had worked assiduously for more than two decades to reduce the potential human errors and other contributions to lowered rearing quality by automating every possible process as this chapter depicts. One of the most significant advancements to mass-rearing technology over the past 50 years, was the adoption of the flash sterilizer (Figure 5.10 (a) and (b) to making insect diet. This apparatus helped automate diet-making to reduce human error that is inherent in diet-making efforts. However, there remained enough human contact with components of the boll weevil rearing system, that the primary cause of failures or degradation of quality was caused by personnel who failed to undertake all sanitation measures (hand washing, use of face masks, proper cleaning of machinery, etc.).

The guiding authority for the Gast facility and the community of researchers in insect-rearing science and technology was late Professor Peter Sikorowski. He dedicated his life and tremendous efforts to communicating the importance of proper understanding of sanitation and basics of microbiology to rearing personnel. He wrote:

> Exclusion of pathogens and opportunistic contaminants that may produce disease in weevil's already stressed by mass rearing conditions is a major consideration in mass production of the insect. The microbial contaminants may cause spoilage of the diet or cause insect mortality directly or indirectly. The boll weevil itself is a complex medium and is subject to spoilage by many species of bacteria and fungi that flourish on it. Spoilage of the diet is caused by the growth of growth in the metabolism of large numbers of microorganisms on the surface or the interior the effect of microbial contaminants on insects depends on the microorganisms involved and, indirectly, on the measures used to control or eliminate them. Obviously, if the losses due to the microbial contamination could be reduced or prevented, the number and vigor of weevils produced per unit of diet would increase.

I was fortunate to have an acquaintanceship with Professor Sikorowski and frequently heard him lament the fact that too often personnel in rearing laboratories violated standards that he tried to teach—fundamentals of basic sanitation. One of the most telling facts of Professor Sikorowski's contention was the fact that most of the contaminants found in weevil cultures, diets, frass, and on various surfaces were microbes associated with humans (Sikorowski et al. 1984; Sikorowski 1984; Sikorowski and Lawrence 1994). Sikorowski demonstrated time after time, that educating insectary workers in fundamentals of sanitation had profound influence on reducing problems with contamination and could even be extended into reduction of problems with pathogens in insect rearing systems (Inglis and Sikorowski 2009a and b).

This diagram (Figure 5.9) places emphasis on the potential impact of personnel (in red rectangle) on most components of diet-making. The diagram could be further elaborated to include almost every other component of the rearing system. I have witnessed actions by personnel that were sometimes explicable and well intentioned toward diligent efforts to improve rearing. However, no matter how well-intentioned actions may be, if they depart from tested, rationale-based SOPs, they can cause profoundly harmful effects on the rearing process. This fact leads to the conclusions that various leaders in rearing system management have reached: education of rearing personnel is of utmost importance. This point is treated in depth in Chapter 12 ("Rearing Education: The Pathway to Improving Rearing Science and Technology") and in several other places in this book. It had been a long-standing commitment of Dr. Sikorowski to use had extensive knowledge of insect/microbial relationships and to try to convey to all the personnel at the Gast facility (as well as to his many students) the importance of sanitation in rearing healthy, reliable, and economically feasible insects. To this end, Sikorowski taught Gast personnel to use Petri plates whose sterile media was exposed to air for given periods of time to help monitor the cleanliness of the various rooms where rearing operations took place (diet-making rooms, rearing rooms, storage rooms, and even rest rooms). Another tool that Sikorowski taught and strongly advocated was the use of RODAC (replicate organism detection and counting) plates. The plates could be set up with various media, depending on the need to count specific types of microbes or general microbial counts (various bacteria and fungi that are non-fastidious and can be cultured).

In the "Conclusion" (GFN11), Griffin et al. 1979 mention the pilot study conducted by the R. T. Gast Boll Weevil Rearing Laboratory (USDA, APHIS) and the Boll Weevil Research Laboratory (USDA, ARS) for the years 1976 and 1977. The target of rearing 5,000,000 weevils per week was met and exceeded in both years, demonstrating the proof of principle that a true mass-rearing system could conclusively be achieved and sustained. In the retrospective provided by Griffin 1984 and Roberson 1984, the problems encountered and the solutions accomplished were discussed with valuable details, especially regarding logistic organization that would enhance the productivity, economics, and sustainability of the program. I feel that the highest value of the publications by Sikorowski et al. 1984 and Griffin et al. 1979 is that the efficacy of developing rearing technology based on scientific principles and practices is unquestionably demonstrated.

5.5.3 Science behind the Boll Weevil-Rearing Technology

The underlying message of these accounts is that the step-by-step technology prescribed in the SOPs was derived from science; and the Gast boll weevil rearing system science principles were well documented in the literature (McIndoo 1926; Vanderzant and Davich 1958; Vanderzant et al. 1959; Vanderzant and Davich 1961; Keller et al. 1962, Gast and Vardell 1963; Everett and Earle 1964; Everett and Ray 1964; Mitlin et al. 1964; Gast 1965, 1966; Gast and Davich 1966; Lloyd et al. 1967; Moore et al. 1967; Walker 1967; Ludvik 1972; Griffin et al. 1974; Griffin and Lindig 1975, 1977; Griffin et al. 1974 Lindig 1979, 1984; Harrell and Griffin 1981; Griffin 1982a, 1982b, 1983, 1984). This long list is not exhaustive as there are dozens of other publications that document in detail the rationale and hypotheses in the works that served as background for virtually every step of the SOPs used in the boll weevil mass-rearing system.

In several publications (Cohen 2001, 2015, 2019) and in my classes, workshops, and symposia presentations, I have emphasized the failing of entomological journals to provide a reliable outlet for rearing research. However, my careful studies of boll weevil rearing to prepare me for writing this chapter have informed me that boll weevil rearing research is a model of how rearing research should be: (1) conducted and (2) treated by authors, editors, and reviewers. I have provided the complete publication by Griffin et al. 1979, which contains the highly detailed SOPs, which can serve as a model of how SOPs should be written. I have found no comparable detailed set of SOPs for insect rearing systems, and I have argued that there is a considerable value in having an outlet for such standards in rearing systems. The outlet for the boll weevil rearing publication was Mississippi Agricultural and Experimental Station. Technical Bulletin, which is a vehicle for publishing work done in the State of Mississippi. The Experimental Station bulletin is not as readily available as would be publications in conventional entomological journals. I have discussed in greater detail suggestions for publishing technology-type rearing papers, which would include SOPs and other technological accomplishment. In Cohen 2018b, I suggested the importance of direct study of rearing system components, including interactions between components. In the model derived from the food science and technology community, human nutrition is part of the focus, but other features of food science such as processing techniques and equipment are equally welcomed in the literature of this discipline. In literature on insect rearing systems, the sole attention is paid to one or more species of insects, but processing science and technology for diets and other rearing system components is largely excluded from technical publications. Only rarely does the entomology community include publications that directly treat inquiry into the nature of the processes and equipment that make many aspects of insect rearing possible. Much of the literature on boll weevil mass-rearing are examples of the noteworthy exception that I hold up as a model here.

5.5.4 Lepidoptera Rearing

Weevils and several species of Lepidoptera were among the most highly automated, highly developed, and well-documented rearing systems. The details of the Gast rearing systems have been documented in several publications culminating in the book, Boll Weevil Mass Rearing Technology by Sikorowski et al. 1984. Supporting this body of information are dozens of shorter publications (spanning the 1960s through the early 2000s) that explain steps or sub-units of the entire rearing process. The Gast Laboratory personnel had the vision to project the concept that the mechanization, automation, and related innovations were applicable across multiple species of insects. In keeping with this vision, Gast scientists and technicians applied the breakthroughs in boll weevil rearing to mass-rearing of several species of Lepidoptera, and conversely, they applied advances in Lepidoptera rearing to boll weevils and later to other insects such as Lygus bugs, lacewings, and other insect species. For example, once the concept of a flash sterilizer, form-fill sealing and automated infestation are developed for one species, it becomes a natural transition to other species.

Two diet kettles (b) were used for mixing diet components, which were then fed as a slurry to the coils in the steam kettle (Figures 5.10(a) and (b) and 5.11(a) and (b)) where diet was heated to 150°C. Hot molten diet was then pumped through the diet feed line to the ceiling line (Figure 5.12), which led into the next room (designated as the "clean room") where the diet was injected into the form-fill apparatus. The form-fill seal apparatus was identical to the one used for planting boll weevil eggs and which is shown in a detailed diagram in Figure 5.8. The R. T. Gast facility was equipped to produce boll weevils and multiple species of Lepidoptera in the millions per week (at least 5,000,000 per week of both weevils and moths), and both the weevil system and the Lepidoptera system were centered in the specially designed flash-sterilizer coupled with a form-fill seal machine.

5.5.5 Equipment Description

Flash sterilizer: Cherry-Burrell No-Bae Uniterm IV sterilizer unit (described by Griffin et al. 1974) with one 30-gallon wash and sanitizing tank and two 30-gallon diet mixing tanks. Barber Coleman 530 series automatic, solid-state temperature control and auxiliary modulating valve for controlling heating and cooling temperatures. Two 1/3 HP, variable speed mixers, 3/8-inch outside diameter stainless steel tubes (Figure 5.5(a) and (b)) and control valves from sterilizer unit to pellet making and larval diet dispensing machines. The tray-forming apparatus is the Formseal® in-line form-fill seal machine. The form-fill machines at the Gast Boll Weevil Laboratory were purchased about 50 years ago, but the technology is readily available today, as it is used for various products such as yogurt cups, condiment cups (both opaque and transparent-type), fruit containers, and many other uses. It should be noted that the shapes and dimensions of the containers are versatile where containers such as the smaller, 32-cell containers vs. larger, flat containers can be made with various dies. The containers can be custom-designed by making dies that are shaped according to the needs of the rearing operations. The USDA, APHIS Methods Development Laboratory in Phoenix used custom-made dies for a variety of rearing container designs, including the containers used for mass-rearing big-eyed bugs (Cohen and Staten 1994).

Once placed in the rearing room, carts with their insects in 32-cell trays were held at rearing room temperature and humidity (27 ± 1°C at 45–55%) in 16:8 light: dark period. After about 8–10 days of incubation, pupae were removed from rearing trays and placed in the adult mating and oviposition cages. Adults were held at 27 ± 1°C at 45–55% and 16:8 light: dark period and fed 10% diluted honey solution and allowed to oviposit on waxed paper suspended in cages. Egg sheets were collected from adult cages daily, washed with 5% bleach solution (50 mL Chlorox [5.25% sodium hypochlorite] in 950-mL sterile distilled water), dried in a laminar flow hood, and placed in planting solution as described above for boll weevil eggs (Figures 5.11–5.16).

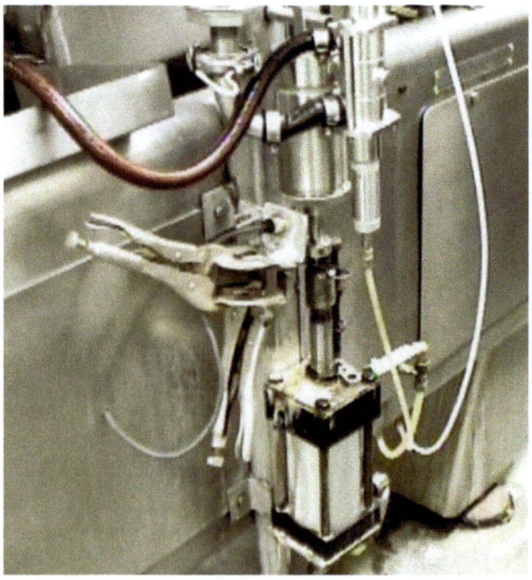

Figure 5.14 Egg planting liquid cylinder and pumping system.

Figure 5.15 (a) and (b) Granular materials dispenser (with corn cob grits). The measured aliquots of corn cob grits are being delivered to the diet surface of a 32-cell tray.

5.6 FUTURE APPLICATIONS OF THE GAST MASS-REARING TECHNOLOGY

Despite the fact that much of the technology described here was developed and refined more than 50 years ago, there is a great deal of potential for application of the mass-rearing technology to current and future mass-rearing operations. Most of the technology in diet-making, diet presentation, integration of insects and diet systems is still extant today with major elements available from the food industry. For example, form-fill packaging of human food items such as condiments is still in active practice as is the food mixing, heating, and other food-processing technology. Even such mass-production technology as extruders and microwave processing of foods remain an important and available part of the food processing industry. Therefore, this technology is readily available for insect-rearing specialists to adopt, especially if they follow the guidelines of the Gast facility for such specialization as oviposition and harvesting accommodation (Figure 5.17).

A modification of a form-fill seal-produced container, made with a Vacuform® machine. In this configuration a wooden die was made with the desired dimensions for a rearing cage for mass-rearing *Geocoris punctipes*. The top of the cage has a cylindrical shape that is made to fit a 3.8 L ice-cream carton lid, which fits perfectly on the cage

Figure 5.16 (a) Shows the open trays about to be covered with Tyvek lidding material that was coated with heat-sensitive glue, and (b) shows sealed trays being fed into the cutter that measured and cut trays into 32-cell units. At this point the trays were ready to be placed by the technician in the white lab coat in slots on the rearing cart and delivered into the larval rearing room.

Figure 5.17 Rearing containers produced by thermoforming.

and allows easy access to the insects and materials inside the cage. The Vacuform machine configuration is potentially useful for laboratories that engage in medium-scale production, rather than true mass-rearing.

5.7 CONSIDERATION OF THE MATERIALS FOR CONTAINERS AND OTHER DISPOSABLE REARING ITEMS

Besides considering the usefulness of different types of plastics for rearing containers, it is important to consider the quality of disposability or whether or not the material can be recycled. The Society of the Plastics Industry (SPI) provides standards and classification for the various types of plastics. Of the seven categories that they describe, only two are of pertinence to the issue of making rearing containers: high-density polyethylene and polypropylene. Both are used for various food containment (milk cartons, yogurt cups, desert or souffle cups, syrup bottles, etc.). Both types of plastic can be recycled, but SPI regards polypropylene as more difficult to recycle than polyethylene. A consideration in future rearing planning is the possibility of re-using the containers, meaning that the materials used can be properly cleaned (Figure 5.18).

Figure 5.18 Commercially produced 32-cell tray. These trays are made of moderately rigid polyethylene plastic, which can be fairly readily recycled. However, it is not durable enough to be easily reused. It should be noted that polyethylene is generally less heat-resistant than polypropylene, and therefore the former is not as resistant to heat treatments in a cleaning process.

In conclusion, the system used in the Gast facility can serve in many ways as a model for a highly automated mass-rearing program. Most of the machinery is from commercial sources, though there were engineering modifications that had to be made to adapt such equipment as form-fill seal machines and flash sterilizers to the custom uses in insect rearing. The processes in the Gast facility could lend themselves to other kinds of diet preparation such as twin-screw extruders, microwave heating treatments, and possibly other heat or radiation sterilization processes. The efficacy of automation was well proven in the several decades of operation of the Gast facility, and the publications that describe the detailed SOPs and the rationale and science behind these SOPs are useful as models for rearing systems of all scopes and scales.

CHAPTER 6

Rearing Systems as Ecological Niches or N-Dimensional Hypervolumes

6.1 INTRODUCTORY COMMENTS

Billions of insects are produced each year in rearing systems that support programs in sterile insect techniques, biological control, as feeders for other organisms, for medical purposes such as wound-cleaning, commodity production, including silk, honey, shellac, and wax, and for research (Cohen 2015). The centerpiece in all these programs is the insect, which is reared under conditions that strive to simulate nature to the extent that they provide all factors necessary for the insect to complete its life cycle in continuous generations. These systems include the insects themselves with all their genetic characteristics and background, appropriate food, environmental conditions, and generally a means of containment that lends itself to keeping the insects in a setting that is suitable to the insect but at the same time convenient to rearing personnel. Rearing systems, in this chapter, are treated as models of processes that were developed on a per species basis for each kind of insect that is being reared successfully; and importantly, the concept is explained that the rearing system simulates the insect's niche in natural or agricultural ecosystems.

6.1.1 The Ecological Niche Concept and Competitive Exclusion Principle

The point of this chapter is to explain how insect rearing systems are artificial insect niches. What insects in nature face in terms of their niche parameters must be met in captivity, and it is important for rearing personnel to understand that the freedom that insects have in nature is, of necessity, replaced by constraints of captivity. Therefore, it becomes the absolute responsibility of rearing personnel to meet the requirements of the insect in terms of providing **all** the niche-defining parameters.

6.1.2 Background Comments about Gause's and Park's Experiments

In a biology graduate seminar in 1968, we studied the concept of the ecological niche concentrating on the G. E. Hutchinson's model of niches as n-dimensional hypervolumes and the competitive exclusion principle (Hutchinson 1957 and 1965) from Gause's treatment of "the struggle for existence." Although much of the thinking behind these principles is grounded in and anticipated by Darwin's writings—based on observations of species in nature, the experimental basis of these lines of thinking was developed through cultures of organisms such as yeasts, protozoans, and insects. In my early exposure to ecology and ecological niches, I did not consider that I would someday be dedicated to understanding how culturing insects (= insect rearing) was largely an outgrowth of the concept and practice of cultivation of microbes. I now realize that insect rearing as it is practiced today has its roots in: (1) silkworm rearing and (2) development of drosophilid (and flesh fly) rearing practices which were directly derived from microbial culture practices.

The concept of ecological niches was articulated by Gause 1932 and Gause et al.1934, and further explored by numerous authors since the original experiments with two species of yeasts and two species of *Paramecium*, *P. aurelia* and *P. caudatum*. Gause et al. had shown that the culture conditions for each species of *Paramecium* would allow continuous generations of each species indefinitely when they were reared alone, but when reared in mixed culture, one species always displaced the other. This displacement was shown to occur with other closely related species competing for the same resources leading to a phenomenon that came to be known as the competitive displacement principle or "Gause's Law." Gause arrived at the concept of ecological niches from his experimentally based enquiries into the "struggle for existence." The niche concept from Gause was explored extensively by Park 1948, 1954, and 1957 in studies of the tenebrionid beetles *Tribolium*, including *T. castaneum* and *T. confusum*. The niche concept became a centerpiece of 20th century ecology as a way of explaining how species with their unique niches interacted in ecological communities and ecosystems. In 1957, G. E. Hutchinson provided a mathematical model of niche complexities as an N-dimensional hypervolume (Hutchinson 1957).

He introduced this revolutionary concept of the niche by assigning the assorted variables that define a species' ecological character using $x_1 \ldots x_n$ for each species in an ecosystem, explaining the model as follows:

> Consider two independent environmental variables x_1 and x_2 which can be measured along ordinary rectangular coordinates. Let the limiting values permitting a species S_1 [*species 1*] to survive and reproduce be respectively x'_1, x''_1 for X_1 and x'_2, x''_2 for X_2. An area is thus defined, each point of which corresponds to a possible environmental state permitting the species to exist indefinitely.

Hutchison goes on to explain:

> We may now introduce another variable x_3 and obtain a volume, and then further variables $x_4 \ldots x_{n,}$ until all of the ecological factors relative to S_1 [*species 1*] have been considered. In this way an n-dimensional hypervolume is defined, every point in which corresponds to a state of the environment which would permit the species S_1 to exist indefinitely. For any species S_1, this hypervolume N_1 will be called the *fundamental niche* of S_1. Similarly for a second species S_2 the fundamental niche will be a similarly defined hypervolume N_2.

This model lends itself to visualization of the niche parameters as shown in Figure 6.1.

The concept of an insect rearing system constituting an ecological niche was explored by Park 1948, 1954, and Park 1957 in a remarkable comprehensive and long-range

Figure 6.2 Red flour beetle *Tribolium castaneum*. The beeltes were kindly provided by Ms. Meredith Favre a graduate student in the laboratory of Professor Marce Lorenzen (Department of Entomology and Plant Pathology, North Carolina State University).

experiments with the flour beetles *Tribolium castaneum* and *T. confusum* (Figure 6.2). In explaining the basis of exclusion or displacement of one species over another, the niche parameters that Park explored included competition for food, tolerances to presence of varying amounts of O_2 and CO_2 as well as tolerances to nitrogenous wastes. Park considered the differential effects of parasitic (pathogenic) responses to the sporozoan (phylum Apicomplexa) *Adelina*

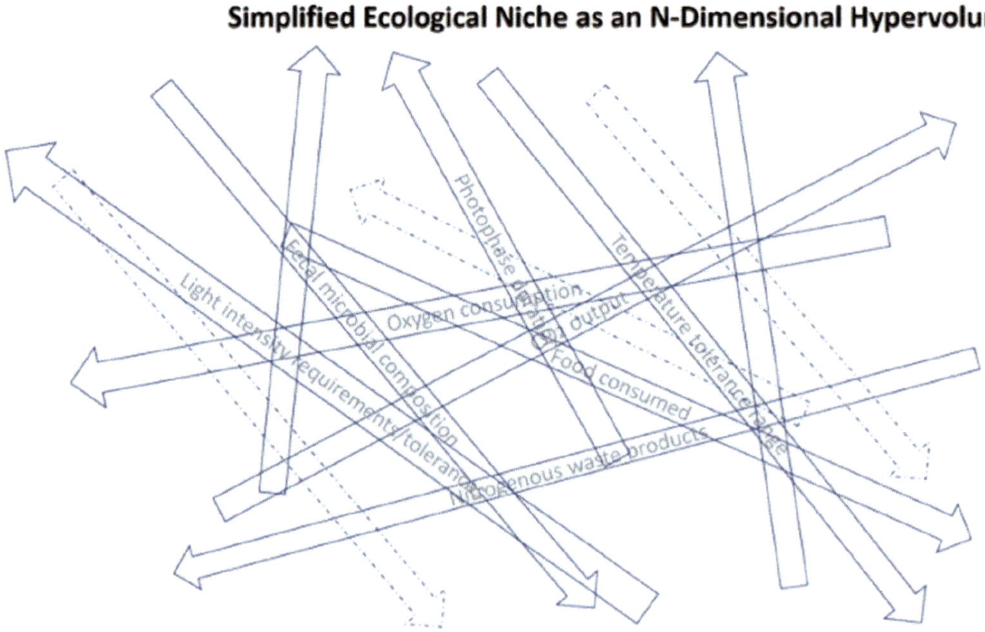

Figure 6.1 Simplified niche parameters expressed in an N-dimensional hypervolume. The figure depicts the idea that there are many facets of niches, represented as intersecting or overlapping vectors. The intersections represent the concept that niche parameters are interactive.

tribolii, which is known to affect insects from various orders of insects. The insights of Park toward subtleties of niche parameters are evident from the following discussion:

1. alteration of the temperature, moisture content, and oxygen concentration within the medium ("micro-climate") from that obtaining in the controlled incubators in such a way as to affect, differentially by species, the natality and/or mortality rates;
2. differential predation in which, say, the larvae and imagoes of one species might consume the eggs and pupae of the other at a rate statistically higher than that directed against their own kind;
3. differential interference with copulation or fecundation; and
4. differential conditioning of the medium (Crombie 1945). This might concern the reduction of available food, the elaboration of environmental poisons, such as excrement and carbon dioxide, or comminution [reduction of particle size] of the flour itself.

6.1.3 Competitive Exclusion in *Tribolium* Species

In Park's 1948 monograph on competition between *Tribolium confusum* (the confused flour beetle) and *T. castaneum* (the red flour beetle, both stored grain pests (Coleoptera: Tenebrionidae), various factors were discussed, including culture temperature, presence or absence of a parasitic coccidian (described by Tanada and Kaya as "protozoans" and classified as follows: Infrakingdom: Alveolata; Phylum: Apicomplexa; Class Conoidasida; Subclass: Coccidia) species *Adelina tribolii*. Park's remarkable extensive longitudinal studies were aimed at defining the niche of the species of Tribolium, and one of the main tools for discerning and defining the niche was competition between these two congeneric species. An example of the definition of the competition is evident from the set of graphs in Figure 6.3.

The diagram in Figure 6.3 represents Hutchinson's concept of the various x-niche parameters. For example, x_1 may represent oxygen consumption, x_2 may represent carbon dioxide output, and x_3 may represent the type of and amount of food consumed and so on through x_n. To meet Hutchinson's expectations, the species filling the niche in Figure 6.1 would be able to exist indefinitely or with stability in the ecosystem in question. Clearly this model suggests a highly complex relationship that is possible for each species and an even more complicated set of ecological interactions in multiple species systems. Figure 6.1 and Table 6.1 describe many (though not all) niche-related parameters in rearing systems.

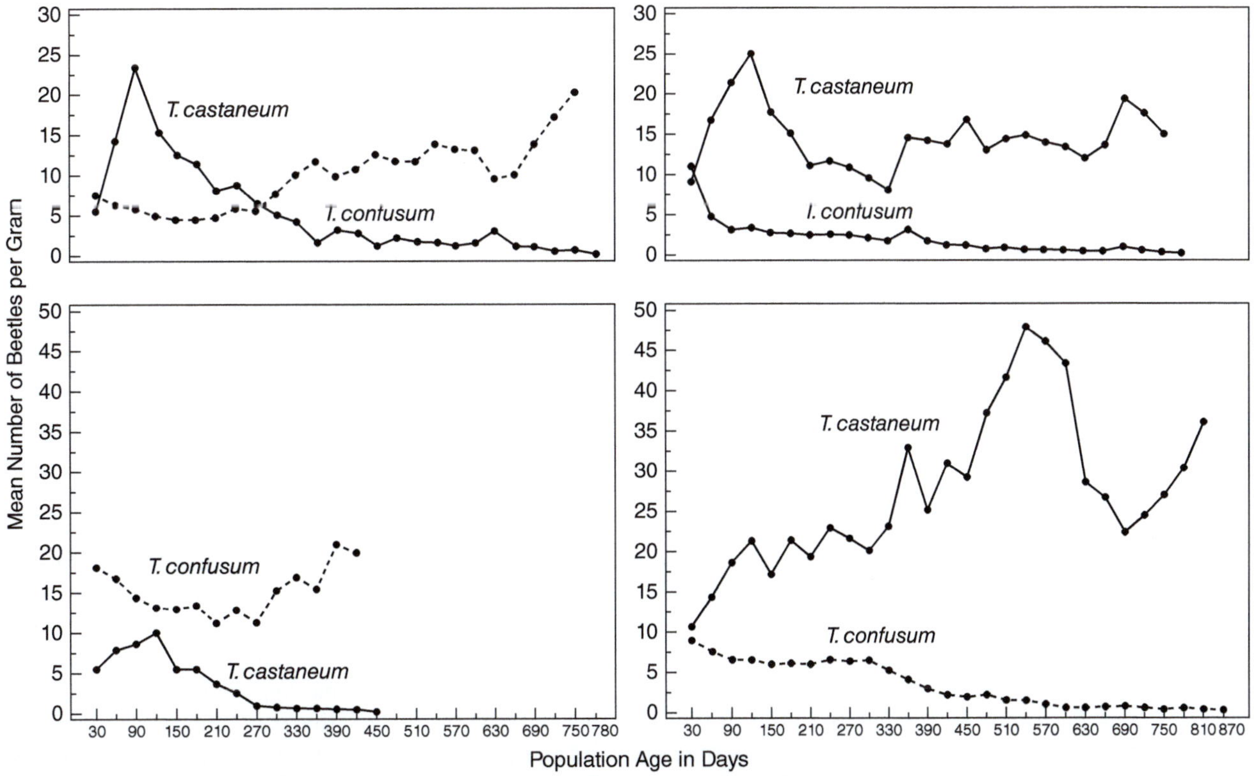

Figure 6.3 From Park 1948 showing the "struggle for existence" or competition between *T. confusum* and *T. castaneum* in experiments where one of the variables was the parasite/symbiont, *Adelina tribolii*.

Table 6.1 Niche Parameters in Rearing Systems

Niche Parameters	Range of Niche Parameters
Starter Insects' Genetic Characteristics	Full genome characteristics, including genetic diversity
Food/diets	Type and amount of food eaten
Waste products	Nitrogenous wastes, moisture, undigested food
Gas exchange	CO_2 output, O_2 intake
Moisture exchange	Water ingested, evaporated Insects' effects on humidity in rearing container through cuticular, respiratory, and fecal water loss, cuticular and respiratory water loss, osmotic values of water in aquatic systems
Light relations	Effects of photoperiod, light intensity, and light quality (including possible other electromagnetic spectrum wavelengths such as UV and IR
Microbial relations	Beneficial microbes, pathogens, commensals, etc.
Population density	How many individuals/unit of rearing space?
Population dynamics	Gregarious vs. solitary, agonistic behavior, competition, group digestion, group thermal regulation, etc.
Effects on substrate (biting, lidding, etc.)	Degrade container by biting holes, webbing to "manage" frass, digging and conditioning soil, pH of water for aquatics, etc.
Mating conditions	Ratios of ♂/ ♀; number of adults, environmental conditions, space for courtship, pheromone plumes, etc.
Oviposition conditions	Materials for egg inserters, conduce conditions for egg-laying and egg development/egg hatch
Populations of competitors (e.g., mites and psocids)	Often on plants used to rear insects and sometimes on artificial diets (see text discussion for details)
Populations of parasites/predators	The large biomass and populations of mass-reared insects offer a target for cryptic parasitoids and predators
Thermal relations	Temperature tolerances and optima influence the success or reared insects and further influence susceptibility to microbial attacks
Silk utilization/dynamics	Multifunctional uses of silk are discussed in the text
Effects of other (than silk) secretions: extra-oral fluids	For example, glandular secretions, oral/salivary secretions, modifying nutritional quality and bioavailability
Scales (as dust hazards, etc.)	Potential for microbial transfers, influencing gas exchange, debris
Waste products	Nitrogenous wastes as toxins, substrates for microbial growth, blocking normal food ingestion by clutter, etc.
Gas exchange	Range of O_2 uptake and CO_2 output on cultured insects, dispersal of O_2 and CO_2 in containers
Insectary soundscape/vibration-scape	Factors that interfere with auditory communication (stray noise) or vibrations that produce stress or induce excessive responses
Factors involved in oxidative stress	Factors in diet or environment that generate free radicals or reactive oxidative species in insects

One of the concluding statements by Park is as follows: "The major conclusions emerging from the above summary are (a) that both species of beetles cannot live together for any extended period, and (b) that, in competition, *T. castaneum* characteristically disappears, with *T. confusum* remaining as the successful organism." This statement, it must be emphasized, pertains to *Tribolium* that are not infected with *A. tribolii*. However, when beetles are infected with the parasite *A. tribolii*, the situation reverses, and *T. castaneum* (graphs B and D in the figure) wins the "struggle for existence." This is despite the fact that both species of beetles, when not parasitized by *A. tribolii*, can continue their normal population dynamics. This means that a subtle but important aspect of the ecological niches of the two species of *Tribolium* is their relative tolerance vs. susceptibility to the parasite.

The "niche as n-dimensional hypervolume" model has stood the test of time and was found to have great explanatory potential for helping us understand organisms' survival and other dynamics, as explained by Pianka 1974 and more recently by Swanson et al. 2015 and Pruitt et al. 2016. In fact, the study by Swanson et al. 2015 shows how the niche or N-dimensional hypervolume can be partially defined empirically by fitting data from stable isotope studies to the mathematical construct originated by Hutchinson. The Pruitt et al. study uses behavioral aspects of species in a spider community to help define the niches of predators that coexist in overlapping spatial relationships.

This raises the questions, what is a niche, and how does the niche concept apply to rearing systems? In the sense used here, the niche is the overall ecological role that populations of a given species play in an ecosystem. The niche includes the foods eaten by the species in question and what uses that species/population as food. The niche further includes effects the species has on the biotic and abiotic factors in the specified system—the reciprocal interactions between the population and the other components of the system. The niche includes physiological parameters like oxygen consumption, carbon dioxide outputs, excretory waste products, and effects of the population on soils or water (such as tunnels in soil or disrupted water flow in microhabitats). It further includes such factors as webbing, or other construction such as patches of soil modified by the organism's digging, building traps (such as pitfall traps by ant lions), and all other physical and biological influences the organism has on its system. A huge part of the niche of an organism is the reciprocal effect of the organism and the microbes in its community. The outcome of this microbial relationship (the microbiome) of the organism's niche has become a subject of intense study recently. Because the scope of this chapter is narrow in terms of ecology, it is important to point out that the complex implications of the ecological niche of a species or population of a species may be narrow or broad, depending upon the complexity of the ecosystem. Also, some organisms have such pronounced importance and influence in their systems that they may be regarded as keystone species—organisms that shape the character of the entire system (Figure 6.4).

This image in Figure 6.4 was captured from a video-microscopy recording of an emerald ash borer feeding on ash phloem wood in the laboratory. In the upper portion, above the feeding site, two mites (unidentified species) can be seen. These mites can enter the rearing system, and with artificial diets can out-compete the emerald ash

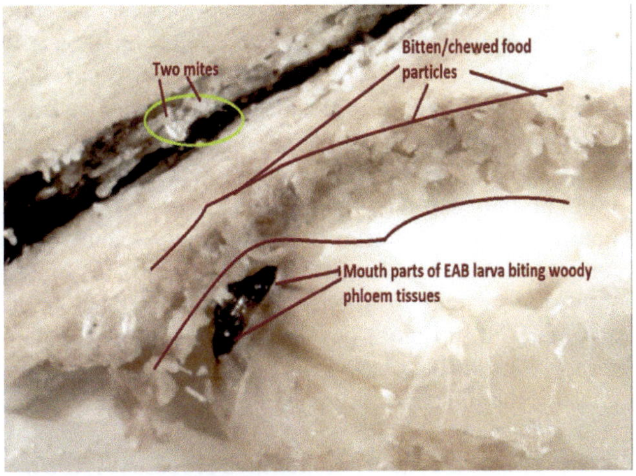

Figure 6.4 EAB larvae with mites.

borers, which are the rearing targets in programs being developed to mass-rear parasitoids of EAB for biological control. This is an example of the kind of complex interactions that can lead to disruptions of our rearing systems, sometimes rendering these system to be out of control.

6.2 THE REARING SYSTEM AS A MICROCOSM OR AN ECOLOGICAL NICHE

In rearing systems for given species of insects, we are dealing with simplified ecological relationships where we generally focus on a single species of insect that we are cultivating for some defined purpose. As pointed out by Cohen 2018, all rearing systems that succeed in supporting production of continuous generations of target insects, there is a complete provision of all the insects' requirements (Figure 6.5). This means that each rearing system, including the environmental parameters, containment, foods, and all other species-related resources, exists as a **microcosm** of the insect's natural world. This can be stated as the rearing system as ecological niche concept. In summary, the concept is articulated by Cohen 2018 as follows:

> Hundreds of billions of insects are produced each year in insectaries or laboratories. These insects are reared for programs of pest control, as feed, for production of various products such as textiles and food, for research and education, among many purposes. The concept is advanced here that rearing conditions are expected to simulate the natural environments from which target insects are derived. Simply speaking, the rearing container and rearing facility are the insect's microhabitat and its overall environment. Therefore, the central premise of this paper is that rearing systems are human-made ecosystems. If our expectations are to be met, that reared insects are healthy, fit, and of high quality—simulating their wild counterparts, rearing personnel must engineer conditions that meet all the insects' needs. In this review, major advancements and progress are documented within the context of the ecology involved in artificial rearing technology. In this review, attention is paid to the food, environmental factors (temperature, humidity or water—in the case of aquatic insects—lighting, gas exchange, and soil—for soil-inhabiting insects), containment, and some special features of rearing systems that were designed to meet special needs of a target insect's biology. Because there are thousands of publications on rearing advancements, the focus is on publications that had the strongest influence on modern day insect rearing systems, as well as papers that represent historically novel rearing concepts. Where the chain of in-formation is missing on how or from where rearing innovations were derived, papers are referenced as documentation of the various rearing system features.

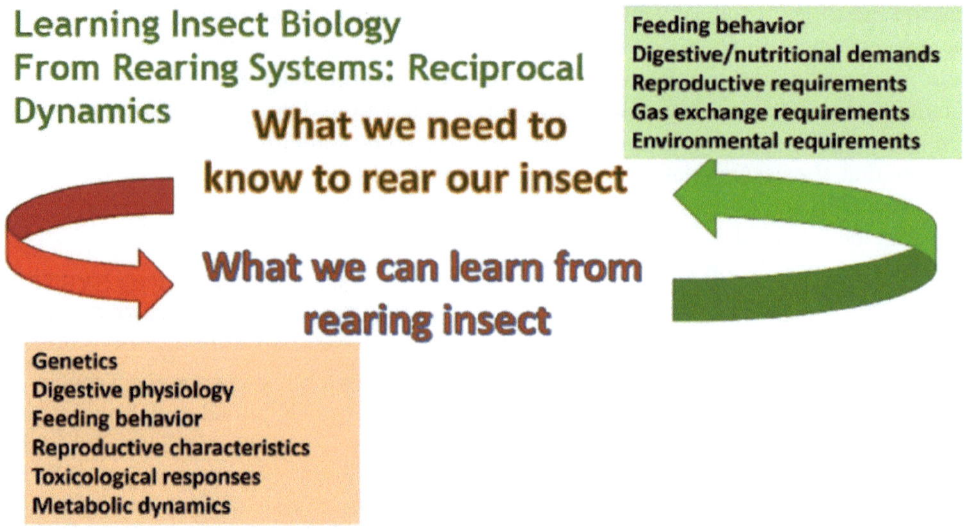

Figure 6.5 Reciprocal relations between development of rearing system and rearing systems as tools of inquiry (from Cohen 2018).

The salient features of this paragraph are that the rearing system is a ***controlled***, ***human-designed*** ecological ***microcosm*** or a human-made version of the insects' ***ecological niche***.

6.3 THE RECIPROCAL INTERACTIONS BETWEEN REQUIREMENTS AND STUDIES OF NICHE PARAMETERS

There is a reciprocal nature of rearing system development and what we can be learn from having a rearing system that allows us to control experiments with target insects. On one hand, in order to develop a working rearing system, we need to know certain minimal information about the nature of our insects; and on the other hand, once we have a rearing system in place, we can conduct experiments to discover further information about the insects that we can rear. The main point of the diagram in Figure 6.5 is that while a base of knowledge was needed to develop a rearing system, there is a rich array of learning outcomes that can be gained from controlled studies of insects in rearing systems. The pioneers in *Drosophila melanogaster* rearing learned early on that in order to study the genetics (and other biological features of biology of *D. melanogaster*), they had to be able to culture it using a reliable, controlled rearing system. To develop such as reliable system, they had to learn to use yeast and well-controlled media; then once they had the flies in culture, they could study the genetics, the nutritional physiology, microbial relations, and countless other aspects of *Drosophila* biology. Thus we have the reciprocity between rearing requirements and studies of the complete ecological niche.

6.4 THE VALUE OF APPLYING THE ECOLOGICAL NICHE CONCEPT TO REARING SYSTEMS

The niche concept entails the ideas: (1) that organisms have complex roles in nature, (2) that these roles include reciprocal interactions with the ecosystem that include the organisms in question, and (3) that the interactions include a great variety of components such as foods eaten, interactions with the physical environment such as soil or water, physiological functions such as gas (O_2 and CO_2) exchange, release of nitrogenous wastes, heat exchange, and myriad other factors. Most of these kinds of interactions can take place, at least in part, with organisms in captivity. For our purposes, insects that we rear can be thought of conveniently and heuristically as undergoing a complex of interactions with the components of their rearing system (i.e., their microhabitat). Most often, when rearing system development is initiated for a given species, there is a basis of knowledge about the target organisms' natural history. Knowledge of some of the features described in Figure 6.2 is already known when rearing is being initiated for the species in question (Figures 6.6–6.8). As the target insect is brought into captivity to begin the rearing process, a heuristic process of trial and correction is undertaken, and the details of the insect's niche begin to emerge into an ever-increasing base of knowledge such as that summarized in Figure 6.3. By knowing about the niche dimensions and characteristics described in Figures 6.2 and 6.3, the rearing personnel can fine-tune the incipient rearing system to make it better fit the target organism's niche. This is inevitably an iterative process (Figure 6.4).

The insects in Figure 6.6 are western bean cutworms (*Striacosta albicosta* [Smith]: Noctuidae: Lepidoptera),

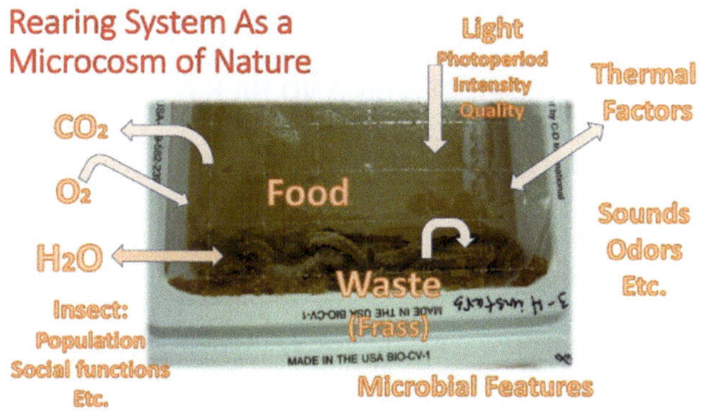

Figure 6.6 Diagram of the concept of the rearing system as a microcosm of nature (i.e., the organisms' niche).

which are being reared on an artificial diet with defined environmental conditions discussed here. The plastic rearing container was designed to permit light, exchange of gases, and protection from stray microbes; however, the uncontrolled factors are omnipresent as phenomena such as stray sounds or vibrations, odors, stray or unintended light, and other factors that influence the rearing process. Under conditions where the insect releases moisture as water vapor during gas exchange or frass water, the moisture content within the container increases to levels of water activity that promote microbial growth.

The diagram in Figure 6.7 shows an algorithmic approach to developing an artificial diet for a generalized insect, starting with trying to use an existing diet for an insect with similar feeding habits. This paradigm of algorithms for trial and error diet development represents the general pattern that had gained historical validation of diet development efforts (Cohen 2015). However, the trial and error approach can be greatly improved upon by using optimization strategies based on design of experiments discussed in detail throughout this book.

Figure 6.8 (a) mold-infested and (b) a healthy cell.

Figure 6.8 (a) shows a portion of a cactus moth (*Cactoblastis cactorum*) rearing system in 8-cell trays where contamination of diet by *Penicillium* sp. is evident while (b) is a healthy cell with healthy larvae (~75) of uniform size thriving. Please note that the 8-cell trays are similar to the tray shown in Figure 6.6, except that the trays in (a) and (b) have had their lidding removed to allow better visualization of the larvae, their diet-medium, and their frass.

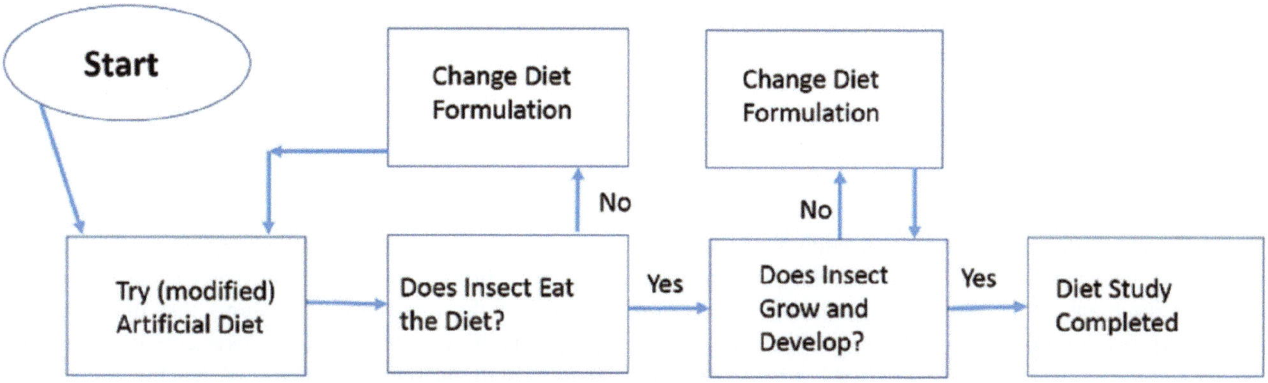

Figure 6.7 A simplified example of an iterative process in development of rearing system components.

6.5 REVIEW OF LITERATURE ON ECOLOGICAL BASIS FOR REARING SYSTEMS

Since the dawn of laboratory rearing systems for insects in the 1st decade of the 20th century, reports have been published to characterize the ecological requirements of target insects. This information started with silkworms (Kelly 1903), a few species of Diptera, including flesh flies (Bogdanow 1906, 1908) and then *Drosophila* (Delcourt and Guyenot 1910).The information presented here is drawn from publications about various aspects of insect rearing. The papers were accessed through *Web of Science*, and the topics presented under subject headings of each section were derived from and explained in the publications. With the exception of some early papers written in German or French, the search was confined to papers written in English or were translated into English such as the works of Hamamura (translated from Japanese). Efforts were made to select papers based on: (a) their historical precedence, (b) their citation record, and (c) their content in relation to other papers that were influenced by the works in question. For the third factor (c), the influential papers were included in this review because of clear statements of ***rationale*** behind experimental factors. Besides information from each paper cited in this review, other details are presented, including the historical position of each topic, the number of citations for several papers, and comments about stated or presumed rationale for each rearing advancement. It is important to mention that the search of the literature does not always reveal papers on the topics expressed in the papers' titles, abstracts, or keywords. As an example, I happen to have knowledge of the contributions of Dr. George D. Butler. However, in my search of "temperature" and "insect rearing" as keywords, Butler's papers were not revealed by the *Web of Science* scope and were brought out only after the "Butler, GD" was included in a search of "Insect Rearing" and "Temperature." Once all three of these terms were fed into the search, 34 papers by Butler on rearing and temperature were found. The only reason that this review's author could find these papers was that he knew that Butler had written several papers on the impact temperature had on rate of development. The point of this comment is that many important advances in and components of established insect-rearing systems have a history that is either: (a) not explained in rearing literature or (b) not easily found by searches conducted through conventional search methods. In fact, it is an unfortunate fact that much of the body of practices used in insect-rearing systems is a matter of an unwritten culture where practices find their way into rearing systems for given species, and tracing many of these practices remains elusive. *This feature of rearing practices lends itself to the undesirable outcome that much of what is practiced in insect rearing is not vetted and not based on science or well-engineered technology.*

6.6 *INSECT-REARING SYSTEMS* DEFINED AND CHARACTERIZED: HISTORICAL PERSPECTIVES AND SILKWORMS

Just as a fundamental unit of chemistry is the atom and a fundamental unit of biology is the cell, the fundamental unit of laboratory production of insects is the ***insect-rearing system***. This all-important phrase connotes **how** the insect is reared: the starter population, food, containers, temperature, humidity, water system characteristics (for aquatic insects), soil conditions (for soil insects), light conditions, and all other special needs the target insect is known to have. This review does not include the incredibly complex rearing systems used in sericulture (silkworms and silk production). Sericulture was begun in China nearly 5,000 years ago (Barber 1992; Hamamura 2001; Vainker 2004; Anonymous 2011), and apparently supersedes all other rearing systems combined in terms of the economic implications. For example, an Indian document about rearing silkworms (Anonymous 2011) states that sericulture employs more than 700,000 families in India alone. Besides their importance for textile industries around the world and other uses of silkworms as food and for medical applications, silkworms are also crucial components of research accomplishments. As of a *Web of Science* search in 2017, more than 9,100 papers (more than 13,000 on "*Bombyx*") were published about silkworms, including a 1903 paper in the *Journal of the American Medical Association* about the use of silkworm gut (not silk) for sutures for kidney surgery! Furthermore, various internet searches with keywords such as "silk production," "economics of silk," and "sericulture worldwide" reveal that throughout many Asian, European, and recently other nations, silk production—especially by the mulberry silkworm (*Bombyx mori*) is of paramount importance. For example, the International Sericultural Commission (https://inserco.org/en/statistics) gives statistics for silk production from 20 countries, showing that the metric tons of silk produced occupy a substantial part of many countries' economic base.

6.7 THE ONTOGENY OF *DROSOPHILA* REARING SYSTEMS

In this review, the main topics of the rearing system concept will be presented, first, as general features of rearing such as containers, temperature, humidity, light, and diet and second through exploration of publications that were influential and involved in the development of several of the mass-rearing systems that underlie major programs. In this chapter, rearing systems for *Drosophila* species will be the centerpiece of rearing these model organisms that are the most thoroughly studied and best understood metazoan.

Besides the thousands of papers written about *Drosophila* genetics since the early 20th century, many more papers have been written about all aspects of *Drosophila* biology on topics ranging from ecology of these flies (Starmer and Fogleman 1986; Starmer et al. 1986; Anagnostou et al. 2011) and further studies on gerontology, nutrition, and countless other aspects of *Drosophila* biology such as feeding choice studies (Ja et al. 2009; Deshpande et al. 2015). The papers by Deshpande et al. 2015 and Ja et al. 2009 cover the important topics of food selection in relationship to fitness of *Drosophila*. The papers of Bridges and Darby 1933 and Bridges 1932, respectively, treat the evolution of diets for *Drosophila* and other aspects of rearing systems such as containers, home-made incubators, and basics of the ecology of rearing *Drosophila* as a laboratory organism. These papers are remarkably in-depth treatments of the progress that took place in the laboratory of Thomas Hunt Morgan. The comprehensive work by Morgan et al. 1913 treats the genetics discoveries with *Drosophila* and other species used as models of Mendelian genetics. The work done in Morgan's lab at Columbia University in the early 20th century shaped the field of genetics and other areas of biology, including insect rearing, to such an extent that many authors have regarded the "Fly Room" where the genetic research of Morgan's group (including, Bridges, Muller, and Sturtevant) became the subject of popular books and a movie called "The Fly Room," discussed in Anonymous 2015.

6.8 *DROSOPHILA* DIET AND REARING SYSTEMS AND THEIR HISTORICAL IMPORTANCE

The two oldest papers on artificial diets were by Bogdanow 1906 and 1908. Both papers dealt with flesh flies and incorporated components that were important in hundreds of diets, including bovine milk casein, purified salt solutions, meridic nutrients such as egg albumins, and even introduced addition of live microbes to condition the diets. The monograph on silkworm culture (Kelly 1903) provides a detailed explanation of all processes required for rearing *Bombyx mori*. However, the paper that lays the foundation of controlled conditions in rearing and diet presentation was the Delcourt and Guyenot 1910 study that explains the importance of controlling all aspects of rearing conditions. This paper and several other pioneering diet studies by Guyenot 1913a, 1913b, 1913c, 1914a, and 1914b made significant advances in *Drosophila* dietetics. For example, in Guyenot 1913a, the subject of the role of live microbes vs. aseptic diets was explored—a topic that had gained extensive attention in insect rearing from the earliest inquiries to the present day when insect gut microbiome evaluations abound. It is valuable to note that in the early days of diet development, the scientific community was in its "Golden Age" of microbiology and genetics, and some of the early studies on insects reflected the contemporaneous interest in the role of microbes in insect nutrition, leading to numerous studies of the essentiality of various species of microbes (e.g., Loeb 1915; Baumberger 1917a, 1917b, 1919; Baumberger and Glaser 1917; Loeb and Northrop 1917; Michelbacher et al. 1932). For historical perspective, it should be noted that Bogdanow 1906 and 1908 had explored among several parameters the importance of defined species of microbes' influence on host insect success.

6.9 DEVELOPMENT OF *DROSOPHILA* DIETS AND DIET-BASED REARING SYSTEMS

Delcourt and Guyenot 1910 explained that in their efforts to study the heredity of *Drosophila* over successive generations, a requisite for accurate determination of virtually all genetic questions was the development of controlled conditions in the rearing process. They explained that the inconsistencies of the foods used to rear the flies, where the qualities and differences in fermentation progress of raisins, bananas, and apples caused variability that was too great to allow interpretation of the genetic results that they were testing. These inconsistencies and difficulties demanded a control of the rearing system—mainly through control of diets. It was also obvious that there was a microbial involvement in *Drosophila* rearing as was addressed by Delcourt and Guyenot 1910; Guyenot 1913a, 1913b, 1913c; Guyenot 1914a, 1914b; Baumberger 1917a, 1917b; Baumberger and Glaser 1917; Loeb and Northrop 1917; and Morgan et al. 1913 (as explained further by Bridges and Darby 1933). The sequence of development of artificial diets for *Drosophila* was: fermented fruit (generally bananas) → mashed over-ripe bananas with yeast added → banana puree extract with yeast and agar → various fruit (raisins, apples, etc.) purees with yeast and agar. Two of the major advancements that were profoundly influential to not only *Drosophila* rearing but also rearing countless other insect species were the addition of yeast (live or dead) and agar to provide a gelled diet.

6.10 THE EVOLUTION OF CONTROLLED DIETS FOR *DROSOPHILA*: MICROBIAL RELATIONSHIPS

From 1910 through 1933, one of the liveliest subjects of inquiry was the relationship between insects and microbes in terms of dietary issues. There was a preoccupation with establishing the role and extent of microbial involvement in the feeding biology of *Drosophila* and other insects, especially other Diptera. A major portion of this progression was explained in elegant detail by Baumberger 1919 who reviewed his own and other researchers' experiments with yeasts vs. bacteria as food sources (as external or internal symbionts vs. nutrients) and live vs. heat-killed microbes. An important and illuminating part of these studies was the series of experiments that sought to tease apart the nutritional requirements that were part of the foods prior to microbial action vs. the foods post-fermentation (meant here as microbial action on media or raw materials) vs. the microbes themselves (Baumberger 1919; Michelbacher et al. 1932). After extensive review of the literature on the roles of microbes in feeding success on natural and artificial diets, it remains difficult to conclude how much of a diet's success in meeting target insects' needs is from direct effects of microbes (changing the nutritional quality of raw materials or media components), the contribution of the original raw materials, or the nutritional value of the microbes themselves (Cohen 2015, 2018). Cohen 2015 concluded that these complex questions about microbe/media/microbe-media interactions must be determined on a per species basis across all insects being reared. In summation, the many diets for *Drosophila* that are currently being used were evolved through a series of refinements of a fruit base with yeast added along with other nutritional and functional components such as corn meal, agar, and antifungal agents. Despite considerable effort to develop completely defined (holidic) diets for *Drosophila* to help establish the basic nutritional requirements for this insect, and in this context, all practical diets for this insect are based on complete foods in conjunction with yeast—mainly baker's or brewer's yeast *Saccharomyces cerevisiae*.

6.11 OTHER ASPECTS OF *DROSOPHILA*

Other aspects of *drosophila* rearing systems are treated in this chapter in separate sections on temperature and humidity, containers, and miscellaneous environmental factors (such as light, radiation, etc.). The section "Environmental Factors in Rearing Systems" is a general review of rearing systems for various species, divided into the major environmental topics. It must be noted that because of the far-reaching influence of *Drosophila* rearing that was developed extensively in the early 20th century, this subject is discussed in several other places in this book.

6.12 ENVIRONMENTAL FACTORS IN REARING SYSTEMS

The principles and fundamentals of insect/environment relationships are so pervasive in understanding the ecology of a given insect that it is impossible to define the niche or to provide the niche-based requirements without thorough understanding of the nature of each environmental factor and the inherent interplay with all other factors. In the insectary, as in nature, there is an intricate interplay between temperature, light, water-factors, diet characteristics, substrates, gases, olfactory components, and potentially many other factors. In Chapter 9 on fitness and stresses, the effects of deviation from the fundamental niche requirements are discussed in detail, but for now, the factors are considered in a simple framework.

6.12.1 Temperature and Humidity

A recent search of the keywords, "Insect rearing" and "temperature" in the *Web of Science*, revealed over 1,000 references. The earliest one was Smith 1931, but careful scrutiny of other papers where rearing conditions are treated reveals that temperature and humidity had been recognized for decades prior to the 1931 Smith paper was published. For example, Kelly 1903 provided excellent detail on the importance of keeping silkworm larvae within a narrow temperature range, and she was even more strident in her emphasis on the importance of temperature and humidity in the synchronization of hatching in silkworm eggs. Bridges 1932 provides an in-depth history of the treatment of temperature in the "Fly Room" at T. H. Morgan's *Drosophila* facilities at Columbia University had been a subject of intense effort to regulate temperature of fly rearing containers. Bridges 1932 wrote:

> DURING the early work on the genetics of *Drosophila melanogaster* the cultures were reared on laboratory tables or on open wall shelving. This method gave irregular results. The cold nights of the winter season made a considerable proportion of the cultures fail to start properly, and also lengthened the generation interval.
>
> In 1913 I built above my laboratory table a large wooden incubator which held all my experimental cultures. This incubator was heated by carbon electric lamps placed in the lowest shelf space. The thermostat was of the "ether-wafer" type. The expansion of the ether forced apart contacts in the heating circuit. A six inch electric fan continuously circulated air down a gap behind the shelving with return through a similar gap between the shelves and the doors.

Elsewhere in this paper and in several other places, Bridges emphasized the importance of temperature regulation in rearing circumstances—that it was not only crucial for optimization and uniformity, but also to help control the very genetic factors that *Drosophila* researchers were trying to explore. For example, he pointed out that linkage/crossover studies are skewed by deviations in temperature. Bridges wrote, "Plough's work on the effect of temperature extremes on crossing-over showed that it was necessary to maintain the temperature constant for genetic reasons as well as to improve culture conditions." Unfortunately, the recognition of the far-reaching effects of temperature and many other facets of rearing conditions is lost to the community of scholars in terms of the consequences of deviations or lack of control of rearing conditions.

A series of papers that were influential in the role of temperature in rearing systems were Smith 1931, Butler 1966, Stinner et al. 1974, and Butler et al. 1983. These papers followed the classic line of inquiry where constant temperatures were the independent variable, and the biological outcomes such as rates of development, survival, and sometimes body weight or reproductive success were observed. Visscher et al. 1979 added the variable of host plant growth temperatures as well as rearing temperatures, showing that temperature, and probably other plant history factors, were influential on rearing outcomes. The more natural condition for insects under field conditions includes variable temperature, humidity, light conditions, and other parameters; however, most insect rearing has been done at a single, constant temperature. But there is a body of literature (51 references identified under the search words, "insect rearing" and "variable temperature"). The earliest papers in this search were from the 1990s such as Manel and Debouzie 1997 where a field-based model is reported, and more recently, Sprugeon and Brent 2016 reported the value of using variable temperature. Manel and Debouzie 1997 used field temperatures to model development. The use of "room temperatures" or slightly above has such extensive use that the original use of these temperatures is no longer cited. However, it remains an interesting and important fact that Bridges 1932 provided a chronicle of the use of temperatures kept constant and slightly above room temperature, and he further explained the importance of the maintenance of stable, well-regulated temperatures in the synchronization of development and in the prevention of genetic or chromosomal alterations of the test insects.

6.12.2 Light Conditions (Photoperiod, Light Intensity, Light Quality, Other Radiation)

Cohen 2015 and 2018 pointed out that the phrase "light conditions" includes photo period, light intensity, and light quality—meaning the wavelengths provided in the rearing system. Although nearly every paper that includes rearing information provides information as an incidental component in rearing conditions, only 49 papers were found with the keywords, "light" and "rearing" in a *Web of Science* search. The range of papers under this category is broad, with papers on different qualities of light through unique papers such as the effects of light conditions on diapause in gypsy moths rearing in a spaceship. The Hayes et al. 1991 paper also includes other useful rearing biology information such as responses of the gypsy moth eggs and larvae to microgravity, cosmic radiation, sub-freezing temperatures, and unusual vibrations. One of the most interesting results of this unusual study is that space travel conditions led to a decrease in diapause from 3–4 months to six days. In fact, the topic of diapause and the environmental factors that induce or influences it is a dominant topic in insect biology. The keywords, "insect rearing" and "diapause," elicited 289 papers in *Web of Science*. When the term "rearing" is removed from the keywords (= "insect" and "diapause"), the search reveals 2,075 papers. The oldest of these papers describes the evolutionary background of what Roubaud 1930 calls, "obligatory larval hibernation" in mosquitoes (*Culex pipiens* L.). The author also mentions the word "diapause" in this paper, so this topic dates back nearly 100 years. Lees was a pioneer in arthropod diapause factor analysis where in 1953 he recognized the importance of light and dark phases in controlling diapause. During that same time frame, C. M. Williams discovered that the induction of diapause involved direct light stimulation to the insect brain, and Harris et al. 1969 demonstrated that in the boll weevil, the quality of light (the narrow wavelengths) were key factors in controlling or contributing to induction of diapause. The Harris et al. group 1969 also demonstrated that infrared radiation is also an important influence on diapause in boll weevils. It is noteworthy that Kelly describes the phenomenon of diapause in silkworm (*B. mori*) eggs, but she does not use the term in her 1903 monograph. Despite not using the term, "diapause," Kelly explains a protocol for breaking it.

In regard to diapause, various light factors are known to influence the onset and termination of this "resting period." These factors include light phases (light period and dark period), light quality (wavelengths), and light intensity. In Figure 6.9(a) and (b), it is evident that the shelf is immediately under the light fixture and therefore has a much higher light intensity than the shelf below. A brief investigation with a light meter in various parts of rearing rooms reveals that there is a considerable spatially based variation in light intensity. Chronic differences in light intensity between insect containers are a possible cause of physiological differences that may lead to unwanted deviations in rates of development of mass-reared insects.

Interestingly, the concept of photoperiod and its influence on biological systems seems to have been developed in the early 20th century in relationship to plants and is first described by Roberts 1927 in the context of growth and fruitfulness of apple trees. In relating photophase to insects, several papers treat various aspects of this subject including the concept of lunar phase (Ito et al. 1993), and the relationship between photoperiod and body size and flight capacity (Gu and Danthanarayana 1992), a biochemical/nutritional basis of response to photoperiod (Claret and Volkoff 1992), and relationship between maternal photoperiod on diapause (Ruberson et al. 1991; Lysyk 1992).

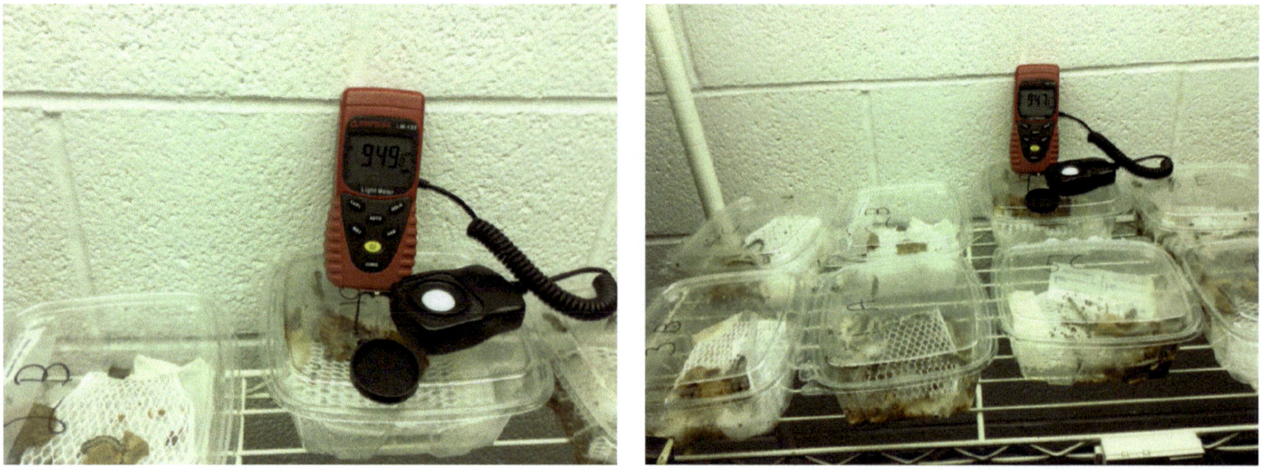

Figure 6.9 (a) and (b) Show a light intensity meter.

The topic of ultraviolet effects on insects has been treated in a few studies such as by McCloud and Berenbaum 1994. The importance of color and light environment as well as hatching rhythm were treated by Connor 2006, Collins et al. 2008, and Tao et al. 2017, respectively. Light intensity has been recognized as a key factor in many aspects of insect biology (Huaijian et al. 2017), and this recognition took place in the earliest studies of rearing *Drosophila* (Bridges 1932). Several laboratories use continuous light, rather than light/dark photoperiods (Patana 1969, 1977; Economopoulos et al. 1987). However, too often rearing facilities use light conditions that are based on conditions that have not been properly documented or vetted on the basis of their target insect's biology. When rearing personnel are trying to develop a rearing system for an insect that is new to them, they simply depend upon conditions that were successful in other laboratories—often this information is passed on by word of mouth (Cohen 2015).

6.12.3 Gas Exchange, Metabolism, and Energetics

While the literature on gas exchange, metabolic dynamics, and energetics is vast, few papers deal directly with these parameters in rearing systems. Reverting to the axiom that the rearing system is our domesticated insects' ecosystem, we are reminded that all the insects' needs must be met, including gas exchange and their related aspects: metabolism and energetics. This means that for terrestrial insects, the containers must meet the O_2 and CO_2 requirements while diets and insects are protected from desiccation due to excessive water loss. This also includes the cases where insects tunnel into diets or are otherwise positioned where insufficient oxygen supplies render the target insects into anoxic states or where excessive carbon dioxide prevents normal metabolic functions, including pH regulation of internal fluids. The intimate relationship between nutrient utilization and gas exchange also underlies the rearing system, leading to a relationship between respiratory quotient (RQ) and the insect's nutritional profile. In this context, it is important to recognize that the nutrient profiles of artificial diets can affect the insects' oxygen requirements, where the insects' inherent nature as a "lipid specialist" vs. a "carbohydrate specialist" can drive the insect to increase its oxygen demand, if there is excessive and unavoidable lipid (especially fats) in the diet (Cohen 1985; Economopoulos et al. 1987; Klok et al. 2016). It is also crucial to recognize that other factors in the rearing environment such as temperature have profound influence on oxygen demands and carbon dioxide dynamics (Cohen 1985; Klok et al. 2016). An excellent explanation of the relationships between metabolic rate variation in tsetse flies is found in Terblanche et al. 2004, and Harak et al. 1998 provide a compelling overview of stress in mealworms and the resulting changes in caloric differences in stress vs. non-stress conditions. It is worth noting that a series of interesting papers have been produced by the Terblanche group over the past two decades, and many of these papers on thermal tolerance, metabolic rates, and related topics are of interest to researchers in the rearing community in the context of food utilization, stress, and evolutionary perspectives of metabolic responses to rearing/environmental conditions (Figure 6.10).

6.12.4 Water (Humidity and Aqueous Conditions)

The subject of water in rearing systems is treated further in Chapter 10 on stress in rearing systems. In nature, water is an all-important, driving force that shapes ecological outcomes. Life, as we know it, cannot survive or thrive without the right allocation of water. Yet water can also be a "deal breaker" in rearing

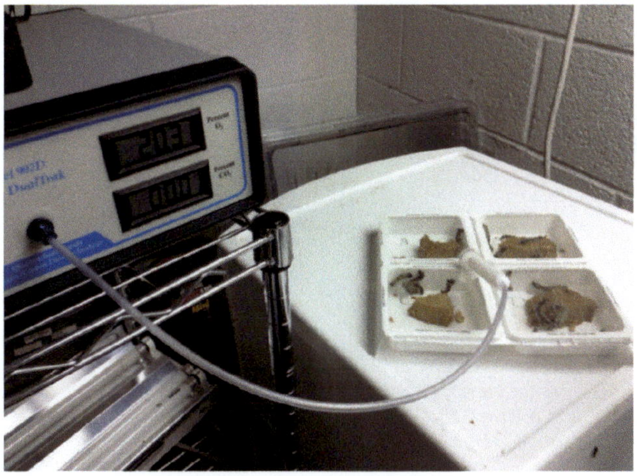

Figure 6.10 Shows simultaneous measurements of % O_2 and CO_2 in rearing containers (8-cell trays) with painted lady butterfly (*Vanessa cardui*) larvae.

situations. Too much water vapor can lead to microbial problems, especially fungi (molds). Temperature balance or regulation can depend upon evaporation of water as a cooling force. Excess water can also cause excretory functions such as production of uric acid or other nitrogenous waste compounds to be improperly regulated. Of course, too little water can lead to desiccation or difficulties with gas exchange and subtle difficulties such as proper shedding of cuticle in what should be normal molting.

6.12.5 Atmospheric Humidity (Terrestrial Conditions)

Often, rearing papers include information about humidity along with the temperature conditions (Smith 1931). However, several papers include studies of specific effects of various humidity conditions (e.g., example Subramanyam and Hagstrum 1991; Cothran and Gyrisco 1966). In general, reports of humidity are nearly as numerous as those of temperature effects in rearing publications.

Using Table 6.2 and Figure 6.11, we can let the two boxes represent a kg or 1.3 m^3 of air confined in a hypothetical box (the left box in Figure 6.11). If we know that the outside air where an insectary is located is 10°C, and the relative humidity is 50%, then this matrix tells us that the amount of water in a kg of air (1.3 m^3) is 3.81 g. If we warm the air in the box to 30°C, we can read the chart at 30°C and 100%, and we see that the same volume (mass) of air can hold 27.17 g of water vapor. Therefore, in the heated box of air, the saturation level is 3.81/27.17

Table 6.2 Absolute Moisture Content of Air (g/kg)

°C/RH	10	20	30	40	50	60	70	80	90	100
2	0.44	0.87	1.31	1.74	2.18	2.62	3.05	3.49	3.92	4.36
4	0.5	1.01	1.51	2.01	2.52	3.02	3.52	4.02	4.53	5.03
6	0.58	1.16	1.74	2.32	2.9	3.47	4.05	4.63	5.21	5.79
8	0.67	1.33	2	2.66	3.33	3.99	4.66	5.32	5.99	6.65
10	0.76	1.52	2.29	3.05	3.81	4.57	5.33	6.1	6.86	7.64
12	0.87	1.74	2.62	3.49	4.36	5.23	6.1	6.98	7.85	8.72
14	1	1.99	2.99	3.98	4.98	5.98	6.97	7.97	8.96	9.96
16	1.14	2.27	3.41	4.54	5.68	6.82	7.95	9.09	10.22	11.36
18	1.29	2.58	3.88	5.17	6.46	7.75	9.04	10.34	11.63	12.92
20	1.47	2.94	4.4	5.87	7.34	8.81	10.28	11.74	13.21	14.58
22	1.67	3.33	5	6.66	8.33	9.99	11.65	13.32	14.99	16.65
24	1.89	3.77	5.66	7.54	9.43	11.32	13.2	15.09	16.97	18.86
26	2.13	4.27	6.4	8.53	10.67	12.8	14.93	17.06	19.2	21.33
28	2.41	4.82	7.23	9.64	12.05	14.45	16.86	19.27	21.68	24.09
30	2.72	5.43	8.15	10.87	13.59	16.3	19.02	21.74	24.45	27.17
32	3.06	6.12	9.19	12.25	15.31	18.37	21.43	24.5	27.56	30.62
34	3.45	6.89	10.34	13.78	17.23	20.67	24.12	27.56	31.01	34.45
36	3.87	7.75	11.62	15.49	19.37	23.24	27.11	30.98	34.86	38.73
38	4.35	8.7	13.05	17.4	21.76	26.11	30.46	34.81	39	43.51
40	4.84	9.69	14.53	19.38	24.22	29.06	33.91	38.75	43.6	48.44

This table (derived from: http://uk.agratechniek.nl/moisture-content-air.html) is useful for determining the direction of water movement in a rearing system. In using this table, it should be remembered that the amounts of water in the various cells are the weight in g of water vapor per kg of air, which is about equal to 1.3 m^3. The highlighted vertical column on the left is the temperature in °C, and the horizontal column on the top of the chart is % relative humidity. For an idea of how this chart works, we can use Figure 6.11.

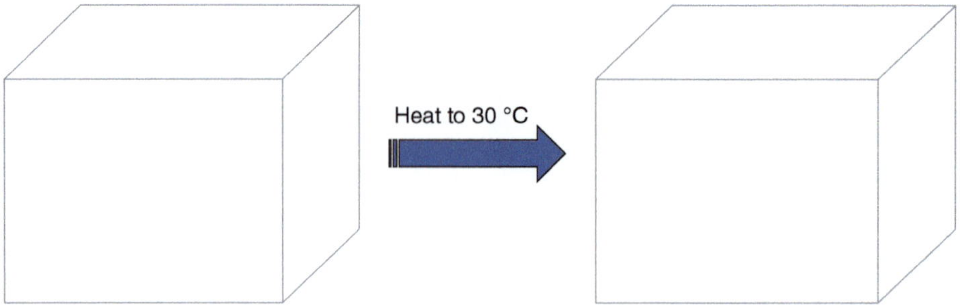

A 1.3 m^3 box of air at 10 °C and 50% relative humidity has 3.81 g of water vapor

The 1.3 m^3 box at 30°C can hold 27.17 g of water vapor; therefore the relative humidity is 3.81 g/27.17 = 0.140 * 100 – 14.0% relative humidity

Figure 6.11 Water vapor-holding capacity of air.

or 0.140 saturated which equals 14.0% relative humidity. A helpful way of looking at the drying potential of the air in the heated box is to think of the warmer air being "hungry" for water that can come from the diet, insects, frass—any material that can give up water to fill the deficit between 3.81 and 27.17. In other words, the 1.3 m^3 mass of air is "looking for" 27.17 minus 3.81 g of water vapor = 23.36 g of water vapor (the two data cells of interest are highlighted for emphasis). This phenomenon of bringing in cool, moist air and turning it into warm, dry air is what I call the "winter desert effect" that takes place in insectaries.

It must be kept in mind that the atmosphere of rearing systems is not the static situation suggested in the example with closed boxes. The air in the insectary is very dynamic in terms of movement such as drafts and influx and efflux of air from outside coming in and air from inside the building going out. There are further dynamics within the containers (cages) and between the containers and the ambient air. Rearing containers have openings for gas exchange, and within the containers, the air is being changed by O_2 uptake by the insects and possibly microbes, CO_2 release by insects and microbes, and water loss from the insects, frass, microbes, and diets. How fast these gaseous materials exchange with the ambient air surrounding the container depends on the biomass of insects and the configuration of the openings. This topic is discussed further in Chapter 10 (Relationships between Fitness, Quality Control, Health, Homeostasis, and Stress), but for now suffice it to say that these dynamics can have profound influence on the fitness and condition of the insects in our systems.

The "winter desert" phenomenon has been recognized fairly widely, and many insectary managers use humidifiers to reduce the potential desiccation stress described above. It must also be recognized that the converse issue takes place in the summer when warm, sometimes humid air is brought into the insectary. Again, considering the values in Table 6.2, if the outside air is 35°C with a 50% relative humidity, and the cooling system reduces the temperature to 25°C, the humidity goes up to nearly 100%. At such high humidity, the risk of mold (and other microbial) growth and condensation cause free water to be present as a potential trap for small insects, especially neonates. Of course, the added moisture from sources within the container also adds to the risks inherent in excessive moisture. In the case of excess moisture, rearing facilities can take corrective action by using dehumidifiers. In geographical areas where temperature and humidity extremes are prevalent, the rearing system must be treated with extra precautions. With the advancement of Global Climate Change, these types of problems must be expected to be much more troublesome. In my own experience, having started my rearing career in Arizona, I know the challenges that accrue from the temperature and humidity extremes that we faced.

6.12.6 Aquatic Insects and Larvae of Mosquitoes

The literature on rearing aquatic insects is too expansive to be covered comprehensively here. In fact, the keywords of "mosquito" and "rearing" produced more than 1,100 references in the *Web of Science* search, dating from Phillips and Swingle 1940 through hundreds of papers in the past few years, alone. The topics in these papers include diets for aquatic insects, and they also contain extensive information on food regimens (Inoda et al. 2009), special containers for water-dwelling species, including papers on simulating moving water (lotic) systems (Keiper and

Foote 1996). Comprehensive reviews of mosquito rearing such as Williams and Williams 2017 and Benedict et al. 2009 are useful for summarizing the most widely used systems of containers, water-handling, and other special features of aquatic systems such as filtering, pH management, lighting, and waste removal. Also, the text book on aquatic insects edited by Merritt et al. 2008 is an excellent compendium on all the special requirements of aquatic insects from various habitats. It is noteworthy that mosquitoes, because of their importance for medical and veterinary domains, predominate literature on aquatic insect-rearing systems.

6.13 CONTAINERS OR CAGES

Using the keywords, "Insect Rearing" and "containers," 83 papers were found as of June, 2019. Under these search terms, the papers go back to 1966. However, it must be noted that hundreds of other papers that deal with topics besides "insect rearing" and "containers" are present in the literature. For example, Bridges 1932, Bridges and Darby 1933, and Morgan et al. 1913, all spend considerable time explaining selection of milk bottles and other containers for rearing *Drosophila melanogaster.* It must be further noted that changing the search strategy to include "insect rearing" and "cages" boosts the number of papers to 359! What may seemingly be a minor part of insect-rearing systems—containers—must be recognized as one of the most consequential parts of the rearing system/ecology concept. A good container must serve the target insect's needs, including thermal features, humidity accommodation, gas exchange, being hospitable to developmental and reproductive needs, and any other microhabitat factors inherent in the insect's biology. The container must also serve the needs of the rearing personnel, including availability of the containers at a reasonable cost, storage accommodation, ease in access to the insects, and cleaning requirements. The comments of one of the earliest papers on rearing Burgess 1908 clearly depicts the recurring theme and basis of insect rearing: to capture natural conditions as much as possible for successful rearing. Here are Burgess' comments:

> One of the serious problems which it was necessary to solve in order to successfully rear the parasitic and predaceous insects which were being shipped from Europe to prey on the gypsy and brown-tail moths was to secure apparatus by means of which these insects, as well as their hosts, could be successfully reared in large numbers. All of the old style equipment in general use by entomologists for rearing work was tested, but in many cases it was found that radical improvements were necessary in order to accomplish the results desired. It was of primary importance to place the insects under as nearly as possible natural conditions and at the same time to keep them in confinement where they could be studied and observed and not allowed to escape from captivity.

In the last paragraph of the paper, Burgess elegantly stated the concept that the rearing conditions must simulate nature—but with conditions that meet the convenience of the investigator (or the rearing system personnel):

> It is perfectly true that breeding devices must be used which are especially adapted to the habits and character of the insects to be studied, and it may be desirable to modify these cages in such a way as to suit the special needs of the investigator.

From these earliest treatments of containers (or cages or rearing devices) throughout the history of insect rearing, a variety of container-types is remarkable. In the long history of silkworm rearing, it is widely seen that no container per se is used in the larger-scale rearing operations (see Figure A2.6 from Appendix 2). As is the case for many rearing innovations, the rearing system for *Drosophila* spp. was often the earliest model. For example, Bridges 1932 wrote about the early rearing system in the laboratory of T. H. Morgan (a famous facility that came to be known as "The Fly Room"):

> In the early work on the heredity of *Drosophila* the cultures were raised in quart fruit jars, in quart milk bottles, and in a very miscellaneous collection of museum jars and other laboratory glassware. The great difficulty was in getting the flies out of these cultures for etherization and examination. By 1913 pint and half-pint milk bottles had become standard, largely on account of their uniformity of mouths, their heavy, strong glass and the ease with which they could be procured.

Bridges 1932 went on to explain in elaborate detail the other advantages of half-pint milk bottles and the rationale behind the refinements, including the importance of being able to clean and even sterilize the milk bottles to prevent carryover of insects from a previous generation as well as for sanitation. For example, he describes the modification of the standard round milk bottle to a more convenient and efficient square shape:

> In 1916 I made experiments to determine the optimum area of culture surface, and concluded that the pint milk bottle was best for the flies, but the half-pint, which was not far inferior in that respect, was greatly superior in ease of handling, etc. Accordingly in 1917 I drew designs and made a wooden model of a special bottle, having the floor area of the pint bottle and the height (135 mm) and other good features of the half-pint bottle.

Bridges 1932 went into further detail about the optimization of container design, including his developing an agreement with a bottle manufacturing company to produce a special, custom-made square version of the half-pint bottle. Many laboratories continued to use the glass milk bottles until the later 20th century when the convenience and low expense of plastic containers overtook glass bottles in popularity.

After Bridges' 1932 paper, there are very few papers that treat containers as a specific, central topic. However, as is the case with many aspects of rearing systems publications, there is a wealth of information about rearing containers that can be found in publications if they are scrutinized. Several papers that treated container specifications and that were influential or that presented novel containment systems among the larger body of rearing papers are Shorey 1963, Shorey and Hale 1965, Ignoffo and Boening 1970, Cothran and Gyrisco 1966, and Singh and Surrey 1980. The Ignoffo and Boening 1970 paper introduced the multiple cell rearing tray which became a tremendous improvement in mass-rearing containment systems, including the innovation of using the food industry's form-fill-seal technology to make various plastic trays such as the 32-cell tray and the 8-cell tray, which greatly reduce handling of individual cups. Several laboratories invested in the purchase of tray-making form-fill apparatus to ultimately reduce costs of tray purchases (Cohen and Staten 1994; Tillman et al. 1997). As Tillman et al. 1997 describe, the multiple-cell trays required specially adapted lidding materials that consisted of a plastic material with a glue adhesive that was either a self-sticking glue or a heat-activated glue. Such lidding was designed to include pre-perforated zones fitted to each rearing unit, allowing adequate gas exchange while reducing the loss of water from the diet and insects. Of course, these gas exchange demands must be met and balanced with protection from excessive water loss through evaporation from the diets and the insects, and for this appropriate containers must be designed such as those described by Tillman et al. 1997.

Some further issues that are inherently important in container considerations are the questions of sufficiency of gas exchange (O_2 uptake and CO_2 release) and water vapor exchange. In years of research in insect rearing and in exchanges of insect-rearing experts over the past 40 years, I have become convinced that the need for appropriate gas and water vapor exchange is of great importance, yet I have not been able to find published information that directly measures the amount of oxygen and carbon dioxide present in the containers during the rearing process. For example, in Figure 6.2 (also in Figures 6.4 and 6.6) where 8-cell trays (Frontier, New Jersey) are used, the lidding material—a thin plastic sheet—has been perforated to allow gas exchange. Similarly virtually all rearing closed rearing containers have some kind of accommodation for gas exchange, and it is generally assumed by performance of the containment systems that gas exchange is adequate to meet the insects' needs.

6.14 REPRODUCTIVE NEEDS IN REARING SYSTEMS

It is clearly a crucial responsibility of rearing system personnel to provide suitable conditions to allow their target insects to perform the reproductive functions that allow them to sustain continuous generations of offspring. These conditions include mating accommodations, suitable oviposition circumstances, and appropriate hatching and developmental requirements. A search of the keywords, "insect rearing" and "reproduction" yielded 539 publications (from 1979 to 2017) with a large range of topics, many including diet quality/nutritional relationships, microbial interactions, or environmental conditions such as lighting and temperature. However, the focus of this topic in the current review is more confined to conditions that affect mating success and oviposition factors. So, for example, the adult cage configuration that was conducive to mating and supply of liquid diet (sugar or honey solution) was presented by Yamamoto 1969 and Bell and Joachim 1976 for rearing two lepidopteran species (*Manduca sexta* Sphingidae: Lepidoptera and *Pectinophora gossypiella* Gelechiidae: Lepidoptera). These papers, as well as Patana 1977, also provide suggestions for oviposition substrates such as waxed paper or paper toweling that lend themselves to convenient substrates for collecting and processing the eggs laid by the various moth species. Similar requirements are met by use of ice-cream cartons with organdy lidding material for lacewing rearing as described by a number of authors, starting with Ridgway et al. 1970. A unique and highly efficient ovipositional accommodation was designed by Patana 1982 who reported using packets made from Parafilm™ covering filled with a carrageenan gel. The packets were laid on top of wire or organdy mesh cage tops, allowing the plant bugs (*Lygus hesperus* Knight, Miridae: Hemiptera) to oviposit into the packets where the eggs could absorb moisture as they would under conditions where the insects ordinarily would oviposit into plant tissues. This system allows the insect-rearing staff to harvest packets without opening the cages or otherwise disturbing the insects, and the packets were reported to contain more than a thousand eggs per 100 cm^2 surface area. The insect-rearing literature contains hundreds of specialized reproduction/oviposition accommodations, too numerous to describe here but all clearly part of the need to meet on a per case basis various insects' needs.

6.15 MICROBIAL ASPECTS OF REARING SYSTEMS (PATHOGENS, SYMBIONTS, AND CONTAMINANTS)

The topic of microbial aspects of rearing systems involves well over a thousand papers on pathogens, symbionts, and contaminants. The study of microbial relationships in insects in rearing systems has been a topic of research focus since the earliest papers by Bogdanow 1906 and Bogdanow 1908. The scope of papers written about "insects" and "microbes" exceeds 1000, with many more papers written about the overlap of these areas but not located by a search of these keywords. More recent attention, starting with a review of insect gut bacteria not of pathogen status, was written by Dillon and Dillon 2004. Since that review, there were 338 papers located as of June, 2019 with the keywords "insects" and "microbiome." Clearly, the topic of microbial aspects involved in rearing systems is too large for this chapter, but it is intended here that the cursory review of microbial involvement in microbial aspects of rearing systems will provide a useful start for investigators to probe this very rich field in more detail. The topic of microbial aspects of insect rearing had been treated extensively by the late Peter Sikorowski, a pioneer and leader in the field of rearing-based microbiology. And two of the most comprehensive, practical, and in-depth reviews are to be found in the two chapters by Sikorowski 2009a and 2009b.

6.15.1 Pathogens in Rearing Systems

As in nature, pathogens can have a huge influence on insect populations in rearing systems, and the close proximity of insects in captivity and conditions dislocated from natural epidemic control renders rearing systems into a status of ripe target of obligate and facultative pathogens as explained in detail by Inglis and Sikorowski 2009a and 2009b. Inglis and Sikorowski 2009a, for example, explain several mechanisms of pathogen transfer in rearing system, with examples from viral, bacterial, protozoan, and fungal microbes. The *Web of Science* search of "rearing" and "pathogens" revealed 1,662 papers over the past 60 years (as of June 17, 2019). A search of "insect" and "pathogens" revealed nearly 10,000 papers, many of which treated the subject of insects held in rearing conditions.

6.15.2 Symbionts in Rearing Systems

The importance of symbionts in successful rearing programs cannot be overemphasized in terms of fitness or even the survival of colonized insects (Cohen 2015). Depending on the taxa of the target insect species, symbionts may be internal or external. Internal symbionts may be part of the insect's gut microbiome (Dillon and Dillon 2004), or they may be held by specialized structures in the insect's hemocoel in mycetocytes or mycetomes (Douglas and Prosser 1992; Douglas 1998; Sandstrom et al. 2001; Nishide et al. 2017, for a more recent review of this topic). External symbionts are found in wood-consuming beetles such as ambrosia beetles (discussed by Klepzig and Six 2004; Six 2012; Biedermann and Rohlfs 2017) where fungus-bearing external structures known as mycangia are adapted to inoculation of fungi into wood galleries. This topic is extremely important in rearing ever-increasing pest populations of forest insects.

6.15.3 Microbial Contaminants in Rearing Systems

Because of the essential nature of rearing systems where highly nutritional diets are kept for extended periods under environmental conditions that are optimal for microbial growth, contamination becomes one of the most prominent challenges to rearing personnel. These conditions are further made problematic by the density of insects in most rearing systems and the proximity of fecal materials (frass) to the insects. The issue of contamination, therefore, has received attention since the earliest years of rearing efforts. For example, Bridges and Darby 1933 discussed the extensive efforts to manage mold in diets used for *Drosophila* rearing in the Fly Lab of T. H. Morgan, dating back to 1913. The control of contamination has included modification of the media (diets) to make them less susceptible to microbial contaminants by using sanitation (Sikorowski and Lawrence 1994; Inglis and Cohen 2004; Inglis and Sikorowski 2009a, 2009b), pH (Funke 1983; Zha and Cohen 2014; Cohen 2015), water activity or use of humectants (Ludemann et al. 1979; Cohen 2015), or antimicrobial chemicals (Ouye 1962; Vail et al. 1968; Singh and House 1970; Ludemann et al. 1979; Funke 1983; Alverson and Cohen 2002; Inglis and Cohen 2004; Zha and Cohen 2014; Cohen 2015).

6.15.4 The Special Case of *Wolbachia*

A special and highly important case of microbial influences on insects in rearing systems is the complex interaction of the bacterium *Wolbachia* spp. with their insect hosts. The search of keywords, "Wolbachia" and "rearing," revealed 194 papers (as of June, 2019), the earliest ones dating back to the mid-1990s such as Sinkins et al. 1995. The comprehensive review by Werren 1997, though now more than 20 years old, still serves as an excellent base of information and overview of the various biological effects of *Wolbachia* such as cytoplasmic incompatibility, influences on sex ratios, feminization of males, induction of parthenogenesis, and various other fitness parameters related to both field and laboratory interactions. The scope of *Wolbachia* as a "manipulator of invertebrate biology" is further explained by Werren et al. 2008. The older review by Werren 1997 is updated by Rani et al. 2009 and Goettel and Inglis 2006, and most recently by Lindsey and Stouthamer 2017. Also, because *Wolbachia* presents such a mixed case of a beneficial effects vs. potential harmful effects of *Wolbachia*, it is valuable to have information about possible management strategies for this bacterium such as those provided by Kyei-Poku et al. 2003 dealing with heat treatment or antibiotic therapy to eliminate *Wolbachia*.

6.16 GENETICS AND EPIGENETICS

First, it should be understood that although genetics is not a component of the ecology of reared insects per se, it must be considered in the context of domestication and selection (deliberate or inadvertent) for traits that relate to ecological parameters such as responses to temperature, light conditions, or nutrients. A search of "insect rearing" and "genetics" revealed 135 results dating from 1976 to 2019; however, this greatly understates the number of papers that deal with various rearing conditions as they pertain to the genetics of the target insects. There are countless "cryptic" implications of how features such as variations in diet components can influence genetic expression of phenotypic characters. Clearly there are thousand more papers on genetics of reared insects dating back to the late 19th or early 20th centuries, especially if *Drosophila* genetics are considered in relationship to various rearing conditions such as temperature effects, diet, or countless other parameters. For example, the basic studies of Morgan et al. 1913 put the foundation of *Drosophila* genetics in the context of controlled rearing conditions into a systematic framework as do countless papers on genetics of colonized insects. This raises an important point emphasized throughout the writing of the late Dr. A. C. Bartlett who was a leading researcher and scholar on the topic of insect domestication and its results (Bartlett 1984, 1994). Some more recent papers such as Thomas 1993 treat the specific interactions between controlled rearing conditions and biological or fitness outcomes in rearing systems for screwworms, and the studies of Bush and Neck 1976 give valuable perspectives of the ecological aspects of genetics in mass-rearing settings. Studies of these kinds of interactions are powerful reminders of the importance of seemingly minor differences in rearing conditions in the biological fitness outcomes and realization of genetic potential of reared insects. Benedict et al. 2009 cover this topic in a systematic review of rearing conditions and mosquito production and quality. The topic of genetics in rearing systems and the interplay genetic and domestication factors is too large to be properly treated in a general review such as the current paper.

The concept of transgenerational effects (epigenetics) in insect-rearing systems is both very important and poorly studied thus far. The effects that all environmental factors may have on our insects' epigenome can translate to the possibility that any and all harmful or stressful events can be recorded by the insects in our colonies! This fact translates to the haunting possibility that every stressful event in our colony's history can make an indelible change in how our insects respond to their current environment. Considering the above paragraph where "cryptic" selection of genetic characteristics can indelibly change the insects we are colonizing, but there are also possible indelible changes to the epigene level of organization that may slant or veer our insects to show phenotypic responses their environment. This topic is discussed further in Chapter 10 on stress in rearing systems.

6.17 CONCLUSIONS

An important point made by Cohen 2015 about rearing systems is that there is a reciprocal relationship between what we need to know to develop a rearing system and what we can learn about our insects once we have an operating rearing system. This point emphasizes the concept of control as a fundamental need in developing a successful rearing system and the fact that controls in experiments with target insects can yield a level of understanding that could not be developed for insects in the field.

In Figure 6.12, the important advancements in insect rearing are noted, including the origins of the concept of control, the introduction of gelling agents, yeast supplements, and antimicrobial agents. Note that silkworm rearing advancements are not noted in this figure. However, the monograph by Henrietta Aiken Kelly 1903 remains one of the most detailed prescriptions for an insect-rearing system. It is apparent that silkworm rearing has been one of the major influences on most aspects of insect rearing as an artificial provision of controlled conditions behind rearing operations.

In a collection of essays on the topic of relationships between ecology and evolution, Hutchinson 1965 explains and richly illustrates how environmental forces interplay with and shape genetic and ultimately evolutionary outcomes in nature. Rearing systems can be regarded as the "ecological theater" as a shaping force for the insects that we rear for various purposes. The rearing specialist who fails to consider the nature and power of the shaping forces in the rearing constraints that we place on our insects subjects is apt to lose control over the products of the rearing system. In setting up an insect-rearing system, entomologists select a highly limited population—a subsample of a much larger population (Mackauer 1972; Bartlett 1984, 1994), violating a major premise of the Hardy-Weinberg Equilibrium Principle, and we subject the insects to intense, albeit often unrecognized selection pressures. The selection parameters and profiles were reviewed in this chapter where temperature, humidity, container configuration, light intensity, diets, and countless other factors were

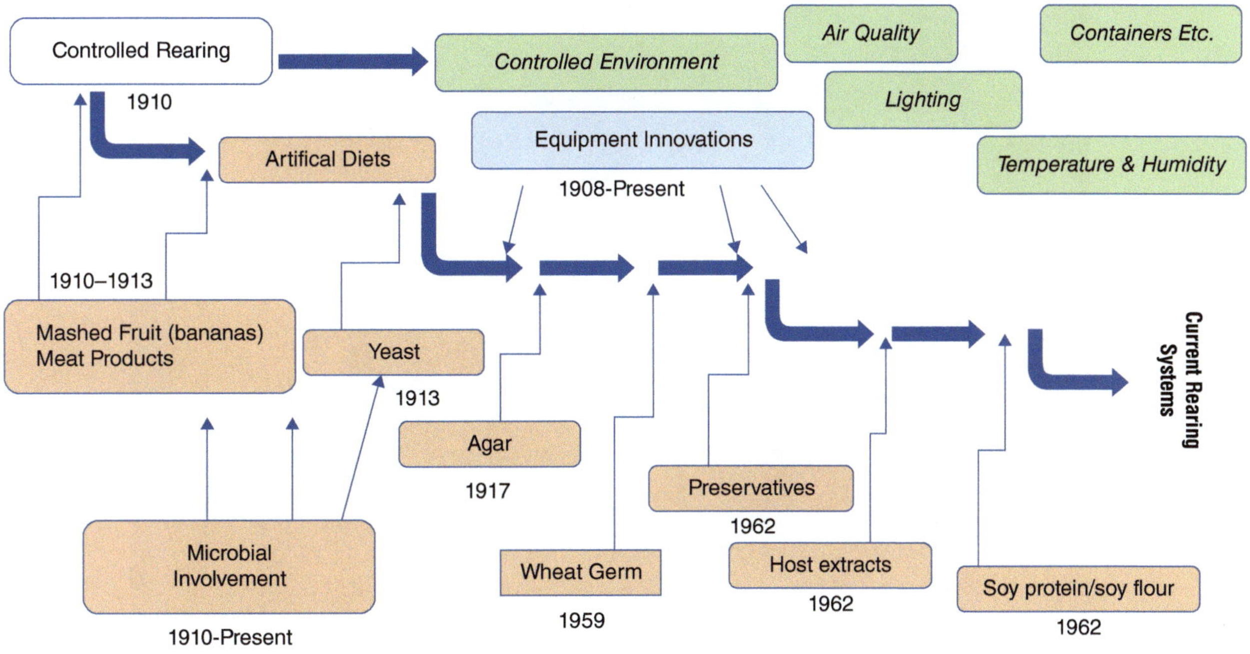

Figure 6.12 Summary of historical events in rearing systems.

imposed upon the insects that are being produced. There is often, in rearing systems, no attention to densities of insects and how density-dependent pressures may affect the outcome of our rearing efforts. An example of the potential importance of the density of insects is currently being investigated in the Cohen laboratory at North Carolina State University where we have discovered that wax moth larvae (*Galleria mellonella* L., Pyralidae: Lepidoptera) for dense aggregations and complex networks of silken tunnels. We discovered that one of the characteristics or consequences of the aggregations was an impressive thermogenesis that raised the temperature of the aggregations by an excess of 10°C (Figure 6.13).

Figures 6.13 (a) and 6.13 (b) show wax worm larvae in glass (Mason) jars. Note that the temperature on the outside of the smaller jar was 12.8°C warmer than the ambient, and note further that the temperature was measured on the outside of the glass, indicating that the temperature inside the jar, in the core of the wax worm aggregation, was even higher. We are discovering that as a result of the thermal aggregations, wax worm larvae are able to complete development much faster than under conditions where thermal aggregations are less dense. We are also exploring other aspects of the thermal aggregations, including various metabolic consequences and implications for microbe/insect interactions.

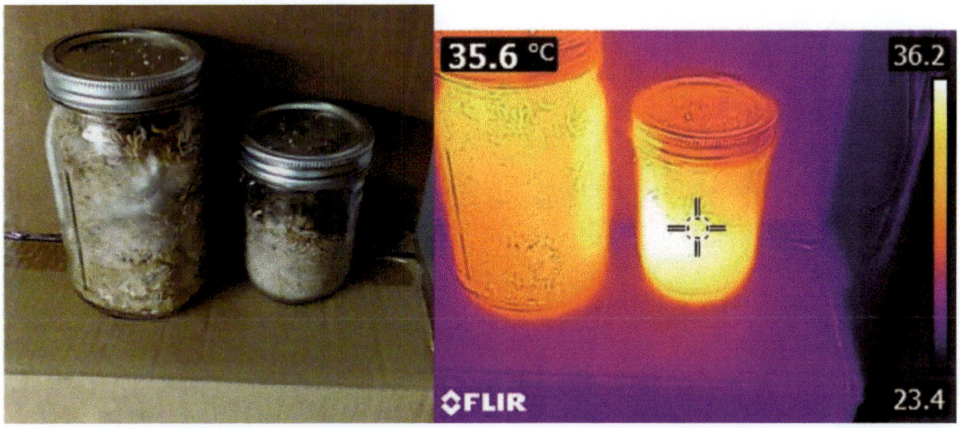

Figure 6.13 (a) Light image and (b) infrared image.

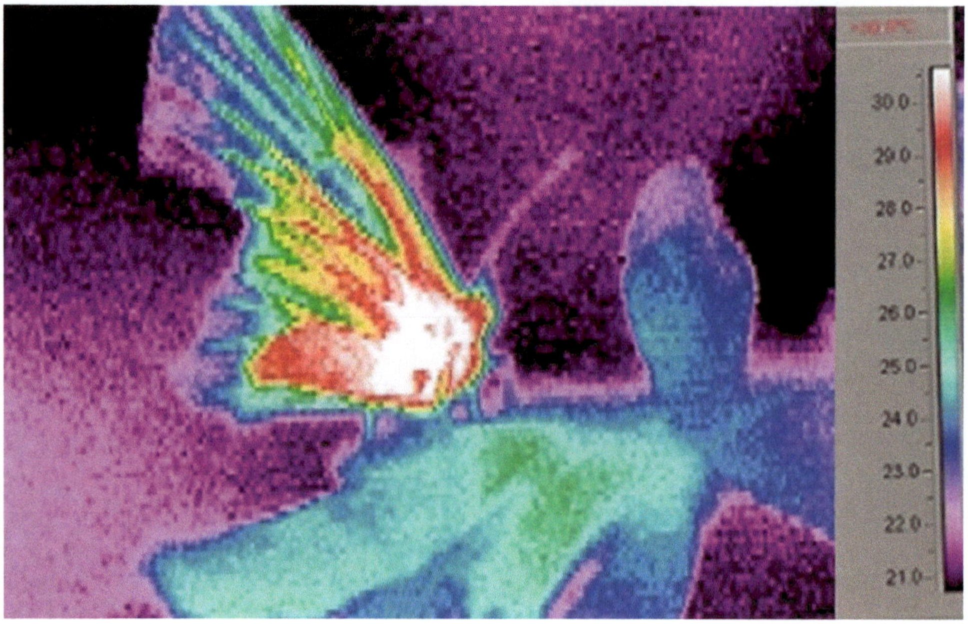

Figure 6.14 Thermal image of moth.

Finally, it was the intention of this chapter to provide a sample of the complex ecological conditions that are present in rearing systems, and it is hoped that this albeit cursory summary of the various conditions will enhance the realization of rearing specialists and the community they serve that the insect-products of rearing systems have been subjected to stresses and selective pressures that may lead to deviation from the expected outcomes. It is therefore incumbent as the responsibility of rearing personnel and rearing system managers to be ever-vigilant to recognize and reduce deviations from a standard of quality in the insect products. This is the subject of the chapters on process control (PC) and quality control (QC) in this book (Figure 6.14).

The moth image in Figure 6.14 was made in early morning in an alfalfa field in early spring near Tucson, Arizona. The moth is resting on alfalfa leaves whose temperature was between 23.5°C and 26.5°C. The background (ambient temperature was 21–22°C, and the core temperature of the moth was 30.5°C, nearly 10°C above the ambient air and more than 5°C above the leaves where it was sitting. Note the gradation of temperatures as we look at the more distal parts of the insect's body. This is an example of a facet of the moth's niche in terms of the temperature dynamics.

CHAPTER 7

Designing Error (Variability) Reduction in Rearing Systems: Reducing Uncertainty

7.1 INTRODUCTION TO ERROR ANALYSIS

This chapter begins with a paragraph from J. R. Taylor's *An Introduction to Error Analysis* (1982):

> Error analysis is the study and evaluation of uncertainty in measurement. Experience has shown that no measurement, however carefully made, can be completely free of uncertainties. Because the whole structure and application of science depends on measurements, the ability to evaluate these uncertainties and keep them to a minimum is crucially important.

It is of paramount importance to recognize that ***error***, ***uncertainty***, and ***variability*** are essentially equivalent (at least for our purposes). "Error," in this statistical sense, is interchangeable with "uncertainty" or "variability." The word "error" does not mean a mistake as the word means in a conversational, informal sense. It is also my aim to raise the readers' understanding of how uncertainty (error or variability) is an enemy of well-controlled rearing systems that give us dependable products (quality reared insects in suitable numbers) that serve their purpose in the best possible way (Pugh and Winslow 1966 and Harrison 2004). This chapter attempts to give rearing specialists a powerful set of tools to help them produce insects that are the best possible components in experiments or applied uses such as biological control agents, subjects in sterile insect technique (SIT) or genetic pest management systems, as feed for other species, or any of the other functions there are for reared insects. Nested in this system of error exploration, are several approaches to recognizing and then alleviating (or reducing) the sources of uncertainty in insect production. Finally, the intention here is to guide the reader through a planning strategy to examine the rearing processes in light of the tenets of statistical process control (PC). This approach will help rearing specialists employ a continuous improvement strategy to produce insects that adhere to quality control (QC) standards.

7.2 THE NATURE OF REARING SYSTEM PROCESSES

One of the fundamental characteristics of rearing systems is that they are process-based: they incorporate multiple processes such as diet preparation, inoculation and harvesting of insects, maintaining environmental components, and every other aspect of our rearing procedures. The fact that processes dominate our rearing operations means that we are obliged to develop PC techniques. The rationale for using PC procedures are that they:

1. Reduce variability, uncertainty, or error
2. Reduce loses or risk of quality (fitness) deterioration
3. Lead to recognition of what needs further study
4. Reduce costs of rearing
5. Provide formalized structure for basis of changes in processes needed to improve quality, production efficiency and reliability

The sense of **variability** here is what Taylor 1982 describes as **error** or **uncertainty**. Many authors (e.g Anonymous 2019c) describe error in experiments as falling into the following general cases:

1. Gross error (also called "operator error" or "human error", or "blunder"): You misread the instrument or recorded the wrong value; someone bumped the table or opened a door during weighing
2. Systematic error: Your balance always reads 0.002 g above zero (can be predicted and corrected)
3. Random error: You get unpredictable variation besides your systematic error. Caused by temperature changes, irregular vibrations, electronic noise in instruments, etc.

At the heart of variability in systems, including insect-rearing systems, is normal distribution of values that describe rearing factors. The two bell-shaped curves (Gaussian distributions) in Figure 7.1 illustrate the point. The outer curve and the inner (narrower) curve share the

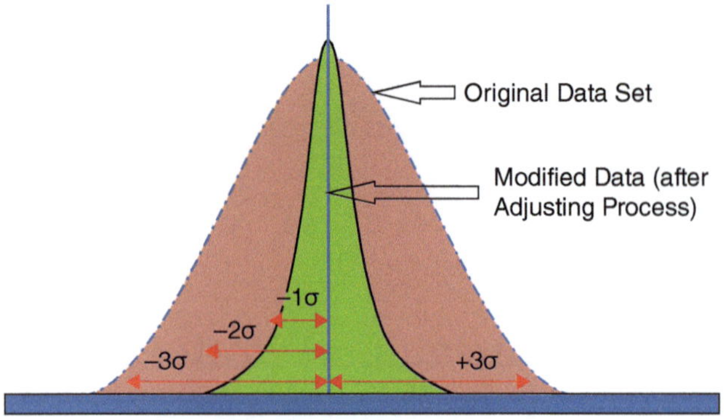

Figure 7.1 Standard bell-shaped curves of normal or Gaussian distributions.

same mean (average) for these distributions, but clearly the outer curve shows a greater variability around the mean than does the inner (green) curve. The statistical description of the variation, the Greek letter σ (sigma), represents the standard deviation of the data around the mean; so for the larger curve, the sigma values represent 68% of the data for +1σ and −1σ, 95% of the data for +2σ and −2σ, and 99.7% of the data for +3σ and −3σ. Clearly, the narrower curve has much lower sigma values, and for our purposes, this would represent a "tighter," more reliable process or rearing system component. An important aspect of error in a designed and controlled system is the concept of the signal being the purpose of the system (such as rearing healthy, reliably available insects) and the noise (the error or variations in the system that detract from obtaining the desired output (Joseph and Wu 2002). This concept will be explored in this chapter and in several other places in this book.

7.3 MEASURING AND EXPRESSING VARIATION

The variation is expressed as positive or negative σ (sigma) or standard deviation of the mean (represented by the central line for both curves). This figure represents standard distributions of a variety of rearing system parameters (including environmental factors such as temperature, humidity, or light intensity) diet factors (including protein content, pH, or gel strength), and insect characteristics that can be measured as continuous distributions (weight or development period, for example). A major goal of rearing personnel is to modify rearing processes so that the variation (error) found initially as the outer bell-shaped curve is narrowed to follow the scope of the inner curve. Such a change would be quantifiable in terms of lower standard deviation values.

Table 7.1 **Moisture Content of Wheat Germ Stored in Containers with and without Silica Gel (Desiccant) Packets**

Day of Test	Without Silica Gel	With Silica Gel
1	4.1	4.1
2	4.8	4.1
3	5.3	4.2
4	5.7	4.2
5	6.2	4.3
6	6.2	4.3
7	6.9	4.3
8	7.2	4.4
9	7.8	4.5
10	8.2	4.5
Mean	6.28	4.3
Variance	1.577	0.024
Standard deviation	1.253	0.155

The package of wheat germ was opened and measured on day 1, then divided into two 100 g aliquots and each placed in a plastic container with or without packets of silica gel (a desiccant). The containers were kept on the laboratory shelf and held at about 20°C with humidity that varied from 50% to 80% over a period of 10 weeks. Containers were opened at one-week intervals to sample the wheat bran for moisture analysis.

Examining Figure 7.1 and Table 7.1 reveals that the distribution for the original data set (in this case representing the force required to penetrate a diet surface with a force gauge) contains much more variation (larger deviation) than the modified data set. For example, the improvement (reduction in variability (error or uncertainty) resulted from developing standards of diet heating procedures where each batch of diet was made with more careful attention to the temperature of agar activation and the duration of the heating process. Therefore, even though the means are identical, the variations around those means are much less in the modified diet-making process. This type of change can constantly be made toward a program of continuous improvement, which involves a deeper understanding of the factors that cause variability. The specific statistical characteristics of the changes are evident in the numerical treatment of this hypothetical situation presented in Table 7.1. Note that all the variability parameters (standard deviation of the population being estimated, standard deviation of the sample, and variability) are all considerably lower in the After Adjustment scenario than they are in the Before Adjustment data set.

It is worth noticing that there are several online discussions of error, often provided for students in physics and chemistry classes. One excellent source of discussion of diagnosis and treatment of error is from a freshman chemistry tutorial by Gerroranal 2020, where the author provides a diagram that illustrates the distinctions between precision and accuracy (Figure 7.2).

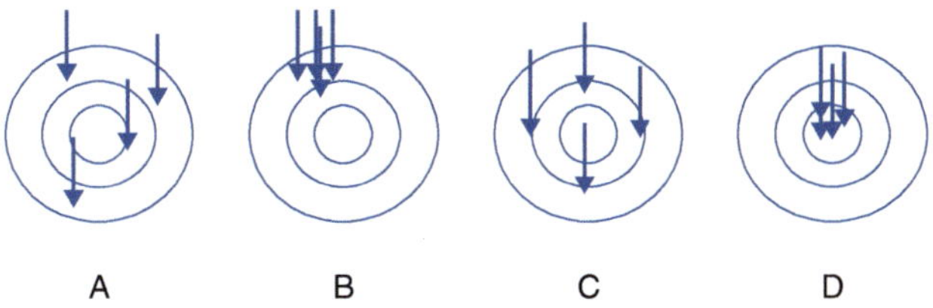

Figure 7.2 **"Accuracy and precision"** is redrawn from Cohen 2015 and is discussed by Gerroranal 2020.

7.4 ACCURACY AND PRECISION EXPLAINED

In Target A, there is neither precision nor accuracy. In Target B, the arrows are close together, but far off the bullseye, so the shooter is precise but not accurate. In Target C, the average of the arrows' position is well-within the bullseye, so they are accurate, but they lack precision. In Target D, the arrows are close together (precise) and they are on the mark of the bullseye, so they are accurate. In rearing systems as in other production systems, the goal is highest possible accuracy and precision. The job of the rearing team is to discover and correct all possible deviations from the highest possible combination of accuracy and precision. In fact, in every step of the rearing process, we must be attuned to every possible way that we can reduce error (variability or uncertainty); which is another way of saying that we are striving for accuracy and precision.

7.5 SOURCES OF ERROR, VARIABILITY, AND UNCERTAINTY

Error can come from every part or component of rearing systems, raw materials, the target insects, fluctuations of environmental factors, personnel, and contaminants, to name a few sources. As we work through the sources of error, we should constantly keep in mind the diagram in Figure 7.3, which is the centerpiece of Montgomery's 1991 and later works on statistical process control. **It is not an overstatement to assert that the single, most important feature of rearing personnel's job description is to increase PC and quality by reducing error, i.e.,** ***to refine processes to make them less variable and to convert, whenever possible, uncontrolled process inputs into controlled inputs.***

7.5.1 Raw Materials in Rearing Systems

Among the most difficult sources of error in rearing systems are the raw materials as they come from suppliers.

The diagram in Figure 7.3 is adopted and modified from Montgomery 2013. Three important statements from Montgomery are: (1) "Quality means fitness for use." (2) "Quality is inversely proportional to variability." (3) "Quality improvement is the reduction of variability in processes and products." These three principles, though simple-sounding, are the basis for QC and PC in all production systems, including insect rearing (Baird 1962 and Montgomery 2013). The topics of QC and PC are really at

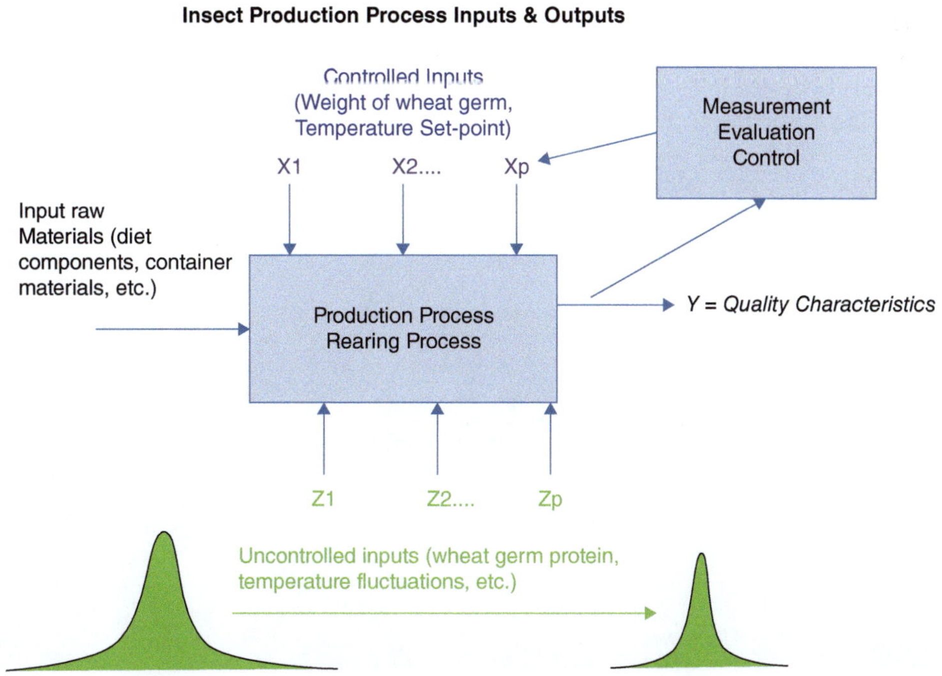

Figure 7.3 Diagram for developing control systems for insect-rearing processes.

the heart of this book, and they are discussed in special detail in Chapters 7, 8, 9, and 10 (error and variability; stress and quality; design of experiments; and working process control: design and operations).

7.6 AN IMPORTANT FORM OF GROSS ERROR (OPERATOR ERROR)

A crucial source of error in rearing systems is inherent in the differences between how personnel perform their functions. In a sense, this form of error is a result of differences between individuals—their personal or physical capabilities (visual acuity, coordination, color perception, training, and even attitude). One rather famous account of the discoveries of mutant *Drosophila melanogaster* eye colors in the laboratory of T. H. Morgan at Columbia University involves one of the students Calvin Bridges extreme visual acuity and color perception. It is said by Bridges' fellow students that he recognized the mutants that the others with less color perception ability missed in examining the miniscule eyes of these tiny flies. Bridges discovered the vermillion eye color mutant, differentiating it from the red (wild) eye; and he noticed the white-eyed mutant just before a bottle containing this important fly was to be discarded. Because the human involvement in experiments and other facets of rearing can be so instrumental in generating error, it is valuable to consider a situation where the comparisons of measurements of two operators can be used to evaluate this type of error.

The measurements in this discussion are tests of gel strength, which were performed by the author and a post-doctoral associate (Dr. Molly Robinett), and the full narrative of these experiments is provided in Chapter 9 in a tutorial on using **mixture design** to optimize responses in a diet texture (rheology) experiment. Briefly, the experiment treated 12 diets, which were identical except for the relative proportions of agar, cellulose, and starch. Once the diets were prepared, they were poured into cups as seen in Figure 7.4. The diets were then tested for gel strength with a fruit pressure tester (to the right in Figure 7.4), and the amount of force required to penetrate the surface was reported as an arbitrary (unitless) value. Each diet was tested six times by each operator (AC and MR). The record of the tests is presented in Table 7.1. The table is divided into diet types (Diet No. 1–Diet No. 12) in the left hand column, with the penetration force recorded in succeeding columns.

On the left, in Figure 7.4, the diets can be seen with penetrometer probe holes, and on the right, the Wagner tester on its test stand is visible. The test stand was used to help standardize operators' measurements to reduce operator error that would occur if the meter were used without the test stand. This is an excellent example of the types of simple changes that can be made in rearing system production processes and in their measurement that can help reduce error in the overall rearing system (Table 7.2).

We performed analytical statistics (ANOVA and pairwise comparisons) to determine thecharacteristics of the different diet. The graphs in Figure 7.5 show AC's and MR's test results.

Both sets of graphs (from Figure 7.5(a) and (b)) show the means and ranges of the penetration tests done by AC (a) and MR (b). Note the similarities in the data for each diet, indicating that two independent operators can get similar results from the penetrometer tests. It is especially impressive that this was the first time that Operator MR had used the Wagner penetrometer. It is of further interest that when comparisons of standard deviation between the two operators were made, MR had higher standard deviations for six diets, and lower standard deviations for the other six diets. This comparison translates to similar precision by both operators. Furthermore, the similar means for each diet measured by the two operators indicate similar accuracies for all the measurements reported here. The comparisons of

Figure 7.4 (a) and (b) Diets prepared according to a "mixture design" from JMP design of experiments (SAS, JMP Version 13).

Table 7.2 Penetrometer Readings for 12 Diets

Diet No.	AC1	AC2	AC3	AC4	AC5	AC6	MR1	MR2	MR3	MR4	MR5	MR6
1	225	235	235	240	235	240	205	215	205	220	215	215
2	220	215	210	220	215	205	195	215	200	215	215	200
3	230	255	255	235	240	215	245	225	240	240	235	240
4	150	155	145	145	155	150	150	140	140	155	170	155
5	255	270	270	260	260	280	195	260	245	275	245	250
6	280	285	285	285	285	280	265	275	265	265	255	270
7	175	180	190	185	190	195	175	170	180	185	185	200
8	100	105	95	100	95	100	90	95	90	90	95	90
9	220	215	220	220	220	215	220	220	215	215	225	225
10	100	100	100	110	115	110	105	110	110	110	100	110
11	135	125	125	135	125	130	135	140	145	140	140	145
12	95	90	105	110	110	110	105	110	100	110	110	110

The results of the penetrometer readings are presented as arbitray values, with the first six readings in each row representing the first operator to do the tests (AC) and the second six readings for each diet by the second operator (MR).

the means for tests of gel strength by the two operatiors (MR and AC) are shown in Table 7.3 (a and b).

7.6.1 Interpretation

The low **within diet** standard deviations for the 12 diets measured by the two operators (Table 7.2) indicate that: (1) there is not a large variability of texture from one part of a given diet to another; (2) the instrument (the Wagner penetrometer) has an inherently high precision or repeatability factor; and (3) the operators were careful in using the instrument to make highly precise measurements.

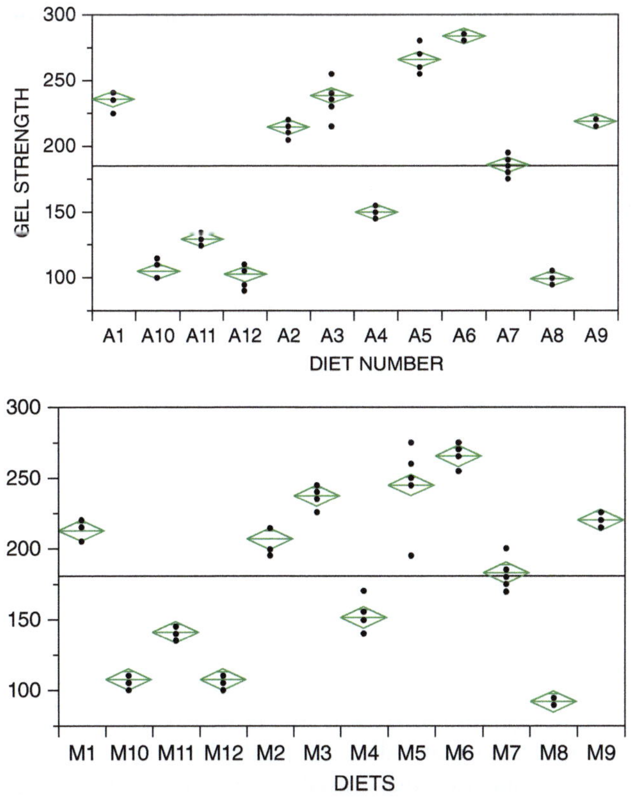

Figure 7.5 (a) and (b) Comparison of gel strength measurements performed by two operators.

Table 7.3 (a) and (b) Comparison of Means of Gel Strength Tests Performed by Two Operators

Level									Mean
A6	A								283
A5		B							266
A3			C						238
A1			C						235
A9				D					218
A2				D					214
A7					E				186
A4						F			150
A11							G		129
A10								H	106
A12								H	103
A8								H	99
M6	A								266
M5		B							245
M3		B							237
M9			C						220
M1			C	D					212
M2				D					206
M7				E					183
M4					F				151
M11					F				140
M10						G			108
M12						G			108
M8							H		92

Note: Levels not connected by same letter are significantly different.

Table 7.3 (a and b) shows the statistical comparisons that were made after the ANOVA test. This test was made with the same set of data that was used for the Mixture Design, and it represents the conventional "confirmatory" statistics, which allow us to evaluate data after the experiment. The Mixture Design is an exploratory approach. Please note that the one-way ANOVA and comparisons of means allows us to determine the diet that gives us the greatest, least, and in-between gel strengths (or weights of pupae, numbers of eggs from adults from each diet, etc.), but it does NOT tell us anything about how cellulose and starch or agar and cellulose interact. The Mixture Design **DOES** let us determine the functions on an **interaction** basis, which we could not do with conventional, confirmatory statistics.

7.7 CASE STUDY OF DEVELOPMENT AND OPTIMIZATION OF *VANESSA CARDUI* REARING SYSTEM

7.7.1 Introduction, Purpose and Background for This Tutorial

This section of the chapter on error, variability, and unpredictability is intended to provide a detailed guide to illustrate how a rearing algorithm/protocol can be coupled with design of experiments for developing and optimizing a complete rearing system for a model insect (the painted lady butterfly, *Vanessa cardui*). The goal is to provide the details of a case study of a rearing system development for an insect that is readily available and that presents fairly few challenges for successful rearing. I selected painted lady butterflies as the subject of this case study. This case study was designed to serve as a practical tutorial for students of insect rearing to provide a user-friendly insect and rearing system. I chose painted ladies because: (1) they are readily available worldwide, (2) I have had considerable experience with rearing them, (3) their development period is fairly short, (4) their relatively large size makes them easy to handle (compared with something like *Drosophila melanogaster*), (5) they are amenable to rear collectively (i.e., they do not display cannibalism as do *Heliothis* species or *Helicoverpa* species), and (6) they are docile, esthetic, and charismatic to have in culture. The downside of *V. cardui* is that while adults will oviposit on several species of plant leaves, they require live plants or freshly cut leaves that can be held in floral vials to keep them fresh for a few days.

7.7.2 Background for Using This Case Study/Tutorial as a Model for Rearing Systems for Other Species

The principles and guidelines used in this case study have broad applications to other insects. The goal of this case study was to provide a model for insectary personnel to follow for the logistics of an experimental design that was complex and required multiple factors to be included. Such features are suggested as mixing the base diet ingredients in a uniform way for those components that would not be varied from one treatment (set of factors) to another. The counterpart logistic approach of this model was to pre-mix the "factors" components, thus reducing the changes of operator error entering the system. A further intention of this case study/tutorial is to demonstrate the stepwise, algorithmic nature of diet development and diet optimization. Finally, what I most wanted to accomplish here was to provide a step-by-step, process-by-process guide to discovery of the sources of variation/error/uncertainty in the rearing system as a whole.

7.7.3 Background for Painted Lady Butterflies

Painted lady butterflies (Lepidoptera: Nymphalidae, *Vanessa cardui* L.) can be minor pests of soybeans and a few other crops (Pedigo 1994); they are important as educational and research subjects as well as being of commercial value for releases during celebrations (https://www.clearwaterbutterfly.com/). Despite their importance in these functions, there seems to be little reported documentation of comprehensive rearing protocols for *V. cardui*. Yet, there are dozens of papers published on various aspects of the biology of painted lady butterflies where an artificial diet was used as part of a rearing system (e.g., Hainsworth and Hamill 1993; Kelly and Debinski 1999; Kollisch et al. 1999, 2003; Kwauk et al. 2014). In the cited references and several others, the diet used to rear *V. cardui* larvae was a proprietary diet offered by the suppliers of the butterflies such as Carolina Biologicals (Burlington, NC) and Wards Science (https://www.wardsci.com). These and other suppliers of *V. cardui* offer instructions and materials for rearing the butterflies, including containers for larvae and adults, handling materials such as paint brushes to gently move larvae, and other components including diets. However, the rearing instructions from suppliers are not intended for production of successive generations of the butterflies, and therefore these instructions do not constitute complete rearing system. In the thesis work of one of my graduate students (Chen Zha) who wanted to use *V. cardui* as a research investigation of the effects of antimicrobial agents in the diet (Zha and Cohen 2014), we had to adopt and modify an existing diet for tobacco hornworms (Yamamoto 1969) to allow us to synthesize a diet and modify the amounts of antifungal agents that we were planning to study. In earlier work in my laboratory, I developed a diet that I called the "YC" Diet, the name indicating that I used the Yamamoto 1969 wheat germ/torula-based diet, which I modified using a trial and error approach. A further-modified form of this diet is presented in Table 7.1.

7.7.3.1 *Vanessa* Rearing System Optimization Logistics

1. Introduction to tutorial on rearing system optimization
2. Starting the colony: eggs
3. Diet (see point 13)
4. Containers
5. Environment
6. Larval growth and development
7. Pupal/chrysalis management
8. Adult emergence conditions
9. Adult feeding
10. Oviposition
11. Egg harvesting/egg management
12. Next generation/starting new colonies
13. Diet optimization

7.7.3.1.1 Introduction to Tutorial on Rearing System Optimization

This portion of the "error chapter" is a tutorial based on an actual case study conducted in the Insect Rearing Education and Research Program at North Carolina State University. In this tutorial, I explain and illustrate how laboratory personnel can proceed to develop and/or optimize a rearing system, using a design of experiments approach to evaluate and reduce error in any rearing system. I selected painted lady butterflies as the model species because of this insect's "user friendly" characteristics in an artificial diet-based rearing system as well as their availability on a cosmopolitan basis. I deliberately and pointedly conducted experiments on painted lady butterflies as part of the development of this chapter so that the narrative would reflect real-world problems in the rearing setting. As a consequence of this approach, I hope to provide a useful "how to-" approach that can be followed and applied by entry level rearing personnel as well as rearing professionals.

7.7.3.1.1.1 Developing and Optimizing a Rearing System: A Case Study with Painted Lady Butterflies — Although they can be minor pests of soybeans and a few other crops (Pedigo 1994), painted lady butterflies (*Vanessa cardui* L. Nymphalidae: Lepidoptera) are important as educational and research subjects as well as being of commercial value for releases during celebrations (https://www.clearwaterbutterfly.com/). They are also widely used as research subjects for a variety of physiological, behavioral, and ecological studies (e.g., Hainsworth and Hamill 1993 studied foraging rules for nectar using *V. cardui*; Herman and Dallmann 1981 used them for studying endocrine biology; Kelly and Debinski used *V. cardui* in studies of effects of food limitation; Kollisch et al. 1999, 2003 used these butterflies to study adipokinetic hormone; Kwauk et al. 2014 used painted lady adults to study morphological dynamics of fluid uptake in the proboscis; Stalter et al. 2000 used painted lady butterfly larvae as models for inquiry into the toxicity of Bacillus thuringiensis; Tilden 1962 studied the ecological aspects of *V. cardui* population movements; and Zha and Cohen 2014 used painted lady in a study of the effects of antimicrobial compounds on larval digestive physiology and nutritional indices). For all these studies, the condition or fitness of the insects is important as a determining factor in painted ladies as an experimental subject. It should be obvious to all researchers in insect biology that the quality of their study and the meaningfulness of their results depends greatly on the biological condition of the experimental subjects.

Therefore, it is important to have a convenient and efficient rearing system for this insect. However, no reports of such a rearing system exist in the literature. This is not to say that there are no existing rearing protocols for *V. cardui*; there are dozens of papers published on various aspects of the biology of painted lady butterflies where an artificial diet was used as part of a rearing system (e.g., list some papers).

7.7.3.1.2 Starting the Colony

In this tutorial I emphasize "***logistics***" in recognition of the need to have well-designed strategies for every aspect of the rearing process. I encourage, in this chapter and throughout this book, thinking of all rearing processes in terms of **algorithms**. For this case study, we can consider as an algorithm any stepwise approach that develops a set of rules for solving problems in our rearing system. For example, once we decide on the species of insect that we are planning to rear, we are faced with the problem (or issue) of establishing a source of those insects. Our problem here is to obtain a starter population of painted lady butterflies, specifically and initially, which life stage should acquire. Are we going to get our insects from the field or from an established colony? There are several practical questions that must be asked including ones about the practicality of the means of acquisition. In the context of the practical decision, a driving force must be the **purpose** of the colony. Clearly, a colony that is to be used for host plant resistance research or microbial ecology (e.g., microbiome analysis) research will have different criteria than a colony to be used for education about insect life stages.

Other practical questions about the source of starter insects include risks of introducing pathogens from addition of wild-collected insects, genetic truncation that accompanies insects from long-established colonies, and the relative advantages of using established colonies as a starter source. In this case study, I elected starting the painted lady colony from a commercial source, Carolina Biological, basing this selection on previous work on this insect in my laboratory. This selection meant that I would have a convenient source of the target insect as eggs that were provided to my laboratory. The eggs were provided in small vials shipped by one-day delivery and hatched within two days post-arrival at my lab. I started the colony with a diet modified slightly from Zha and Cohen 2014, and after about three weeks of larval growth and pupal development, I had a set of about 50 adult *V. cardui* to use for the development of a rearing process described in this chapter.

7.7.3.1.3 The Diet

The major emphasis in this case study is on diet manipulation and optimization. The diet used in these experiments was derived from a formulation reported in Zha and Cohen 2014. The basic diet formulation is presented in Table 7.4, and the rationale for the experimental manipulations is explained in the narrative on the design of experiments protocol.

7.7.3.1.4 Containers

Because larval *V. cardui* are fairly gregarious, they can be reared communally in groups of five or more (Figures 7.6–7.12). We used four kinds of containers for communal rearing, but these different containers were not compared with one-another for rearing efficacy. This leads to the issue that is central to this book on design, operation, and control: how can we use data-based procedures to vet or validate each part of the standard operating procedures that we develop and continuously improve? For the purposes of this case study, I chose to confine attention to the diet optimization. However, in future studies, the issue of containerization should be part of the test-based optimization procedures.

For preparation of the containers shown in Figure 7.6, a dissecting needle was used to make the holes. In this setup, ten eggs were placed in the container with ca. 2 × 3 cm sections of diet. As larvae developed past 2nd and 3rd instar, the diet was placed on top of a mesh

Table 7.4 Basic Diet Formulation

Modified "YC" Diet for Painted Lady Larvae	Weight (g)	High Amounts	Low End of Weight Range		High End of Weight Range	
Wheat germ	20	20	19.4		20.6	
Alfalfa powder	9	9	8.73		9.27	
Sucrose	5	5	4.85		5.15	
Torula yeast	6	6	5.82		6.18	
Vanderzant vitamins	2	2	1.94		2.06	
Beck salts	1	1	0.97		1.03	
Ascorbic acid	1	1	0.97		1.03	
K sorbate	0.5	0.5	0.485		0.515	
Methyl paraben	0.5	0.5	0.485		0.515	
Na propionate	0.5	0.5	0.485		0.515	
Agar	7	7	6.79		7.21	
Water	260	260	252		268	
			Range for low amounts		Range for high amounts	
			Low	High	Low	High
Sweet potato flour (LO/HI)	5 (+/–0.15)	10 (+/–0.3)	4.85	5.15	9.7	10.3
Rice bran extract (LO/HI)	3 (+/–0.09)	6 (+/–0.18)	2.91	3.09	5.82	6.18
Citric acid (LO/HI)	1 (+/–0.03)	2 (+/–0.06)	0.97	1.03	1.94	2.06
Casein or whey (Biozzz) or soy protein (note: there are three alternative components here)	9 (+/–0.27)	9 (+/–0.27)	8.73		9.27	

Note: That the "base diet" components are presented in unshaded, and the variable components are presented in shaded. In Figure 7.7, the diet formulations suggested by Custom Design from JMP are presented. The design is a full-factorial type where all combinations/all "levels" of sweet potato flour, rice bran extract, citric acid are specified, and the three protein sources (casein, soy, and whey) are also presented. Note that the components in unshaded are the ones whose amounts are fixed values for the entire experiment. The components in shaded are the variables (called "factors" in the JMP format) in this experiment. Note further that the table is sub-divided to guide the technician to the acceptable limits discussed in the following paragraph. Therefore, the target weight for wheat germ is 20 g. The allowable boundaries dictated by the "3% rule" are a low of 19.4 g and a high of 20.6 g. Note that for the smaller amounts such as 1 g of Beck salts, the balance used must be capable of reading to the nearest hundredths of a gram in contrast with the larger amounts such as wheat germ, which can be read on a balance that reads to tenths of a gram. For the set weights below one gram, the balance must be capable of being read to thousandths of a gram in order to assure weights that are within 3% of the target amount.

Figure 7.6 This container is a plastic Ziplock storage box, with 20 holes punched into the lid for air exchange.

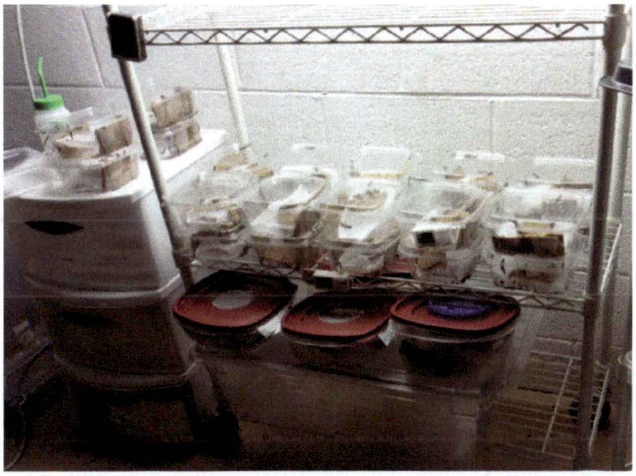

Figure 7.7 The distribution of the 24 test diets in 24 test containers.

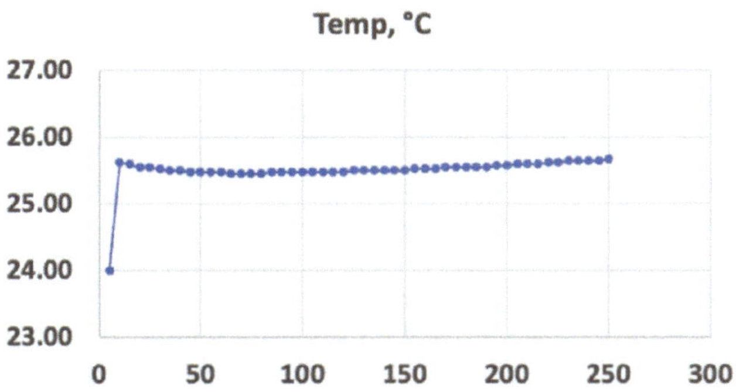

Figure 7.8 Temperature over a four-hour interval in painted lady rearing room.

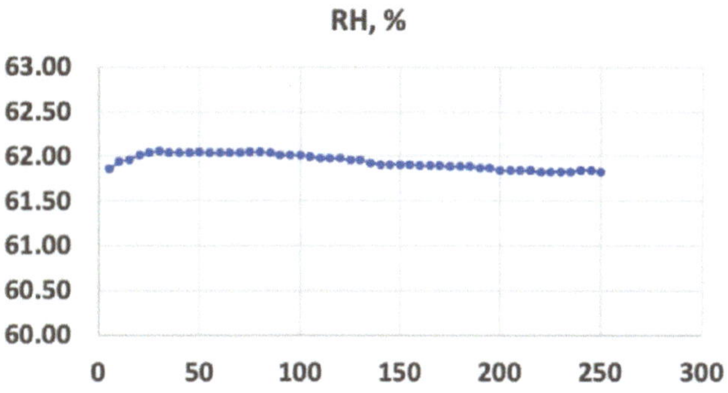

Figure 7.9 Relative humidity over a 4-hour interval in painted lady rearing room.

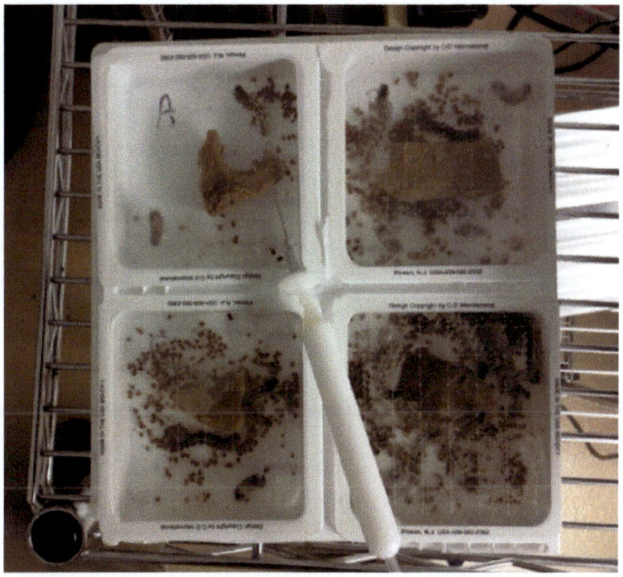

Figure 7.10 O_2 measurements vs. number of larvae in *Vanessa* rearing: closeup view.

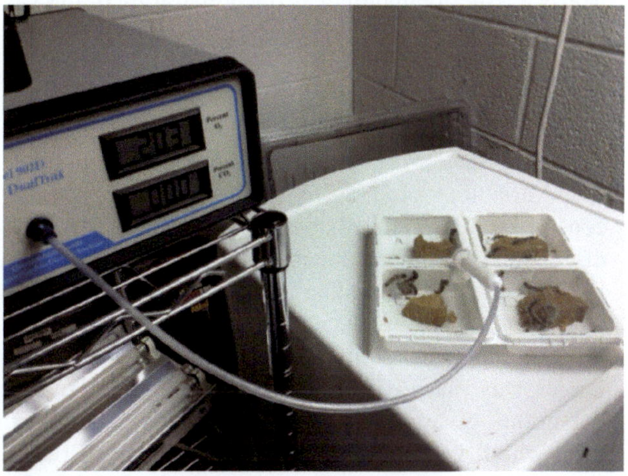

Figure 7.11 O_2 measurements vs. number of larvae in *Vanessa* rearing 2.

Figure 7.12 Measurements of light reaching rearing containers in different parts of the rearing room.

covering over a square 6.5 × 6.5 cm Petri dish. The mesh allows air flow and is especially intended to permit frass droppings to fall way from the diet to reduce diet/frass contact and contamination. During the early stages of the rearing system development, several containers were tested, including condiment cups (30–150 mL) and storage cups (450 mL). The 12 × 12 × 3 cm plastic containers shown here were selected because of their capacity to hold groups of ten larvae. The eight-cell trays (shown in this figure and Figure 7.5) were also used in early trials, but the space was to confined to permit prepupae to form their chrysalis, allowing it to hang from the top of the container (Figures 7.4 and 7.8).

The containers seen in Figure 7.7 were placed in different positions every day. Changes in positions included alternation of the top vs. bottom container, and alternating the horizontal position of the containers. The logistics of randomization of container position proved to be nearly as time-demanding as the preparation of the various diets.

7.7.3.1.5 Environment

In this section, various factors that may contribute to error/variability of larval growth and development are considered. The point that is constantly underlying the discussion of each component is the question, how is the factor in question promoting or retarding differential rates of growth and development?

The first data point was recorded before the data logger was placed in the rearing room. The remaining points give an accurate sampling of the temperature, which varied within 0.5°C for the entire rearing period. The temperature was measured with data loggers that had been placed within the rearing containers (Figure 7.7). The recorded temperatures for the measurement period described in this figure were between 25.0 and 26.0°C. However, it should be noted that temperatures have been shown to vary from container to container, depending upon the position of the containers with respect to one-another, their distance from the light source, and their position on the shelves. Considering growth-influence from temperature in relation to Q_{10} (Chapters 6 and 10) even slight differences in temperature from one position to another could help account for variation in developmental progress and biomass of individual larvae.

The relative humidity within rearing containers (as indicated by the data logger placed in one of the rearing containers) was close to 62% for the entire measurement period (Figure 7.9). At this relative humidity, mold growth is almost entirely absent (Cohen 2015). It noteworthy that the size and number of pinholes in the container were verified to be appropriate for moisture loss from the diet strips, larvae, and frass.

In the test shown in Figures 7.10 and 7.11, larvae were place in numbers varying from 2-8 in the compartments of the 8-cell trays (from C.O. International, New Jersey) with perforated lidding. The intention behind the perforations in the tray lidding is to provide adequate gas exchange for the insects in each cell. However, the assumption that there is adequate gas exchange in rearing containers has not been documented on a per container basis. The effects of stress generated by reduced oxygen atmospheres and increased carbon dioxide are well-documented for example by VandenBrooks et al. 2018, Hodkinson 2005, Harrison et al. 2009 and 2010, and Harrison and Haddad 2011. For example, VandenBrooks et al. 2018 showed that Drosophila melanogaster reared under hypoxia conditions (12% O_2 levels) showed significant changes in the structures of the tracheolar system and the distribution and number of mitochondria. This topic is further discussed in Chapter 10, but for the current discussion it is sufficient to consider the question of how many larvae per cell would constitute an oxidative stress (hypoxia) and carbon dioxide (hypercapnia or hypercarbnia)?

Figure 7.11 shows the O_2/CO_2 analyzer, with its readouts for % O_2 and % CO_2. Note that the readings of O_2 and CO_2, respectively, are 20.3% and 0.00%. However, in the cells where eight larvae were kept, with no accommodations for increased gas exchange, the O_2 levels were measured as 18.2%-19.3%, and CO_2 levels reached more than 1% (equaling 10,000 ppm) on a chronic basis. These levels of lowered oxygen and elevated carbon dioxide were not further evaluated to determine whether they constituted significant stress, but they were noted and they should be considered as subjects of further study in the domain of potentially chronic stress (hypoxia and hypercapnia) in reared insects, in general.

7.7.3.1.6 Larval Growth and Development

The growth of larvae was measured once during the prepupal period after about 14 days of rearing from neonates to late larval stage. More frequent measurements would have given more precise indication of progress on the various diets, but the consideration of additional stress from handling the insects prevented the evaluation of running progress in growth. Because the larvae were held in communal containers, it was not possible to measure development through each larval instar, because shed cuticles could not be attributed to a given larva. One feature that prevailed in this rearing system and is typical of rearing systems universally is the non-uniform size of larvae within each container (Figure 7.6).

If Figures 7.12 and 7.13 the light meter (Amprobe LM 120, Everett, WA) is seen, registering 947 Lux. The light meter used here is sensitive enough to allow differences to be detected between locations just centimeters apart. The light measurements also indicated the large differences between light intensity from shelf to shelf, and of course the light intensity was always much lower in the rearing room than it was in outside light, especially in full sunlight where the light intensity could be at least ten times as great as it was even under direct light from an artificial light source. Our instrument did not measure light quality (wavelength), but this can possibly be another source of variability in rearing conditions.

7.7.3.1.7 Pupal/Chrysalis Management

A key feature to mass-rearing of *V. cardui* is providing appropriate sites for the larvae to proceed into the chrysalis stage. An important part of this process is assurance that the larval containers have a lid or ceiling that encourages normal, healthy attachment and pupation. Figure 7.13 (a–c) shows the conditions that were developed to accommodate pupation/chrysalis formation.

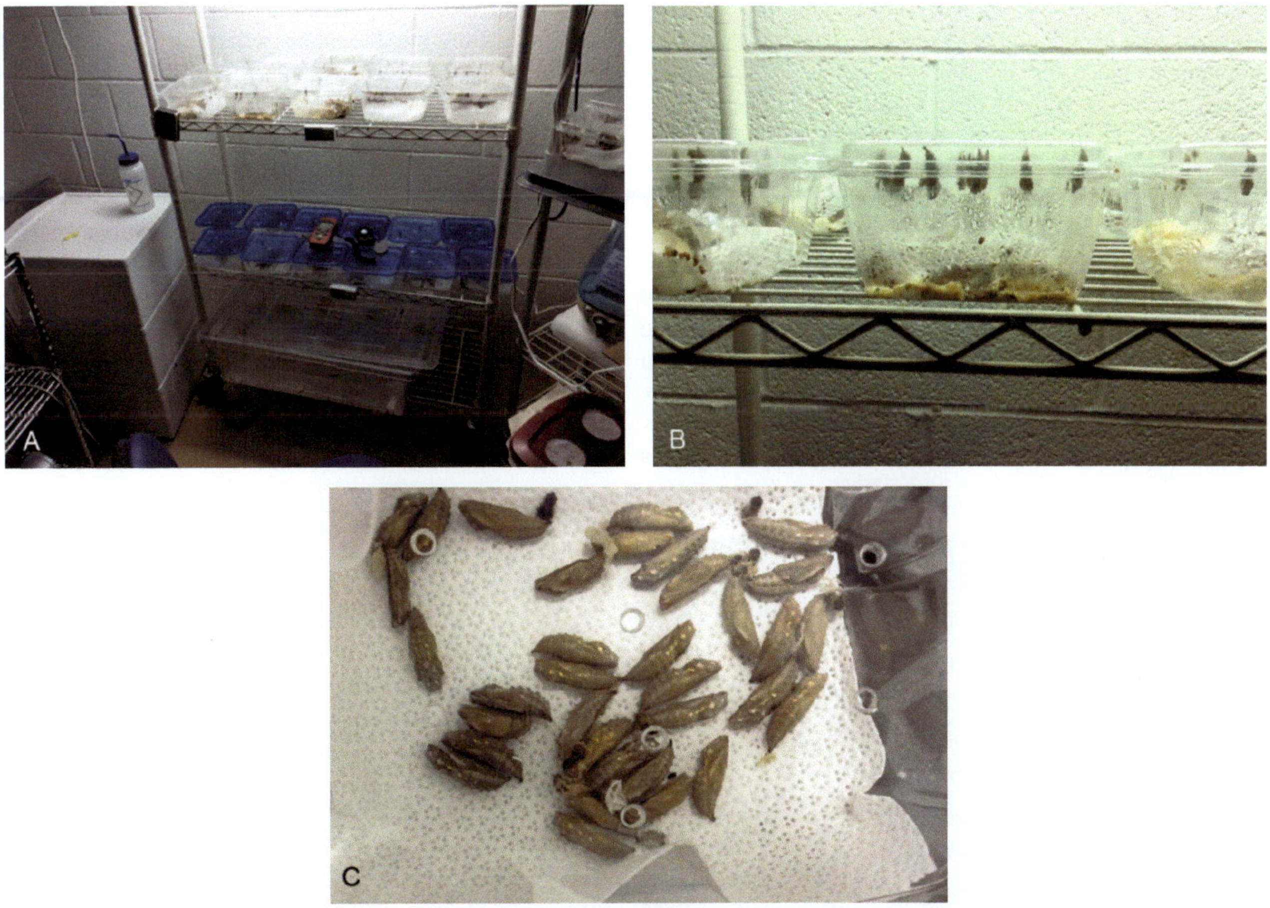

Figure 7.13 (a–c) Rearing setup for factorial tests of Vanessa diets. Besides factorial diet tests, this image also depicts the measurements of light intensity in different positions of rearing containers. Figures (a) and (b) show two kinds of containers (ziplock with blue lids and dart clear containers, both about 1 L) where larvae readily moved to the top to start chrysalis attachment and formation as is evidenced by the healthy chrysalises in figure (c). As a quality control measure, samples of 5–10 pupae per 100 were weighed and checked for normal adult emergence.

Figure 7.14 Adult *Vanessa cardui* in oviposition cage with mallow plant.

7.7.3.1.8 Adult Emergence Conditions

Nearly 100% of the chrysalises (chrysalids) were formed on the lid of the container. This made it convenient to place the lids into the adult cages to allow them to emerge without disturbance to the pupae, which responded to handling with rapid vibrations. The successful emergences of adults from chrysalis-casing were in excess of 90%, except for those reared on inferior diets (Figure 7.14).

Adults were allowed to emerge from chrysalis undisturbed. Honey (undiluted) was streaked on the mesh of the container, and a water wick (dental wick fitted into water vials) was placed inside the cage to allow ad libitum access to water and artificial nectar (made of a 5% honey solution with 0.1% potassium sorbate as a mold inhibitor). Small potted mallow plants were placed in cages as oviposition sites. The potassium sorbate was found to be both acceptable to the adult insects and an effective mold inhibitor.

7.7.3.1.9 Adult Feeding

In preliminary experiments (not reported here), adults were separated into three feeding systems consisting of: (1) undiluted honey and free water on cotton wick, (2) diluted honey and free water on cotton wick, and (3) diluted honey without water. The adults in system (1) had the highest egg production and longevity.

7.7.3.1.10 Mating and Oviposition

The cages used for adult emergence and feeding were used for mating and oviposition sites. The cages were the Carolina Butterfly Sanctuary (Carolina Biologicals, Burlington, NC), which are free-standing mesh cages that are easily cleaned mild (1%) bleach solution with water spray to rinse the bleach solution. The honey and meconium (reddish patches in Figure 7.11) are easily removed, allowing the next generation to have clean, sanitary cages. The mesh also permits water spray to be administered with squeeze water bottles, which are also used to water the potted mallow plants (*Malva sylvestris*) (Figure 7.15a). Also, semi-artificial oviposition sites were used to help reduce the amount of potted plant materials used for oviposition (Figure 7.15b). The semi-artificial oviposition site (Figure 7.15b) was made with excised leaf portions held to paper towel by a window of Parafilm. The leaves could be kept somewhat moist by wetting the paper towel, which allowed water diffusion to keep the leaf section from drying out excessively. It is evident from Figure 7.15b that this oviposition site was acceptable to the butterflies.

Figure 7.15 (a) and (b) Oviposition on mallow leaves on plant (a) and excised leaves (b).

7.7.3.1.11 Egg Harvesting/Egg Management

Eggs were attached to the leaf surface with a cement that holds the eggs firmly. Efforts were made to free the eggs with sterile tap water where leaves were immersed in a beaker, which was sonicated for 5 minutes. This method freed few eggs, so a 0.5% solution of bleach (sodium hypochlorite) was used, which freed most of the eggs. Finally, a soft paint brush was used to mechanically remove the remaining eggs, which were placed on a sheet of Whatman # 3 filter paper in a Buchner funnel where the eggs were washed with sterile tap water to remove the excess bleach solution and to partially dry the eggs. This method did not seem to damage the eggs, as there was more than 90% hatch with this method. Note that ways were devised of using excised leaves without placing the whole mallow plant into the adult cage (Figure 7.12(b)). The excised leaves and portions of leaves were kept moist with damp paper towel, and leaves were surrounded with Parafilm to preserve moisture. Note that some eggs were laid on the Parafilm.

7.7.3.1.12 Next Generation/Starting New Colonies

Eggs collected as described in section "Egg Harvesting/Egg Management" were placed in rearing containers like the ones seen in Figures 7.3 and 7.4. Diet aliquots were added to the containers, and newly hatched larvae (neonates) were allowed to find the diet and begin feeding. Fresh diet was added to containers every third day. The remainder of the rearing process was as described in section "Introduction to Tutorial on Rearing System Optimization" (**7.7.3.1.1**) through section "Egg Harvesting/Egg Management" (7.7.3.1.11).

7.7.3.1.13 Diet Optimization

Table 7.4 presents the ranges for each diet item based on a plus 3% and minus 3% error. It is a standard in the Cohen laboratory to set error boundaries of +/– 3% of each diet ingredient. This standard, though admitedly is arbitrary, gives technicans and other rearing professionals a common ground and a uniform set of expectations. It is important to note that this seemingly simple standard has far-reaching implications in terms of error reduction (potential for reproducibility, PC, and ultimately QC). The implications of this standard also help provide guidelines for decisions about instruments and other measuring devices. For example, we weigh the larger masses of diet materials on a high capacity balance that reads to grams. For water measurements where, in this case, 260 g of water are the target, a reading of 260 g on the balance (after the tare container is placed on the balance and set to "zero") may actually mean that the water weighs from 259.500 to 261.500—all readings in this range registering

Figure 7.16 Shows the diet component rice bran being weight on a balance readable to tens of mg.

"260 g." Because our standard for weighing error for water is 3% above 260 and 3% below, the error that we described here is well-within the set standard. However, if we tried to weigh the sucrose, which is targeted at 5 g, on the same balance, what reads as "5" g might be 4.5 or 5.5 g of sucrose—amounts that are in excess of our 3% standard. Therefore the ranges from Table 7.1 indicate to the technician that she or he must use a balance that has finer (more precision) readability than is allowed by readings to the nearest gram (Figures 7.16–7.18).

Note that the weight shown in Figure 7.16 is 3.03 g of stabilized rice bran, which is well-within the acceptable range of 2.91–3.09 g (which equals the target weight of 3.00 g, ± 3%). Note also that the plastic container to the right of the balance is used to hold rice bran so that it will not be returned to the source container seen in the background. This practice of using a sub-aliquot to allow addition or deletion of material being weighed is an

Figure 7.17 Logistics of multiple diet mixing are illustrated here.

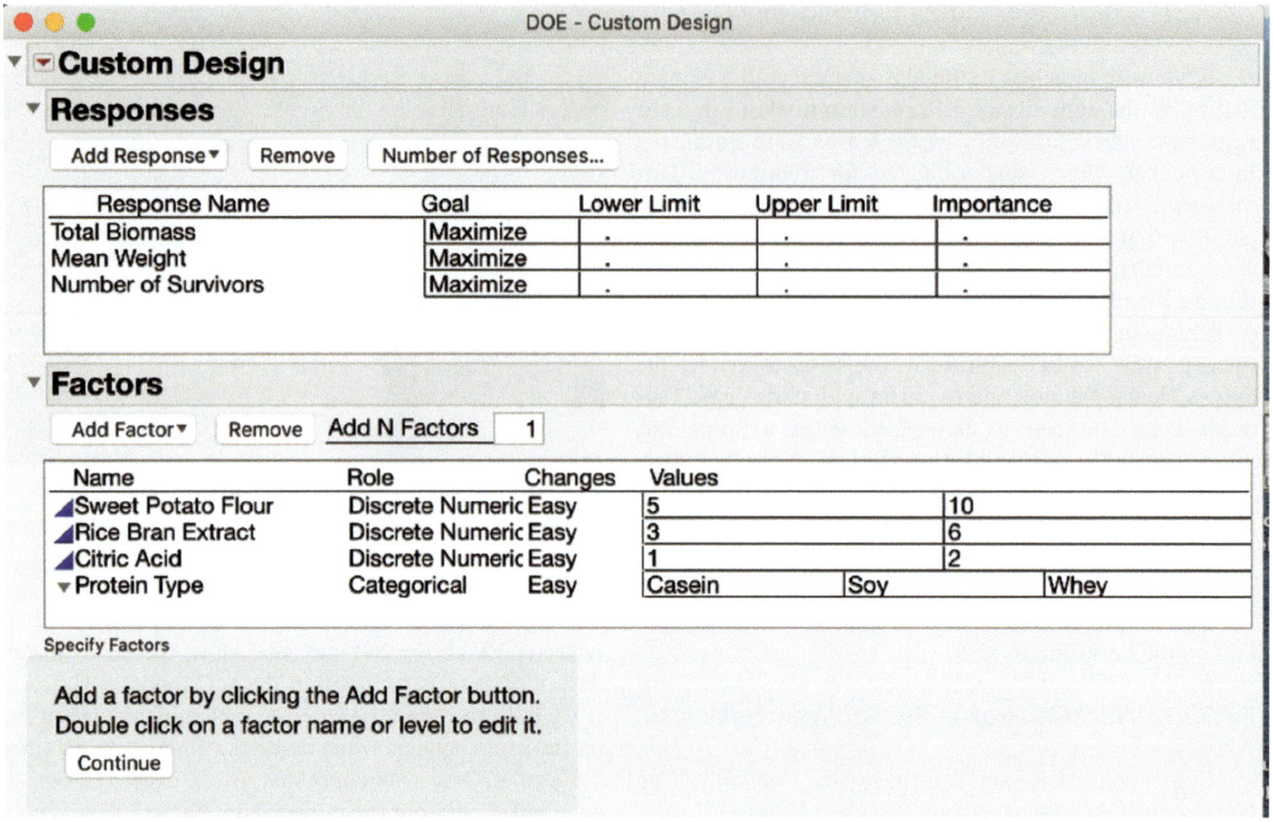

Figure 7.18 Initial setup of Custom Design.

effective way of preventing contamination, and it is convenient to use the open plastic beaker to allow repeated adding or removal of material. It is important to note that the amount of material that will be discarded constitutes an economic cost of reducing error. The reduction in error must be considered in relation to the differences between compromised quality in insects vs. a frugality that may compromise the whole purpose of the rearing process.

There are 24 different diets being made whose formulations are listed in Table 7.4. Figure 7.17 shows the dry ingredients being assembled in the full-factorial diet. The logistics recommended here are to mix all the dry ingredients that are part of the standard diet (the portion that is not being modified shown in black font in Table 7.4); then the variable components are added. Note that the component, stabilized rice bran, is being measured in Figure 7.16. The target amount was 3.0 g. The balance reads to hundredths of a gram, and this reading is 3.03 g, which is 0.03 g above the target amount. Calculating the error in weighing, we divide 0.03 by 3.00 giving us an error of 0.01 = 1%. Our standard (discussed above) was set at +/– 3% error, so we decided that the component shown here is within the boundaries of our target error. If the weight of the rice bran were 3.11 (for example), the error would be 0.11/3.00 = 0.037 (rounding off to 0.04 or 4% error). This would have been deemed unacceptable. For the technician performing the task of diet preparation, the 3% standard would translate to 3.00 g + 0.09 or 3.09 g maximum and 2.91 g minimum (Table 7.4).

Note that in Figure 7.18 under "**Response Name**," the outcomes that we were testing were specified: larval weight, pupation number, and mortality (after seven days). Note further that the Goals for each, respectively, are "Maximize," "Maximize," and "Minimize." Our knowledge of rearing/diet phenomena informs us that the greater the weight of larvae, the higher the fitness of the insect; likewise, the larger the number of larvae that successfully pupate, the greater the fitness with respect to developmental success. Conversely, fewer deaths of developing larvae are to be taken as an affirmation of the rearing system components' success. In the next section, "Factors" are listed, the factors in this experiment were amounts of "sweet potato flour" (5 and 10 g), "rice bran extract" (3 and 6 g), "citric acid" (1 and 2 g), and finally "protein" type "casein," "soy flour," and "whey." Please note further that in the Custom Design program in JMP, we can mix treatments, in this case, amounts of certain ingredients vs. kinds of another ingredients (kinds of proteins).

This point is worth discussing in greater depth. The analytical power of this type of experimental design is enormous. We can determine, using this type of design of experiments, whether sweet potato flour as a novel ingredient is an asset, liability, or neutral factor in the rearing of painted lady larvae. We can ask the same question of rice bran extract, and the same, again, for citric acid. By using different amounts of the various test materials, we can get a better understanding of how much each factor influences rearing quality (fitness) outcomes. We can further determine the answer to the question, "which type of protein (casein, soy, whey) is best for the insects' fitness?" Furthermore, we can not only make the determinations of the fitness contributions of each of the factors, but we can also gain an understanding of the **interactions** between the different components. The power of being able to evaluate the contributions of multiple variables is of vast potential in development and improvement of rearing systems! This is one of the most powerful features of this approach to experimental design. It must further be added that we can modify the design by asking other questions about simultaneous variables such as temperature, light conditions, cage factors (size, number of insects per container, openings in the container permitting gas exchange, etc.). Note in Figure 7.19 (after selecting "Continue"), there are choices in Model. The choices are shown in Figures 7.19 and 7.20. Note that Figure 7.20 shows the setup of the Custom Design dialog after "Continue" was selected. Please note, further, that the choices include "Main Effects" and the "Interactions" that we wish to test, and all of the estimabilities are by default deemed

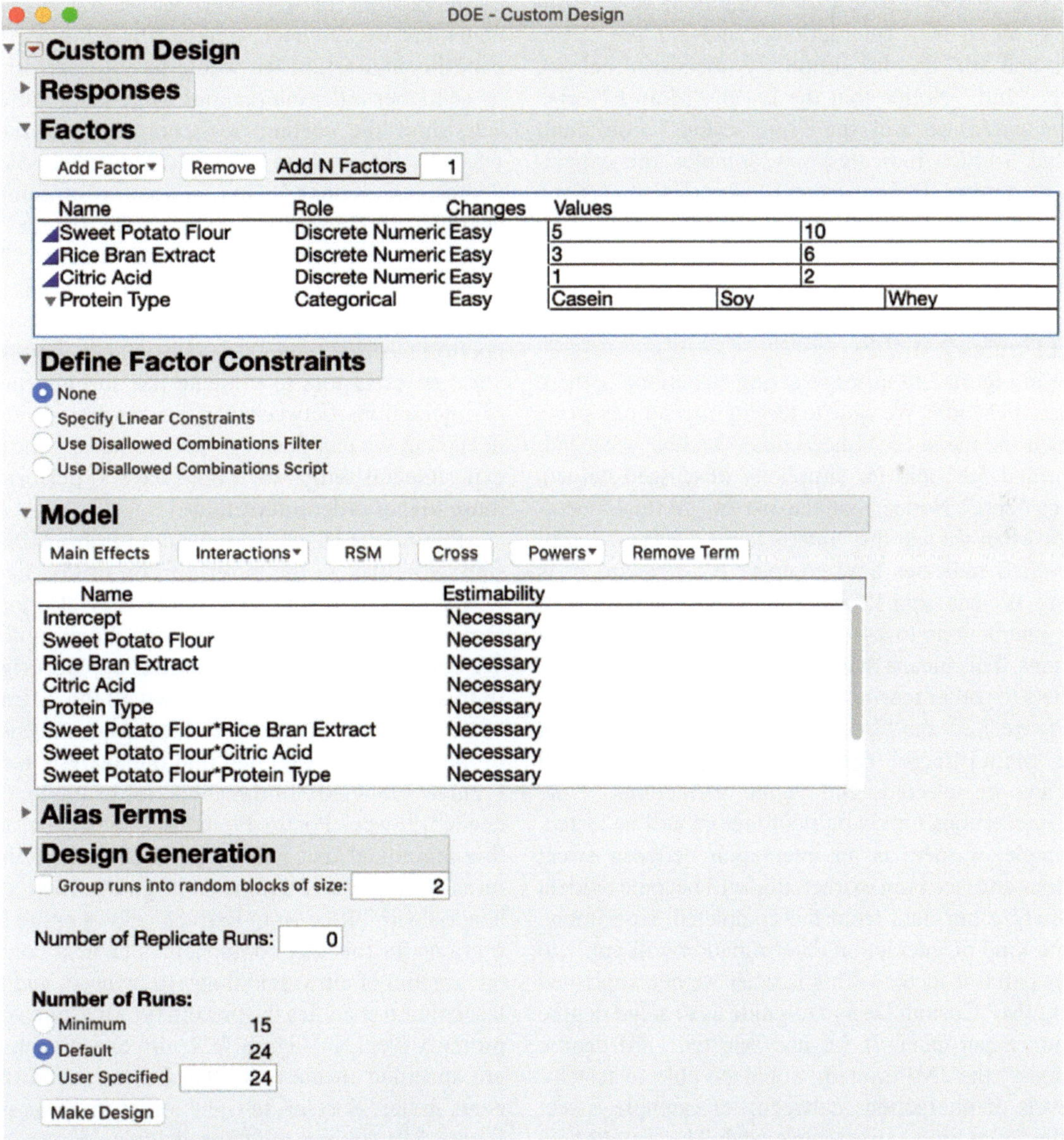

Figure 7.19 The Custom Design dialog where Factor Constraints and Model parameters are chosen.

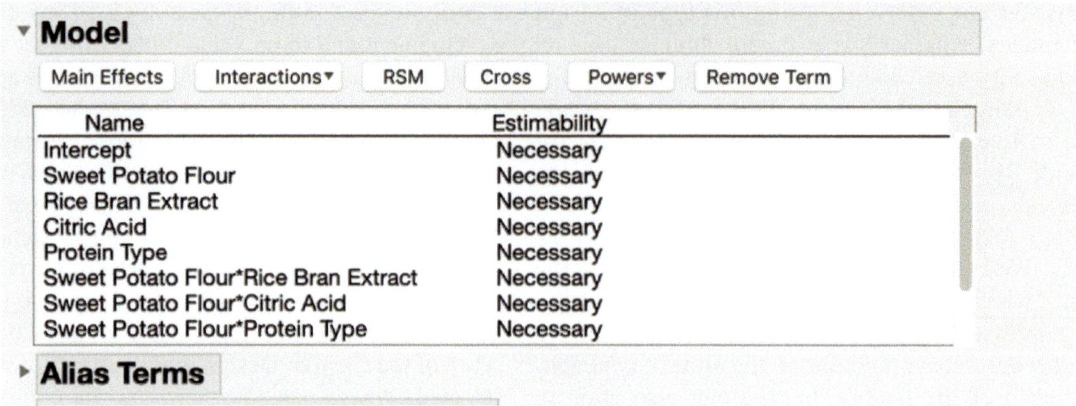

Figure 7.20 Isolated portion of Custom Design dialog showing parameters to be tested.

"Necessary." We want to test for interactions so we drop down the menu for "Interactions" and we see "2nd, 3rd, 4th, and 5th" and for simplicity (discussed below), we select "2nd." Notice that the list of "Main Effects" increases when we add the "Interactions" command with "2nd," which indicates how complex the experiment is becoming. It went from 12 experiments for the minimum with no interactions to 24 experiments with 2nd degree interactions. This means that we will have to set up 24 different diets (or other rearing factors that we wish to test) The remaining flow of analytical steps are found in Figures 7.21–7.29.

Note in Figure 7.19 (after selecting "Continue"), there are choices in Model. We want to test for interactions so we drop down the menu for "Interactions" and we see "2nd, 3rd, 4th, and 5th" and for simplicity (discussed below), we select "2nd." Notice that the list of "Main Effects" increases when we add the "Interactions" command with "2nd," which indicates how complex the experiment is becoming. It went from 12 experiments for the minimum with no interactions to 24 experiments with 2nd degree interactions. This means that we will have to set up 24 different diets (or other rearing factors that we wish to test).

In Figure 7.20, the "Interactions" are specified along with the "Main Effects." This is what the JMP system will test because we selected "2nd degree interactions." Note that the interactions for six pairs of factors will be tested. For example, if there is an interaction between sweet potato flour and rice bran extract, this will become evident as we analyze our data from the completed experiment. The same kind of interaction determination will apply to the other pairs of factors. This is what we accomplished by asking JMP Custom Design to guide us to a 2nd degree interaction experiment. If we had selected "3rd degree interactions," the JMP system would be able to test for three levels of interactions between, for example, sweet potato flour/rice bran extract/citric acid. This would be a very powerful tool to evaluate the all-important concept of interactions between various rearing system components, but we can become very limited by the number of experimental setups we would have to perform to detect these higher order interactions. Note that Figures 7.19 and 7.20 show the portion of Custom Design dialog shows what the JMP system will test because we selected "2nd degree interactions." This is what we accomplished by asking JMP Custom Design to guide us to a 2nd degree interaction experiment. If we had selected "3rd degree interactions," the JMP system would be able to test for three levels of interactions between, for example, sweet potato flour/rice bran extract/citric acid. This would be a very powerful tool to evaluate the all-important concept of interactions between various rearing system components, but we can become very limited by the number of experimental setups we would have to perform to detect these higher order interactions.

Figure 7.21 shows the the output from JMP's "Design" step. Note that in the lower portion of this figure, under "Data Table Options," we made the selection in "Run Order" of "Sort left to right," from the drop down menu. We could have sorted "Randomize" or "Sort right to left." In this box, the JMP user is shown all the 24 unique combinations that make up the full factorial analysis called for in the inputs that were fed into the analysis factors (from Figures 7.18–7.20). In this table, for example, the column headed "Sweet Potato Flour" shows all the amounts of this ingredient that will be put into each of the 24 mixtures of "factors": 5 grams or 10 grams. Next, the column headed with "Rice Bran Extract" shows either 3 grams or 6 grams for this diet component. The next column shows the amount of citric acid that will be used, and finally, the last column specifies the protein type (casein, soy, or whey protein). Next, in Figure 7.22, the experimental logistics are specified in the table that has arranged the components in the "Sort left to right" order that was specified in Figure 7.21, giving the "Sweet Potato Flour" column 12 combinations that contain 5 grams and 12 combinations

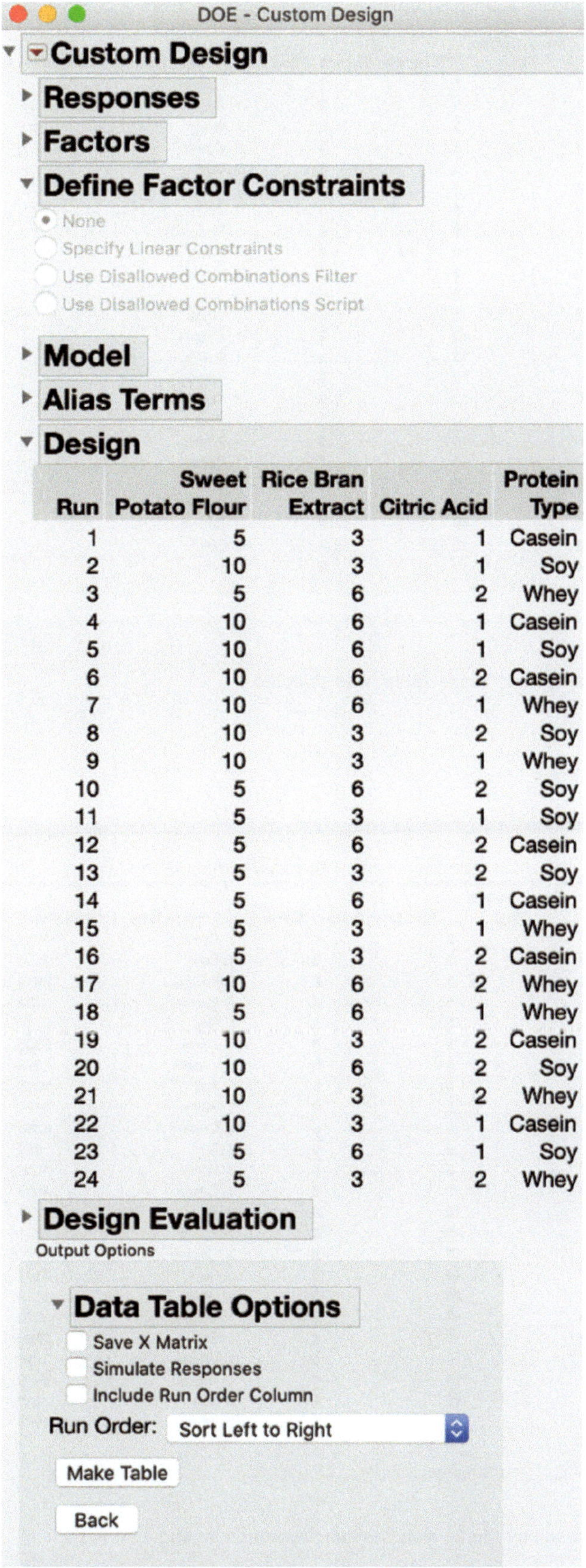

Figure 7.21 Custom Design dialog for the painted lady butterfly diet.

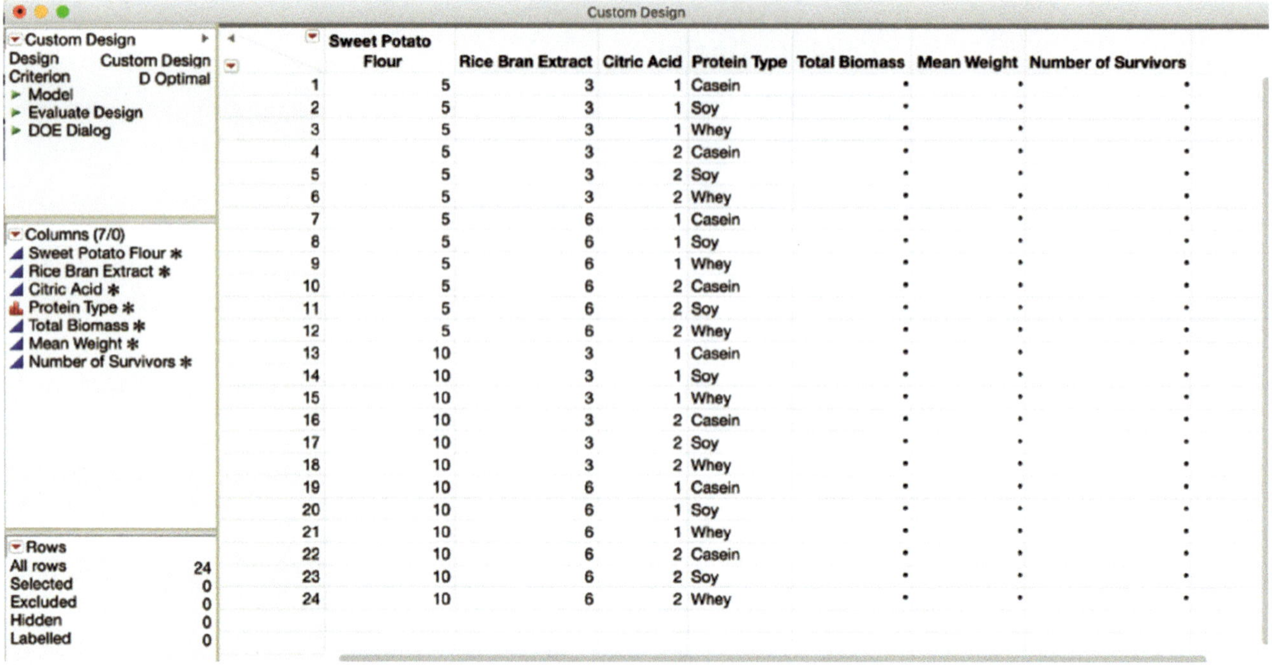

	Sweet Potato Flour	Rice Bran Extract	Citric Acid	Protein Type	Total Biomass	Mean Weight	Number of Survivors
1	5	3	1	Casein	•	•	•
2	5	3	1	Soy	•	•	•
3	5	3	1	Whey	•	•	•
4	5	3	2	Casein	•	•	•
5	5	3	2	Soy	•	•	•
6	5	3	2	Whey	•	•	•
7	5	6	1	Casein	•	•	•
8	5	6	1	Soy	•	•	•
9	5	6	1	Whey	•	•	•
10	5	6	2	Casein	•	•	•
11	5	6	2	Soy	•	•	•
12	5	6	2	Whey	•	•	•
13	10	3	1	Casein	•	•	•
14	10	3	1	Soy	•	•	•
15	10	3	1	Whey	•	•	•
16	10	3	2	Casein	•	•	•
17	10	3	2	Soy	•	•	•
18	10	3	2	Whey	•	•	•
19	10	6	1	Casein	•	•	•
20	10	6	1	Soy	•	•	•
21	10	6	1	Whey	•	•	•
22	10	6	2	Casein	•	•	•
23	10	6	2	Soy	•	•	•
24	10	6	2	Whey	•	•	•

Figure 7.22 Custom Design dialog after "Make Table" was selected.

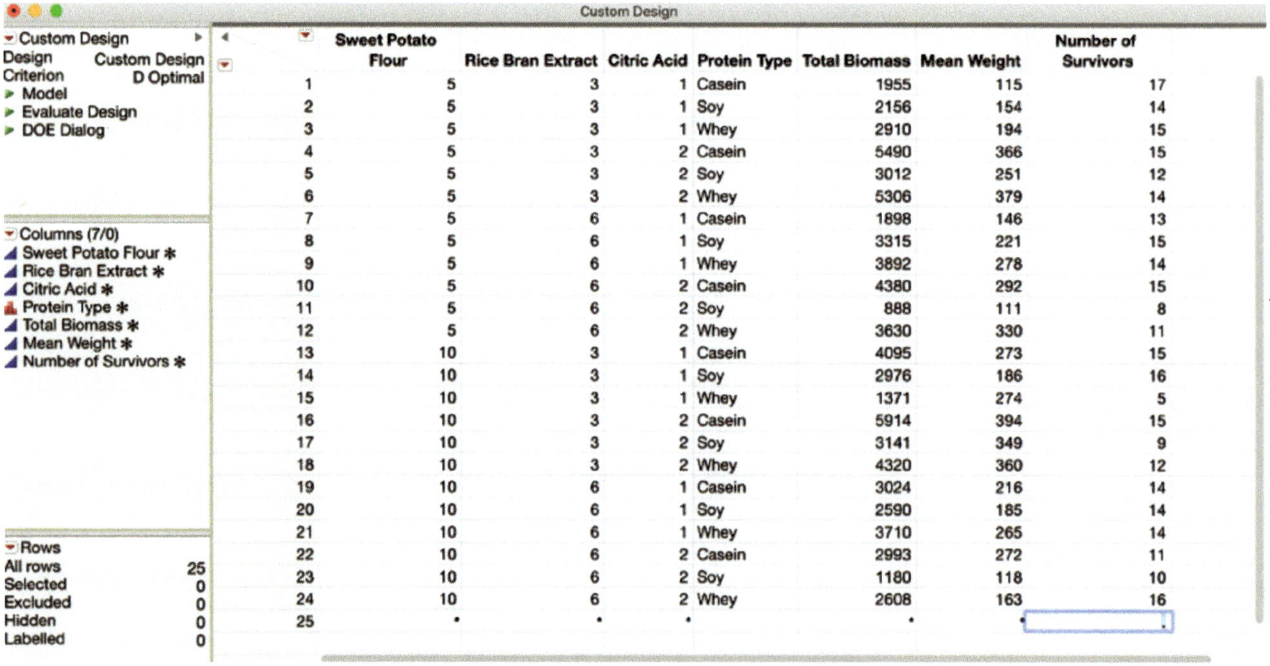

	Sweet Potato Flour	Rice Bran Extract	Citric Acid	Protein Type	Total Biomass	Mean Weight	Number of Survivors
1	5	3	1	Casein	1955	115	17
2	5	3	1	Soy	2156	154	14
3	5	3	1	Whey	2910	194	15
4	5	3	2	Casein	5490	366	15
5	5	3	2	Soy	3012	251	12
6	5	3	2	Whey	5306	379	14
7	5	6	1	Casein	1898	146	13
8	5	6	1	Soy	3315	221	15
9	5	6	1	Whey	3892	278	14
10	5	6	2	Casein	4380	292	15
11	5	6	2	Soy	888	111	8
12	5	6	2	Whey	3630	330	11
13	10	3	1	Casein	4095	273	15
14	10	3	1	Soy	2976	186	16
15	10	3	1	Whey	1371	274	5
16	10	3	2	Casein	5914	394	15
17	10	3	2	Soy	3141	349	9
18	10	3	2	Whey	4320	360	12
19	10	6	1	Casein	3024	216	14
20	10	6	1	Soy	2590	185	14
21	10	6	1	Whey	3710	265	14
22	10	6	2	Casein	2993	272	11
23	10	6	2	Soy	1180	118	10
24	10	6	2	Whey	2608	163	16
25	•	•	•		•	•	•

Figure 7.23 Shows an interim result for the 24 diets that we have been testing.

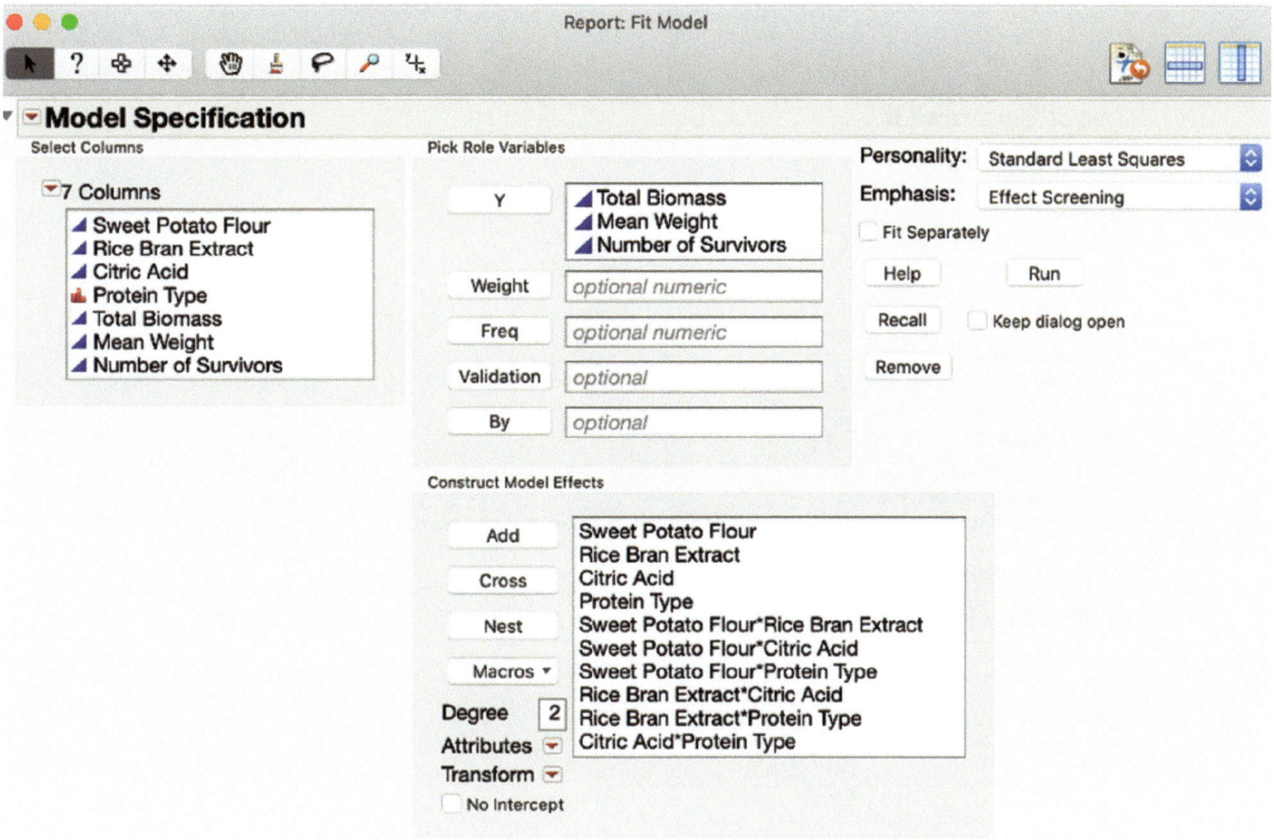

Figure 7.24 Shows the dialog interaction after "Model" was selected from the upper left-hand corner of Figure 7.28.

with 10 grams of the sweet potato flour. Next, the column with the rice bran extract label specifies that the first 6 diets with 5 grams of sweet potato flour will have 3 grams of rice bran extract while the other 6 diets with 5 grams of sweet potato flour will have 6 grams of rice bran extract. Considering only these two factors (sweet potato flour and rice bran extract), this experiment has four unique combinations of the two mentioned components.

Next, the third column specifies the amounts of citric acid (one gram or two grams) for each combination of factors. Finally, the fourth column specifies the three protein types that will be combined with the other three factors that are arranged in their eight different combinations. The three kinds of proteins mixed with the various combinations of the other three factors, provides the unique $8 \times 3 = 24$ completely unique diets that will be mixed with the standard diet, which is described in Table 7.4, including the wheat germ, alfalfa meal, and other components that are common to all the diets. One more step towards the analysis is seen in Figure 7.22, the columns labelled "Total Biomass," "Mean Weight," and "Number of Survivors" are specified but with missing data that is to be filled in as the results of the experiments are tabulated. Once the experiment is completed where the larvae are fed the diets, and weighed (collectively and individually), and the number of survivors is recorded for each treatment (each of the 24 unique diets), these data can be entered into the spreadsheet indicated as Figure 7.22. This leads to the arrangement seen in Figure 7.23.

In Figure 7.23, we can see that the selections that we made, as described in Figures 7.18 through 7.22, which led to the experimental design where there are 24 unique combinations of diet components. In this figure, we have the results for "total biomass" that we obtained by collectively weighing the living larvae in each container (1–24). We calculated the mean weight by dividing the total biomass by the number of surviving larvae. Note that in Figure 7.20, we can see that the selections that we made, as described in Figures 7.1–7.5, led to the experimental design where there are 24 unique combinations of diet components. Half the diets contain the lower level of sweet potato flour (5 g), while the other half (other 12) contain the higher level (10 g). This is shown in the column labeled "Sweet Potato Flour." In the next column, labeled "Rice Bran Extract," we see that the first six diets contain the lower levels of rice bran extract, as do 13–18. For the citric acid component, the lower levels are found in 1–3, 7–9, 13–15, and 19–21. Finally, the protein sources are ordered as "casein, soy, and whey," each sequence repeated eight times so that each diet in the 24 is unique

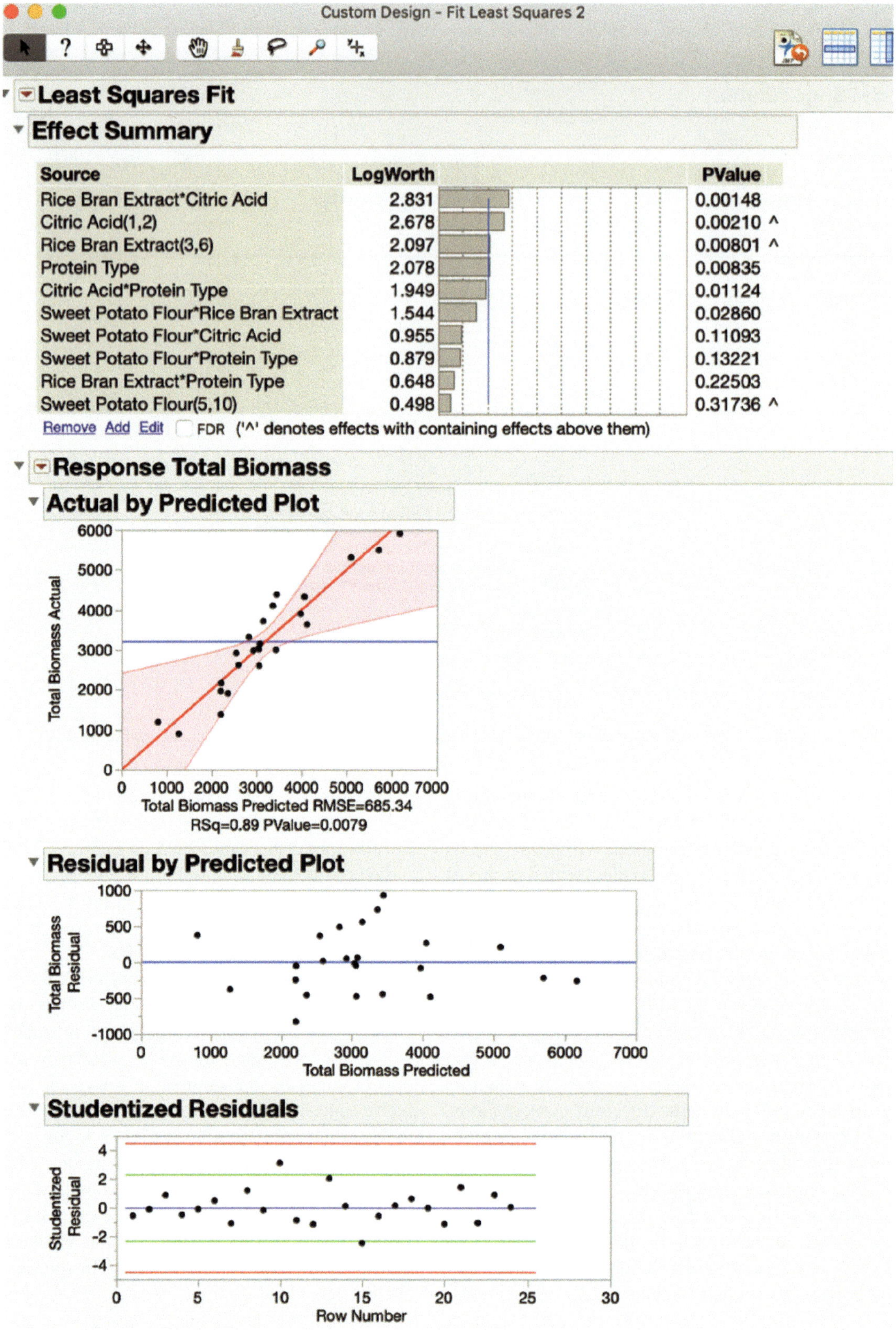

Source	LogWorth	PValue
Rice Bran Extract*Citric Acid	2.831	0.00148
Citric Acid(1,2)	2.678	0.00210 ^
Rice Bran Extract(3,6)	2.097	0.00801 ^
Protein Type	2.078	0.00835
Citric Acid*Protein Type	1.949	0.01124
Sweet Potato Flour*Rice Bran Extract	1.544	0.02860
Sweet Potato Flour*Citric Acid	0.955	0.11093
Sweet Potato Flour*Protein Type	0.879	0.13221
Rice Bran Extract*Protein Type	0.648	0.22503
Sweet Potato Flour(5,10)	0.498	0.31736 ^

Figure 7.25 Shows the ten most significant factors of the full factorial diet experiment.

Scaled Estimates

Nominal factors expanded to all levels

Term	Scaled Estimate	Std Error	t Ratio	Prob>|t|
Intercept	3198.0833	139.8947	22.86	<.0001*
Sweet Potato Flour(5,10)	-37.91667	139.8947	-0.27	0.7925
Rice Bran Extract(3,6)	-355.75	139.8947	-2.54	0.0316*
Citric Acid(1,2)	373.75	139.8947	2.67	0.0256*
Protein[Casein]	520.54167	197.8409	2.63	0.0273*
Protein[Soy]	-790.8333	197.8409	-4.00	0.0031*
Protein[Whey]	270.29167	197.8409	1.37	0.2050
Sweet Potato Flour*Rice Bran Extract	-120.25	139.8947	-0.86	0.4123
Sweet Potato Flour*Citric Acid	-174.5833	139.8947	-1.25	0.2435
Sweet Potato Flour*Protein[Casein]	325.79167	197.8409	1.65	0.1340
Sweet Potato Flour*Protein[Soy]	102.41667	197.8409	0.52	0.6172
Sweet Potato Flour*Protein[Whey]	-428.2083	197.8409	-2.16	0.0586
Rice Bran Extract*Citric Acid	-602.9167	139.8947	-4.31	0.0020*
Rice Bran Extract*Protein[Casein]	-289.125	197.8409	-1.46	0.1779
Rice Bran Extract*Protein[Soy]	-58.25	197.8409	-0.29	0.7751
Rice Bran Extract*Protein[Whey]	347.375	197.8409	1.76	0.1130
Citric Acid*Protein[Casein]	601.875	197.8409	3.04	0.0140*
Citric Acid*Protein[Soy]	-725.75	197.8409	-3.67	0.0052*
Citric Acid*Protein[Whey]	123.875	197.8409	0.63	0.5468

Figure 7.26 Shows the interactions for the response, "Total Biomass."

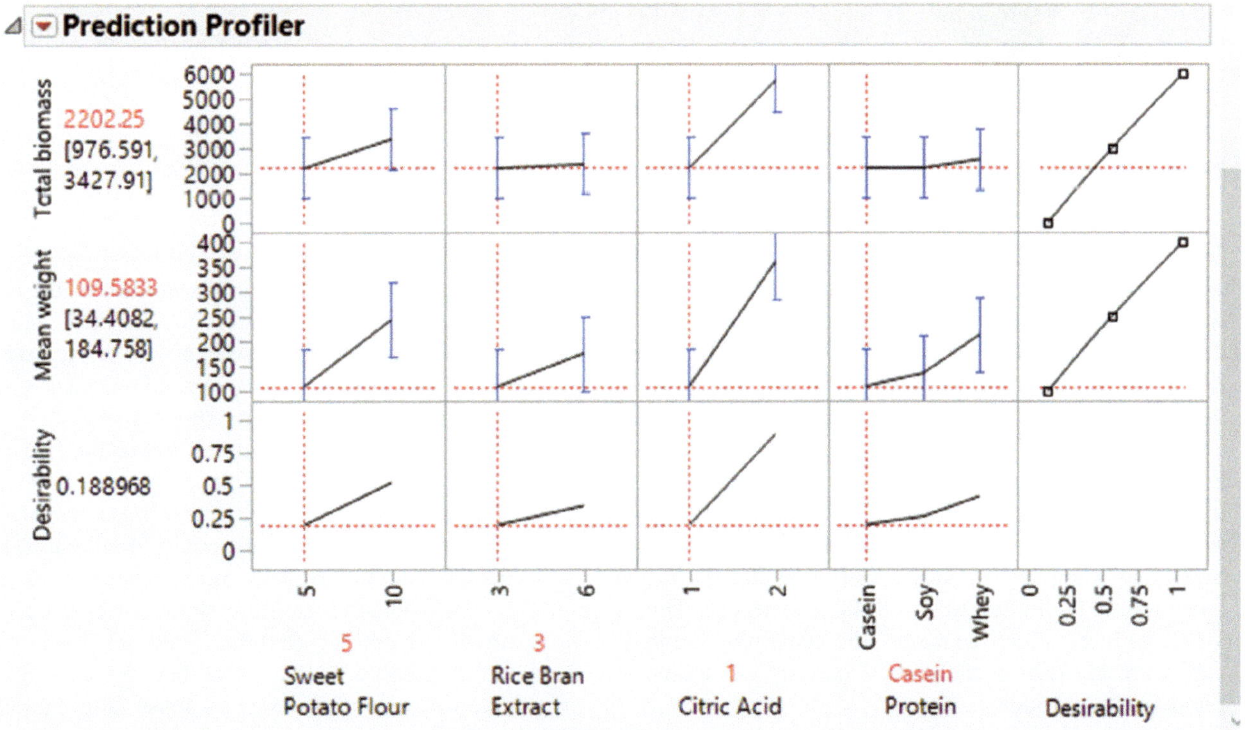

Figure 7.27 Shows the prediction profiler provided by JMP and showing graphically the trends of the experiment for individual components.

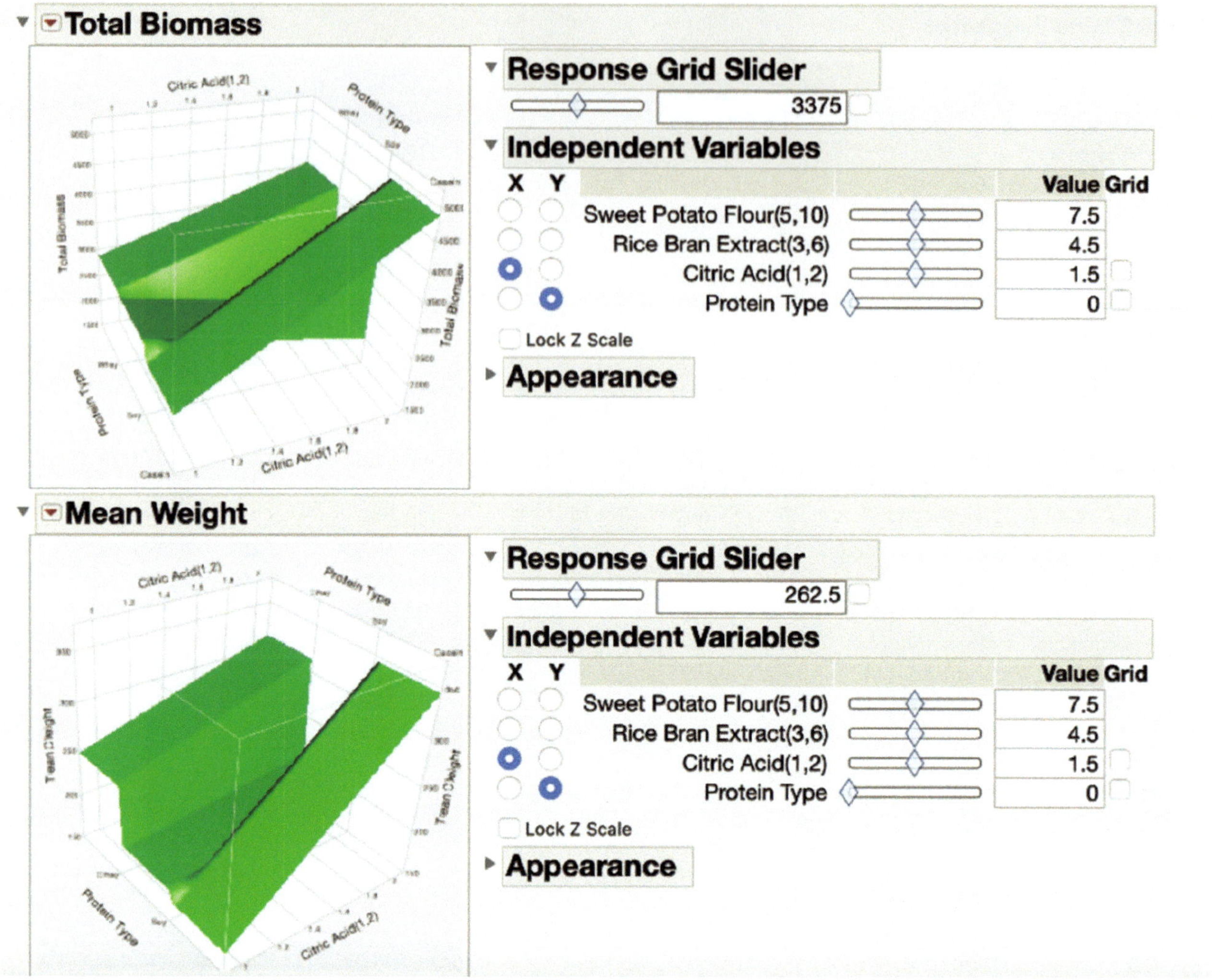

Figure 7.28 Surface plot of "Total Biomass" vs. "Citric Acid" and "Protein Type."

from all the other 23 diets. For clarity, the results of the experments are shown in Table 7.5. A step that is not shown here is the selection of the term "Model" found in the upper left corner of Figure 7.23. That selection led to the appearance of the box seen in Figure 7.24. Here, all the factors, which are present as defaults are presented, and for simplicity, the default values for "Pick Varable Roles," "Weight," and others are accepted. Also, in the upper right portion of Figure 7.24, the "Personality" (= "Standard Least Squares") and "Emphasis" (= "Effect Screening") are left in place, and immediately below these terms, the button titled, "Run" is selected. This is a crucial step, which generates the remaining outputs, which are the results of the data analysis, all shown in Figures 7.25 through 7.29.

In Figure 7.25, it should be noted that the lower the PValue, the more significant the factor. Included in this figure are plots of residuals, which we will not treat here because the statistical implications are beyond the range of this discussion. However, it should be mentioned that the flexibility of the JMP programs allows a full range of user expertise, from beginners to statistical analysis to well-trained statisticians who can gain many insights into the statistical basis of their efforts to reduce error and enhance the reliability of their rearing systems.

Note from inspection of Figure 7.26 that the JMP program picks the statistically significant factors and interactions ($P < 0.05$) by presenting the **Prob > t** in red font. Note also that the factors (sweet potato flour, citric acid, etc.) and the interactions (rice bran extract*citric acid, citric acid*protein[soy], etc.) are shown in graphic form with bars whose polarity (left vs. right) and size indicate the nature of the interactions—whether they are negative or positive in their influence on response (total biomass). This gives us a perspective of, for example, which protein (casein, soy, or whey) was best for contributing to the total biomass gain. Casein, for example, was shown to have a significant positive effect on biomass, but soy protein was

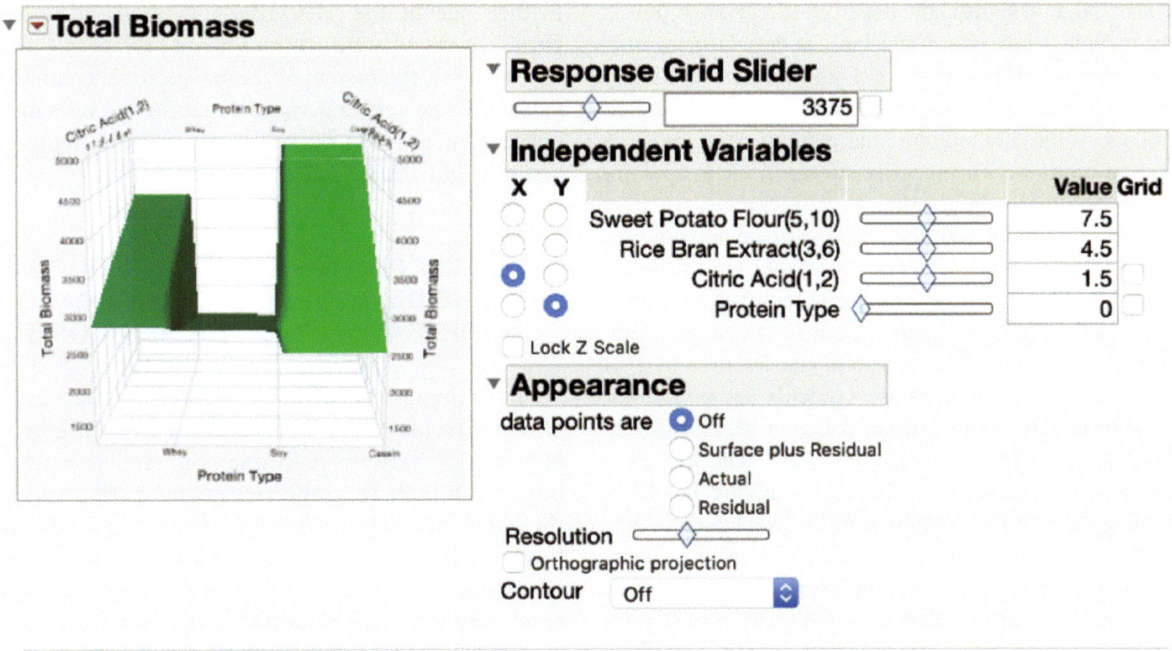

Figure 7.29 Surface plot of "Total Biomass" vs. "Citric Acid" and "Protein Type."

Table 7.5 **Interim Results for Number of Survivors and Composite Biomass ("Total Biomass") Collected Eight Days after Neonates Were Added to Each Diet**

Diet Number	# Survivors	Total Biomass in mg	Mean wt.
1	17	1,955	115
2	14	2,156	154
3	15	2,910	194
4	15	5,490	366
5	12	3,012	251
6	14	5,306	379
7	13	1,898	146
8	15	3,315	221
9	14	3,892	278
10	15	4,380	292
11	8	888	111
12	11	3,630	330
13	15	4,095	273
14	16	2,976	186
15	5	1,371	274
16	15	5,914	394
17	9	3,141	349
18	12	4,320	360
19	14	3,024	216
20	14	2590	185
21	14	3710	265
22	11	2993	272
23	10	1180	118
24	16	2608	163

It is **important** to note that the test results provided below are from the response (outcome) presented as total biomass (**not** "number of survivors" or "mean weight"). It should further be noted that the other two parameters, "number of survivors" and "mean weight," track the total biomass and give similar results when they are run as the experimental responses.

shown to have a significant negative effect on biomass. Whey protein showed a numerical tendency to improve biomass accretion, but it was not significant statistically. However, citric acid, though it consitutes far less than half of 1% of the diet biomass, has a significantly positive impact on biomass. The specific effects of citric acid are, however, especially interesting as it interacts very positively with casein (P = 0.0140) and very negatively with soy (P = 0.0052). This effect of citric acid with different proteins suggests some fascinating nutritional and possibly behavioral implications that merit further exploration.

The prediction profiler is one of the features of JMP statistical analyses that helps users quickly and unambiguously determine the factors that are of greatest consequence in the designed expermeents (Figure 7.27). The top set of graphs shows the trends for individual components' effects on total biomass. The higher level of sweet potato flour trended toward higher total biomass and mean weights while the higher level of rice bran extract showed a slight positive trend in mean weights. The higher level of citric acid (2 g vs. 1 g) showed a strong positive trend. Whey appeared to trend slightly toward higher weight gains than soy or casein. The prediction profiler is a standard function built into a wide variety of design of experiments and artificial neural network programs in JMP. This very graphic, very visual form of data interpretation is very useful to those of us who find visual treatment of statistical analyses easier to grasp than tables. What cannot be presented in the static images of a page is the most crucial feature of the prediction profiler: its dynamic interactive capability. Each variable factor in the experiment can be manipulated by the user placing the cursor on the red dotted lines and dragging the lines to different concentrations. This allows the user to use the statistical power of the program to predict the outcomes of different concentrations or other factors such as temperature, mixing speed, etc.

In Figures 7.28 and 7.29, the surface plots are shown for total biomass (top) and mean weight (bottom). The surface plots were drawn by JMP when the drop down menu was selected by clicking the red triangle next to the least squares fit seen at the top of Figure 7.22. Other profiler options include cube plots, and contour profiler (as well as the surface profiler seen here). In these images, the Z axis is the responses (total biomass and mean weight), while the X axis is citric acid, and the Y axis is protein type. In studying the graphs, it becomes apparent that there are three sections for the protein, the one closest to the reader (casein), the middle one (soy), and the farthest from the reader (whey).

Figure 7.29 presents a view of the same surface graph as the one in Figure 7.28 (showing total biomass in relation to protein type and citric acid concentration). Here it is evident that the experiments where casein was the protein source yielded the highest biomass, with less biomass with whey, and no value in using soy protein. The prediction from this figure is that increasing citric acid would further benefit the *V. cardui*'s biomass accumulation. However, we must be cautious about project this trend too far; however, the results of these experiments attest to the value of citric acid and suggest that further tests of higher concentrations of citric acid could lead to higher yields and possibly greater fitness outcomes.

7.8 EXPERIMENTAL/REARING LOGISTICS AND DEVELOPMENT OF STANDARD OPERATING PROCEDURES (SOPs)

It is important to recognize that while the design of experiments (DoE) approach is powerful as a tool to help rearing systems researchers to resolve which components in their rearing systems are most important in the quest for reliable and efficient rearing optimization, there are tradeoffs that must be considered in developing experimental plans. First, these approaches, while conducive to testing multiple factors in the same experiments, the number of test setups may become unwieldy. For example, to perform a full-factorial analysis of two levels of three factors, it takes six experimental setups. To perform three levels of three factors, it takes nine experimental setups. To perform three levels of four factors, it takes 12 experimental setups, and so on. In the present experiment, four factors were treated (sweet potato flour, rice bran extract, citric acid, and protein source), and the first three were treated at two levels (amounts) and the third was treated with the same weight of three different protein sources, totalling 24 unique experimental setups. If the experimenter wished to replicate the experiments, for example three times, this would demand 72 experimental setups, i.e., making up 72 diets and running the experiment, including measuring the results (e.g., weighing the insects, counting the number of survivors, etc.) 72 times.

Second, the rearing team must decide on which parameters are most likely to be important for the development/optimization of the rearing system in question. The experiences of the team members and a search of the relevant literature are of great importance here. If a rearing system is already in place, the existing procedures and SOPs can be used as starting points and as control protocols for the experiments that are to be conducted as DOE approaches. If the rearing system is not in place in the laboratory where the work is being done, the researchers will need to consult the literature or other authoritative sources. The bottom line is that while the statistical optimization procedures are very powerful tools, they are still dependent on the expertise of the researchers and the background literature or knowledge base that exists for the insects being investigated. However, when little or no background information is available, the statistical processes and DOE-based optimization can be powerful

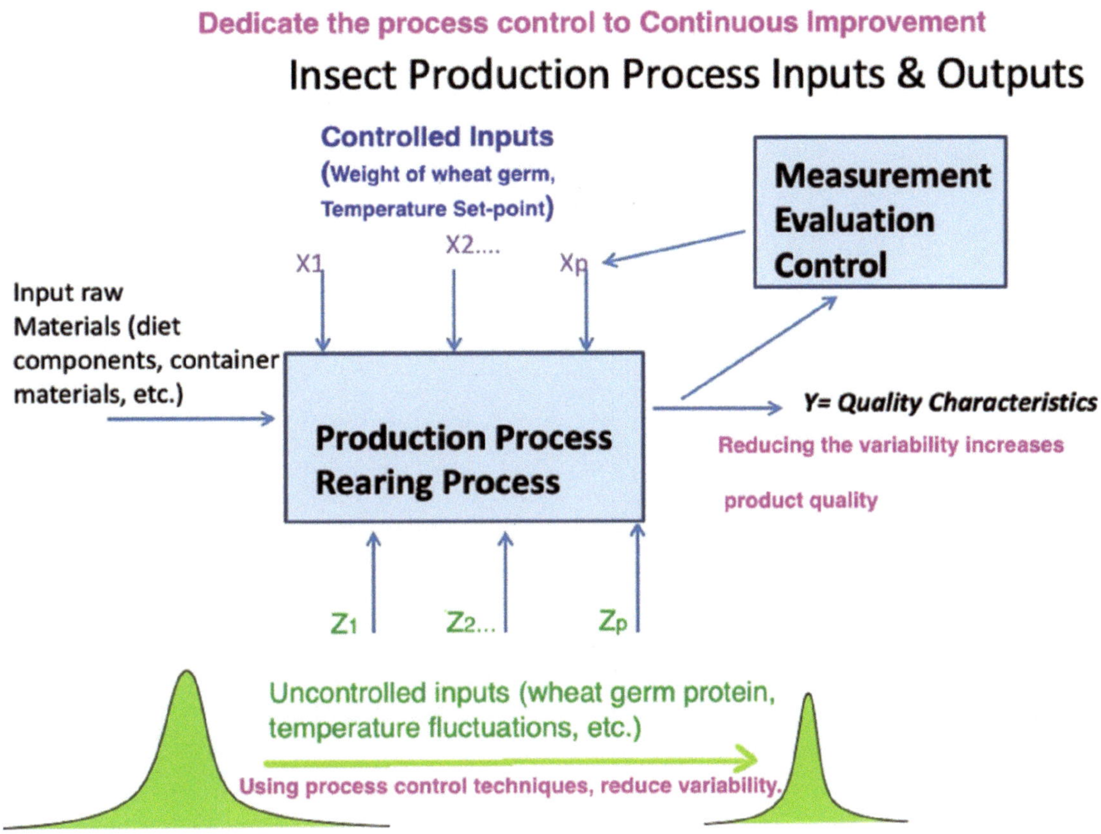

Figure 7.30 Revisited Process Control Chart.

tools to learn about the target insects in a stream-lined and strongly data-driven manner.

7.8.1 Conclusions Regarding Diet/Rearing System Optimization Studies

As a result of these studies, the diet presented in Table 7.4 was determined to be the optimal diet up to this point, and the variables that were adoped were: (1) higher amount of sweet potato flour, (2) higher amounts of citric acid, (3) lower amounts (or total exclusion) of rice bran extract, and (4) casein (in the sodium form, which is more soluble than other forms of casein such as technical grade). These tests also deemed the plastic containers (750 mL; 25 × 15 × 8 cm) were suitable for rearing ten larvae per container, with about 20 g of diet added every three days and placed on square Petri dishes covered with plastic mesh to allow frass pellets to drop away from the diet. The temperature range was 25–26 +/– 1°C; the humidity 60 +/– 8% were suitable. Light intensity of 650 Lux +/– 150, with a photoperiod of 16:8 light: dark (Figure 7.30).

All the topics in this chapter were underpinned by the concept of evaluating and then reducing the sources of error, i.e., reducing variability and the concomitant uncertainty that eminates from variablity.

The control chart in Figure 7.30 summarizes the revisions of the chart in Figure 7.3. It reflects the intentions of the chapter, and really this entire book, to dedicate the efforts of all rearing personnel to use process control in the form of reduction of error to move our rearing programs toward the philosophy of **continuous improvement**. The rearing team (be it a group of 50 rearing professionals or a single person doing part-time rearing) is to be ever vigilant to notice deviations in the rearing conditions (temperature, humidity, air circulation, lighting conditions, diet texture, presence of microbial contaminants, behavioral responses of the insects, etc.). To elevate the detection of deviations from the set-point conditions, rearing personnel must resort to data collection and then data analysis. Merely collecting data does not, in itself, promulgate improvements, but analyzing the data such as the nature of any (preferably many) features of the input of materials and other system components. As these data collections are made, the rearing personnel sort the inputs into the categories of controlled inputs (X_1, X_2, X_p) and uncontrolled inputs (Z_1, Z_2, Z_p). Then the task is to reduce the variability of the controlled inputs, and further to convert as many of the uncontrolled inputs into controlled inputs. Through this coarse tuning then fine tuning of the inputs, higher quality characteristics will result.

CHAPTER 8

Building a Process Control System

8.1 INTRODUCTION TO PROCESS CONTROL IN REARING SYSTEMS

In keeping with the title of this book, the term "design" indicates the concept that rearing systems are products of an engineering phenomenon. Although the point is often overlooked, researchers and other "rearing professionals" employ design techniques to develop and improve rearing practices. As rearing personnel carry out rearing processes, they are executing rearing "operations," and as a key factor in the need for continuous improvement, the personnel are (or should be) in a continuous quest to gain **control** over all possible aspects of the rearing system's components. The key to control is the estimation and reduction in **variation** (error = uncertainty) in the existing system. At this point, a statement from Montgomery 1991, 2009, and 2013 emerges as being of paramount importance: "**The primary objective of quality engineering is the *systematic reduction of variability* in the key quality characteristics of the product**." (My bold type, Montgomery's italics).

Figure 8.1 is adopted from Montgomery, and it shows "Distribution A" of process/product qualities as a very broad bell-shaped (standard distribution) curve. Distribution B has less "spread," and distribution C has the narrowest spread, with each successive distribution having lower standard deviation. Montgomery's point in this diagram is that quality engineering strives to take our production system from the preliminary high variability to increasingly lower variability. Figure 8.2 shows the distribution of standard deviations around the mean.

8.2 SPECIFIC CONCEPTS OF CONTROL IN RELATION TO VARIATION DISTRIBUTIONS

It is noteworthy that: (1) the 0.3% of data points that would lie outside the three standard deviations would indicate that the points are "out of control" (a concept discussed further in the quality control discussion, and (2) the basis of the Six Sigma system of quality control and product assurance is grounded in the distribution of a product's attributes within 3 sigma (three standard deviations) above and below the mean. The goal of reducing variability (indicated in Figures 8.1 and 8.2 by standard deviation) is equivalent to stating that we are trying to reduce the uncertainty that accompanies products (= insects) of unacceptable fitness or quality.

8.3 EXAMPLES FROM A CASE STUDY OF A PLANT BUG REARING SYSTEM

Montgomery 2013 also makes the point that as the production system "matures" into greater and greater control, it achieves this goal by applying routine process control and further refines operations by applying design of experiments techniques. An example of the kind of sampling that would lead to the distributions in Figure 8.1 would be a distribution of life history factors of insects (such as western tarnished plant bugs, *Lygus hesperus*) in a production system. Table 8.1 shows results from an experiment with *L. hesperus* fed diets with different kinds and different amounts of antimicrobial agents (from Alverson and Cohen 2003a), and from this table we can consider several choices as to biological (life history) outcomes such as survival rate, biomass accumulated, eggs per female (mean fecundity), rate of development, and/or duration from egg to onset of oviposition. Clearly, when comparisons are made between the control (in Table 8.1), it is evident that nearly all fitness/life history parameters for *L. hesperus* treated with various antimicrobial agents reflect adverse effects of the agents.

For example, if we select the parameter, eggs/female/day, the mean (μ) fecundity for the control was 26.77 ± 1.95σ, meaning that 99.7% of the females would have produced between 20.92 and 32.62 eggs per day, provided that the females were from a population that was conforming to the "normal" established and expected fecundity values. By comparing the values for fecundity of the other fecundity measurements in this experiment, we find a number of departures from the "within tolerance" range of oviposition rates. For example, most of the

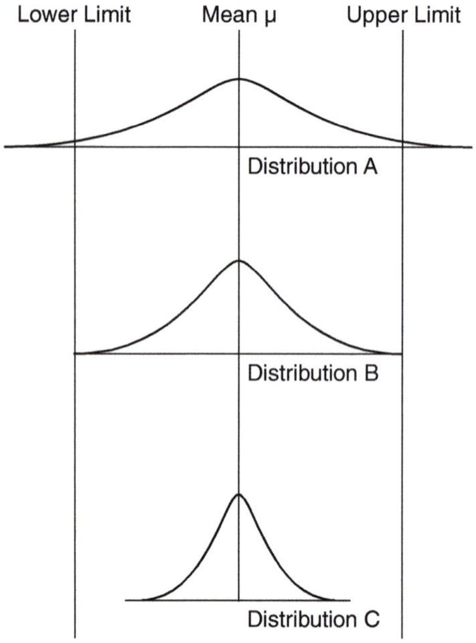

Figure 8.1 Reduction of variability illustration.

L. hesperus fed 1,000 ppm formalin would be detected as being "out of tolerance," and all the *L. hesperus* fed any methyl paraben (from 1,000 to 2,000 ppm) would be out of tolerance. It is noteworthy that the Alverson and Cohen experiments were designed to test the effects of additives that were being considered for use in colonies that were threatened by mold contamination. However, often, we find in our colonies that deviations take place from "within tolerance" ranges for substances or conditions that are not being controlled. Therefore, it becomes clear that having the baseline established for various biological, chemical, and physical factors can provide a sort of experimental control for our colonies—a control that simulates the type of control that we utilized in the antimicrobial agent experiments.

The statement about developing a baseline in the above paragraph can become a driving force for the remainder of the process control commitment. If we revisit the *Lygus hesperus* rearing system discussed in the previous paragraphs, we discover that the rearing system was originally developed after Debolt 1982 reported the first artificial diet that would sustain successive generations of *L. hesperus* and that a gel packet made of carrageenan/water gel would stimulate oviposition. The colony had been reared for about 50 generations on the Cohen 2000c diet, and the control population from Table 8.1 was used to establish the baseline for comparison of various antifungal components that were being considered for use to replace formalin due to risks to rearing personnel associated with this material's carcinogenicity. The type of data collection reported in Table 8.1 could be used not only to determine changes in means of various fitness parameters but also the possible reductions in variability that we are continually trying to reduce. The model of variability reduction in

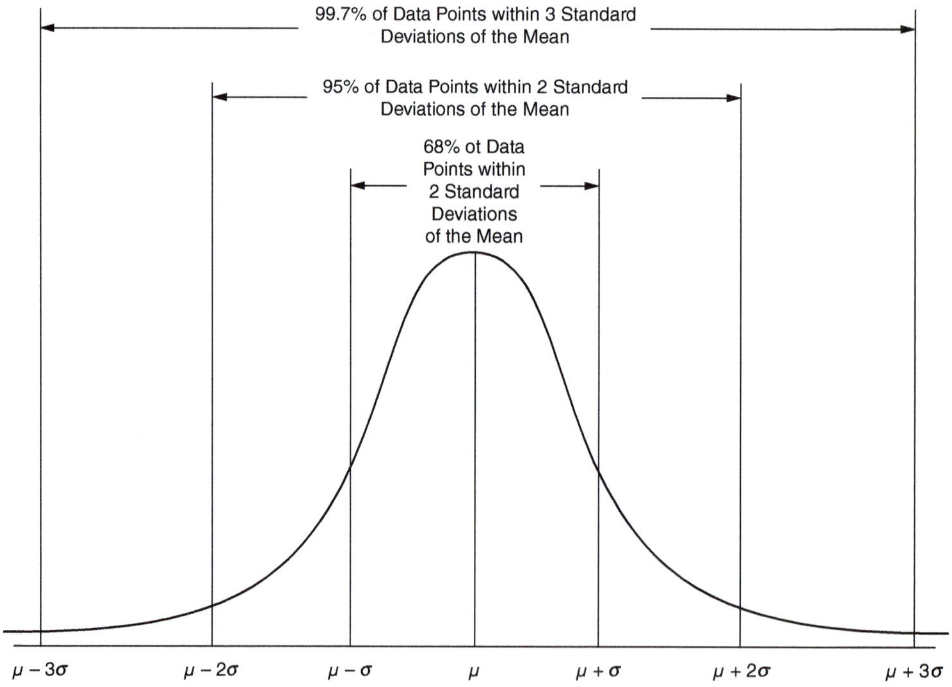

Figure 8.2 Distribution of points in a standard distribution, "bell-shaped curve."

Table 8.1 Biological (Fitness) Parameters in a Rearing System for the Western Tarnished Plant Bug, *Lygus hesperus* (Data from Alverson and Cohen 2003a)

Antifungal Agent	Concentration (ppm)	Total No. of Survivors (+/– SE)	Total Biomass (Dry Wt. mg) (+/– SE)	Eggs per Female per Day (+/– SE)	Days to Adult Emergence (+/– SE)	Days to Oviposition (+/– SE)
Control		147 (3.0)a	834 (48.4)a	27 (1.95)a	16.3 (0.7)d	19.0 (0.6)a
Benzoic acid	2,000	137 (4.2)a	704 (51.2)abc	21 (2.58)abc	17.0 (0.0)cd	20.0 (0.0)a
	3,000	146 (3.0)a	773 (31.1)abc	20 (2.90)abc	17.0 (0.0)cd	20.3 (0.9)a
	4,000	119 (19.8)a	643 (70.9)abc	18 (3.43)abcd	16.7 (0.3)cd	20.3 (0.7)a
Formalin	500	123 (23.0)a	671 (120.7)abc	17 (3.39)bcd	17.3 (0.3)cd	20.7 (0.7)a
	1,000	123 (16.8)a	627 (64.8)bc	16 (5.29)cd	17.7 (0.7)cd	20.7 (1.2)a
	1,500	130 (3.5)a	681 (44.4)abc	16 (2.03)cd	17.3 (0.3)cd	20.3 (0.9)a
Methyl paraben	1,000	117 (9.2)ab	581 (24.3)c	11 (1.46)d	17.7 (0.3)bc	21.0 (0.0)a
	1,500	78 (25.0)b	303 (114.0)d	1 (0.92)e	18.7 (0.7)b	25.0 (1.5)b
	2,000	23.3 (8.7)c	91.9 (54.3)e	0 (0.00)e	21.7 (0.3)a	26.0 b
Propionic acid	800	117 (14.7)ab	685 (67.1)abc	25 (1.98)ab	16.3 (0.7)d	19.7 (0.7)a
	1,200	127 (12.3)a	706 (52.3)abc	21 (3.08)abc	17.0 (0.0)cd	20.0 (0.6)a
	1,600	126 (1.5)a	696 (11.7)abc	24 (3.15)abc	17.0 (0.0)cd	19.7 (0.3)a
Sorbic acid	800	137 (9.6)a	781 (57.6)ab	22 (4.19)abc	17.0 (0.0)cd	20.0 (1.0)a
	1,200	129 (14.3)a	773 (65.8)abc	22 (1.69)abc	17.0 (0.0)cd	20.7 (0.3)a
	1,600	116 (18.1)ab	657 (104.1)abc	20 (5.29)abc	17.0 (0.0)cd	20.7 (0.9)a

Lower-case letters following means and standard deviations indicate that means with the same letters are not different at the 0.05 level of significance.

Figure 8.1 can be used as the goal of the process/quality control personnel in insectaries as a whole (Figures 8.3 and 8.4).

Figure 8.3 shows the configuration of the Lygus mass-rearing system used at the USDA, ARS Biological Control Mass-Rearing Research Unit (BCMRRU) at Mississippi State, MS, to rear western tarnished plant bugs and tarnished plant bugs *Lygus hesperus* and *L. lineolaris*, respectively. About 20,000–40,000 plant bugs (*Lygus* spp.) were kept in the mesh cages with packets of artificial diet and oviposition media (= carrageenan gel). The rearing system originally developed by Debolt 1982 and Patana and Debolt 1986 and modified by Cohen 2000c and 2003a involved the feeding of the insects through the cage's mesh and through a layer of Parafilm composing a diet packet. The adult females were also able to oviposit into the gel packets through the mesh and Parafilm layer. The Debolt diet for *L. hesperus* was found to be very complex and expensive to produce, and it did not adequately support production of *L. lineolaris* prompting studies of diet improvements that would alleviate these problems. The Cohen 2000c and 2003a diet incorporated full-fat soy flour and several other components that allowed the diet to be formulated without expensive purified chemicals such as linolenic and linoleic acid, salt mixtures, and casein as well as other simplifications. The Cohen-led research team at the USDA, ARS lab at Mississippi State also optimized an oviposition packet formulation consisting of 4% carrageenan (= Gelcarin 812), that included potassium sorbate salt that reduced mold infestations in the egg-collection packets and enhanced the gelling properties of the carrageenan. These optimization steps permitted the BCMRRU to step up the mass-rearing system to a production of about 100,000 adult *L. lineolaris* per week, serving as the basis for biological control and other experimental procedures aimed at control of this pest. It must be noted, however, that the improvements in the mass-rearing of tarnished plant bugs and western tarnished plant bugs were made in incremental steps because we were hindered by not knowing (at that time) the benefits of using multi-factorial design of experiments engineering procedures.

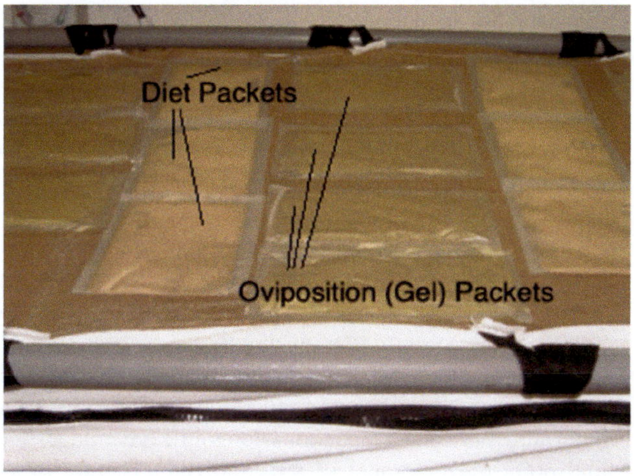

Figure 8.3 *Lygus* mass-rearing system.

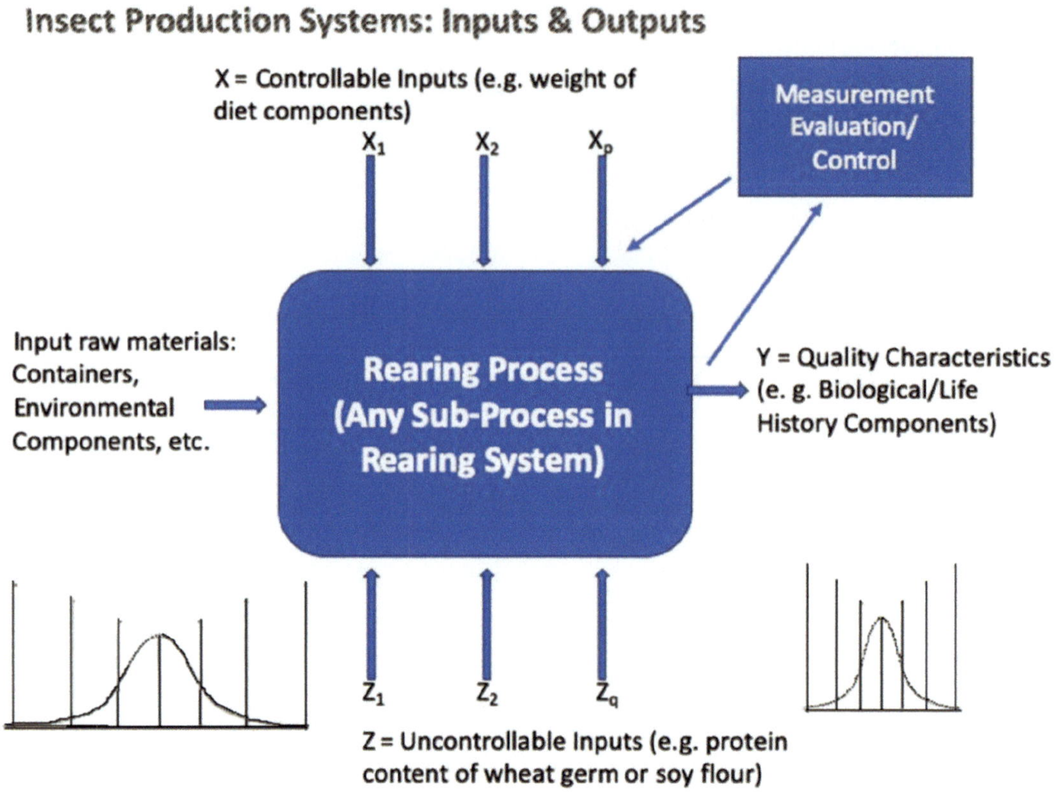

Figure 8.4 Inputs and outputs in rearing production systems.

The efforts were what we would now consider basic and modest advancements in process control.

8.4 WHY DO WE INVEST SO MUCH EFFORT IN PROCESS CONTROL?

If the recommendations in this chapter and throughout this book are followed, it would mean a considerable expenditure of time, effort, and other resources collecting data, performing experiments according to the demands of design of experiments, meetings to interpret and discuss the measures taken toward PC-development. It would be natural for the reader to ask if these efforts and costs are worthwhile. The following matrix is an effort to answer the questions about the value of PC:

1. What we accomplish by using process control
2. Reduce variability (i.e., reduce error or uncertainty)
3. Reduce losses or risk of quality deterioration
4. Develop key inflection points for problem-solving/trouble-shooting
5. Recognize what needs further study
6. Reduce costs of rearing
7. Provide formalized structure for basis of changes in processes needed to improve quality, production efficiency, production volume (or numbers), and reliability (i.e., this is part of a continuous improvement strategy)

These seven advantages of using process control are the reasons or rationale for any or all statistical process control efforts, and the main focus is on reduction of variability that is ubiquitous and found in every system and process. The most basic underlying tenet of process control is measurement or ascertaining the nature of the process in ways that allow repetition, replication, and evaluation. The general tenet translates to the imprecation made by Montgomery 1991 (and Shewhart and Deming before him): collect data on every possible and practical aspect of the system. In a rearing system, this data collection could mean measuring and keeping records (data sets) of environmental conditions, background and current characteristics of the target insects, the diet components and diet characteristics, features of the containers, and the array of features that may influence the rearing process (such as noises, odors, changes in personnel, etc.).

It should be understood that in the reduction of variability, the other advantages automatically accrue, reduced losses of quality (reduced deterioration), spotlighting the points in the system that need more attention (further study) and approaches to problem-solving, and along with the increase in quality and reduction of wasted time and materials, there is a tremendous economic advantage. The final point (development of a formalized structure or set of protocols) is best characterized as establishment of infrastructure.

Some of the most basic contributions to building infrastructure in insect-rearing systems came from the pioneering work on Drosophila where such authors as Delcourt, Guyenot, Baumberger, and Morgan established the importance of careful control of rearing conditions. Some of the most explicit statements about control in rearing infrastructure come from one of the young researchers in T. H. Morgan's lab, Calvin Bridges who wrote:

> In 1913 I built above my laboratory table a large wooden incubator which held all my experimental cultures. This incubator was heated by carbon electric lamps placed in the lowest shelf space. The thermostat was of the "ether-wafer" type. The expansion of the ether forced apart contacts in the heating circuit. A six- inch electric fan continuously circulated air down a gap behind the shelving with return through a similar gap between the shelves and the doors.
>
> Shortly after this, two other large incubators were equipped for other workers. In these the thermostat contacts controlled a relay which turned the lamps on and off. ***Plough's work on the effect of temperature extremes on crossing-over showed that it was necessary to maintain the temperature constant for genetic reasons as well as to improve culture conditions.*** (Bold and italic font added by Cohen)

Bridges 1932 and Bridges and Darby 1933 wrote about the importance of using controlled conditions both for the benefit to the colony and for the "genetic reasons" where what the researchers were trying to explore in terms of a genetic basis could be altered or influenced by environmental conditions such as temperature. Controlled experiments on genetics (or any other biological feature of our target organisms) mean that all factors are controlled so that we can know that the results we observe are due to the factors that we are testing, rather than some stray factor (which we can now regard as "special causes," which are discussed in detail later in this chapter) (Figures 8.4 and 8.5).

The major goals of process control are illustrated in the diagram in Figure 8.4, which was modified from Montgomery 1991 and 2009. The goals are to convert uncontrolled inputs (Z factors) into controlled inputs (X factors). The diagram indicates that through the process of converting what had previously been uncontrolled inputs and sub-processes to controlled ones will lead inexorably to the improvement of the system's quality output. It must also be noted that part of this system of improvement is the decrease in error or variability of each step in the overall rearing process. This means that components and process steps that had some degree of control are continuously being refined to have less variability or greater control. The steps can be defined as follows:

1. Changing uncontrollable inputs into controlled inputs
2. Eliminating **special causes** (= assignable sources of error or production deviations)
3. Reducing the variability of common causes
4. Supporting efforts at continuous improvement with data-driven, hypothesis-based improvements in rearing programs

How do we proceed through developing process control procedures in our rearing system?

1. Determine the most important parameters on a per system basis
2. Make measurements of parameters **while the rearing system is working well**
3. Start developing control charts with upper and lower control limits (3σ above and 3σ below the mean)
4. Determine instances when process is going out of control (i.e., determine **special causes** of deviation while also developing **common causes** of deviation)
5. Make corrections of special causes
6. Continuously make adjustments to reduce the variability of parameters (going from "Distribution A" in Figure 8.1 increasingly toward "Distribution C" in that figure)

Rewards for applying process control in rearing systems (large and small systems):

1. Avoidance of problems such as colony collapse and decline in quality
2. Reduction of waste and incumbent costs of waste
3. Greater control over the product (numbers and quality or fitness of insects)

8.5 BACKGROUND FROM PIONEERS OF QC/PC: SHEWHART AND DEMMING

These points further reinforce the central concept of this book and of process and quality control systems in general: reducing the amount or scope of variability (error) will inherently increase the quality and reliability of the system. The increase in quality and reliability is inextricably connected to an increase in the system's value or the value of the products.

In discussing the kinds of error/variability/uncertainty in process and systems several authors such as Shewhart and Demming have provided an overview that relegates the causes as **special** or **common**. Common causes are often thought of as random, natural, regularly occurring variation while special causes are unusual and unpredictable. Special causes are "surprises" that may result from a power failure during a process, a contamination problem that is connected to an unusual weather phenomenon (as I write this, I am dealing with mold problems in my lab associated with Hurricane Florence), or other unpredictable errors. Special cause uncertainty (or deviation from the rearing norm) is often large-scale or catastrophic. Therefore, it is important to determine special cause error

Table 8.2 Dealing with Special Causes

Examples of Special Causes	Procedures to Remedy Problems
1. Climate control system goes out	Repair/replace system
2. A batch of diet is ruined by over-heating (or under-cooking)	Make a new batch of diet
3. A mold outbreak sweeps through the insectary, related to surge in humidity	Determine cause of humidity surge and repair it (leaks, etc.) Install dehumidifiers
4. A diet component (wheat germ, soy, yeast) arrives in a deteriorated (extremely low-quality state)	Replace component with fresh, vetted material
5. A new pathogen enters the rearing system	Identify the pathogen; determine its mode of entry; correct or block that mode of entry
6. A new material (container, lidding, oviposition substrate, etc.) is defective (poor air exchange from lidding, escapes)	Identify/isolate the problem and replace the defective material
7. A routine procedure is violated by a new staff member	Identify the problem and re-train staff member

and eliminate it. Here are some examples of special cause sources of error and the measures to correct or eliminate them (Table 8.2).

8.6 DEALING WITH SPECIAL CAUSES IN A REARING SYSTEM

Although many of these special causes may have cropped up unexpectedly, having carefully developed standard operating procedures (SOPs), well-trained personnel, and an alert process control person/team can often anticipate these and similar problems (surprises). I offer here some measures that can be put in place as part of the rearing system's SOPs. For example, **Special Cause 1** may have been averted with regular checks on the climate control such as vibration/noise checks on compressors, air pumps/air movement systems. Routine cleaning of ventilation system, checks on cooling fluid levels, and regular replacement of air filters (or replacement of the cooling/heating system itself) can avert the catastrophic losses of an entire colony. Many rearing facilities have the luxury of having extra incubators or rearing rooms that are redundant, and some rearing programs have systems of alarms that notify (via computer alarm capabilities) key personnel that a failure is underway. The responsible personnel can take action of correcting the climate failure before it has caused the colony to die off.

Special Cause 2 can be noticed before the diet is presented to (inoculated with) the target insects by using established range of temperatures and cooking times, and this can be further safeguarded against by having, for example, a texture determining step at the end of the diet-making procedure. A properly prepared gelled diet will have a narrow range of textures that are not too hard or too soft, and this can be determined with an appropriate rheological device such as a penetrometer or texture analyzer. The texture test (or pH test or some kind of chemical test such as an enzyme activity test for peroxidase or phenoloxidase, for example) can provide the diet synthesis personnel key information about whether the diet was made properly.

In my experience, **Special Cause 3**, an unexpected mold outbreak, is one of the most important and prominent causes of mass die-offs in rearing systems.

Special Cause 4, where a standard component of a diet or some other part of the rearing system is defective upon arrival, is another frequent occurrence. Often a supplier that was deemed an acceptable source for rearing materials changes a material or allows the material to become defective, and the rearing personnel who acquire the material are caught unaware that the material was defective.

Special Cause 5 (unanticipated appearance of a pathogen) has derailed rearing programs that were otherwise productive and reliable. **Case Study X**: For example, in our rearing program in the Biological Control and Mass-Rearing Research Unit at Mississippi State University (BCMRRU, USDA, ARS), we had an outbreak of a microsporidian in the *Lygus hesperus* culture. The colony was originated from a stock colony originated in USDA, ARS Biological Control Laboratory in Tucson, AZ (Debolt 1982) and was adopted for a program established to mass-rear *Lygus* parasitoids. The pathogen was of undetermined origin, but it could have come from the original colony, or it could have come from *Lygus lineolaris* individuals that were derived locally from alfalfa fields near the rearing facility at Mississippi State's Gast Mass-Rearing Laboratory.

Special Cause 6, a new material (container, lidding, oviposition substrate, etc.) is defective (poor air exchange from lidding, escapes). Especially when upscaling the numbers of insects in developing a mass-rearing system from a lower level of production, containers and their associated materials often lead to problems that require optimization of gas exchange, access to diet, and prevention of cannibalism or other population density-based interactions.

Special Cause 7, a routine procedure is violated by a new staff member: as is the case with many failures from special causes, there is little published documentation of this phenomenon. However, in my experience, this is a widespread occurrence in rearing systems at all scales from laboratory bench colonies to mass-rearing facilities. In the USDA, ARS Biological Control Laboratory in Tucson and in the BCMRRU lab in Mississippi, we kept careful records on the productivity of our colonies.

8.7 DATA COLLECTION

One of the most perplexing and far-reaching questions that students of rearing system process control ask, is what kinds of data do we collect? The follow-up questions pertain to treatment of the data, once it is collected.

8.7.1 What Kinds of Data Do We Collect?

It is important to have a firm knowledge of the raw materials in diets and in other portions of our rearing systems. Because there are so many raw materials that go into our rearing systems, it is essential to be selective about which components merit the most attention. The decision about whether to pay more attention to a component such as wheat germ or soy flour vs. a component such as sugar must be based on the likelihood (= probability) that the component is likely to cause error in the system and that the error would be of greater significance than that caused by components of lesser importance. Personnel with knowledge about diet components would know that wheat germ and soy flour are: (1) major nutrient sources in a diet, (2) very complex in their biochemical (nutritional) composition, (3) likely to vary from source to source, and (4) relatively unstable. In contrast, sucrose sugar is generally very pure (table sugar off the shelves of grocery stores is considered to be one of the purest foods available). Sucrose is very stable, so storage tends to be safe in terms of product deterioration or contamination. Other components that are simple (rather than complex mixtures of nutrients and feeding stimuli such as wheat germ, wheat bran, rice flour, soy flour, etc.) may pose other problems that would make them subject to data collection and scrutiny.

For example, a purified protein such as casein (from bovine milk) or purified soy proteins can have multiple variations from product to product or from batch to batch. Taking casein as an example, far too often diet formulations such as Yamamoto 1969 list casein without any further specification. However, as pointed out by Damodaran 2008 and Cohen 2015, the word "casein" can be very ambiguous. There are several types of bovine milk phospho-proteins called caseins (αS1, αS2, β, κ) which include different fractions of milk whose solubilities and pk indicate different physical, chemical, and nutritional properties as well as different nutritional values. The associated moieties such as phosphate groups, sugar, lipids, and vitamins confer important differences in the nature of caseins in insect diets (Cohen 2015). To further complicate our grasp of caseins in diets, there are potassium and calcium salts of casein molecules (sometimes known as "caseinates") that can complicate the role and value of casein in a given diet. The point of this discussion is that data should be maintained about which casein is being used in our diet.

8.7.2 Undertaking Data Collection

The kinds of data must fit the following requirements:

1. They must be objectively measured
2. They must be easily measured
3. There must be appropriate timing of data collection
4. The data must be meaningful (relevant to your rearing situation)
5. Data collectors (personnel) must be educated in how to properly collect the data

Descriptions such as weight in grams, linear dimensions, wavelengths, light intensity, temperature, relative humidity, concentration, etc. are examples of objective measurements. Descriptive terms such as "heavier, brighter, longer, etc." are subjective and inappropriate for the statistical analysis process that is our ultimate goal. As much as possible, the types of data that we collect should be easily measured. The weight of a wheat germ sample is easier to measure than the percent protein or percent lipid of the sample. The amino acid concentration of the protein in wheat germ is even more difficult to measure, and while it is very pertinent to the rearing situation, it is generally unrealistic for most rearing laboratories to be able to routinely make measurements of the amino acid concentrations. The third character of good data measurement systems is that what is being measured is related to the outcome. Establishing the relationship between measurements of processes and components that occur early in the rearing cycle with the outcomes and collective endpoints of those processes is a key feature of a good data collection system. That is to say for example, if a treatment of adults laying eggs early in the rearing cycle predisposes late instar larvae, pupae, or adults to respond abnormally, it is important to tie the early adult/egg treatment to the later outcome. The following diagram represents a 15-step rearing cycle representing the rearing process for painted lady butterflies (described at length in Chapter 7). Beginning with adults feeding on artificial nectar (e.g., diluted honey), Step 1, in Generation I, would be the process of feeding adults with adequate amounts of nutrients in appropriate amounts. Step 2 would be the allocation of nutrients to the eggs being loaded with yolk proteins and other nutrients, and Step 3 would be the oviposition of the eggs on the surface of the mallow leaves. Step 4 would be the egg handling conditions prior to hatch; and Step 5 would be the hatching of the eggs and their inoculation onto the diet. The steps proceed through larval development, pupation (formation of the chrysalis), and finally, the emergence of the adult that is starting Generation II:

$1 \rightarrow 2 \rightarrow 3 \rightarrow 4 \rightarrow 5 \rightarrow 6 \rightarrow 7 \rightarrow 8 \rightarrow 9 \rightarrow 10 \rightarrow 11 \rightarrow 12 \rightarrow 13 \rightarrow 14 \rightarrow 15$

However, if the artificial nectar were somehow deficient because of excess dilution, contamination by microbial factors, deterioration because of excess light or heat, then the nutritional allotment of the adults might be inadequate to support to the full reproductive potential of the adults that are starting Generation II. This is a matter of maternal allocation of nutrients to eggs, and inferior allocation of some of the eggs could cause the eggs that were undernourished to have a cascade of problems such as deformities, sub-standard size, or disproportionate mortality (Figures 8.5–8.7).

Starting at the top of the chart (Figure 8.6), your rearing/QC/PC team must first identify the problem areas or the rearing issues that you want to address ("identify the problem areas"). For example, you might set the topic as increasing the fitness of your *Lygus* (tarnished plant bug) colony. Upon further discussion, you may decide to focus on: (1) increasing biomass of the adults per rearing container, (2) decrease mortality between neonate stage and adult stage, and (3) increase the number of eggs per female (see Table 8.1). In working on these problems with a team, it is often useful to use one of the total quality management (TQM) tools known as "mind-mapping."

In Figure 8.7 "Mind-Mapping," is illustrated to show how a process control team has been doing brainstorming about what factors can be modified to improve rearing outcomes or to reduce problems in the colony. Starting with the insects in the center of the diagram, someone suggests that the culture's history may be of great importance. Someone else suggests that the diet is also a possible source of departures from optimal fitness, while others suggest environmental factors are bound to play key roles in quality and fitness. In a well-run brainstorming and mind-mapping session, initially all ideas are recorded, possibly with yellow sticky notes that can be added to or moved as the

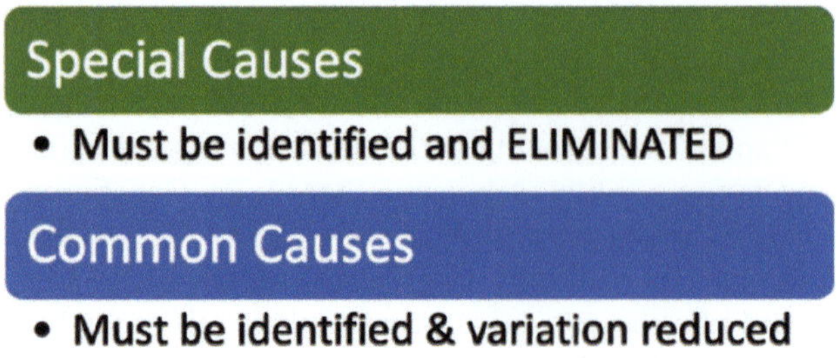

Figure 8.5 Identifying and eliminating special causes and common causes.

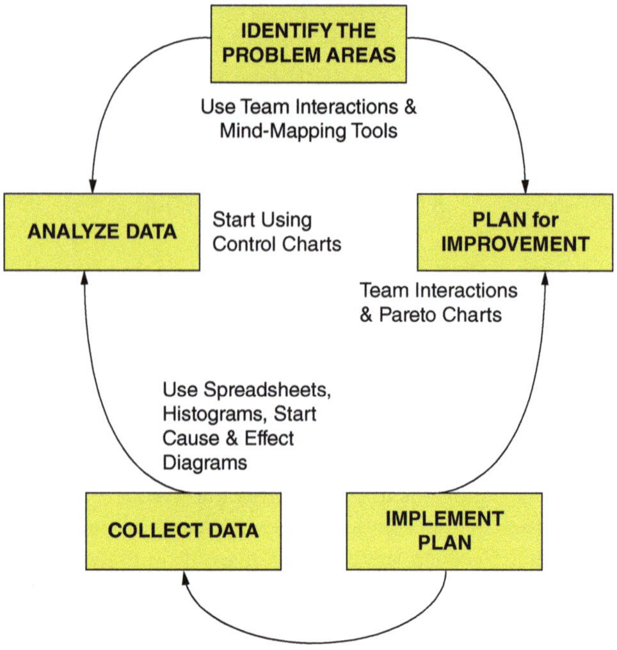

Figure 8.6 A plan-do chart.

group focuses in on the major issues that they perceive. No idea is rejected or treated with disrespect; and this rule of brainstorming allows everyone to feel comfortable participating in this productive practice. Someone with a little background in nutrition may suggest that the lipids in the diet are bound to be very important, and if problems with growth or molting are observed, possibly this is due to some kind of lipid deficiency. It is possible that at this early point the lipid (or protein or vitamin) suggestion may be ahead of its time, but the idea goes up on the whiteboard or table where the mind-mapping is being done.

8.7.3 Cause-and-Effect Diagrams (Fishbone Diagrams)

One of the most effective total quality management (TQM) tools for developing process control systems is the Fishbone Diagram, also known as a "cause-and-effect diagram" or an "Ishikawa diagram." To create an interactive form of the diagram, various online sources offer templates, and some statistical systems such as JMP (from SAS, Cary, NC) provide templates. However, we have found in my laboratory that a highly versatile template can be created in PowerPoint, with an example of the diagram provided in Figure 8.8 (cause-and-effect diagrams I, II, and III). After the brainstorming described above and the mind-mapping, the PC/QC team can convert their mind-map into a preliminary cause-and-effect diagram, such as the one here where the main components that could cause failures are identified as diet factors, containers, microbial factors, environmental factors, personnel factors, and

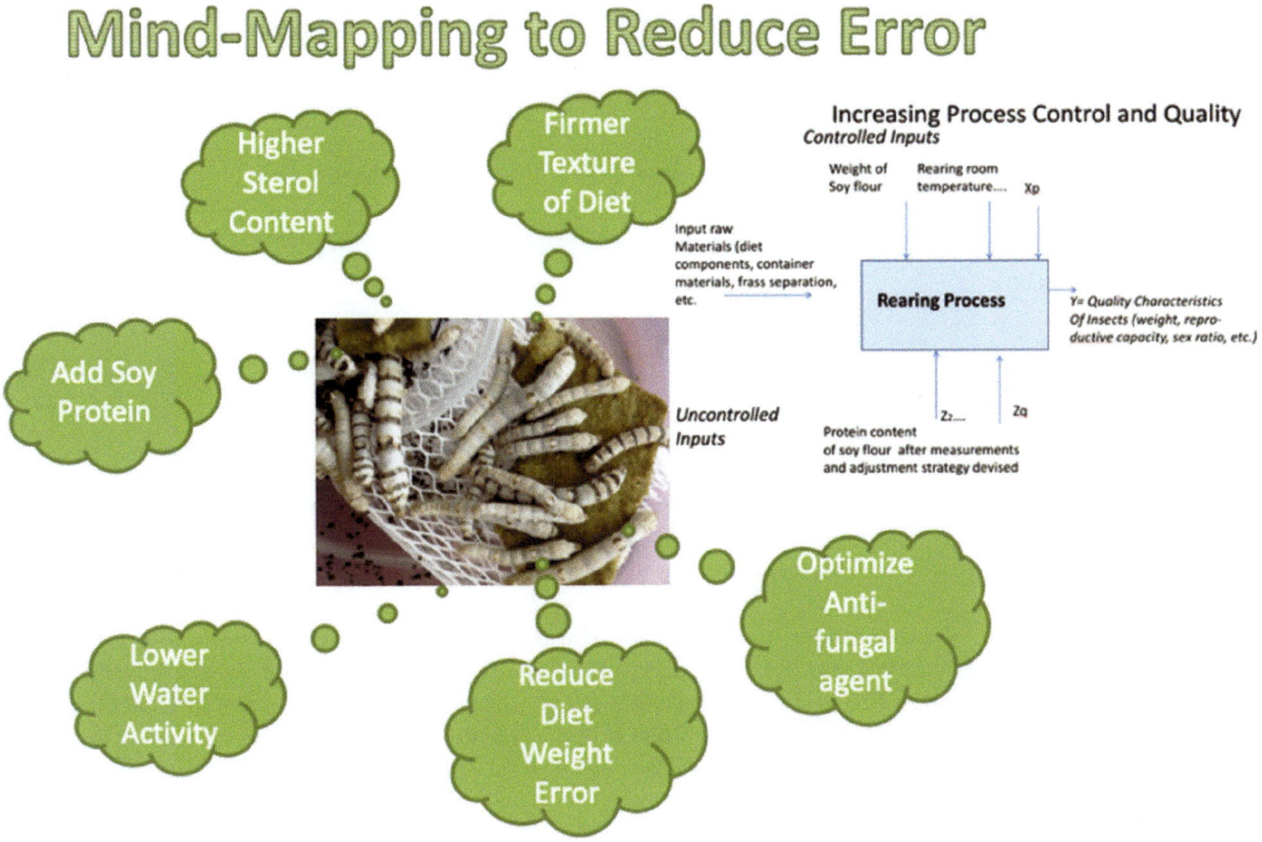

Figure 8.7 Mind-mapping technique for improvements of fitness.

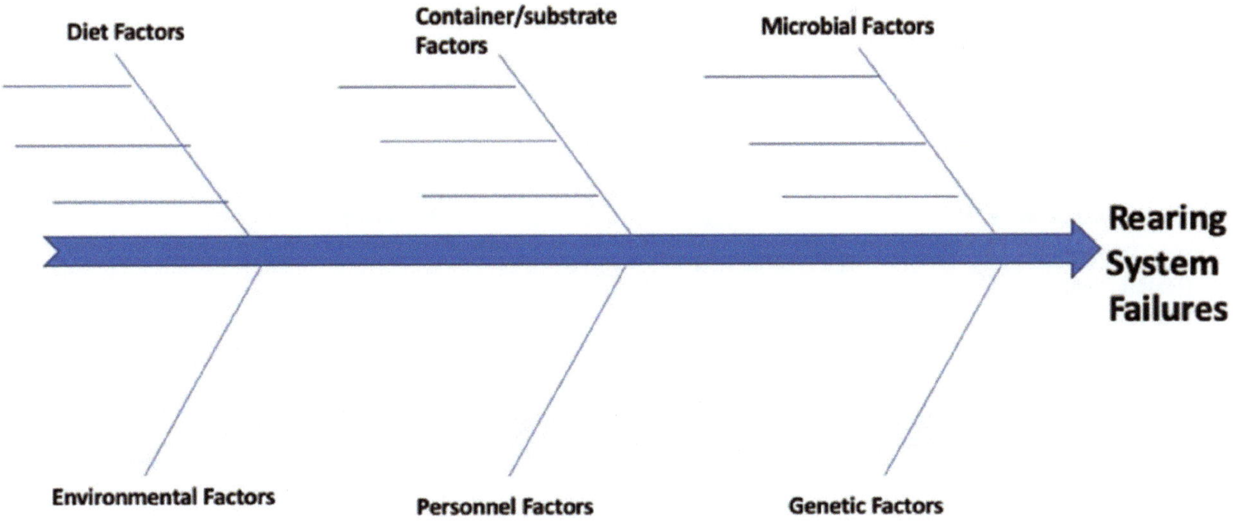

Figure 8.8 "Cause-and-effect I" shows a preliminary diagram designed to help insectary personnel identify causes that lead to the result: rearing system failures. As the "process/quality group" works on the causes of failure as they currently understand them, they present the relationships as sub-groups that lead to the "backbone": rearing system failures depicted with the main arrow. As the PC/QC team further discusses possible sources of rearing system failures, the topic, "Diet Factors," may be broken down further into such possible culprits as component freshness, problems with dietary sources (such as the supplier of soy flour or agar having to be replaced), or possible diet processing factors (changing mixers or heating devices). These sub-topics of "Diet Factors" can be listed as the "fish skeleton" articulations listed in Figure 8.9 under "Diet Factors."

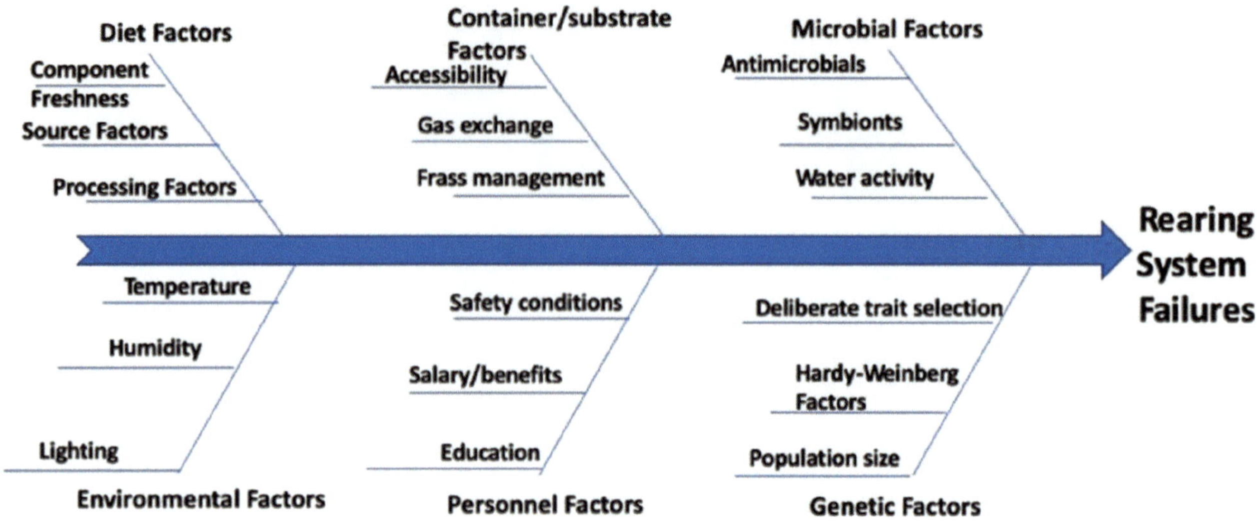

Figure 8.9 More advanced cause-and-effect diagram for "rearing system failures" analysis.

genetic factors. It should be noted that the mind-mapping and cause-and-effect diagramming are tentative. They can be further explored or rejected without disrupting the process of rearing system exploration. The more that participants know about the process(es) in question, the more likely the direction of the process exploration will be productive toward problem-solving. However, even a naïve team can reach higher points of understanding their system by using these processes efficiently (Figures 8.9 and 8.10).

In Figure 8.9, what had been begun in Figure 8.8 was developed further as a result of brainstorming and/or further data collection. At first, the suspected causes of a rearing system failure were proposed to be diet factors, container/substrate factors, microbial factors, environmental factors, personnel factors, and/or genetic factors. However, upon further inquiry, the six factors specified here were expanded to include several sub-sets of each factor. The value of this type of analysis and the diagram that serves the analysis is that as new causes are discovered, they can be added to the diagram. The additional causes can then be further analyzed with Pareto analysis to help the quality/process control team refine the system. For example, when the environmental causes were shown to be of central importance in rearing system failures (Figure 8.10), this factor can be treated as a main cause, and its underlying causes or factors can be further investigated until the system is completely stable and under control.

In Figure 8.10, the PC/QC team may have delved in depth into the environmental failures, and then that topic can become a cause-and-effect diagram in itself. The

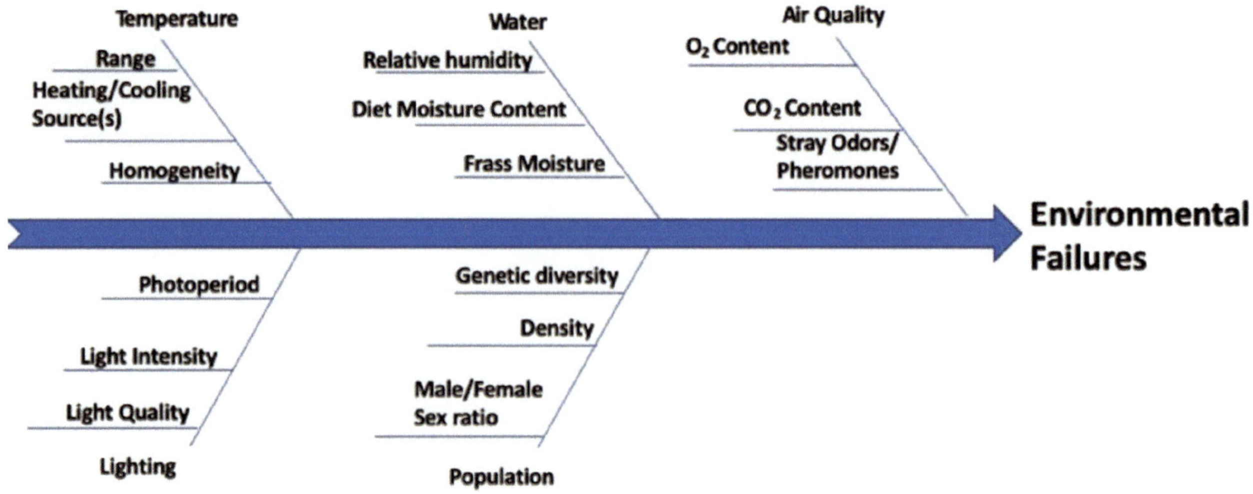

Figure 8.10 Amplification of the environmental causes of rearing system failures.

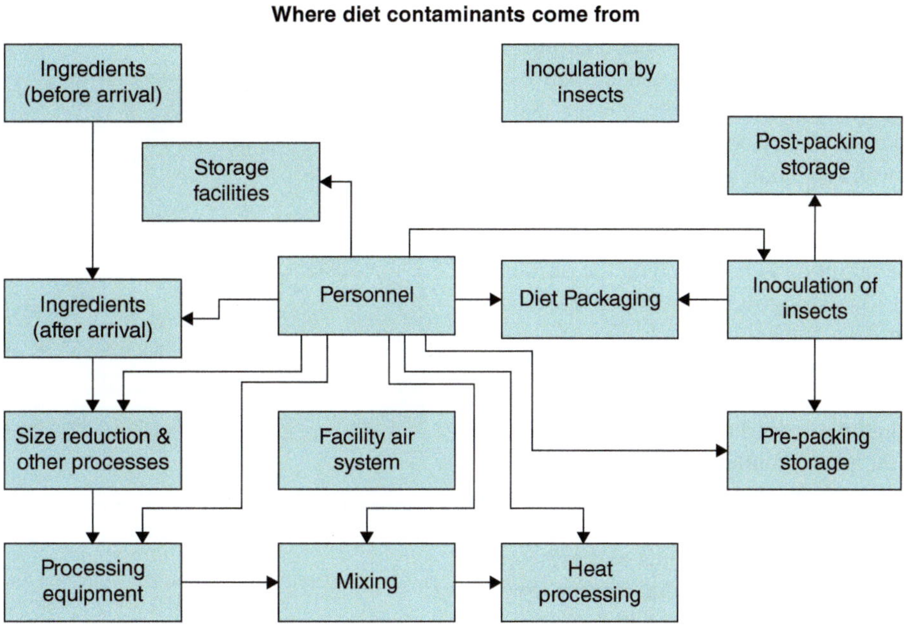

Figure 8.11 Flow diagram of possible sources of microbial contaminants in insect diets.

environmental factor, temperature, can be broken down into the range of temperature: possibly the range is inappropriate to the insect in question (too great a range or too narrow). The heating and cooling sources may be problematic in terms of evenness of temperature throughout the rearing room. Under "Lighting," various light-related issues can be problematic such as the photoperiod (light: dark phases); or light intensity may be inappropriate (too bright or too dim); and quality (color spectrum) of the light source may be an issue. These topics are discussed in more detail in Chapters 6 and 10. In summation, the various sources of failure, once identified, can then become topics of study, beginning with data collection. For example, for the environmental sources of failure, the team can obtain (usually inexpensively) instruments for collecting data on environment, light meters, recording thermometers, and hygrometers (such as data loggers). The team can also employ techniques for determining population density, oxygen and carbon dioxide concentration, and many other environmentally pertinent facets of data.

8.7.4 Flow Diagrams as Tools for QC/PC

Along with mind maps and cause-and-effect diagrams, **flow diagrams** are a useful way of dealing with planning infrastructural principles. In Figure 8.11, a flow diagram (adapted from Cohen 2015) is presented to help track the sources of microbial contaminants in rearing systems in general and specifically in diets. The flow diagram shows that personnel are considered here to be the most prominent source of microbial contamination, where they directly handle ingredients, perform the various steps (handling new raw materials, grinding, sieving, mixing, etc.) in diet production. The other identified sources of contamination include the insects, the air-handling system, and the raw materials before they arrive in the rearing facility. Figure 8.12 shows mold growing on a bag of soy protein (Nutrisoy) at the Gast Facility (USDA, ARS, BCMRRU, Mississippi State) in a walk-in refrigerator kept at about 4°C. This is an example of

Figure 8.12 Mold-contaminated bag of Nutrisoy (from ADM, Archer Daniels Midland) at the USDA, ARS Gast Facility. This diet component had been stored for several years in the walk-in refrigerator seen in this photograph. The mold on the outside of the package raises questions as to the quality of the product within the packaging. A question for the process control/quality control personnel is whether or not this material should be used or discarded.

microbial contaminants coming from the storage facility, as depicted in Figure 8.11. The recommendations from most manufacturers/suppliers for products such as soy protein, soy flour, wheat germ, vitamins, etc. are that they are to be kept under cool, dry conditions.

The walk-in was kept cool, but not dry enough to prevent opportunistic mold from forming on the package, and possibly in the package as well. Because the soy protein's arrival date at the facility was not noted on the packaging, we had no idea of how long the product had been stored there prior to the development of a mold bloom—a basic demonstration of how important it is to date all incoming materials (Cohen 2015). It is also not clear as to whether any of the mold had penetrated the packaging and started growing within the paper container. However, the potential of raw materials to develop problems with microbial contaminants and other kinds of degradation (moisture uptake, oxidation, infestation by rodents and arthropods, etc.) increases with the duration of storage and with other factors such as climate and sanitary conditions. In one of the rearing facilities where I had served, I once found several bags of wheat germ that someone had discarded into a dumpster. Curious about why bags of an expensive material had been discarded, I reached into the dumpster to try to sample the material (one bag having been torn open). I found that the opened bag and several others that were unopened contained wheat germ that had the consistency of a clay brick! I wondered: (1) how long the wheat germ had been stored, (2) what kind of changes had occurred in the nutritional and palatability qualities of the hardened material, and (3) how often do rearing personnel use such materials, even though they required being chiseled out of the storage container and pulverized with a hammer. I had experienced a similar phenomenon with a package of brewer's yeast, which I used in a diet for wax worms with negative results!

The airborne mold spores that started the colonies visible in Figure 8.13 are illustrative of the portion of Figure 8.11 that depicts "the facility air system" as a source of microbial contamination. I am including here under "facility air system" the potential for mold or other microbial contaminants to originate from various surfaces such as walls, ceiling tiles, floors, and ventilating ducts. The laboratory where the photograph from Figure 8.13 was taken had a chronic problem with mold in the ventilation system. This problem did not cease until the system was completely replaced with new ductwork. The number of colony-forming units (viable mold spores) was reduced by extensive cleaning and air filtration, but the problem would always recur especially when the filtration system (Figure 8.14) was not in active use. Besides counting of airborne culturable microbial contaminants, another useful method for insectary sanitation is known as the RODAC system (RODAC = Replicate Organism Detection and Counting) and can be used for microbiological control of

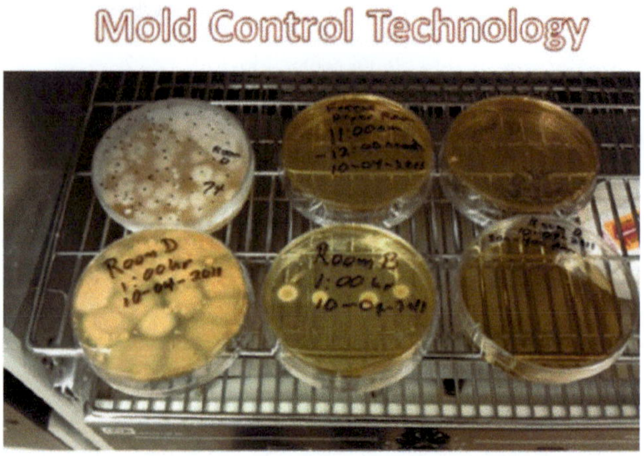

Figure 8.13 Control of airborne mold in two rearing rooms. Petri plates with Sabouraud's agar were exposed to the rearing room air for one hour, then incubated for three days at 25°C. When the two rooms were initially occupied, they presented severe mold problems (left two plates). After thorough cleaning of walls, floors, ceiling tiles, and counter tops with 2% bleach solution, there was a reduction in counts of mold spores (indicated by the fewer colony-forming units in the center of two Petri plates). However, it was only after introducing an in-room HEPA filtering system that nearly all airborne spores were eliminated (indicated by the nearly mold-free dishes on the right).

Figure 8.14 A portable Honeywell air filtration system. This system was used to successfully and dramatically reduce airborne mold spores in a 450 cubic foot rearing room.

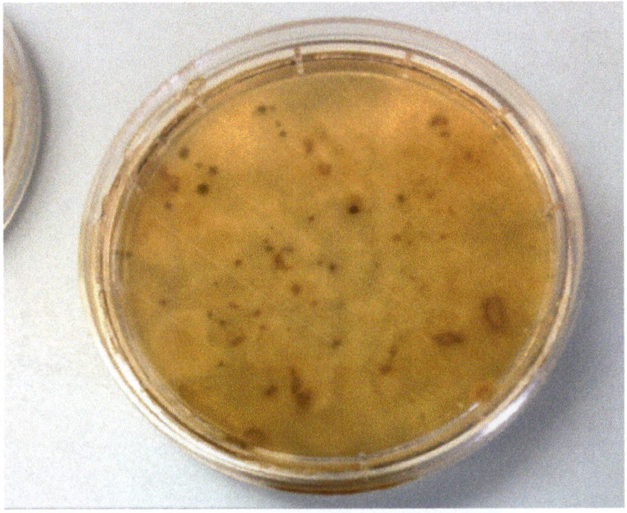

Figure 8.15 A RODAC plate. This RODAC (Replicate Organism Detecting and Counting) plate was used to determine the effects of various microbial amelioration procedures. It was recommended by Sikorowski et al. 1984 for an important part of the process control systems for reduction of microbial contamination in rearing systems. Because the plate readily lends itself to counting colony forming units per unit surface area, trends in increase, decrease, or stable populations of surface microbes can be measured and incorporated into a continuous improvement system of microbial control.

all surfaces. The plates can be purchased from various suppliers of microbiological media, or they can be made up in the lab with techniques described by Sikorowski et al. 1984, Bruch and Smith 1968, or McGowan 1985. Most sources recommend the use of trypticase soy agar with lecithin and Polysorbate 80, a generalized medium for a broad variety of bacteria and fungi with two agents that neutralize disinfectants (lecithin and Polysorbate 80) (Figure 8.15).

8.7.5 Using Pareto Charts

Pareto charts are a PC/QC tool that are based on the "80/20 principle": the idea that a single factor can account for 80% of all outcomes. If, for example, 80% of all larval deaths stem from diet problems, then it stands to reason that the best way to reduce larval mortality is to discover and correct the diet problem(s). The Pareto concept is a means of focusing the production team's attention to what really matters—what priorities should be put on the top of the "to do" list. Clearly, Pareto analysis will not always lead to an 80/20 ratio, and it should be understood that in complex systems such as insect-rearing systems, there are often multiple causes for various outcomes. In the hypothetical constructs for "causes of rearing system failures" (Figures 8.9–8.11), we posited diets, microbes, environmental issues, genetic factors, container issues, and personnel as possible causal agents of the failures (Figure 8.16).

The plot on the left in Figure 8.16 shows the list of causes thought (or hypothesized) to be responsible for rearing system failures. The plot on the right in Figure 8.16 shows an initial, pre-data-collection plot based on the thoughts and discussions of the PC/QC team. If any or all the team members have experience with rearing successes and failures, their thoughts of which causes may be most important can be very valuable. However, the next steps are to set up tests as objectively as possible to measure the components contributing to failures. This raises a highly crucial point: how do we determine actual causes and effects?

An excellent example of a robust effort to solve a rearing problem is the program undertaking by a group of scientists under the leadership of Dr. Tom ODell of the US Forest Service. The story is told by ODell 1992 and Keena 1993 (and later, the story was told in technical detail by ODell et al. 1997 and by Keena et al. 1998). In her 1993 essay, Keena explained the abnormal performance syndrome in gypsy moths (an occurrence that was taking place in the late 1980s at the USDA, APHIS Methods Development Laboratory at Otis Airbase in Massachusetts). The syndrome consisted of higher than normal mortality and irregular development of larvae and asynchronous advancement to pupal stage by an inordinate number of *Lymantria dispar* L. (Lepidoptera: Erebidae). The malady came to be known as "straggler syndrome" or "abnormal performance syndrome" or APS. Keena 1993 wrote, "In October of 1989, the team invited expert witnesses from around the United States to a strategic planning session in Hamden, CT. For two days we discussed and developed a plan of action. A multitude of possible suspects to follow up on emerged. APS would be a tough case to solve. The team met a couple of months after the meeting and selected eight primary suspects to investigate." The topics ("suspects") for investigation were: (1) diet, (2) rearing environment, (3) egg treatment, (4) larval density, (5) container issues, (6) inbreeding, (7) phenotype, and (8) pathology.

The narrative about the symptoms of the abnormal performance syndrome is a good starting place for the systematic investigation of causes mentioned above. Keena 1993 described the symptoms of APS-affected "families" (= common members of APS-egg clusters) as: (1) reduced egg-hatch, (2) inability of the neonate to fully chew its way out of the egg, (3) neonates that do escape eggs fail to implant in the diet, (4) larvae that do accept diet develop asynchronously, (5) between neonate and adult stage, there is an exceptionally high mortality, and (6) besides the straggling and high mortality, there is also a high rate of deformity, especially in pupae. The APS team systematically attacked the problem by addressing each aspect of the topics ("suspects") such as pathology.

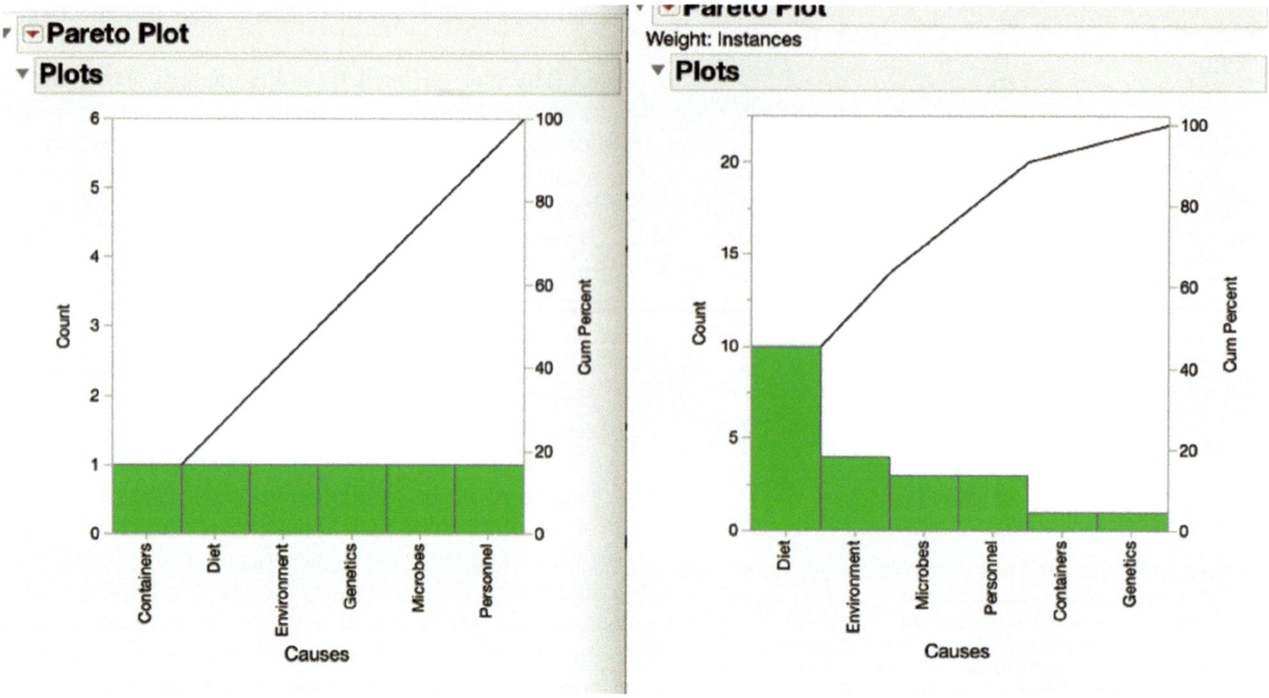

Figure 8.16 Developing Pareto plots.

The team concluded that there was no direct evidence of bacterial or other possible pathogens such as nuclear polyhedrosis virus fungus infection.

They further ruled out cryptic bacterial pathogens by using antibiotic therapy (oxytetracycline) and mild heat treatment, which is known to destroy some pathogens—all to no avail in curing the syndrome. The narrative goes on, explaining the reasons for rejecting the various potential causative agents, often with the fruits of deeper scrutiny of the rearing conditions revealing information about problems with containers and outcomes related to inbreeding (consanguinity), but these factors were not at the heart of the APS problem. The possibility that diet factors were the problem was investigated (I had been one of the consultants present at the October, 1989 meeting on APS in Hamden, CT), and the recommendation was to consider the most unstable portions of the diet. The composition of the diet was wheat germ, casein, methyl paraben, sorbic acid, vitamin mixture, agar, Wesson salt mix, and water. Tests were made of the degree of deterioration of fatty acids (which are subject to peroxidation), oxidation of various labile components such as proteins and their component amino acids, and of course ascorbic acid, which is known to deteriorate due to oxidation-reduction reactions with various cations—reactions accelerated by heat. However, all these factors were ruled out by analytical testing procedures. The conclusion that the APS team reached after a few years of in-depth investigations, especially into diet-related issues, was that the mineral mixture—a common diet ingredient called Wesson salts—had a change in the form of iron phosphate (ferric phosphate).

Regarding an interaction between diet and genetics, Keena writes:

> When the female parent was reared on diet containing crystalline ferric phosphate, the progeny had APS, but when the female parent was reared on diet containing amorphous ferric phosphate, the progeny did not have APS, regardless of the male parent's diet. This indicated that there was a significant maternal effect involved in APS. In addition, in families where APS was expressed in the progeny, there was a significant paternal effect in addition to the strong maternal effect. We also found that genetic differences in susceptibility to iron availability existed both between and within families. This indicated that the phenotype of each individual (whether it had APS or did not have APS) was partly determined by an underlying genetically determined sensitivity to dietary iron availability.

It is an interesting question about the gypsy moth diet: how much iron is contributed by other components in the diet such as wheat germ. The USDA Data Base (https://fdc.nal.usda.gov/fdc-app.html#/food-details/445010/nutrients) lists the iron content of 100 g of wheat germ to contain 5.54–6.55 mg of iron (depending on whether it is raw or stabilized wheat germ). Though the other diet components do not contain significant iron contents, the wheat

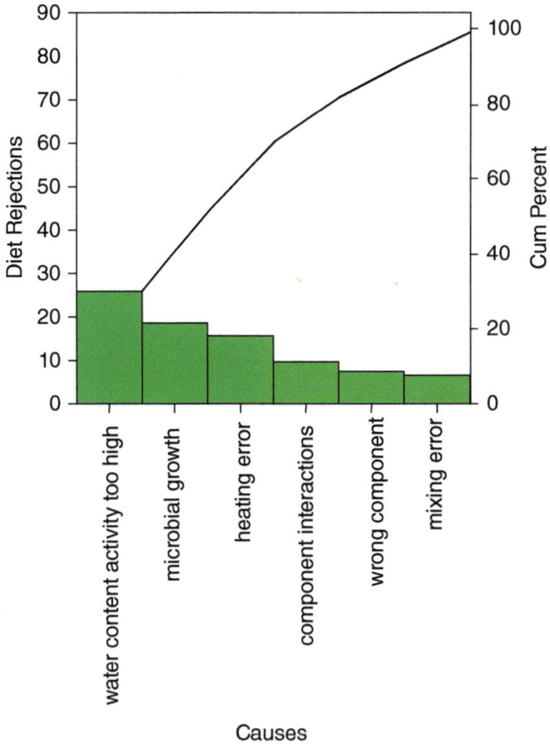

Figure 8.17 Pareto chart of causes of failure in an insect diet. Early in the PC-development process, we develop Pareto charts based on an iteration of rationale → hypothesis → test → modified hypothesis → test →→→ ....

germ can be regarded as a considerable source of iron. The iron content of leafy materials such as raw cabbage contains lesser amounts of iron per unit of dry weight compared to various wheat germ products. Therefore, the question of why the character and content of the iron from Wesson salts merits further attention.

All this being said, the highly organized procedures used by the Gypsy Moth APS team stand as models of using Pareto-type thinking and testing causes-and-effects with scientifically-based inquiries provides a format of systematic approaches to insect-rearing system infrastructure (Figures 8.17–8.19).

In Figure 8.18 (a), the data are presented for moisture analysis of a soy flour that was stored for 15 months under room temperature conditions. The moisture balance used to determine moisture content of samples from the soy flour is seen in Figure 8.18 (b). Beginning with freshly acquired and newly opened soy flour, monthly readings of moisture content were made to determine how moisture content may have varied under room temperature storage conditions where no moisture absorbent such as silica gel packets were used. Data from Figure 8.18 (a) were analyzed in JMP 14 "Control Chart Builder," resulting in the control chart presented in Figures 8.19(a) and (b).

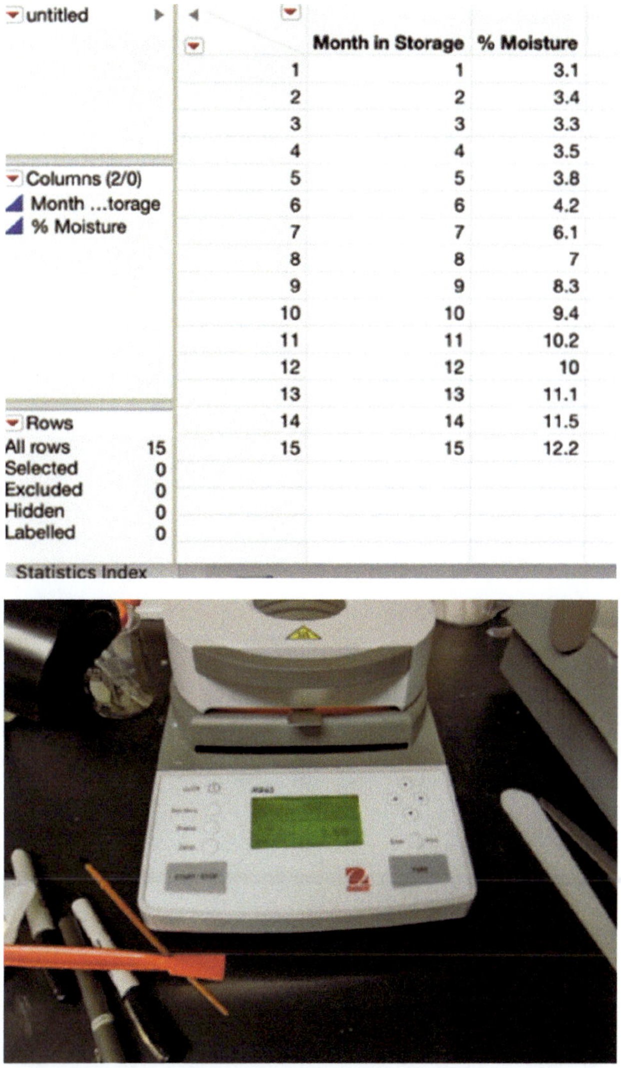

	Month in Storage	% Moisture
1	1	3.1
2	2	3.4
3	3	3.3
4	4	3.5
5	5	3.8
6	6	4.2
7	7	6.1
8	8	7
9	9	8.3
10	10	9.4
11	11	10.2
12	12	10
13	13	11.1
14	14	11.5
15	15	12.2

Figure 8.18 (a) and (b) Moisture Analysis of soy flour.

Figure 8.19 shows that there was a progressive increase in moisture content as the soy flour was stored and utilized (opened) periodically. In the C chart, all points are within control limits indicated by the upper control limit and the lower control limit (UCL and LCL, respectively). It should be noted that these limits are set at the mean (μ) +/− 3σ, making this chart a standard control chart. In the Individual chart, the changes from month to month are also clearly indicated, with the first six points being below the LCL (bottom red line), and the last six points indicate that the moisture content was above the UCL (upper red line). This chart is a Shewhart variables configuration, which means that the JMP system "noted" a trend where variation was not random, but rather it was following a trend—in this case a tendency for points' values to increase through time. The terms "variable" and "attribute" refer to the treatment of data as measured data (length, weight, absorbance, % water, etc.)

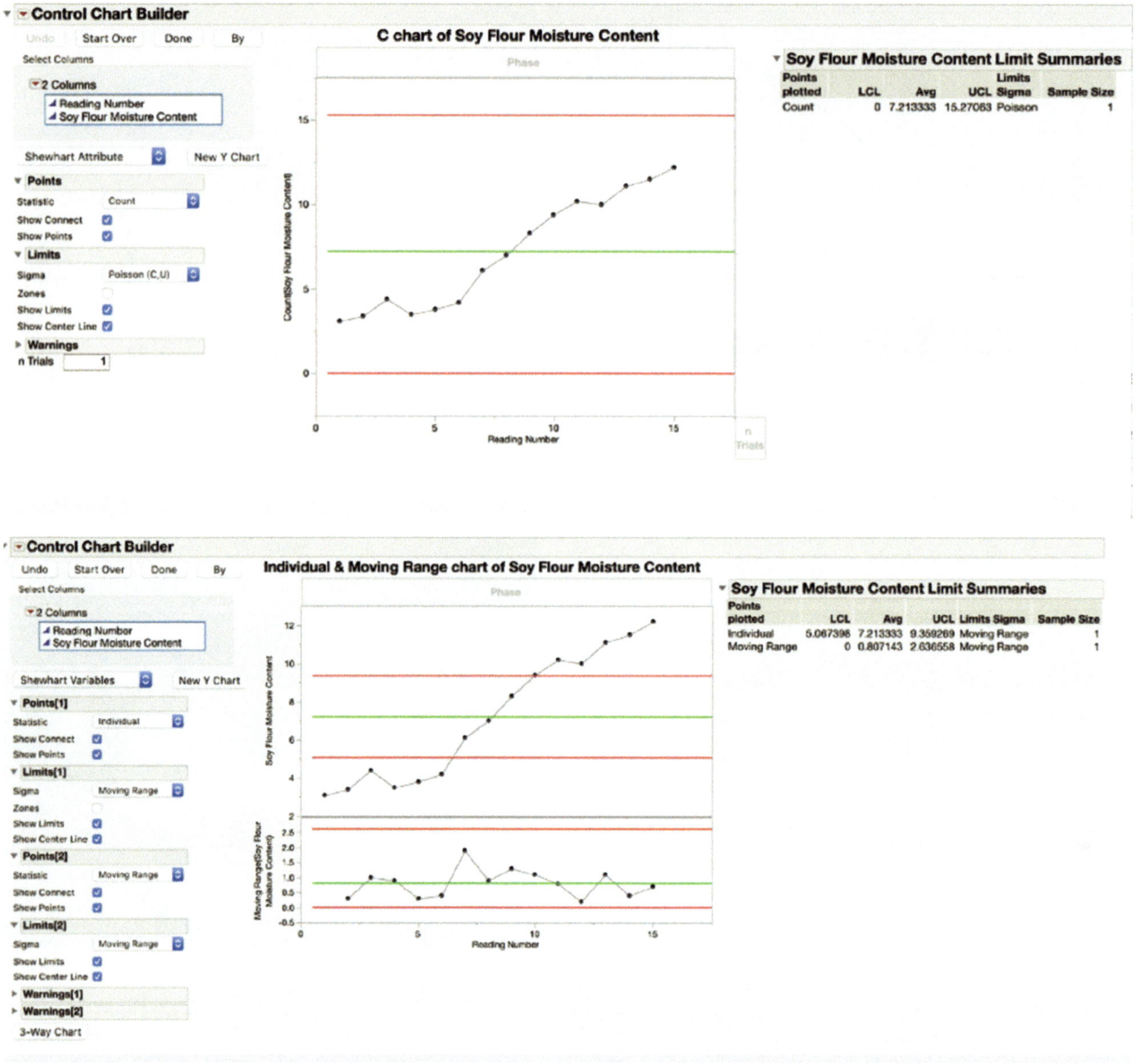

Figure 8.19 Control charts (C chart in [a] and individual and moving range for [b]) for moisture content of soy flour.

vs. counted data (number of survivors, number of males, etc.). The UCL and LCL were set more narrowly than how they were set in the C chart above. The mean (μ) for the individual range is indicated by the green line (=7.21), which is the same for both the C chart and the Individual chart.

The power of the control charts is that for C charts, the rearing personnel concerned with control of their rearing system (with any and every facet of the system) can be visualized at a glance. The same data set that generated both the C chart and the Individual chart tells us two important stories, in this case about the soy flour storage history: (1) the actual moisture content of the flour is under control (within the bounds of the UCL and LCL); however, (2) there is a trend in the data from Figure 8.18a is substantial and continuous moisture gain throughout the storage period. The first finding is to be learned from the C chart, and the second finding stems from the Individual chart. It should be noted that while these statistical tests are helpful in showing trends to rearing personnel, they must not be thought of as an over-riding force that replaces or trumps experience and scientific knowledge. The moisture content of the soy flour was 3.1% when we opened the package that was freshly shipped from the supplier. As is the usual case, a freshly opened package of various raw materials has a low moisture content that had been protected from the environment by proper packaging and possibly packages of desiccant in the container. The fact that the soy flour continued to pick up moisture and ended up at more than 12% moisture should raise our attention to the question about quality of the material. For one thing, if the formula for our diet calls for 100 g of soy flour, we are actually giving our insects only 97 g of soy flour and 3% moisture with the fresh material. However, when we use 100 g of the older soy flour, we are giving our insects a diet with only 88 g of soy flour and 12 g of water. This means that insects fed the diet with the 12% moisture are getting nearly 10% less soy flour than they were supposed to get. The next issue is that if the diet components contain more water, are moisture-related activities such as microbial growth, non-enzymatic reactions such as oxidative damage, and enzymatic reactions deteriorating the quality of this raw material? And finally, is the increased moisture content a kind of indicator of various kinds of deterioration not related to moisture content or water activity?

The following section of this chapter deals with another use of control charts to help guide or evaluate interpretations of temperature and humidity in a rearing room. Data were collected using a data logger (Hobo), and the output for several days is presented in Figure 8.20. The logger was set to record temperature and humidity at five-minute intervals for 24 hours/day for 13 days, totaling nearly 4,000 data points for each microclimate factor. Even with high-capacity statistical systems, these data sets are so complicated that I have presented a small window of data collection (less than two hours) for SPC-based data analysis.

In Figure 8.20, the temperature (lower lines in black) and the relative humidity (upper lines in blue) are indicated over a 13-day period. Note that the fluctuations in both temperature and relative humidity seem to undergo large, exaggerated swings, but the extent of change is largely within a three-degree F range, and the humidity fluctuations are mainly between 4% and 6% with a mean of about 62%. However, it should be noted that between September 18 and September 21, there was a considerable drop in both temperature and relative humidity. This is an excellent example of how important it is to collect data as extensively as possible. The low region for temperature was during the day of September 19 where the temperatures ranged from 75.5 and 78.5°F; and the humidity ranged from 52% to 55%. The reason for this change was not clear, but we noted that rate of development had declined during this period.

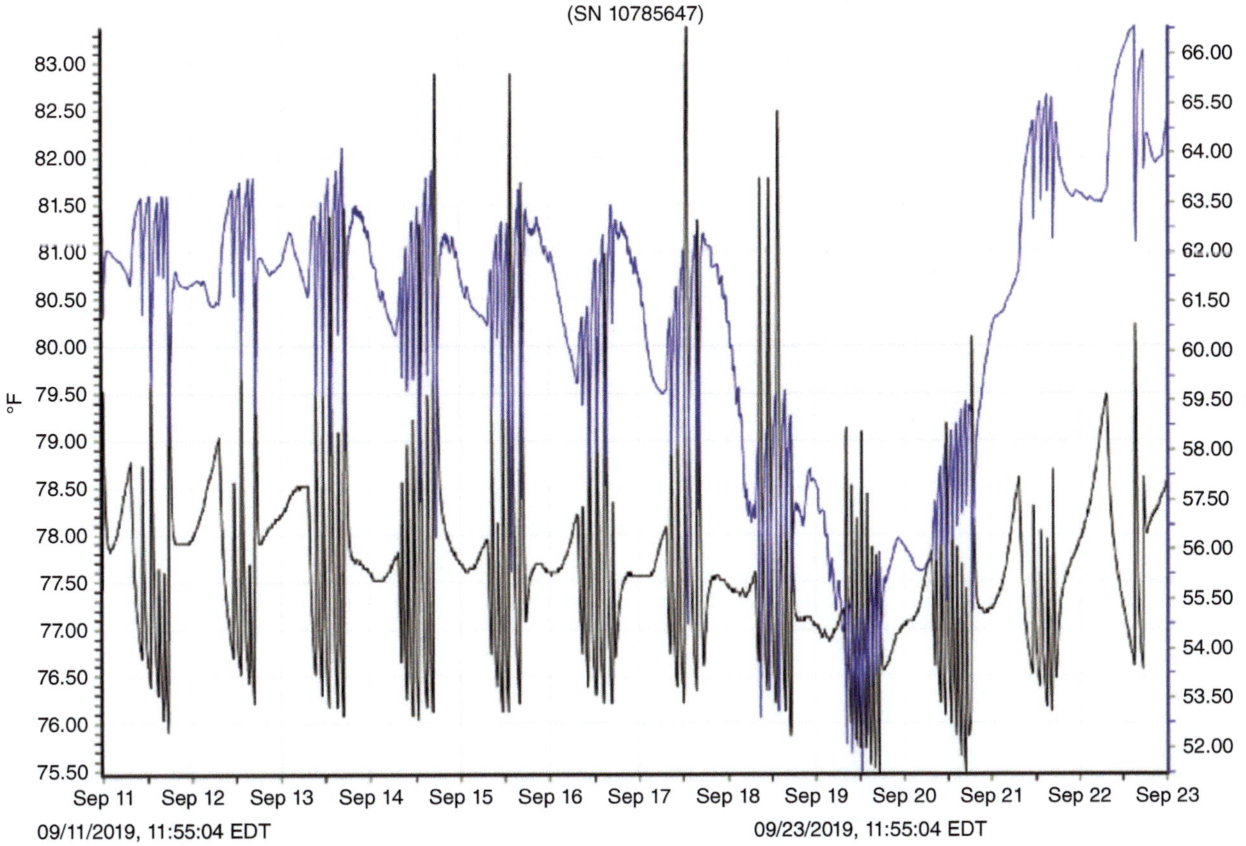

Figure 8.20 Data logger readout of rearing room temperature and relative humidity.

Table 8.3 Hobo (Data Logger) Readings of 20 Temperatures and Humidities in a Rearing Room

Reading Number	Temperature °F	Relative Humidity
1	79.6	36.6
2	80.0	37.7
3	80.5	39.2
4	80.9	40.3
5	81.2	41.3
6	81.4	42.1
7	81.6	42.8
8	81.6	43.4
9	81.7	43.9
10	81.8	44.3
11	81.9	44.7
12	81.9	45.0
13	82.0	45.3
14	82.0	45.5
15	82.0	45.6
16	82.0	45.8
17	82.0	45.9
18	82.0	46.0
19	82.0	46.2
20	82.1	46.3

Table 8.3 represents the actual temperature data taken from the spreadsheet that generated Figure 8.20, with only 20 consecutive points in time represented, for reasons of simplicity and clarity. If insectary personnel considered only the data table, it could be interpreted that there was a continuous increase both in temperature and humidity from Reading Number 1 through Reading Number 20. However, the points are not conspicuously changing, and the graphic presentations in Figure 8.21 (a) and (b) (temperature measurements) and (c) (humidity measurements) are more clearly grasped in the graphic representations. The application of a control chart (easily obtained from the JMP Chart Builder for Control Charts) shows the mean of the individual data points for temperature and for humidity, and the control charts (b and c) show the upper and lower control limits under conditions where the temperatures and the humidities were changing in a pattern that was of sufficient magnitude that the majority of the data points were deemed out of control. This trend was emphasized by the Control Chart Builder which selected "variables" to emphasize the trend of continuously rising temperatures. However, from a strict perspective of calculating the mean and standard deviation, the Control Chart Builder also is able to present the data points' variation around the mean and shows them to be well within the six sigma limits, which represent three times the standard deviation (3σ) above and 3σ below the mean.

The lower charts in Figure 8.21(b) and (c) represent the moving range—a way of clearly depicting shifts in data values that are abrupt and of considerable magnitude. Using both types of control chart (individual and moving range) is useful to the rearing system personnel who are trying to track the well-being (quality = fitness) of their rearing processes. The individual points can relate to the personnel which specific portions of their system may be

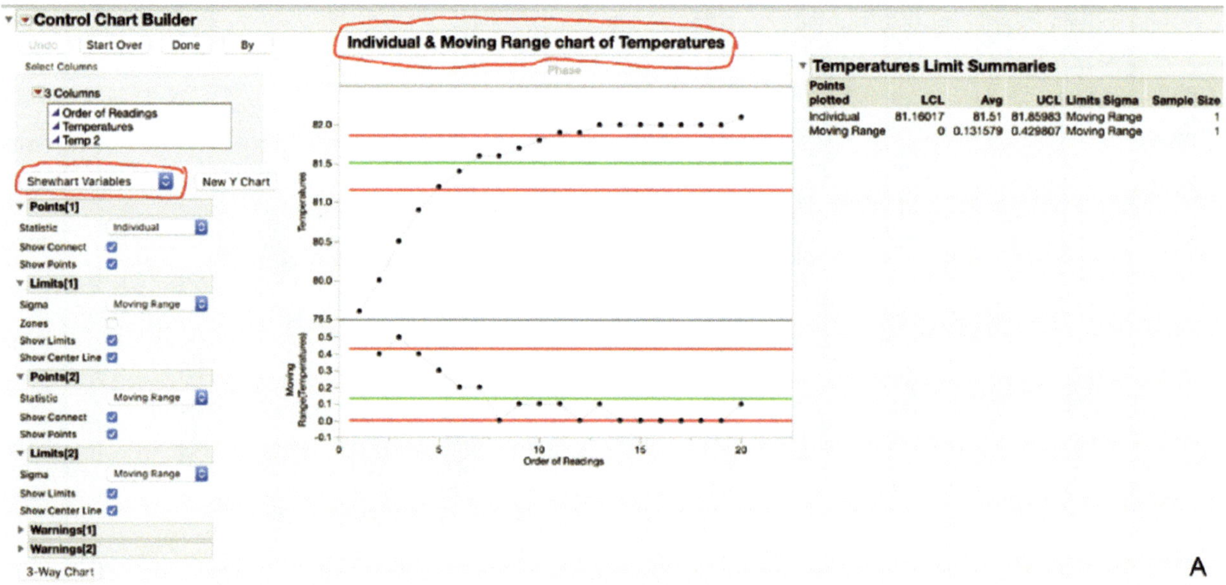

Figure 8.21 (a) An individual and moving range chart of temperature expressed as Shewhart variables, (b) a C chart of temperature expressed as Shewhart attributes, (c) an individual and moving range chart of humidity expressed as Shewhart variables.

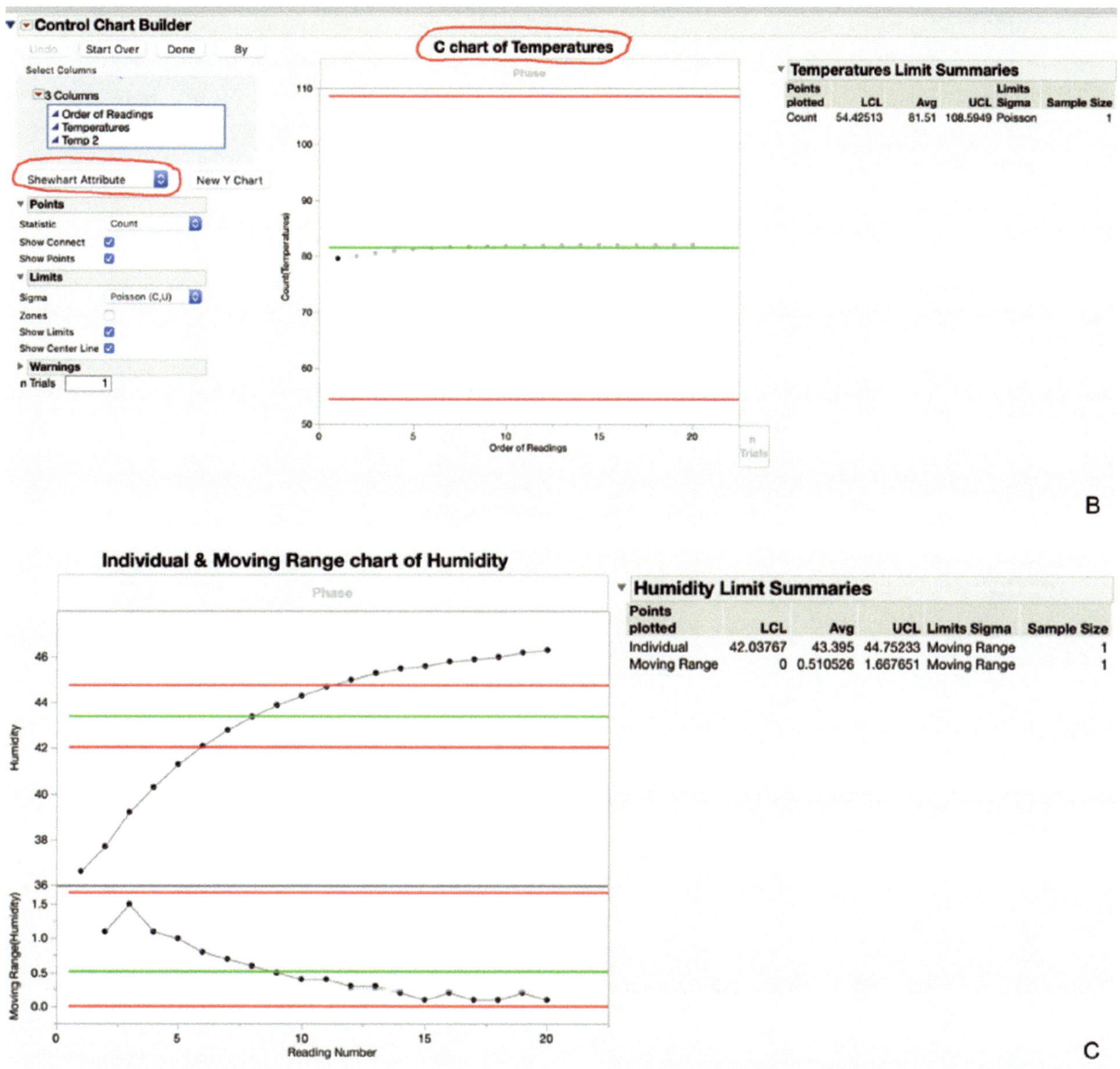

Figure 8.21 *(Continued)*

out of the boundaries that we call "control." In control chart (b) of Figure 8.21, the temperature was out of the range of the mean 81.5°F (from 81.16 to 81.86) in 13 out of 20 readings over the two-hour period where data are represented here. It is possible that if a biological problem were detected in the insects, the cause of the problem could be attributed tentatively to the "out-of-range" temperatures. By collecting data such as we did with the data logger, or as is possible with other measurements in the system, we begin to build a basis for objectively approaching the problems with data-based hypotheses. A final point is to be made here: the relatively small magnitude of the deviation in temperature and humidity shown in Figure 8.20 (a–c) would make it difficult to interpret a major problem being related to temperature or humidity deviations (Figure 8.22).

One more control chart will help us understand the potential power of control charts to establish the baseline for trends and tendencies that deviate from the trends that are "in control" (Table 8.4).

In Table 8.4, actual measurements of water activity were taken and recorded from 1 to 18. The data were entered into a JMP Control Chart Builder program, and the bottom plot was obtained. Note that all points indicated that the wheat germ water activity in all samples were ***in control***. However, what if the 19th sample measured 0.5? The top chart shows that a reading of 0.5 would be ***out-of-control***. Note that with the first 18 samples, the

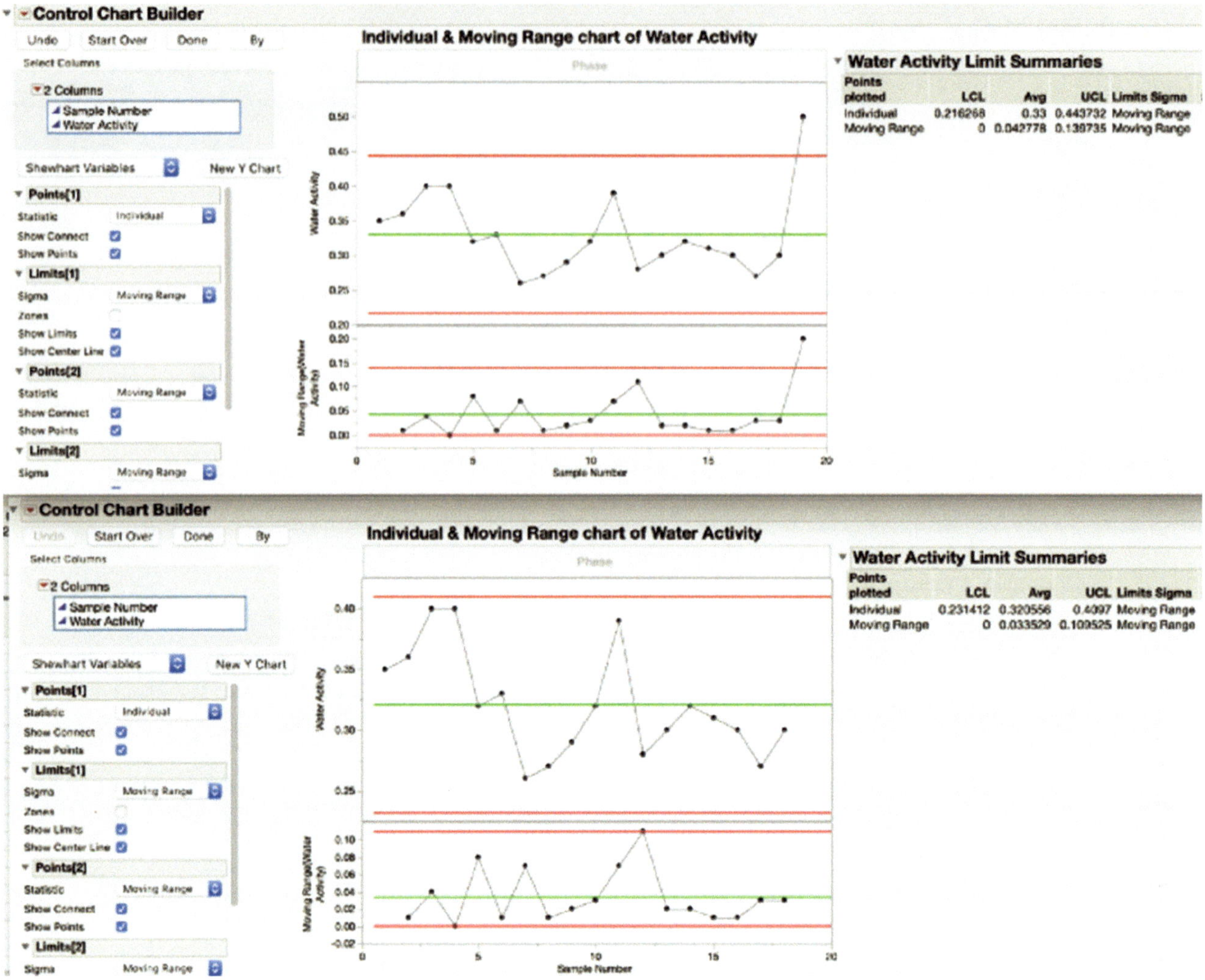

Figure 8.22 A data set where the water activity of wheat germ samples was tested after storage.

Table 8.4 Measurement of water activity of soy flour

Reading Number	Water Activity
1	0.35
2	0.36
3	0.40
4	0.40
5	0.32
6	0.33
7	0.26
8	0.27
9	0.29
10	0.32
11	0.39
12	0.28
13	0.30
14	0.32
15	0.31
16	0.30
17	0.27
18	0.30
19	0.50

mean water activity was 0.32, and in the data set with 19 samples, the mean water activity went up slightly to 0.33. However, the jump to 0.5 in only one data point was enough to raise the upper control limit (UCL) from 0.4097 to 0.4437, an indication of the sensitivity of standard deviation to the large water activity value in the final data point. It is of interest to recognize that on a statistical basis the interpretation that the 0.5 point was out-of-control is also in agreement with the physical reality that "something is going on" to have caused this increase in water activity. If the cause of the elevation in water activity was a storage problem or a result of some type of problem with the supplier or other source of change, it is important that the quality/process control team recognize that something needs to be done to correct the problem so that it does not repeat itself.

The bottom line—the important take-home message—is that the statistically based approach (statistical process control = SPC) to insect-rearing systems is a very helpful discipline that imposes or suggests careful attention to all relevant facets of the rearing system, but SPC in itself does not fully inform rearing personnel how to make decisions. Even with the powerful tools of SPC, rearing specialists must use judgment as to which factors best relate to the quality of their rearing processes. A helpful way of regarding these relationships is to consider rearing specialists as scientists who understand the biology of their insects in the context of the complexities of the rearing system. This concept is well aligned with the concepts discussed at length in Chapter 6 on "Rearing Systems as Ecological Niches or N-Dimensional Hypervolumes." Rearing specialists must be constantly aware that their insects are of a genetic character that has been brought into the laboratory just as an organism in the wild may have been genetically isolated from its conspecific counterparts. This is the facet of rearing studies and rearing assumptions linked to a kind of island biogeography where barriers to mating and genetic exchange with the other members of the species are limited by the rearing barriers. The thinking at this level must also include the tenets of the Hardy-Weinberg law dealing with selection, geneflow, and restricted population size. Besides the genetic facets of the rearing complexities, there are also the environmental factors where the needs for "normal" gas exchange, temperature effects, humidity-relationships, lighting features, and all other environmental features must be understood and practiced. The other special issues that must be considered are those pertaining to diet of an insect that has no choices as they might in nature and a virtually gradient-free microenvironment that is especially accentuated by the containers that simply do not exist in nature. Finally, and very importantly, the contact with personnel must be considered where there is a nearly constant artificial interaction with humans, with their forces of selection (deliberate or inadvertent), exposure to diseases, odors, sounds, etc. that are completely alien to the insect as it would live in its natural world.

In the following figures, there are some real-world examples of components of processes that can be fitting subjects of reduction in variation such as is treated in Figure 8.4. Figure 8.23 shows three balances with

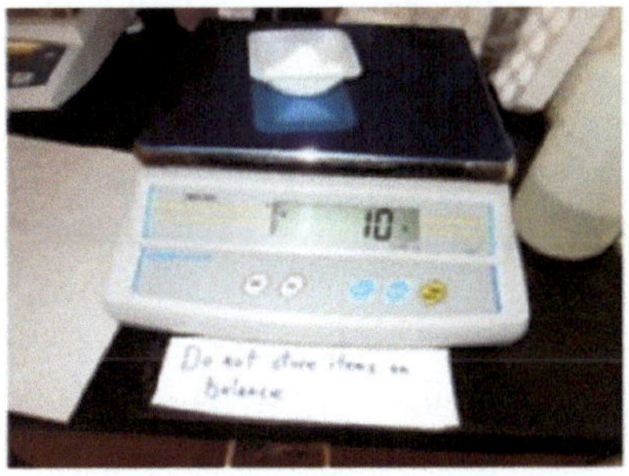

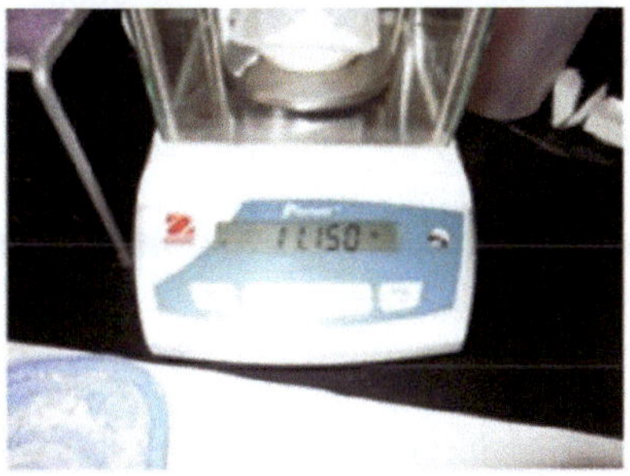

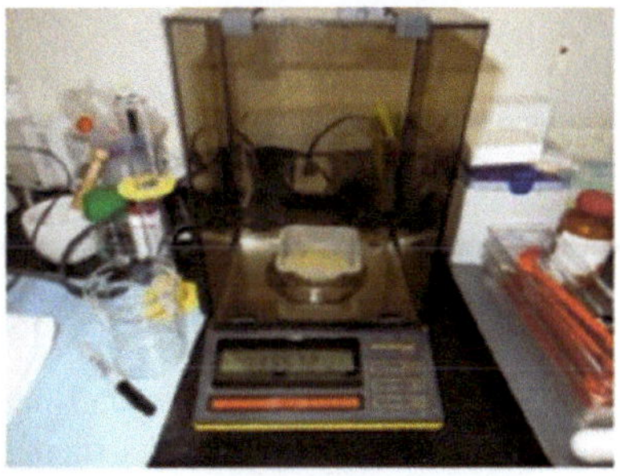

ADAM balance OHAUS Pioneer Sartorius

Same soy flour sample weighed on 3 different balances: note that there is about 3% more soy flour than the 1st balance indicates.

ADAM balance	OHAUS Pioneer	Sartorius
12	12.362	12.3582
12	12.364	12.3577
12	12.364	12.3582
12	12.365	12.3579
12	12.364	12.3579
12	12.365	12.358
12	12.364	12.3579
12	12.364	12.3581
12	12.365	12.358
12	12.364	12.3579

Soy flour measured from 3% to 12% water content.

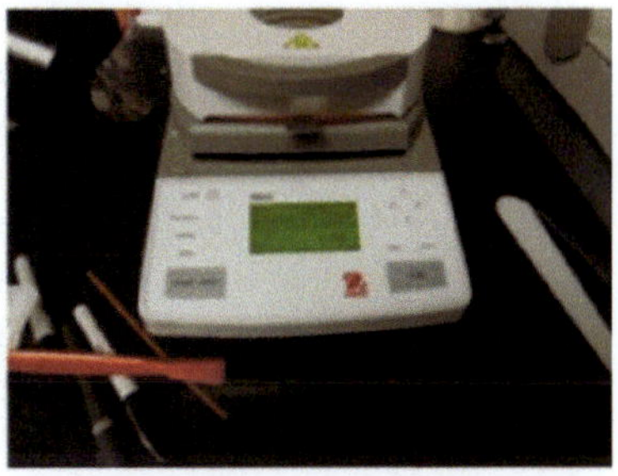

Moisture Balance

Figure 8.23 Sources of error and reduction of precision and accuracy.

readability of grams (the Adam balance), milligrams (the Ohaus balance), and tenths of milligrams (the Sartorius balance). The three balances were tared with the weigh boat, then a sample of about 12 g of soy flour was added, and the sample plus weigh boat was placed on each balance ten times with the results recorded. The least sensitive balance reported exactly 12 g for each weighing (giving a mean of 12, 0 standard deviation, and 0 difference between readings); the second balance gave the readings seen in Figure 8.23 with a mean of 12.3639, a standard deviation of 0.001 g, and a difference of 0.003 g between the highest and lowest readings, and the most sensitive balance gave a mean of 12.3580 g, with a standard deviation of 0.0002, and an 0.0005 g difference between the highest and lowest readings. These differences in precision of the balances may be important in various experiments where outcomes may be close, and the readability and precision of the balance (or other measuring instruments) may obscure differences (or similarities) of the true values. This is discussed further in Chapter 9 (Section 9.2.3) where weights of insects on different diets were measured with balances of different readabilities. However, the most important point here is that the target weight of 12.0 g seemed to be met with the least sensitive balance, and it appeared that the same sample weighed exactly 12 g. But when the other two balances were used, it was found that the true weight of the sample was closer to 12.360 g, resulting in an error of about 30% more soy flour than was intended to be added to the diet. The lack of sensitivity/readability of the Adam balance could have gone the other way, with sample weighing about 30% less than the intended amount. The take-home message of this discussion is that the reduction in variability that is a primary concern in increasing control of rearing processes includes a wide variety of measures, including selection of appropriate precision equipment and the techniques in using this equipment (such as proper calibration of balances and volumetric equipment, leveling of balances, proper placement of balances on stable tables or lab benches, protection of sensitive balances from vibration of their tables and avoidance of air drafts), careful and meticulous reading of the equipment (i.e., avoiding human error), and proper education of personnel in procedures such as cleanup, not moving balances and other delicate equipment unless proper procedures are followed, and establishing ranges of tolerance for measurements. Within this framework, other sources of variability and quality of rearing processes can be related to the handling of components such as how they are stored, including environmental conditions of storage, duration of storage, and containers where storage is being done (Figure 8.24). A simple issue such as storing in a frost-free freezer can result in destruction of the components' nutritional quality and palatability (Cohen 2015).

The issue of tolerances for reduction of error is discussed at length in Chapter 7 in the discussion of Table 7.2 where the "+/− 3% tolerance rule" is discussed. Briefly, the practice of developing, as part of the SOPs in a rearing system, the target is established, and the acceptable range of tolerance is explicitly specified. In Table 7.2, the target of 1.0 g of ascorbic acid, for example, was specified as 0.97–1.03 g, and of course this zone of tolerance demanded that a balance that reads to hundredths of grams was required. The same kinds of tolerance zones can be set for every other measurable component and process in the rearing system (temperature, humidity, light intensity, etc.). This practice, which is under the strict control of the rearing personnel, can lead to a desirable **system of high precision rearing**.

This topic (Figure 8.24) is covered extensively by Cohen 2015, but a few comments are appropriate here. Two poorly defined but undoubtedly important sources of variability in rearing systems are the sources of materials and handling of those materials, especially storage. What is meant by source is the supplier, brand, batch or any indication of the nature and origin of materials—especially diet components. There are ambiguities to names such as wheat germ, yeast, soy, oat bran, etc. Some materials are simple enough that the brand, supplier, or origin is probably not important. Other materials have characteristics that make them very likely to cause variability in rearing procedures.

8.7.5.1 Low-Risk Components

Sucrose is an example of a low-risk component. Sucrose or common table sugar (sometimes labelled "granulated sugar") is one of the purest substances readily available in the commercial market. Careful chromatographic analysis of sucrose, even a product obtained from a grocery store shelf, is close to 100% pure sucrose, with traces of glucose and fructose (Cohen unpublished data on gas chromatographic analyses of sugars using silylation derivatives). Sucrose is stable and can be stored for years in its original packaging at room temperature with little negative consequence; though compromised packaging can result in absorption of moisture and possibly infestation of pests. Other low-risk components include other sugars, most free amino acids, and many antifungal compounds.

8.7.5.2 High-Risk Components

Materials that are entirely or primarily complex foods such as yeast products, wheat germ, soy flour, wheat bran, oat bran, eggs of fowl, meat products, and many dairy products. The first problem with most of the products mentioned here is that the names are ambiguous. When "yeast" is specified, it may be active or inactive. It may be derived from one of the most commonly used species, *Saccharomyces cerevisiae* (often called baker's yeast in

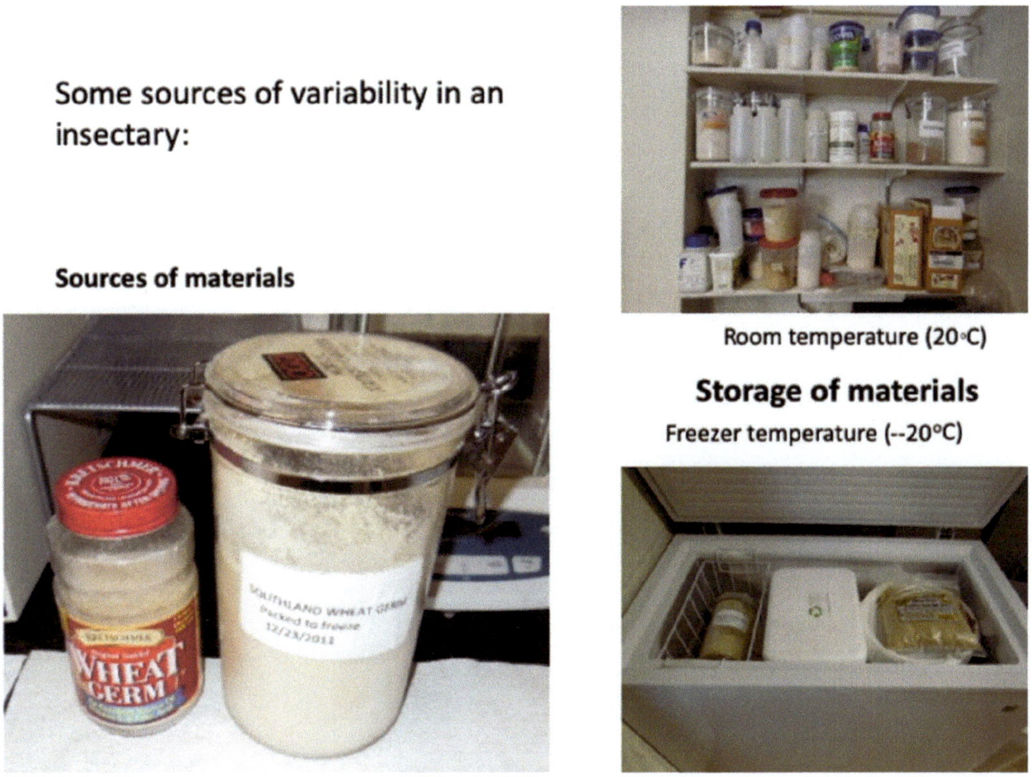

Figure 8.24 Sources and storage of materials (low-risk and high-risk components).

its active form and brewer's yeast in its deactivated form). Another ambiguity regarding some *S. cerevisiae* products is that they are made from the cultivation of the yeast on molasses or some other medium, and the yeast product is inactivated by heat and is called "nutritional yeast." In the truest sense, if the product is to be called brewer's yeast, it should have been derived as a by-product of brewing, especially beer, and this product must be put through a debittering process to make it palatable (by removing the flavors of the hops). Other commonly used yeast products from *S. cerevisiae* are known as yeast autolysates or yeast extracts. Without further discussion of these seemingly subtle differences in *S. cerevisiae* products, it must be emphasized that there are great differences between the nutritional value, palatability, bioavailability, and stability of these materials, and confusion of one material with another can lead to drastic problems in the diets of which they are a part. Furthermore, once these differences are understood, it must be realized that there can be considerable differences between yeast products with seemingly the same name, such as "brewer's yeast," depending on the brand.

CHAPTER 9

Tutorials on Using Design of Experiments with SAS, JMP Protocols

9.1 INTRODUCTORY COMMENTS ABOUT DESIGN OF EXPERIMENTS (DOE)

An important advantage of using a systems approach to design and define operations in rearing systems is that a pre-existing engineering framework is in place and has proved itself as useful for developing and improving countless types of processes and production systems. The points behind these statements regarding design of experiments (DOE) and systems are made by Montgomery 1991 and 2013. In his 8th edition, Montgomery 2013 makes the following statement about use of his book in conjunction with software products:

Design-Expert, JMP, and Minitab Software

> During the last few years a number of excellent software products to assist experimenters in both the design and analysis phases of this subject have appeared. I have included output from three of these products, Design-Expert, JMP, and Minitab at many points in the text. Minitab and JMP are widely available general-purpose statistical software packages that have good data analysis capabilities and that handles the analysis of experiments with both fixed and random factors (including the mixed model). Design-Expert is a package focused exclusively on experimental design. All three of these packages have many capabilities for construction and evaluation of designs and extensive analysis features. Student versions of Design-Expert and JMP are available as a packaging option with this book, and their use is highly recommended. I urge all instructors who use this book to incorporate computer software into your course.

In his Preface, Montgomery 2013 also makes important points about the **empirical model**, which he points out lends itself to manipulation (with the appropriate engineering tools such as surface plots or contour plots) that leads to ***mechanistic understanding*** of the system in question. Montgomery writes:

Empirical Model

> I have continued to focus on the connection between the experiment and the model that the experimenter can develop from the results of the experiment. Engineers (and physical, chemical and life scientists to a large extent) learn about physical mechanisms and their underlying mechanistic models early in their academic training, and throughout much of their professional careers they are involved with manipulation of these models. Statistically designed experiments offer the engineer a valid basis for developing an *empirical* model of the system being investigated. This empirical model can then be manipulated (perhaps through a response surface or contour plot, or perhaps mathematically) just as any other engineering model. I have discovered through many years of teaching that this viewpoint is very effective in creating enthusiasm in the engineering community for statistically designed experiments. Therefore, the notion of an underlying empirical model for the experiment and response surfaces appears early in the book and receives much more emphasis.

An excellent explanation for how DOE is applicable to countless systems is provided by Sall et al. 2017. The authors discuss the fact that "experimentation is the fundamental tool of the scientific method." Given this premise and that the point of this chapter and of this book is to provide a framework for treating insect-rearing systems with the scientific method, it behooves us to examine the relationship between the statistical methods of DOE. The introduction of this point about application of DOE to scientific inquiry is best expressed by Sall et al. 2017:

> In an experiment, a *response* of interest is measured as some *factors* are changed systematically through a series of *runs*, also known as *trials*. Those factors that are not involved in the experiment are held (as much as is practical) at a constant level. Then any variation or effect produced by the experiment can be attributed to the changes in the design factors and natural variability. The goal is to determine whether and how the factors

affect the response. Experimental design addresses this goal—it allows us to learn the most about the relationship between the response and factors for a given number of runs. Compared to ad hoc trial and error, experimental design saves money and resources by reducing the number of necessary trials.

Key Concepts

DOE, and statistics in general, enable us to make decisions and develop product and process knowledge in light of random variation. Experimentation is a true journey of discovery.

Throughout this chapter and in other parts of this book where SAS JMP®, Version 14 procedures are applied and explained, we will be examining the ***responses*** in our rearing systems. We will also be working with the ***factors*** that are the *variables* in our experimental (therefore scientific) approach to insect-rearing systems. To achieve the **rigor** (discussed in detail in the Preface and in Chapter 14) that we strive to incorporate into our rearing system development, we will find that carefully designed experimentation and proper analysis of the experiments lead to high standards of scientific rigor. The responses, which can be considered the **outcomes** of our experiments, may include any of the biological factors such as development time, survival, weight, physical dimensions, colors, fecundity, digestive or metabolic efficiency, immune system characteristics, and any other measurable outcomes of our experiments. Responses may also be economic factors (such as lower costs of production or low labor intensity), and they may include customer satisfaction with our rearing products (a key element of quality control).

We get to the responses in our experiments by the variables that we set up—known here as the factors that we are testing. Is one complex diet component such as wheat germ superior to another complex food such as soy flour or rice bran? Is a rearing system component from one supplier superior to another such as brands of rearing containers? Does heating the diet to 100°C produce a diet that leads to greater biomass in our insects than does heating the diet to 90°C? Does 14: 10 light to dark cycle produce adult silkworms with high fecundity than does 16: 8 photoperiod? All the questions of kinds of diet components, types of rearing containers, diet preparation temperatures, and photoperiods are factors, while the biomasses, fecundity, survival, etc. are responses to the manipulation of the factors (variables). We can see from this how our experimentation is a "journey of discovery:"

Experimentation Is Learning

The word *experiment* has the same root as the words *expert* and *experience*. They all derive from the Latin verb *experior*, which means to try, to test, to experience, or to prove. Typical experimentation is a familiar process of trial and error. Try something and see what happens. Learn from experience. Learn from changing factors and observing the results. This is the *inductive method* that encompasses most learning. Designed experiments, on the other hand, add to this method by giving us a framework to direct our experiences in a meaningful way.

Sall et al. point out that the DOE experiments that we perform are learning experiences where we are deepening our understanding of the cause and effects relationships in our rearing systems. Both from the perspectives of developing rearing systems (R & D efforts) and rearing systems optimization, the use of DOE experiments helps us with developing working rearing systems and programs of continuous improvement.

Controlling Experimental Conditions Is Essential

We are easily fooled unless we take care to control the experimental conditions and environment. This control is the critical first step in doing scientific experiments. This is the step that distinguishes experimental results from observational, happenstance data. You can obtain clues of how the world works from non-experimental data. However, you cannot put full trust into learning from observational phenomena because you are never completely sure why the response changed. Were the response differences due to changes in the factor of interest, or some change in uncontrolled variables? The goal is to obtain knowledge of cause and effect.

9.2 TUTORIAL CASE STUDIES USING JMP

The following case studies are examples taken from various projects in the Cohen laboratory (Insect Rearing Education and Research Program at NCSU). The intention in presenting these tutorials is to give readers an opportunity to follow step-by-step protocols where several JMP procedures are presented, representing actual rearing experiments. It should be noted that JMP provides a number of tutorials in the form of webinars and online presentations, as well as in publications such as Sall et al. 2017 book on JMP. It should also be noted that besides JMP (from SAS), several statistical packages are available for use in DOE studies. These include but are not limited to Minitab (from Minitab, Inc.), S-Plus (Mathsoft, Inc.), and Design-Expert (Stat-Ease, Inc.). I have focused here on the JMP procedures out of habit and convenience, and because this is the system with which I am most familiar, I have limited the tutorials to the JMP format. It is my intention that by careful attention to the powers of DOE and related statistical matters, students of insect-rearing systems will be motivated to find a statistical package with which they can become comfortable (after a learning curve is satisfied), and they can proceed to treat all their rearing investigations (research, development, and optimization) using DOE and related statistical issues. *I conclude these remarks by repeating the concept that if I could redo my 40 + year career in insect diets and rearing systems, I would have conducted **all** my research in the framework of DOE and other statistically related treatments of my experiments.*

9.2.1 Case 1: Using the Taguchi Method

Using the Taguchi Method to help determine how several diet factors may affect the gel strength of our diet, we start off by possibly having noticed that some diets for a target insect seem to invite more robust consumption than other diets. Possibly an alert technician noticed that some diets seem mushy or overly soft, and these diets are not as attractive to the insects as ones that seem firm and crisp. After discussing this with the staff, the rearing team decides to try to discover which diet components detract from the firmness of the diet (the gel strength), and they decide to perform a texture analysis using a firmness-testing instrument such as a penetrometer (pictured later in this chapter in the section on mixture design).

The starting point for the Taguchi Method is shown in Figure 9.1 where the "Click Category:" is used to select "DOE." The starter contains the major categories (custom design, definitive screening, classical, and design diagnostics). "Taguchi Arrays" is the interactive box at the bottom

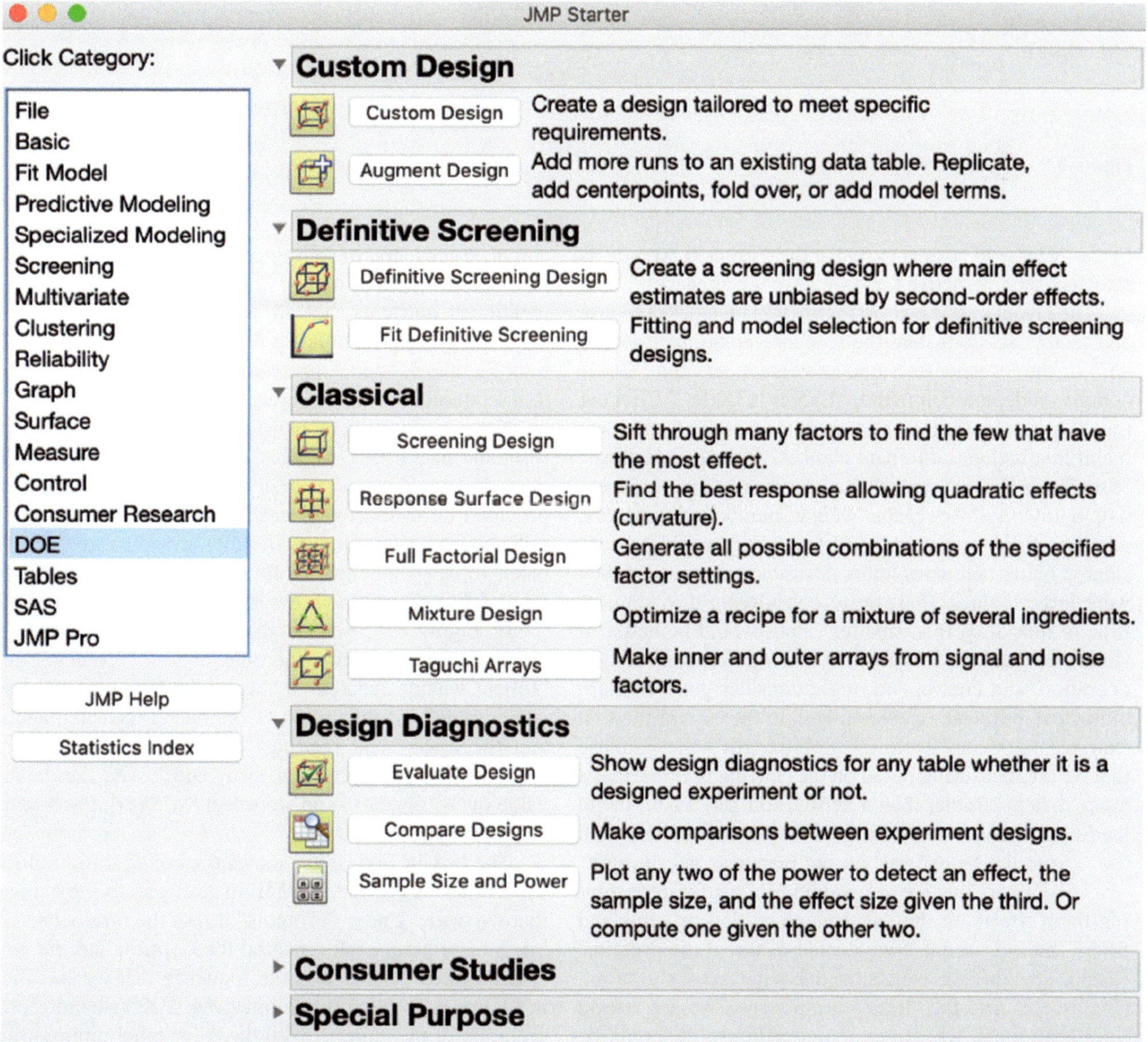

Figure 9.1 Shows the JMP starter menu.

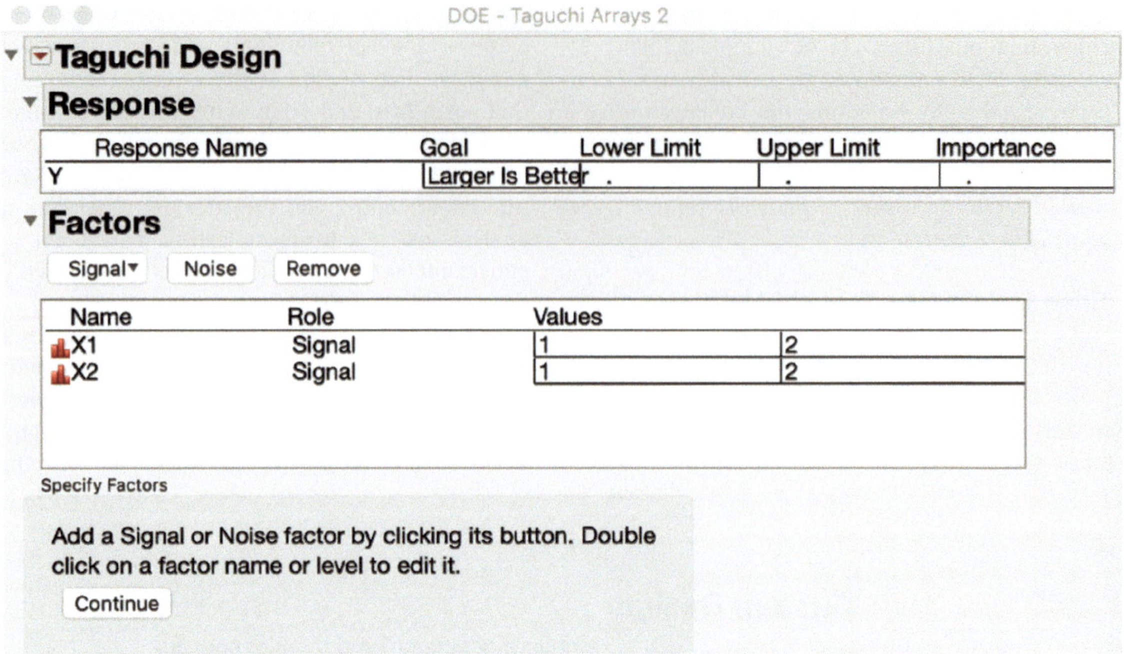

Figure 9.2 Taguchi design (arrays).

of the "Classical" category. Once the "Taguchi Arrays" is selected, the interactive box in Figure 9.2 appears.

The layout of this box (in Figure 9.2) includes response and factors sections, and the response contains a default "Y" in the "Response Name" category, and the "Goal" contains a default command, "Larger is Better." Clicking the "Larger Is Better" statement causes a drop down menu that includes alternate choices, "Nominal Is Best," "Smaller Is Better," and "None." From a rearing perspective, many responses (pupal weight, number of survivors, amount of diet consumed, and fecundity) would be considered better outcomes (more desirable outcomes) if they were larger values. If we were considering development time or mortality, then smaller values would be better. It should be realized that for our rearing systems' design, operation, and control, we might consider goals that are biological, physical, or economical. In the case of the first example that we are exploring, gel strength is the response that we are measuring based on the starting premise that a crisp, firm gel (rather than a mushy, soft gel) would invite feeding responses of our target insects; so we start with the "Larger Is Better" goal for our response: gel strength.

The Factors that we are wanting to test for determining their effects on the gel strength of diet proteins and fillers (bulking agents), we decide to test these proteins: casein, soy protein isolate, and bovine serum albumin (Figure 9.3). For the "filler" components, we are testing lima bean meal, wheat germ, and cellulose. It should be noted that both lima bean meal and wheat germ are not only fillers or bulking agents, but they are also sources of complex nutrients, including proteins, carbohydrates, lipids, vitamins, and minerals. Cellulose is void of any of the mentioned nutrients, but all three components certainly have the quality of bulking or texture-influencing function. Figures 9.3 and 9.4 provide the names that we added to the interactive box in Figure 9.2.

Note in Figure 9.3 that when we chose to add factors (proteins and filler), we also chose "3 levels" from the drop-down menu, rather than the "2 levels" choice. This provided three spaces for each factor. All of these terms will show up in later interactive dialog boxes and in results boxes to be seen below. At this point the "Continue" button is selected, and the box for Figure 9.4 appears.

In Figure 9.4, we see that the two factors specify "Inner Array" because the role of the "Proteins" and "Filler" was designated as "Signal," rather than "Noise" in Figure 9.3. At this point, "Continue" is selected again, and the dialog box reading "how many times do you want to perform each inner array run?" And the default value of "4" appears; and we select "4." Next, the box for Figure 9.5 appears.

The box in Figure 9.5 shows the configuration of the experiment suggested by JMP in response to the choices that we made. Under "Proteins" it lists the three different kinds of protein, each repeated three times, and the next column titled "Filler" has the sequence that we specified in Figure 9.3. The 3rd column gives a short-hand representation of the pattern, with the "-" symbol representing the 1st level specified, the "0" symbol representing the 2nd level specified, and the "+" representing the 3rd level

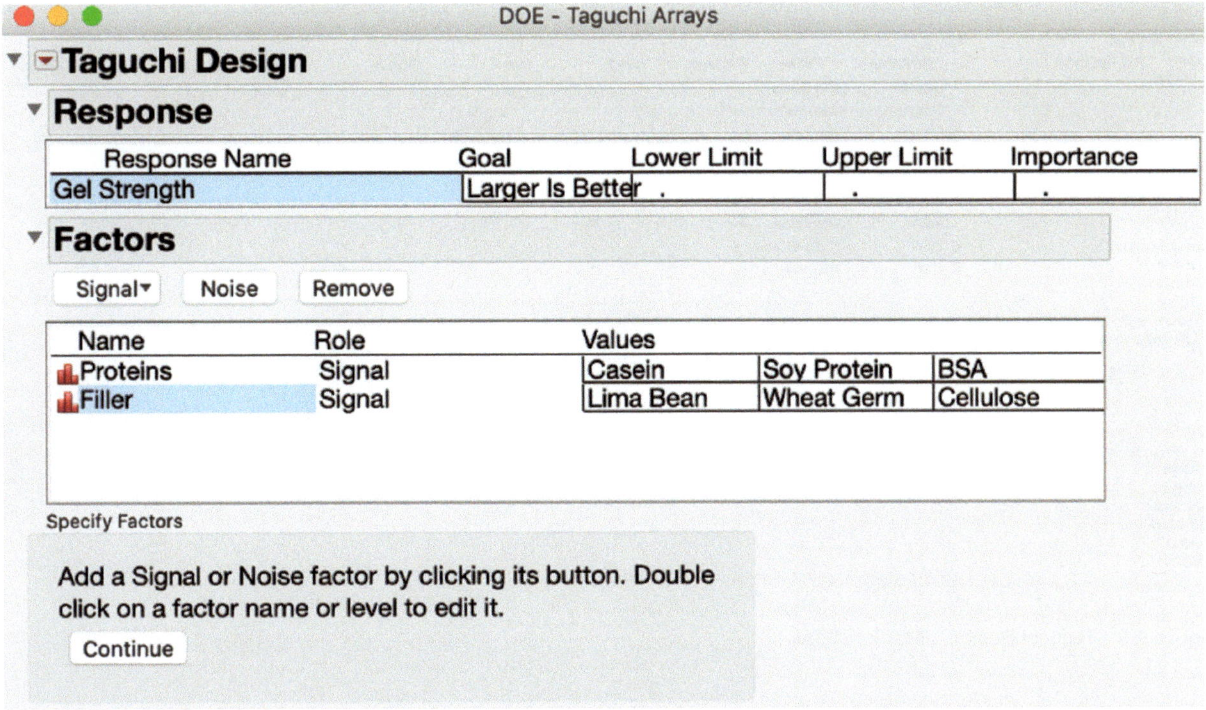

Figure 9.3 The interactive box with the response and factors filled in.

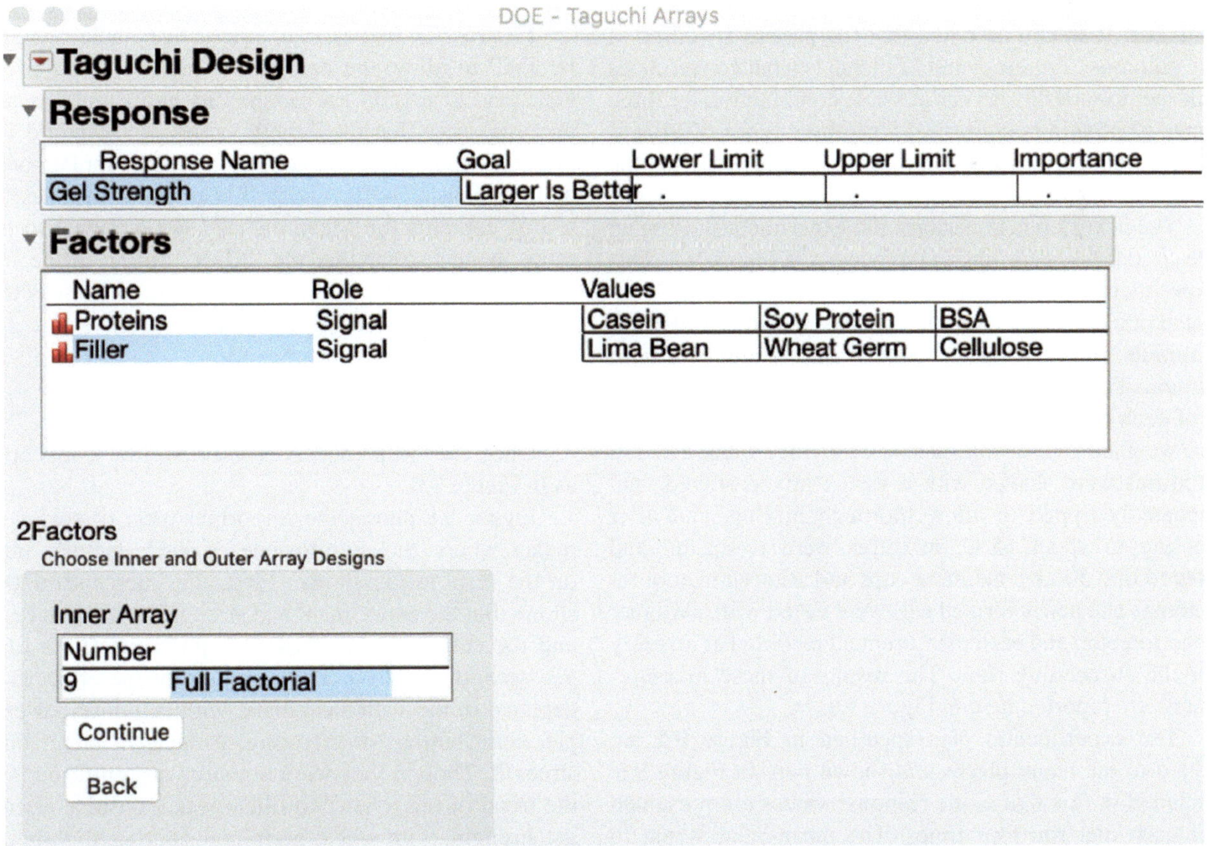

Figure 9.4 The interactive box for Taguchi Design after "Continue" was selected.

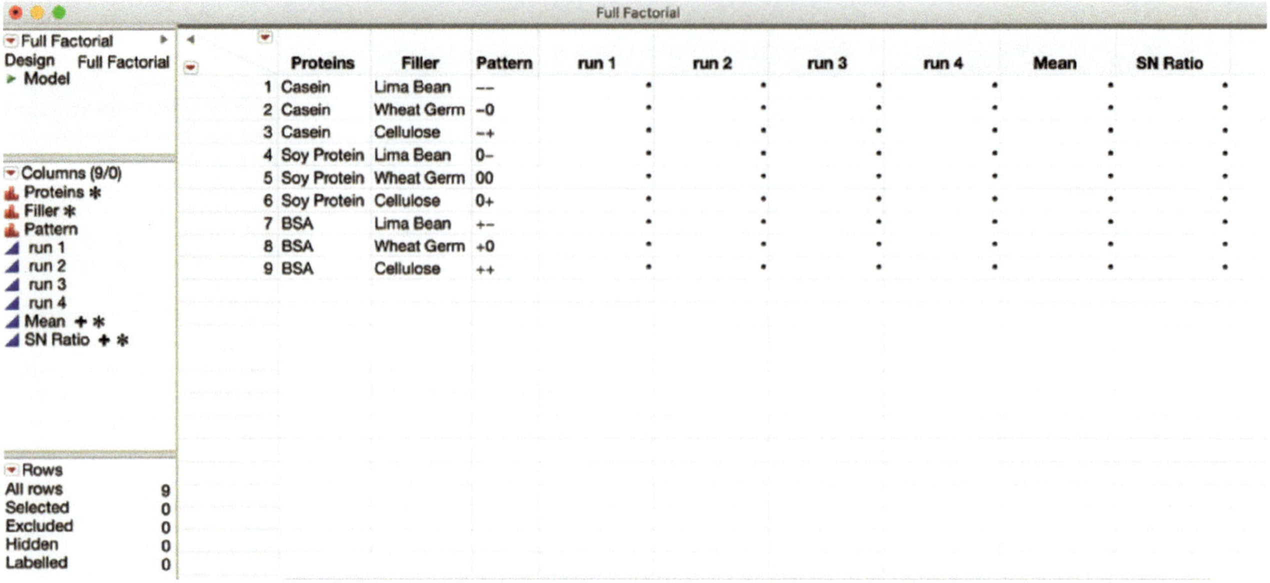

	Proteins	Filler	Pattern	run 1	run 2	run 3	run 4	Mean	SN Ratio
1	Casein	Lima Bean	--	•	•	•	•	•	•
2	Casein	Wheat Germ	-0	•	•	•	•	•	•
3	Casein	Cellulose	-+	•	•	•	•	•	•
4	Soy Protein	Lima Bean	0-	•	•	•	•	•	•
5	Soy Protein	Wheat Germ	00	•	•	•	•	•	•
6	Soy Protein	Cellulose	0+	•	•	•	•	•	•
7	BSA	Lima Bean	+-	•	•	•	•	•	•
8	BSA	Wheat Germ	+0	•	•	•	•	•	•
9	BSA	Cellulose	++	•	•	•	•	•	•

Figure 9.5 Taguchi Design full factorial box.

specified in Figure 9.3. The next four columns are blank at this point, with a dot to indicate that no values have been entered yet. The next two columns are the mean and signal to noise ratio to be generated by the experimental outcome. It should be noted that the pattern specified in the columns "Proteins" and "Fillers" is what constitutes a full factorial design. Every possible combination for three types of protein being mixed with three types of filler is covered here. Each combination is unique from all the rest.

The next step is to perform the experiment, setting up the conditions according to the pattern in Figure 9.5. This experiment was a simplified setup where the "diets" consisted of a 1.5% agar solution that was heated to 100°C for 1 minute, then adding 4% of each filler and 2% of each protein. The hot agar solution (47 mL) was added to the 3 g of each combination of filler and protein that had been pre-weighed in a 50-mL sterile centrifuge tube. The hot mixtures were shaken with a vortex mixer, stirred, and repeatedly tipped to allow thorough mixing, and after cooling to about 55°C, mixtures were re-shaken and poured into 50-cc condiment cups and allowed to cool for 2 hours. The newly formed gels were tested with a Wagner penetrometer, and each measurement recorded as an entry for the succeeding runs. The results of these measurements are reported here in Figure 9.6.

The experimental plan specified in Figure 9.5 are added to the blank places and shown here in Figure 9.6. Included in this figure, the response values are presented for each diet run four times. The means and signal to noise ratios were used by the JMP program to generate the results shown below. Figure 9.7 shows the report: fit model box, which includes the names of the columns ("Proteins," "Filler," etc.), the roles of the variables, the "personality" and "emphasis" and the choice of running the data analysis ("Run").

Figure 9.7 shows the interactive table generated by JMP to allow the operator to pick the roles of variables and to specify the Personality and Emphasis of the analysis. Note that the default values of "Standard Least Squares" and "Effect Screening" are listed for Personality and Emphasis. Also, under "Pick Role Variables," the Y box contains the Mean and SN Ratio (signal to noise ratio). Note further that the "RUN" button in the upper right portion of the figure is now ready to be selected to command the JMP analysis to go forward.

9.2.1.1 *Results*

Once the "Run" button is selected, the results appear as in Figure 9.8.

Figure 9.8 shows the important table of scaled estimates where the significance of each factor's impact on the response is shown. First, the intercept (<0.0001) shows that the experiment had strongly significant results, and the cellulose also showed significant effects on the gel strength (0.0264). The direction of the effect is also depicted in the scaled estimate, with cellulose, for example, contributing (or effecting) a negative effect on gel strength. Though they were not quite at significant levels, the trend of the other two fillers was a positive effect on gel strength, with wheat germ contributing slightly more than lima bean meal to gel strength. Interestingly, the proteins had different trends from one another, with casein

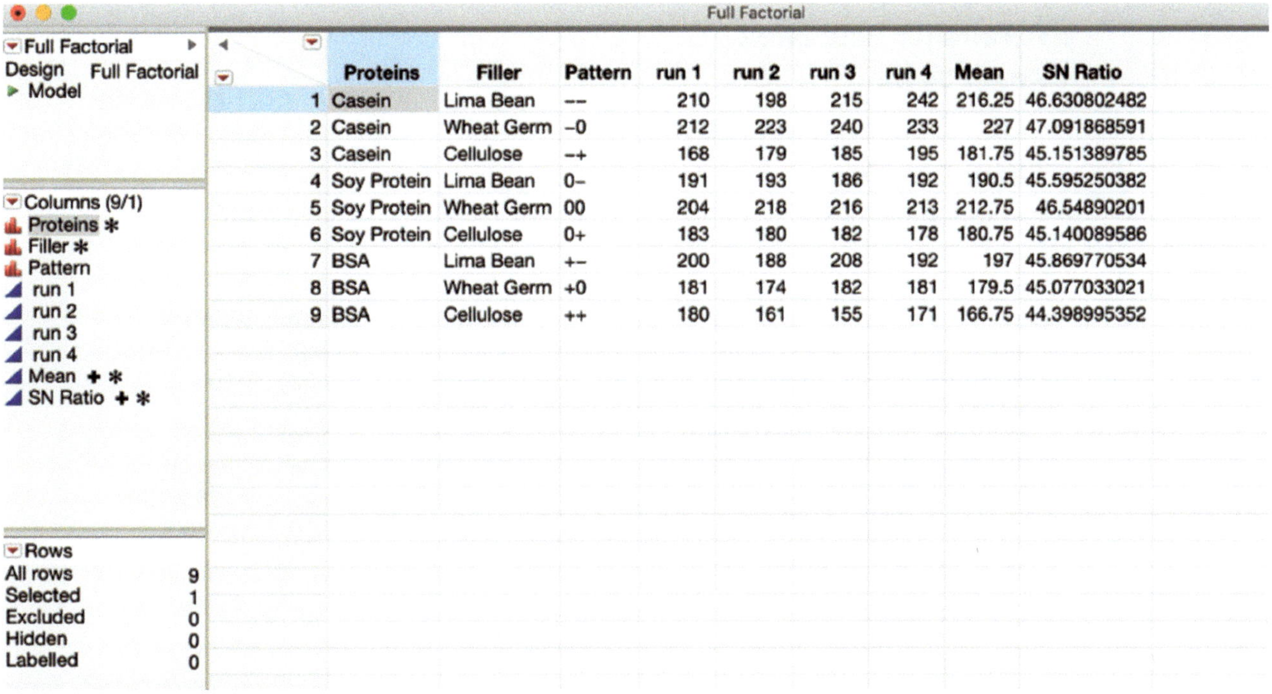

	Proteins	Filler	Pattern	run 1	run 2	run 3	run 4	Mean	SN Ratio
1	Casein	Lima Bean	--	210	198	215	242	216.25	46.630802482
2	Casein	Wheat Germ	-0	212	223	240	233	227	47.091868591
3	Casein	Cellulose	-+	168	179	185	195	181.75	45.151389785
4	Soy Protein	Lima Bean	0-	191	193	186	192	190.5	45.595250382
5	Soy Protein	Wheat Germ	00	204	218	216	213	212.75	46.54890201
6	Soy Protein	Cellulose	0+	183	180	182	178	180.75	45.140089586
7	BSA	Lima Bean	+-	200	188	208	192	197	45.869770534
8	BSA	Wheat Germ	+0	181	174	182	181	179.5	45.077033021
9	BSA	Cellulose	++	180	161	155	171	166.75	44.398995352

Figure 9.6 Results of the experiments with nine unique combinations.

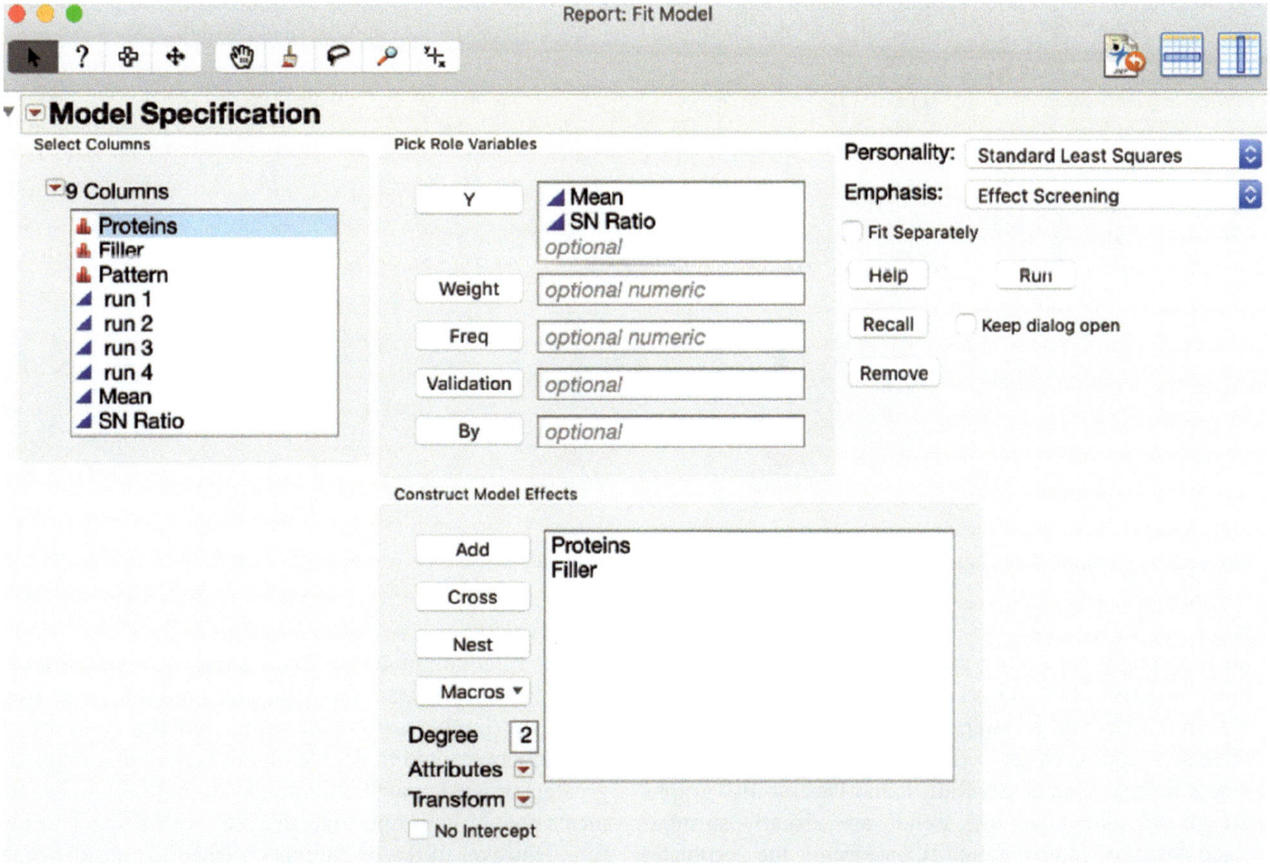

Figure 9.7 Taguchi analysis model specification.

▸ **Parameter Estimates**

▸ **Effect Tests**

▸ **Box-Cox Transformations**

▸ **Effect Details**

▾ **Scaled Estimates**

Nominal factors expanded to all levels

Term	Scaled Estimate	Std Error	t Ratio	Prob>|t|
Intercept	45.722678	0.169928	269.07	<.0001*
Proteins[BSA]	-0.607412	0.240314	-2.53	0.0648
Proteins[Casein]	0.5686756	0.240314	2.37	0.0771
Proteins[Soy Protein]	0.038736	0.240314	0.16	0.8798
Filler[Cellulose]	-0.825853	0.240314	-3.44	0.0264*
Filler[Lima bean]	0.3092632	0.240314	1.29	0.2675
Filler[Wheat germ]	0.5165899	0.240314	2.15	0.0980

▾ **Prediction Profiler**

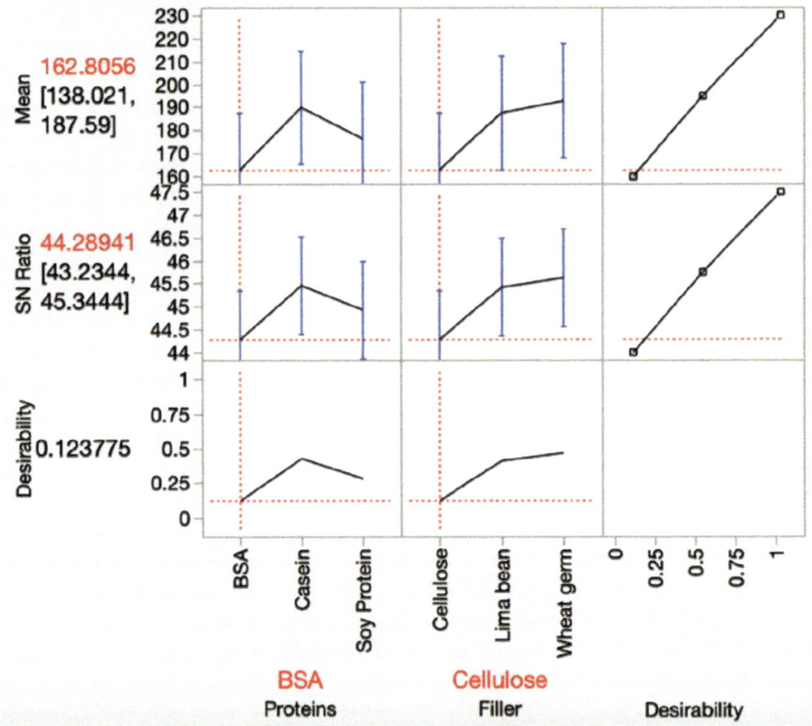

Figure 9.8 Results of the Taguchi analysis for gel strength vs. proteins and fillers.

fairly strongly enhancing gel strength, while BSA had a fairly negative effect on gel strength, and soy protein had almost no effect on gel strength. The visualization of these effects is enhanced by viewing the prediction profiler, where means, Means, SN Ratios, and Desirability graphs are shown and where the trends are visually summarized for easy interpretation. Considering the responses (gel strengths), the highest values were obtained from the casein (for the protein components) and the wheat germ (for the filler components). The weakest gels (or the greatest amount of detraction from gel strength) were from the BSA (bovine serum albumin) and the cellulose. If researchers wanted to decide on the factors that make the greatest or least contribution to gel strength, these experiments would help optimize the diet with respect to texture. However, if the researchers wished to consider other factors such as feeding stimulation, nutritional value, bioavailability, or diet stability, they would have to design experiments where they measured other responses such as how much diet was consumed per unit of time or biomass accumulation of larvae. However, regardless of the questions being asked, if the experiments were designed with a DOE procedure such as Taguchi Arrays, the researcher would find the prediction profiler and the accompanying analyses helpful, especially if they follow the guidance of statisticians such as Sall et al. 2017 who explain:

> The settings of each factor are connected by a line, called the *prediction trace* or *effect trace*. You can grab and move each vertical dotted line in the Prediction Profile plots to change the factor settings. The predicted response automatically recomputes and shows on the vertical axis, and the prediction traces are redrawn. The Prediction Profiler lets you look at the effect on the predicted response of changing one factor setting while holding the other factor settings constant. It can be useful for judging the importance of the factors.

One last feature of the JMP program for Taguchi Arrays is to be seen in the strongly visual representation of the experiment treated here shown in Figure 9.9, which shows the various levels of response to the factors

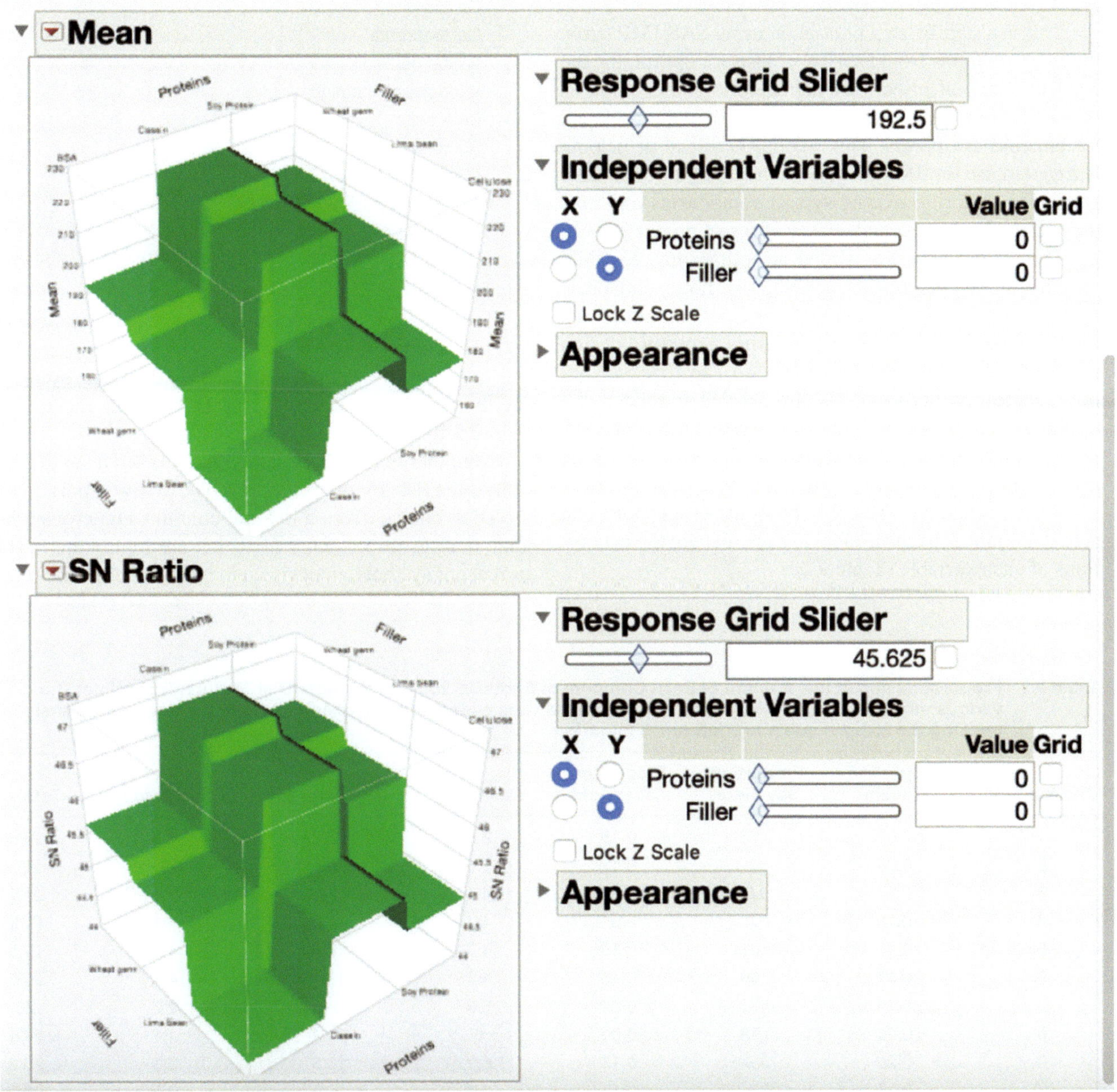

Figure 9.9 Three-dimensional representation of gel strength means and signal/noise ratios in relationship to types of protein and types of filler in agar-gel diets.

tested here. Clearly the diets with casein and wheat germ gave the highest gel strength response "platform," and the cellulose/BSA combination gave the lowest response. Such three-dimensional structures when used with several diet factors and several responses to those factors can become a powerful tool for rearing system development and optimization. This attests convincingly to the power of the statistically based exploration possible with DOE.

9.2.2 Step-by-Step Tutorial on Using JMP Mixture Design to Optimize an Insect Diet's pH

This is a step-by-step tutorial on using SAS JMP (from the JMP programs from DOE) as a basis for optimizing the concentrations of insect diet ingredients. The example is taken from one of my research projects on diet development for emerald ash borers. This tutorial is aimed at helping learners to use the JMP program, which has a tremendous power for exploring rearing system components and how they affect the process and quality of insect production.

This experiment was aimed at optimizing pH while varying three components, casein, soy flour, and citric acid. At first, it would seem that citric acid would be the key variable for pH set points. By using a mixture design for the three mentioned variables, we can discover not only the desirability (in terms of a pH that simulates that of phloem wood) but also the possible interactions of components that are what Cohen 2015 calls "cryptic buffering agents." Below (Table 9.1) is the total diet, with the three variables in shading. Table 9.1 shows the proportions and actual amount of each variable in each diet are the proportions of each variable (Table 9.2).

Table 9.1 Proportions of the Three Variables (Casein, Soy Flour, and Citric Acid) in the 12 Diet Mixtures. (Diet modified by Cohen and based on Keena et al. 2015)

Cohen modified from Keena et al. 2015 Diet Components	Weight (g)
Water	150
Agar (Frontier)	6
Potassium sorbate	0.3
Calcium propionate	1
Beta sitosterol	0.1
Starch (potato)	6
Cellulose	25
Ca ascorbate	1.1
Wheat germ oil	1
Vanderzant vitamins	2.5
Beck salts	1
Sucrose	6
Casein	5
Soy flour (Southland)	20
Citric acid	0.25

Table 9.3 shows the proportions of diet components and pH outcome (response).

To begin the tutorial in JMP® 14.2, begin by opening the program and finding the DOE in the menu as in Figure 9.10.

Note that in the "File" list, "DOE" is 4th from the bottom of the list (Figure 9.10). The blue arrow is pointing to the DOE choice (from a list that contains numerous other analytical features, many of which are highly applicable to rearing systems analysis and optimization). The menu

Table 9.2 Proportions and Actual Amount of Each Component Added to 200 g of the Agar/Diet Mixture. Table Note: The various mixtures of casein, soy flour, and citric acid were added to the cooked diet from the steam kettle after cooking the 200 g aliquots to 95oC for 10 minutes

Diet Number	Proportion Casein	Proportion Soy Flour	Proportion Citric Acid	Casein (g)	Soy Flour (g)	Citric Acid (g)	Total (g)
1	0.09	0.76	0.15	2.5	21	4.1	27.6
2	0.09	0.8	0.11	2.5	22	3	27.5
3	0.15	0.8	0.05	4.1	22	1.4	27.5
4	0.15	0.8	0.05	4.1	22	1.4	27.5
5	0.17	0.68	0.15	4.7	18.7	4.1	27.5
6	0.17	0.68	0.15	4.7	18.7	4.1	27.5
7	0.19	0.71	0.09	5.3	20	2.7	28
8	0.21	0.73	0.05	6	20.1	1.4	27.5
9	0.25	0.6	0.15	6.9	16.5	3.5	26.9
10	0.25	0.6	0.15	6.9	16.5	3.5	26.9
11	0.28	0.628	0.092	7.7	17.3	2.5	27.5
12	0.28	0.67	0.05	7.7	18.4	1.4	27.5

Table 9.3 **Proportions of Diet Components and the pH Outcomes From the Various Test Mixtures**

Diet Number	Casein	Soy Flour	Citric Acid	pH
1	0.09	0.76	0.15	3.96
2	0.09	0.8	0.11	4.19
3	0.15	0.8	0.05	4.66
4	0.15	0.8	0.05	4.82
5	0.17	0.68	0.15	4.01
6	0.17	0.68	0.15	3.88
7	0.2	0.71	0.097	4.19
8	0.218	0.731	0.05	4.66
9	0.25	0.6	0.15	4.04
10	0.25	0.6	0.15	3.99
11	0.28	0.628	0.092	4.28
12	0.28	0.67	0.05	4.58

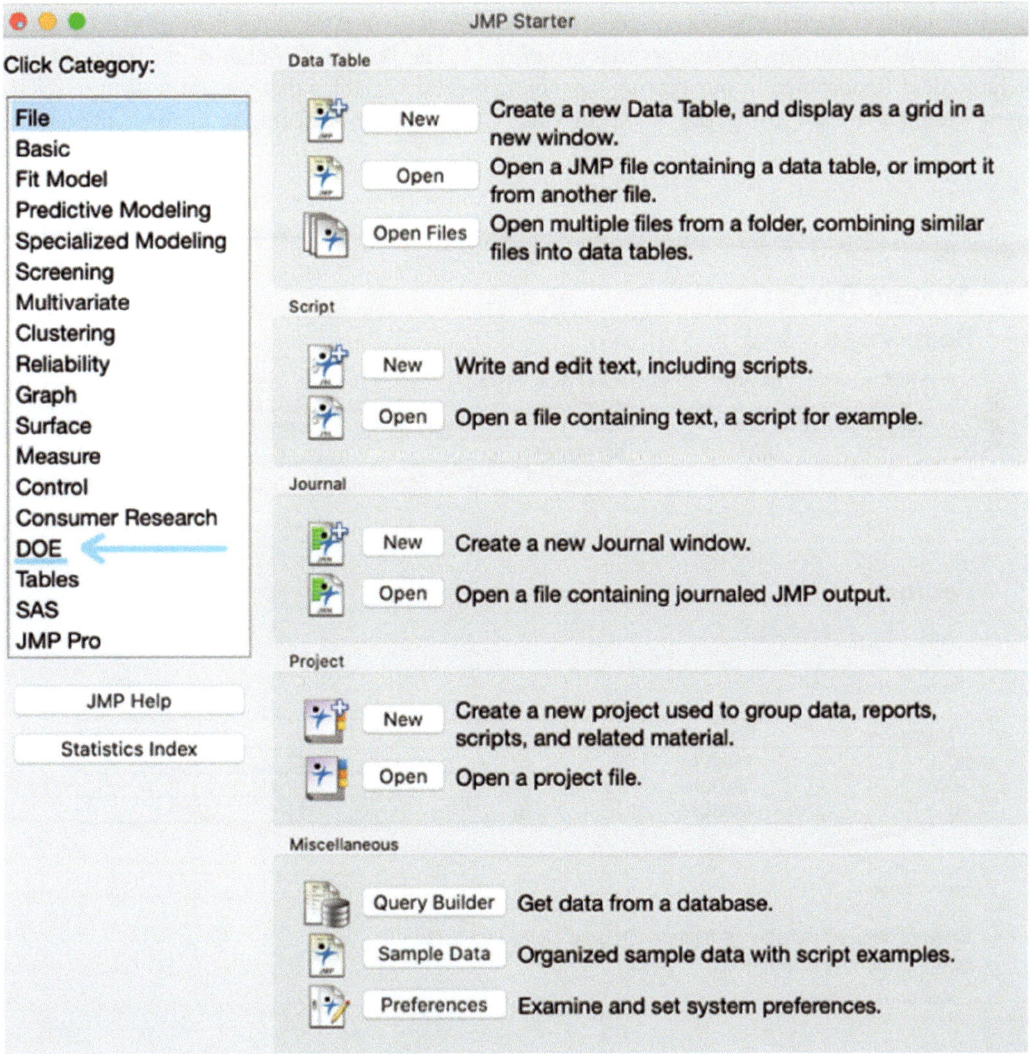

Figure 9.10 Sarting the JMP program.

for various programs in DOE are listed under the major categories, "Custom Design," "Definitive Screening," "Classical," "Design Diagnostics," "Consumer Studies," and "Special Purpose" programs. Readers are encouraged to explore these programs and sub-programs as they provide statistical guidance that has great potential for using statistical support for rearing system inquires.

For the current tutorial, once the "Mixture Design" is selected, the box in Figure 9.11 appears. The red numbers, 1, 2, 3, and 4 are noted to help explain the step where the "Responses" and the "Factors" are being considered.

The dialog under "Mixture Design" in Figure 9.11 includes to main sections: **Responses** and **Factors**. The **responses** are the outcomes of the experiments that we are performing in our DOE procedures. Typical responses of interest in rearing systems may include weight of the insects, mortality (or survival), duration of a developmental stage, number of eggs laid, number of hatched eggs, length of wings, amount of a diet consumed, hardness (or other rheological properties) of a gel, days until mold forms, number of adults that could fly out of a container, and many, many more measurable parameters that are of physical or biological importance in our rearing system. In the current example, we are using pH of the diet as our response to a mixture experiment. Elsewhere in this chapter, we will be using gel strength (firmness) as the response to diet mixture experiments.

In circles 1, 2, and 3, in Figure 9.11, the dialog for response name (1), the goal (2), and the lower and upper limits (3) are to be specified by double-clicking on the item (such as "Y") and filling in the name, then selecting the goal (maximize, minimize, or match target); then the numerical limits are specified. The specification of the numerical limits (upper and lower) takes some experience with the components being tested. The experience may be from the investigator (researcher or QC person) having previously performed the rearing component being tested or it may be the experience presented in a publication. In the current study, the response of pH was approximated by the investigator by measuring pH of several diets that we were working with in my laboratory, and it was determined that most of the diets' pH values fell between 3.5 and 5.0. This knowledge helped us avert doing extraneous inquiries with pH values in the strongly acid range (below 2.0) or far about neutral.

The **factors** (number 4 in Figure 9.11)are the experimental variables that we are testing (circled X1, X2, and X3). Double-clicking the factor names, we can substitute

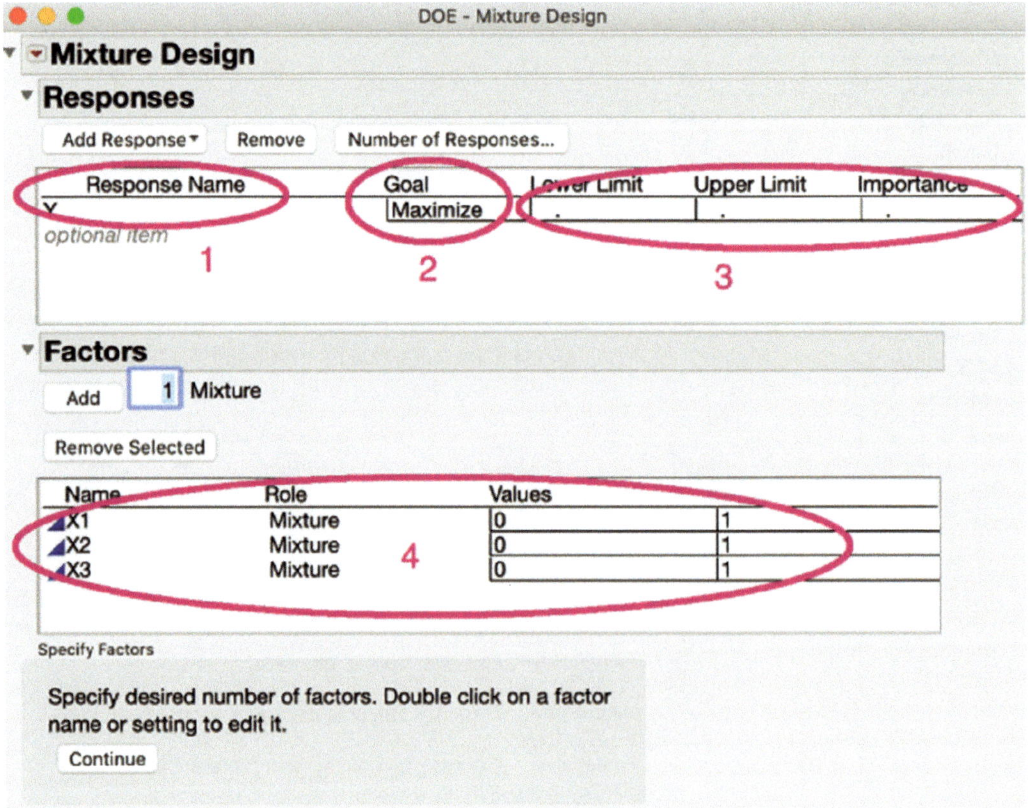

Figure 9.11 Mixture Design dialog box.

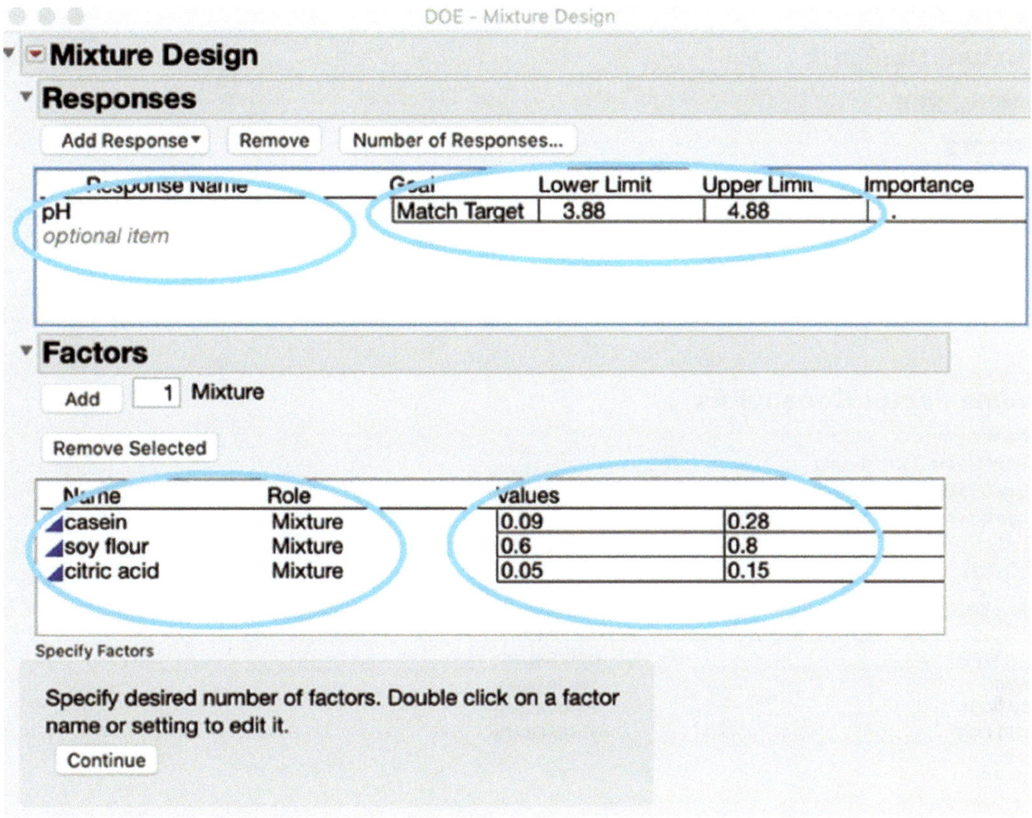

Figure 9.12 Mixture Design interactive box, filled in with desired values for the current experiment.

our terms, "casein," "soy flour," and "citric acid." Next, also in circle 4, we fill in values that are within a reasonable range for each factor. By "reasonable," I mean what our experience or that of previous researchers has taught us (Figure 9.12).

Notice that the default box in Figure 9.12 has been filled in with "pH" for Response Name and "Match Target" chosen for the Goal. Note that the lower and upper limits were specified (3.88 and 4.88 for pH responses), and the names of the factors have been entered, as well as the lower and upper range of desired values. At this point, the "continue" box is selected, and the interactive box as in Figure 9.13 is present (Figures 9.14 and 9.15).

It is important here to note for the Mixture Design box that the number of replicate runs is selected in the box near the bottom (under "Design Generation."). This box (Figure 9.13) allows important features to be added to the analysis, including the specification of the Main Effects and the Interactions. I feel that for designing and optimizing rearing systems, the Interactions options are especially important because one of the most complex features of rearing systems, including artificial diets, is the nature and extent of interactions between system components. In my own past 40 years of experience with artificial diet development and rearing system optimization, I have found the interactions between components to be the most evasive, yet important, features of how the system works. The evaluation of interactions is the most far-reaching ways of arriving at the mechanisms of rearing systems and the basis for a building a body of predictions about how various diet and other rearing system components will perform.

In this example (in Figure 9.14), the "Optimal" was selected. The other models, under linear constraints, can be useful in highly sophisticated mixture experiments. The simplest one (optimal) was selected here for ease of starting for novice users of these engineering optimization procedures. It is noteworthy that the outcome that we were seeking in using the mixture procedure was to find the combination of diet components that gave us the closest fit to a pH that was considered most desirable.

For the first test diet specified in the table in Figure 9.15, the proportion of casein was 0.09, soy flour 0.76, and citric acid 0.15 (which totalled equals 1.0). The respective proportions of casein, soy flour, and citric acid for the second test diet were 0.09, 0.80, and 0.11 again totalling 1.0, and so on for all 12 diets. Once the diets were all mixed,

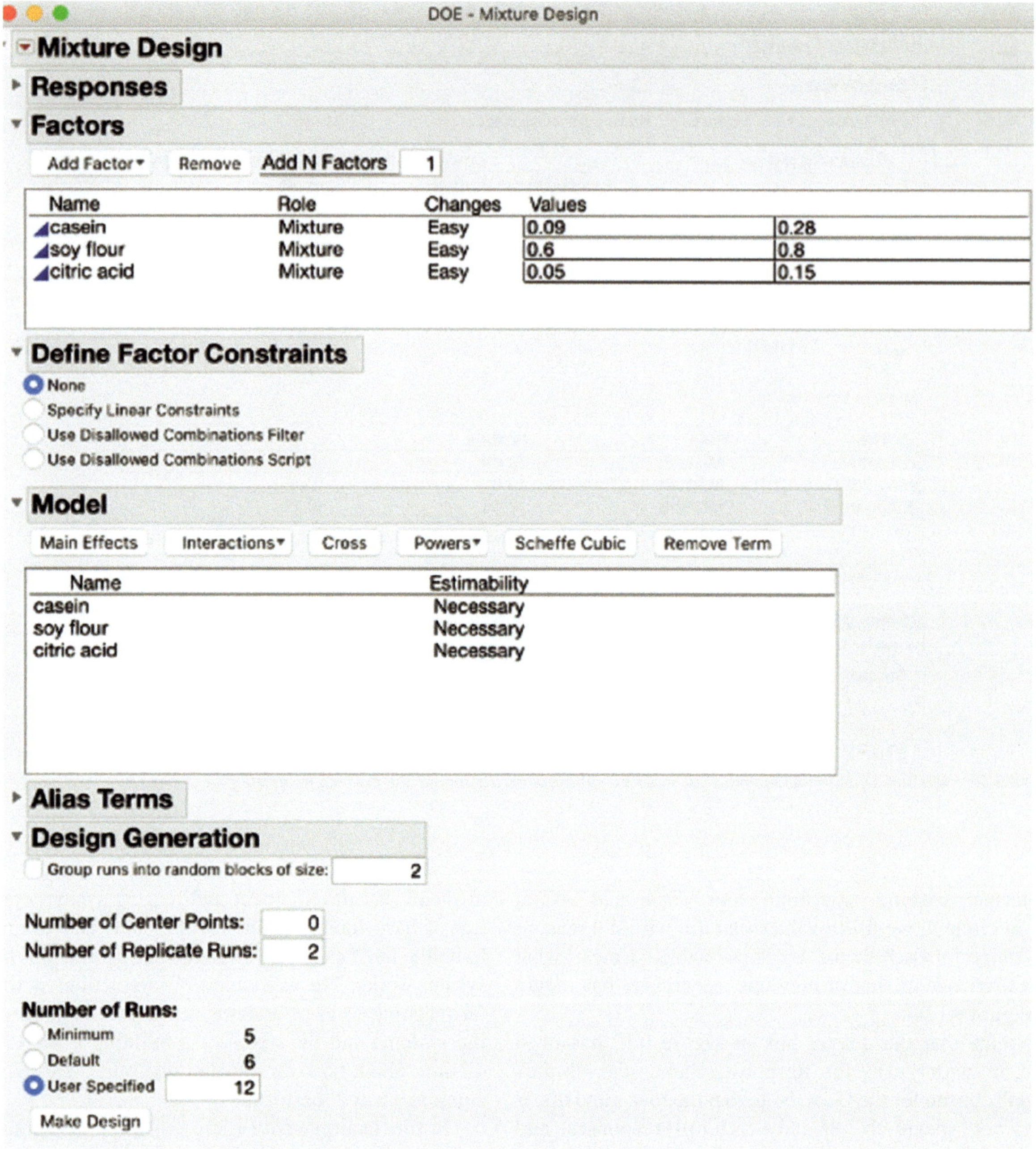

Figure 9.13 Mixture Design dialog box with interactive design generation table.

heated, poured into cups and allowed to cool, pH measurements were made, and the values in the pH column were obtained and recorded in the pH column next to the respective mixtures.

At this point "Model" is selected from the terms displayed in the upper left portion of the figure, and the interactive dialog box in Figure 9.16 appears.

The box in Figure 9.16 is the command box that starts the analysis, once all the factors have been entered. As with other DOE processes in JMP, the run button on the upper right is selected at this point, and the following reports appear.

The "Effect Summary" in Figure 9.17 shows the P values or significance of contribution of each factor, with soy flour being shown to have the most impact on pH response. It shows further that there is an indication of interaction, the most interaction taking place between soy flour and citric acid. The "LogWorth" along with the "P value" are the indicators of magnitude of the interactions and components' contribution.

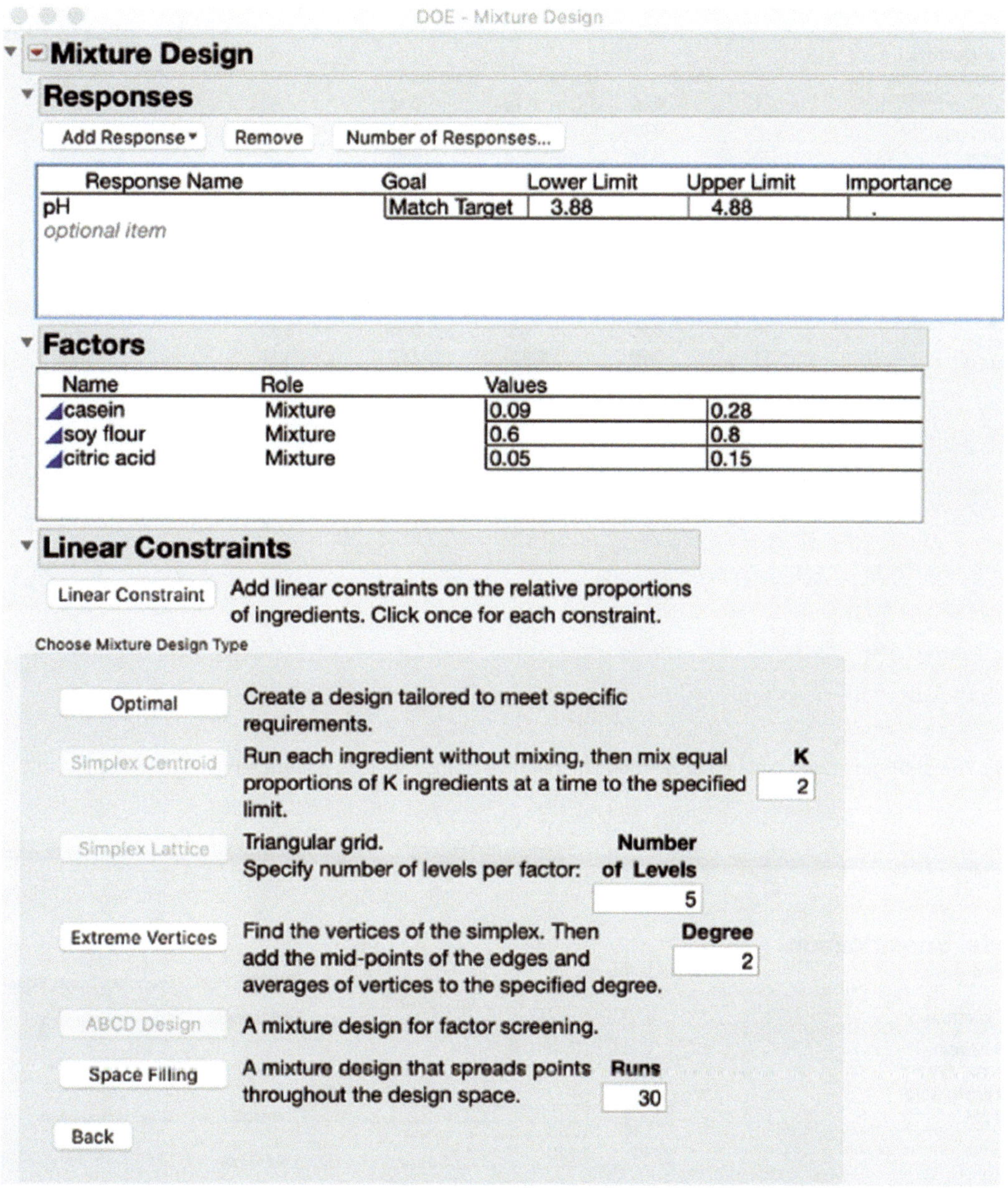

Figure 9.14 Dialog box for Mixture Design with "Choose Mixture Design Type" dialog.

It should be noted that in all likelihood, the strongly significant soy influence is related to the very large amount of soy present in the mixture, compared to citric acid or even casein.

For rearing personnel who are not statisticians, the prediction profiler is one of the most helpful representations of the experimental outcomes. The profiler (Figures 9.18 and 9.19) shows with visual ease, that of the three components, citric acid (despite its very small mass) contributed most to the pH outcomes, going from a pH of about 4.6 at the lowest concentration, to about 3.9 at its highest concentration. While the difference in concentration was 1.0% from lowest to highest, the difference in pH was 0.7 pH units. At the same time, soy flour influenced a pH change of 0.4 pH units over a range of 20% of the mixture.

The prediction profiler in Figure 9.19 has an additional feature of allowing interactions between components by allowing the user to select a factor's window, and move the vertical, dotted line (in this case to a lower concentration of casein) to show how the other components and outcomes will be affected. Compare Figure 9.19 with Figure 9.20 to see this effect. This visual tool can help rearing system researchers make projections based on the

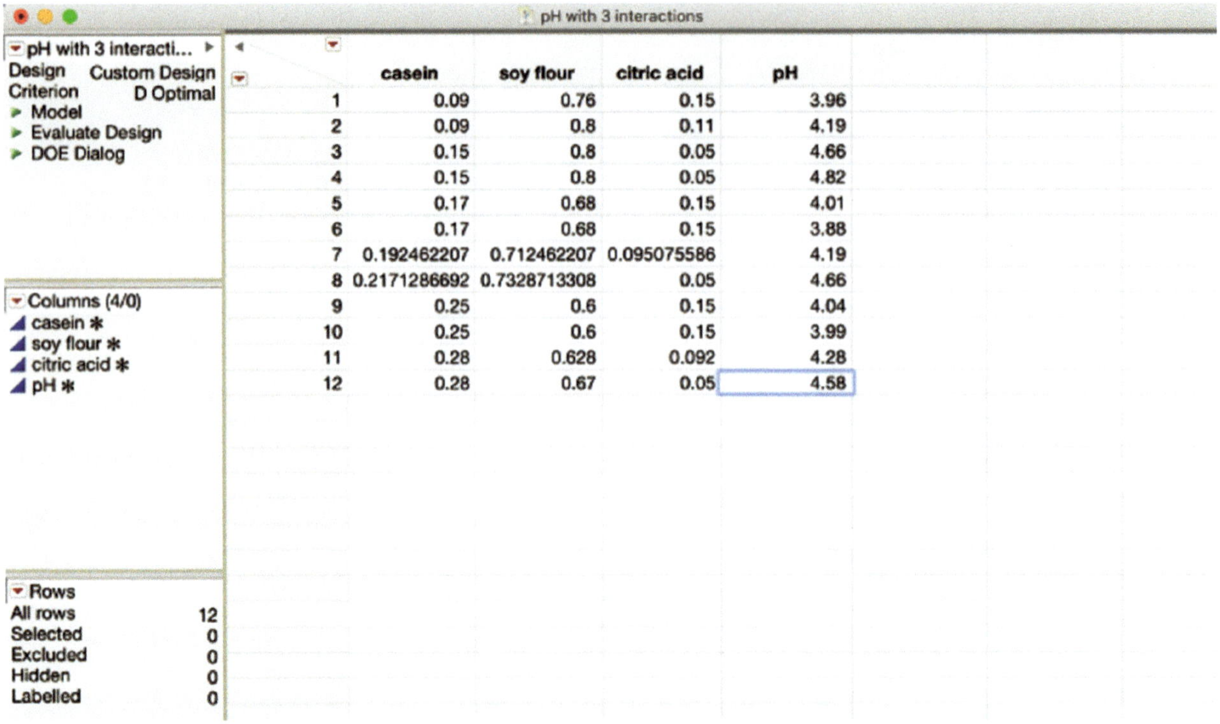

	casein	soy flour	citric acid	pH
1	0.09	0.76	0.15	3.96
2	0.09	0.8	0.11	4.19
3	0.15	0.8	0.05	4.66
4	0.15	0.8	0.05	4.82
5	0.17	0.68	0.15	4.01
6	0.17	0.68	0.15	3.88
7	0.192462207	0.712462207	0.095075586	4.19
8	0.2171286692	0.7328713308	0.05	4.66
9	0.25	0.6	0.15	4.04
10	0.25	0.6	0.15	3.99
11	0.28	0.628	0.092	4.28
12	0.28	0.67	0.05	4.58

Figure 9.15 The JMP-generated table for proportions of the three components to generate the mixture to be tested.

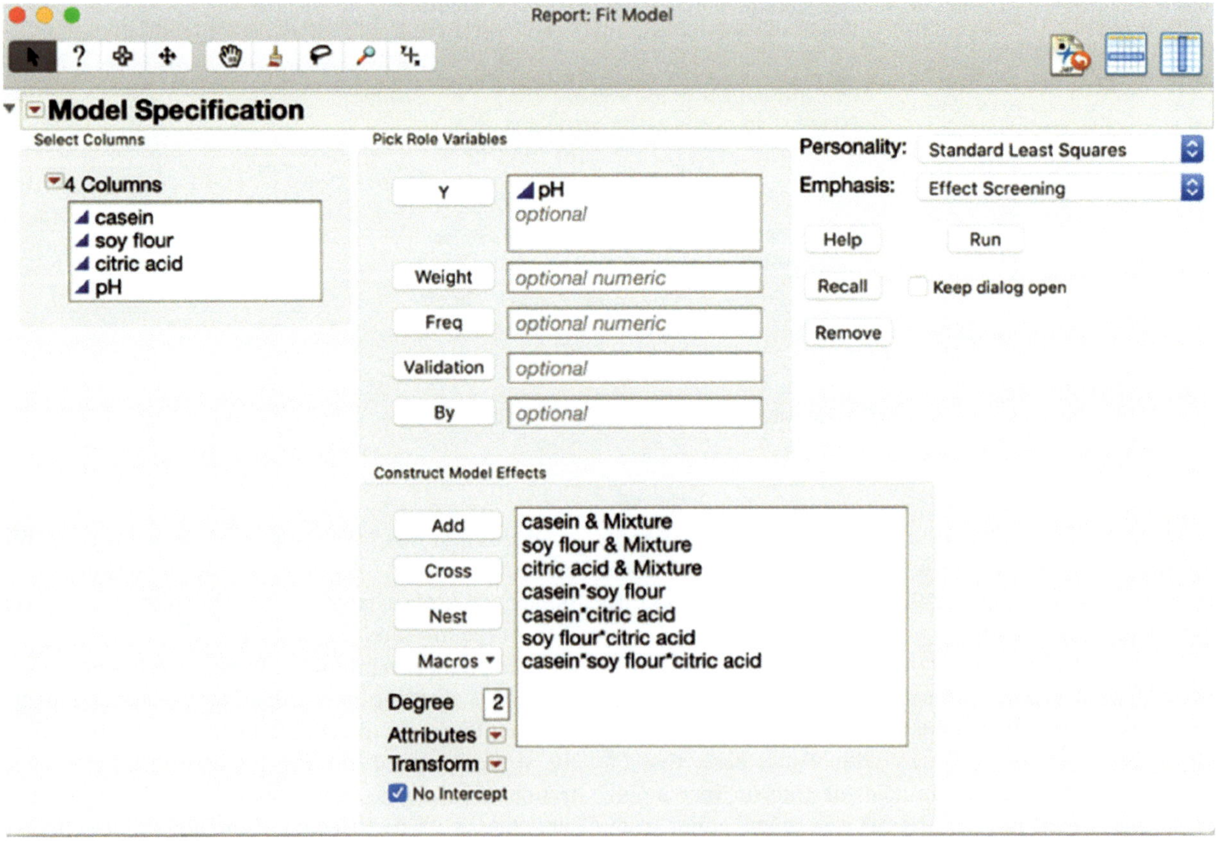

Figure 9.16 Report for "Fit Model."

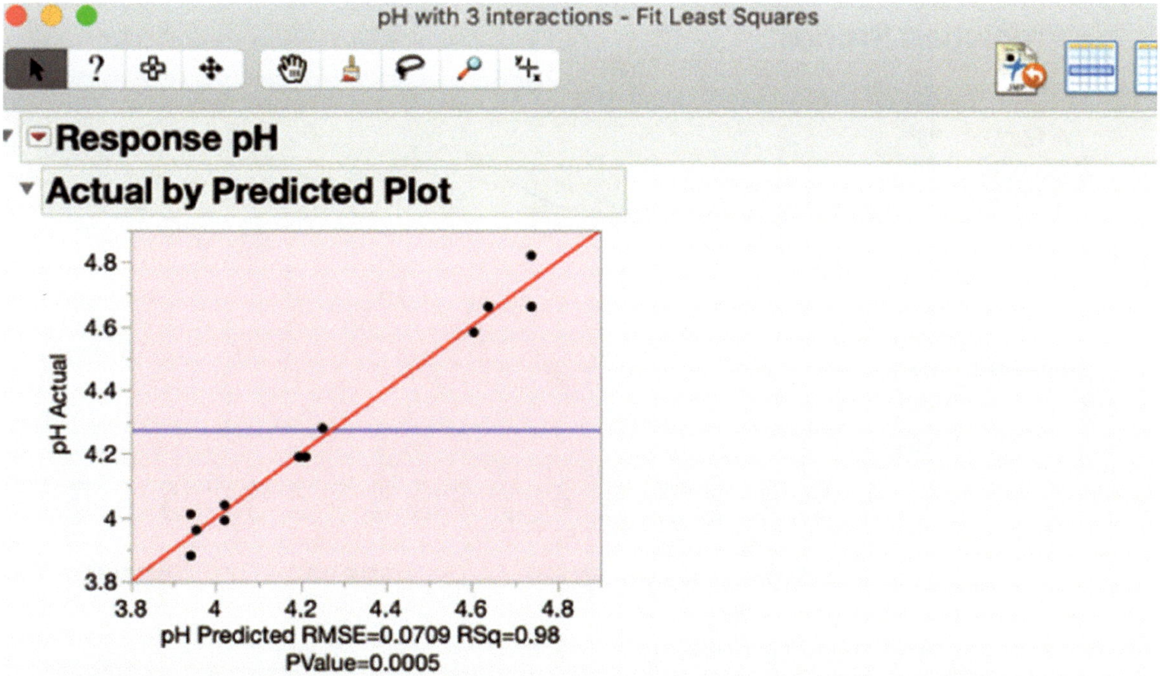

Effect Summary

Source	LogWorth	PValue
(soy flour-0.6)/0.26	5.705	0.00000
(casein-0.09)/0.26	5.104	0.00001
(citric acid-0.05)/0.26	2.116	0.00765
soy flour*citric acid	0.846	0.14247
casein*citric acid	0.552	0.28060
casein*soy flour	0.198	0.63323
casein*soy flour*citric acid	0.005	0.98783

Remove Add Edit FDR

Lack Of Fit

Source	DF	Sum of Squares	Mean Square	F Ratio
Lack Of Fit	2	0.00260121	0.001301	0.1734
Pure Error	3	0.02250000	0.007500	Prob > F
Total Error	5	0.02510121		0.8487
				Max RSq
				0.9806

Figure 9.17 Actual vs. Predicted pH results and effects summary.

data that they obtained from their experiments to project outcomes that are statistically tied to or associated with the actual outcomes.

In Figure 9.20, the red lines that are interactive in the JMP analysis were manipulated to show the effects on other components by manipulating one component. The manipulation/projection was done with the red cross-bar selection tool. The interactive, predictive nature of the cross-bars makes the prediction profiler tool extremely useful in optimizing mixtures. This is an example of how the experiments with variation of three components in 12 diet combinations can allow predictions that preclude the mixing of dozens to scores of different diets. This illustrates the predictive power of design of experiments.

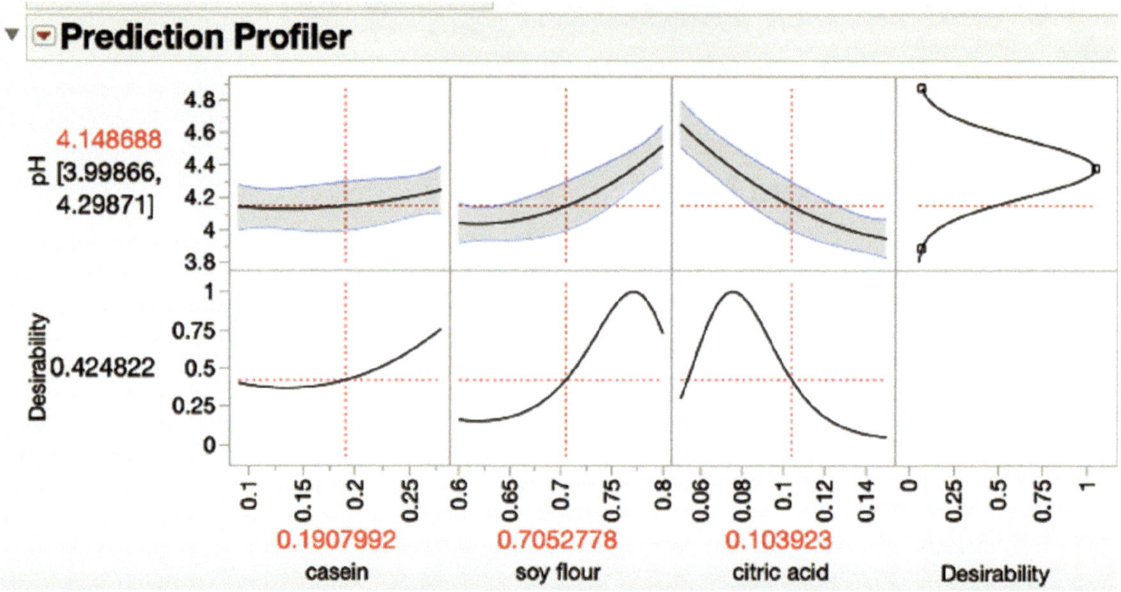

Figure 9.18 Prediction profiler for pH optimization excercise.

The above two profiler reports (Figures 9.20 and 9.21) are based on identical data (the same mixtures of variables—casein, soy flour, and citric acid in the same diets for both profiles). However, the upper profiler predictions are based on constraints settings of pH 4.50–5.50, and the lower profiler was based on constraint settings of pH 4.1–pH 4.9. It is evident from these profiles that casein makes a very slight positive contribution to higher pH and soy flour makes a more substantial positive contribution to raising pH, but citric acid makes a negative contribution (or the more citric acid, the lower the pH). The last statement would be expected (that citric acid would lower the pH of a diet, but the effects of the soy flour are not expected), and this type of mixture experiment can allow us to tease apart the effects of "cryptic" pH influences, "cryptic" buffering capability, and most important of all: the interactions between components can be brought to light.

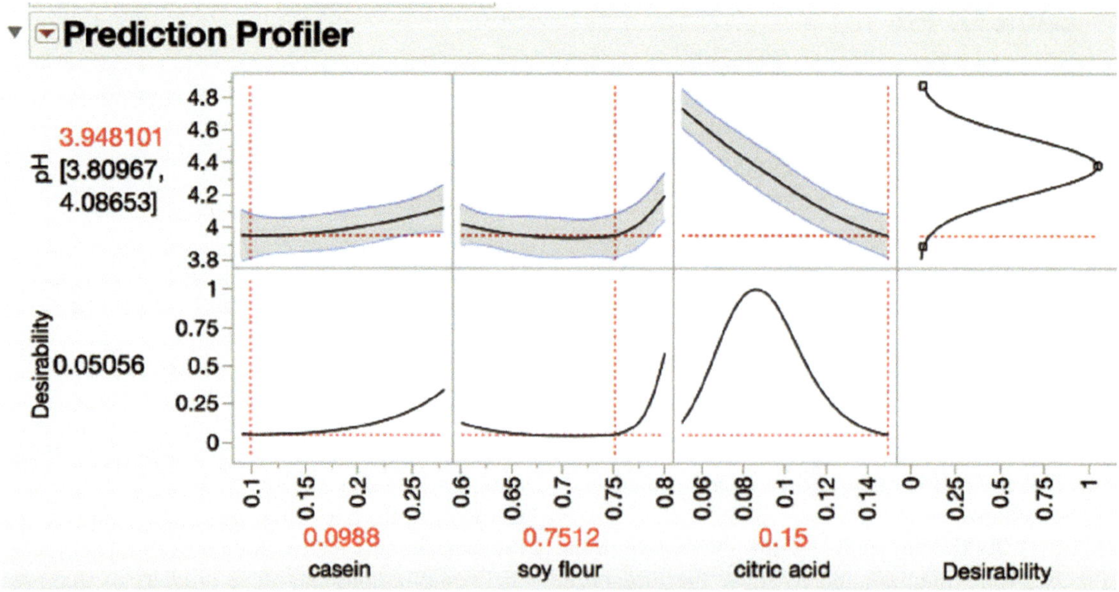

Figure 9.19 Prediction profiler showing interactive dialog.

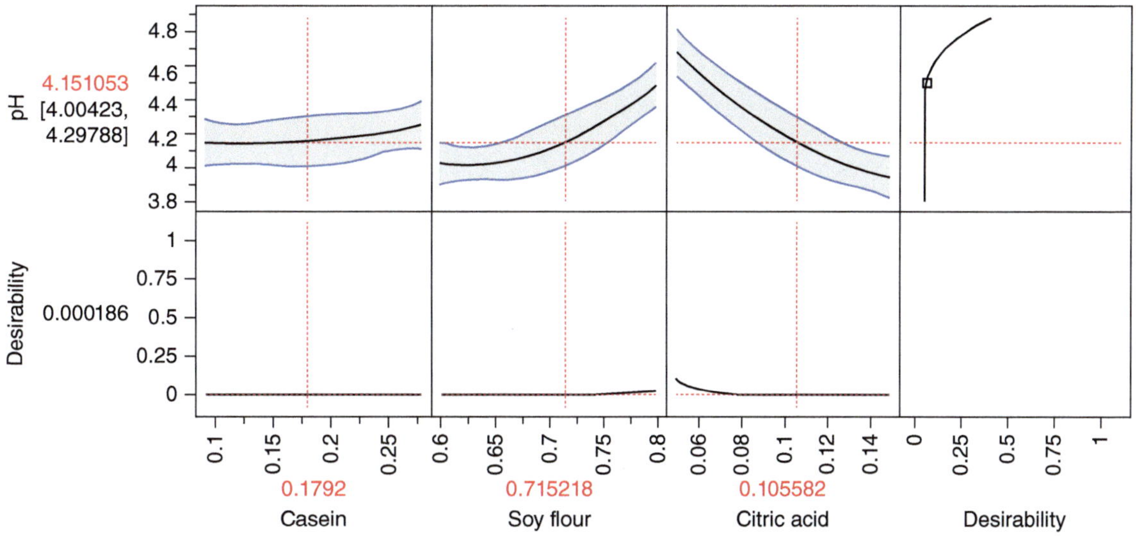

Figure 9.20 Prediction profiler with different amounts of casein, soy flour, and citric acid from those in Figure 9.19.

The profile in Figure 9.22 shows the zones or concentrations of the three components that contribute to the desirable pH range. For visual orientation, this diagram is helpful in showing that the centerpoint of the pH outcome is achieved with a ratio of the casein, soy flour, and citric acid are about 0.2, 0.7, and 0.1, respectively. It is also to be noted that the mixture triangle has an interactive predictive tool that allows the researcher to manipulate the data for projections of the effects of varying one of the three components in relation to how the other two components are affected in terms of the final pH outcome.

The plot in Figure 9.23 shows the relative contributions (as orthogonal estimates) in a Pareto-type diagram. As is the case with Pareto plots discussed throughout this book, this type of plot helps researchers decide which components have the most important effects on the experiments conducted here.

Figure 9.24 (a–c) shows three-dimensional plots of soy flour and casein vs. pH. As an interactive component, the 3-dimensional plot can be turned to allow the visualization of the surface to help illustrate the relationships between components. The interactive dialog allows the user to switch to the other parameters (such as casein and citric acid vs. pH).

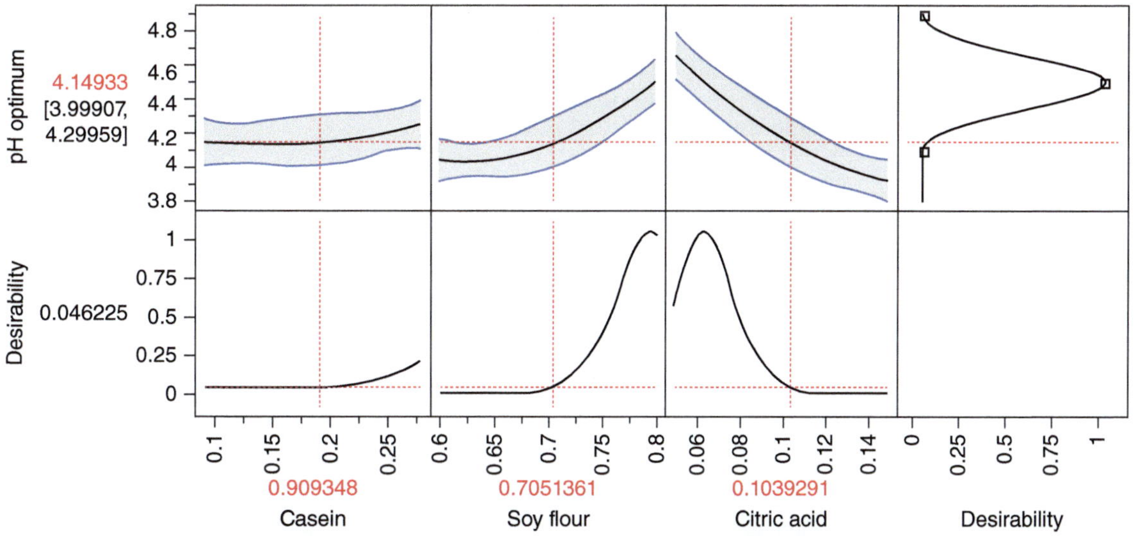

Figure 9.21 Prediction profiler with constraints settings of 4.1–4.9 (rather than 4.50–5.50).

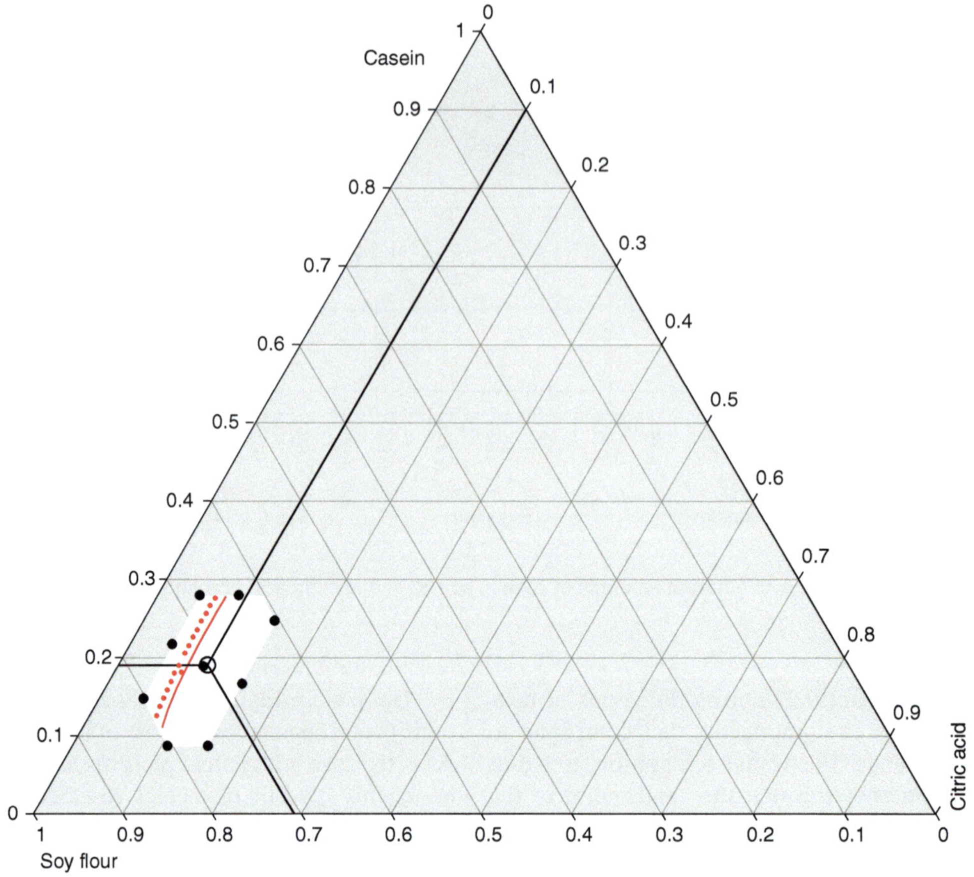

Figure 9.22 Mixture profiler triangle.

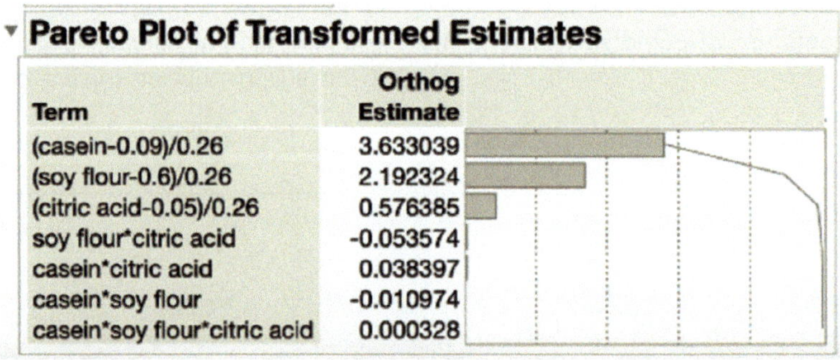

Pareto Plot of Transformed Estimates

Term	Orthog Estimate
(casein-0.09)/0.26	3.633039
(soy flour-0.6)/0.26	2.192324
(citric acid-0.05)/0.26	0.576385
soy flour*citric acid	-0.053574
casein*citric acid	0.038397
casein*soy flour	-0.010974
casein*soy flour*citric acid	0.000328

Figure 9.23 Pareto plot of the contributions of the three components.

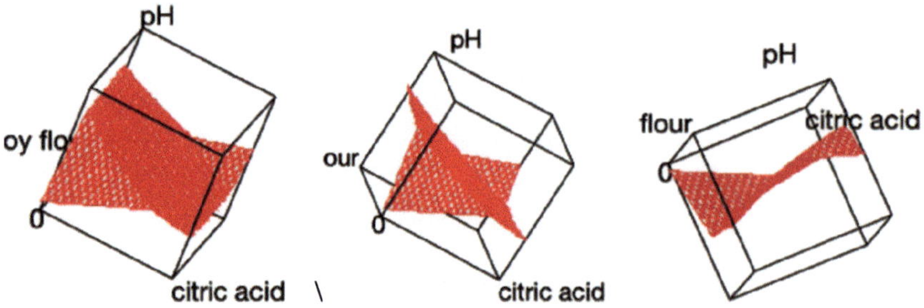

Figure 9.24 (a–c) Contour profiler plot.

Table 9.4 Components in the Wax Worm Diet Used for Full-Factorial Test

Components	Weight (g)	% of Each Component
Beeswax	32	5.5%
Bran (wheat, oat, rice)	136	37.4%
Yeast (brewer's, torula)	44	5.5%
Water	32	8.8%
Honey	68	18.7%
Glycerin	64	17.6%

9.2.3 Tutorial on a Full-Factorial Design for Optimizing Components of a Wax Worm Diet

9.2.3.1 *ANOVA and Comparison of Means Analysis of Wax Worm Diets' Effects*

The following example is taken from an experiment that was done in a rearing course that I taught recently. The three graduate students participating in this class (Ms. Mauri Hickin [= "M"], Ms. Tabatha Komar [= "Ta"], and Mr. Mario Simon Pinilla-Gallego [= "S"]) were assigned a rearing experiment using wax worms, with the assigned task to test diets that contained two different yeast products (torula yeast and brewer's yeast) and three different grain products (wheat bran, oat bran, and rice bran). The diets were prepared for the students, and the students were provided with 60 2nd instar wax worms to use in six treatments (different diet formulations). The diets had the composition specified in Table 9.4.

The components that were selected as **factors** to be tested for their effects on the **response** (weight gain of larvae) were the cereal or bran component. The types of bran included wheat, oat, or rice. The other variables or **factors** were the type of yeast: torula or brewer's yeast. The experiment was designed to answer the question about which factors gave the best **response**, which was measured as the weight of the larvae after a prescribed feeding period. The response chosen here was one of the important life history traits, which are often measured in diet and rearing system evaluations (for example, weight, development rate, mortality, and fecundity).

The purpose of the experiment was to give the students experience in handling insects in a rearing experiment and to familiarize them with some statistical (DOE-based) methods used for evaluating and optimizing diets. The students (M, Ta, and S) infested each of the six diets with ten wax worms, and reared the insects for ten days in their laboratory. At the end of the experiment, students removed the insects and weighed them on balances in their laboratory. The students provided their individual data sets for analysis by JMP, and obtained the following results from a one-way ANOVA test that showed that there were significant effects of diet formulation and were further analyzed with an each pair Student's t Test at P = 0.05 (see Figures 9.25–9.30).

Figures 9.25 and 9.26 show the results of the ANOVA and of the Each Pair Student's t test analysis by student Ta, respectively. In these tests, the diets with rice bran (R) and torula yeast (T) were shown to support the greatest amount of larval growth.

Note that student S had used a balance that reads to tenths of a gram, so all weights fell into three categories,

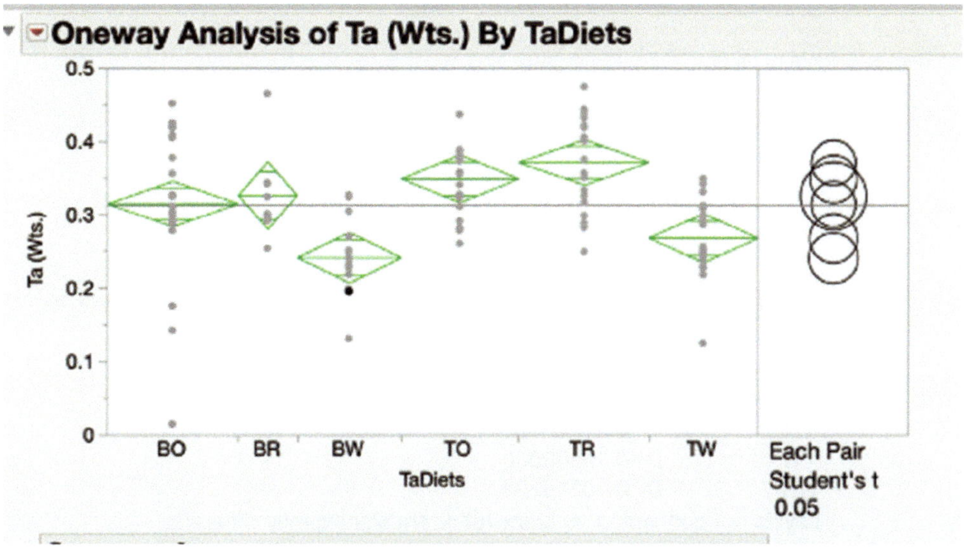

Figure 9.25 Ta ANOVA.

Connecting Letters Report

Level				Mean
TR	A			0.37126000
TO	A	B		0.34913333
BR	A	B		0.32623333
BO		B		0.31477200
TW			C	0.26821765
BW			C	0.24139375

Levels not connected by same letter are significantly different.

Figure 9.26 Ta comparison of means.

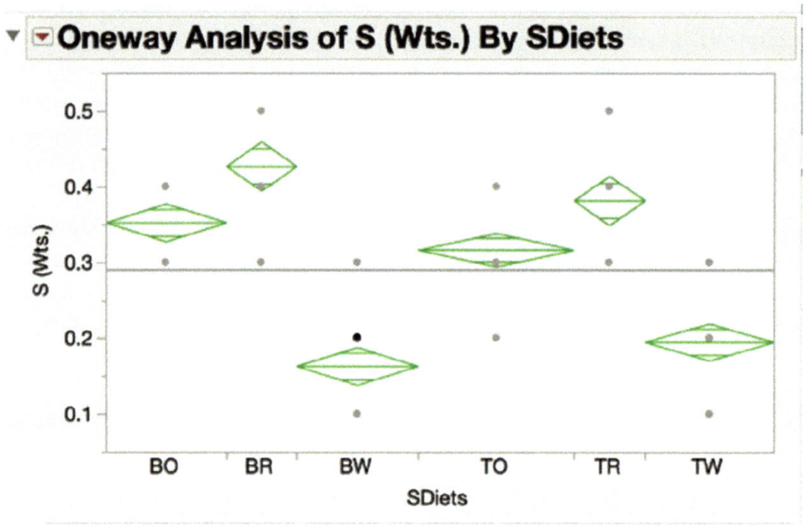

Figure 9.27 Student S-ANOVA of diets.

Connecting Letters Report

Level					Mean
BR	A				0.42727273
TR	A	B			0.38181818
BO		B			0.35263158
TO			C		0.31666667
TW				D	0.19500000
BW				D	0.16315789

Levels not connected by same letter are significantly different.

Figure 9.28 Student S's connecting letter report.

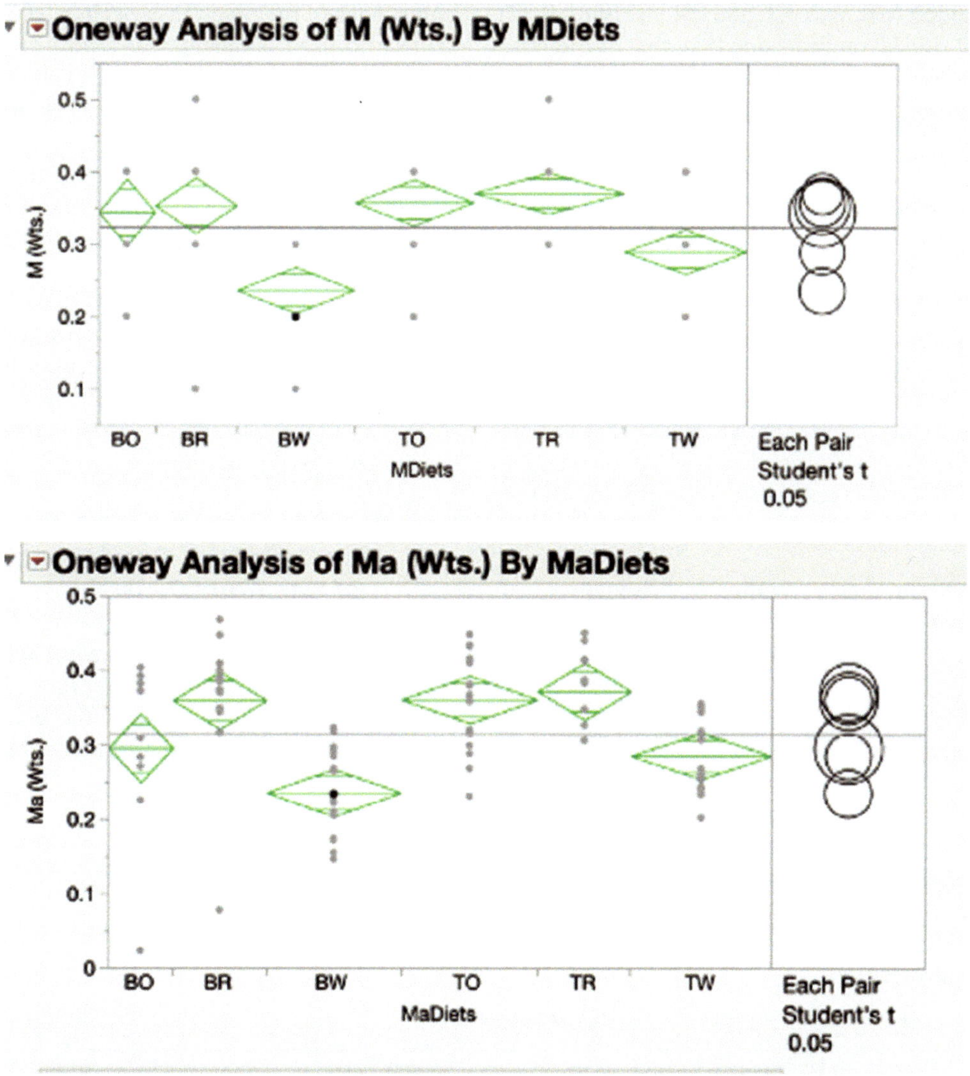

Figure 9.29 (a and b) Student Ms. Mauri Hickin's results of diet trial showing mean weights (a) values were rounded to nearest 0.1 g and (b) read at 0.001 g.

Connecting Letters Report

Level				Mean
TR	A			0.37083333
TO	A			0.35789474
BR	A			0.35384615
BO	A	B		0.34444444
TW		B		0.29000000
BW			C	0.23684211

Levels not connected by same letter are significantly different.

Connecting Letters Report

Level				Mean
TR	A			0.37184615
TO	A			0.36068421
BR	A			0.35946154
BO		B		0.29555556
TW		B		0.28525000
BW			C	0.23563158

Levels not connected by same letter are significantly different.

Figure 9.30 (a and b) Connecting Letter Reports for Student M's Analysis at two different levels of balance precision.

ones that rounded to 0.2, to 0.3, and 0.4 g (Figure 9.27). The effects of this lesser precision than that of the other two students (Ta and M) can be seen in the comparison of the figures representing the means and standard error.

Figures 9.27 and 9.28 show that in Student S's experiment, the diets containing rice bran, like those of Student Ta, supported the greatest larval weight gain.

From the data set submitted by "Student Ma": ANOVA treatments of the diet optimization experiments. Figure (a) shows the analysis of data that had been rounded off to tenths of g, while figure (b) shows the original data measured in thousandths of g. It is noteworthy that as a result of the rounding, the means were changed. This is especially evident in the Connecting Letters Report (based on the Each Pair Student's t test) where diets BR and BO are separated from one another when the more precise data are analyzed.

Figure 9.30 (a and b) compares "Student Ma's" report based on the Each Pair analysis/comparison of means. Note: the two connecting reports in (a) and (b) are derived from the tests visualized in Figure 9.29 (a and b, respectively). Besides the findings regarding the diet components' effects on larval growth, these student reports serve as an interesting example of the importance of precision in experiments. Clearly, the students who used balances that read to thousandths of a gram gave greater precision and resolution of the ingredient effects than did the lower precision balance, which read to tenths of a gram.

It should be noted that this part of the data analysis is the conventional (non-DOE) analysis of the experiment. The ANOVA analysis made it clear that there was a significant effect of the diet components, and the a posteriori analysis/comparison of the means showed which means reflected the diets that were conducive to the weight gain of the larvae. However, despite the usefulness of the conventional statistical analysis, the full power of the predictive values of the data analysis is not manifested unless we use a DOE-based procedure.

9.2.3.2 Full-Factorial Design Tutorial on Optimizing Components of Wax Worm Diets

What follows is the application of DOE's full-factorial design used by students in the rearing class to optimize the diet by experimenting with the cereal components (brans) and the yeast components of the existing diet (Figures 9.31–9.36).

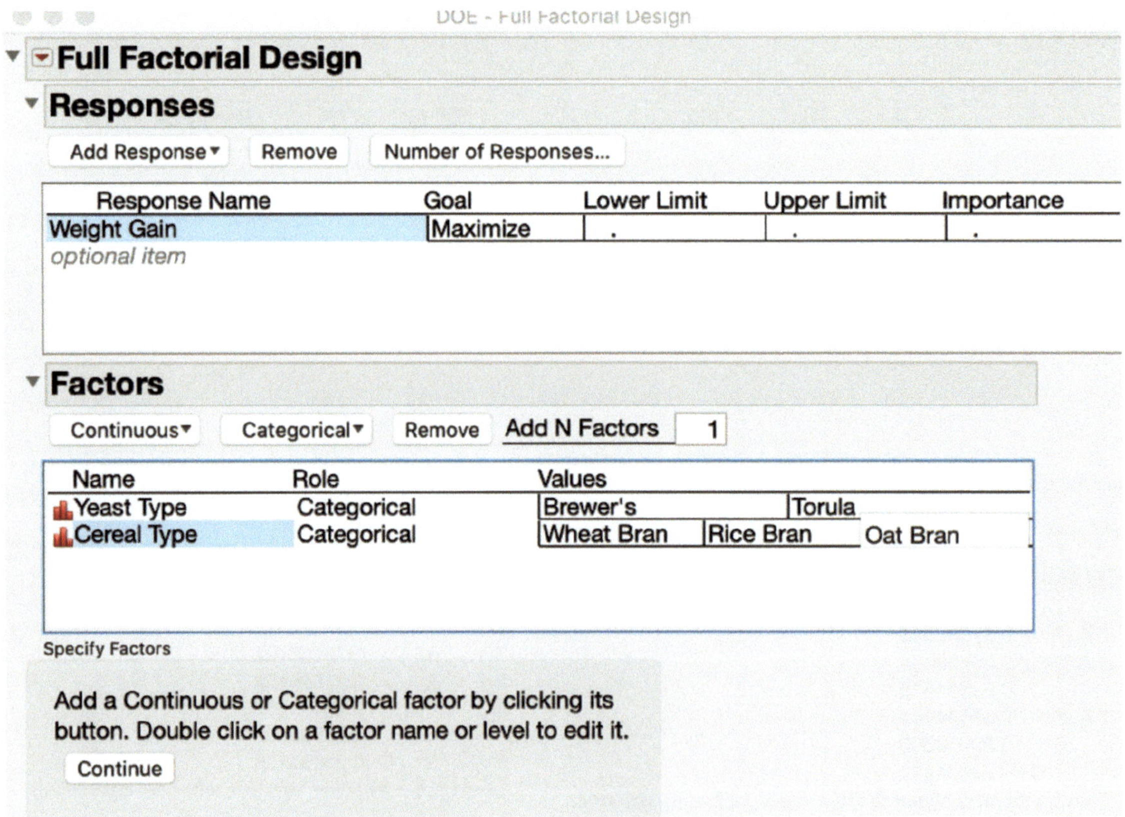

Figure 9.31 The initial interactive dialog after the responses and factors were filled in.

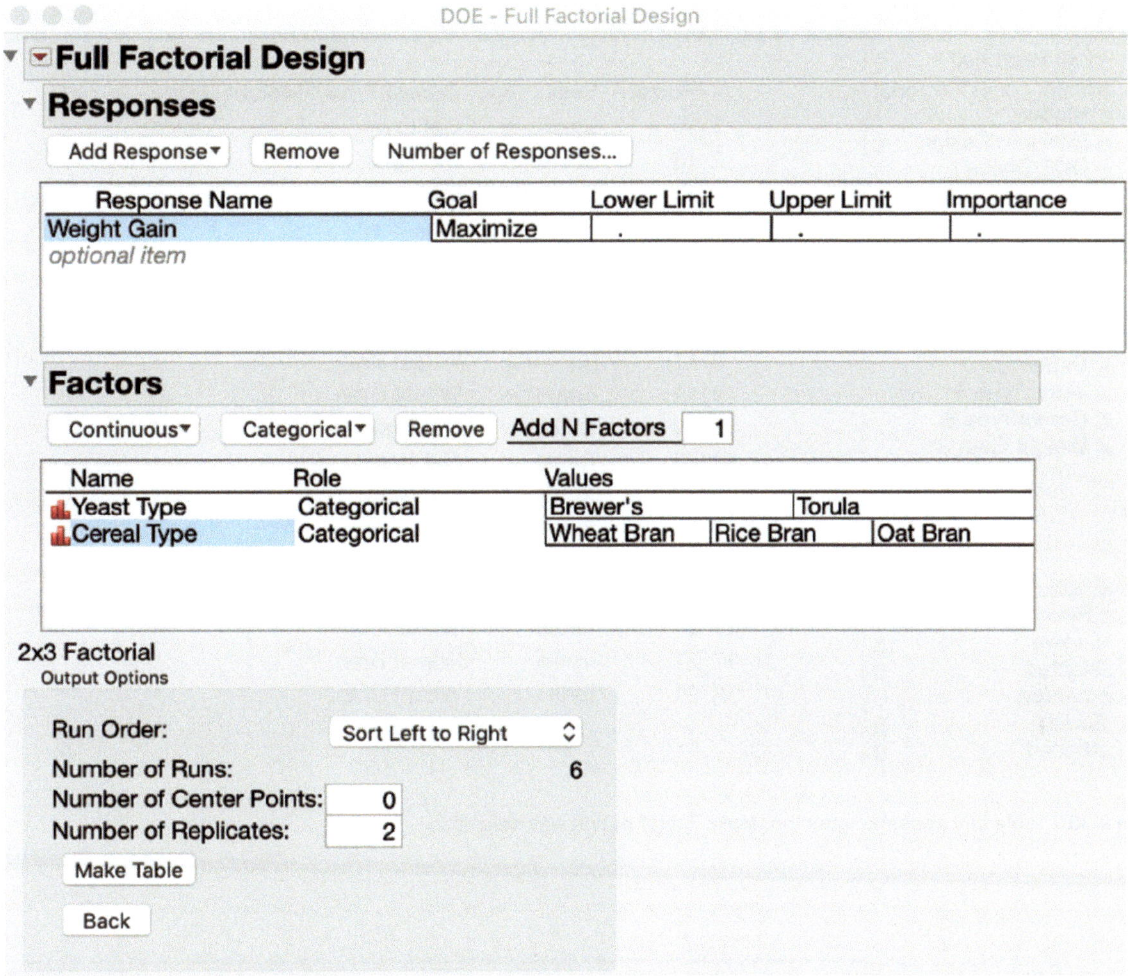

Figure 9.32 Design of the diet components that were varied in the wax worm experiment. Note that the yeast component had two variables (brewer's yeast and torula yeast = B and T, respectively), and the cereal type had three components (oat bran, rice bran, and wheat bran). In the interactive dialog used here (Figures 9.31 and 9.32), the number of replicates was specified as 2 (Figure 9.32). The resulting prescription from the JMP format generation was six different diets, BO, BR, BW, TO, TR, and TW. Note that in the original plan, the assignment was to weigh the larvae after ten days, then allow the larvae to pupate so that pupation rate could be determined, and the number of emerging adults could be measured so that the third column in Figure 9.33 could be completed. However, there were logistic problems that prevented the completion of the last two parts of the experiment.

At this point, the "Continue" button had not yet been selected. In Figure 9.32, the "Continue" button had been selected, and "Run Order" and the "Number of Replicates" had been filled in.

Note that the response, which is for "Weight Gain" in Figure 9.33, is left blank, awaiting the results of the experiment to be entered as is seen in the following figure.

Note from Figures 9.33 and 9.34 that each pattern (Brewer's yeast/Oat bran = 1-3, etc.) is repeated three times, once for each student's response in each category. Note also that the values used in the "Weight Gain" response category were the means of weight gains in each student's experimental outcomes. It is important to note that in Figure 9.34, the experimental responses for each student's tests were reported in the "Weight Gain" column. The next step in performing the analysis is to select the "Model" interactive button seen in the upper left portion of the figure under "2 × 3 factorial."

Below these choices seen in Figure 9.35, the interactive "Run" button appears, and selection of this button causes the analysis to begin. The results of the analysis are presented in Figure 9.36.

This (Figure 9.36) is where the value of the DOE approach becomes apparent. The "Terms" in the scaled estimates shows the factors and the potentially interacting factors. First, the "Intercept" term is shown to be strongly significant (with a <0.0001 probability that the experimental outcomes, overall, were due to chance).

	Pattern	Yeast Type	Cereal Type	Weight Gain
1	13	Brewer's	Oat Bran	•
2	13	Brewer's	Oat Bran	•
3	13	Brewer's	Oat Bran	•
4	12	Brewer's	Rice Bran	•
5	12	Brewer's	Rice Bran	•
6	12	Brewer's	Rice Bran	•
7	11	Brewer's	Wheat Bran	•
8	11	Brewer's	Wheat Bran	•
9	11	Brewer's	Wheat Bran	•
10	23	Torula	Oat Bran	•
11	23	Torula	Oat Bran	•
12	23	Torula	Oat Bran	•
13	22	Torula	Rice Bran	•
14	22	Torula	Rice Bran	•
15	22	Torula	Rice Bran	•
16	21	Torula	Wheat Bran	•
17	21	Torula	Wheat Bran	•
18	21	Torula	Wheat Bran	•

Figure 9.33 Table that appeared after the "Make Table" button was selected.

	Pattern	Yeast Type	Cereal Type	Weight Gain
1	13	Brewer's	Oat Bran	296
2	13	Brewer's	Oat Bran	315
3	13	Brewer's	Oat Bran	353
4	12	Brewer's	Rice Bran	360
5	12	Brewer's	Rice Bran	326
6	12	Brewer's	Rice Bran	427
7	11	Brewer's	Wheat Bran	235
8	11	Brewer's	Wheat Bran	241
9	11	Brewer's	Wheat Bran	163
10	23	Torula	Oat Bran	361
11	23	Torula	Oat Bran	349
12	23	Torula	Oat Bran	317
13	22	Torula	Rice Bran	372
14	22	Torula	Rice Bran	371
15	22	Torula	Rice Bran	382
16	21	Torula	Wheat Bran	285
17	21	Torula	Wheat Bran	268
18	21	Torula	Wheat Bran	195

Figure 9.34 Shows the weight gain (in mg) of the larvae from the three students' experiments.

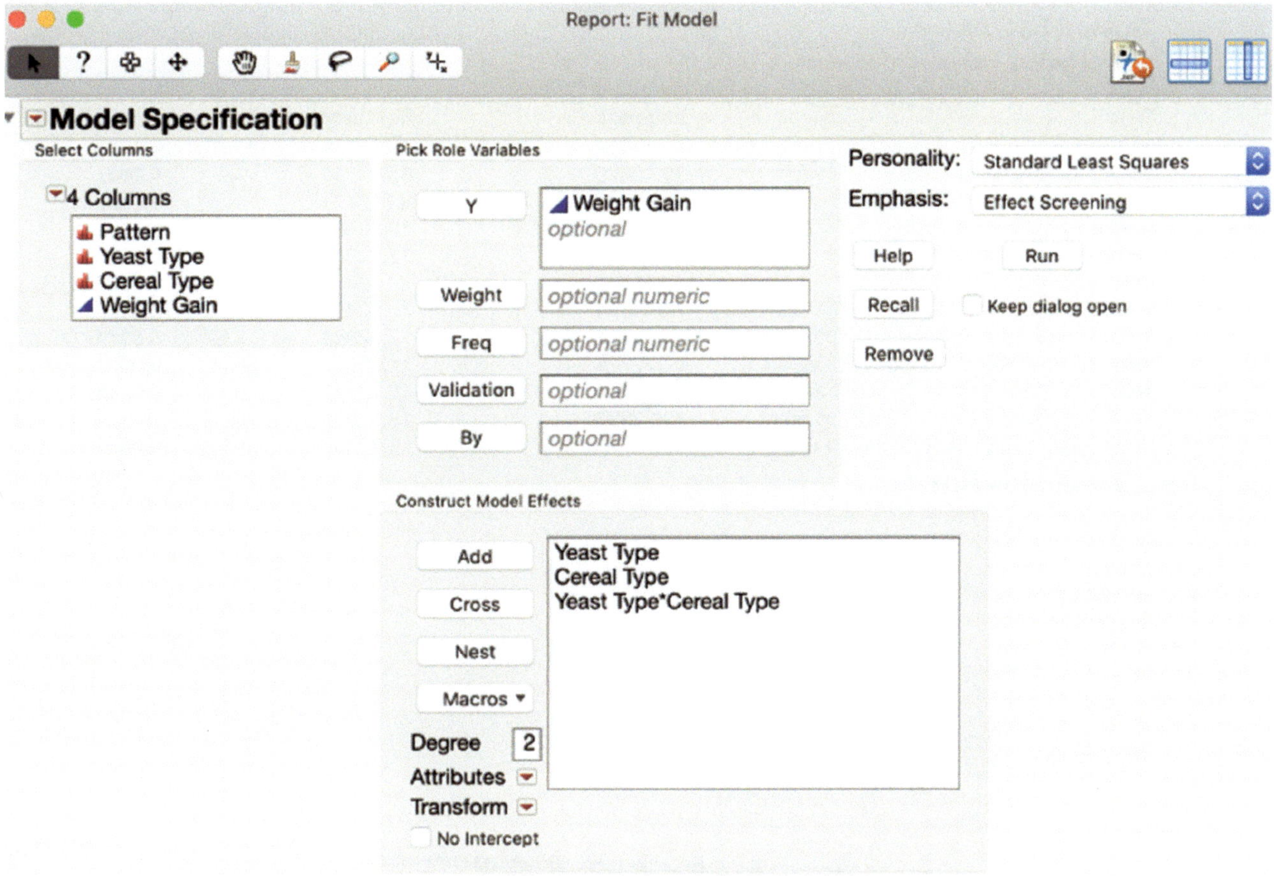

Figure 9.35 Shows the "Fit Model" dialog where the "Personality" and "Emphasis" appear with the default values of "Standard Least Squares" and "Effect Screening," respectively.

The factors that have significant outcomes are shown in red in the column labeled Prob>|t|. The next significant outcome is attributed to Cereal Type [Wheat Bran]. Attention to the multiple graphs under the scaled estimate heading shows the negative (left) direction of the effect and the positive (right) direction of the effect. The wheat bran shows a strongly negative effect, indicating that wheat bran was deleterious to the weight gain of the larvae. By contrast, the next cereal factor, rice bran showed a significant positive effect on the weight gain (P = 0.0003). Although none of the other factors or factor * factor interaction was significant, there are clearly trends that can give clues to the possible positive or negative effects from those factors. For example, oat bran had a trend toward a positive effect on larval weight gain. The prediction profiler helps provide a visual interpretation of the results, with the graph of the weight gain vs. cereal type showing that rice bran is strongly better than wheat bran, and oat bran is somewhat better than wheat bran. Also, the trend seems to be that torula yeast is superior for wax worms' weight gain than is brewer's yeast. Clearly, these trends would lend themselves to further experimentation with the components.

9.2.4 Tutorial on a Mixture Design Optimization of Texture in an Insect Diet

Here is another actual case performed by a research associate (Dr. Molly Robinett, a former postdoctoral research associate at the USDA, APHIS Lab in Brighton, MI and pictured above, making diets) and me. Pictured above, the 12 diets made as mixtures of an EAB larval diet that we are testing. We made the diets according to the profiles of cellulose, starch, and agar suggested by the JMP Mixture Design, and we tested them with a penetrometer, as a blind experiment with Molly performing one set of tests and Allen doing another set of tests, recording data from six penetrations per cup over the 12-cup range (Table 9.5).

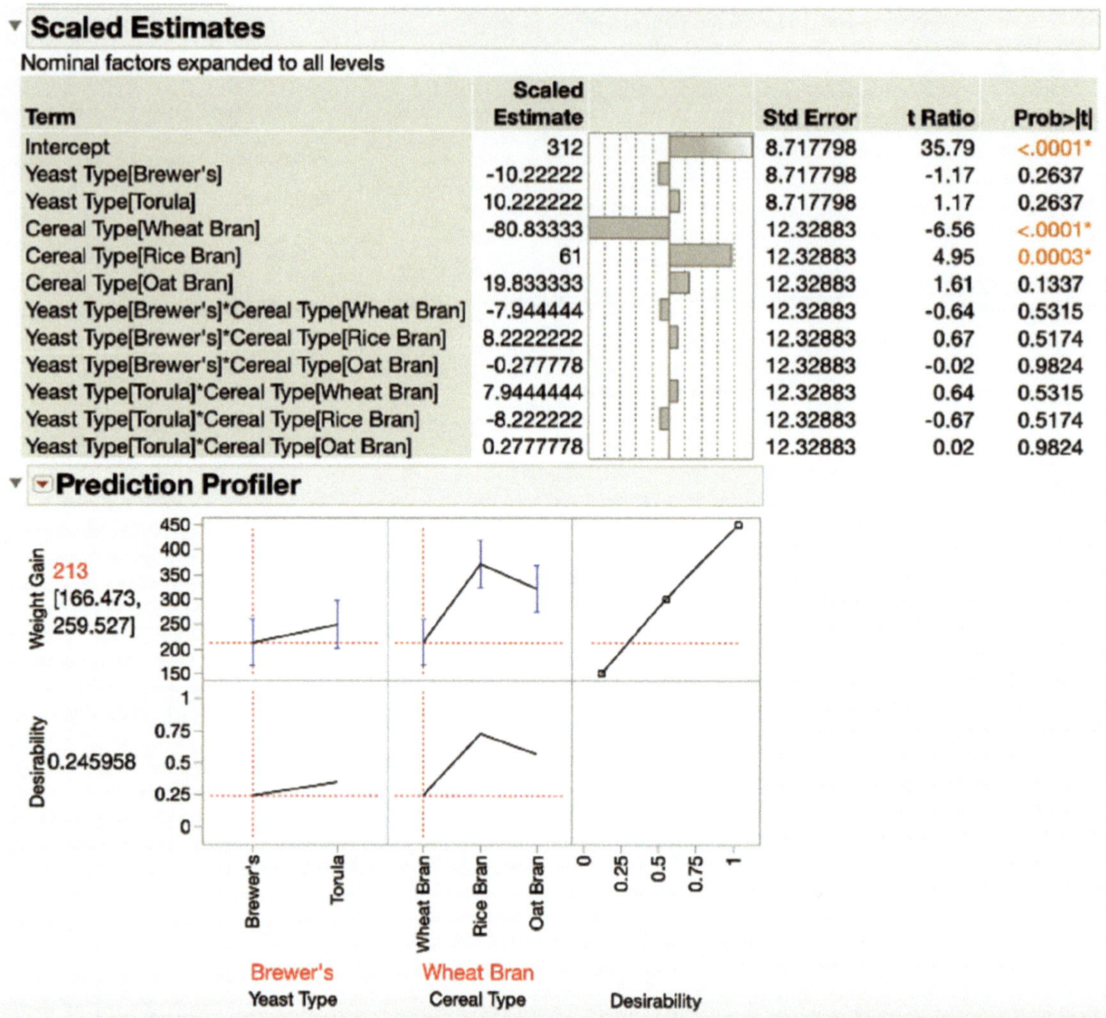

Figure 9.36 Figure showing partial results of the DOE full-factorial design analysis, including the scaled estimates and the prediction profiler.

Table 9.5 The Penetrometer Data Collected by Molly (MR) and Allen (AC)

Diet No.	AC1	AC2	AC3	AC4	AC5	AC6	MR1	MR2	MR3	MR4	MR5	MR6
1	225	235	235	240	235	240	205	215	205	220	215	215
2	220	215	210	220	215	205	195	215	200	215	215	200
3	230	255	255	235	240	215	245	225	240	240	235	240
4	150	155	145	145	155	150	150	140	140	155	170	155
5	255	270	270	260	260	280	195	260	245	275	245	250
6	280	285	285	285	285	280	265	275	265	265	255	270
7	175	180	190	185	190	195	175	170	180	185	185	200
8	100	105	95	100	95	100	90	95	90	90	95	90
9	220	215	220	220	220	215	220	220	215	215	225	225
10	100	100	100	110	115	110	105	110	110	110	100	110
11	135	125	125	135	125	130	135	140	145	140	140	145
12	95	90	105	110	110	110	105	110	100	110	110	110

The data entries for the mixture analysis were composite means of both operators' data, but further below, we provide comparisons of the data collection by two independent operators (in the ANOVA and paired t-tests).

We used the Mixture Design from JMP in the way as displayed in Figures 9.37-9.42.

We used for the three diet constituents that we were trying to mix to give us "desirable" gel strength qualities of an insect diet that we are trying to develop for rearing emerald ash borers (noted in Figure 9.38). The diet consists of several other components that we did not vary, and in all diets, the total for the three components was 23 g of cellulose + starch + agar. The proportions in the "Factors" (= constraints) box were provided by Cohen, based on his experience with these materials (his knowing that too little agar would make a goo, rather than a gel, but too much would be impossible to mix; too little [or no] starch would not be nutritionally acceptable to the insects, and so on for physical and biological rationale).

Please note that these proportions in each row in Figure 9.39 all equalled 23 g in our diet preparations. The last column in Figure 9.39 was determined by our making multiple penetrometer measurements to allow us to calculate a mean gel strength for each diet; so 223.75 is the mean of 12 penetration tests.

Note that the model shown in Figure 9.40 allows development of such factors as the interactions. Note also that under "Number of runs," the user specified was selected and 12 runs were entered. The more runs that are performed, the higher the resolution and probability that interaction trends will be demonstrated; however, conducting a greater number of runs also becomes costly in terms of resources such as time and materials. To perform the next step, the "Make Design" selection is made.

In Figure 9.41 the role of variables is specified, and the selection of standard least squares and effect screening stipulated for personality and emphasis, respectively, selected. Next, the "Run" selection is made to complete the analysis; and the following outputs are presented in Figures 9.42–9.45.

The effects of the varying proportions of the three mixture components on gel strength are shown in Figure 9.42. Recalling that we specified "Maximize" for the "Goal" of the experiment in the early steps of the process, we can see that increasing amounts of cellulose caused a reduction of gel strength; increasing concentrations of potato

Figure 9.37 (a–d) Preparing diets and performing gel strength tests.

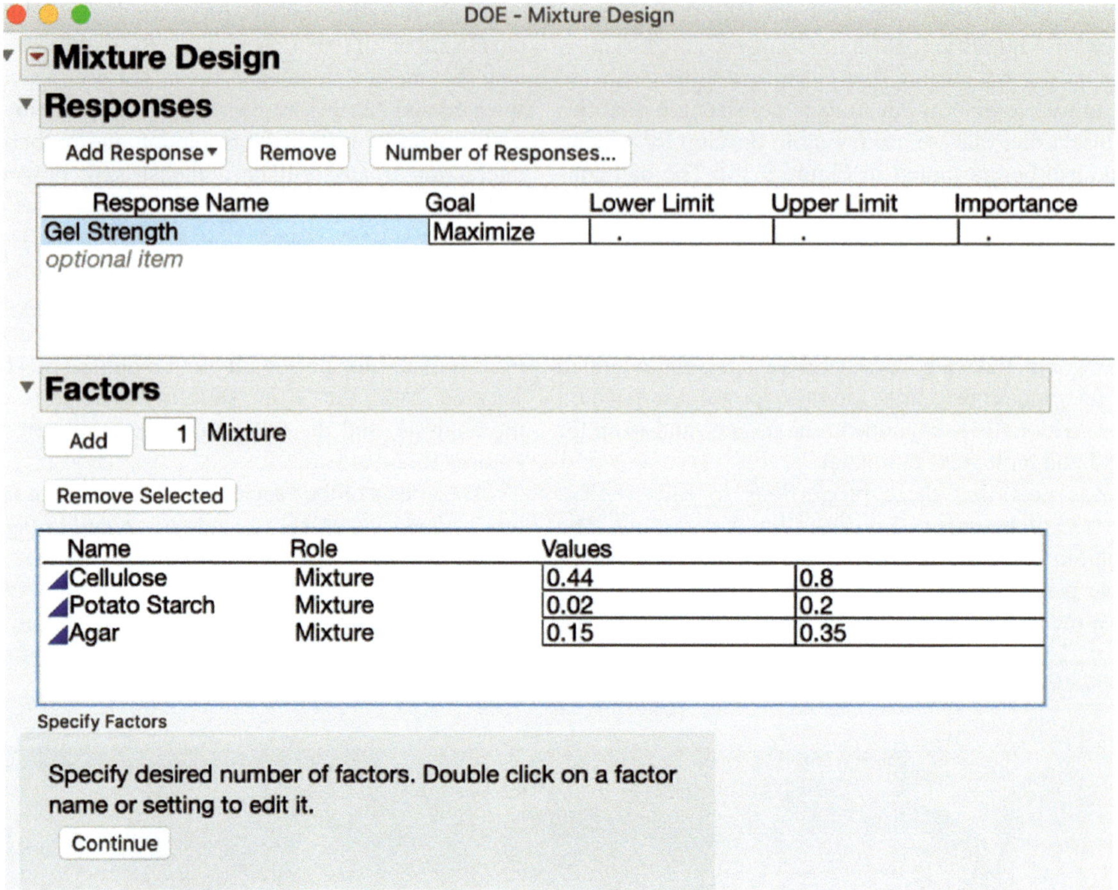

Figure 9.38 Shows the constraints for proportions of cellulose, potato starch, and agar for the diet mixture experiments.

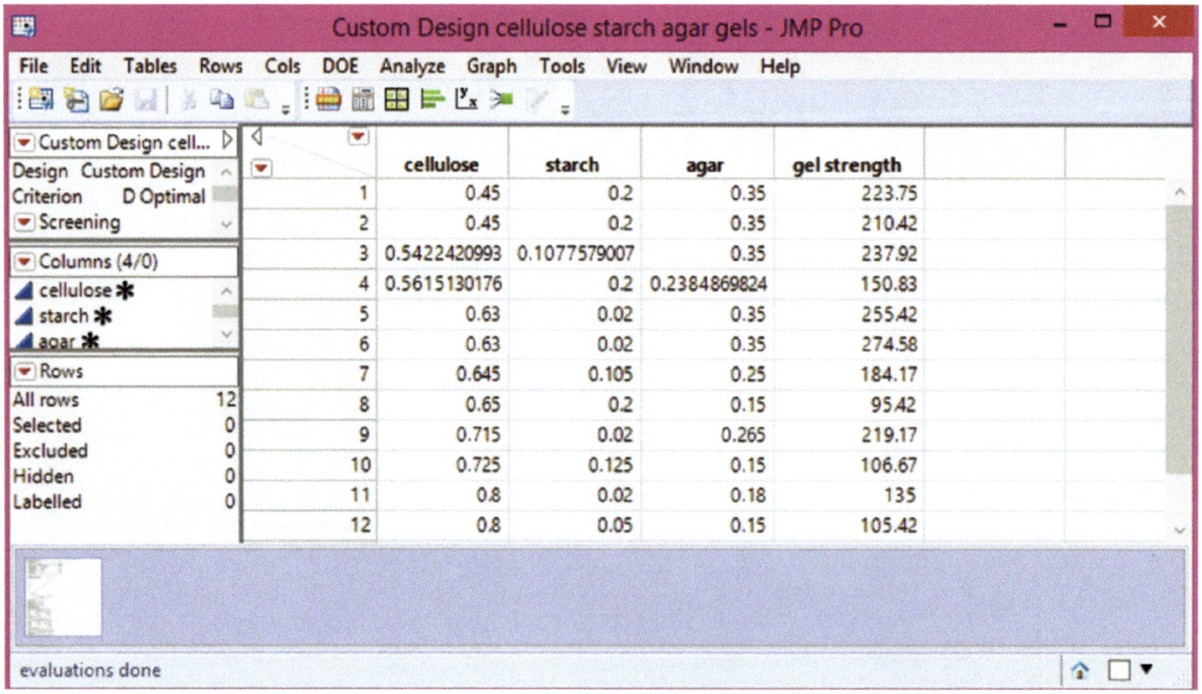

	cellulose	starch	agar	gel strength
1	0.45	0.2	0.35	223.75
2	0.45	0.2	0.35	210.42
3	0.5422420993	0.1077579007	0.35	237.92
4	0.5615130176	0.2	0.2384869824	150.83
5	0.63	0.02	0.35	255.42
6	0.63	0.02	0.35	274.58
7	0.645	0.105	0.25	184.17
8	0.65	0.2	0.15	95.42
9	0.715	0.02	0.265	219.17
10	0.725	0.125	0.15	106.67
11	0.8	0.02	0.18	135
12	0.8	0.05	0.15	105.42

Figure 9.39 Shows the proportions of cellulose, starch, and agar that JMP told us to use in the 12 diets that we prepared.

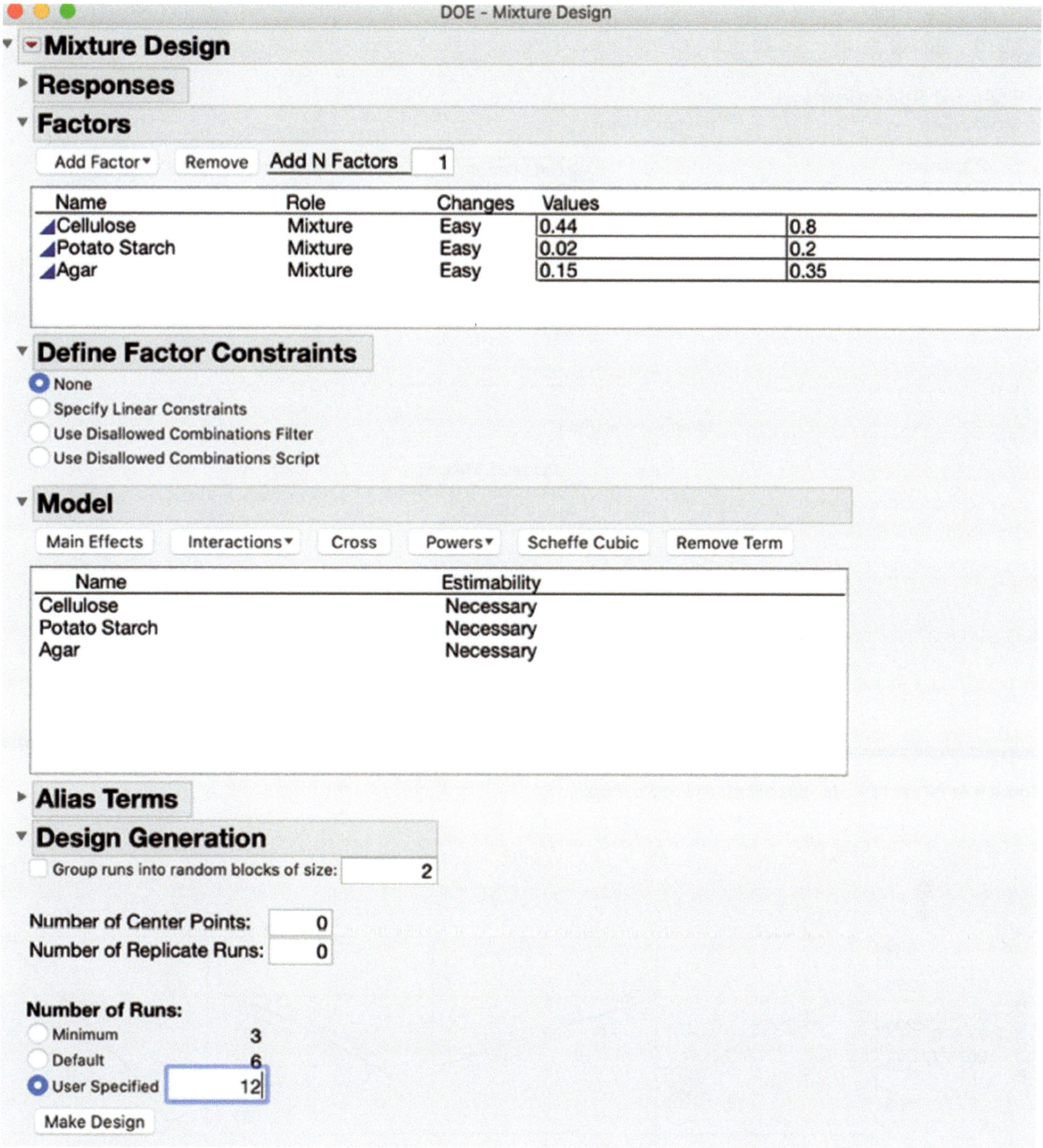

Figure 9.40 Mixture design model.

starch had a negative effect on gel strength, and increasing amounts of agar had strongly positive effects on gel strength.

9.2.4.1 *Interpretation*

If we were deciding on the inclusion and amounts of the three components in a given diet where we wanted to increase gel strength, this experiment would indicate that cellulose should not be used because of its strong detraction from strong gels; and starch was of no value. On the other hand, increasing the amount of agar would be beneficial. However, with our broader scope of understanding, we realize that cellulose and starch have other roles in the diet; therefore, we would have to consider those factors before discarding cellulose and starch. For example, there is evidence that for some insects such as silkworms, cellulose stimulates feeding (Hamamura 2001)

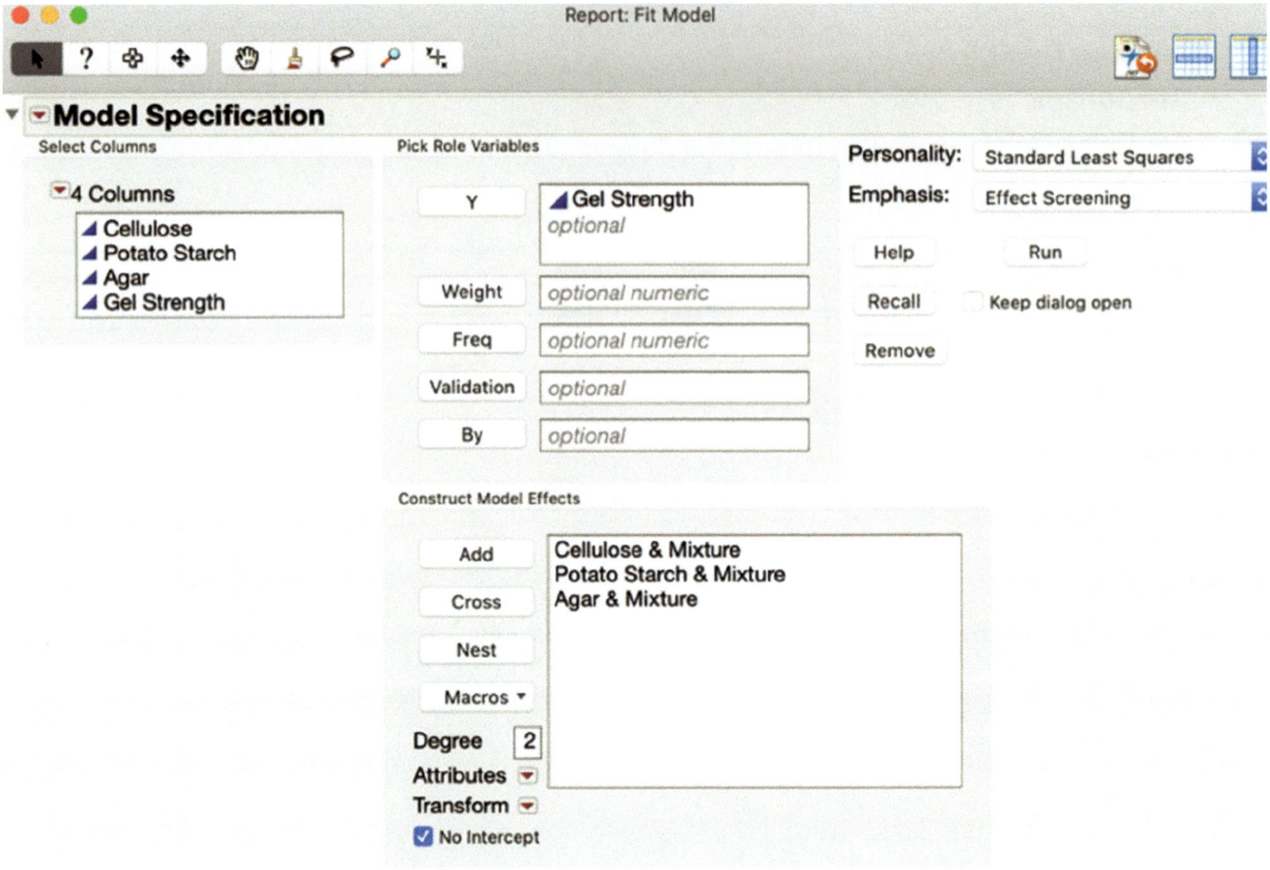

Figure 9.41 The "fit model" specification dialog is shown.

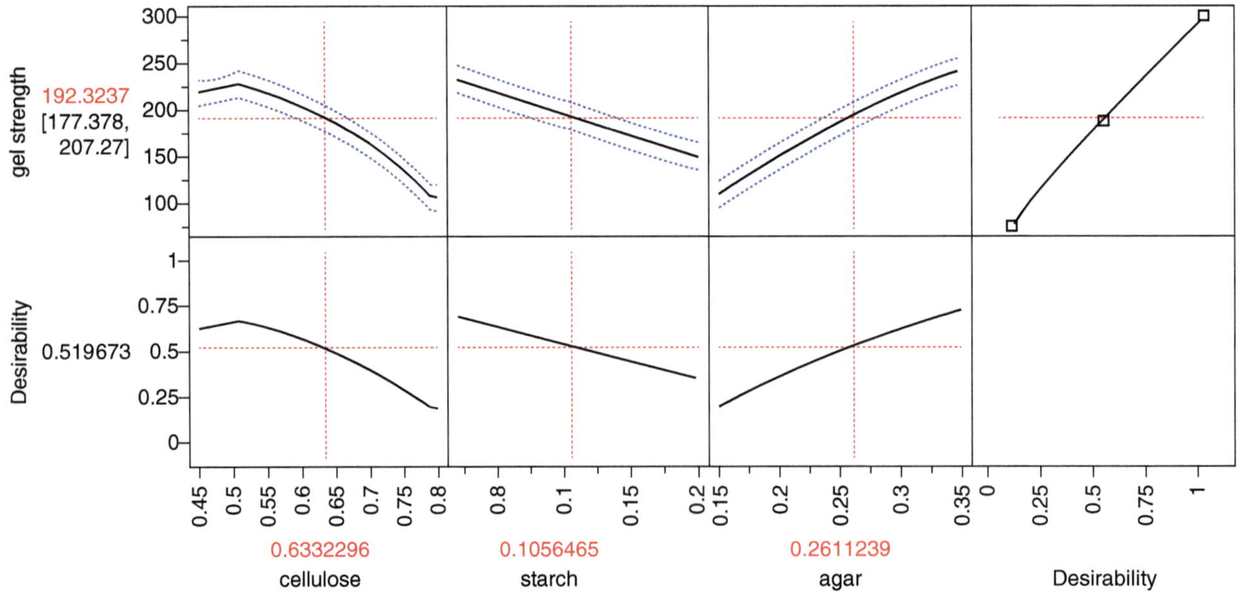

Figure 9.42 Shows the prediction profiler for the mixtures.

Below: "Sorted Parameter Estimates":

Term	Estimate	Std Error	t Ratio	Prob>|t|
(cellulose-0.45)/0.38	99.476534	9.180433	10.84	0.0001*
(agar-0.15)/0.38	263.94583	53.69383	4.92	0.0044*
cellulose*agar	320.4236	118.7706	2.70	0.0429*
cellulose*starch*agar	-596.5189	324.8753	-1.84	0.1258
starch*agar	348.5142	218.997	1.59	0.1724
cellulose*starch	202.72343	139.4704	1.45	0.2058
(starch-0.02)/0.38	-17.98629	70.91145	-0.25	0.8099

Figure 9.43 Values for sorted parameter estimates.

and it also stimulates peristalsis and digestive efficiency (Cohen 2015). We also know that starch is a souce of energy and materials for various metabolic processes, so before discarding it from the diet, we would be well advised to consider its purposes besides the contribution to gel strength. However, the importance of the strategy of using DOE to determine various effects of all diet (and for that matter all rearing system) components, it is helpful to understand what would otherwise be cryptic effects of certain ingredients. When one line of inquiry suggests that we, for example, increase the cellulose in a diet, understanding the implications of such an increase would have on diet texture could prevent unanticipated problems (Figures 9.43–9.45).

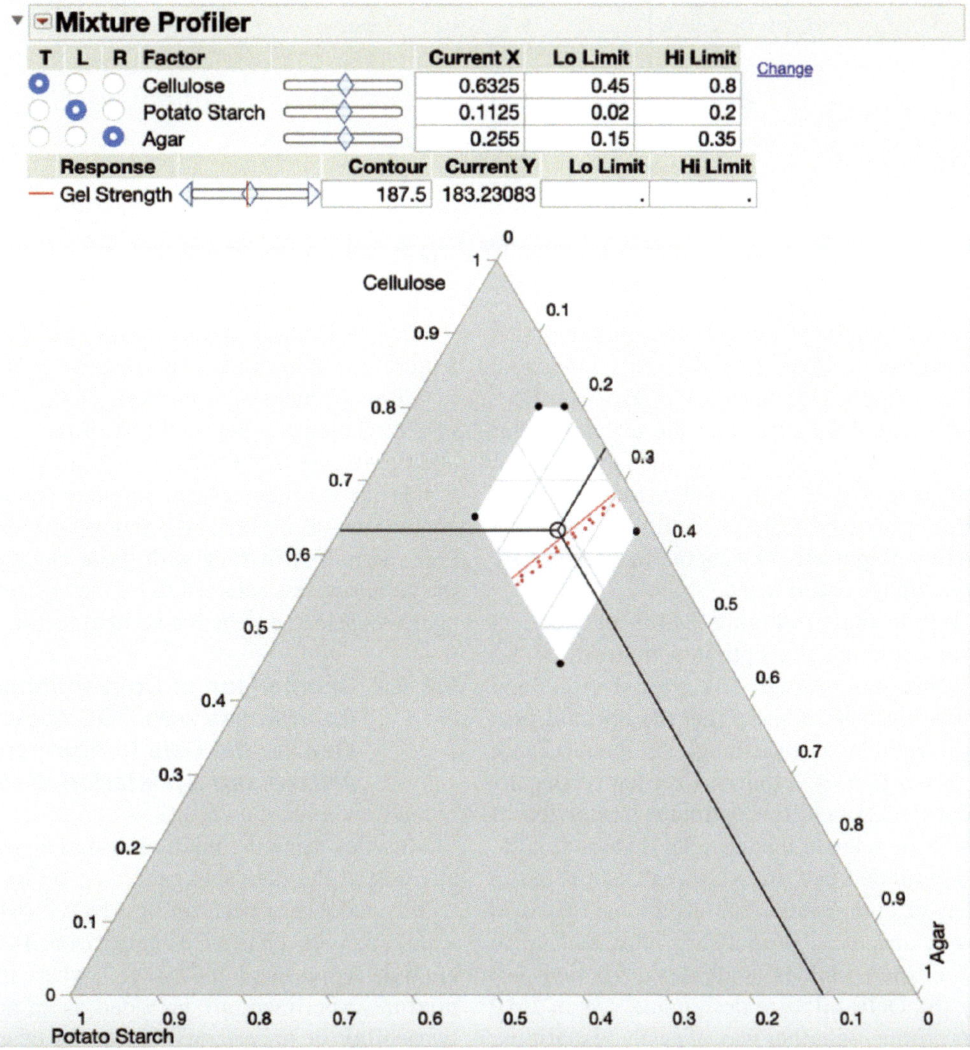

Figure 9.44 The response triangle showing the optimal values for each factor.

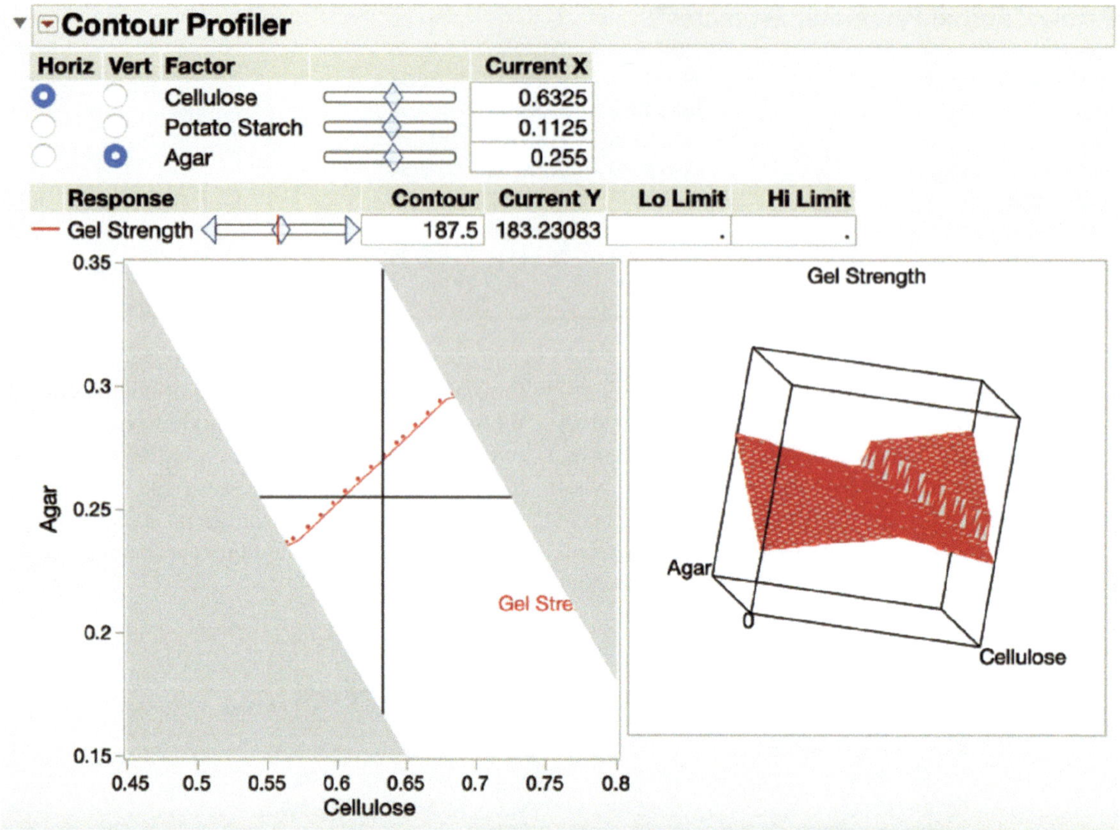

Figure 9.45 A contour profile of two factors ("Cellulose" and "Agar") plotted against "Gel Strength".

This table in Figure 9.43 presents the estimates that allow the researcher to determine the most influential factors and their significance toward the response. Here it is evident that cellulose, agar, and the cellulose*agar interaction had the most influence on gel strength, but the trend suggests that a 3-way interaction between cellulose*starch*agar is possible. It should be noted that the JMP program graphically depicts the factors that are significant with orange or red fonts.

The response triangle in Figure 9.44 is an excellent visual tool for depicting the optimum mixture of the three components being tested. The interior circle and the lines eminating from it show that the optimal proportion of agar (right leg of the triangle) is about 0.2550; the optimal proportion of cellulose (left leg of the triangle) is about 0.6325; and the optimum proportion of potato starch (bottom leg of the triangle) is about 0.1125. The reader is reminded that the mixture design is based on proportions of components that are being varied in the experiment, and these proportions must add up to 1.00 or 100% of the variables' total mass. Another tool generated by the DOE mixture design analysis in JMP is the contour profile—another way of easily visualizing the relationships of the data in a complex, multiple factor experiment.

Figure 9.45 is a contour profile, which is an interactive feature that allows the experimenter to turn the the plot in a three-dimensional manner, enhancing visualization of the relationship between two of the variables (factors) and the response. The interactive character of the profiler is that the exerimenter can simulate changes in the concentrations of the factors to determine how the response (gel strength) will vary with those changes. In this case the cellulose was selected to be the horizontal factor, and agar was selected to be the vertical factor.

9.2.4.2 Comparison of Conventional Statistical Approaches with DOE Approaches: Viewing the Data Interpretation with ANOVA and a posteriori Statistical Tests

Besides using the mixture design as shown above, we also tested the data sets generated by the two operators (Molly and Allen), performing ANOVA (which was significant) and then a pairwise comparison. The resulting data analysis is presented in Chapter 7, where the results of the mixture experiment are provided as an analysis of error (variability or uncertainty). In terms of error reduction, the estimation and evaluation of error in terms of standard deviation or standard error is an essential component in reducing variability and uncertainty. As we inspect every possible component and process in our rearing systems in a highly detailed and granular way, we have the data base to guide us through the process control measures explained in Chapters 7 and 8. This scrutiny for sources of error and the subsequent reduction of the error is in accord with a major theme of this book, depicted in Figures 7.3 and 7.30. The reduction of error is at the heart of **design** and **control** of insect rearing systems' **operations**.

9.2.5 Artificial Neural Networks: Background and Case Study with *Vanessa cardui* Diets

In dealing with the extreme complexities of rearing systems and analysis of variability and uncertainty, statistical systems such as artificial neural networks can be very useful tools to help us understand the deviations and departures in our system's reliability and predictability. The literature on the statistical basis of rearing related topics is sparse, with some notable exceptions such as several papers by Bianconi and his associates. For example, Bianconi et al. 2010 applied the concept of artificial neural networking to help understand the interplay and dynamics of blowfly populations in a rearing system. They wrote:

> The complexities of the nutritional ecology of blowflies could be clarified and detailed by the deployment of appropriate modelling techniques such as artificial neural networks, which are mathematical tools widely applied to the resolution of complex biological problems. A notable feature of artificial neural networks is their independence from any assumptions about the theoretical distribution shape of the data used. Also, the performance of artificial neural networks over linear or non-linear regression-based statistical models is at least interesting, as the dimensionality and/or the non-linearity of the systems increase (Schultz and Wieland 1997; Haykin 1999; Schultz et al. 2000, Wasserman 1993).
>
> Neural networks algorithms are founded on the construction of models that consist of a great number of simple processing units called neurons (or nodes) that possess several connections between them and are lined up in layers. The number of neurons in the input layer corresponds to the variables that will be used to feed the neural network and should be the most relevant variables to the problem in question (Jang 1993; Haykin 1999).
>
> Artificial neural networks were conceived with the aim of imitating the human brain functionality. Thus part of the terminology used in the area of artificial neural networks, namely neurons, synapses, learning, layers, etc., is due to such a fact.

We used *Vanessa cardui* as a test subject to bioassay diets with combinations of agar vs. gellan (hydrocolloids or gelling agents), wheat germ vs. soy flour as main diet staples, brewer's yeast vs. torula yeast, and casein vs. whey protein. The rationale for using *V. cardui* was that their growth rate was rapid, and this would allow us to perform replicated experiments in a faster time frame than would be possible for testing diets with EAB larvae. The experiment was a learning/training technique to help us adopt the artificial neural network for further work with EAB larvae and adults. The diet variants are listed in the table (spreadsheet) here (see Figures 9.46–9.50).

The factors tested as shown in Figure 9.46 were gelling agents (agar vs. gellan), complex nutrient material

	Agar/Gellan	Wheat Germ/Soy	Brewers/torula	Casein/whey	Weights Rep 1	Weights Rep2	Weights Rep 3
1	agar	WG	B	C	111.8	103.3	147.6
2	agar	WG	B	W	155.7	144.1	112.8
3	agar	WG	T	C	58.2	53.8	52.5
4	agar	WG	T	W	143.3	143.2	131.2
5	agar	S	B	C	152.7	123.7	120.9
6	agar	S	B	W	111.4	91	60.2
7	agar	S	T	C	133.4	99.3	142.5
8	agar	S	T	W	186.7	146.4	142.4
9	gellan	WG	B	C	33	20.7	21.7
10	gellan	WG	B	W	50.7	18.8	19.2
11	gellan	WG	T	C	61.8	26.9	41.3
12	gellan	WG	T	W	7.7	4.1	2.8
13	gellan	S	B	C	100.4	36.4	195.4
14	gellan	S	B	W	106.8	137	69.4
15	gellan	S	T	C	70.8	37.5	54.3
16	gellan	S	T	W	154.5	170.1	101.2

Figure 9.46 A table of variables or factors and the responses in a study of painted lady butterfly larval diets.

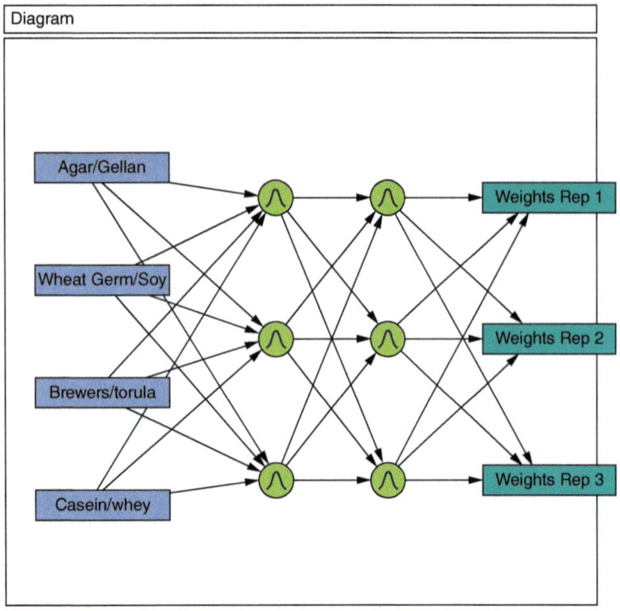

Figure 9.47 Neural network diagram of weights (per replication) vs. factors (gelling agents, main nutrient sources, types of yeast, types of proteins)

(wheat germ vs. soy flour), yeast type (brewer's vs. torula), and protein source (casein vs. whey).

The diagram in Figure 9.47 represents the structure/design of the artificial neural network, with the middle light green circles representing the inner ("neural") layers. The darker green rectangles to the right represent the outcomes (= responses) of the insects in terms of pupal weights.

The tables report in Figure 9.48 includes the statistical values derived from the neural network analysis using a Gaussian assumption for this model. With appropriate study of the various standard deviations (or other error terms), the rearing system researchers can determine the extent to which each component (such as agar or soy flour) contributes to the "wobble" in the system, i.e., the inherent error. If a factor such as soy flour is found to be associated with a large error (variation or standard deviation), then the properties of that soy component can be explored further for homogeneity within the batch of material or the degree to which it contributes to mixing error. In my laboratory, for example, the error or variability of mold formation in patches of diet led to our questioning the degree to which the antifungal agent methyl paraben was being mixed in our painted lady diet. This led to adoption of mixing the methyl paraben with high-lipid wheat germ, by a trituration process, which far better disbursed the methyl paraben than did simple addition of the lipophilic compound to the dry ingredients of the diet. This discovery and problem solution led to our routinely mixing lipophilic compounds with high lipid, high biomass components such as wheat germ, re-fatted soy flour, and wheat germ oil.

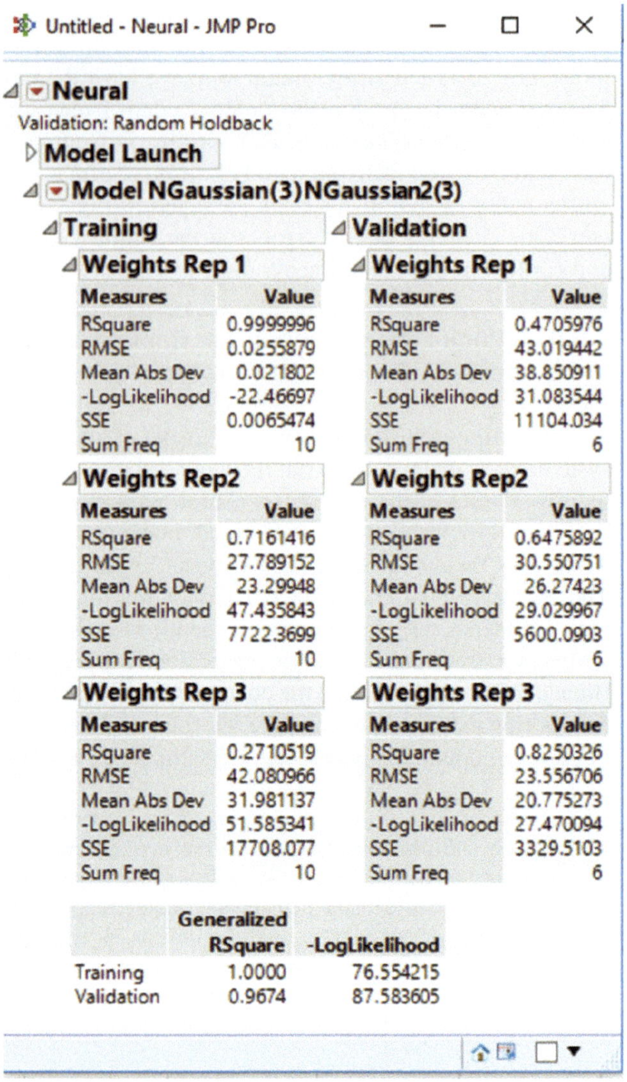
Untitled - Neural - JMP Pro

Neural

Validation: Random Holdback

Model Launch

Model NGaussian(3)NGaussian2(3)

Training

Weights Rep 1

Measures	Value
RSquare	0.9999996
RMSE	0.0255879
Mean Abs Dev	0.021802
-LogLikelihood	-22.46697
SSE	0.0065474
Sum Freq	10

Weights Rep2

Measures	Value
RSquare	0.7161416
RMSE	27.789152
Mean Abs Dev	23.29948
-LogLikelihood	47.435843
SSE	7722.3699
Sum Freq	10

Weights Rep 3

Measures	Value
RSquare	0.2710519
RMSE	42.080966
Mean Abs Dev	31.981137
-LogLikelihood	51.585341
SSE	17708.077
Sum Freq	10

Validation

Weights Rep 1

Measures	Value
RSquare	0.4705976
RMSE	43.019442
Mean Abs Dev	38.850911
-LogLikelihood	31.083544
SSE	11104.034
Sum Freq	6

Weights Rep2

Measures	Value
RSquare	0.6475892
RMSE	30.550751
Mean Abs Dev	26.27423
-LogLikelihood	29.029967
SSE	5600.0903
Sum Freq	6

Weights Rep 3

Measures	Value
RSquare	0.8250326
RMSE	23.556706
Mean Abs Dev	20.775273
-LogLikelihood	27.470094
SSE	3329.5103
Sum Freq	6

	Generalized RSquare	-LogLikelihood
Training	1.0000	76.554215
Validation	0.9674	87.583605

Figure 9.48 Neural network "Model Launch"

The graphs "Profiler" graphs in Figure 9.49 show the outcomes for the various pairs of inputs. For example, in the lower left graph, the weights of pupae were higher with agar than they were with gellan gum. The next frame shows the slight advantage of using soy flour (S) over wheat germ (WG). The next frame to the right shows the slight advantage of using brewer's yeast vs. torula, and the final bottom panel shows the considerable advantage of using casein vs. dairy whey protein. These findings that were determined simultaneously in an experiment where 16 different diets were tested simultaneously over three replications show the contributions of each rearing component and with the proper analysis of the sources of variation, experiments such as these can help the research team determine the sources of variation that contribute most to important rearing outcomes such as the factors that determine the fitness (quality) of the insects in our rearing system.

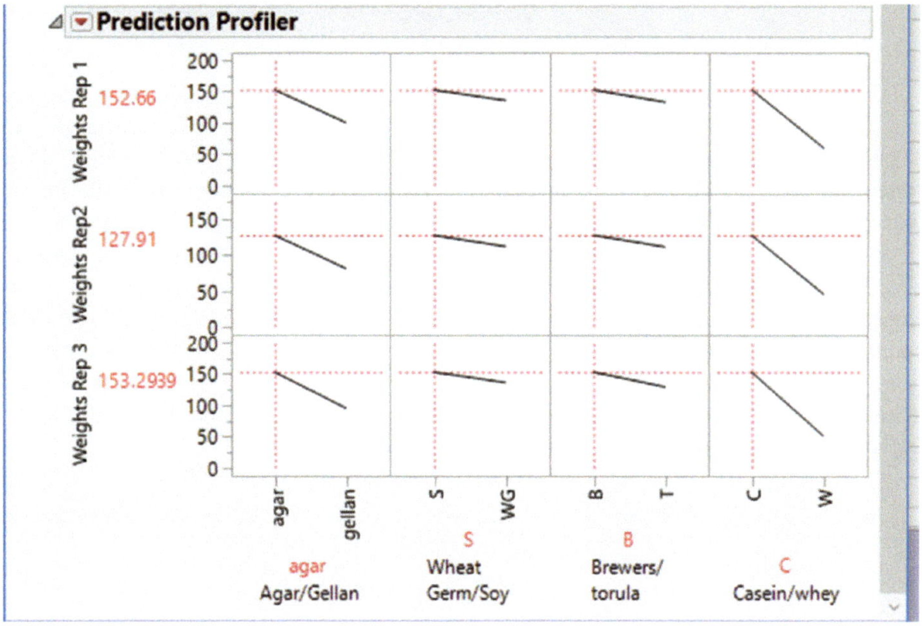

Figure 9.49 Prediction profiler for the neural network.

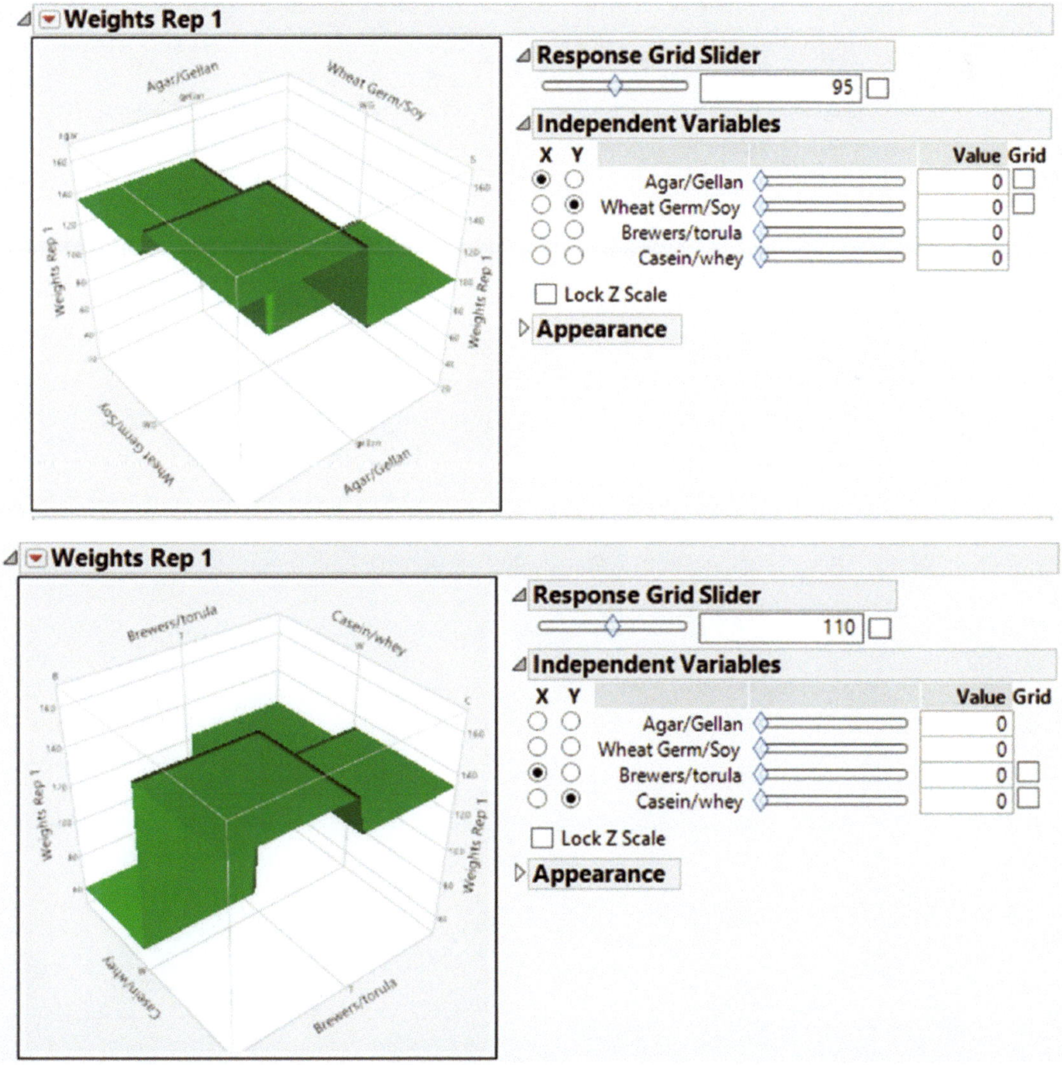

Figure 9.50 (a) and (b) Surface plots of factors analyzed in the painted lady butterfly neural network experiments.

Figure 9.50(a) shows the representation of the responses of pupal weights to agar/gellan and wheat germ/soy flour and the brewer's vs. torula yeast and casein/whey. This type of representation, especially with its interactive and dynamic manipulation potential, is an excellent tool for optimization and amplified research and development strategies. In Figure 9.50(b), the variables were changed to the yeast and protein tests, showing the relative responses to these components.

In conclusion, this chapter can be used as a guide for students of insect rearing who are not educated as engineers or statisticians. The intention was to give, as much as possible, step-by-step instructions for addressing the syntax and format of the JMP system. The chapter is also realistic for rearing applications, as all the examples were tested in the laboratory with actual rearing experiments. It is emphasized that this treatment is not as in-depth as the JMP system will allow once the student of design of rearing experiments becomes more familiar with the JMP (or other statistical package) system. Two references that can take a rearing researcher to the full depths possible are the Montgomery 2013 book and the SAS 2018 book, both available online.

CHAPTER 10

Relationships between Fitness, Quality Control, Health, Homeostasis, and Stress

E. F. Knipling 1979 described his vision of insect rearing as follows:

> Mass rearing of insects is still a young science. With the help of insect geneticists, insect nutritionists, and insect behaviorists, insects might be reared under conditions that will make them equally, if not more, vigorous and more adaptable to the environment than the wild population. These improvements are likely to occur after further experiences and research.

I used Knipling's quote in two *American Entomologist* Forum articles Cohen 2001 and 2019, and I believe it remains relevant in the current book.

10.1 OVERVIEW OF RELATIONSHIP BETWEEN FITNESS, QUALITY, HOMEOSTASIS, AND STRESS: FITNESS

As rearing practitioners, we can adopt the vision of our rearing program as "sustainable production of **healthy** insects." A well-established and useful way to define the health of our insects is in terms of their fitness. **Fitness then is at the heart of quality control and is a biological basis of quality control.** Herbert Spencer coined the expression "survival of the fittest" in his *Principles of Biology* 1864. Spencer used "fitness" to explain the basis of natural selection pertaining to the basis of Darwin's theory of evolution. Fitness has several levels of meaning that can be taken as a general concept or health/well-being or as a carefully defined mathematical concept of population genetics. For our purposes, fitness means the genotypic and phenotypic characteristics of an organism that favor survival and reproduction under a given set of conditions. In a rearing context, we must be aware of the concept of fitness in nature vs. fitness in our rearing setting and in the environment into which our reared insects may be released or the laboratory situation where the insects are to be studied. Fitness can be considered as how well an organism meets the constraints of its ecological niche, and for our purposes, the niche can be thought of as the artificial conditions that we provide to our populations of reared insects. This "rearing system as ecological niche" concept is discussed elsewhere in this book. For now, it is useful to treat the rearing conditions as the composite environment that we force upon the insects that we rear. Keeping in mind the considerable differences between our artificial rearing environments and nature, we must be mindful that we place our target insects under severe selective pressures that are often different from the forces of selection in nature.

In the context of fitness and quality control, it is useful to consider the **health** of our reared insects to be a set of characteristics that are related to **homeostasis** or the physiological state of our insects in a kind of sustainable equilibrium ("steady state"). And, in turn, homeostasis can be thought of the adaptive capacity of our insects to respond to and overcome the effects of **stress**. Conversely, stress can be defined as any of the forces that challenge homeostasis. These concepts can be developed into a model of genetic potential, the organism's adaptations (or capacity) to maintain its homeostatic state and reach its full reproductive potential in the conditions of stress. This is illustrated in Figure 10.1.

Figure 10.1 attempts to convey the concept of "health" in terms of the various biological potentials that the insect (organisms, in general) can realize. For example, if under ideal (optimal) conditions, the insect can produce 300 fertile eggs, a healthy female insect can be expected to produce all of those eggs, and if the conditions were sustained, all of those eggs would hatch and develop into adults making up the next generation. The concept is illustrated in the upper right (reproductive potential) where the equation $\Delta N/\Delta t = rN\ (K - N/K)$ where for a given species under optimal conditions, ΔN the change in numbers (population growth) is over Δt a change in time (specific increment of time) equals the reproductive potential (r) times the instantaneous population times the quantity of the carrying capacity or K minus the number in the instantaneous population divided by the carrying capacity.

Under this concept of genetic potential and actualization of each component of the insect's biology, all the

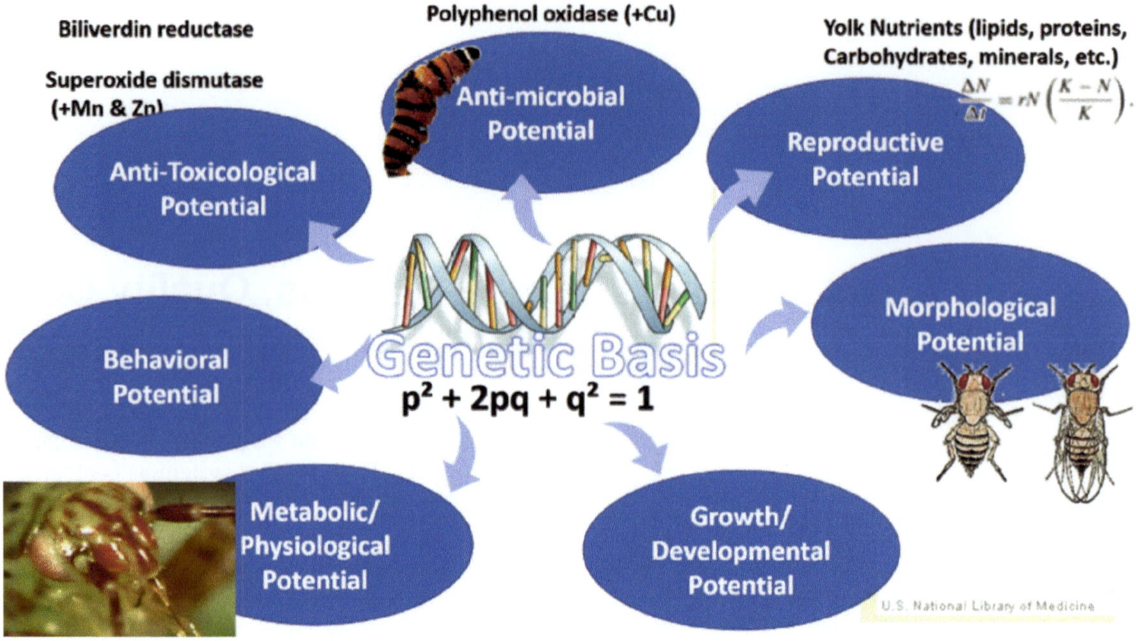

Figure 10.1 Diagram of the interplay between an insect's genetic basis and biological characteristics leading to a balanced, sustainable homeostatic state.

components would work in accord with each other, harmoniously cooperating in allowing the insect to reach its full biological potential in terms of finding and using nutrients, optimizing growth and development rates, overcoming diseases, avoiding or solving problems with environmental threats (such as desiccation or over-heating), find mates, etc. all adding up to a population of that insect that is in harmony with its environment, i.e., it is in a state of health. These healthy insects are generally considered as being of high quality, and the rearing conditions that lead to this healthy, highly fit state are controlled to allow the rearing program to continue production of insects with the ascribed qualities.

10.2 ARE FITNESS AND QUALITY CONGRUENT OR MERELY RELATED?

It is important to recognize this point about insects in our rearing systems: while these terms are related, fitness is not a synonym for quality. **While fitness connotes the degree to which an organism is adapted to its environment, quality is the degree to which the organism in question meets the needs or expectations of the users.** As discussed in several places in this book, quality control is intricately connected with the uses for insects which are the products of our rearing systems. In fact, the quality of insects produced in our rearing systems can be defined as the degree to which our insects meet the standards or expectations for which the insects are being produced. For some uses, specimens that are close to death could meet the needs of certain stakeholders (e.g., where the insect is to be used for a collection or if the insect's mating behavior [Figure 10.2], external features—such as mouth part anatomy [Figure 10.3] or antenna ultrastructure—are being studied]). However, for most purposes that require insects to be reared, only healthy insects are suitable; or in other terms, the healthier the insects, the better for most uses. For example, in sterile insect technique (SIT) programs, the sterile insects must be competitive with the wild counterparts, and for biological control the predators and parasitoids must be fit enough to not only survive in the field but also display adequate search, capture, and voracity characteristics. The literature on SIT and biological control contains scores of papers that address the issue

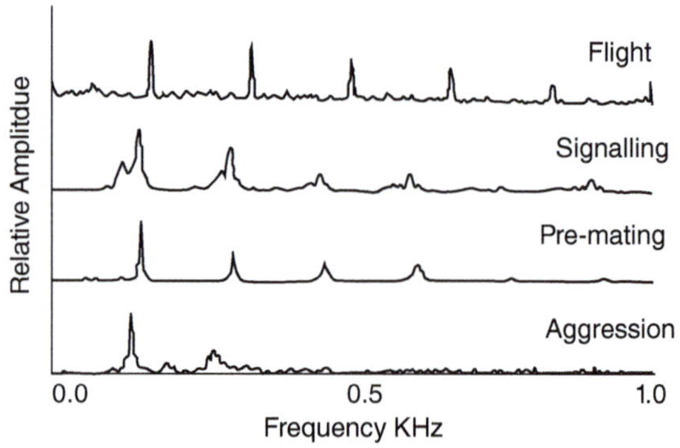

Figure 10.2 Sound frequency signatures of male *A. suspensa*.

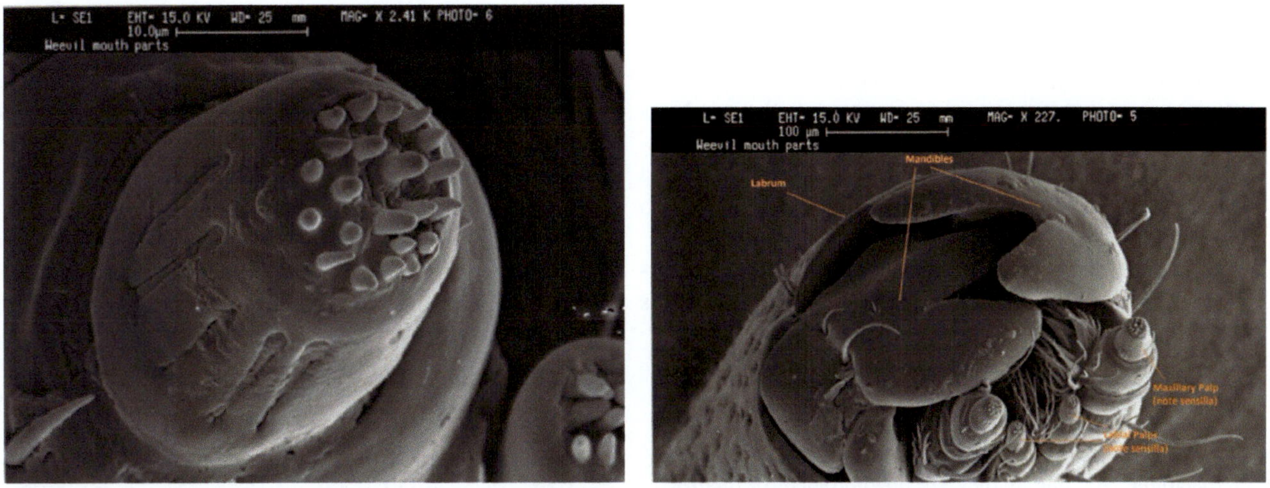

Figure 10.3 (a) and (b) Mouthparts of boll weevil reared on Lindig 1984 Diet at Gast Mass-Rearing Facility.

of fitness (quality) of insects and the success or failure of programs that utilize mass-reared insects. Other uses of reared insects demand standards of quality: for example, where feeder insects are utilized the nutritional value of the insects is of utmost importance. Also, where insects are used for research, it is axiomatic that the quality of the insects is directly proportionate with the quality of the research conclusions. The old axiom from computer science, "garbage in → garbage out" (GIGO), certainly applies to insects that are derived from rearing systems. Failings in the quality of insects used by researchers result in failings of the research quality.

10.3 HOW DO OUR CONCEPTS OF FITNESS AND QUALITY STEM FROM THE PROCESSES ESTABLISHED IN DEVELOPING OUR REARING SYSTEMS?

As rearing systems (colonization practices) are being developed and various components (diet, temperature, humidity, light conditions, etc.) are being vetted, the test of the newly developed components is often FITNESS of the insects, with fitness expressed in various terms such as rates of survival, fecundity, fertility, weight, development rate, pupal eclosion rate, etc. In the section that follows, the background of quality control (QC) in rearing systems formally introduced by Chambers in 1977 will be examined with considerable attention paid to the background papers cited by that author to help build a context and argument for how and why the industrial and engineering concept of QC can be applied to insect rearing, especially mass-rearing. The topic of QC's history and applications is reviewed by Cohen 2015 in the chapter on "Fundamentals of Process Control, Quality Control, and Problem Solving."

10.4 REVIEW OF QUALITY CONTROL

10.4.1 Why Was the Concept of Quality Control Applied to Rearing Systems?

A recent article "An Introduction to Quality Control in Manufacturing" (at newcastlesys.com Tuesday, March 6, 2018) relates the history of QC in the Japanese auto industry. The anonymous article points out that Edward Deming was the person who influenced the Japanese to use standards and principles of QC in post WWII to turn around the auto industry and make it the standard of automobile quality and economics. The article points out that for his contributions to the Japanese economy, Deming was considered "a god." In a Web of Science search of the literature using the keywords, "quality control" and "mass-rearing," as of February 1, 2020, there were well over 300 papers listed. The earliest reference in that search was the *Annual Review of Entomology* paper by Derrell Chambers 1977, which cited 106 publications that were considered as relevant to the groundwork for advancing the application of QC to insect rearing. To convey the scope of the earliest paper to treat insect rearing as a production process that like most (all?) production systems with the need for QC, I point out that most of the publications cited by Chambers were what were then recent papers (1970s) with ten publications from the 1960s and one from 1937, where Needham 1937 provided "handy hints" on insect rearing. In the same time frame as the Chambers' *Annual Review* paper was published, an important monograph on QC in mass-rearing by Boller and Chambers 1977 appeared. At first, the response of the entomological community was slow to take up the application of production engineering's practice of QC. The flow diagrams, process control charts, quality charts, Pareto diagrams, and statistical terminology and equations were alien to entomologists. However, the logic of the pioneers was recognized as an extremely useful tool in improving the products and processes of insect-rearing systems, and an increasing number of papers with direct reference to QC and many others with indirect references to QC began to appear.

The Chambers paper provided an excellent and far-reaching road map for insect rearing, especially with this prescient statement:

> Nevertheless, I emphasize the importance of critical studies on insect colonization to entomology. Rearing and rearing research must have the image of a credible discipline. As long as they are down-graded, or quality research pertinent to the subject is euphemistically called physiology, life history, or some other more esoteric term, the topic will not receive the attention it needs or deserves.

I take the liberty here to amplify some key features of Chambers statement without meaning to put my words into his pen. By "critical studies on insect colonization," I take Chambers to be calling for direct and scientific approaches to developing and improving insect-rearing systems. I have spoken of this need in my writing (Cohen 1992, 2001, 2015, 2019, for example). I suggest that "critical studies" contain explicit statements of rationale and experiments designed to render the scientific basis of the incipient rearing procedures. The thorough and competent science behind rearing advancements would help rearing research be (or become) "a credible discipline." The last statement is one to which I am very sympathetic—about euphemistically calling the discoveries "physiology, life history ...," etc. and Chambers' expression that rearing researchers have apparently felt the need to disguise or adorn rearing studies with esoteric terms resonates with me and with other rearing researchers who have had problems publishing their results of rearing studies. I discuss this at length (Cohen 2019), but suffice it to say here that in my experience I had no problem getting accepted papers on biochemistry of digestive enzymes, studies in physiological ecology, behavioral ecology, and morphology—all of which were undertaken to develop rearing systems for various insects. But upon "admitting" that the works were undertaken to help develop or improve a rearing system, I had problems getting the papers published. Over the past 40+ years that I have been in the rearing research business, I have heard similar experiences from dozens of frustrated colleagues.

In leading up to this statement, Chambers explained that there were many papers dealing with what he included under the umbrella of "quality," papers where the characteristics of insects produced in controlled studies were used as criteria for assessing the fitness of insects reared under the conditions being tested. This is to say that before QC had become a standard for mass-rearing systems, the research referred to by Chambers generally reported results of studies that were inherently mechanistic and/or cause and effect inquiries. Factors such as diets/diet components, genetics, and environmental conditions were investigated and evaluated as to how well they suited the fitness of the insects being reared. In studies such as diet or feeding dynamics (e.g., Shorey and Hale 1965; Keiser and Schneider 1969), behavior (e.g., Baumhover 1965; Bruell 1967; Sharp et al. 1983), and genetics (e.g., Alley and Hightower 1966; Chabora 1969). Soon after the Chambers 1977 and Boller and Chambers 1977 publications appeared, the USDA, Animal and Plant Health Inspection Service (APHIS), the International Atomic Energy Agency (IAEA), the World Health Organization (WHO), and the Food and Agriculture Organization of the United Nations (FAO) all became leading supporters of the concept and practice of QC in insect mass-rearing systems. As we realize the value of QC in mass-rearing and smaller-scale rearing systems, the next issue is defining what we mean by "quality" and "control."

10.4.2 How Is "Quality Control" Defined?

Chambers 1977 provided the following definitions and comments about "quality control:"

> Quality is in more critical need of definition. Although it is of increasing concern to those rearing insects in large numbers and is the objective of considerable research, it often exists only in the eye of the beholder. As I have noted previously (Chambers 1975) quality has, by dictionary definition, three components-skill, relativity, and reference-that is, quality is the degree of excellence in some trait(s) or skill(s) relative to a reference. Thus, for the purposes of this paper, skills are the performance requirements for achieving the objective, relativity has to do with ranking (comparing) the degree of excellence in performing the skills, and **references are the standards, needs, and constraints against which the skills are compared**.
>
> The dictionary definition of **control includes checking, verification, comparison, regulation, periodicity, and restraint**. In its congruence herein with quality there are also implications of input and feedback mechanisms, cross-check comparisons, decision making, and developing protocols for the imposition of control upon quality. Protocols are badly needed in insect production, and their development and application to decision making should receive major attention. However, developing the means for making comparisons has been an objective of the greater part of recent research, and such research has

tended to stress the effects of inputs. I try to make the point herein that ***feedback mechanisms and information should form a critical portion of the data used in establishing and refining protocols because feedback effects occur throughout the process of production and have an impact upon its efficiency and effectiveness***. Unfortunately, these impacts are often overlooked or accepted quid pro quo. **[Bold and italics are Cohen's addition to Chambers' statement.]**

In defining quality, Chambers cautions us that in some part, quality is subjective ("in the eyes of the beholder"). He further describes it in terms of the skill of the "performance requirements" (which could mean the performance of the personnel or the automated/robotic equipment that process the steps of rearing operations). In general usage, the word quality can mean something of high value or superior and it can mean the characteristics of something such as a product obtained through a process. For QC, I suggest the definition of **quality as definable characteristics** of an entity. With this definition, we can consider the end product of any process as having definable characteristics, and we can then use those definable and measurable characteristics as **standards**.

In the sense of quality as measurable characteristics, the subjective portion of the definition is important to consider. If we are producing sterile tephritid flies for an SIT program, from the grower's perspective, high-quality flies are the ones that reduce the pests well-below economic threshold. The flies that most efficiently eliminate or reduce damage to the grower's crop are the ones of desirable quality. Whatever biological characteristics are possessed by the flies, as long as they do the job of protecting the crops, they are of desirable quality. For the production staff at the mass-rearing facility, the end-goal of reducing pest damage in the field is not measurable. Therefore, the burden of the QC team at the rearing facility is to **translate** characteristics (qualities) of flies at the facility to predictions or projections of performance in the field. The QC team must find objectively and reliably measurable characteristics of flies produced in their lab that correlate highly with desirable field performance, and this had been a Holy Grail sought by various rearing teams charged with mass-rearing **consistently high-quality insects**. The characteristics that had been sought by rearing researchers are discussed further in this chapter and elsewhere in this book.

The second part of the QC discussion by Chambers is the ***control*** part where this term means some kind of feedback system that includes comparison and regulation. This could include a post-production screening to vet the insect-products or to screen and eliminate the insects that do not fit the standards. Alternatively, an elaborate and sensitive feedback system in developing the production process can include an "early warning" system that would relate undesirable deviations in the process to the inferior quality of the product, and the error-causing deviation would be corrected early in the process. For example, if a low-protein wheat germ was found to cause reduction in the vigor of the insects in a rearing system, checking the protein content of the wheat germ ***before*** it was added to the diet and replacing substandard material with wheat germ of vetted characteristics would help assure the circumvention of the vigor problem (Figures 10.2 and 10.3(a) and (b)).

The explanation by Chambers explicitly calls for "feedback mechanisms..." that "occur throughout the process of production and have an impact upon its efficiency and effectiveness."

Regarding laboratory adaptation of insects being reared, Bush 1975 wrote:

> Difficulty has also been encountered in getting wild screw-worm flies to oviposit and mate under laboratory conditions [Hightower et al. 1972]. The solution to this problem in the past has usually been to use the few flies that mate under artificial conditions as progenitors of the next generation. After five or six generations, "normal" fertility and fecundity is restored. However, these "normal" flies are the very ones that are most likely to be genetically aberrant and may be useless for release purposes.

In Figure 10.2, the sound analysis of *Anastrepha suspensa* (the Carrabean frutit fly) display the following behavioral traits: flight; signaling; premating (after mounting the female but before copulation); and aggression (attempted territorial competition). (Redrawn from Chambers [in Bush] 1975.)

The weevil mouthpart SEM images (Figure 10.3) are useful for morphological study of the feeding apparatus of *Anthonomus grandis*, including detailed views of the mandibles and chemo-sensilla. However, the image does not reflect that the weevils were shown to have a visual impairment due to an issue of β-carotene in its diet as demonstrated by Reinecke 1991. In the tables that follow, Reinecke 1991 reports that weevils with no added β-carotene were recaptured at a rate of 6.4%, significantly less than those with 160 mg/kg added to their diet, where there was a 15.8% recapture rate, and those with the intermediate amount of β-carotene had an intermediate % of recaptures. It is important to note that the experiment reported in this table involved the delivery of β-carotene supplements to adult weevils, demonstrating that visual acuity that was lost in larval undernutrition (with respect to β-carotene) could be somewhat corrected in adults. See Table 10.1 from Reinecke 1991.

Reinecke 1991 stated, "Recapture in pheromone-baited field traps of boll weevils fed 3 concentrations of

Table 10.1 Boll Weevil Vision as Indicated by Trapping Success

Beta-Carotene (mg/kg diet)	Number Released	Mean % Recaptured[a]
160	1,857	15.8a
16	1,943	11.9ab
0	1,934	6.4b

[a] Treatment replicated three times. SEM estimated to be 1.4 based on the mean square error derived from ANOVA ($F = 12.36$; $df = 2, 4$; $P < 0.01$). Means followed by the same letter are not significantly different ($P < 0.01$)

beta-carotene for 5 d after adult emergence." In his concluding statement, Reinecke 1991 made the following comment:

> The effect of the adjusted diet on boll weevil behavior reported in my work may be of interest to workers who have used the Gast Rearing Laboratory weevil in the past for other studies and to those who rear boll weevils on artificial diets that may be deficient in beta-carotene.

This statement pertaining to the data in Table 10.1 should be considered by the stakeholders who use reared insects in their programs, including the many research studies where insects from rearing systems (rather than from nature or field populations) are used as research subjects with little or no consideration (or mention) of the vagaries of reared insects' quality.

10.5 HOMEOSTASIS AND STRESS

A key assumption of this discussion is that the well-being of the insects in our rearing system (as is the case for organisms in nature) is directly connected with the organism's ***steady state*** or ***homeostatic state***. ***Stress*** can be defined in terms of the forces in an organism's internal and external environment that counter or challenge homeostasis. At the center of the **stress → homeostasis** "tug of war" are the forces or processes whereby an organism can meet all (or most) of its **genetic potential** in terms of utilization of resources and accomplishing a reproductive potential. Biologists recognize that the conditions of life are contrary to the equilibrium conditions that nature imposes on all systems—conditions that are driven by forces of entropy (Schrodinger 1944)—forces that must be countered by the organism to keep its intricate organization. I am suggesting that Schrodinger's entropy is a driving force that we consider as stress. Maintaining homeostasis demands that the organism expend energy to obey or actualize the genetic programming that underlies the structural and metabolic demands of living systems. As a simple example, we can consider a silkworm. Starting with an embryo of a silkworm within an egg, we can devise a simplified algorithm of the processes it must undergo to complete its development, to complete its life history, and ultimately to succeed in reproducing.

10.5.1 Developing an Algorithm for Tracking Stress

Hatch → Prepare Body (firm cuticle, produce digestive enzymes, etc.) → Find Food → Eat Food (making decisions about the feeding stimuli) → Digest Food→ Absorb Digested Food → Circulate Absorbed Materials → Cells Absorb Specific, Appropriate Nutrients → Cells Grow and Divide → Tissues Form Until Pre-molting Conditions Are Met → Successive Molts Are Undertaken → Pupation → Adult Stage Reached → Mating Takes Place → Eggs Are Laid → Eggs Hatch →→→ {The cycle is perpetuated for successive generations.}

In this highly oversimplified algorithm, there are countless steps, systems, subsystems, and other interacting components that form the web of metabolic systems, morphological events, and behavioral expressions of the genetic potential. Many of the steps such as metabolic pathways involved in cuticular color patterns are somewhat flexible (Dahlman 1969; Schmidt and Skerra 1994; Lee et al. 2008; Nguyen et al. 2019) as to whether the genetic expression will be fully realized while other components or subsystems are obligatory for the organism's survival (such as essential nutrients in mulberry leaves as described by Hamamura 2001). In two systems related to the findings of Reinecke 1991, the color of the cuticle in many larval insects and vision are related by the connection between a complex of pigment molecules such as bilin-binding protein in *Pieris brassicae* (Schmidt and Skerra 1994) and the so-called biliproteins in *Rhodinia fugax* (Lepidoptera: Saturniidae) described by Saito 1998. The bilins are various pigment molecules related to porphyrins (such as hemoglobin and chlorophyll). The functions of bilins are multifold, but in our discussion their function as pigments is the focus. In the two insect species mentioned here as well as many others, the bilins are transported in the hemolymph and are deposited in the cuticle. The bilins in both cases described here are blue, and with a carotenoid (with a yellow or orange hue), the complex of bilin, carotenoid, and protein provide a green molecule, which when deposited in the cuticle confers a green color to the insect. Interestingly, and very important to the quality and characteristics of our reared insects, many components or factors can be in-between absolutely essential and fully optional. For example, Reinecke 1991 found that the standard boll weevil diet for mass-rearing produced adult weevils with visual impairment due to a suboptimal amount of β-carotene. For years, the weevils reared in the Gast facility (discussed at length elsewhere in this book) lacked full visual acuity—a defect that was detected with colored field traps. The fact that the weevils seemed to pass quality standards of weight, reproductive potential, and other life-history traits did not reveal that there was a failure of the lab-reared weevils to meet their biological/genetic potential. **This raises a question of how often such undetected failures occur in our rearing systems and the insect products of these systems, and how much difference does this short-coming make in the quality of our research or programs that depend upon the insects in question.**

This leads to the discussion of factors that translate to the quality/fitness of the insects being produced in rearing systems. When we ask about the sources of failures of quality characteristics, the most probable factor is ***stress***. It is logical and useful to consider any factors that influence an organism's departure from a well-balanced homeostatic state can be defined as stress factors. Stress, in a biological context, includes environmental, nutritional, microbial, and genetic factors. Nutritional stress factors include lack of required nutrients, excess of some nutrients that results in imbalance of key nutrients, and any factors that reduce the bioavailability and stability of nutrients. Environmental stress factors include conspicuous parameters such as temperature, humidity, light conditions, and gas exchange, but we must also consider the less conspicuous factors such as vibrations, sound disruptions, insect scales, air movement, and even saturation or distribution of semiochemicals (Figure 10.4).

10.5.2 Nutritional Stress

This is a complex and far-reaching phenomenon where the absence or excess of a single nutrient or suite of nutrients can cause malnourishment or under-nourishment resulting in departures from life-history regularities, poor growth and development, reproductive failure, immunological problems, metabolic disorders, and other physiological, morphological, and/or behavior abnormalities.

Figure 10.4 "Blue *Manduca sexta*" showing biliverdin (or "biliproteins") conferring bluish coloration rather than green coloration (see hornworm in inset) which is seen in *M. sexta* that had been feeding on plants rich in carotenoids and xanthophylls (orange and yellowish pigments).

Jaumann and Snell-Rood 2019 demonstrated that adult life-history features and behavioral traits (choosiness of oviposition sites in cabbage leaves with or without conspecific models) in cabbage white butterflies (*Pieris rapae*) were strongly affected by quality of larval diet. The dietary factor that these authors used was amount of protein (casein and torula yeast) in larval diets, and they found that "Overall, adult nutritional stress decreased investment in multiple reproductive traits, including a behavioral trait, but larval stress increased investment in flight, potentially to disperse away from nutritionally poor environments." The authors summarized their study in part as follows: "This experiment and others … suggest that poor nutrition in insects may have adverse effects on multiple reproductive traits simultaneously, for example, fecundity and behavioral traits such as choosiness." And they conclude that, "Poor nutrition during the adult rather than larval stage seemed to have larger effects in cabbage white butterflies, especially for reproduction. Adult nutrition may be particularly important for these butterflies because adult food can compensate for poor nutrition during the larval stage." The implications of these findings (which are in keeping with numerous references cited in this paper) are that there is a phenotypical plasticity in insects that can manifest itself as underperformance in terms of expression of genetic potential. This is important in insects both in nature and in our insect cultures. The topic of nutrient availability is treated extensively by Boggs 1986, 1987, 1997, 2003, 2009, and by Boggs and Gilbert 1979. In nature, the vagaries of nutrient availability can result in compensation that allocates nutrients to such structures as flight muscles in lieu of allocation of limiting nutrients to reproductive scope. This type of situation is the fodder for natural selection of insects that develop strategies that optimize the interplaying factors that either enhance reproductive potential or survival of the individual. With limitations on the "expected" levels of nutrients (that is the types and proportions of nutrient profiles that the organism has evolved to require), the organism has to make behavioral, morphological, or metabolic choices as to how to utilize a single limiting nutrient or cluster of nutrients.

The concept of **limiting** nutrients leads to a discussion of the concept of "**limiting factors**" or **Liebig's Law of the Minimum** (discussed also in the Chapter 6 on "Rearing Systems as Ecological Niches or N-dimensional Hypervolumes"). Liebig's law has classically applied to explaining why a given species or population does not occupy or thrive in a given habitat, despite the fact that most of the habitat's characteristics seem to be suitable for the organism's survival. In fact, Liebig advanced his concept of limiting factors to help explain plant nutrition (and other plant-related issues of surviving and thriving such as adequate sunlight, soil texture, warmth, etc.) (Figure 10.5).

The model in Figure 10.5 is derived from Justus von Liebig (1803–1873) who proposed that the distribution of plants could be explained by the presence of adequate amounts of the limiting factors such as sunlight, water, and various nutrients. The concept was later expanded to explain the dimensions of ecological niches where limiting factors included or precluded the success of a given species in an ecosystem. Here, we are applying the concept to explain the relative success of an insect in our rearing system.

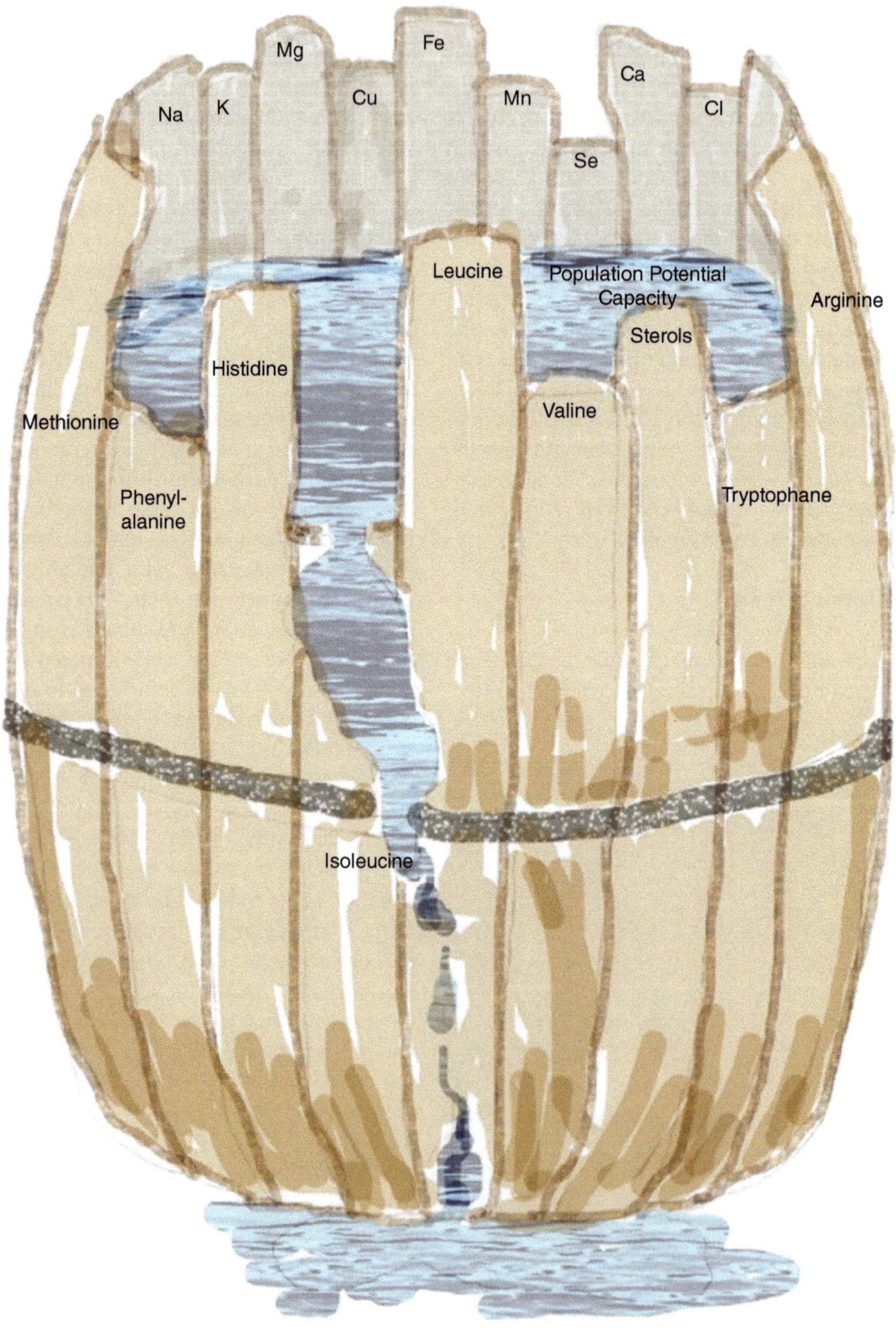

Figure 10.5 "Liebig's Barrel" (drawn by Jackie Cohen).

Assuming that our rearing system is providing all of the nutrients that our target insect requires or is adapted to utilize, except for one, the one missing or inadequate quantity of one nutrient will limit the insect in some way. The limitation may be in one metabolic, behavioral, or morphological function—one aspect of the insect's complete biology. But if the function that is being thwarted by the missing nutrient, the function will not occur as it normally should, and the insect will fail in our rearing system. It is important to recognize that Liebig's law may apply simultaneously for several subunits of the insect's biology

Figure 10.6 summarizes the nutritional basis of feeding homeostasis that is taking place in these healthy silkworms. The macronutrients in the diet which contains about 10% mulberry leaf powder include the proteins and other macronutrients (larger amounts of each nutrient class) from the mulberry source and the supplementary components from soy flour, starch, and other carbohydrate and lipid sources. The micronutrients include the vitamin and salt mixtures that Hamamura 2001 and others have shown to be conducive to enhanced growth in silkworms, as well as enrichment of ascorbic acid. The phenolics (including flavonoids and related secondary compounds) are supplied by the mulberry leaf. All these factors interplay in helping make the artificial diet meet the sensory stimulation features demanded by the highly fastidious silkworm larvae (this includes a suitable texture that gives the "bite-feel" of crisp, fresh mulberry leaves and the token stimuli that induce biting and swallowing). When all these factors are satisfied, the insect's digestion, absorption, and metabolic pathways are stimulated to function in an over-all homeostatic, positive growth fashion (Anderson and Anton 2014).

10.5.3 Hormonally-Based Sugar and Lipid Homeostasis in a Nutritional Framework

Lee and Park 2004 demonstrated the activity of a genetically based peptide: adipokinetic hormone (AKH), which is a mediator of hemolymph sugar and lipid concentrations and also drives a behavior of food seeking in *Drosophila melanogaster.* Lee and Park point out that the hormone AKH is present in insects in diverse taxonomic groups, and the thinking is that the gene for this peptide or its homologs is present throughout Insecta though the patterns of hemolymph sugars, and lipid profiles may vary among the various species. Besides its role in sugar and lipid homeostasis, AKH has been shown to have multifunctional effects in various species of arthropods, including cardio-acceleration, migration of retinal visual pigments, induction of cytochrome *P450*, inducing or activating the gene for fatty acid binding protein in flight muscles, intervention with glycogen breakdown and synthesis, and excitation of motor neurons—all summarized by Lee and Park 2004. The relevance of this far-reaching discussion of the multiple functions of a seemingly simple neuropeptide (AKH) is that in various rearing situations, the carbohydrate and/or lipid composition of the diet can have unexpected and unanticipated consequences

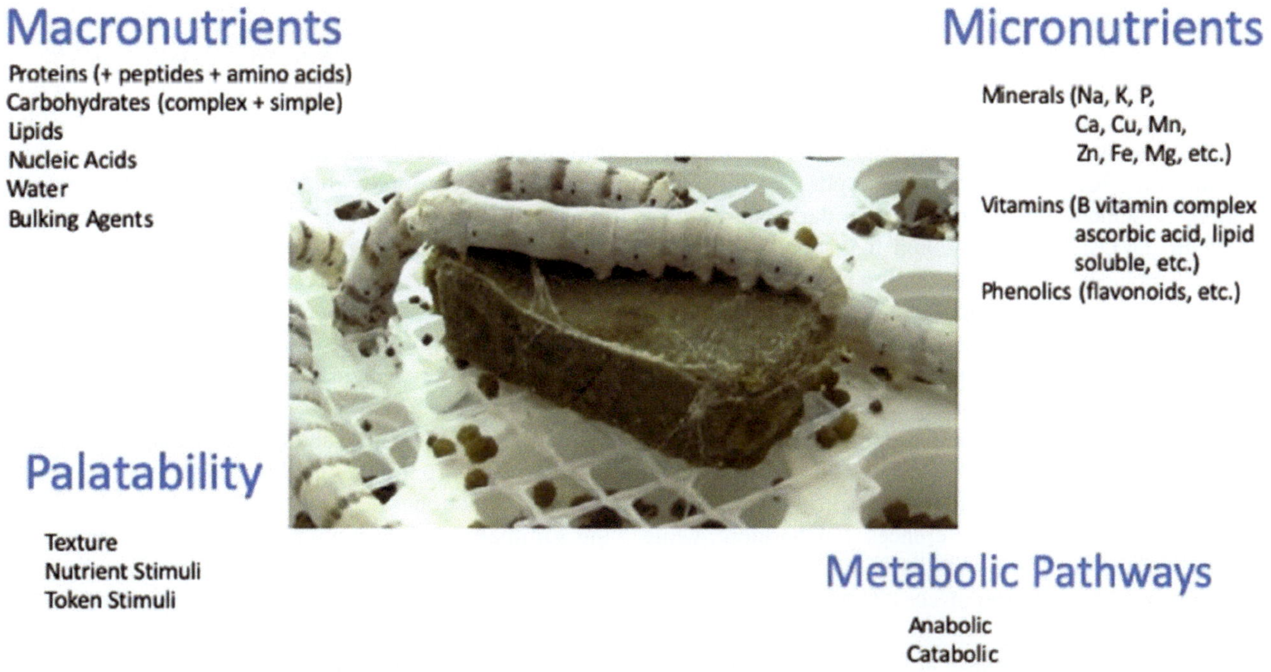

Figure 10.6 Relationships of feeding dynamics and nutrition.

in various systems in our reared insects. For example, the stress of having too little food or diet that is not balanced in terms of sugars or fats can lead to a variety of symptoms that may result in insects that are abnormal in terms of some feature or ultimately death. Any rearing factor that causes over-production of AKH can lead to a cascade of effects, possibly starting with over-activity of our insects in search of food and ending with any number of abnormalities connected with lower than normal blood sugar concentrations.

10.5.4 Environmental Stress

This includes all factors in the insect's ambient setting including temperature, water/moisture, light conditions (photoperiod, light intensity, light quality), oxygen/carbon dioxide content, waste product toxins such as nitrogenous wastes, pH of diet or aqueous media, physical characteristics of the container, etc. A convenient model of environmental stressors can be derived from the discipline of environmental physiology or physiological ecology. Physiological ecologists have conventionally studied the interplay between organisms' environment and the biological/physiological responses of organisms, with an especially careful consideration of organisms in extreme environments such as deserts, arctic habitats, high mountain zones, or aquatic habitats with extreme salinity. In extreme environments in nature, organisms must survive conditions with temperature extremes, very low humidity, very high salt content, or chronic bombardment with ionizing ultraviolet (UV) radiation.

10.5.4.1 Light Conditions and Stress

Light conditions are among the highly important components of the environmental factors in rearing systems rearing systems (Lindroth et al. 1993; Hori et al. 2014; Scholl et al. 2015; Zhang 2018). The phrase, "light conditions," includes photoperiod (light: dark phases of a 24-hour day), light intensity, and light quality (= wavelengths included in the artificially provided light spectrum). In literature on plant responses to light conditions, there are hundreds of reports with specific attention to all the specified light conditions but the search terms "light conditions" and "insect rearing" yielded only 115 references in a Web of Science search as of February 15, 2020, while a search of "light conditions" and "plant physiology" yielded 690 references, and "light conditions" and "plant propagation" yielded 495 references. Furthermore, while the plant references dealt with light intensity, quality, and photoperiod in depth, the insect references touched very little on these three terms.

A few studies included attention to the effects of factors such as constant light, constant darkness, and mixed light/dark conditions. For example, Bebas and Cymborowski 1999 found that constant light, especially in late larvae and pupae, caused a reduction in fertile (eupyrene) sperm bundles in *Spodoptera littoralis*. This finding is an indication of the possible far-reaching effects that a seemingly innocuous rearing factor such as photophase can have sweeping and dire consequences in a rearing system. It raises questions of what light intensity threshold is required for production of fully normal, viable (fertile) sperm.

Kim et al. 2019 reported an experiment where they determined that exposure to green light can have far-reaching effects on the physiology and behavior of some insects. The context of the study by Kim et al. is that light traps that emit green wavelengths (about 520 nm) can reduce the JH titer in larvae and pupae that have been exposed to the light source for prolonged periods (see Figure 10.7). I have treated the subject of light conditions in greater depth in Chapter 6 on "Rearing Systems as Ecological Niches or N-dimensional Hypervolumes."

10.5.4.2 Thermal Relations and Stress

For all insects that are subjects of insect rearing, it is important to remember that there are two factors that shape the insects' responses to thermal conditions genetic potential and acclimation. In studying desert organisms' responses to very high temperatures, for example, physiological ecologists ask how the survivors deal with temperatures that would cause non-desert organisms (mesic-adapted organisms) to have their metabolic pathways to run amok, and on a more short-term basis, the organisms can reach lethal maxima or critical thermal maxima (Cohen and Pinto 1977; Edney 1977): temperatures at which the organism dies (with chronic exposure = **lethal** maximum) or goes into a heat-induced torpor (with short-term/acute exposure = critical thermal maximum). We also learn from studies of physiological ecology that metabolic pathways—all governed by enzymes—are highly influenced by shifts in temperatures (Figure 10.8). That is, the enzymes become denatured or distorted to the point where they no longer function normally. For short and mild periods of excess heat, enzymes (and proteins, in general) can undergo a reversible change in conformation (three-dimensional structure) causing the proteins to deviate in their functional efficiency. Figure 10.9 illustrates the summation of phenomena that influence an insect's response to a broad range of temperatures, starting with lower temperatures where the insect has not reached the threshold of activity—typically a basis of dormancy or cold-inactivity sometimes referred to as "cold torpor." As the body temperature rises, two factors are at work,

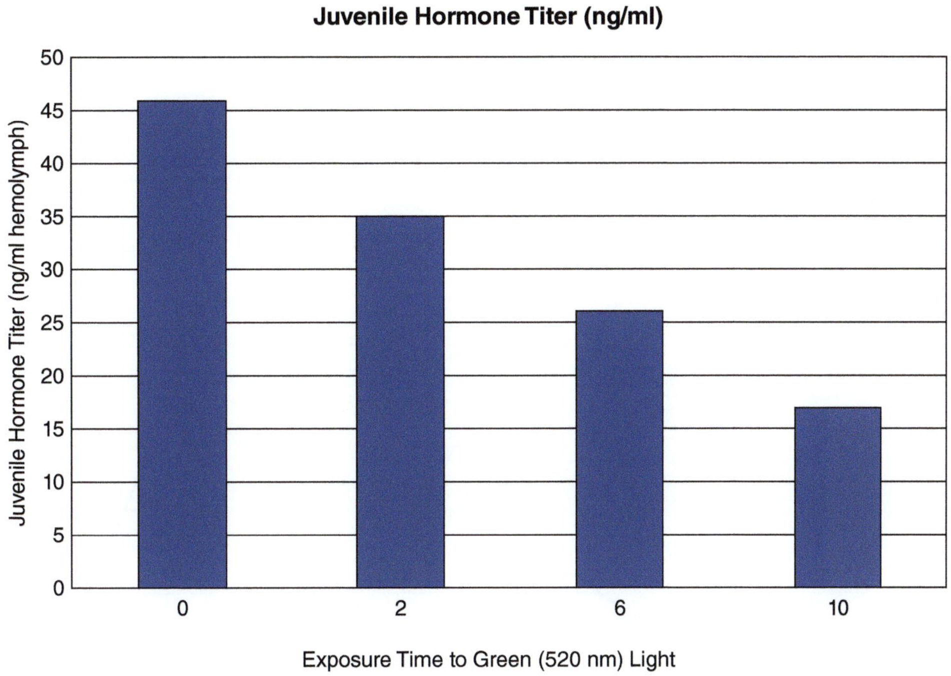

Figure 10.7 Effect of green light on JH titer in *Mythimna separata* larvae. (Redrawn from Kim et al. 2019).

the purely physical phenomenon where chemical reactions accelerate with an increase in molecular activity relating to heating. Simultaneously, the insect's enzymes are changing shapes as temperature increases, and there is an increasing efficiency of enzyme function with the effect of accelerating the metabolic activities up to the point of the optimum temperature being reached indicated graphically by the peak in the bell-shaped curve (a Gaussian distribution). Figure 10.9 further shows that the decline in metabolic rate as the temperatures go past the optimum temperature is driven by departure of the enzymes in the insects' metabolic pathways due to

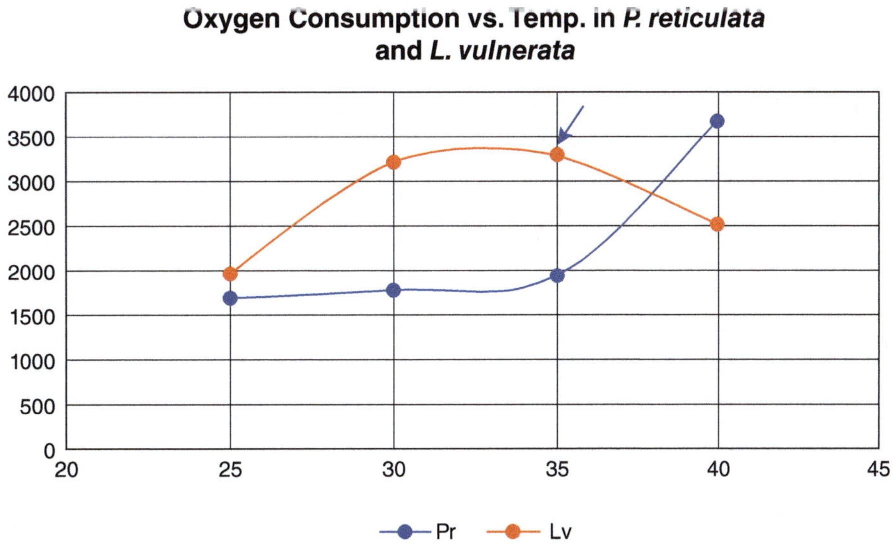

Figure 10.8 O_2 consumption per g body weight/h in *Pleuropasta reticulata* and *Lytta vulnerata* (two species of blister beetles, Meloidae).

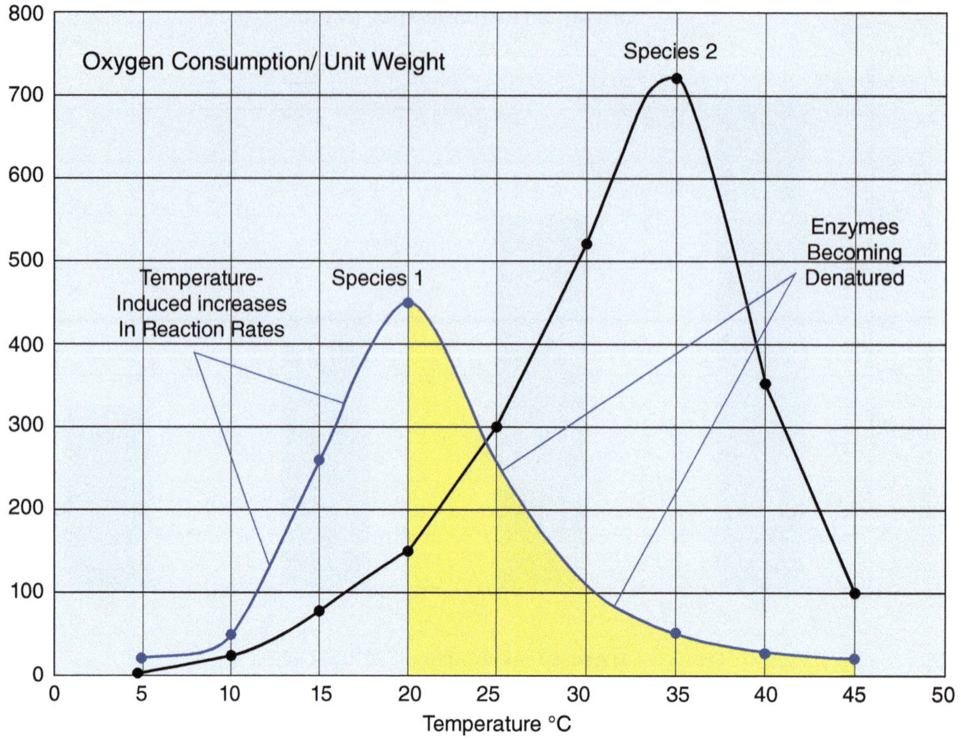

Figure 10.9 HMR–Hypothetical metabolic rate vs. temperature.

denaturation or departure from optimum shape. Through the forces of the Arrhenius relationship between physical processes and temperature, the chemical reactions would follow more of a parabolic curve than a Gaussian one.

The outcome of this push and pull of metabolic and purely physical forces is that insects in our rearing systems are subject to very deterministic physical and biological forces that shape the well-being of the insects in our rearing systems. Clearly, because our insects' growth, development rate, nutrient utilization efficiency, behavior, etc. are all driven by ambient temperature, we are obliged to furnish the correct thermal environment to allow our insects to thrive in their "rearing niches." This is an axiom of rearing (that temperature is an all-important component of the rearing environment), and **any** departure from the optimum temperature for our species of insect (and the population subsets that we are culturing) constitutes **stress** (Figures 10.8–10.11).

Note in Figure 10.8 that for *Lytta vulnerata*, which is active in cooler temperatures than is *P. reticulata*, the peak of oxygen consumption is between 30°C and 35°C, while in the more thermophilic blister beetles (including *Phodaga alticeps* and *Cysteodemus armatus* shown in Figure 10.11), the peak oxygen consumption temperatures are considerably higher. The lesson from these measurements of rates of metabolism as indicated by O_2 consumption is that metabolic potential (potential for growth and development rate along with all other metabolic/physiological processes) is in accord with the adaptation temperatures, including both genetic potential and shorter-term acclimation. The "take home" message for rearing personnel is that peak performance of their insects can be expected from information about the adaptations of our insects to their environments of origin. Too often, this concept is ignored when rearing systems are designed and established.

Figure 10.9 is a conceptual overview of the effects of temperature on metabolic rate of two species of insects, species 1 being adapted to cooler climate, species 2 being a warm climate-adapted insect. In species 2, the optimum temperature is higher than that of species 1, and the peak metabolic rate is also higher as expressed in units of oxygen consumed per unit of body weight per unit of time. It is also noteworthy that for species 1, the change in metabolic rate for a 10°C change in body temperature (the Q_{10}) is about 9.0 between 10°C and 20°C, while for the same temperature differential the Q_{10} is about 6.

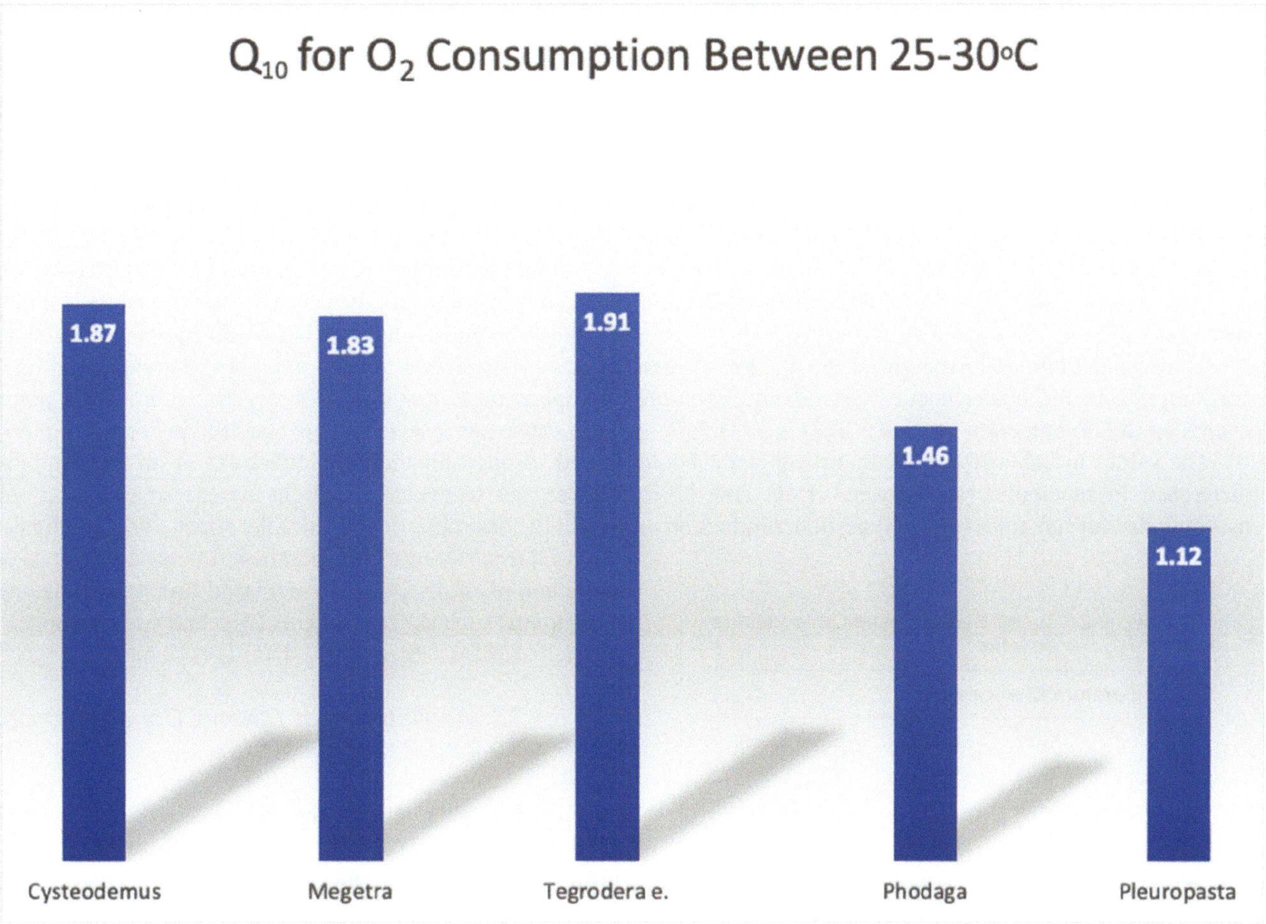

Figure 10.10 Blister Beetle Q_{10} between 25°C and 30°C. In this figure (based on a study by Cohen and Pinto 1977), the increase in metabolic rate for the blister beetles *Cysteodemus armatus*, *Megetra cancellata*, *Tegrodera erosa*, *Phodaga alticeps*, and *Pleuropasta reticulata* are shown in terms of oxygen consumption rates (µL/g/h).

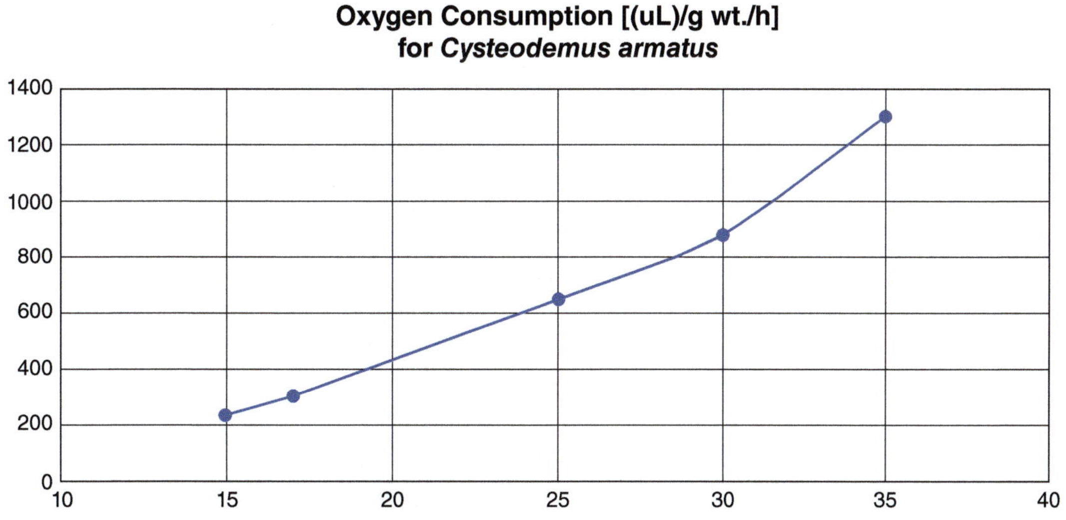

Figure 10.11 O_2 consumption by *Cysteodemus armatus* between 15°C and 35°C.

The concept of Q_{10} is important as a basis for metabolic and physiological activities of insects in nature and in rearing systems. The Q_{10} is defined as follows:

$$Q_{10} = \left(\frac{R2}{R1}\right)(10 / T_2 - T_1)$$

https://www.physiologyweb.com/calculators/q10_calcuType equation here.lator.html

Log Q_{10} = $10/T_1–T_2$ log O_2 consumed at T_2/O_2 consumed at T_1.

Cohen and Pinto 1977 measured the Q_{10} for oxygen consumption in the blister beetle *Cysteodemus armatus* over six pairs of temperatures (Table 10.2).

The values in Table 10.2 show that there is a 3.4-fold difference in metabolic rate between 15°C and 17°C, meaning that all functions (rate of digestion, production of ATP from fuels, excretory metabolics rates, etc.) changed dramatically over the slight increase in temperature. Meanwhile, the change in metabolic rate between 25°C and 30°C involves only a 1.87 shift in metabolic rate. Therefore, this table is pointing out dramatically that Q_{10} changes over a range of temperatures are not linear throughout the temperature range. The point of this dramatically non-linear change in metabolism as a function of temperature is that in our insectaries, what may seem to be only slight deviations from set points (intended temperatures) can have a huge effect on all the metabolic functions of our insects. An unintended and unnoticed drop in temperature below one of these strong inflection points can cause our insects to slow essential processes and proceed through all their life functions at an unacceptably lower rate. Conversely, a slight increase may have a profoundly desirable or undesirable effect with unintended consequences. For example, a huge increase in metabolic rate can result in a greatly increased threat from damaging toxins such as free radicals (FR) and from pathogens, which are also subject to Q_{10}-based phenomena.

Table 10.2 Q_{10} for Blister Beetle *Cysteodemus armatus*

Temperature Changes (°C)	Q_{10}
15–17	3.38
17–25	2.60
15–25	2.74
25–30	1.87
25–35	2.02
30–35	2.16

Note: the Q_{10} calculations in this table were made with a program provided by physiologyweb.com. In this table, Q_{10} values are shown for several pairs of temperatures. It is noteworthy that the 2 degree differential between 15°C and 17°C shows the steepest Q_{10} of 3.38, meaning that the metabolic rate, as indicated by O_2 consumption, had increased by more than threefold between 15°C and 17°C. This suggests that the onset of activity for this population of beetles collected from host plants *Larrea divaricata* (= creosote bush) was close to 15°C, which is in keeping with the observations in the field as discussed by Cohen and Pinto 1977. The beetles had been recently collected from the field in late-March 1974 where daily air temperatures ranged from 10°C to 40°C near plants in full sunlight, and surface temperatures were in excess of 50°C in full noonday sunlight where beetles were actively crawling from plant to plant. Beetles were observed to be dormant early in the morning when temperatures were under 15°C prompting the researchers to measure oxygen consumption starting at 15°C and going into the mid-30s for other measurements of respiration. I am making this point at length here because it is important in a rearing setting to match the field conditions with the rearing conditions and especially to recognize where rearing temperatures fit into the profile of adaptations (genetic predisposition and acclimation).

10.5.4.2.1 Applications of Thermal Considerations and Rearing Insects

Far too often, temperature regimes in rearing systems are established with an arbitrary basis. Furthermore, the often-observed phenomenon of non-synchronous development of laboratory populations is neglected in terms of the causes of this undesirable outcome. While several reports and discussions of problems from non-uniform temperature conditions appear in the literature (such as ODell 1992), too often problems stemming from thermal heterogeneity and other temperature-related issues are underappreciated (Figure 10.12).

10.5.4.3 Cold Stress in Rearing Settings

Although heat-stress has received more attention in rearing research, it is also important to consider the effects of chilling on the fitness of insects from rearing settings. Sometimes the application of cold is deliberate for storage or pre-conditions to other treatments such as sterilization (e.g., Diallo et al. 2019). Diallo et al. 2019 found that the quality of tsetse flies was adversely affected in terms of survival, adult emergence from pupae, and flight capability. The authors found that there was an adaptation to the cold-stress conditions where flies that had been routinely stress (by chilling and being subjected to vibrations involved in travel) demonstrated improvement in their fitness, indicating that there was an adaptation to these stresses. The authors also improved the methodology of reducing the time of exposure to stressful conditions, which also improved the fitness of the flies and their usefulness for the SIT program.

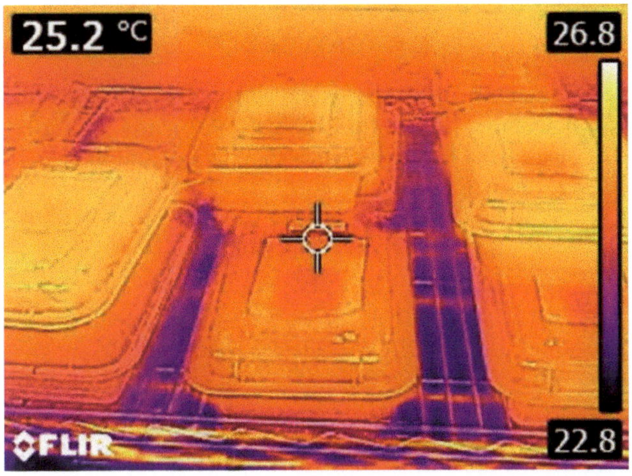

Figure 10.12 Thermal inconsistencies in rearing rooms. This is a thermal (infrared) image of rearing containers. Note the temperature range of 4°C in this small area of the rearing room, where temperature is partially dependent on the position of the containers in relation to light.

10.5.5 Toxicological Stress (Toxins, Free Radicals/Oxidative Factors Requiring Antioxidant Compounds and Enzymes) and Microbiological Stress (Diseases/Contaminants/Disruption of Symbionts)

Recent studies of nutritional ecology have shown that an important aspect of nutritional homeostasis is the support of immune functions in insects (Cotter et al. 2011; Ponton et al. 2011). Cotter et al. 2011 demonstrate that proper proportions of protein to carbohydrate are crucial in optimizing immune functions against pathogens. These authors also point out that there are hierarchical tradeoffs between immune functions and life-history characteristics influenced by protein/carbohydrate ratios.

10.5.5.1 How Do Insects Deal with Stress from Reactive Oxygen Species (ROS)?

Along with the many potential microbial stress factors, insects face an even more ubiquitous threat to their ability to maintain homeostasis: the onslaught of FR or reactive oxygen species (ROS). The concepts of immunity against microbial threats and responses to threats from ROS sometimes merge because many aspects of the two seemingly disparate threats often become congruent. That is, one source of ROS may be pathogenic microbes, and at the same time, one part of insect defenses may be the generation and utilization of ROS to help attack and kill microbes. The complexity and seeming ambiguity here is in the ubiquity of free-radical generation in so many processes such as ATP-production through oxidative phosphorylation along with countless other routine metabolic processes. Metabolic pathways abound where superoxides (O_2^-) and hydrogen peroxides (H_2O_2) are formed (Figure 10.13).

The diagram in Figure 10.13 shows several of the defenses, all based on genetic potential and nutritional/

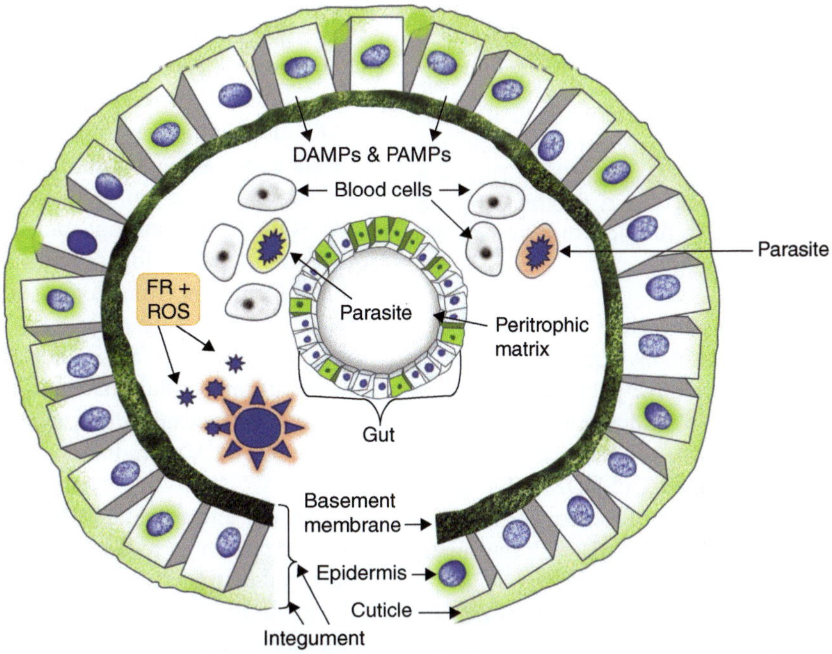

Figure 10.13 Diagram of immunological and anti-toxicological features of insects.

metabolic resources. The figure shows several of the types of interactions with pathogens and toxins that illustrate molecular systems such as DAMPs (damage associated molecular products) and PAMPs (pathogen associated molecular products). Note in this diagram that there are also free radicals (FR) and reactive oxygen species (ROS) illustrated. Note also that blood cells and epidermal cells and blood cells are shown to generate defensive components (DAMPs and PAMPs). Also, it is important to note that this diagram is highly simplified and does not illustrate molecular signals of damage nor molecular signals of damage recognition. Components such as calcium and certain proteins such as apolipophorin and lipids such as prostaglandins and eicosanoids are key intermediates in the damage/pathogen recognition and initiation of action. This topic is treated in excellent detail in a review of insect immunity by Dubovskiy et al. 2016. In the following discussion, I am drawing from Dubovskiy et al. 2016 and from another excellent review by Perry et al. 2010. rFom these two reviews, I assemble the underpinnings of defenses against pathogens and toxins in insects. Close scrutiny reveals that these two defense systems have many common key factors, especially with respect to FR- or ROS-based damage.

Perry et al. 2010 wrote the following about one of the most important families of molecules that protect cells and their parent organisms from ubiquitous oxidative stress:

> The ubiquitous superoxide dismutases (SODs) catalyze the disproportionation of superoxide to molecular oxygen and peroxide and thus are critical for protecting the cell against the toxic products of aerobic respiration. As noted by Nick Lane in his book *Oxygen: The Molecule that Made the World*, the discovery and naming of superoxide dismutase activity is "... in the opinion of many, the most important discovery of modern biology never to win a Nobel prize." This discovery and its implications transformed research on oxygen free radicals and anti-oxidants and their biological implications: There are greater than 60,000 scientific papers published on the superoxide free radical and its functions in more than 100 human pathologies.

Perry et al. 2010 point out further that superoxide (as shown here) is generated by many life processes including aerobic metabolism, oxidative phosphorylation, photosynthesis, and a plethora of pathogen-host interactions. Box 10.1 contains a summary of reactions of a copper/zinc-type superoxide dismutase (CuSODZn) (EC 1.15.1.1) showing that the oxidation state of the copper is affected by the interaction of the SOD molecule with the superoxide radical.

It should be noted that the superoxide radical (O_2^-) shown in Box 10.1 has a highly reactive electron that is very aggressive in attacking various cell components, including DNA, cell membranes, and other components as well as damage that superoxide does to other molecules outside of cells, such as components in insects' hemocoels. The production of hydrogen peroxide (H_2O_2) reduces the threat to biological systems (intracellular and extracellular components), but because hydrogen peroxide is also potentially toxic, it must be safely eliminated by a second enzyme in the "free-radical quenching system," catalase (EC 1.11.1.6). The action of catalase is to convert toxic hydrogen peroxide to harmless water and oxygen.

BOX 10.1

Reactions of superoxide dismutase (SOD) in quenching superoxide radicals (O_2^-) and catalase converting hydrogen peroxide to water and molecular oxygen.

$$Cu^{++}SOD - Zn + O_2.^- \rightarrow Cu^{+}SOD - Zn + O_2 \rightarrow Cu^{+}SOD - Zn + 2H^+ \rightarrow Cu^{++}SOD - Zn + H_2O_2$$

$$2H_2O_2 + \text{catalase} \rightarrow 2H_2O + O_2$$

From the perspective of stress in rearing systems, the microbial invaders and toxins from various sources are threats to the well-being (health) and homeostasis of the insects in our colonies. Therefore, it is desirable for us to either remove or preclude the threat of the potential pathogens or toxins or to assure the capability of our insects to fight-off these agents of stress. Since we cannot remove all sources of pathogen and toxin-related stress, we are obligated to assure that the insects have the resources to protect themselves such as making sure that the raw materials are present and available to allow the insects' homeostatic systems to execute their functions. This means that for the two enzymes discussed in this section, SOD and catalase, we must provide high-quality protein in adequate amounts, and we must also assure that the cofactors such as iron for catalase and Zn, Cu, and Mn for SOD are present and bioavailable to our insects. To further emphasize this point in rearing systems, I provide this information from Nojima et al. "There are two major factors contributing to cellular stress in organisms. One comprises internal factors, such as respiration, production of energy, anti-inflammatory, and redox system. The other comprises external factors, such as environmental factors, ultraviolet (UV) irradiation, infection, temperature, and humidity."

Often, antimicrobial compounds are added to insect diets to prevent or reduce microbial stress, especially from bacterial or fungal contaminants. However, antimicrobial compounds are, by their very nature, toxins to the target microbes but possibly also to the insects that we are trying to protect. Too often, the potentially harmful effects of antimicrobial agents are neglected in considering the expectations of the reared insects. Various biological symptoms of the toxicology of antimicrobial substances were discussed originally by Ouye 1962, and then treated extensively by Singh and House 1970a, 1970b, Funke 1983, Alverson and Cohen 2002, and Zha and Cohen 2014. These studies demonstrated diminution or compromise of fitness in insects challenged with various levels of antimicrobial compounds. In earlier studies, gross, overall effects of antimicrobial compounds include reduction in biomass, delays in normal development times, and increased mortality in a compound-dependent/dose-dependent way. However, studies that focus on more specific effects reveal that toxicity of antimicrobial compounds has far-reaching effects on factors such as digestive efficiency, metabolic efficiency, and other physiological outcomes (Zha and Cohen 2014).

It is important to note that the effects of the various antimicrobial agents varied from species to species, with substances such as methyl paraben being harsh with some species of insects and fairly benign to other species. It is also noteworthy that the manifestations of the toxicity were varied, and these studies suggested that the toxicity may be expressed in ways that may not be suspected. An example of the unexpected kinds of toxicity is provided in a recent study by Büyükgüzel et al. 2020 who reported a far-reaching phenomenon that they described as "fluctuating asymmetry" that they demonstrated in Drosophila melanogaster in response to feeding on diets with neomycin, which is used in some insect diets as an antibacterial substance. Büyükgüzel et al. showed that there was a dosage-dependent response in terms of asymmetry of the larval anal papillae and adult wing venation. The authors discussed the ramifications of such asymmetries relating to potential malfunctions of structures that are normally bilaterally symmetrical.

10.5.6 Microbial Stress

10.5.6.1 General Microbial Stress Issues and the Insect Immune System

Studies of insect immunity abound, including hundreds of papers on the relationship between the immune system and rearing conditions, especially diets and nutritional aspects of homeostasis. Their immune system protects insects from attacks by viruses, bacteria, fungi, and parasites with what Lemaitre and Hoffmann 2007 call innate defense reactions, including encapsulation, phagocytosis, coagulation, melanization of potential pathogens. Lemaitre and Hoffmann 2007 also explain the roles of antimicrobial peptides (AMPs) and other microbial stress-related defenses from the fat body, hemocytes, and gut of Drosophila, which these authors present as a model system for insect immunity in general. The various kinds of microbial stress can be related to: (1) the presence of pathogens, (2) the presence of contaminant-microbes, (3) the absence of a mutualistic symbiont, and (4) harmful interactions between microbes (such as competition or displacement of symbiotic microbes) or interfering interactions between microbes and non-microbial factors (such as presence or absence of chemicals that adversely affect an otherwise beneficial symbiont). The complexity of the potential interactions of microbes and insects is vast, but a few examples will suffice.

Lee et al. 2008 provide a foundation of how several seemingly disparate aspects of insects' biology are related to microbial stress and the ability of insects to combat stress. They reported that *Spodoptera littoralis* life-history traits (larval mortality, development rate, and pupal weight), antimicrobial characteristics (lysozyme and polyphenol oxidase activities), and degree of melanization were influenced by protein quality of the larval diet:

> Insects respond to microbial infection by the rapid and transient expression of several genes encoding potent antimicrobial peptides. Herein we demonstrate that this antimicrobial response of *Drosophila* is not aspecific but can discriminate between various classes of microorganisms. We first observe that the genes encoding antibacterial and antifungal peptides are differentially expressed after injection of distinct microorganisms. More strikingly, *Drosophila* that are naturally infected by entomopathogenic fungi exhibit an adapted response by producing only peptides with antifungal activities. This response is mediated through the selective activation of the Toll pathway. (Note from Wikipedia: the Toll-like receptor is a leucine-rich curved, membrane bound receptor that plays a key role in the innate immune system).

In the diagram in Figure 10.14, the structures diagrammed as protruding from the cell wall of the gram-positive bacterium are described by Lemaitre and Hoffmann 2007 as a "lysine-type peptidoglycan." This bacterial surface "signature" is what binds or docks with the Toll-type receptor which is located on the hemolymph side of the fat body cells' membrane.

This recognition of invading Gram-positive bacteria is a precursory step to the signaling of **antibacterial peptides** (AMPs) whose sequence is specified in genes for defenses against this type of pathogen (Figure 10.14). The cascade of events described here in the case of Gram-positive bacteria, the predominant AMP is known as Defensin (a product and consequence of Toll readout). Lemaitre et al. 1997, Vodovar et al. 2005, and Lemaitre and Hoffmann 2007, and further describe (in addition to the Toll-like receptor) the Imd pathway is involved in production of AMPs. An important conclusion that the model organism, *D. melanogaster*, used in scores of studies of immune response to microbial and parasite challenges has the ability to discriminate between types of pathogens and parasites in terms of the production of protective/

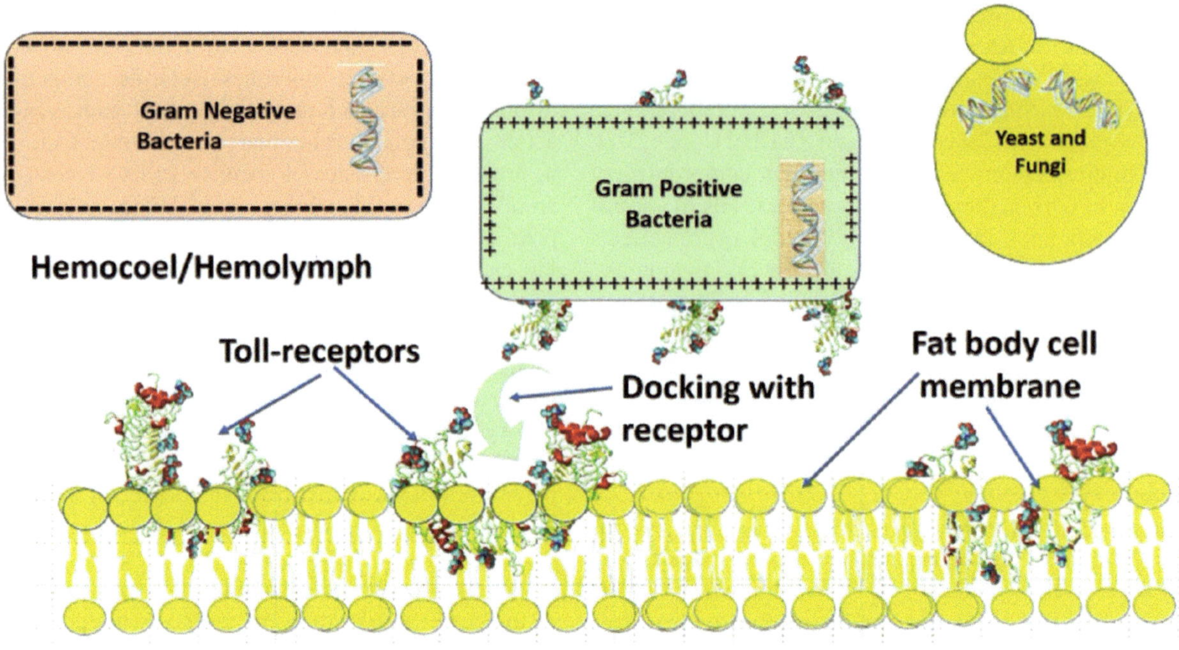

Figure 10.14 From Lemaitre and Hoffmann 2007 shows in highly abbreviated form, the specific response of the immune system of *Drosophila melanogaster* to different kinds of microbial invaders (pathogens).

defensive AMPs. It is also important to note that besides AMPs, insects have the various defenses summarized in Table 10.3.

It should be noted that Table 10.3 is a greatly simplified expression of the complexity and refinement of the immune system in *D. melanogaster*, which is the insect whose defensive responses have been most thoroughly studied. While some of the specifics may be peculiar to *D. melanogaster*, the defensive strategies generally apply to all insects. **An important take-home message of this**

Table 10.3 A Summary of the Defense System Described by Lemaitre and Hoffmann 2007

Defense	Comments about Defense	Site of Defense Origin
Antimicrobial peptides		Fat body, hemocytes, epithelia?
Epithelial immunity	Includes local AMP expression, mechanical barrier, + ROS-generation	Gut, cuticle
Hemocyte generation (hematopoiesis)	Various defenses related to hemocytes made possible. Coagulation, regulation of minerals related to defenses such as iron manipulation; includes plasmatocytes consuming and destroying invasive agents	Lymph gland, hemocytes (crystal cells, plasmatocyte—for phagocytosis—and lamellocytes for encapsulation)
Melanization	Includes oxidation of phenolic compounds in the production of melanin which helps kill microbial invaders and is associated with wound-healing onset; involves sequestration of microbes;	Damaged tissues (cuticle, etc.) and crystal cells—containing pro-polyphenol oxidase, which they release into hemolymph upon disruption
Coagulation (clotting)	Reduces "bleeding," immobilizes microbial invaders, helps promote destruction of invaders, helps promote healing mechanisms. Special proteins such as *homolectin* are involved in the clotting reaction.	Damaged cuticle
Phagocytosis	Involves several types of receptor proteins—some referred to as entry receptors specific to certain pathogens, also include DNase II which destroys invaders' DNA	Plasmatocyte-type hemocyte
Encapsulation	Involves destruction of parasites by lamellocyte-produced toxins, including ROS-generation. Also lamellocytes become sticky and cohesive to form layers around invasive parasites (which build the capsule), formation of actin-protein also involved	Lamellocyte-type hemocyte

It should be noted that this summary is an oversimplification of the system presented by Lemaitre and Hoffmann and their many references.

Table 10.4 **Antimicrobial Peptides (AMPs) from Drosophila melanogaster as Reported by Lemaitre and Hoffmann 2007**

Peptide	Main Target of Activity
Diptericin	Gram-negative bacteria
Attacin	Gram-negative bacteria
Drosocin	Gram-negative bacteria
Cecropin	Gram-negative bacteria
Defensin	Gram-positive bacteria
Drosomycin	Fungi
Metchnikowin	Fungi

discussion is that while there is exquisite genetic organization and complexity behind defenses of insects, there is also the fact that the raw materials such as amino acids and minerals must be present in adequate amounts and in suitable proportions for these defenses to become manifested against microbial stress factors. It is also important to note that the environment in which these immune pathways are manifested must be considered to be influenced by other rearing system factors such as oxygen availability, temperature, and management of waste products. In other words, to realize our goals of producing healthy, fit, high-quality insects in the face of microbial stress, our rearing systems must provide conditions conducive to healthy defensive systems' operations (Table 10.4).

Vodovar et al. 2005 demonstrated that the responses of *D. melanogaster*'s AMPs and other antimicrobial defenses are very much influenced by the type of microbe involved in the tests of immune system challenge. They showed that *Pseudomonas entomophila* (*Pe*), which is strongly pathogenic to *D. melanogaster*. The ingestion of *Pe* by *D. melanogaster* larvae and adults leads to massive destruction of gut cells and that infection by this Gram-negative rod-shaped bacterium. Along with the gut cell damage, the bacterium induces a persistent change in the larval genes related to immunity.

Vogelweith et al. 2015, using European grape berry moths as models, discuss the costly nature of immune functions in terms of tradeoffs to achieve fitness. Vogelweith et al. discuss the various immunological challenges from microbes and eukaryotes, many of which activate different immunological pathways, which the authors call "arms" of their immune system. They call the "choices" of different arms of the immune system tradeoffs. They conclude that the tradeoffs are influenced primarily by the microbe/parasite load and especially by the nutritional value of the food. They write:

> Inherent to the cost of immunity, the immune system itself can exhibit tradeoffs between its arms. Phytophagous insects face a wide range of microbial and eukaryotic parasites, each activating different immune pathways that could compromise the activity of the others. Feeding larvae are primarily exposed to microbes, which growth is controlled by antibiotic secondary metabolites produced by the host plant. The resulting variation in abundance of microbes on plants is expected to differentially stimulate the insect antimicrobial immune defenses. Under the above tradeoff hypothesis, stimulation of the insect antimicrobial defenses is expected to compromise immune activity against eukaryote parasites. In the European grape berry moth, *Eupoecilia ambiguella*, immune effectors directed towards microbes are negatively correlated to those directed towards eukaryotic parasites among host plants. Here, we hypothesize this relationship is caused by a variable control of the microbial community among host plants by their antibiotic metabolites. To test this hypothesis, we first quantified antimicrobial activity in berries of several grape varieties. We then measured immune defenses of E. ambiguella larvae raised on artificial diets in which we mimicked levels of antimicrobial activity of grape berries using tetracycline to control the abundance of growing microbes. Another group of larvae was raised on artificial diets made of berry extracts only to control for the effect of nutrition. We found that controlling microbe abundance with tet- racycline in diets did not explain variation in the immune function whereas the presence of berry extracts did. This suggests that variation in immune defenses of *E. ambiguella* among grape varieties is caused by nutritional difference among host plants rather than microbe abundance. Further study of the effects of berry compounds on larval immune parameters will be needed to explain the observed tradeoff among immune system components.

The immune systems that Vogelweith et al. 2015 investigated were hemocyte concentrations (counts), polyphenol oxidase activity, and AMPs. Some important points are provided by this study regarding assessment of immune function as an indicator of fitness and quality in rearing systems: (1) the conditions of the rearing system, including microbial load (= potential microbial stress) and nutritional quality of the diet (natural or artificial) will have strong bearing on the insects' choice of responses to microbial stress; (2) the presence of certain compounds in the diet, especially phenolic compounds such as chlorogenic acid, caffeic acid, and other secondary compounds such as some flavonoids can be very influential as to the anti-microbial response system the insect puts into play; and (3) it is important to select the most appropriate system(s) to evaluate depending on the insect, the types of microbial organisms (or parasites) causing the stress, and the rearing conditions—especially dietary nutritional value. Defenses against microbial stress include nodule formation by insect

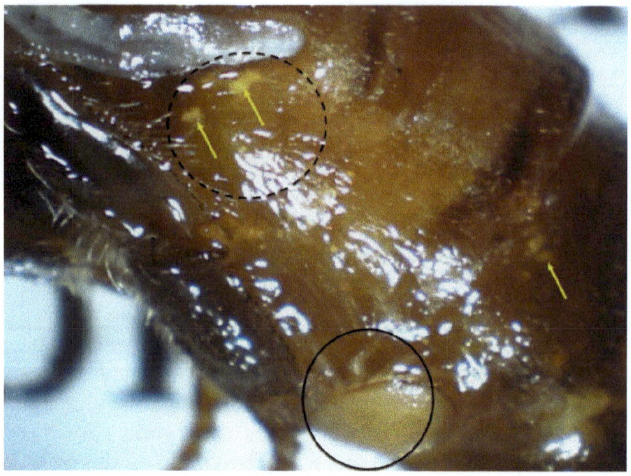

Figure 10.15 Honeybee nodules. This figure shows the formation of tyrosine nodules in a honeybee infested with Varroa mite.

hematocyes (Figures 10.15–10.18), a process that can demand considerable resources and imposes further stress on the insect.

10.5.6.2 How Are Wolbachia Species a Special Case of Microbial Stress in Rearing Systems?

Along with our considerations of defensive systems against stresses imposed by potential harmful microbes, we must also recognize that in insects that have evolved into a relationship with certain symbiotic microbes, the insects will not reach their full potential unless they are allowed to incorporate symbiotic microbes into their systems. An example of such a microbe is the bacterial genus, *Wolbachia*. More than 20 years ago, Werren 1997 made the following statement:

> *Wolbachia* are a common and widespread group of bacteria found in reproductive tissues of arthropods. These bacteria are transmitted through the cytoplasm of eggs and have evolved various mechanisms for manipulating reproduction of their hosts, including induction of reproductive incompatibility, parthenogenesis, and feminization. *Wolbachia* are also transmitted horizontally between arthropod species.

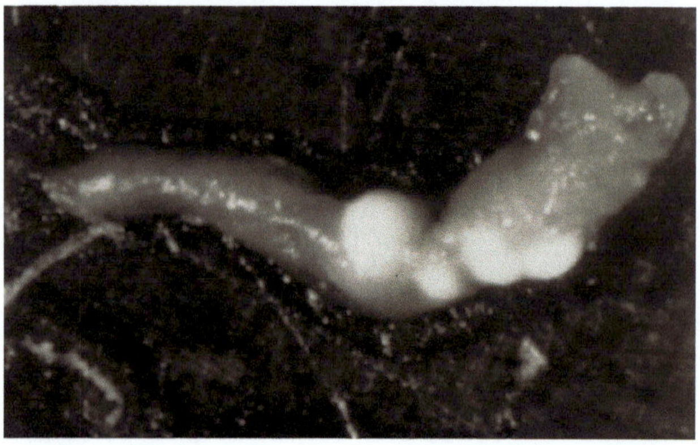

Figure 10.16 Honeybee tyrosine nodules in gaster of a bee that came from a Varroa-infested cell.

Werren 1997 also described the following outcomes of Wolbachia infection:

- Male killing
- Feminization of infected males
- Parthenogenesis
- Rapid speciation
- Cytoplasmic incompatibility

As Werren 1997 points out, these factors can lead to both positive and negative impacts on the insects affected by *Wolbachia*, and the all of the outcomes have profound influence on the insects that we are interested in colonizing. Clearly such factors as male-killing or feminization of males can have considerable bearing on issues of normal reproduction and our colony's genetic diversity. Clearly, too, the possibility that the *Wolbachia* effects have both positive and negative implications for our colonies.

Since the 1997 Werren summary of *Wolbachia* which earned 1,000 citations as of February 3, 2020, more than 4,000 studies have been published about the effects and importance of this bacterial group throughout Insecta, including the several species of insects being mass-reared. Mascarenhas et al. 2016 further describe the ubiquitous occurrence of *Wolbachia* spp. in Insecta. They described horizontal transfer of *Wolbachia* in colonies of tephritid fruit flies and some parasitoids of the flies. The Mascarenhas et al. 2016 paper is also useful as a review of about 100 papers published since the Werren 1997 review.

10.5.6.3 Microbial Contamination Issues

Although pathogens have gained considerable attention, as have mutualistic symbionts, one of the greatest and most impactful aspects of insect/microbe relationships in rearing systems is microbial contamination. Generally, when we speak of microbial contaminants, we are focusing on environmental viruses, bacteria, fungi, and protozoans. The concept of "environmental microbes" includes the vast numbers and species of universally present microorganisms that are present on/in foods, surfaces, the air, insectary personnel, and on or in the insects themselves.

Top left and bottom left: room D before treatment; top right and bottom right: room D after thorough cleaning of

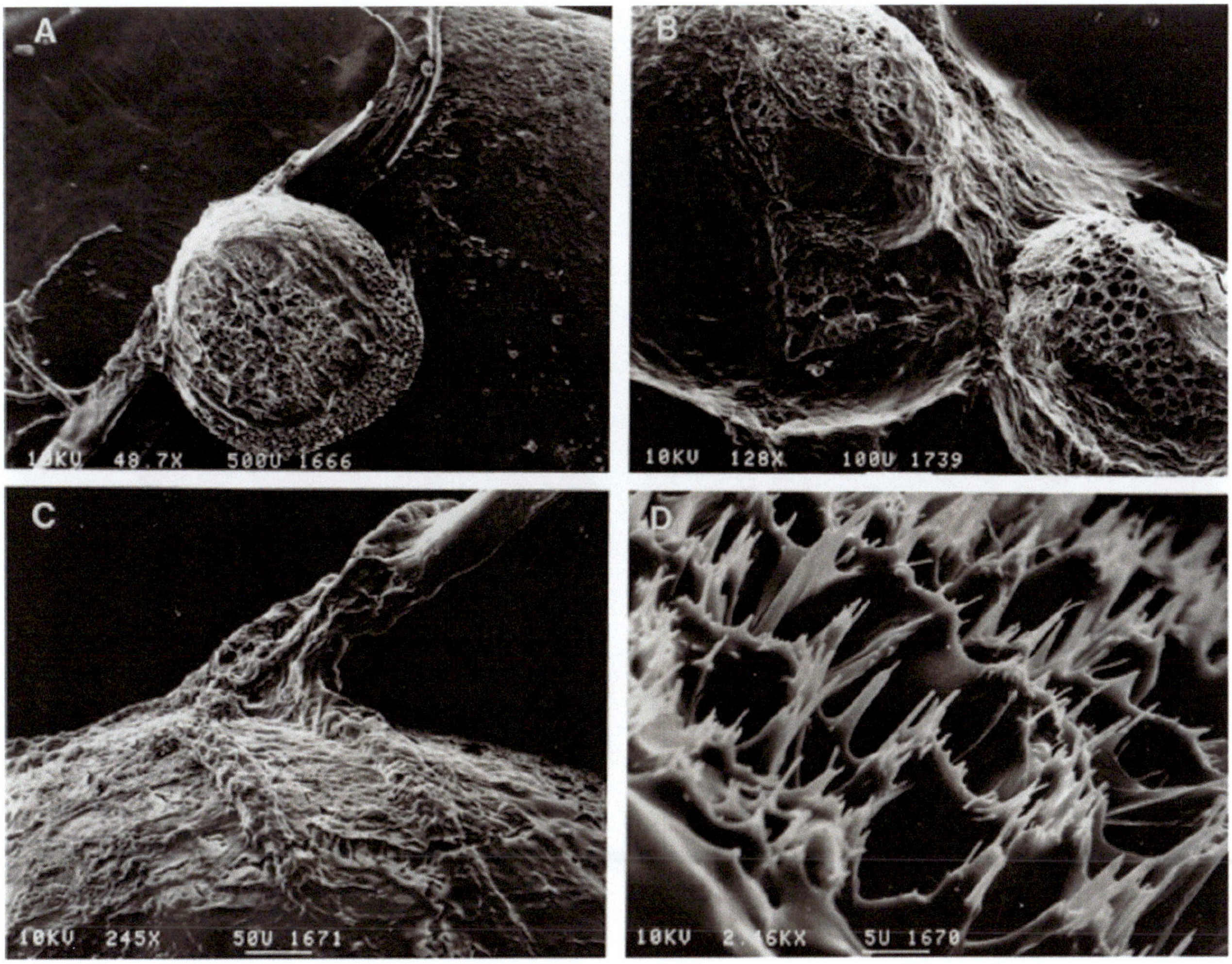

Figure 10.17 Tyrosine nodules in honeybees infested with Varroa mites. These images show the gut (above) showing four conspicuous nodules and (below) scanning electron micrographs of nodules covered with hemocytes (Erickson et al. 1997).

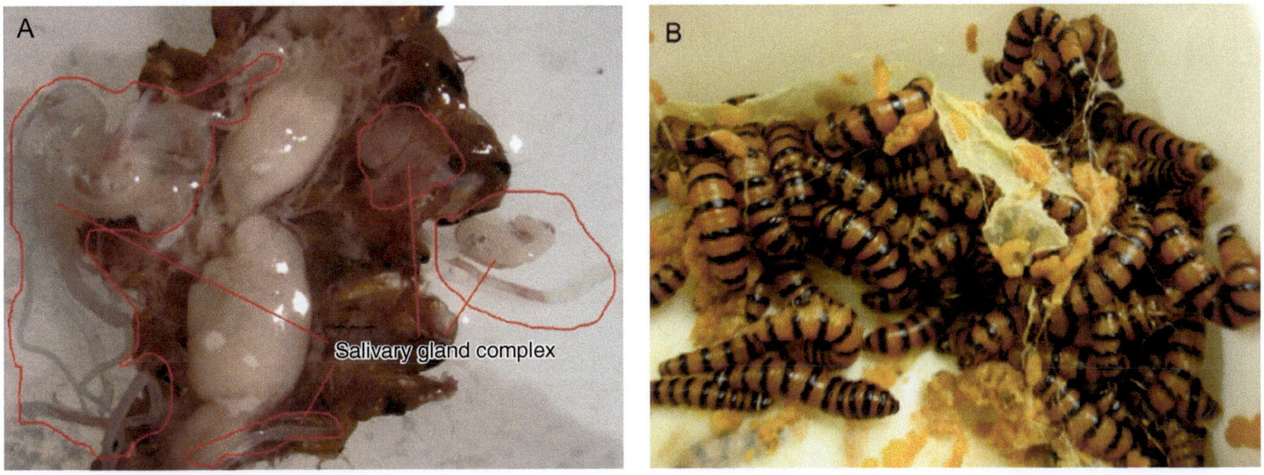

Figure 10.18 (a) Microsporidium-infected cactus moth larva. (b) Healthy cactus moth larvae.

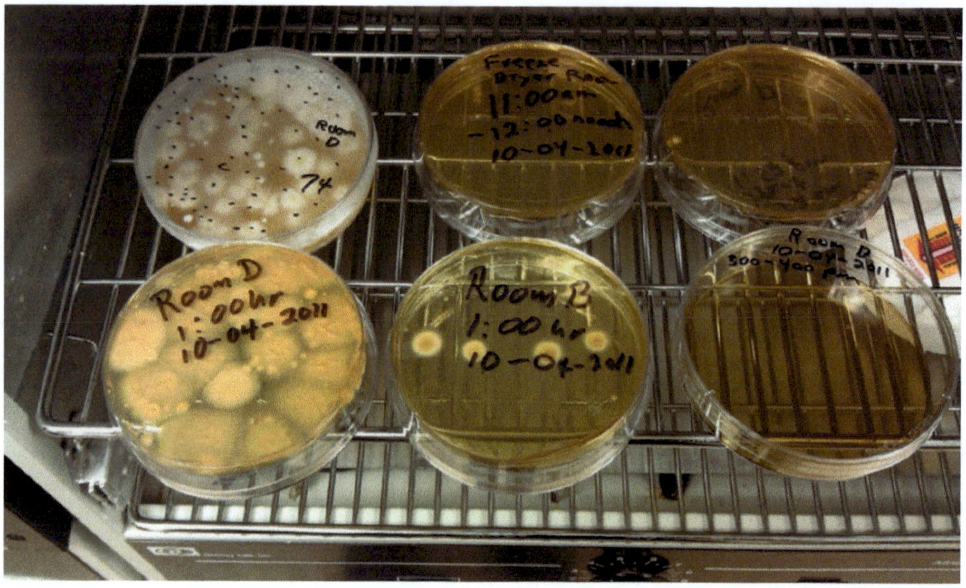

Figure 10.19 Airborne contaminants in rearing rooms.

walls and floors and employment of portable air filtration system (Figure 10.19).

The air filtration system shown in Figure 10.20 is capable of removing more than 99.97% of particles that are 0.3 µm. This means that most fungal spores would be eliminated by a single pass through this filtration system, based on the observation that most fungal spores range from about 4 or 5 µm, while viral particles are generally under 1.0 µm, with Corona virus being about 0.1 µm. With this HEPA filtration running constantly in a relatively small space, it can scrub the air several times per hour. For hospital operating rooms, the standard is 20–25 air changes per hour (ACH), according to the following site:

https://www.hfmmagazine.com/articles/3246-studying-airflow-in-the-or.

Figure 10.20 Honeywell portable air filtration device.

The site named Pure Air (author of the site's writeup is Don Musilli at https://www.pureairsystems.com/2012/07/15/calculating-air-exchange-rates-for-hepa-filter-systems/) which provides the following description of how to calculate room area (actually volume) to help estimate whether the air scrubbing system is adequate for a rearing room or clean diet prep room:

> You need to determine the area of the room in cubic feet, so you first need to measure the length and width of the room. Let's say it is 15 feet long and 20 feet wide. Multiply 15 × 20 and you get 300 square feet. Now multiply that number (300) times the height of the room. We will make in 10 feet high. So you multiply 10 x 300 and you get 3,000 cubic feet. This is all it takes to determine the area of the room.

Using this procedure, and checking the cubic feet per hour the air purifier can handle, you can determine the expected rate of hourly exchanges the air purifier will deliver. This can be used as part of the process control improvement program, and coupling this with air contamination counts such as those illustrated in Figure 10.19, the issue of reducing microbial contamination is on its way to being properly addressed. Coupling this with diet sanitation procedures and cleaning surfaces, the possibility of microbial outbreaks can be greatly reduced (Figures 10.20 and 10.21). **Finally, and most importantly,**

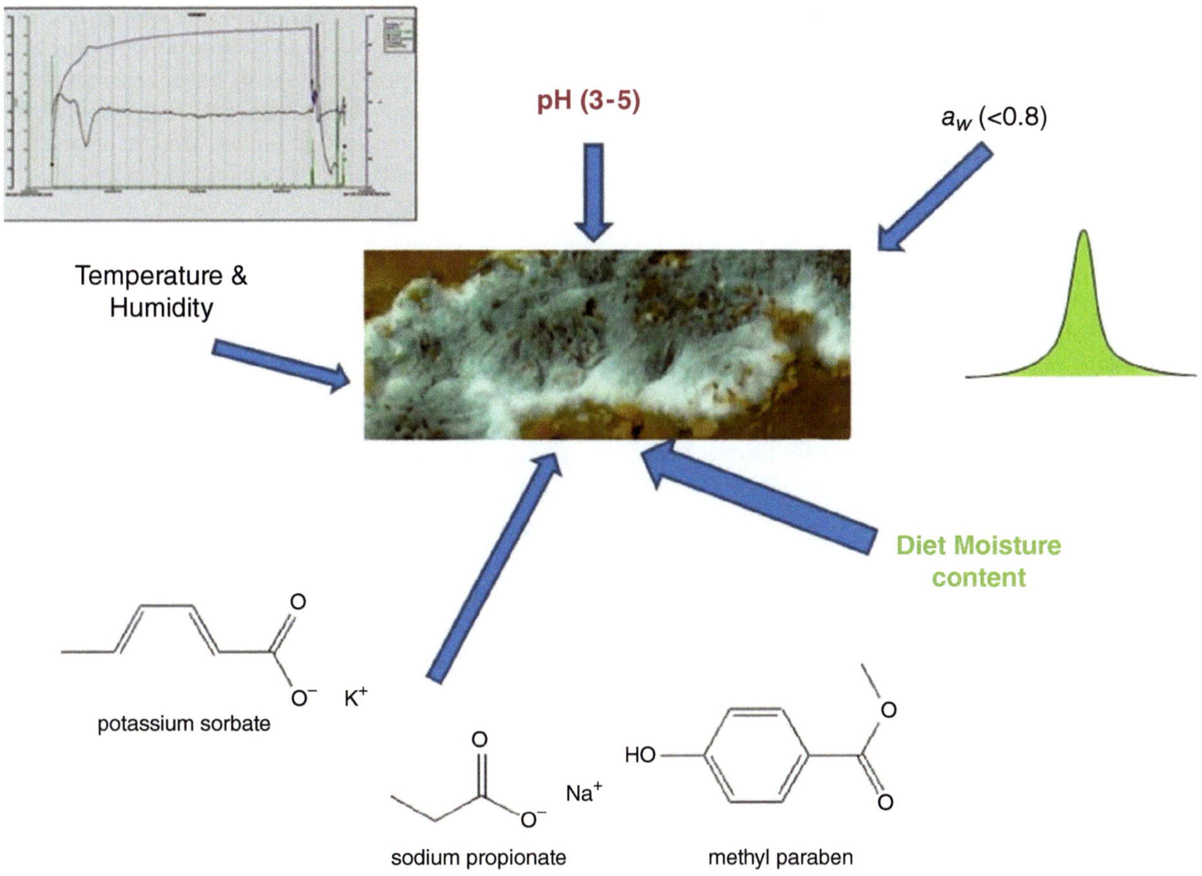

Figure 10.21 Microbial contaminant stress factors.

if the personnel are educated about sanitation (wearing masks, gloves, lab coats, hand washing, etc.), **the risks of microbial deterioration of the rearing system can become a thing of the past.**

In Figure 10.21, factors that contribute to or are related to stress in colonized insects are presented. In the upper left inset, the temperature and humidity conditions are represented. Microbial growth, especially mold contaminants such as the Penicillium species seen here, are encouraged by typical rearing temperatures and humidity. The container's humidity and the water activity of the diet (*aw*) are the chief contributors to mold, and these factors are exacerbated by the microbial load, which is increased by airborne spores and chronic exposure to the insects' frass. Both microbial contaminant stress and pathogen stress are further exacerbated by conditions of crowding and improper treatment of diet, such as over-cooking or under-cooking where nutrients and protective (prophylactic) agents are destroyed or microbes are not fully killed. An important means of demonstrating microbial load that insects can be exposed to in our rearing systems is through the use of RODAC plates for demonstrating microbes on surfaces and Petri plates opened to laboratory air to demonstrate airborne contaminants (discussed further in Chapter 5 on "Case Study of the USDA, ARS R. T. Gast BCMRRU: Boll Weevil and Lepidoptera Mass-Rearing").

10.5.7 Genetic Stress (Genetic Truncation, Selection for Low Fitness Traits)

In the body of literature about the characteristics and consequences of genetics in insect rearing, there is a compendium of information about the importance of considering and recognizing that there is an interplay between the quality and character of genetic influence on the target insects. Starting with earlier treatments of genetics in rearing by pioneers such as Mackauer 1972, Bartlett 1984, and Hoy the special treatment of genetics in insect rearing systems has received much needed attention.

Sorensen et al. 2012 write:

> Mass- and laboratory-rearing of insects is a key component of several integrated pest management strategies. **However, mass- or laboratory-rearing can have dramatic effects on insect performance via laboratory adaptation, inbreeding depression, inadvertent selection or through direct rearing effects (e.g. crowding and artificial diets).** Thus, when rearing insects intended

for field release, or when research is intended to reflect wild population performance, there is a need to establish quality under these conditions. **In reality, however, much of the focus on quality control of mass- or laboratory-reared insects concerns maintenance of facility output (i.e. numbers) and, to a much lesser extent, individual trait effects (e.g. mating competitiveness or field flight performance).** This is due to both practical issues (time or budget constraints), but also to limited understanding of the relationship between laboratory and field assays and the basic biology of the species. Regardless, trait performance is widely acknowledged as an essential factor in integrated pest management success. Recent studies on basic insect biology have made significant progress in understanding several aspects of field performance, such as the ability for thermal treatments to mitigate poor performance via phenotypic plasticity or the effects of artificial selection, and have lessons which may transfer to applied quality control for pest management. Here, we review the key challenges faced in insect rearing when focusing on quality control for field performance, some major recent advances in understanding field performance of insects in relation to environmental factors (e.g. thermal adaptation), and **the potential synergies between basic biology and quality control research foci.**

10.5.7.1 General Genetic Considerations

The points made in the above paragraph by Sorensen et al. 2012 are a highly cogent and valuable discussion of the relationship between genetics and rearing or more broadly between the general biology of the target insects and the rearing procedures. The paragraph could serve both as the introduction to the chapter on stress, quality, and rearing procedures, but it could also be the conclusion. If we search back to the original publications and discoveries in insect rearing—the evolution of the *Drosophila* rearing system, we quickly come to realize that the purpose of developing a rearing system for this insect was to gain knowledge of genetics. Though this topic is covered in more detail elsewhere in this book (Chapters 3 and 6, for example), it is appropriate to characterize the kinds of issues in genetic study led to some of the most far-reaching rearing advancements, including the use of highly stabilized thermal environments, refined practices of containerization, utilization of agar, yeast, and numerous additives for stabilization and nutritional enrichment. The origin of the concept of rearing insects in controlled (***defined***) environments is voiced in the 1910 paper by Delcourt and Guyenot. These authors expressed that the difficulties in discerning phenotypic expressions and accurately inferring the genetic basis of these expressions could only be alleviated by controlling the foods and environmental conditions of the rearing process. Delcourt and Guyenot described the chaos of phenotypic outcomes when different fruit purees were used, along with other chaos-causing factors such as microbial involvement, moisture content and other variables that had crept into the rearing process that the early *Drosophila* geneticists tried to wade through (see other papers by Bridges 1932, Bridges and Darby 1933, Baumberger 1917a, 1917b, and Guyenot 1914a).

10.5.7.2 Epigenetics/Transgenerational Effects and Rearing Stress (Also Includes Heat Shock Proteins)

Figure 10.22 is a simplification of the arrangement of histone proteins associated with DNA and arranged in a relationship of regulation of the genes via a mechanism of deactivation or activation of genes by modification of methyl (CH_3) or an acyl group to particular regulatory sites on the histone protein. In this diagram, a methyl group (symbolized by an orange star) was added to the central histone protein, deactivating it and shutting down the gene regulated by that histone. Depending on the organism and the conditions, this epigenetic (transgenerational) process may go on with: (1) a short-term deactivation or activation of genes with methylation or acylation reactions (or de-methylation or de-acylation reactions); and (2) long-term, transgenerational activation or deactivation of

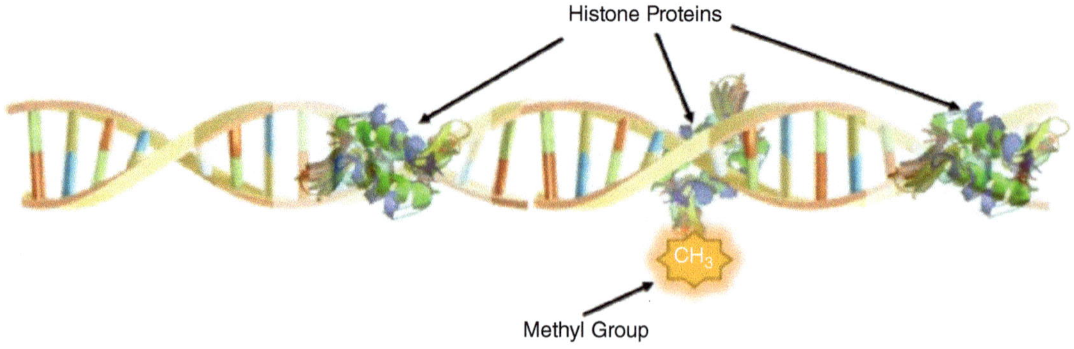

Figure 10.22 Histone regulation of gene expression.

genes. The key concepts here are that the genes are not the only factors in organisms' responses to their environments: epigenes are also important factors. An intriguing and compelling example of an epigenetic phenomenon in humans is described by Mukherjee 2016. In late 1944, the Dutch were subjected to extended and harsh famine by having much of the Netherlands cut-off from ports where food could be delivered to the people. The starving Dutch people resorted to such non-nutritive food that they were malnourished for more than a year (until the end of WW II). The phenomenon described by Mukherjee was where the children of the starved parents suffered from eating disorders where they severely overate, and had the various diseases associated with extreme overeating (obesity, increased rates of cancer, heart disease, and diabetes). Even more strange was the fact that the grandchildren of the original starved population also showed signs of the extreme reaction to starving showing the same propensity for over-eating and its associated ailments—all this despite the availability of adequate food and education about nutritional well-being. This story, Mukherjee explained, was an example of the power of transgenerational changes—not resulting from genetics but rather epigenetics.

The power and potential subversive character of these epigenetic phenomena lead me to conclude that possibly one of the most sweeping and influential factors in insect-rearing systems is the phenomenon of transgenerational effects on our cultured insects. The phenomenon was introduced to the entomology community in 1974 by Cohet and Bouletre who discussed the influences of growth temperature on the copulation reactivity of female Drosophila. The concept of epigenetics has been discussed for the past 75 years, with varying degrees of documentation, but there is no question that a solid base of information confirms that there the transgenerational impact of experiences on organisms' phenotype is something real (Figures 10.22 and 10.23). For example, a recent study showed that effects of maternal diet and age can be passed on to future generations of the water flea, *Daphnia pulicaria*, a freshwater crustacean (Goos et al. 2018). The authors wrote:

> Our results indicate that transgenerational effects of maternal age and environment persist across multiple generations, and vary between successive generations. Further, our examination of the interaction network between life-history traits of each generation suggest that the mechanisms for such persistent transgenerational effects are highly dependent on maternal condition.

Also using *Daphnia* as a model, Garbutt and Little 2017 tested the relationships between rate of food consumption, body size (biomass), and resistance to disease finding that the offspring (F_1) generation from larger *Daphnia* were more resistant to infection from a bacterial pathogen than were smaller *Daphnia*, and the effects of faster feeding F_0 Daphnia were to produce offspring (F_1s) that were more disease-resistant than offspring of slower-feeding F_1s. Thus, there was a clear transgenerational advantage to the coupled characters of fast feeding rates and larger body size in terms of lower susceptibility to pathogens. Because these experiments were conducted only over two generations, there was no evidence of the persistence of the advantage.

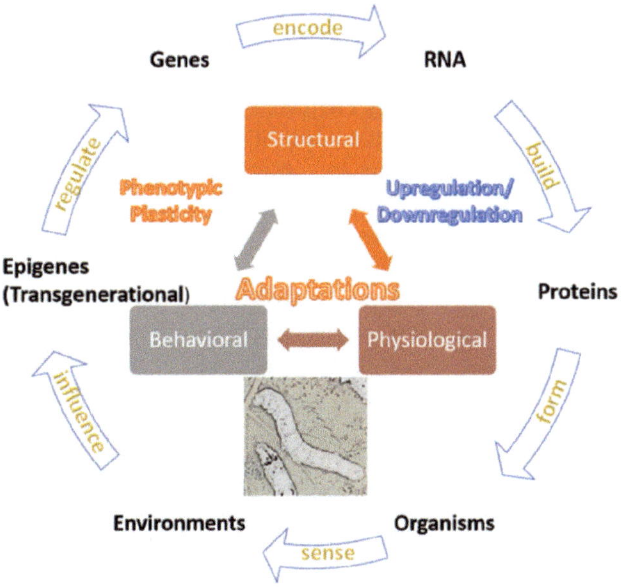

Figure 10.23 Organization of "Environmental Genetics/ Epigenetics". A helpful way to read the diagram is to start with "Genes" and say "Genes encode RNA (which) build Proteins (to) form Organisms (which) sense Environments (which) influence Epigenes (which) regulate Genes...." Centered in the diagram are adaptations, which can be summarized as behavior, structural (morphological), and physiological. The adaptational complex responds to and interplays with the environment, and the adaptive changes in the organisms (such as the silkworms in the inset) can be upregulated or down regulated so that between the genetic and epigenetic components, the organisms express or demonstrate phenotypic plasticity. The phenotypic plasticity or elasticity is in the sense that was discussed by Hildebrandt et al. 2018. (Adapted and redrawn from Mukherjee 2016.)

Importantly, some studies are showing that some species escape transgenerational effects of nutritional depravation, at least in a portion of their responses. For example, Mbande et al. 2020 found that the beetle (*Neolema abbreviata*) could be "rescued" from the adverse effects of parental nitrogen deprivation with larval access in the F_1 generation of higher nitrogen diets. However, it must be noted that there were some factors in this species and in these tests that showed some conflicting responses, indicating that some factors are susceptible to "rescue" from poor nutrition in parents while other life-history factors continue to suffer from transgenerational nutritional stress. These conflicting results within the same study indicate some of the complexity of epigenetic factors, seemingly telling us that some of the epigene switches are more readily turned on and off than others.

Zizzari et al. 2016 explain that among the various environmental stresses faced by insects and other organisms, food shortage is of prime importance. They include food shortage or low nutrient quality as major influences on "life-history trajectories," and they include numerous references documenting this major stress factor. In their study with springtails, they showed that parents reared on a poor diet "primed their offspring for more efficient resource use." As an interesting component of their study, which divided maternal vs. paternal influences on efficiency of resource uses, they found:

> … a mis- match between maternal and offspring food environments generated sex- specific adverse effects: female offspring born to well-fed mothers showed a decreased flexibility to deal with low-food conditions. Notably, these maternal effects of food availability were not observed in the sons. Finally, we found that the relationship between age and size at maturity differed between males and females and showed that offspring life-history strategies in *O. cincta* are primed differently by the parents.

The discovery of gender-related transgenerational effects adds an important twist to the complexity of the various aspects of epigenetic phenomena.

In an important departure from studies of transgenerational effects, Triggs and Knell 2012 showed (in *Plodia interpunctella*, a pyralid = Indian meal moth) that the immune competence of offspring is strongly dependent upon the food quality of the progenitors. In these experiments, two diets were provided to the insects, a standard (high-quality) diet consisting of 10:1:1 parts of wheat bran: brewer's yeast: glycerol vs. a poor diet consisting of 20:1:1 wheat bran: brewer's yeast: glycerol. The fitness of the late 5th instar larvae was evaluated in terms of weight, polyphenol oxidase activity (PO), and mean hemocyte counts. The researchers found that "The most dramatic effects on immunity were seen in the PO data where a poor parental diet reduced offspring PO activity by c. 11% on average if the father received the poor diet, by around 15% if the mother received it and by around 26% if both parents received the poor diet." It is notable that while having both parents on a poor diet caused the most dramatic decline in immune characteristics, the maternal contribution was more pronounced than was the paternal contribution to offspring immune competence.

All of these studies, and many more not cited here, indicate the important potential of transgenerational (epigenetic) degradation of insect fitness/quality in our rearing systems. Therefore, along with our development of quality control, quality assessment procedures are essential for our evaluation of genetic fitness of our reared insects; knowledge of the possible decline of our colonies due to epigenetic effects "recorded" in our insects' epigenome is also worth considering.

An extremely important group of stress-related proteins have been labeled "heat-shock proteins," though now it is understood that their effects are far more inclusive than merely being related to heat-stress. Indeed, they have been demonstrated to confer protection against osmotic stress, oxidative stress, and other unhospitable conditions. Concerning heat shock proteins, Chen et al. 2019 write "Heat shock proteins (Hsps) are conserved molecular chaperone proteins that are ubiquitous in both eukaryotic and prokaryotic organisms. In addition to mediating the folding and assembly of proteins, Hsps perform critical functions in protein homeostasis and also modulate protein localization and degradation." The topic of transgenerational regulation of heat shock proteins is described eloquently in the abstract by Chen et al. 2019 where the authors characterize their finding about red flour beetles *T. castaneum*:

> The regulatory function of histone acetylation in the expression of genes encoding heat shock proteins (Hsps) has been documented in *Drosophila melanogaster*; however, knowledge of the role of acetylation in modulating Hsps in other insect pests is limited. In this study, two full-length cDNAs encoding inducible Hsp70 (designated TcHsp70) and heat shock cognate 70 (TcHsc70) were isolated and characterized in the red flour beetle, *Tribolium castaneum. TcHsp70* and *TcHsc70* cDNAs were 2256 and 2132 bp and encoded 1941- and 1893-bp open reading frames, respectively. The deduced TcHsp70 and TcHsc70 proteins contained 646 and 630 amino acids, respectively, and contained sequences typical of the Hsp70 family, including the EEVD motif for cytoplasmic localization. Expression patterns after heat shock indicated that *TcHsp70* was strongly heat-inducible, whereas the expression level of *TcHsc70* remained unchanged under heat shock. RNA interference-mediated knock-down of three genes encoding class I histone deacetylases differentially influenced both basal and heat shock inducible expression of *TcHsp70* and *TcHsc70*, suggesting the involvement of histone acetylation in epigenetic regulation of *Hsp70* transcription in *T. castaneum*.

The regulation of heat shock proteins in the family known as TcHsp70 and TcHsc70 is controlled by modifications of histones by an acetylation reaction, which is another important means of histone "switching on" and "switching off" as is histone regulation by methylation. Chen et al. 2019 further clarify the epigenetic controls as follows: "The best-known epigenetic processes include histone post-translational modifications (PTMs), DNA methylation, and small RNA interference. Acetylation is a form of histone PTM, which is modulated by the antagonistic activities of histone acetyltransferases (HATs) and histone deacetylases (HDACs)." The details and complexities of these controls are beyond the discussion of this book. However, the key point and relevance of this complex process of gene regulation (epigenetic regulation) is that in our rearing systems, we constantly have events, often unintended or non-deliberate, which can significantly but cryptically modify our insect-rearing products. **The question that is ever-present is to what extent does a rearing condition trigger an epigenetic outcome that may have far-reaching and long-term even transgenerational effects on the nature and quality of our insects?**

Hildebrandt et al. 2018 discuss and demonstrate fitness distinctions between epigenetic modifications and acclimation in their treatment of *phenotypic plasticity/flexibility*. They define this flexibility or phenotypic plasticity as "the ability. To alter the individual phenotype (without any changes in the genetic background) with changing environmental conditions...." They write:

> Most animal species have the genetically determined ability to live and reproduce in environments that differ in various abiotic and biotic conditions (salinity of the medium, temperature, presence or absence of predators, availability of food etc.). The way and the extent a certain trait changes in response to an altered environmental condition within the critical values that limit survival or fitness is termed reaction norm. The reaction norm represents a function that is used to quantitatively describe the immanent changes in life cycle parameters (growth rate, fecundity, etc.) in their dependency on changes in environmental conditions.

The examples that Hildebrandt et al. 2018 use are from studies of responses to osmotic stress in hydroids and mosquito larvae, and they make distinctions about the responses to this stress in terms of life-history responses to the possible variations in osmolality of their aqueous environment. They explain that the ability to avert excessive expenditure of energy to regulate their internal osmotic potential under conditions of environmental challenge (too low or too high osmolality of the surrounding water) is a measure of the organisms' fitness to the existing environmental conditions. If they must divert energy from growth, fecundity, or other life-history factors to osmoregulation, then they are considered poorly adapted to the environmentally stressful condition. Hildebrandt et al. 2018 raise the important issue of "reversibility of plastic changes," i.e., whether the organism can return to its normal condition after the stress factor has been relaxed or return to the original values. Within this discussion, the authors differentiate between ***phenotypic plasticity*** (where the changes are fixed within the organism under stress) and ***phenotypic elasticity*** (where the changes can go back and forth with the exertion and relaxation of the stress factors). In both cases, there is a demonstrated ability to change the adaptive character in response to stress, but the difference is in how flexible the organism's response is to continuing changes in the stress factors. These questions are very pertinent to our purposes in insect rearing. If any of the stresses that we impose inadvertently in our rearing system—deviation in temperatures, nutritional quality, microbial stress, oxygen deprivation, excess waste materials (such as CO2 or nitrogenous wastes), or water deprivation—how phenotypically plastic is our organism? **This is to ask, once we have imposed the stress on our target organisms, how prone are they to return to their normal state?**

VandenBrooks et al. 2018 report deep structural modifications in the gas exchange systems of *D. melanogaster* reared under conditions of: (1) normal atmospheric oxygen concentration (= 21%), (2) higher than normal O_2 (31%), and (3) lower than normal O_2 (11%). In response to these extremes, the tracheoles had greater diameters, more branch points, number of tracheoles in low oxygen-reared flies and just the opposite in high-oxygen-reared flies. Also, as seen in Figure 10.24, the ratio of numbers of mitochondria to tracheolar area, the insects in the low O_2 environment responded to O_2 stress by enhancing the metabolic structures and processes to match the demands of their environments. The important message of this study is that as organisms respond adaptively to various forms of stress, they must make tradeoffs that are likely to reduce fitness (quality) in other aspects of their

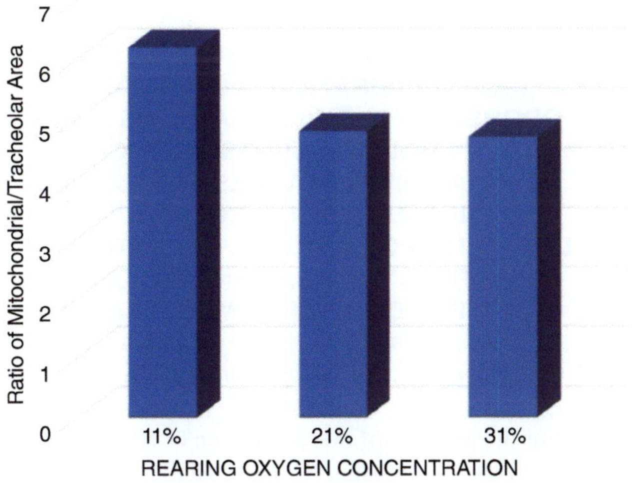

Figure 10.24 "Oxygen stress". The ratio of mitochondrial/tracheolar volume in *Drosophila melanogaster* reared at three levels of atmospheric oxygen. (The figure is redrawn from VandenBrooks et al. 2018.)

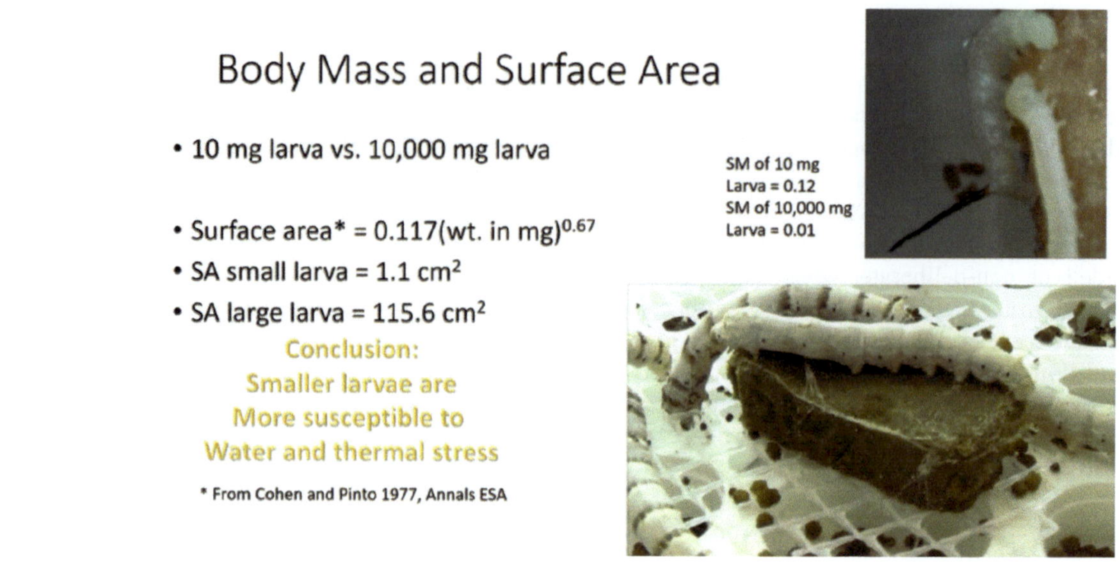

Figure 10.25 Stress from environmental factors.

morphology, physiology, and/or behavior. There is a significant cost of making more mitochondria and greater surface area and biomass of tracheoles (Figure 10.24).

Figure 10.25 shows the relationship between biomass (body mass) and surface area of insects subject to exchanges of heat and water with their environment. The smaller the insect, the greater the surface-to-mass ratio, which translates to heightened stress for smaller insects. The stresses can be from water loss or excessive heat transfer. The concept of stress relationships between surface-to-mass ratios is a pervasive theme in insect rearing, explaining many instances of mortality or reduction in fitness.

10.6 MISCELLANEOUS FORMS OF STRESS (SOUND/VIBRATION "POLLUTION"), ODORS AND CHEMICAL FACTORS, LACK OF NATURAL GRADIENTS, RANDOM OR REGULAR DISTURBANCES, DISRUPTION OF NORMAL SOCIAL CONDITIONS, DISRUPTION OF NATURAL CUES FOR LIFE HISTORY ACTUALIZATION, AND OTHER DISRUPTIONS OF HOMEOSTASIS

10.6.1 Disruption of Sounds/Vibrations

For the purpose of this discussion of sound issues in the insectary, I am using the expression, "soundscape." I am adopting the soundscape concept from several sources that treat sounds in natural habitats such as those described by Krause and Farina 2016 and Pijanowski et al. 2011. These publications deal with the concept that there is a complex presence of sounds that are part of the natural ecosystems of the world. Upon learning of the concept that there is an ecological importance to the soundscape's characteristics in given habitats, I began to think about how our rearing rooms have a variety of sounds, noises, and vibrations that are alien to the organisms that we extract from nature to rear in our insectaries or other artificial environments. This raises questions about how much the various species that we rear may be more or less affected by sounds.

Although one of the most familiar associations that we have with insects pertains to the sounds that they make (buzzing flies: Emily Dickinson's poem, "I heard a fly buzz when I died," the buzzing of bees, the chirping of crickets, etc.), little attention has been shown to the sounds of insects in a rearing situation. With this inattention comes the possibility that what might be a highly important component of our insects' homeostasis and normal functioning comes the potential stress or disruption of the normal biology that we inadvertently impose on our colonies of insects. What is brought to my immediate attention as I discuss this issue are the lacewings that were an important part of my own rearing research. In my efforts to develop diets and other rearing components for mass-rearing the lacewing *Chrysoperla rufilabris* (Neuroptera: Chrysopidae), I had not understood the importance of the mating calls discovered in various species of lacewings. In fact, the various species of lacewings from the genus *Chrysoperla* are so similar that the only non-genetic means of differentiating between them is by analysis of the mating calls. The calls are ultra-low-frequency vibrations called thrumming, a sound by abdominal movements of lacewing adults, calling to their mates. The sounds and the sonograms that visualize the sounds can be found on a site posted by Charles S. Henry, Dept of Ecology & Evolutionary Biology, University of Connecticut, CT and presented here: http://hydrodictyon.eeb.uconn.edu/people/chenry/Cryptic_songs.html#Top. Also, the details of the sound-based species identification are discussed by Henry et al. 1999a, 1999b among several papers on the topic of species-specific calls by lacewings. In the website that presents these amazing sound patterns, Henry writes:

> Below are the known song species of the *carnea* group of *Chrysoperla*, which are morphologically very similar to one another. Beneath each species name is an oscillograph and sonograph of the typical song of the species, recorded at 25 degrees C. The time axes are seconds, and the vertical axis of the sonograph (the lower of the two graphics) is in Hertz (NOT kilohertz, as in more familiar insect songs!). Click on the large-print, highlighted species name or any part of either graphic, and you can hear what these songs "sound" like (although you would have to put your ear almost on top of the singing insect, in order to perceive anything remotely like this in the field or laboratory). These substrate-borne signals are of very low frequency, so do not expect to hear much unless you use headphones, or have a subwoofer hooked up to your computer.

We have been aware of sound/vibration "pollution" in the insectary, and using a simple sound meter (SPER Scientific), we had measured baseline decibel readings of 40–50 dB in rearing rooms where fans, heaters, and other equipment items were turned off. However, when fans, heaters, and other standard lab equipment items were turned on we measured values as high as 65–75 dB. We did not measure vibrations reaching rearing shelves and containers, but it is likely that there are considerable vibrations that can be felt by the insects. There are various vibration-detecting devices on the market, ranging in price from $100 (US) to $10,000. Some of these devices may be used to develop QC ranges to help determine the degree of vibrational noise that may be long-term stress factors. Clearly, more study of this subject is needed—a topic covered further in the Chapter 13 (on "Future of Insect Rearing"). Recent evidence of the capability of sounds or vibrations to disrupt normal insect behavior is demonstrated by Avosani et al. 2020. These researchers showed that playbacks of mating calls (vibrations) could disrupt normal mating activities of *Bactericera cockerelli* (Triozidae: Hemiptera). This work, which seeks to deliberately disrupt normal behavior with vibrations or sounds, implies that inadvertent disruption or distraction from vibrations could affect mating or other aspects of behavior in laboratory-reared insects. This is certainly an example of an inadvertent stress factor that has not been considered adequately in rearing situations.

10.6.2 Disruptions or Stresses in the Odorscape

Conchou et al. 2019 present an excellent overview of the odorscape in nature:

> Olfaction is an essential sensory modality for insects and their olfactory environment is mostly made up of plant-emitted volatiles. The terrestrial vegetation produces an amazing diversity of volatile compounds, which are then transported, mixed, and degraded in the atmosphere. Each insect species expresses a set of olfactory receptors that bind part of the volatile compounds present in its habitat. Insect odorscapes are thus defined as species-specific olfactory spaces, dependent on the local habitat, and dynamic in time.

This abstract from Conchou et al. 2019 excellently summarizes the emerging discipline of insect odorscapes. I suggest extending the concept in the first sentence to include the importance of olfaction to include volatile and contact chemoreception in the whole ecosystem where sources of chemo-stimulation may be dead and decaying tissues from animals, for carrion-feeding insects, and various volatiles and contact stimuli for vertebrate blood feeders, etc. That being said, Conchou et al. are correct in expressing the wealth of and diversity of volatile compounds that fill the air in various terrestrial ecosystems. In fact, the concepts of the odorscape can be extended to the aquatic environment in a sort of "chemoscape" where dissolved chemicals can serve as attractants or other chemical signals for aquatic insects, as well as other aquatic organisms. I am borrowing the term, "chemoscape" from Villalba et al. 2019, who use the term to include nutritional compounds and plant secondary compounds in the setting of natural chemical effects, especially on ruminants. In the last sentence in the quotation above, the authors point out that insects, on a species-specific basis, express sets of olfactory receptors that bind the volatile organic compounds (VOCs) that are signposts of that insect's environmental cues—specific "signature" molecules that tell corn earworm adults that they are closing in on appropriate hosts for their oviposition activities. Likewise, boll weevils have receptors that are tuned to cotton plant volatiles, and mile-a-minute weevils are tuned to mile-a-minute plants—appropriate sites for their host plant selection for oviposition. The concept of odorscapes is most helpfully presented by Conchou et al. 2019:

> Manipulations of pest-insect odorscapes are a promising approach to answer the strong demand for pesticide-free plant-protection strategies. Moreover, understanding their olfactory environment becomes a major concern in the context of global change and environmental stresses to insect populations. A considerable amount of information is available on the identity of volatiles mediating biotic interactions that involve insects. However,

in the large body of research devoted to understanding how insects use olfaction to locate resources, an integrative vision of the olfactory environment has rarely been reached. This article aims to better apprehend the nature of the insect odorscape and its importance to insect behavioral ecology by reviewing the literature specific to different disciplines from plant ecophysiology to insect neuroethology. First, we discuss the determinants of odorscape composition, from the production of volatiles by plants"... "to their filtering during detection by the olfactory system of insects.... We then summarize the physical and chemical processes by which volatile chemicals distribute in space... and time... in the atmosphere. The following sections consider the ecological importance of background odors in odorscapes and how insects adapt to their olfactory environment. Habitat provides an odor background and a sensory context that modulate the responses of insects to pheromones and other olfactory signals. In addition, insects do not respond inflexibly to single elements in their odorscape but integrate several components of their environment. We finally discuss existing methods of odorscape manipulation for sustainable pest insect control and potential future developments in the context of agroecology."

The remainder of the quotation from Conchou et al. 2019 is devoted to explaining how a more granular and comprehensive understanding of the insect/odorscape interactions can be put to practical use in manipulating cropping systems (and possibly natural ecosystems) for the protection of those systems in a more benign manner that has been historically practiced such as with overuse of pesticides. In the remaining section of the abstract the authors discuss the practical advantages of studying and grasping the complex relationships of the insects' using their exquisitely fine-tuned detection systems to meet the olfactory map of the environment that the olfactory signals constitute. I have included the details of this abstract and further suggest careful reading of the entire paper for rearing specialists who seek to determine how much their rearing systems are departures from the natural highly complex olfactory environments to which the insects had evolved for tens of millions (or more) years. In reading the Conchou et al. paper and touching on the many papers in the references in this paper, it becomes compelling to me to think about how much we have abstracted the environment of the insects we rear in terms of the olfactory simplification we provide in our rearing systems. Not only have we set up systems that delete the olfactory complexities of the natural world and agroecosystems, but we have also replaced the odorscape of nature with an artificial odorscape of diet components that our insects would never encounter in the wild. While we try our best to give the insects food that is acceptable to them, we are missing a huge complexity of the natural environment and selecting insects that can survive in the unnatural world we have created for them (Cohen 2019). While we need to develop rearing systems as simplifications of nature, it would be valuable to delve as deeply as possible into the ways that we may have perverted the natural complexities and nuances of the insects we rear, and it would be helpful for the customers of our rearing system products to better appreciate how much our rearing process may be producing insects that do not reflect their natural counterparts—wild insects. A final thought about this subject: some insects seem to resist our efforts to simplify their existence and sustenance for our convenience. Even after 5,000 years of domestication, the mulberry silkworm (*Bombyx mori*) holds out in relinquishing its taste for mulberry. In all the efforts to develop artificial diets for silkworms, the only ones that work, even moderately well, are those formulations that contain at least token components of mulberry—specifically white mulberry (*Morus alba*).

10.7 CONCLUDING REMARKS ABOUT STRESS AND HOMEOSTASIS

In this chapter, the points that stress and homeostasis are interrelated in intricate ways. A way of looking at fitness is that it is the organism's ability to retain its homeostatic capabilities in the face of stress. The thrust of the chapter was to indicate the many facets of stress-response occur in insects in their natural context and in the rearing system environment. Furthermore, the fitness discussed throughout this chapter is inextricably woven with the quality of the insects. A challenging aspect of these realizations is for rearing specialists to dissect or parse the components or process steps in their rearing systems to determine how departures from optimal conditions induce stress, which, in turn, challenges the insect products' quality.

CHAPTER 11

Publishing Rearing Papers

11.1 SUMMARY AND PURPOSE OF THIS CHAPTER

This chapter suggests approaches and standards for insect-rearing papers, to make them more suitable and effective for publication. Guidelines are provided for publishing two kinds of rearing papers: (1) insect rearing-technology papers and (2) insect rearing-science papers. The technology papers offer rearing methodology or practical aspects of rearing system development and optimization. The rearing-science papers would include outcomes of rearing experiments developed with controls, variables and are based on explicitly stated rationale and explicit hypotheses. This paper also suggests welcoming papers based on discoveries of various aspects of feeding biology and treating scientifically rearing systems' components and properties. It is suggested that the treatment of insect-rearing systems should be modeled after the extensive and successful practices of the food science and food technology community (Cohen 2015). The food science model includes welcoming papers that do not necessarily involve living subjects. It is argued here that the outcomes of following these suggestions will greatly enhance the progress of insect-rearing inquiry, and this progress will support a much greater practical and basic science-based understanding of insect biology.

11.2 OVERVIEW AND HISTORY

The complexity and intricate ecological, physical, and biological relationships in insect-rearing systems suggest an excellent topic for scientific exploration. Insect-rearing systems are intricately interwoven networks or webs of materials that have complex interactions between components and within the framework of the target insects' biology (Cohen 2001, 2015, 2018a, 2018b, 2019). In fact, in this book and in Cohen 2018a, 2018b, rearing systems have been described as human-designed or engineered ecosystems or microcosms of an insect's world. This means diet, all environmental requirements (temperature, water, light, air, soil, and all other environmental factors), reproductive accommodations (mating space, oviposition sites), and social structure accommodations must be provided. Also, the rearing system context must include containment, and access for personnel to conduct their care in providing food, cleaning, harvesting, and any other special functions that must be engineered into the rearing process. In an early milestone paper Delcourt and Guyenot 1910 presented the foundation of insect-rearing systems: control. In fact, the silkworm rearing systems (described in the silkworm chapter in this text) are sterling examples of ways to achieve control in rearing systems. Throughout the development of rearing systems over the past century, the concept of control has been the driving force behind nearly all published rearing advancements. The rearing system concept entails an engineering basis, and all aspects of rearing systems lend themselves to control strategies that are in accord with the principles of systems engineering.

Over the more than 100 years since the Delcourt and Guyenot 1910 paper on control of insect rearing appeared, thousands of publications have been produced regarding the biology of reared insects. These papers contain various aspects of rearing systems detailing the tactics of providing micro-environments that provided temperature, moisture (humidity), light conditions, gas exchange, reproductive needs, protection from contaminating microbes, and various other requirements that are met in nature but must be deliberately provided for captive insects. Despite the large number of papers that treat some aspects of insect rearing, it is apparent to long-time practitioners of insect rearing, that the publications represent only a "tip of the iceberg" of all that is practiced in rearing facilities. There are several reasons for the gaps in the body of published rearing information: (1) in some rearing operations, innovations are proprietary. (2) Some organizations that develop and use rearing procedures are structured to solve problems but do not have a publication "culture." (3) In many cases, rearing innovations are considered tangential or incidental to the main purpose of a project, so the rearing part of a program gets neglected. (4) The infrastructure of journals and other publication outlets does not contain a formal structure for communicating rearing information. For Item (1), it is clear that for organizations that profit from efficient rearing of an insect, divulging their techniques would possibly threaten their advantage over competitors. This could include operations where the rearing is to produce insects as a product or where the insects are used for research for product testing. For Item (2), organizations such as the USDA, APHIS or USDA, Forest Service are charged with solving problems such as control of invasive pests, and their mission does not include transfer of technology or scientific understanding.

Regarding Item (3), development or adoption of a rearing system is generally considered an incidental part of the main program such as biological control, sterile insect techniques (SITs), or some other pest control process. Similarly, scientists who use insects as research subjects often regard the rearing process as a necessary but trivial inconvenience as an avenue to doing research with such disciplines as host plant resistance, genetic pest management, insect physiology/biochemistry, and numerous other research purposes. In my experience, many researchers who are diligent about experimental details such as electrophoresis gel characteristics or chromatographic solvent composition fail to consider their target insect's stress levels. The thinking of these researchers seems to be, "if my insect, then it is suitable for the experiments I am performing." Anyone familiar with the intricacies and details of insects' biological characteristics can recognize that this attitude is naive as to the many factors that can be manifested in an insect's biology, especially in light of the many potential stress factors that occur in rearing systems.

Item (4) (clearly expressed standards) is one of the most serious barriers to improved communication of rearing advancements: it is the lack of a clearly defined set of standards for rearing publications. Lacking such standards, researchers must design their studies according to their interpretation of papers they choose to use as models or they must invent novel designs, and in turn, editors and reviewers must judge the merits of the papers without clear guidelines or standards (Cohen 2001, 2015, 2018a, 2018b, and 2019). Authors, editors, reviewers, and readers are therefore left with questions as to whether a rearing contribution must be completely successful, novel, and of sweeping scope to be worthy of publication. By "completely successful," I am saying that many reviewers consider a paper unworthy of publication unless it presents a complete, de novo system of mass-rearing a species that has not been reared previously. This is in contrast with "**incremental accomplishments**," which offer improvements in existing systems or steps toward a completely developed mass-rearing system. The advancements in completely successful systems can be considered **saltatory progress** vs. **incremental progress**. Although there could be a problem with overloading the literature with trivial progress reports, there is another problem with overly stringent expectations that filter out accomplishments that may appear to be minor but actually have far-reaching potential (Cohen 2001, 2015, 2018a, 2018b, 2019).

The centerpiece of rearing systems is an appropriate diet. Diets must meet requirements of palatability, nutritional value, bioavailability, and stability (Cohen 2018a, 2018b). They are greatly influenced by nuances such as types of materials used in them and the processes used to make them. For example, Cohen 2015 has discussed at length important differences in the quality of diets produced with different wheat germ sources or differences in heating processes. Diets are technological constructs that may have been developed sometimes by heuristic procedures or sometimes by detailed engineering approaches (Cohen 2015). Since Bogdanow 1906 and 1908 reported an artificial diet for blowflies, thousands of insect diet papers have been published, and trillions of insects have been reared on artificial media. Diet-based mass-rearing programs include SITs for tephritid flies, screwworms, and pink bollworms, as well as tens of billions of feeder insects and insects for biological control. Also of great importance is the extensive research made possible by diet-based rearing techniques; this point cannot be overstated. Considering the successes of diets as centerpieces of rearing systems and the dependence of science and society on the capability to produce healthy insects on diets, it would seem self-evident that diet research would have a strong standing in the scientific community. However, despite the many successes in diet-based rearing systems, there remain many barriers to the research and reporting of diet accomplishments. Although several of the earliest papers—dealing with *Drosophila* diets—were published in our most respected journals (*Science*, *Nature*, the *Proceedings of the National Academy of Science*, 1910–1925), a cryptic trend emerged where insect diet papers and other rearing reports were not welcomed by journals. Over four decades, I have tried to understand why this trend developed and what could be done to correct it. These points are discussed extensively by Cohen 2019.

Based on my studies of rearing/diet papers and acquaintanceship with rearing/diet specialists, I conclude a major problem is that we lack clear standards upon which researchers can model their studies and that reviewers and editors can use to evaluate the reports. The problems are in four areas: (1) many diet papers lack a sound scientific or technological basis. (2) Many useful findings "fall through the cracks" and are never published or they appear only as cryptic information that is extraneous to diet and rearing advancements. (3) The audience or stakeholders of diet research accomplishments regard the findings **simplistically** and without respect for the difficulties of diet research and the potentially rich intellectual understanding which diet/feeding biology studies can provide. (4) The audience of diet research is chronically concerned that incremental accomplishments or negative data will be published and will somehow dilute the quality of entomology journals. The outcome of these concerns is that a great deal of information fails to be published, sometimes because of the threat of rejection, what I consider a "flinching" on the part of researchers who know in advance that if they try to publish a study that does not have sweeping success to report, the effort to publish will be treated with condescension and rejection (Cohen 2019).

11.3 CHANGING CURRENT PARADIGMS ABOUT INSECT REARING

The scientific community would profit from adopting some changes in how rearing publications should be conceived by rearing researchers and their audience, including editors and reviewers of rearing papers. Therefore, I suggest here that insect-rearing systems should be regarded as: (1) tools of inquiry about insect biology and (2) tools for study of rearing system components and their interactions.

1. **Tools of inquiry about insect biology**. This point is based on the reciprocal interactions between basic science of insects' biology and applied, technological accomplishments with rearing systems. For example, knowledge of feeding biology such as properties of peritrophic matrices or enzyme characteristics in extra-oral digestion inform diet development researchers about how incipient diets are being utilized by insect targets. A compelling example of the connection between the basic and applied science (technology) is to be found in the teachings of Vincent G. Dethier on the reflex of fly proboscis in response to stimulation be certain chemical stimuli, leading to technological outcomes such as development of diets for a wide variety of insects and a tremendous advancement of our understanding of complex sensory stimuli and feeding responses (Dethier 1976). The works of Hamamura 1959 and 2001 on silkworm diets are a wonderful testimony to the basic science/diet technology outcomes as are numerous works by Fraenkel, Beck, Schoonhoven, Chippendale, Vanderzant, and others who worked seamlessly and productively on basic and practical aspects of what I call "feeding biology" or "feeding science" (Cohen 2015).
2. **Tools for study of component and process interactions and effects** (i.e., study of the **physical/chemical properties** of diets as well as their **biological properties**. This potentiates studies of insect diets in parallel approaches to the highly productive studies of human foods and feeds for livestock and pets. This opens a place for formal studies of the science and technology of rearing systems. **Rearing system science** is the approach to understanding mechanisms, cause and effect relationships, and the discovery of the nature of rearing system components. **Rearing system technology** is the practical application of the scientific principles; it is the application and invention that stems from problem-solving inquiry. Although the relationships of or distinctions between science and technology can be philosophical and abstract, a working framework ascribes to science inquiry, causal relationships, predictions, and understanding of the principles of nature. Technology encompasses invention and practical ability to construct useful products—in this case, insect-rearing systems. Although commonly the terms "science" and "technology" are used as indistinguishable pairs, there are important commonalities and distinctions for our purposes in dealing with insect-rearing systems. In the process of inquiry about rearing system science and technology, the rationale behind a specific inquiry is important in both domains. Because science is concerned with discovery of cause and effect relationships, science-based rearing system inquiry should be hypothesis driven. However, for rearing-technology advancements (inventions), we are more concerned with what works for our practical purposes. Often, but not always, understanding of the scientific basis of a technological breakthrough helps researchers and inventors reach practical, heuristic outcomes; the science behind an outcome may not be testable and may not be required to use rigorous, repeatable techniques in developing a useful rearing system (or subsystem).

11.4 RECOMMENDED STANDARDS FOR REARING PAPERS

1. **Rearing-science and rearing-technology papers**. Rearing papers of two kinds should be welcomed: (a) rearing technology and (b) rearing-science. Rearing-technology papers would include descriptions of rearing-related processes, including innovations in diet synthesis procedures (size-reduction, mixing, heating, post-synthesis conditions, etc.), diet presentation procedures (forming diet configurations, packaging, preservation, etc.). The types of advancements made in deriving mass-rearing systems from laboratory scale systems should be welcomed here. An excellent example of this is the development of the twin-screw extruder system used for mass-rearing pink bollworms. This important innovation was published in an engineering journal (Edwards et al. 1996), rather than an entomological outlet. This point is further developed below in the **specific recommendations for the technology and science papers**. The role of the extruder as a revolutionary innovation in mass-rearing is discussed at length in this book in the chapter on Insect Rearing Systems (Chapter 3). Rearing-science papers are based on discovery of or amplification of our understanding of insect biology. The connection with rearing is that the discoveries of basic science phenomena and issues are derived

from controls and variables that are manipulated in the rearing system. Admittedly, this is a potentially vast topic. It's origins were seen, for example, when researchers such as Thomas Hunt Morgan and his laboratory team of researchers (A. H. Sturtevant, H. J. Muller, and C. B Bridges, known as the "fly boys") discovered phenomena such as mutations, chromosome structure, and gene linkage.

2. **Explicit rationale and hypotheses**. Both rearing-technology and rearing-science papers should be based on a clearly stated rationale, and science papers should be further based on clearly stated hypotheses. Four examples of clear statement of rationale are found in early papers on Drosophila rearing. Baumberger 1917a explained that because yeasts were found to be the principle natural food of *Drosophila* spp., he included them in his artificial diet for these flies. He also explained a rationale for using agar in diets because of the convenience of counting fly eggs in a semi-clear medium (Baumberger 1917b). Similarly, Guyenot 1913 explained the basis of using controlled conditions for rearing Drosophila spp., and he also explained his rationale for using live yeast as a rearing component, based on his observations that the fly larvae appeared to consume substantial amounts of yeast, thanks to his observations of the low population density of yeast cells in cultures of Drosophila with high populations of fly larvae (Guyenot 1914). It is also exemplary that although Baumberger had made similar observations of the larva/yeast relationship, Baumberger 1917a cited the earlier paper of Guyenot 1914, which is a proper treatment of credit for discoveries by previous authors.
3. **Physical/chemical/biological properties of rearing system components**. Papers should be welcomed without necessarily involving direct insect experimentation. A paper that deals with a diet's physical or chemical characteristics can be fully valuable without direct inclusion of insects. This characteristic is fully subscribed to in the human food community. Hundreds of food science and technology papers are published weekly that contain no direct involvement of human subjects. Food processing methods that protect known nutrients, improve texture (or other sensory and rheological qualities Cohen 2015), protect foods from microbial contamination, and other features are not subjected to tests with human (or other animal) subjects. The same should be applicable to insect diets, and the potential progress in dietetics would be immense. An important subset of this concept is that while we can gain a great deal from using food science and food technology literature, there are important aspects of insect diets that are unique to insect rearing. For example, insect diets are often unique mixtures of components that do not appear in human or livestock foods; furthermore, insect diets are often expected to last (remain stable) at rather high temperatures, high humidity, and for extended periods of time (often weeks). Indeed, it is common that rearing systems such as those devised for pink bollworms and tobacco budworms are expected to last for a 2–4 weeks development period (Cohen 2015). In extreme cases such as with tree-inhabiting species such as emerald ash borer, target insects are expected to feed on a single (unchanged) diet for more than six weeks (Keena et al. 2015). In addition to these issues, insect diets often serve as living places for target insects (with tunneling, use of diets and frass for refugia for pupation, or where frass and diets are closely associated). All these features lend themselves to the demand for (or clear benefit from) direct study of the characteristics and dynamics of insect diets, as well as other aspects of rearing systems.
4. **Incremental vs. saltatory accomplishments publishable**. Often important rearing discoveries appear to be incremental rather than saltatory advancements. Many editors and critics of insect-rearing papers have reasonable skepticism about the value of repeating previous works with only minor nuances of difference between the current work and what has already been thoroughly tested. However, it is often difficult and subjective to decide between trivial increments and important advancements. An excellent example of this issue is illustrated by the report of wheat germ in a diet for boll weevils (Vanderzant et al. 1959), which showed that except for a slight positive effect on oviposition, wheat germ did not seem to hold much promise as a diet ingredient. However, the next year when it was used in the diet of pink bollworms (Adkisson et al. 1960a), wheat germ proved to be a highly useful diet component, which, in fact, revolutionized diets for phytophagous insects.
5. **Papers with negative results**. Related to the points about incremental results are studies that offer negative results. Often negative results are not derived from flawed experimental designs, and the relentless denial of publication of papers with negative results has long been at odds with a major purpose of publication: the communication of important information that has been obtained through careful study. Automatic denial of papers that demonstrate only negative results not only hides important information from future researchers, but it also encourages authors to disguise or exaggerate results so that they can be viewed as positive. This point is discussed in detail by Cohen 2001.
6. **Utilization of exploratory statistics in addition to confirmatory statistics**. Statistically based strategies have long been used by engineers and food scientists to improve their systems, and recently diet researchers have utilized this approach productively (Lapointe et al. 2008; Assemi et al. 2012). As I discussed in a summary on development and improvement of diets (Cohen 2015), this approach has great potential for using statistical thinking to efficiently optimize diet development. Also, experimental packages such as

design of experiments (DoE) in SAS and JMP systems can be used for process control and quality control, and if these approaches are used early in the process of diet development they can become an excellent "package" of empirically derived, well-vetted process control for diet production in mass-rearing systems. This topic is covered at length in this book in the Chapters 8–10.

7. **Microbiomes and microbially-processed diets**. Increasingly, procedures for determining the populations of microbes in insect microbiomes have been enhanced with molecular techniques. These techniques lend themselves to highly productive studies of the roles of various microbes in domesticated vs. wild (or feral) insects' natural biota and the changes in this microbiome that result from artificial diet treatments. Furthermore, the literature on the involvement of various microbes in processing diet ingredients is ever-increasing and promising for greatly improved diet components (Cohen 2015, 2018a, 2018b). Many robust techniques have become available in the last decade where molecular approaches make the identification and quantification of microbiota possible, often without the demands of culturing microbes that are fastidious and had previously defied identification (Cohen 2015).
8. **Various aspects of palatability, nutritional value, bioavailability, and stability of diets**. These topics offer a rich potential for inquiry into what makes standard and "exotic" diet components useful and potentially enhancing to incipient diets. This certainly includes such topics as the character of diet matrices, structural properties of useful diets, and especially bioavailability. Once we have come to understand that failures in rearing are often attributable to problems with palatability, bioavailability, and stability, we can gain useful perspective about the relative contributions that nutrition vs. these other three factors make in rearing successes and failures.
9. **Unusual or novel subjects such as using insects for treatment of wounds or papers that advocate using insects as human food**. As treated in the chapter on "Who's Who in Insect Rearing?" several insectaries are now established to produce clean, available, and carefully packaged fly larvae for wound cleaning (maggot therapy). Also, a growing body of information is emerging in refereed publications about production of insects intended for use as human food. Such papers could be useful if they are based on sound science and technology principles such as the study by Lund and Parrella 2015. Furthermore, other novel uses of insects may include such applications as using insects to detect explosives, insects as production systems (bioreactors) for novel pharmaceutical products such as human proteins (Cohen 2015 reviews several such papers). Other uses for insects such as decorations and celebrations are in some subcultures a considerable enterprise.

11.5 REARING TECHNOLOGY VS. REARING-SCIENCE PAPERS: SPECIFIC RECOMMENDATIONS FOR TYPES OF PAPERS THAT WOULD BE VALUABLE TO THE SCIENTIFIC COMMUNITY

11.5.1 Rearing-Technology Papers

Although several rearing publications are of the type that I am calling "technology," too often very valuable technology (sometimes considered "methods") information is never published or is hidden so that the advancements are unavailable to our community. Over my 40 years of study of diet development, I have become aware of the loss of valuable information where researchers had ideas, often based on valuable experience, but were discouraged by editors and reviewers from expressing the rationale behind innovations or improvements in insect diet and other aspects of rearing techniques.

11.5.1.1 Examples of Rearing-Technology Topics That Merit Publication and Wide-Spread Attention

11.5.1.1.1 Studies of Biotic and Abiotic Rearing System Components

Cohen 2001, 2003, 2015, and 2019 has argued that it would greatly benefit the entomology community if journals treated insect-rearing papers the way food science and food technology papers are treated. That is, in food science and technology, many papers are published that do not use human subjects to validate the research being reported. For example, hundreds of studies per year appear in journals dealing with human food science and technology. Papers abound in topics such as effects on texture or nutrient stability of various processes and additives. Processing practices (such as extrusion and microwave heating) are studied routinely, with what the food science community considers publishable work. Effects of various processes on ascorbic acid or other vitamin content are perennial topics of interest in food science and technology journals. In contrast, in the entomology literature, there is a huge gap in what can be called "abiotic" studies—research on characteristics of diet or other rearing system components. I present here a few examples of topics that would be of great value to explore with scientifically-based research but would likely be rejected by entomological publications' editors and reviewers.

11.5.1.2 Characteristics of Gelling Agents (An Example of Inquiry into Abiotic Characteristics)

Nearly everyone who has prepared gelled diets must have noticed that the nature of the gelled diet is often different from batch to batch. Sometimes, the diet texture is crisp and firm. Other times, the diet comes out mushy. Those diet-makers who are very observant often notice that there are degrees of diet texture outcomes that are somewhere in-between very firm and too mushy to properly be called a gel. For our current discussion, we can dismiss the question of "does the firmness/mushiness continuum affect the insect?" The question here is "what influences or causes the gel to be more or less firm?" Using the rheological tools (borrowed from food science), we can do simple tests of gel firmness such as penetrometers discussed throughout this book (for example, in Chapter 9). To start an inquiry into the factors that cause differences in gelling characteristics in insect diets, we can propose several questions. **Question 1**: Is the manufacturer (or supplier) of the gelling agent—starting with agar for this example—an influence on gel strength? If the rationale is to use a penetrometer to determine gel strength, and the hypothesis is that gels derived from different manufacturers will display different gel strengths, we can begin our testing. For the current experiment, I used five agar products that I had in my laboratory from five suppliers. I will not report the manufacturers' name in association with the products, as the purpose of this discussion is not to provide a critique of different agar products but rather to show that there are differences that need to be understood. The tests were conducted where a 1% mixture of agar from each supplier was heated to boiling to activate the agar; then cooled set gels were analyzed with a Wagner penetrometer. The brands were labeled "A" through "E." The results of this analysis are presented in Figures 11.1 and 11.2.

In related questions, the type of gelling agent could be expected to influence the texture (gel strength in the present discussion) of the diets in question. Three common gelling agents, agar, carrageenan (kappa type), and gellan gum, are all what we can call for purposes of this discussion, primary gelling agents. We also can include in an inquiry about gelling characteristics the effects of possible gelling **synergists** and **antagonists** (discussed by Cohen 2015). Briefly, various substances that are used in diets commonly, and other substances, not conventionally used in artificial diets may act in ways to enhance the gelling characteristics of the primary gelling agents (*synergists*); while other substances may interfere with the gelling potential of primary gelling agents. We call the latter substances *antagonists*. This topic is the subject of numerous studies in food science journals (Tang et al. 1996; Mao et al. 2000; Lee and Lucey 2004, 2010; Mende et al. 2013; Sun et al. 2018).

In an experiment, using the DoE approach, a series of diets being developed for emerald ash borers were tested with the goal of maximizing the hardness as measured by gel penetration force (Figures 11.3–11.5). The diets contained the three primary gelling agents, agar, carrageenan (Gelcarin 812), and gellan gum. Each of these gelling agents was combined with one of four hypothesized gelling synergists, locust bean gum, xanthan gum, alginate, and guar gum. The diet formulations are shown in Table 11.1.

The results of this profile show clearly that the strongest gels in terms of penetration force were ones made with agar, while the weakest ones were based on gellan gum. The best synergist was locust bean gum, with alginate and guar gum being similarly weak, and xanthan gum being the least effective synergist. These two examples (testing the brand of gelling agent = agar) and testing kinds of primary gelling agents vs. potential synergists suggest only a few of the questions that could be asked of various rearing-system components—in this case, diets. In each of the cases considered here, ***abiotic*** components of the rearing system were tested without recourse to bioassays. Other kinds of discovery about the core nature of our rearing systems can be exploration of such system components as microclimate, materials treatment of oviposition

Table 11.1 Diets Tested for Penetration Force (Synonymous with Gel Strength)

Diet Components	Diet 1	Diet 2	Diet 3	Diet 4	Diet 5	Diet 6	Diet 7	Diet 8	Diet 9	Diet 10	Diet 11	Diet 12
Soy flour	10	10	10	10	10	10	10	10	10	10	10	10
Saw dust	18	18	18	18	18	18	18	18	18	18	18	18
Starch	3	3	3	3	3	3	3	3	3	3	3	3
Agar	1.5	1.5	1.5	1.5								
Carrageenan (Gelcarin 812)					1.5	1.5	1.5	1.5				
Gellan gum									1.5	1.5	1.5	1.5
Locust bean gum	0.75				0.75				0.75			
Xanthan gum		0.75				0.75				0.75		
Alginate			0.75				0.75				0.75	
Guar gum				0.75				0.75				0.75
Water	70	70	70	70	70	70	70	70	70	70	70	70
Sucrose	3	3	3	3	3	3	3	3	3	3	3	3

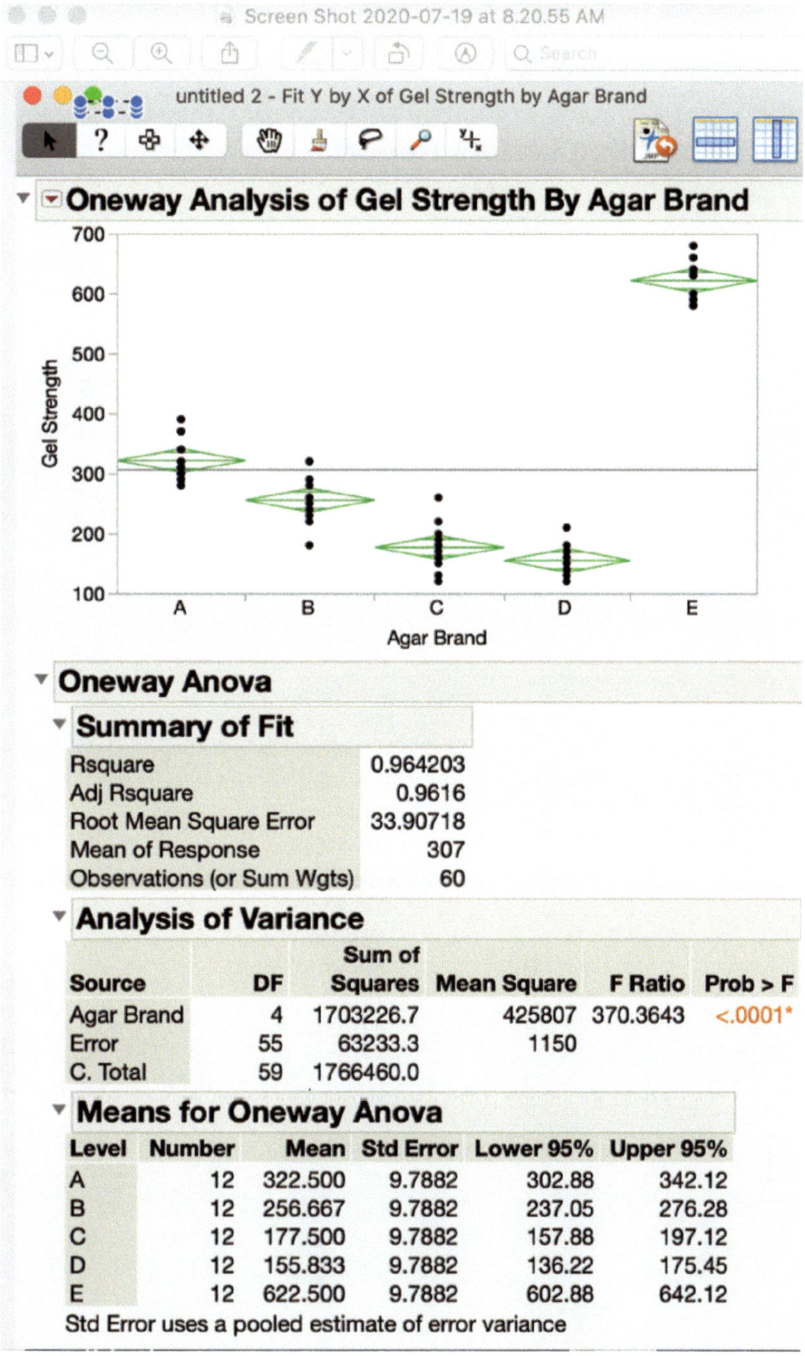

Summary of Fit

Rsquare	0.964203
Adj Rsquare	0.9616
Root Mean Square Error	33.90718
Mean of Response	307
Observations (or Sum Wgts)	60

Analysis of Variance

Source	DF	Sum of Squares	Mean Square	F Ratio	Prob > F
Agar Brand	4	1703226.7	425807	370.3643	<.0001*
Error	55	63233.3	1150		
C. Total	59	1766460.0			

Means for Oneway Anova

Level	Number	Mean	Std Error	Lower 95%	Upper 95%
A	12	322.500	9.7882	302.88	342.12
B	12	256.667	9.7882	237.05	276.28
C	12	177.500	9.7882	157.88	197.12
D	12	155.833	9.7882	136.22	175.45
E	12	622.500	9.7882	602.88	642.12

Std Error uses a pooled estimate of error variance

Figure 11.1 Analysis of variance (ANOVA) of mean gel strengths for five brands of agar.

components, and countless diet materials and diet processing steps. The model of human food science and technology should stand as an example of how highly beneficial knowledge can be gained without the necessity of having insect subjects. In cases where the rearing research and optimization is directed at microbial components of the rearing system (with microbes that may be symbiotic, commensal, parasitic, or contaminants), there is what we can consider an intermediate organism (the microbes) that exist in between or in association with the insects in their abiotic setting. There exists in the literature on food science, a wealth of information about "food safe" (= GRAS or generally regarded as safe) antifungal and other antimicrobial materials (see Fellows 2009, Stark and Tan 2003, Dijksterhuis et al. 2013, Silva and Lidon 2016, and Leyva Salas et al. 2017) that can have useful applications for rearing systems, especially insect diets and container-based frass microbial management. All of these compounds and

Screen Shot 2020-07-19 at 8.22.22 AM

Connecting Letters Report

Level					Mean
E	A				622.50000
A		B			322.50000
B			C		256.66667
C				D	177.50000
D				D	155.83333

Levels not connected by same letter are significantly different.

Figure 11.2 Connecting letters report for mean gel strengths of the five brands of agar. Clearly, brand E is far superior in gel strength than any of the other brands, and brand A is the second best agar in terms of gel strength. Brands D and C were less than a third as effective at forming firm gels as brand E. If gel strength is important to a given species of insect being fed in a rearing system, then the brand of agar could be of significant consequence to the rearing operation.

Level										Mean
d1	A									5092.9625
d5		B								3442.4654
d12			C							2994.4733
d4				D						1933.4395
d3				D						1791.9683
d6					E					1343.9762
d8						F				1123.9099
d7						F	G			990.2983
d2							G	H		856.6866
d10								H		789.8808
d9								H	I	697.5315
d11									I	624.8310

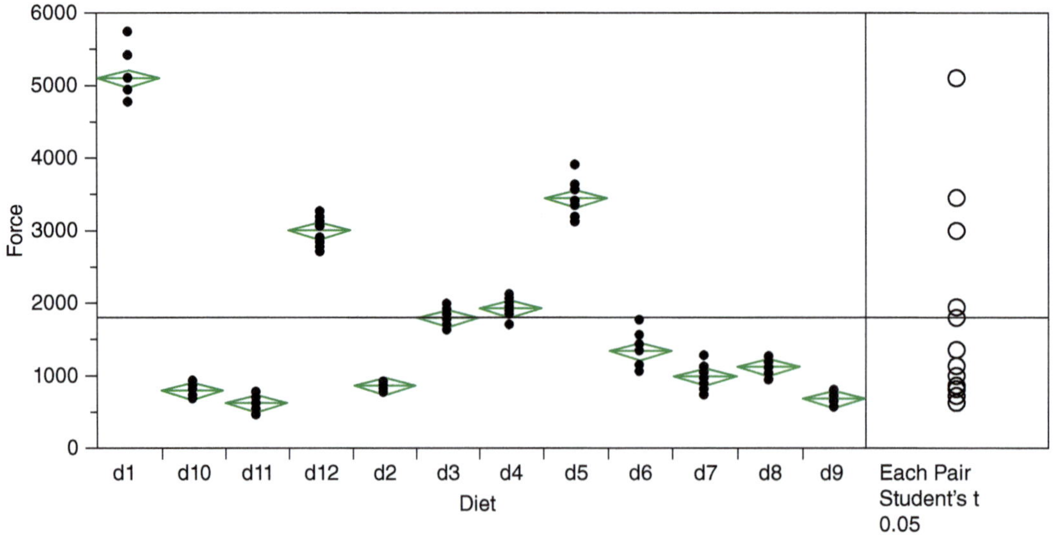

Figure 11.3 Comparison of penetration force in 12 diets whose primary gelling agents were agar, carrageenan, and gellan gum, and whose putative synergists were locust bean gum, xanthan gum, alginate, and guar gum. The results of the ANOVA are shown below, and the relationships between diet means are shown in the connecting letters matrix shown above. To help clarify the complex results of these tests, the full-factorial test was done as shown in Figure 11.4.

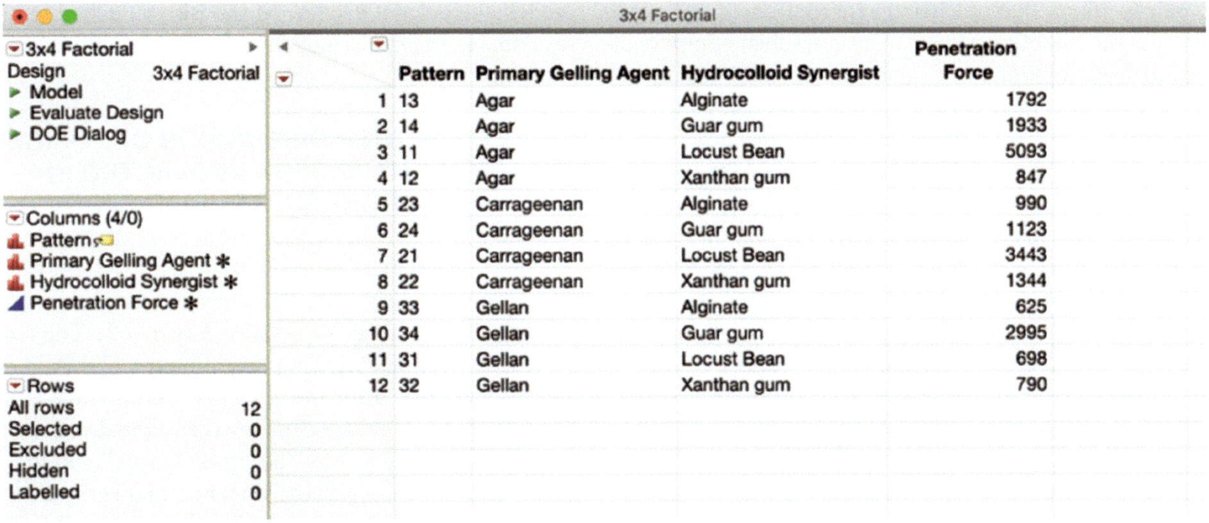

	Pattern	Primary Gelling Agent	Hydrocolloid Synergist	Penetration Force
1	13	Agar	Alginate	1792
2	14	Agar	Guar gum	1933
3	11	Agar	Locust Bean	5093
4	12	Agar	Xanthan gum	847
5	23	Carrageenan	Alginate	990
6	24	Carrageenan	Guar gum	1123
7	21	Carrageenan	Locust Bean	3443
8	22	Carrageenan	Xanthan gum	1344
9	33	Gellan	Alginate	625
10	34	Gellan	Guar gum	2995
11	31	Gellan	Locust Bean	698
12	32	Gellan	Xanthan gum	790

Figure 11.4 Table for full-factorial analysis of gelling agents.

Scaled Estimates

Nominal factors expanded to all levels

Term	Scaled Estimate	Std Error	t Ratio	Prob>|t|
Intercept	1806.0833	.	.	.
Primary Gelling Agent[Agar]	610.16667	.	.	.
Primary Gelling Agent[Carrageenan]	-81.08333	.	.	.
Primary Gelling Agent[Gellan]	-529.0833	.	.	.
Hydrocolloid Synergist[Locust Bean]	1271.9167	.	.	.
Hydrocolloid Synergist[Xanthan gum]	-812.4167	.	.	.
Hydrocolloid Synergist[Alginate]	-670.4167	.	.	.
Hydrocolloid Synergist[Guar gum]	210.91667	.	.	.
Primary Gelling Agent[Agar]*Hydrocolloid Synergist[Locust Bean]	1404.8333	.	.	.
Primary Gelling Agent[Agar]*Hydrocolloid Synergist[Xanthan gum]	-756.8333	.	.	.
Primary Gelling Agent[Agar]*Hydrocolloid Synergist[Alginate]	46.166667	.	.	.
Primary Gelling Agent[Agar]*Hydrocolloid Synergist[Guar gum]	-694.1667	.	.	.
Primary Gelling Agent[Carrageenan]*Hydrocolloid Synergist[Locust Bean]	446.08333	.	.	.
Primary Gelling Agent[Carrageenan]*Hydrocolloid Synergist[Xanthan gum]	431.41667	.	.	.
Primary Gelling Agent[Carrageenan]*Hydrocolloid Synergist[Alginate]	-64.58333	.	.	.
Primary Gelling Agent[Carrageenan]*Hydrocolloid Synergist[Guar gum]	-812.9167	.	.	.
Primary Gelling Agent[Gellan]*Hydrocolloid Synergist[Locust Bean]	-1850.917	.	.	.
Primary Gelling Agent[Gellan]*Hydrocolloid Synergist[Xanthan gum]	325.41667	.	.	.
Primary Gelling Agent[Gellan]*Hydrocolloid Synergist[Alginate]	18.416667	.	.	.
Primary Gelling Agent[Gellan]*Hydrocolloid Synergist[Guar gum]	1507.0833	.	.	.

Prediction Profiler

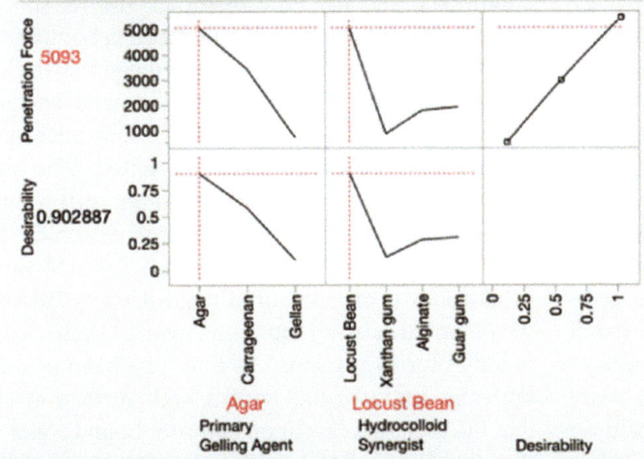

Figure 11.5 Prediction profiler (from SAS, JMP 14.0).

countless others that are being reported daily and coming from the literature on novel antimicrobial compounds derived from a plethora of organisms must be tested: (1) for their effectiveness against the microbes being targeted for control and (2) for their safety with the insects that we are trying to protect. This leads to the following point about the ultimate need to test innovations in all aspects of rearing with insect subjects. Note: in this discussion, for simplicity, gel strength (measured as pressure required to puncture the gel or to make it fail) has been the focus. However, in many studies of food rheology (such as Tang et al. 1996), other characteristics were also considered important such as brittleness, elastic properties, extensibility, and other rheological (textural) qualities. All of these qualities deserve further study with insect subjects to establish, on a per insect basis, the importance of these sensory qualities of insect diets. This topic is discussed further in Chapter 13 on future issues in insect-rearing systems and was the subject of a chapter on rheology in Cohen 2015.

11.6 BIOTIC CHARACTERISTICS AND BIOASSAYS

Clearly, at the heart of all of our rearing systems is the insects that we are culturing. Therefore, once the abiotic explorations reveal the nature of the factors in question, further explorations are in order to discover how the new knowledge and understanding apply to the target insects. For example, once the differences in texture are discovered in rheological terms such as resilience, hardness, viscosity, gumminess, cohesiveness, and stringiness, the path to discovery of the acceptable and optimal range of these qualities to target insects can be established. As discussed by Cohen 2015 and 2019, how much of a diet's palatability is determined by the sensory properties of texture vs. those of chemical sensory stimulation? If ascorbic acid is a feeding stimulant, can a diet that is mushy vs. crisp override the stimulus effects of the ascorbic acid? In humans' sensory responses to various foods, it is clear to everyone that a sausage with the texture of yogurt would be considered unpalatable. However, the range and depth of studies that would shed useful light on the nature of rearing system components suggests a rich array of studies that should be publishable in entomology journals—without the use of insect subjects. Throughout the history of publications on insect rearing, the convention has been to publish papers with insects as the primary focus as research subjects. It is obvious that the "insect-centered inquiry" remains a crucial part of potential future lines of study and discovery. However, to restate a central point, once we know that a given nutrient or other rearing system factor (such as adequate oxygen supply or appropriate temperature) is established, exploration of rearing materials and processes does not need to be slavishly attached to bioassays during the open-ended process of novel discovery.

11.7 FURTHER DISCUSSION OF TOPICS THAT HAVE BEEN NEGLECTED IN INSECT-REARING PUBLICATIONS

The development of the mass-rearing system for pink bollworms (*Pectinophora gossypiella*) was used in the successful SIT (discussed in detail in Chapter 3). The rearing system development accomplished by the USDA, APHIS and cooperators from the growers' community and the State of California Department of Food and Agriculture. The basic diet that was the "grandparent" of the currently used medium was the wheat germ diet described by Adkisson et al. 1960a (with roots in several earlier efforts to rear pink bollworms on artificial diet). In a chapter on pink bollworm rearing, Stewart 1984 described the rearing technology that the USDA, APHIS facility in Phoenix, AZ, USA had been using for a SIT program for 14 years prior to this chapter. In that chapter, Stewart mentioned that the diet that facility used was essentially "from Ouye 1962." However, examination of the Ouye paper reveals that several components were added to the 1984 diet. It is also important to recognize that the Ouye paper was the pioneering introduction of deliberate use of antimicrobial compounds to control bacterial and fungal infestations in diets. Careful reading of the Stewart 1984 chapter reveals many topics that were worked out in the development of the "pinkie" rearing system but whose basis in terms of rationale is not explained. This means that when other researchers are trying to solve the same kinds of problems, they must "reinvent the wheel" with each problem.

In line with this meticulous reporting of diet treatment procedures, factors had to be established for precise replication. This included such factors as the heating temperatures for making diet, the duration of heating, mixing conditions, and the type of mixer and its speed. A diet mixed at 1,000 rpm in a steam kettle is bound to have different characteristics than a diet blended at 10,000 rpm in a high-speed blender. Likewise, the particle size of solid components of diets such as lima bean meal and wheat germ had to be established and vetted. The size reduction was accomplished with a hammer mill where either a 10.0-μM sieve was used, making an extremely fine flour. This was in contrast to the use of a 500-μM sieve, which produced a coarse-textured meal. Cohen 2015 discussed the outcome of using lima bean meal of these two extremes in a diet for *Lygus* spp. The fine lima bean powder milled with the 10.0 μM had such a high surface to mass ratio that the diet made from this powder bound water so tightly that the diet had to be diluted with extra water to make the diet capable of being poured into diet feeding packets. The original diet 2,000c was made with lima bean meal that was made from grinding dried lima beans in a household coffee grinder. The modification made with an industrial capacity hammer mill was initially processed with a 10.0-μM sieve until it was discovered that the diet failed to elicit fitness levels previously established for *Lygus* bugs 2,000c. Once the research team learned that the culprit in the rearing system deterioration was the particle size of the lima bean meal, the 500-μM sieve was substituted, and rearing production numbers soared (Cohen 2015). This is an example of how a seemingly trivial aspect of a rearing system can drift inadvertently and unnoticed into unacceptable processing conditions. In well-established and highly productive research systems, hundreds of such changes or corrections must be made as was the case for the pink bollworm rearing described by Stewart 1984, documented below.

The rearing technology described by Stewart 1984, including the diet formulation information, remained essentially unchanged until the "game changing" paper by Edwards et al. 1996. The Phoenix pink bollworm team, led by Mr. E. Miller and the Edwards team from a USDA, Agricultural Research Food Technology Laboratory in Albany, CA, joined efforts to mass-produce pink bollworm diet using state-of-the-art technology from food science: the twin-screw extruder. The Edwards et al. 1996 paper describes the engineering modifications that were developed specially to produce a high-moisture product (something unusual in the conventions of food processing) and the final version of the diet (which continues to be used today).

The Edwards et al. 1996 paper is an excellent example of a diet-technology publication. It is significant that this paper was published in an engineering journal, rather than an entomology journal. It is of further interest that the paper has been cited only three times since its publication 20 years ago, and only one of the citations is in an entomological context (the other two in a food engineering context). This paper did not include experiments with insects, but instead the physical and chemical properties of the diets and the conditions for processing the diets are presented in excellent detail. For example, this paper provides key information about the effects of the extruder processing on thiamine content, microscopic structure of diet, texture, moisture content, and various other diet characteristics in association with processing steps. It also provides specifications of temperatures, pressures, mixing, and flow rates of diets being processed by the new extruder technology. The information, as presented, is adequately instructive so that researchers can adopt the techniques to apply to their own diet development objectives, for other insects, for example.

Over nearly 40 years of pink bollworm diet development and improvement, only a sketchy treatment of the process can be found in the literature, and many of the major innovations remain undocumented or untraceable. In fact, I argue that the adoption of soy flour and the application of extruder technology are two of the most important innovations or improvements of mass-rearing systems. The topic of microbial management in pink bollworm rearing is treated in several papers including the Ouye paper and another one by Fred Stewart et al. 1976, but the other aspects of rearing, especially ones regarding diets, remain obscure. It is also important to note that the *Challenges in Insect Rearing* book edited by King and Leppla (where Stewart's chapter appears) is no longer in print, and as a USDA publication never was as available to the scientific community as are periodical publications such as ESA journals. A further fact of relevance is that the King and Leppla book is one of several government publications where rearing technology is characterized, and though the publication of diets and other rearing accomplishments helps meet some of the needs to make available discoveries that help furnish the scientific community with useful information, the recourse to only "in house" documents that lack the public attention and often lack the vetting that refereed journal provide hinders the potential of important discoveries to reach appropriate audiences. This is not to say that the "in house" USDA publications did not do a valuable service, but the restrictions in reach to a world-wide audience must be recognized as drawbacks to expectations that the valuable information that I describe here would greatly serve the scientific community. It is also noteworthy that along with the government publications described here, there are a few monographs produced and distributed by no-government organizations such as the United Nations, FAO, and IAEA. Some of these publications present excellent explanations of rearing technology. One of the truly outstanding monographs in terms of technology of diets and other rearing components is the book on codling moth rearing by Dyck 2010. Authors would be well served to model after the Dyck monograph their explanations of diet technology. However, for diet-technology papers to be highly available for users in the rearing community, these documents need to be shorter than the nearly 200 pages of the Dyck codling moth book 2010.

11.8 FURTHER SUGGESTIONS FOR TREATMENT OF REARING-TECHNOLOGY PUBLICATIONS

This leaves the specifications of how rearing-technology papers should be treated by authors and reviewers. I suggest as a model of technology papers, the format of (a) process control and (b) standard operating procedures. A framework that includes the steps in the diet-making procedure might include the protocols and standards for each step. One useful organizational scheme could be the list and specifications for the equipment and supplies used in the diet. The other part of a diet-technology paper would be a presentation of the process steps from the beginning to the end of diet production. I realize that the details of this kind of approach can become excessive or they can be too sparse to be useful. However, a realization of the possible error-inducing outcomes makes me suggest more specification rather than less can be useful to the paper's readers. Drawing from the chapter by Stewart 1984 on pink bollworm production, I will illustrate this point regarding diet processing here: (1) diet ingredients are listed in a generic form. It would be useful to provide specific information about the specifications of diet components. For example, the Pink Bollworm Facility has developed an elegant gel strength testing process where specific amounts of agar mixed and heated in water are measured routinely whenever a new batch of agar is being considered. The Facility has a strict range of acceptable standards in terms of grams of penetration force per cm^2 required to puncture the gelled agar under standard measurement (temperature) conditions. Batches of agar that do not meet their standards (fall within the specified range) are rejected. The second ingredient listed by Stewart 1984 is toasted soy flour. However, there are many soy flour products that can be considered to be "toasted," but they may vary in terms of the amount of toasting, mesh (= particle size), whether they are defatted or re-fatted, and even protein content may vary from one brand to another. Even within brand variation is possible, and it would be helpful if the specifications were explained and if a range could be given for each possible variable. Similar comments apply to the wheat germ mentioned by Stewart as component 4 in the diet.

Some suggested examples of topics for technology papers:

1. Physical outcomes of processes (effects of heating on texture, mixing and particle distribution, solubility and distribution of diet components, shaping diet)
2. Chemical outcomes of processes (effects of heating or storage conditions on antioxidants, effects of heating, cooling, storage on stability of diets and diet components, interactions of diet components such as chelating agents and mineral availability). Examples of this type of paper are Edwards et al. 1996 and Cohen and Crittenden 2004
3. Container/packaging characteristics and diet qualities: gas exchange and water loss, light penetration and diet stability, containers or packaging with antimicrobial effects such as chitosan membranes for diet coating, waxes for diet coating. Excellent examples of this type of research can be found in a series of papers and technical monographs that came from laboratories such as the USDA, ARS laboratories at Tifton, Georgia (USA), Mississippi State University, and the Stoneville, Mississippi laboratory. A very influential example of these techniques is described by Tillman et al. 1997
4. Comparisons of conventional vs. novel materials in diets (chemical and physical comparisons of diet components such as raw vs. toasted soy flour; defatted, full-fat, re-fatted soy flour)

11.9 CONCLUDING COMMENTS ON REARING-SCIENCE PAPERS

There are two major categories of "rearing-science" papers that should be pursued: (1) rearing discoveries that were arrived at by scientific approaches and (2) basic science papers, with a rearing basis, that are applicable to the base of knowledge helpful in our enhanced understanding of the insects' biology. In the first category, the use of sound and explicitly explained rationale and hypotheses in the formulation of new or novel or improved rearing systems should be welcomed. The second category includes a wide range of scientific inquiries about such topics as insect feeding biology, reproductive dynamics (mating conditions, intricacies of the reproductive system under controlled conditions, and oviposition), developmental biology, and features that control diapause as a few examples of how rearing system development support basic biology studies. As an example of the type of scientific outcomes resulting from studies using insect-rearing systems, the famous geneticist and evolutionary biologist T. Dobzhansky and his research team conducted scores of studies using *Drosophila* species—papers that explored the genetic basis of evolution. One such study that stands as a most important demonstration of the interaction of genes (genotypes) and environmental factors is a paper by Dobzhansky and Spassky 1947 where the authors used *D. pseudoobscura* collected from various habitats. They selected varieties (races) of *D. pseudoobscura* where there were clear genetic markers and partial chromosomal mapping available. They placed the insects in containers where there were mixtures of the different genotypes, and they held some of the containers at standard rearing temperature and other containers at lower temperatures. After several generations of the mixed rearing at "normal" and cold-stressed temperatures, the authors found that there were shifts in the population numbers of the different genotypes. This demonstrated that under controlled rearing conditions, evolutionary shifts (micro-evolution or changes in gene frequency) could be made to occur. Furthermore, a wealth of papers on genetics discoveries are discussed above in the section above on "Recommended Standards for Rearing Papers." Many papers that were based on scientific principles were an essential part of the development of basic ecology knowledge—a concept discussed at length in the chapter on "Rearing Systems as Ecological Niches or N-Dimensional Hypervolumes."

Diet formulation papers are the practical or applied side of insect dietetics science. The papers in this domain report formulations (often including processing practices such as use of twin-screw extruders or flash sterilizers in diet preparation) that are useful in rearing target insects. An example of the type of paper that exemplifies this approach is the Adkisson et al. 1960a paper where wheat germ was introduced for rearing pink bollworms. The Adkisson et al. 1960a paper meets at least some of the standards of a "scientific paper" by using a control diet (the established casein diet) and introduces wheat germ as a replacement for casein and several other components. The authors collected data on rearing outcomes with the control diet, two variable-containing diets, and even some outcomes with natural host plant. While the importance of this paper cannot be over-emphasized, I might add that it would have been even more helpful to the entomological community to have a clear statement of the **rationale** for using wheat germ. The authors mention that the new wheat germ diet contained fewer added ingredients than the standard: casein diet. They also point out that the wheat germ diet gave results that were equal to or superior to the standard, control diet based on casein. However, they did not explain the expectations (**hypotheses**) regarding wheat germ. Retrospectively, several researchers (including Erma Vanderzant) explained many of the qualities of wheat germ that helped explain its power as a keystone insect diet component, but an explicitly stated rationale and the hypotheses stemming from the rationale (within the original paper) would have been very informative and would have helped shape mechanism-based inquiry about wheat germ in its various forms (raw, rolled, toasted, coarsely ground, finely ground, etc.) and possible substitutes for wheat germ such as various other grain products.

An important difference between diet-technology papers and what I am calling rearing-science papers is that the technology papers report methods that are of interest because they "work." That is, they are practical, and the techniques may have been developed, for example, by heuristic practices (Cohen 2015). This is to say, that they are empirically vetted, but they do not necessarily stem from application of the scientific method. They may have been derived from procedures that did not necessarily have controls and variables as would be the case with scientific discoveries. Furthermore, technology techniques should reflect data-based treatments of procedures or components that can serve as and lead to explicit process control standards. Generally, the importance of this type of treatment is overlooked or treated minimally. Even when product specifications, including brands, are reported, there is often so much variability or ambiguity in diet and often other rearing system components reporting that it becomes difficult or impossible to repeat the conditions that shaped the original findings. Many examples of this are available, but one of the most striking is in ambiguities of the common insect diet components such as soy products.

11.10 THE MODEL OF INQUIRY INTO SOY PRODUCTS IN INSECT DIETS

The use of soy components in insect diets mentioned in Western literature seems to have begun in the 1960s (e.g., by Redfern 1963 and then by Shaver and Raulston 1971). The use of soy in Asian diets for mulberry silkworms (*Bombyx mori*) had been used for at least two decades (summarized by Hamamura 2001) before it was adopted by Western researchers. It should be noted that there is a great deal of ambiguity about the terminology for soy components in insect diets (soy protein, soy protein isolate, soy flakes, soy flour, soy meal, defatted soy flour, full-fat soy flour, toasted soy flour, and raw soy flour). Even a seemingly unambiguous product such as Nutrisoy actually represents several different soy foods, including a brand of tofu and tempeh from Australia, a soy milk from Singapore, and a series of flours and meals from Archer, Daniels, Midland Company. The ambiguity of this terminology is a serious problem as the products mentioned here have vastly different qualities and will perform very differently if one is substituted for another. This soy issue reflects other diet components that can have vast differences nutritionally, in terms of palatability, bioavailability, and possibly in other ways in our insect diets. Such materials as yeast components, wheat germ, and even gelling agents can have significantly different behaviors in insect diets. The key point of all this is that diet papers should have very clear statements of the range of specifications for such potentially ambiguous components. The name of the company that is the source of the components can be helpful, but often this information retains an ambiguity. Therefore, a well-chosen set of technical specifications for potentially ambiguous or variable components should be part of the rearing paper.

The characteristics of soy nutrients with respect to insect diets are discussed extensively by Cohen 2015. However, for the current treatment of the topic of rearing system components, the following paragraph explains the kinds of concerns about which rearing system personnel would benefit.

For example, staying with the topic of a diet component such as soy flour, what can be listed and explained would include source of the material (company that supplied it, batch information, possible changes in new versions of a product—in cases where the supplier is forthright and explicit about changes in their processing or production stream). Also factors that can be considered are: percent of protein, fats, fiber, sugar, ash, and if available mesh (range of particle size), and whether it is raw or heat-treated should be mentioned. I further suggest that other factors can be useful topics for publications. Such questions as, how influential are storage conditions should be asked? What are the effects on stability when diet components are stored at room temperature, storted in a refrigerator or freezer, kept in a vacuum-packed container or kept with desiccant and oxygen scavenging packets? Other worthwhile questions include how long was the allowable shelf-life for the component in the facility where the research was being done, and how do different storage containers influence shelf life of diet materials?

11.11 BACK TO BASICS: SOURCES OF VARIABILITY OF REARING SYSTEM COMPONENTS

One of the realities of insect-rearing systems in the context of suppliers is that the chain of product handling from the origin to the final user is that there are many complexities that can be relevant to the success or failure of the component in the rearing system. Continuing the use of soy flour (or related soy products) as an example, the start of the soy flour production process is in the field where the soybeans are grown. We know that even for what is seemingly the same component, from batch to batch, soy beans can vary because of their genetic strain and their growing conditions (including soil, watering conditions, fertilizers, etc.). A company that obtains its soybeans from various regions and growers then processes the soybeans in potentially different ways. These different ways depend on the company's practices, which may include different kinds of ripening and drying of the beans, cracking, extracting lipids, heat-treating, further milling (grinding the beans that have been cracked, dehulled, extracted by solvents to remove most of the lipids). In some processes, the bean flour or meal is modified in a process called "re-fatting," where the lipids that had been removed in de-fatting have been added back to the flour/meal to improve texture and nutrition from unsaturated fatty acids and phytosterols.

11.12 CONCLUDING WITH SUGGESTED EXAMPLES OF TOPICS FOR REARING-SCIENCE PAPERS

1. Scientifically-based rearing systems
   a. Topics are based on explicitly stated rationale and hypotheses
   b. Experimental designs are grounded in comparisons of rearing components, subsystems, and processes
2. Science behind rearing system development and evaluation
   a. Application of DoE approaches to rearing system optimization (engineering-based statistical quality/process control)
   b. Nutritional characteristics and value of foods and diet components (classical nutritional studies)
   c. Nutrient self-selection and related studies
   d. Feeding behavior (responses to sensory stimuli, behavior of mouth parts, etc.)
   e. Digestive or other feeding-related enzymes
   f. Analysis of insects' foods (chemical and physical analysis)
   g. Bioavailability determinations of various natural and artificial foods
   h. Factors that contribute to diet stability (antimicrobial materials, antioxidants, emulsifiers, etc.)
   i. Effects of lighting features (intensity, quality, photoperiod) on diapause, growth profiles, and other biological factors
   j. Studies of micro-environment in rearing systems (thermal and humidity gradients, gas transfer in rearing containers, and insectary design modeled after nature
   k. Microbial relationships in insects from rearing systems (microbiome analysis in relation to rearing conditions)
   l. Genetic relationships to rearing conditions: studies of deliberate and inadvertent domestication
   m. Other studies of diet component and processing functionality

11.12.1 Eleven Suggestions for Enhancing Rearing Publications from Cohen 2001

In 2001, Cohen discussed several approaches to establishing a formal status for insect rearing. He included several suggestions for improvements in rearing papers. The suggestions from the earlier paper are in italics, and my updated suggestions follow:

1. ***Whenever possible, studies should be experiment-based, centered on hypothesis/control/variable methodology***. This statement is the basis of proposing the insect-rearing papers should be science-based. Too many papers on insect rearing are anecdotal or not rationale- and hypothesis-driven. Important factors in making rearing papers better contribute to scientific understanding are attention to mechanisms behind experimental outcomes, properly designed and controlled experiments, and attention to interactions of rearing systems' components—especially factors that are part of the rearing experiments that are the subjects of intended publications. This implies that observations and measurements are included in the experiments, and the data from these measurements is a powerful tool in evaluating the experimental outcomes and sources of error that can be reduced to improve rearing outcomes. These topics are important

components in the following suggestions about experimental design and statistics, quality and process control, and explicit statements of rationale.

2. ***Appropriate experimental design and statistics should be used.*** This point goes hand-in-hand with and is an outgrowth of Suggestion 1. In today's scientific world, Design of Statistics procedures are available to every researcher, with some statistical packages being available free, while other statistical systems are available on institutional levels. Currently, exploratory statistics are becoming a well-established tool, and these statistical approaches help researchers with less sophisticated statistical backgrounds to take advantage of elegant experimental designs and interpretational algorithms. This includes, for example, mixture designs, full-factorial and partial factorial designs, and other approaches that allow screening designs, response surface designs, and space-filling designs. These packaged experimental designs include quality and process control approaches that lend themselves tremendously to development and improvement of insect-rearing systems (Cohen 2015).
3. ***Quality control or quality assessment should be a clear-cut and integral part of the reports.*** Quality control (QC) and process control (PC) have become common factors in mass-rearing systems where millions to billions of insects are produced routinely (Cohen 2015). Statistically based QC and PC standards have been adopted in large-scale insectary operations to make quality and process standards part of continuous improvement programs and a basis of meeting needs of customers of the rearing products. However, these standards are adopted late in the process of mass-rearing system development, and it would not only be useful to more seamless adoption of quality/process standards, but preliminary (or early) research works would also be enhanced by adoption of data acquisition and data analysis that helped researchers determine the sources and extent of variation in their incipient research system. For example, a control chart showing the upper and lower control limits of temperature or protein content of a diet component could be a powerful and highly documentable measure of the quality of a foundational experiment.
4. ***Proper citation of past progress should be made along with a clear statement of how the paper in question advances the previous state of knowledge.*** Far too many rearing papers treat background as a "given" that does not need to be discussed or documented by reference to past works. Admittedly, this is a difficult issue for authors and reviewers to deal with, as there are so many facets of rearing studies that are involved in the rearing process. If a rearing component is conventional, it is reasonable to omit specific, part-by-part references. However, authors of rearing papers should be careful to review the background for the processes that they include in their rearing studies, and if a process or material is central to the experimental questions that the paper is exploring or if the experiments adopt a novel concept from a published source, the author(s) of that publication should be cited.
5. ***Clearly and deliberately stated rationale should be made for every novel facet of the work.*** This is closely related to Suggestion 1, both being concerned with mechanistic-driven, science-based studies. This point relates to the rationale for the experimental components of rearing research projects. Rearing papers would be far more valuable if they were always predicated upon explicitly stated rationale. Most of the rearing innovations that have proved to be successful have a scientific, mechanistic reason for their success. While it is true that many rearing factors are so complex or otherwise difficult to unravel, the quest to understand **why** things work should always be a goal of rearing researchers. However, too often rearing papers are treated by authors, reviewers, and editors as being only of practical or heuristic value. This adds a black-box or mystical character to rearing advancements. These two suggestions (4 and 5) bring into focus the value of stating rationale for rearing advancements as well as citation of past sources. When the author of introduced a suite of techniques that made mass-rearing of tobacco hornworms possible (Yamamoto 1968), he stated rationale for the types of containers, diet-holding platforms, and various other physical features of the rearing system that allowed healthy reproduction by the captive hornworm adults. When the same author introduced a diet for larvae of the tobacco hornworm, Yamamoto 1969 introduced a diet for these insects, he revolutionized the rearing of these and many other insects. However, two of the main components of the diet, wheat germ and torula yeast, remain an enigma. The author did not cite the original source of the wheat germ in the hornworm diet, though the paper published by another group of authors (Adkisson et al. 1960a) was most likely the original source of the idea of using wheat germ. It would have been helpful to the readers to know that the addition of wheat germ was originated as an extension of the earlier paper, and it would have added further stature to the paper and its authors who clearly deserved citation. The other issue, the use of torula yeast, is an enigma to me because this material clearly works well (as proved by the hundreds of generations of hornworms reared on this diet), yet there is no stated rationale for why torula was chosen in light of the several other papers published prior to where brewer's yeast was used as a nutrient source (Yamamoto 1968, 1969).
6. ***Editors, reviewers, and other people who judge rearing accomplishments (e.g., promotion and tenure committees) should be made aware of the painfully long and high-risk nature of these advancements, and they should judge them accordingly.*** This point addresses the tendency of the audience of rearing papers to expect that all accomplishments worthy of publications to be saltatory, rather than incremental, advancements. When the early papers in *Drosophila* rearing appeared, they were inherently novel,

ground-breaking studies. In the early 1900s, the use of controlled rearing systems and artificial diets (Bogdanow 1906, 1908; Delcourt and Guyenot 1910; Guyenot 1913; Guyenot 1914; Baumberger 1917a, b), the introduction of yeast to *Drosophila* diets (Delcourt and Guyenot 1910; Guyenot 1913; Baumberger 1917a), and the introduction of agar-based diets (Baumberger 1917b) were entirely novel, and the early rearing systems described in those papers (cited here) were clearly saltatory advancements. When wheat germ was added to Lepidoptera diets (Vanderzant et al. 1959; Adkisson et al. 1960a), this was a major advancement that sent ripples throughout the domain of insect rearing. However, when wheat germ was included for the first time in a diet for the tobacco hornworm, *Manduca sexta* (Yamamoto 1968, 1969), this was by its inherent nature an incremental accomplishment, but it had huge ramifications in influencing rearing systems for other insects and for the vast amount of research and production based on the *M. sexta* rearing technology.

7. ***A clear distinction should be made between insect dietetics and nutrition. Works in both domains should be welcomed.*** Because nutrition has long been an established science with inherently mechanistic and experimental basis, when applied to insects, it was automatically respected and regarded as a true science. Conversely, because the word "diet" has had use in day-to-day discussions of foods in the context of cooking and other uses where there was not necessarily scientific rigor, insect dietetics has been relegated to a position of lesser respect. Even the often-used term for diet formulations as "recipe" induces the users of the term and the audience to regard diet formulation as a non-scientific process. One often hears diet specialists and their audience referring to "tweaking" an established diet, which connotes making minor, often trivial changes in a diet or other features of a rearing system. My argument is that nutrition studies deserve respect and a place in the publication of science and technology studies, but dietetics should also be treated scientifically and should, likewise, have a respected place in literature on insect rearing.
8. ***The problems incumbent in insect rearing as a service field should be recognized and addressed.*** This is not especially a standard for publication, but it is instead a background concept that needs serious attention. The history of rearing systems development has shown that the practical goal of establishing parameters that Delcourt and Guyenot developed in 1910 involve control over the production of insects for various purposes. This is the **heuristic concept** of insect rearing as contrasted with the **scientific concept** where one seeks to understand the rationale behind and interactions between rearing system components. **This contrast is tantamount to establishing a rearing system in order to produce insects vs. the discovery-based system of understanding the insects through manipulation and controlled rearing inquiries.** There are many examples that illustrate the concept here, but the emergence of the science of genetics stemming from the practical, heuristic field of plant or animal breeding makes the point. People were developing strains or breeds of plants and animals (such as domesticated wheat or dog breeds) long before the science of genetics emerged in the late 1900s. The development of genetics as a science did nothing but enhances the breeding efforts and allowed remarkably useful domestic organisms on a continuing basis. So it would be with insect rearing, if rearing experts were given an "open license" to treat rearing studies with scientific principles and rigor.
9. ***Specific formal education in rearing science and technology should be established.*** Since I entered the field of insect rearing around 1980, I have been trying to establish a system of formal education to help existing and emerging rearing specialists to have better tools to use for inquiry and discovery of rearing system components. I have been teaching classes, helping to establish workshops, and writing papers to help improve the formal education infrastructure. I have pointed out that thus far, all of us who consider ourselves as rearing experts or rearing professionals have had to learn most of our knowledge and understanding through trial-and-error, on the job training, and somewhat haphazard experiences with insect-rearing literature. While this can be a complex discussion, suffice it to say that a well-designed formal system of rearing education would greatly expedite all aspects of rearing, including the quality and scope of insect-rearing publications.
10. ***A specific, special section in insect-rearing science and technology should be added to at least one of the ESA [Entomological Society of America] journals.*** This is discussed in greater detail by Cohen 2019; however, briefly, it would be a great improvement for rearing specialists and those seeking rearing papers to have in at least one major entomology journal an established section dedicated to insect rearing. Such a section would have leadership from a section editor with considerable expertise in insect rearing as well as a list of at least 30–50 reliable and knowledgeable reviewers in the topic of insect rearing. This type of publication infrastructure would also be characterized by a set of explicit standards for publications—standards that authors, editors, and reviewers would have available to guide all these parties to higher quality papers that would improve the field of insect rearing. It should be noted that recently, the *Journal of Insect Science* (now and ESA journal) routinely publishes insect-rearing papers; however, rearing publications remain sporadically treated by the other ESA journals. It should also be recognized that the opinions on treatment of rearing publications are my opinion, and I realize that this is a controversial point. However, good science and good scientific publications should be open to airing differences in opinion.
11. ***A formal sub-discipline encompassing insect food science and technology should be established with a place for publications in this special field in one***

of the ESA journals. These two points (Suggestions 10 and 11) are aimed at the largest international association of entomologists and one of the longest-standing sources of insect-rearing publications. I discuss elsewhere (Cohen 2019) how the ESA publication (*Journal of Economic Entomology*) has been one of the leading publishers of papers on insect rearing since the journal's inception in 1908; however, at the same time, it has been very restrictive in its treatment of insect-rearing papers. My argument in Cohen 2019, 2015, and 2001 is that if ESA journals and various other publications (especially open-access publications) would establish a systematic and clear-cut set of standards for rearing papers such as the ones offered here, the quality of rearing publications and the research behind these publications would be greatly improved. I further argue that with the more open attitude toward rearing studies, the field would blossom and would undergo a renaissance of learning and enlightenment about insects rearing under the controlled conditions of insect-rearing systems.

11.12.2 Cases that Exemplify Significant Rearing Contributions

Many of the earliest rearing papers that helped launch the field of insect rearing provide excellent models for the kinds of papers that should be published to report advances in the science and technology of rearing. The inauguration of the design concept from the 1910 paper by Delcourt and Guyenot 1910 was already discussed, as were *Drosophila*-rearing papers by Baumberger 1917a, b, and Guyenot 1913 and 1914. Other papers that are excellent models of rationale-based rearing research on *Drosophila* and other Diptera include Baumberger 1919 who provides an excellent review of papers in the first two decades of the 20th century dealing with the possible roles of microbes in the natural history and rearing dynamics of insects. In an extensive thesis on nutritional issues in flesh flies, Michelbacher et al. 1932 take up the issues raised by Bogdanow 1906 and 1908 regarding the interplay between nutritional ecology of insects (especially Diptera) and the microbial associations that are intimately associated with insects from this order. Both Baumberger 1919 and Michelbacher et al. 1932 discuss the rationale for experimentally teasing apart the questions such as the roles of live vs. dead microbes and the changes in the nutritional composition rendered by the microbial actions. The questions of the nutritional value of the media (such as flesh or fruit) vs. the nutritional value of the microbes vs. the metabolic (fermentation) products of the microbes and the media remain a complex and incompletely understood relationship today.

Other cases of highly informative discussions of rationale for various refined experimental details are found in the papers of Burgess 1908 and Bridges 1932 and Bridges and Darby 1933. Burgess 1908 wrote a pioneering paper on various devices used for rearing insects. Apparently, many of the cages, lighting systems, and other rearing system components were in use by the time Burgess' paper was written, but these methods had not been documented in the literature and Burgess was the first to do so, though the detailed methods for materials, equipment, and design remain largely neglected in the literature. It should be mentioned that an even earlier publication by Kelly 1903 was published on silkworm rearing, and in that monograph, Kelly provided explicit details on home-made incubators and other equipment for rearing *Bombyx mori*; the details of Kelly's monograph are presented in more detail in Chapter 4.

In terms of valuable, detailed rationale for rearing system engineering or design, one of the most impressive treatises is from Calvin Bridges, who is known for the work he did in the pioneering genetics studies in the T. H. Morgan genetics laboratory (see Morgan et al. 1913) where A. H. Sturtevant, H. J. Muller, and Calvin Bridges worked in the famous "Fly Room" at Columbia University through the early decades of the 1900s until the 1930s. For example, to explain the details of a prototype incubator, Bridges 1932 wrote:

> DURING the early work on the genetics of *Drosophila melanogaster* the cultures were reared on laboratory tables or on open wall shelving. This method gave irregular results. The cold nights of the winter season made a considerable proportion of the cultures fail to start properly, and also lengthened the generation interval.
>
> In 1913 I built above my laboratory table a large wooden incubator which held all my experimental cultures. This incubator was heated by carbon electric lamps placed in the lowest shelf space. The thermostat was of the "ether-wafer" type. The expansion of the ether forced apart contacts in the heating circuit. A six-inch electric fan continuously circulated air down a gap behind the shelving with return through a similar gap between the shelves and the doors.

Elsewhere, Bridges explained the importance of temperature regulation in rearing circumstances—that it was crucial for uniformity, and to help control the very genetic factors that *Drosophila* researchers were trying to determine. He pointed out that linkage/crossover studies are skewed by deviations in temperature. Bridges wrote, "Plough's work on the effect of temperature extremes on crossing-over showed that it was necessary to maintain the temperature constant for genetic reasons as well as to improve culture conditions." Unfortunately, the recognition of the far-reaching effects of temperature and other facets of rearing is lost in terms of the consequences of lack of control of rearing conditions. These arguments by Bridges and others in the early works on *Drosophila* make powerful cases for the importance of control in

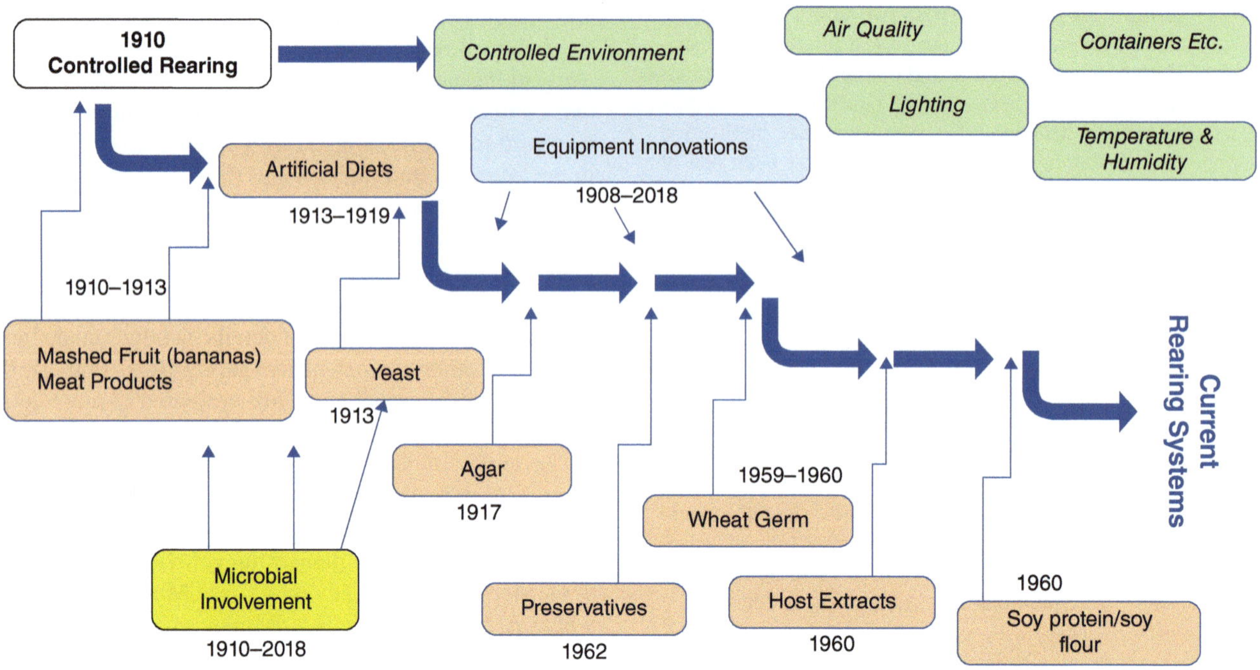

Figure 11.6 Historical background for current rearing systems and functions of insect-rearing systems.

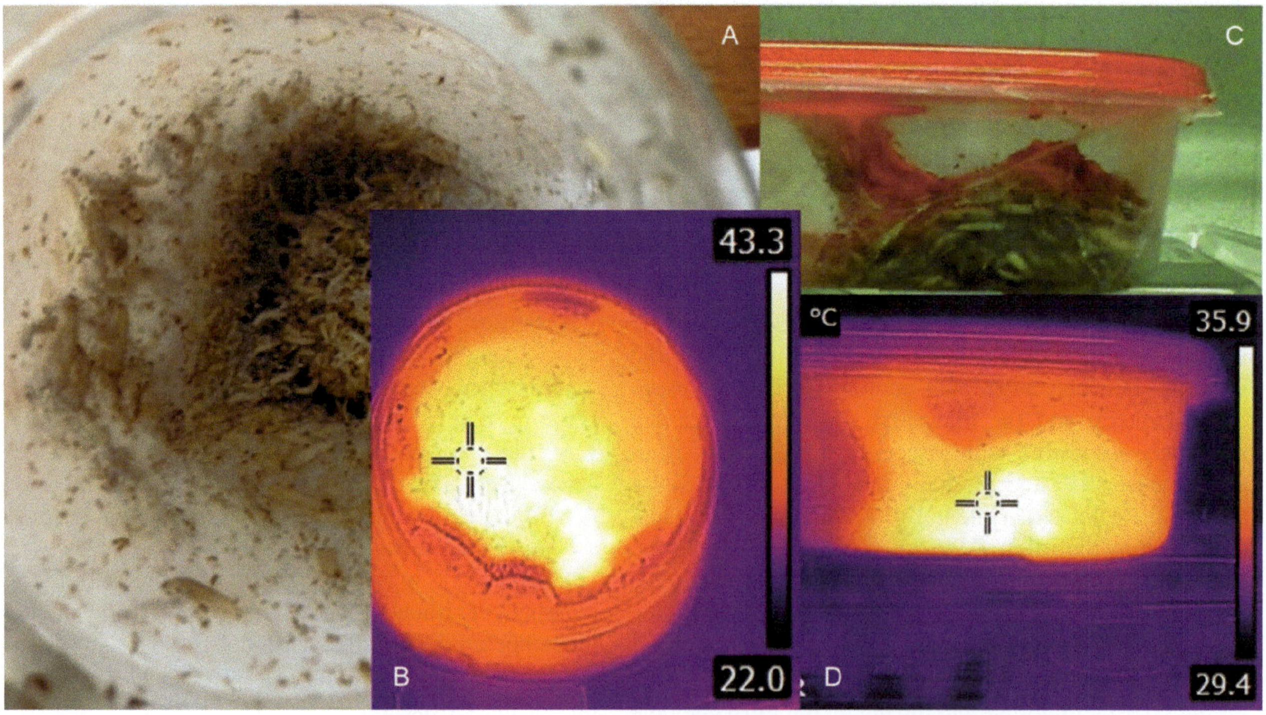

Figure 11.7 Wax moth thermogenesis in rearing containers. Figures (a) and (b) are views through the top of a 2-L glass jar ([a] being a visual light image, [b] being a thermal image, showing the elevated temperatures generated by larval masses). Figures (c) and (d) are images from a plastic (Rubbermaid™ 3-L container). Figure (c) is a visible light image, and (d) is a thermal image showing the webbing and frass-covered mass teeming with larvae.

insect-rearing systems. They further make the case for the hidden (cryptic) sources of deviation, disorder, and detriment to quality that can take place as threats to the products that we intend to gain from all our rearing efforts (Figures 11.6 and 11.7).

Each innovation represented here evolved into a fundamental part of modern insect-rearing technology. Many contributions such as the initial introduction of wheat germ or adoption of preservatives were viewed as an incremental contribution. However, many seemingly incremental advancements became giant leaps in the progress of rearing science and technology. An important take-home message from this historical perspective is that the current state of rearing systems is an outgrowth of systems and system-improvements from the past. To maintain an infrastructure that supports rearing high-quality insects with efficient and economic methods, there must be a clear set of publication standards that assure future researchers and rearing system managers that the techniques that they are using are vetted and based on solid science.

Throughout their history and development insect-rearing systems have been used initially for tools of discovery for various basic biological sciences and later as the bases of various programs where mass-rearing serves purposes such as control or sources of insect products.

The organization of the larval colony is typical of how the larvae spin webbing as scaffolding leading to the top of the container giving them access to fresh air. This is an example of a discovery of a fundamental science phenomenon that was not only derived from a rearing situation, but it could not have been recognized outside of a rearing situation due to the cryptic nature of the wax moths in their natural habitat: inside honeybee hives. It should be noted that the topic of diet-induced thermogenesis and its biological consequences is a current topic of investigation in the Insect Rearing Education and Research Program at NCSU (Cohen, personal observation).

CHAPTER 12

Rearing Education

The Pathway to Improving Rearing Science and Technology

This chapter is a comprehensive treatment of insect-rearing education. It is intended as a starting guide for potential teachers/instructors who wish to further education in rearing for the ever-increasing number of people entering the profession of rearing. It contains a considerable amount of first-person narrative based on my experience as an educator, in general, and as an instructor in the topic of insect-rearing systems. It is derived from my 5+ decades of experience in teaching and my 4+ decades of experience as an insect-rearing specialist. In all cases, the successes and failures at teaching rearing are derived from real-life situations.

12.1 GENERAL BACKGROUND ON REARING EDUCATION

In virtually every field, education is fundamental to progress and professionalism. In ancient and medieval times, learning a profession was achieved mainly through apprenticeship or some form of on-the-job training. As knowledge bases increased, especially during and after the Renaissance, it became a reality that more efficient learning systems had become essential. For example, the son learning from his father, a farmer, how to gather and plant seeds and how to use manure or fish heads as fertilizer and how to harvest the crop. In the primitive systems of agriculture and other fields such as cloth-weaving, arrow-making, or baking, there was more religion or superstition involved in the processes than science-based technology and complex cause and effect-based practice. These generalizations about the importance of formal education in the development of all professional fields certainly applies to rearing education—especially formal education.

Until the year 2000, there was virtually no such thing as formal education in insect rearing. Two small exceptions to this statement are a couple of graduate seminars that I taught at the University of Arizona in 1980 and again in 1990. Other than those classes, everyone entering the field of insect rearing had to depend upon on-the-job training and self-teaching. Although several thousand papers were published about various aspects of insect rearing (Cohen 2018b, 2019), from the years 1910 through 2000, there was no curriculum established for rearing science and technology. This left emerging rearing specialists to try to find an order in rearing practices by trial and error, on-the-job training from rearing specialists who themselves were self-taught, or searching a large body of literature on entomology where rearing papers were stochastically distributed. In my review of insect-rearing system literature (Cohen 2018), I point out how most papers on aspects of insect rearing are difficult to find through key word searches of terms that directly relate to rearing information.

In the late 1990s, my friend and USDA, ARS colleague Dr. Frank Davis and I discussed and lamented the issue of lack of formal training for insect-rearing specialists. As two late-career rearing specialists stationed at the ARS labs at Mississippi State University (MSU), Frank and I discussed this issue during our coffee breaks in Frank's lab where we were often joined by the late Peter Sikorowski and Doug Inglis (who succeeded Peter as insect pathologist in the MSU Entomology Department). During that period, my laboratory (the Biological Control and Mass-Rearing Research Unit = BCMRRU, headed by Don Nordlund) was studying formally and deliberately the concepts and practices of insect mass-rearing. One of our important insect targets of our research was the emerging pest, the coffee berry borer (CBB), and through Don Nordlund's and Clarence Collison's stimulation and ARS support, we were asked to host a three-day workshop on mass-rearing CBB. In that workshop, about 70 researchers from nearly every coffee-growing country in the world attended, and we offered a condensed curriculum that included environment, genetics, pathology/microbial issues, diet, and automation/equipment aspects of rearing systems in general and for CBB, specifically.

It was from the model established with the CBB workshop that Frank Davis, Doug Inglis, and I presented to the MSU Department of Entomology the idea of offering yearly workshops in the principles of insect rearing. After considerable discussion within the Department, we offered, in the year 2000, the first workshop on Principles and Practices of Rearing Quality Insects, and we attracted 30 participants who came from all over the world to increase their knowledge of insect rearing. We decided on 30 participants as a target number because of practical limitations in room size and logistics of handling that many people. A further, more complete history of the MSU workshops, which have become a yearly institution since the first workshop, is covered in an *American Entomologist* paper by Professor John Schneider et al. 2018.

After my involvement in the first few workshops at MSU, I started my own approach to rearing education, initially as part of a private research company (Insect Diet and Rearing Research, LLC) that I had established, first in Tucson, AZ, then at North Carolina State University's (NCSU) innovative incubator on Centennial Campus. For the past ten years, I have been coordinating the Insect Rearing Education and Research Program as part of my duties as research professor in NCSU's Department of Entomology and Plant Pathology. In this position, I have been leading research on rearing systems for various insects and teaching on-site (face-to-face) courses in insect rearing and an Online course in fundamentals of insect rearing.

Cohen 2015 and 2019 has estimated that exclusive of silkworm rearing and honeybee husbandry (where **millions** of people make all or part of their living by rearing insects), there are thousands of people worldwide who make their living by rearing insects. And it remains an unfortunate fact that there is little formal education for these "rearing professionals." We know from study of the rise of professionalism in business, engineering, medicine, teaching, and other fields that professional standing and education go hand-in-hand. Prior to the emergence of modern medicine, surgeons were barbers and did not begin to enjoy the respect that they receive as physicians today until their profession was subjected to standards of academic learning, including sanitation, anatomy, physiology, and other related subjects (McGrew 1985; Himmelmann 2007). Similarly, before John Dewey recommended rigorous education, teachers were not considered professionals until systematic formal standards and courses were developed. Agriculture had been practiced from the dawn of civilization in nearly every human culture, but until it became formalized through scientific practices and meticulous technological inquiries, it remained a matter of mystery and even superstition. Religious practices were often associated with restoring fertility to the soil, and various sacrifices (such as the Fisher King Myth described by Sir Frazer in the Golden Bough) were made, until it was learned that returning dead and decaying organic matter to the soil, helped replace nutrients depleted by crops. The magic and superstition involved in breeding domestic plants and animals finally gave way to the scientific principles of heredity revealed by Gregor Mendel and Charles Darwin. Despite the fact that most trades and disciplines practiced through the ages were learned by on-the-job training in guilds and apprenticeships.

12.2 REARING EDUCATION AND TRAINING

In the same vein, most of learning in the rearing community is on-the-job training, sometimes self-taught and sometimes taught by another person who was self-taught. Such self-teaching is not as efficient nor comprehensive as a well-vetted, cross-disciplinary education can be. Furthermore, the conventional practices of rearing in a given rearing system are usually derived from haphazard trials. Many of the practices in a given rearing system were derived from other existing rearing systems (the practices of confluence and derivatization), too often without recourse to disciplined, meticulous, and controlled scrutiny. The haphazardness of the adoption of novel, emerging rearing technology is often repeated in the teaching of that technology to entry-level personnel. These critical statements are not intended to be insulting to rearing personnel, including those who develop or greatly modify and adopt rearing systems. My criticisms are intended to help develop an educational infrastructure by setting the foundation of standards of both rearing system development/adoption and the educational approaches to spreading information and knowledge about rearing systems. My thinking is derived from the concept that can be paraphrased as, "If I give a person a fish, he/she will have a meal; if I teach him/her to fish, he/she will have a lifetime of fish to eat." If we teach the established practices in a given rearing system, our entry-level person will be able to rear the insect, but if we teach the basis of the practices (their **scientific foundation**), they will be able to apply the principles to problem-solving and developing new or improved rearing systems. This leads to the concepts of **training** vs. **education**. Training is teaching someone **how** to do something; education is teaching a person **how** and **why** they are to do things. Good training is rigorous and meticulous. Good education is at once rigorous, meticulous, **and flexible**. In the highly useful matrix of Bloom's Taxonomy (discussed below), the lower levels of the hierarchy dwell in training while the higher levels are characterized as education: flexibility of thinking and striving to understand the "why's" as well as the "how's." The projects and tests from my students of insect rearing, presented in Appendix 4, are examples of how students applied the scientific principles taught in the lecture part of my courses to the hands-on practical execution of those concepts. The tests especially reflect the achievement of the higher levels of scientific, critical thinking implied in the Bloom model.

12.2.1 Learning by Doing (Hands-On Rearing Education)

Figures 12.1–12.6 show students from on-site insect-rearing courses at NCSU from 2010 through 2020. The

Figure 12.1 (a and b) Students in insect-rearing courses at North Carolina State University (NCSU) (2010–2020). (a) Upper left: Pei-Shan Wu and Fu-Chyun Chu; lower left: NCSU rearing workshop attendees; upper middle: Mauri Hickin (left), Mario Simon Pinilla-Gallego and Tabatha Komar (right); lower middle rearing class of 2011 (first for-credit course at NCSU): bottom row (left to right): John Hanley, Jonathan Cammack, Alana Jacobson, Rick Santangelo, Kelly Oten; top row (left to right): Allen Cohen, Andrew Ernst, Amy Lockwood Murillo, Michelle Meck, Nancy Maxwell Brill, Heather Moscrip, and Micah Gardner; upper right: Folukemi Adedipe; lower right: Chen Zha. (b) Lower left: Nancy Maxwell Brill; lower right: the 2015 rearing class; upper: Micah Gardner Kelly Oten; upper right: Mr. Chris Pernell showing Universal Texture Analyzer procedures to Arun Babu (Note: nearly all of these students of the NCSU rearing classes are in professional positions now in the field of entomology and related disciplines).

Figure 12.2 Students in 2019 insect-rearing science and technology class learning about mosquito rearing in Dr. Michael Reiskind's laboratory. Left to right: Mauri Hickin, Tabatha Komar, research technician Anastasia Figurskey, Mario Simon Pinilla-Gallego, and Dr. Kristen Hopperstad. Figures key is explaining the rearing system following Dr. Hopperstad's lecture on an overview of mosquito rearing systems.

most effective learning situations were those where students got to reinforce lecture principles with hands-on, learning-by-doing experiences and then were obliged to critically interpret the meaning of the science they were accomplishing. It was the intense student involvement in testing, question and answer sessions in class, and laboratory work with interpretations that made up the best learning practices. Interspersed with all the experiences were the group activities based on problem-based learning philosophy, which taught students to collaborate and yet be responsible for individual rendering of their work (Figure 12.7–12.9).

Figure 12.3 Students in A. C. Cohen's 2019 insect-rearing science and technology class learn to measure antioxidant/free radical scavengers in insect diets. Left to right: Mauri Hickin, Mario Simon Pinilla-Gallego, and Tabatha Komar.

12.3 BACKGROUND: MY INTRODUCTION TO INSECT REARING

I offer a brief biographical narrative here as an example of one of the many entomologists who learned insect rearing by on-the-job training. Until recently, all

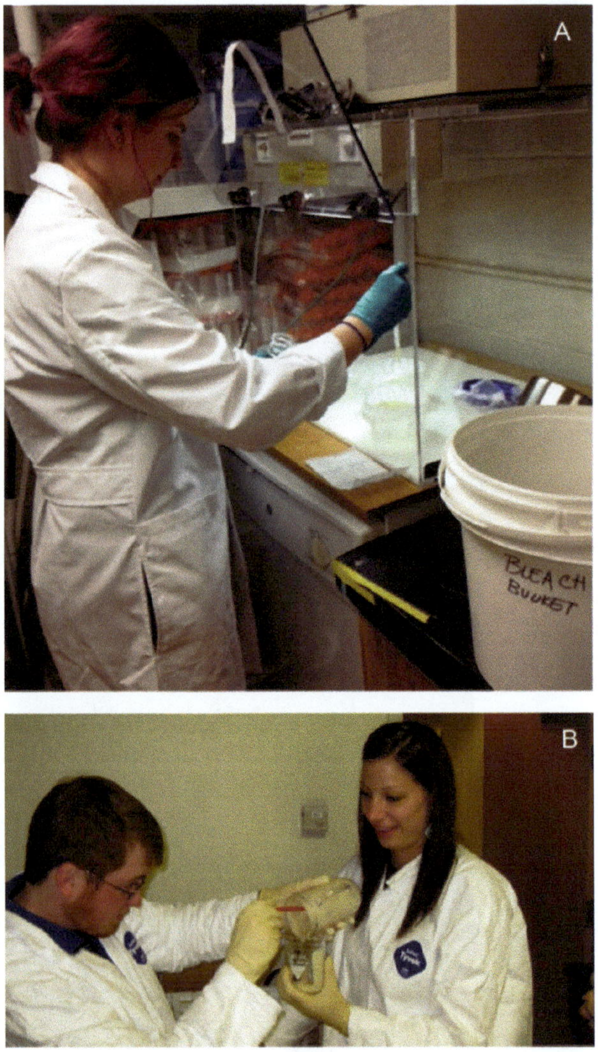

Figures 12.4 (a–d) Recent insect-rearing course students taking part in hands-on experiences. (a) Former Student and Technician Katie Youngs. (b) Kelly F. Oten and Micah Gardner. (c) Folukemi ("Kemi") Adedipe. These former students are pictured here working on their rearing class assignments exemplifying the "learn by doing" or "hands-on" approach to rearing education. (d) Former insect-rearing class (2011) students Nancy Brill Maxwell and Alana L. Jacobson (right) (now, Drs. Brill Maxwell and Jacobson) preparing artificial diets thrips for her class rearing project.

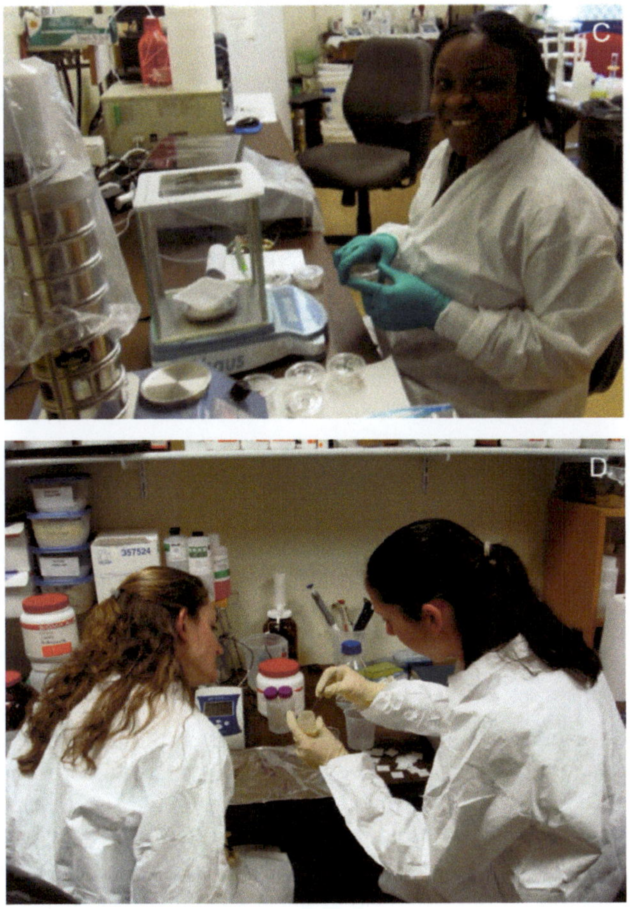

Figures 12.4 (*Continued*)

Figure 12.6 Former rearing students from 2013 rearing course (now Drs. Terri Billeisen in foreground and Diane Silcox Reynolds in background) performing ELISA tests as part of their doctoral projects. They are performing this research in the laboratory of Dr. Ann Black (Immunoreagents Inc.) under the guidance of Mr. Brad Heidinger, cooperators with Insect Rearing Education and Research Program, North Carolina State University (NCSU).

Figure 12.5 Recent picture of two of the top students (Meredith Favre, left and Nicole Gutzmann, right) in the insect-rearing course (2017), here performing analyses of total protein in their research subjects, *Tribolium castaneum*. The two students are completing their graduate studies under Dr. Marce Lorenzen in the Genetic Pest Management Program at North Carolina State University (NCSU).

Figure 12.7 (a–c) Insect-rearing course student (Andrew Ernst) working with rheological instrument to measure texture differences in a tobacco hornworm diet with and without added corn oil. The rheological (textural) differences are seen in (b) and (c).

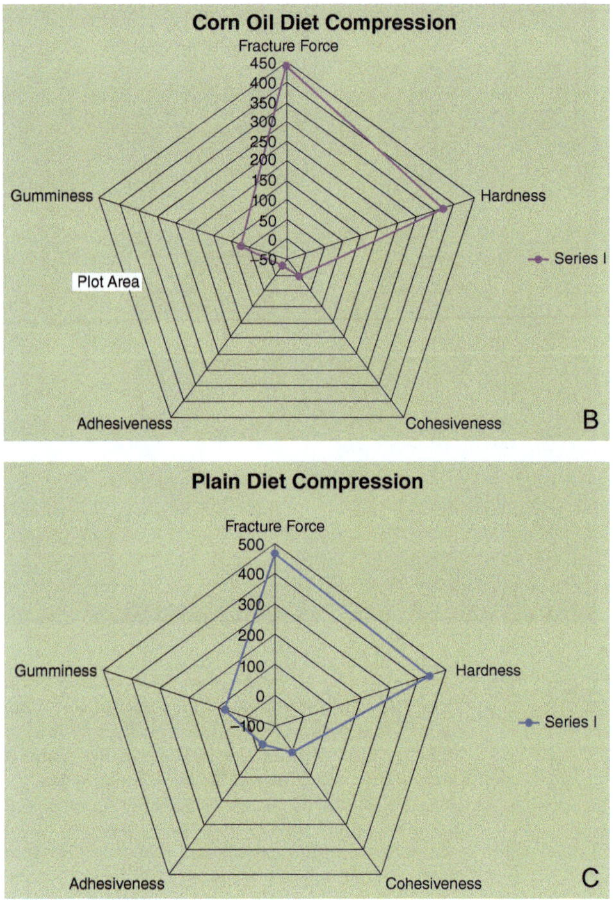

Figures 12.7 (*Continued*)

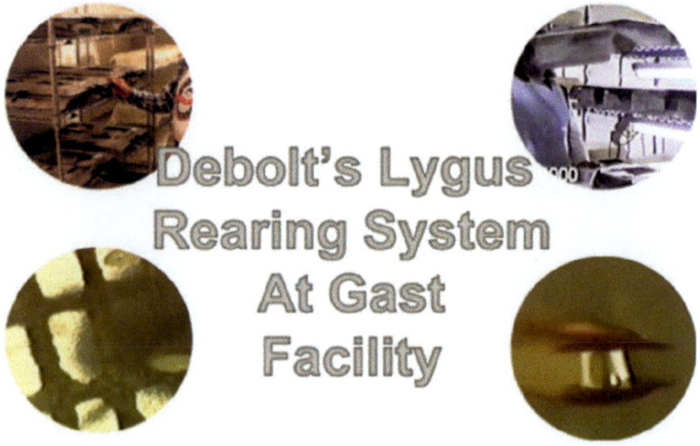

Figure 12.9 Dr. Jack Debolt visiting rearing room for *Lygus hesperus* at USDA, ARS, BCMRRU, at Mississippi State University (MSU) circa 2002. Dr. Debolt was one of my most valued colleagues at the USDA, ARS Biological Control of Insects Laboratory at Tucson, AZ and someone from whom I learned much of my understanding and appreciation of insect rearing. Working with him and Mr. Ray Patana at the Tucson laboratory were remarkable experiences for a novice like me in insect rearing.

rearing education had been informal on the job training. In 1980 and again in 1989, as an adjunct professor in the Entomology Department of the University of Arizona, I offered one-unit graduate courses in insect feeding biology and rearing, where about 20 students were exposed to a semi-formal treatment of the issues in insect rearing, as I then understood rearing science and technology. I tried to use an approach that treated rearing as it was derived

Educating the Next Generation of Insect Rearing Professionals: Lessons from the International Insect Rearing Workshop, Mississippi State University, 2000–2017

Topic	Instructor	Time Frame
Production System	Dr. Frank Davis Dr. SenSeong Ng	2000-2017 2014-2017
Environmental Biology	Dr. John Schneider	2000-2017
Nutrition and Diet	Dr. Allen Cohen Dr. Muhammad Chaudhury	2000-2002 2004-2017
Quality Control	Dr. Norman Leppla	2000-2017
Facilities Management	Dr. William Fisher Mr. Peter Ebling	2000-2015 2016-2017
Population Genetics	Dr. Michael Caprio	2000-2017
Facilities Design	Dr. William Fisher	2001-2017
Health and Safety	Dr. John Reinecke Dr. SenSeong Ng	2000-2011 2014-2017
Microbial Management	Dr. Douglas Inglis Dr. Louela Castrillo Dr. Roberto Pereira	2000-2002 2004-2016 2017

Figure 12.8 Dr. Frank Davis a leader in insect-rearing education and world authority on Lepidoptera rearing retired from USDA, ARS, and active in developing and teaching insect-rearing workshop program at Mississippi State University (MSU) from 2000 to 2019. The workshops at MSU are discussed by Schneider et al. 2018.

from empirically-derived knowledge. At the time I was offering these courses, most of my understanding of rearing was based on my experiences in the Biological Control of Insect Laboratory (USDA, ARS) where Mr. Raymond Patana, Dr. Jack Debolt (Figure 12.9), and Dr. C. Glenn Jackson were our rearing experts under the supervision of Dr. Harry Graham. Between these scientists, there was well over 100 years of rearing experience with dozens of species of insects. The laboratory, like many USDA facilities, had a culture (using the term "culture" in the social sense, rather than biological cultures) of rearing processes that made our lab a remarkably rich environment for me to learn about rearing from the masters. Interestingly, these rearing masters were mainly self-trained with some smattering of input by earlier specialists who, in-turn, learned rearing on the job.

My own professional background before I became an entomologist and researcher was in education (initially as an English teacher, then as a biology instructor at a community college). So I was sensitive to teaching/learning situations and best practices of teaching (Figure 12.9). Part of my background in education taught me the importance of the students' involvement in their education—that if they did not have "buy-in" to their learning tasks, their success as learners was far less than where they were involved and motivated to learn the depths and breadth of their subject. The most compelling part of this Socratic concept was "learning by doing."

12.3.1 The Initiation of Formal Training and Education in Insect Rearing

In the mid-1980s, Professor Brooks 1984 expressed the need for formal education in rearing-related topics (courses in genetics, physiology, pathology, biochemistry, and engineering), but it was not until 2000 that workshops in insect rearing were offered as formal education opportunities (Cohen 2001, 2019; Schneider 2013; Schneider et al. 2018). In each of these papers cited here, it is mentioned that the impetus to provide these educational workshops on insect rearing was Dr. Frank Davis. I must add to this that in the late 1990s, Frank Davis, Doug Inglis, and I had conversations about the worldwide need for insect-rearing education. At the time both Dr. Davis and I were scientists with the USDA, ARS at MSU, and Dr. Inglis (who is now a scientist at Ag Canada in Lethbridge, Alberta, Canada, currently studying gastro-intestinal pathogens in livestock) was an entomology professor at MSU and the Departmental pathologist having filled the position vacated by Professor Peter Sikorowski, the long-time insect pathologist. After considerable brainstorming and planning, we provided a plan of offering workshops in insect rearing. When we approached the Entomology Department at MSU, our proposal of offering a Departmentally sanctioned workshop in rearing, the idea was met with skepticism that a sustainable group of potential attendees would attend, especially when we suggested a registration fee of about $1,000 US, for five days. At the same time period, Mr. Donald Nordlund had mentioned our idea to the administrators of a program on mass-rearing of the CBB (*Hypothenemus hampei*); Coleoptera: Curculionidae Scolytinae, a worldwide and most serious pest of coffee crops. The mass-rearing program was partially conducted at the USDA, ARS, Stoneville and at Mississippi State facilities (in the Gast Facility), and the program administrators were anxious to sponsor a three-day workshop that we provided at MSU. Seventy-five attendees from 15 to 20 countries attended the workshop (Figure 12.8).

12.3.2 Progress in Rearing Education via Workshops

The robust attendance helped convince the Entomology Department personnel and MSU administrators that provision of insect-rearing workshops would be not only feasible but also potentially profitable and even something that would bring favorable attention to the University. We offered the first workshop in the general topic of insect rearing in September of 2000 where 30 participants from several countries attended. After much discussion, we decided to limit participation to 30, although there were more participants who had to be put on a waiting list. The workshops were held in the Clay-Lyle Building (which housed the Department of Entomology) and included lectures on fundamentals of insect rearing, insect diets, ecology of insect rearing, genetics of reared insects, insect pathology, quality control in mass-rearing, and safety in the insectary. The initial group of instructors of these topics were Dr. Frank Davis, Dr. Allen Cohen, Dr. John Schneider, Dr. Michael Caprio, Dr. Doug Inglis, Dr. C. O. Calkins, Dr. Norm Leppla, and Dr. John Reinecke. The updates on the workshops at Mississippi State are presented by Schneider et al. 2018. As Schneider et al. point out, the workshops have been an institution in insect rearing with yearly offerings for nearly two decades (Figure 12.8).

12.3.3 Departure from Rearing Workshops

I discontinued my participation in the MSU after my retirement from the USDA, ARS in 2003, but I continued to offer workshops from my new locations in Tucson, AZ and later at NCSU. I had formed a private company to allow myself to pursue diverse interests in insect-rearing research and rearing education, first working out of Tucson, then as an incubator company on the NCSU Centennial Campus where I was able to do research and teaching. Soon after coming to NCSU, I started an adjunct position with the Department of Entomology where I was able to teach rearing courses and serve on graduate student committees, mentoring students with interests in insect rearing and insect feeding biology. Much of my teaching and research was in association with Professor Fred Hain and his forest entomology group, and in 2010 I left my company and joined the Department of Entomology as a Research Professor. In my functioning as Research Professor, I was able to receive grants, make research service agreements, and teach classes as a member of the NCSU Graduate Faculty. The first class that I taught at NCSU was a graduate seminar in insect rearing, taught in 2010, where each of about 20 students presented seminars on various topics in insect rearing (insect feeding biology, insect/microbial interactions, dealing with microbial contaminants, the environments of rearing systems, etc.). The course was so popular, that I was asked in 2011 to teach a course in the Special Topics in Entomology series where 11 graduate students took the course as a three unit offering in Insect Rearing Science. The topics and course expectations for this course and the four offerings that followed are presented in Table 12.1.

It is important to note that the first time I taught the on-site insect-rearing class, I did not test the students, but instead based grading on class participation and on class projects (some of which are presented in Appendix 4). The course was predominantly lecture where I covered the topics in Table 12.1, and the students were given opportunities to work with various cooperating laboratories on campus, including the Food Rheology Laboratory in the Department of Food, Bioprocessing, and Nutritional Science, Materials Science, etc. where they could get experience with techniques and equipment not conventionally used by entomologists (see Figure 12.7(a) where student Andrew Ernst is working with rheology equipment to measure texture differences in *Manduca sexta* diets with and without corn oil).

These workshops have been offered at MSU under the leadership of Dr. Davis for the past 17 years as five-day sessions where instructors provide lectures and demonstrations in rearing-related topics (insect diets, rearing ecology, pathology, quality control, special rearing equipment, and safety). I was an instructor in the first three workshops, until my retirement from USDA, ARS in 2003. After my retirement and between 2003 and 2011, I offered seven rearing workshops in Tucson, AZ and at my current location at NCSU. More recently, as a member of the Department of Entomology at NCSU, I offered a graduate seminar in insect rearing as a one-semester course; and then at alternate odd years from 2011 through 2019, I offered an on-site, for-credit courses in the principles of insect-rearing science and technology. My changing

Table 12.1 Course Syllabus for Insect Rearing (Special Topics in Entomology 591/791), from NCSU Department of Entomology and Plant Pathology

		Insect Rearing/ENT 591/791	
Week	**Date**	**Topic**	**Due**
1	January 7	Introduction to Rearing	
2	January 14	The Insectary/Feeding Biology	
3	January 21	Feeding Biology/Diets	
4	January 28	Diet Chemistry	
5	February 4	Diet Physics/Rearing Equipment	Written Project Outline
6	February 11	Diet Physics and Diet Development	Test 1
7	February 18	Statistical Process Control	
8	February 25	Microbial Aspects of Rearing	
9	March 4	Spring Vacation	
10	March 11	Microbial Aspects of Rearing	
11	March 18	Analysis in the Rearing Laboratory	Written Project Draft
12	March 25	Analysis and Bioassays	Test 2
13	April 4	Rearing Genetics	
14	April 11	Domestication and Rearing	
15	April 18	Special Topics/Soils, Aquatics	Final Project
16	April 25	The Insectary/Design and Safety	
17	May 1	Final Exam	
		Grading for ENT 591/791	Points
		Attendance (5 pts/class)	150
		Project Outline	25
		Project Draft	75
		Project Final	150
		Constructive Participation	100
		Test 1	150
		Test 2	150
		Final Exam	200
		Total Points for Course	1,000
		Extra Credit for "field/lab" trips	100

teaching emphasis from workshops to actual courses was based on several practical and pedagogical reasons.

These courses were taught as special topics in entomology offerings, and they included lectures (three hours per week) and lab exercises where students gained hands-on experience with rearing and related topics such as microscopy, diet-making, chemical tests of protein content, antioxidants, texture (rheology) analysis, and other aspects of laboratory background. Also, I have been offering an online course in insect rearing for the past three years and have had about 80 people (mainly rearing professionals) complete this course. Finally, a series of symposia or meetings of experts has been provided at various professional meetings, especially at ESA annual meetings. I suspect that I have missed a number of such meeting-based teaching/learning situations, for example, where specialized information was presented on topics such as silkworm rearing.

12.4 REARING EDUCATION AND REARING RESEARCH

I named the current discipline at NCSU where I am a research professor, The Insect Rearing Education and Research Program. I deliberately separated yet combined the terms "rearing education" and "rearing research." The reason that rearing education needs a special place in the entomological community is that there are many special facets of rearing that need to be deliberately and consciously addressed. The need to address these topics specially stems from several realities: (1) students or learners in insect rearing are from various backgrounds and come to rearing learning at a wide range of academic background; (2) there are many important aspects of rearing that require special attention or highly tailored explanations or rationale; and (3) the gaps between important underlying principles of rearing and the practical procedures of rearing need to be **explicitly** addressed.

1. **The diversity of workers in insect rearing**. I have been teaching aspects of insect rearing for 40 years, and while much of my teaching has been at the graduate level with people holding bachelor's or master's degrees in entomology (or closely related fields), many of the people whom I taught were strictly entry-level in science. In most of my workshops, for example, I would have at least one or two people who had never had a course in entomology, many of whom never had a course in biology. In those same workshops, I often had people with advanced degrees, including a PhD in entomology. Some workshop participants had more than 20 years in insect rearing; yet some of those people came to the workshop not knowing the difference between sorbic acid and ascorbic acid (though they had used both in diets that they made routinely). I worked with students and participants who could be expert in pipetting and those who had no idea of how to properly use a pipette nor a balance either for that matter. I worked with engineers who knew thermodynamics "inside out" but had little grasp for critical thermal maxima or lethal maxima of body temperatures of insects. In fact, it is one of the most challenging things about insect rearing that by its nature it draws from people with diverse backgrounds and varied levels of background education. Yet once these diverse people step into a role in an insectary, there are things that they need to understand—things for which their backgrounds did not prepare them. A mechanical engineer who has highly sophisticated understanding of torque, friction, and quantum mechanics needs to understand the delicate and nuanced effects of the wrong photoperiod in initiating diapause in a gypsy moth. A biochemist who knows the structure and function of all the Krebs cycle enzymes may be totally unaware of the properties of stainless steel vs. glass in heat transfer during diet-making. And people of nearly all and every background (or lack thereof) need to learn the special vulnerability of neonate larvae with huge surface to mass ratios hatching in conditions of high temperatures and low humidity.
2. **The specialized nature of protocols in rearing systems**. So many procedures in rearing systems are alien to most people of most backgrounds. Even food chemistry experts are challenged by the especially complex mixtures in many (most?) insects' artificial diets. Commonly, more than 10–20 ingredients are mixed in diet-making processes; then the diets are mixed, heated, and otherwise processed in ways that human foods are not treated; then, at the end of the processing, the diet must remain stable, palatable, and nutritious for days or even weeks under warmer than room temperature conditions while being eaten by an insect, which is apt to transfer microbes that contribute to diet spoilage. The various diets and other rearing components lead to special treatments of rearing conditions that are unlikely to appear in other domains and therefore require specialized explanations and understanding. For example, the many kinds of diet mixing, heating, and packaging necessitate rearing personnel to learn to use steam kettles, autoclaves, extruders, and other diet preparation machinery. People with training in microbiology are likely to be familiar with autoclaves and special protocols to assure sanitation or sterility, but the microbiology background does not pre-train people to use properly steam kettles, flash sterilizers, or twin-screw extruders. Furthermore, specialized packaging with materials such as Parafilm or other diet-containing materials requires training that does not exist in many (or any) other disciplines.
3. **The gaps between underlying principles and established practices**. This is a complex point to understand and a most essential one if rearing education is going to meet the needs of insectary personnel in various jobs. Often, the scientific or mechanistic

principles that underlie routine insectary protocols are not clearly understood by rearing personnel. I will provide several examples to support and explain this generalization.

i. **Case 1: Sanitation**. Insectary workers are told that they should practice basic hygiene when handling insects, diets, containers, and other rearing items. However, the majority of insectary workers have never had a course in microbiology or any other formal training in sanitary techniques. While standard operating procedures (SOPs) call for diet sterilization (usually with heat), once an insectary worker exposes the diet or the insect being transferred to the diet to unfiltered laboratory air, the worker's breathing or talking-projected droplets, the worker's hands or even an un-sanitized paint brush, the rearing container's sanitation has been compromised. The difficulty of "selling" insectary workers the importance of hygiene is amplified by my decades of experience with personnel who insisted on removing or improperly using their protective equipment (masks, gloves, lab coats, and aprons) while handling their rearing materials. I am writing this section during the heart of the COVID-19 pandemic, and I see the continual struggle that medical authorities and leaders have convincing people of using masks and gloves and reducing social contact, not to mention meticulous hand-washing and other practices that would help us save one-another's lives. This makes prospects of gaining full compliance from insectary workers in helping to protect their insects from microbial contaminants. However, our main and possibly only hope is that thorough education will improve our workers' care to apply proper sanitation. The video that showed (using laser light and special cameras) the mist of droplets from coughing or even talking is an impressive educational aid. Also, by having workers trained with culture plates (such as RODAC plates discussed by Sikorowski 1984, Bruch and Smith 1968, and McGowan 1985) where they are instructed to press sterile plates to various laboratory surfaces to determine the counts of microbes that are present in their various laboratory environments (computer keyboards, rearing room walls, bathroom sink and toilet handles, etc.). If workers are trained to use the plates carefully so that they do not contaminate them with non-target surfaces or sources, they can see after two–three days of incubation at rearing room temperatures how many microbes are present on these surfaces, and they can even do experiments where they compare cleaned/sanitized surfaces with such surfaces prior to sanitation. This can be a very compelling lesson in sanitation, which most well-meaning workers will find enlightening and encouraging toward their development of higher levels of sanitation. This discussion of the value of sanitation can be one of the fundamentals for improving performance of workers at all levels of insectary duties. Once the rigors of the microbial basis of sanitation are established for personnel, the other duties of rearing can have the foundation needed for higher quality job performance.

ii. **Case 2: Elements of error in measurements**. The concept of error is discussed in Chapter 7; however, the role of teaching/learning the statistical and physical basis of error (or variation) can be another basic example and model of rigor in rearing duties. For example, a pipette (such as a 1,000-µL pipette) and a sensitive balance (one that reads in thousandths of a gram or ten thousandths of a gram) can be used to teach the concepts of accuracy and precision. The learner (trainee) can be asked to deliver 1000 µL of water to a clean plastic weigh boat. If the balance is properly set up (leveled and the weigh boat pre-tared), the reading should be 1.000 g. If the learner repeats the measurements ten times, the readings should be the same for all ten readings **if there is no error**. If the learner is encouraged to try to use steady, well-timed uptake and delivery of water vs. trying out a hasty uptake and delivery, he/she will learn quickly from the readings how poor pipetting technique can lead to large deviations in the measurement of volume.

12.5 APPROACHES AND METHODS OF TEACHING INSECT-REARING SCIENCE

12.5.1 On-the-Job Training/Education

I have been a teacher for 55 years. My teaching experiences include insect-rearing workshops, seminars, online and on-site courses, and, of course, on-the-job training (not to mention courses in English and many aspects of biology). I taught graduate students in entomology, professional entomologists, and rearing facility personnel, using a variety of teaching techniques, including individualized assignments of readings and lab-projects where students/learners were extensively on their own in terms of discovery. I often had recourse to teaching individuals whom I was training/educating to perform rearing operations where I used a one-on-one interaction, which I refer to as the "Socratic Method" (mentioned above). In this method, the teacher engages the student/learner in a dialogue based on questions and answers that aim to develop in the learner (and really in the teacher, too) a higher level of **critical thinking**. As conveyed in the discussion of Bloom's Taxonomy, science, technology, and other higher education practices must include critical thinking that is in a continual process of developing the discipline in question to a higher level of reasoning. Insect rearing flourishes when it is treated with constant efforts of the practitioners to find rationale and deeper understanding of the system and its components. There is a simultaneous requirement in rearing system practices for processes to be performed with *exactness*, *uniformity*, and *rigorous adherence to established principles and practices* **and** to be *questioned* as to **why** we do the prescribed steps and **how** could they be improved. The well-trained, educated insectary worker learns early in her/his career what is the appropriate time and place for adherence to SOPs and questioning those SOPs. This leads to the practice of **continuous improvement in the rearing system**. A hallmark of this balance between the status quo and continuous improvement is discipline. **The disciplined mind is at once obedient to rigorous adherence to protocols and prepared to challenge and improve those protocols but only after the basis of the protocols is mastered**.

One of the best technicians that I supervised (and there were several really good ones who helped my career in every success that I accomplished) was an undergraduate student from the University of Arizona, Nina Crowe-Urias. She came to my lab with a major in dairy science and a background as a farm-hand who tended dairy cows from an early age and had no compunction to follow routine hard work. At the time when I was trying to develop an artificial diet for the big-eyed bug, *Geocoris punctipes* Say (Hemiptera: Geocoridae, formerly Lygaeidae), Nina was charged with keeping our *Geocoris* colony going (feeding them coddled beet armyworms and green beans) and with assaying the various artificial diets that we were testing. A major point that I came to understand with my research on diets was that the success of the rearing efforts was only as good as the competence of the rearing staff. The best diet formulation in the world will fail, if the technician making the diet does not follow SOPs properly. With Nina, I had an assistant who would carry out the experiments to the letter of the SOPs that we had been establishing. Prior to Nina's entry into my lab, we had tested hundreds of diets, all of which were total or nearly total failures. Once I had discovered that the feeding paradigm that we were adhering to was erroneous (that hemipteran predators were hemolymph-feeders), and once we revised our feeding paradigm to an extra-oral digestion model, we were on our way to developing truly successful diets (Cohen 1985a and 1985b).

The new diet model predicted that using semisolid food with high concentrations of nutrients, we would better meet the needs of an arthropod that injects prey with digestive enzymes to liquify the internal contents and ingests a nutrient-rich slurry. This led to our using various blended meat products to simulate the internal contents of prey, and it began our course of developing meat-based semisolid diets (Cohen 1985a; Cohen and Urias 1986). This approach revolutionized the field of predator and parasitoid dietetics. But the questions of how to formulate the diets, package the diets, and all other rearing matters such as replacement of plant materials as a water source and proper oviposition sites remained to be resolved (Figure 12.10).

A second technician who helps personify the ideal professional who was trained/educated on-the-job to become one of my all-time best lab workers was Ms. Lisa Smith. She came to my lab for a temporary job rearing predators, especially lacewings, in the Phoenix, Western Cotton Laboratory (USDA, ARS). She had an undergraduate major of education with a respectable science background, but no courses in entomology, food science, engineering, or any of the other courses directly applicable to insect rearing; so Lisa's development as a rearing specialist was entirely on-the-job.

On-the-Job-Training: Positive points: very hands on (learn by doing), very immediate, direct contact, quick feedback from instructor to learner, convenient in terms of learner not needing to travel. Negative points: limited in scope to immediate system (no broad educational concepts), expertise of instructors often limited.

12.5.2 Workshops

For the most part, besides a few on-site and online courses offered by NCSU (see the next section), the primary form of insect-rearing education has been workshops offered by MSU over the past 20 years and a few workshops offered at NCSU over the past 15 years. The format of the workshops is largely a "top-down" form of instruction with little opportunity for exchange of ideas during the workshop activities. Having conducted a number of workshops, I arrived at a position critical of the learning environment where participants are expected to listen to instructors for about eight hours a day (with some breaks and short visits to nearby facilities). The opportunities for participants to raise issues that are pertinent to their own rearing situations and experiences are limited to brief questions and short answers by the instructors. In my experience, the workshops consist of about 90+ percent lecture and passive learning and less than 10% active learning. Also, a deficiency of workshops is the lack of feedback that comes from testing and student/participant-based projects. Finally, I am critical of the compressed time frame of the workshops where about 40 hours of teaching/learning are to take place within a five-day period. There is little chance for the ideas and new learning to "incubate" in the minds of the learners, and little opportunity for later follow up once the workshops have ended. All this criticism aside, the workshops have been pioneering in the quest to provide some kind of formal rearing education, and the interactions between students with one-another and some (albeit limited) contact with authorities on insect rearing are of value.

Figure 12.10 Ms Lisa Smith preparing artificial diet for green lacewings and other predators; and cover of 1997 *Agricultural Research Magazine*, showing lacewing larva (*Chrysoperla rufilabris*) that was reared on the Cohen and Smith 1998 artificial diet.

12.5.3 On-Site Courses

12.5.3.1 On-Site vs. Online Courses

I have been teaching both on-site and online courses in insect rearing over the past decade (from 2010 through 2020), and I conclude that on-site (face-to-face) courses allow more dynamic, immediate, and tangible educational experiences. In contrast, however, online courses allow instructors to have a larger audience that can "attend" classes remotely thus potentially reaching a worldwide audience. In the current times of the COVID-19 pandemic, there has been a revolution in the delivery of online materials with vehicles such as *Zoom* and *Google Meet* adding to existing systems such as Microsoft's *GoToMeeting* allowing both synchronous and ***asynchronous*** classes. Briefly, the ***synchronous*** meetings are live and must be coordinated as to times when the classes can meet. While this is a drawback from having meetings/classes with participants in different time zones, it has an immediacy that comes close to face-to-face interactions. I have even used these formats with young children (e.g., second grade classes) showing them videos of silkworms and holding lively discussions of the subject. The access to the desktop of the computer allows a huge variety of media to be utilized. However, various problems inherent in these systems make it difficult to do longer sessions such as would be desirable for all-day workshops. The various platforms have features such as raising hands and other interactive capabilities that enhance learning. In asynchronous classes, the presentation is much more "top-down" where the presenter talks to the class with no immediate chance for questions or discussion. The convenience of asynchronous online classes is unparalleled, and depending on the instructor's and students' agreement, the classes can go on with students having complete access to the materials for days, weeks, months, or indefinitely. The students can pick up a lesson/module at any time, work on it during a lunch break or after work, even in the middle of the night. The student can also repeat the lesson without restriction, so that difficult points can be reviewed until the students feel that they have learned everything.

12.5.3.2 Testing in Online and On-Site Classes

Both types of online instruction can be enhanced with testing, and for some universities there are arrangements where students can go to a local center to take tests under a proctor's supervision. In my classes, all testing is take-home, so there is no need for supervision. This leads to the point that I have found, over my career in teaching, that while objective tests are easier to grade than essays or even short answers, objective tests are limiting in terms of the scope of understanding that they can test. Even really capable students can be boxed into performing poorly on objective tests. This point becomes clearer in Appendix 4 where I have provided tests from several students, and the reader can see the degree of critical thinking and exploration of the various levels of Bloom's Taxonomy that is displayed.

12.5.3.3 On-Site Courses

On-site teaching lectures provide the immediacy of student/teacher and student/student interactions in a real-time basis. Often, the teacher presents a statement, question, diagram, or image that elicits in at least one student a response that can be stimulating to other students or that can elicit in other students a fresh idea or perspective. While a well-designed online course can also involve other students in one student's response, but this does not take place with the immediate reaction possibilities inherent in on-site conditions. Besides the dynamics of student-student/student-teacher interactions, on-site teaching allows the extra dimension of live and immediate interplay. The many photographs in this chapter and in Appendix 4 show students performing hands-on operations that allow observations by the instructor (and other students for that matter) and especially comments by the instructor (such as "you forgot to level your balance," "your pipetting technique is nice and smooth: class, notice Ms. X's pipetting precision," and "you have the ovarioles really clearly showing the insect's reproductive capability"). These comments (positive and negative) are highly informative when they are shared with the class both in lab and demonstration situations; and even in lectures, the interactions between instructors and students at various levels of ability in the subject make for a wonderful learning situation. Live teaching also allows humor and subtlety that can enhance the learning situation (Table 12.1).

The following is an example of the projects that can be done especially easily in on-site courses. The example here is a composite of posters that students provided as the vehicles for presenting their class project results:

1. Laboratory ("Hands on") Experiences
2. Reviewing Relevant Literature
3. Synthesizing and Analyzing Rearing Science Content (Critical Thinking about Rearing Issues)
4. Communicating Rearing Science Inquiry and Discovery

The topics of these posters (Figure 12.9; also see Appendix 4) included tests of free radical scavenging activity of diets of lacewings, levels of methyl paraben tolerance by imported cabbage worms, effectiveness of different yeasts in lacewing diets, rheology (texture analysis) of hornworm diets, and tolerance to surface sterilization by scarab beetles. These studies represent group research projects performed in the manner of Problem-Based Learning. The groups of two or three students did three projects for the semester, where they were charged with doing research on rearing various insect pertaining to the students' interest but integrating various techniques taught in the course. The techniques included various analytical approaches (protein content, lipid content, free-radical scavenging capacity, rheology [texture] analysis, and various diet-making as well as other rearing system techniques). One part of the assignment was to develop and explain explicit rationale and hypotheses that guided the research. Some of the students' research was conducted in their own laboratories, and some of the research was conducted in the Insect Rearing Education and Research Laboratory under Dr. Cohen's supervision. Students also had access to the Food Science Rheology Laboratory and several other campus facilities (Figure 12.11). Students were encouraged to interact with one another, as well as with several science programs that were related to insect-rearing science (for example with the electron microscope facility and the BTEC facility for bioreactor facilities).

Besides the students' access to various laboratories (such as Food Science Rheology and bioreactor facilities, etc.), students are expected to visit my lab several times per semester to see firsthand what kind of rearing activities is being done and especially to get "mini-tours" of the rearing and rearing-related analytical equipment we use. For example, we have extensive video microscopy, various grinders, HPLC, spectrophotometers, lipid extraction equipment, water activity meters, moisture balances, and freeze drying equipment that the students get to experience

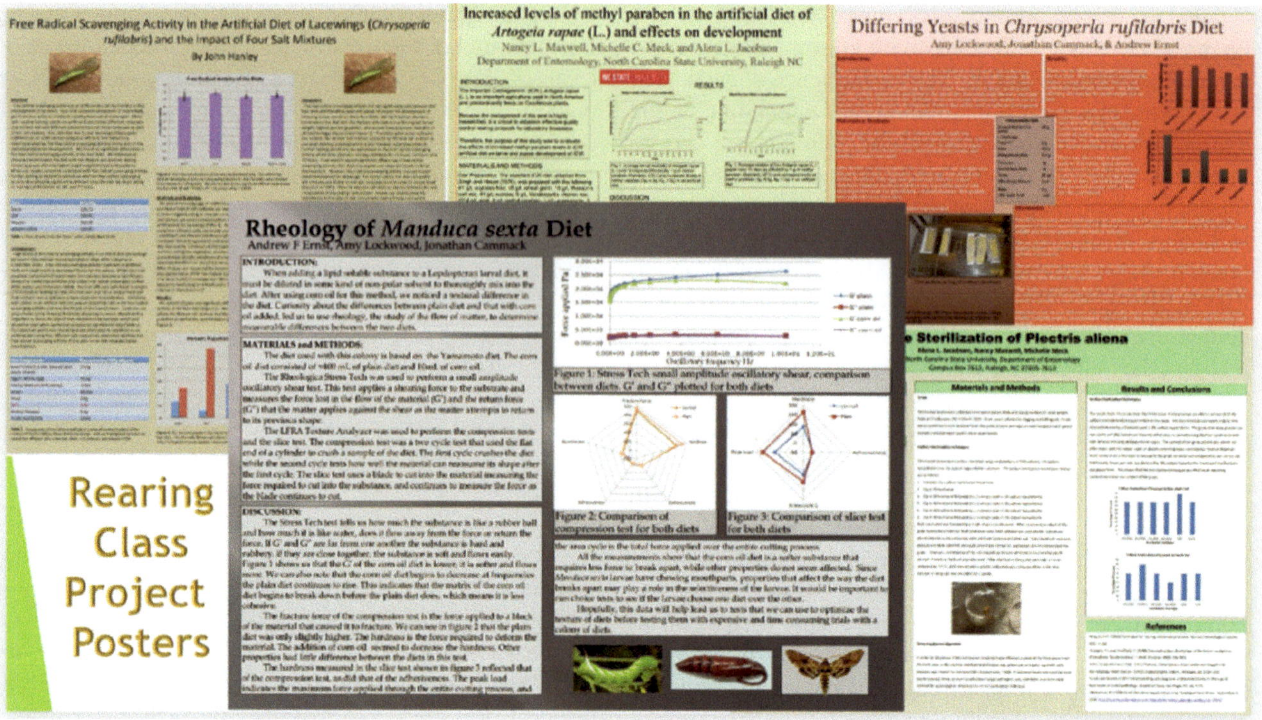

Figure 12.11 Composite image of posters produced by students in the insect-rearing course taught by Allen Carson Cohen in the Spring of 2011 (posters authored by John Hanley, Nancy Maxwell, Micelle Mech, Alana Jacobson, Amy Lockwood, Johnathan Cammack, and Andrew Ernst).

firsthand and are encouraged to use for their projects (see Figures 12.1–12.5).

12.5.4 Online Courses

The items that follow are taken from my current online course.

Thus far, more than 100 participants have completed this course (Figures 12.12 and 12.13). The course is offered online in an asynchronous format, meaning that there is no set meeting time; the participants are able to work their way through the units (called "modules" here). The **advantages** of the course are that while the participant is enrolled (we allow a six-month access period), she/he can repeat the modules at will, stop during the midst of a module, and otherwise deal with the materials at her/his own pace. A further advantage is that participants have access to the instructor for the full enrollment period, and they can ask questions via email contact. The course is presented in a Moodle format (moodle.org) and provides the versatility in terms of participants having access to various Moodle components such as quizzes, access to PDFs pertaining to given subjects, and tutorial videos. Something that I have found to be both an advantage and a disadvantage of this course and asynchronous courses is that the participant is responsible for her/his own learning. This course is currently offered as a non-graded learning experience, and though there are short, objective (mainly True/False) quizzes, the depth of learning that would be desirable is not tested by this method. In my past five years of experience with this online course, I have found that most students have few to no questions and therefore have virtually no learner-teacher contact. In a few cases, I have found that highly motivated participants asked probing questions about their own rearing experiences, and these persons were able to profit greatly from the course, in terms of having the basics laid out in an organized, systematic way, and at the same time having access to the instructor to have him explain in-depth concepts that they did not understand or content that they wished to have treated in further detail.

Some examples of effective learning are discussed here. A few participants commented on the large number of people who rear insects as their occupation or profession. That challenged me to further probe numbers of rearing personnel and the types of rearing programs that were in operation in our current world. My search revealed that there were hundreds of thousands of people worldwide who made all or most of their living rearing silkworms, especially in Asian counties. I found that several dozen people reared medical maggots for maggot therapy; and there are several hundred workers worldwide who rear insects as pets, biological

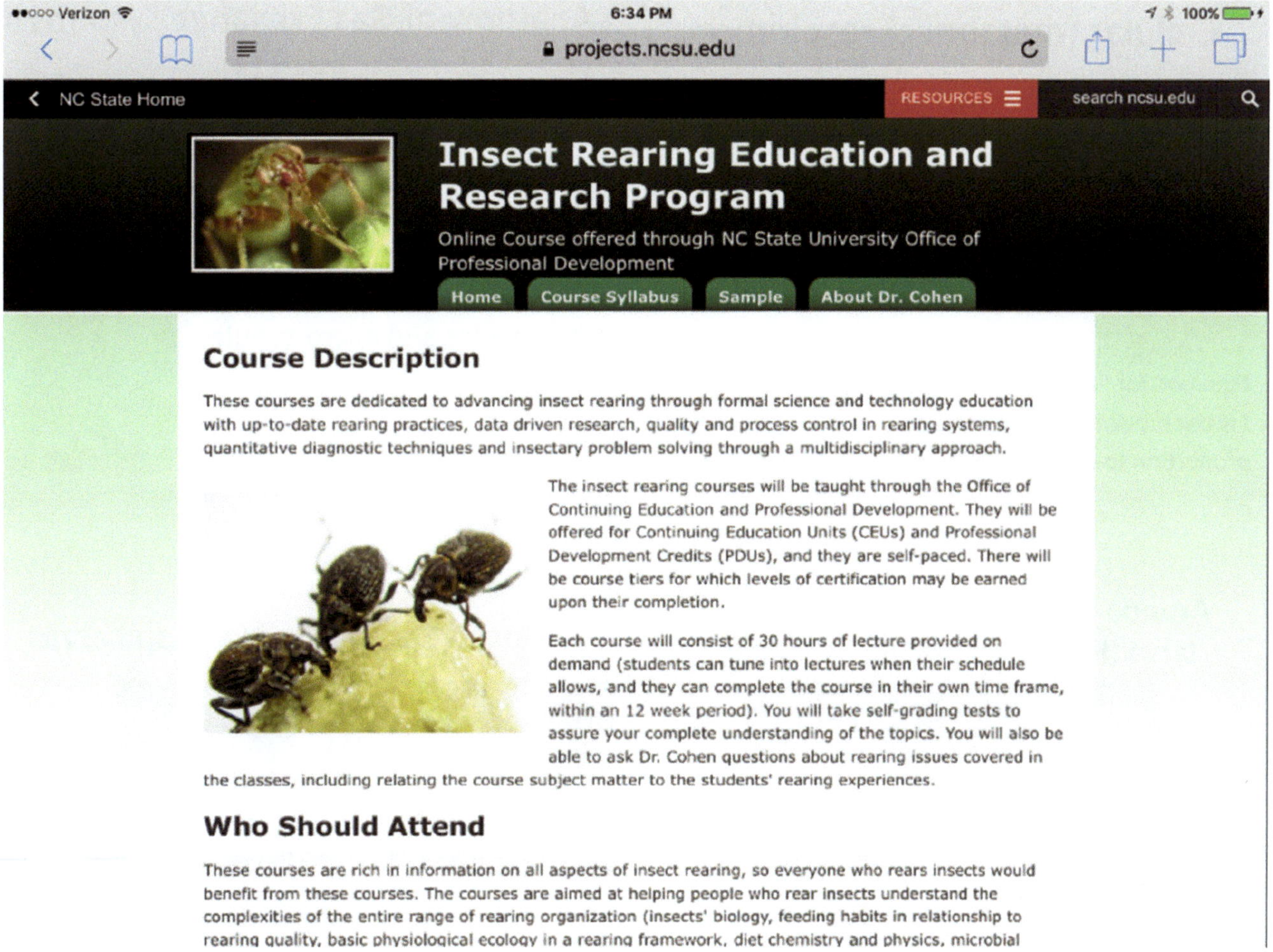

Figure 12.12 Online course announcement.

control agents, and for celebrations (such as butterfly releases at weddings). Several hundred people in the United States alone rear wax worms for testing toxins and microbial virulence.

Below is an example of a presentation (Gas Exchange from Module 12 of the online insect-rearing course). The intention behind this presentation is to provide potential rearing class instructors with a format and script for

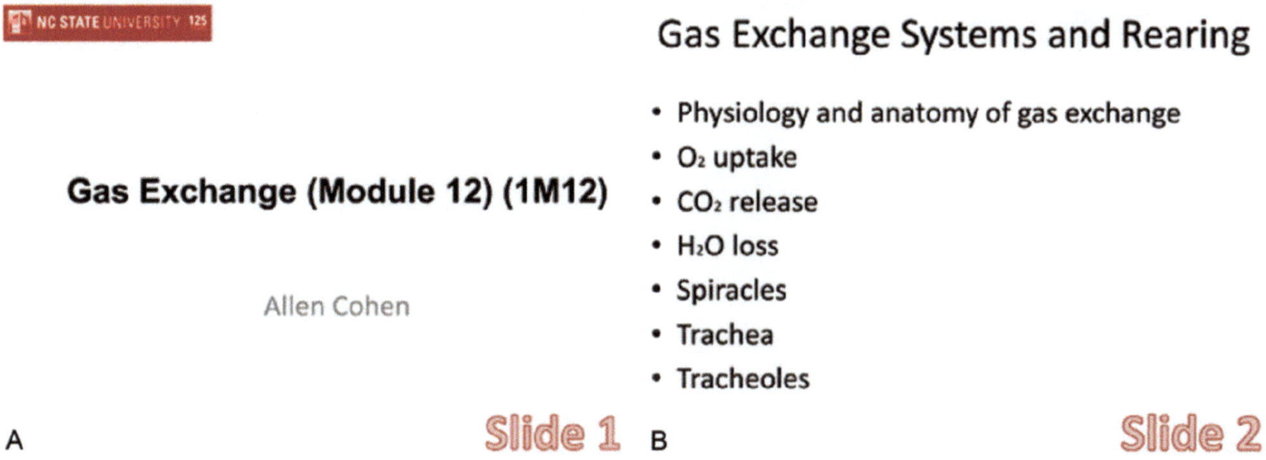

Figure 12.13 (a–l) PowerPoint slide show for gas exchange in insect rearing systems.

(Continued)

Cellular (metabolic) respiration

- Fuels + $O_2 \rightarrow CO_2$ + water (+ energy transfer ATP production)
- Fuels include carbohydrates and lipids and sometimes amino acids
- Virtually all biological processes in your rearing system are ATP dependent!

$$J = -\chi \frac{\partial C}{\partial x}$$

Equation for Ficke's 1st Law where J is the movement of material in proportion to concentration and distance travelled

Slide 3

C

Anatomy of Gas Exchange Systems in Insects

- Spiracles
  - Open and close to permit gas entry and exhaust
  - Must not be blocked by diet, fluids, etc.
- Trachea
  - Flexibility and bellows action vs. diffusion
- Tracheoles and micro-tubules

Slide 4

D

Arundo gall wasp *Tetramesa romana* larva (Hymenoptera: Eurytomidae)

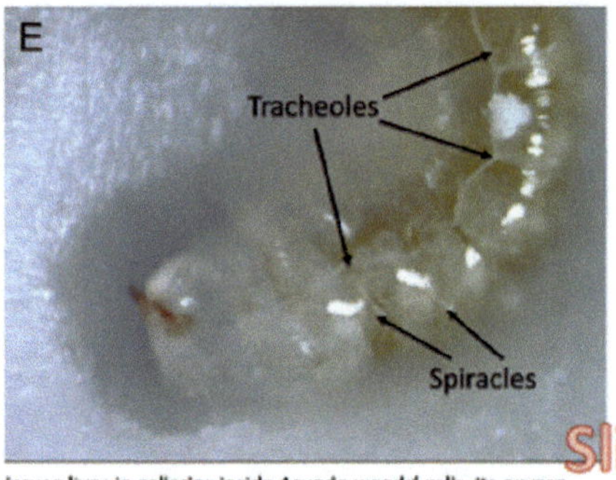

Slide 5

This gall wasp larvae lives in galleries inside Arundo weeds' galls. Its oxygen supply is limited, resulting in selective pressure to have highly efficient gas exchange structures.

Closed Spiracle in *Vanessa* Larva

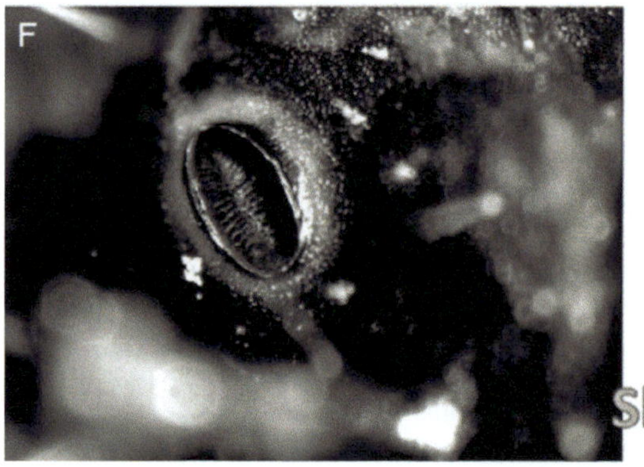

Slide 6

Larvae digging galleries in diet

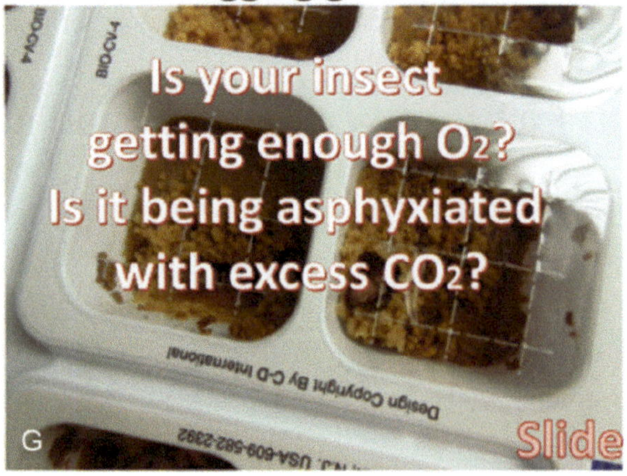

Slide 7

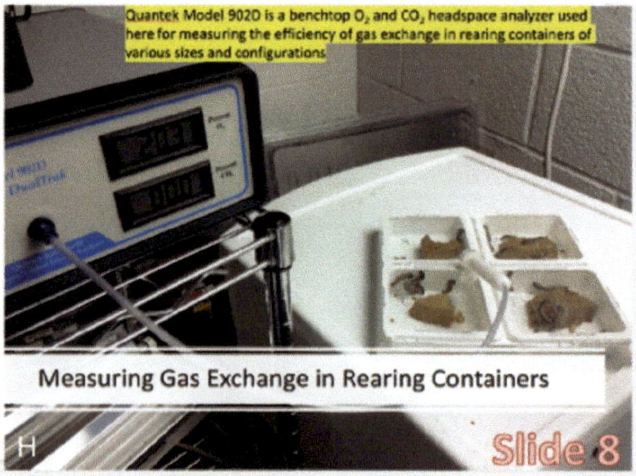

Slide 8

(Continued)

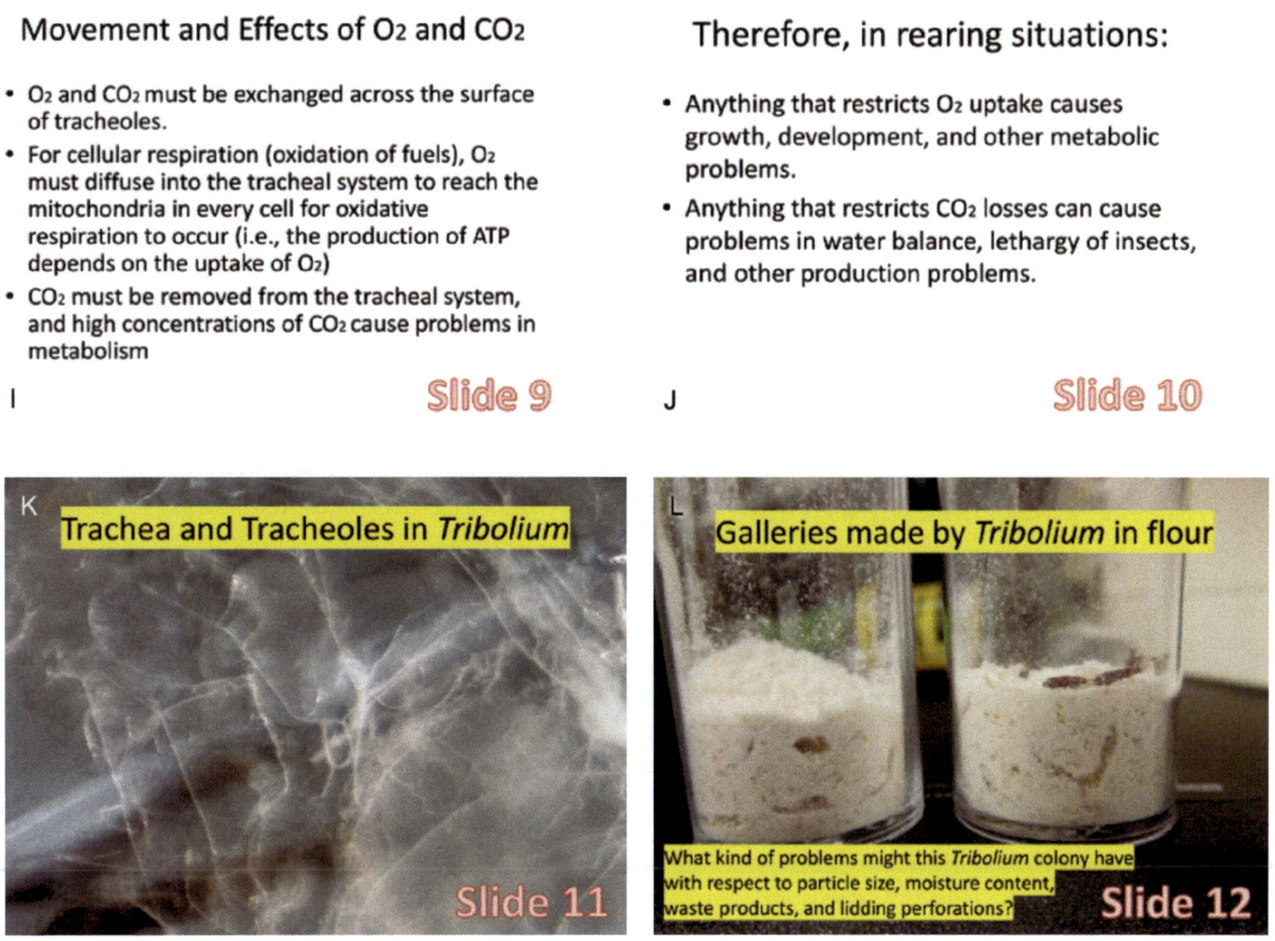

Figure 12.13 (*Continued*)

a typical learning module that is present in the online course. If the reader checks the outline in Table 12.2, she/he will find that virtually all aspects of rearing are treated, including the basic aspects of the insects' biology and the physical characteristics of rearing system processes.

This figure depicts a slide show of a lesson in an online insect-rearing science and technology course. While some students of insect-rearing courses have little or no background in insect physiology, including such specialized topics as gas exchange, others have more extensive and in-depth knowledge of the anatomy, physiology, and biochemistry of gas exchange systems of insects. Such knowledge might include the nature of spiracles in oxygen uptake and carbon dioxide release, control of spiracular openings, means of mass movement of gases in the tracheole system, including bellows movements by muscular exertion, diffusion of gases following Ficke's "laws" of diffusion $\left(J = -\chi \frac{\partial C}{\partial x}\right)$ and the biochemical aspects of respiration, including such cellular phenomena as glycolysis, Krebs cycle, beta oxidation of lipids, deamination of amino acids, and the electron transport chain. The degree of sophistication of a lesson on gas exchange and respiration can be complex enough to fill several levels of lecture detail or it can be simple where it is explained that oxygen is necessary to process fuels and produce cellular energy "currency" known as ATP, and carbon dioxide must be removed to keep the system from "backing up." However, regardless of the sophistication of the students, there is a need to connect the biological nature of O_2/CO_2 exchange and the special features of rearing systems that enhance or impede the all-important gas exchange process. While it is not possible in this book to present the entire online course in Fundamentals of Insect Rearing Science and Technology, I provide here a sample of the presentation using a module (lesson) on gas exchange in rearing systems.

The narrative of the PowerPoint lesson on Gas Exchange in Rearing Systems (Module 12 in the online rearing course) is presented here. The title slide identifies the topic (**Slide 1**), and the items to be covered are listed in **Slide 2**, where the narrator would explain that the lesson aims to depict the structures (anatomy) of

Table 12.2 The Detailed Course Outline for the Introduction to Insect-Rearing Science and Technology

Introduction to Insect-Rearing Science and Technology
a. Overview (Who Rears Insects and Why Do They Rear Them?)
1. Course Philosophy (Module 1) (1M1)
2. Who Rears Insects? Why Do They Rear Them? Where Are They Reared? (Module 2) (1M2)
b. Insectary Structure
1. Insectary Layouts/Floor Plans (Module 3) (1M3)
2. Insectaries As Environments (Module 4) (1M4)
3. An Overview of Rearing Containers (Module 5) (1M5)
4. Basics of Insectary Sanitation (Module 6) (1M6)
5. Basics of Insectary Safety (Module 7) (1M7)
c. The Biological Characteristics and Requirements of Insects
1. Insect Life Cycles in Rearing Settings (Module 8) (1M8)
2. Insect Feeding Systems I: Nutrition (Module 9) (1M9)
3. Insect Feeding Systems II: Mouthparts (Module 10) (1M10)
4. Insect Feeding Systems III: The Digestive System (Module 11) (1M11)
5. Gas Exchange (Module 12) (1M12)
6. Movement and Internal Communication (Module 13) (1M13)
7. Rearing-related Components of Insect Development (Module 14) (1M14)
8. Reproduction (Module 15) (1M15)
9. Circulation and Transport (Module 16) (1M16)
10. Social Biology (Module 17) (1M17)
11. An Overview of "Healthy" Insects (Module 18) (1M18)
d. The Domestication Process (what are the steps in and consequences of domestication?)
1. Insect Genetics: A Brief Overview (Module 19)(1M19)
2. Domestication: An Overview (Module 20) (1M20)
3. Domestication: How It Affects Rearing and Insect Quality (Module 21) (1M21)
e. Understanding Insect Feeding Systems as They Pertain to Rearing (Module 22) (1M22)
f. Understanding Insect Diets (what's in diets and what do the components do?)
1. Types of Diets & Diet Terminology (Module 23)
2. Components of Diets & What They Do (Module 24)
3. More About Components of Diets (Module 25)
4. Interactions of Insect Diet Components (Module 26)
5. Stability of Insect Diets & Components (Module 27)
6. Water in Diets (Water Content and Water Activity) (Module 28)
7. Texture of Insect Diets I (Module 29)
8. Texture of Insect Diets (Module 30)
9. Handling and Treatment of Diets and Diet Components (Module 31)
10. Important Issues in Diet Presentation (Module 32)
g. Essentials of Insect Reproduction (mating, oviposition, and diapause in rearing systems)
1. Mating in Rearing Systems (Module 33)
2. Oviposition in Rearing Systems (Module 34)
3. Diapause in Rearing Systems (Module 35)
h. Microbial Management
1. Kinds of Microbes Relevant to Rearing (Module 36)
2. Kinds of Microbial Interactions (Contaminants, Symbionts, Pathogens, and Commensal Microbes) (Module 37)
3. Managing Microbes I (Module 38)
4. Managing Microbes II (Module 39)
5. Managing Microbes III (Module 40)
i. The Insectary as Environment
1. Temperature and Its Biological Importance (Module 41)
2. Humidity and Its Relationship with Rearing Issues (Module 42)
3. Light (Photoperiod, Intensity, and Quality) (Module 43)
4. Gas Exchange (Module 44)
5. Noise and Vibrations: the Insectary Soundscape (Module 45)
6. Air Quality (Module 46)
7. Special Cases I (Soils) (Module 47)
8. Special Cases II (Aquatic Systems) (Module 48)
j. Introduction to Statistical Process Control (SPC) in Rearing
1. What Is Statistical Process Control? (Module 49)
2. What Are the Sources of Variation in the Rearing Setting? (Module 50)
3. Using Fishbone Diagrams (Module 51)
4. What Do You Measure and How Do You Measure It? (Module 52)
5. Using SPC Tools for Problem Solving and Process Improvement (Module 53)
6. Using Pareto Charts and Pareto Thinking (Module 54)
7. Using Fishbone Diagrams (Module 55)
8. Developing and Using Process Control Charts (Module 56)
k. Managing Processes in Rearing Systems (Instruments and Equipment)
1. Basic Laboratory Instruments for Rearing I: pH and Weighing (Module 57)
2. Basic Laboratory Instruments for Rearing II: Water Activity and Water Content (Module 58)
3. Basic Laboratory Instruments for Rearing III: Volumetric and Spectrophotometric Measurements (Module 59)
4. Basic Laboratory Instruments for Rearing IV: Microscopes (Module 60)
5. Diet Instruments I (Module 61)
6. Diet Instruments II (Module 62)
l. What Every Rearing Specialist Should Know: Developing Education and Academic Background for Self-Improvement as a Rearing Specialist
1. Further Coursework (Module 63)
2. Reading Books and Journals (Module 64)
3. Networking and Tapping into Other Disciplines (Module 65)
4. Using the Internet as a Resource (Module 66)

the gas exchange system and some pertinent matters of the physiology of gas exchange—such as why adequate amounts of oxygen must be available to the insect if it is going to achieve its full growth potential with the most efficient possible utilization of fuel molecules to support the insect's production of ATP, which is the currency of virtually all cellular activities. The nature of oxygen uptake and carbon dioxide release is touched on, including the fact that for the gas exchange to take place, wet surfaces are required, therefore adding the issue of water loss to the complex properties of gas exchange. Also, it is here that special matters of gas diffusion are introduced (including Ficke's first law of diffusion). These points are further explained in **Slides 3 and 4**, and in the voice-over narrative, the point is emphasized that under conditions where the insect is oxygen-starved and CO_2-overloaded, it must resort to anaerobic metabolism, which is about 1/20 as efficient at ATP production, and all normal growth and activities are hindered. The photographs in **Slides 5, 6, and 11** reinforce the understanding of the tubular nature of the tracheoles, and in **Slide 6**, the spiracle from a painted lady larva is seen in its closed position. The narrator can also explain here that an excess of CO_2 can lead to keeping spiracles open (gaping) so that water is lost excessively from the respiratory (gas exchange) system. Such a loss translates to symptoms in the rearing system as lethargic insects that are not eating robustly, slowed growth, slowed development, and various problems with deformities and loss of immune function (meaning an increase in rates of infection). These problems can be exacerbated by factors such as insufficient ventilation in lidding of rearing containers, failure of diet structure (collapse of tunnels, for example) due to gelling agent failures or particle sizes of diets such as flours and meals for *Tenebrio*, *Tribolium*, or other stored products feeders, or over-crowding of insects per rearing container. These potential failures of the gas exchange accommodations are illustrated in **Slides 9 and 12**. An excellent means of detecting problems in gas exchange can be developed around the use of an O_2 and CO_2 measuring system such as the Quantek Model 920D benchtop headspace analyzer (Quantek Instruments, Inc. 183 Magill DriveGrafton, MA 01519). It is worth noting that the instrument used in our laboratory is equipped with a Luer-lock hypodermic syringe needle that we use to penetrate lidding. The system is designed for measuring small volumes of gas in headspaces from packages of dried foods and pharmaceutical products. We have found the instrument to be sensitive to differences in the O_2 and CO_2 present in 30-ml rearing cups, as well as larger rearing containers such as the 8-cell trays seen in **Slides 7 and 8**. There is an overview/take home message to be conveyed from **Slides 11 and 12**. The tracheal system seen in **Slide 11** is highly branched, with trachea splitting into smaller (capillary-sized) tracheoles, which deliver and remove O_2 and CO_2 to and from individual cells. Since EVERY cell in the insect's body requires such delivery and removal, it is crucial that the presence of these two gases is as rich in O_2 and as low in CO_2 as possible (i.e., as close to 21% O_2 and about 0.04% CO_2). This means that in a colony of *Tribolium* (**Slide 12**), there is a dependence of the insects on gas exchange capability that allows O_2 and CO_2 to diffuse adequately from the outside air to the beetle within the tunnels in the flour. Factors such as particle size that is too small or moisture content that is too high can influence the permeability of the flow—for example, if the flour is caked from moisture and insect waste products or if the batch of flour had an exceedingly high portion of fine particles.

Finally, it should be noted that the PowerPoint presentation and the script demonstrated here can also be used as the approach to lecturing in on-site (or face-to-face) classes. It should further be noted that because of space limitations in this chapter, this is merely an example of the kind of presentation that can be made for each of the topics listed in the online course, which consists of 66 Modules about various topics of rearing.

12.5.5 Teaching Methods

Teaching procedures can be divided into informal and formal education, and within these categories, there are many subdivisions. Informal education includes independent rearing, on-the-job training, and trial and error learning. All of us who entered the field of insect rearing have found ourselves with questions about how or why to perform various rearing practices, and after asking colleagues the questions about rearing techniques, often we would turn to the literature. The literature on rearing goes back to foundational papers from the early 1900s (Cohen 2015, 2018), but the information is scattered and often difficult to access. The rearing information in the majority of publications is cryptic and can only be found with scrutiny of papers and books that have titles, abstracts, and keywords that provide no clue that there are valuable contents about rearing (Cohen 2018).

12.5.5.1 The Curricula for Two Courses Are Presented in Tables 12.1 and 12.2 and a Sample of a PowerPoint Lecture from the Online Course Is Presented in Figures 12.13 (a–l)

12.5.5.1.1 Critical Thinking and Testing in Rearing Education

One of the central goals of all education, including insect-rearing studies, should be to prepare rearing personnel to think critically about rearing processes/rearing systems. This is where rationale interfaces with the better understanding and more powerful approach to insect rearing. This is where the questions of how and why things work or fail to work become central. This is the mechanistic approach to rearing. If a diet component such as agar is important to the success of the diet, what features of the agar and its interactions with other diet components and the insects connect to the success? Likewise, all other components of diets and the other parts of the rearing system should be targets of understanding by the rearing professional.

12.5.5.1.2 Using Bloom's Taxonomy to Deal with Critical Thinking

Appendix 4 contains exams that I had given in recent on-site courses that I taught in insect rearing. In these test exams and in all tests that I administered, I strived to employ the recommendations and guidelines from Bloom's Taxonomy. Briefly these concepts were presented by Bloom et al. 1956 and were revised and updated by Krathwohl 2002. The summary of these concepts is presented here in Table 12.3 derived from a website by Armstrong 2018. Basically, Table 12.3 shows in ascending order the levels of thinking sophistication, going from the simplest form of learning or dealing with knowledge from remembering → understanding → application → analysis → evaluation → creation. A rearing student might deal with remembering that carbohydrates, proteins, and lipids are major classes of insect nutrients; he/she may understand that some nutrients are essential rather than non-essential. The student may realize that a diet may be failing because it lacks an essential nutrient (application); then he/she may research the composition of the diet's components to find which ones provide or fail to provide the essential nutrients (analysis).

The levels of Bloom's Taxonomy and a synopsis of each category (from Armstrong). Note, the levels of thinking complexity increase as we go to the higher levels of this table. The bottom step is the least complex or most basic level of learning. It is essentially pure memorization without necessarily understanding or being able to apply what is learned. For example, the student is required to repeat the names of the three main sections of an insect's body (head, thorax, and abdomen). Or the student must draw from memory the structure of a glucose molecule. As we ascend the taxonomic ladder (or hierarchy), we go into the deeper levels of thought: for example, the glucose molecule has several polar groups (–OH groups, for example) so it is compatible with water.

The application of Bloom's Taxonomy has become a standard of teaching and evaluating learning at the full range of learning and critical thinking from the most basic levels to the most sophisticated ones. Often the Bloom model is used in education classes to help guide teachers to encourage more sophisticated thinking about any given subject. In the current case of insect-rearing education, writing questions and assigning projects that demand (or encourage) the students to go beyond just remembering and understanding, but for them to apply what they have learned, analyze the outcome (data interpretation or analysis of rearing system components), and then go on to evaluate (for example several diets' effects on the life history traits of their insects) and finally to create new levels of improved rearing systems or scientific models that had not previously existed. Within this entire Bloom's framework, students of insect rearing are most likely to enhance the practice of rearing and the knowledge that stems from the experiments by using the scientific method to achieve their goals.

Table 12.3 Simplified Hierarchy in Blooms Taxonomy

Level of Taxonomy	Example
Create	Produce a new or original work
Evaluate	Justify a position
Analyze	Draw connections between ideas
Apply	Use information in new situations
Understand	Explain ideas or concepts
Remember	Recall facts and basic concepts

12.5.6 Testing

Examples of the tests used to reinforce the learning of insect-rearing systems from the Bloom's Taxonomy framework and the rationale-driven, hypothesis-based science are provided in Appendix 4.

12.5.7 Other Ways of Learning Rearing (Symposia, Professional Meetings, Websites)

There is a growing body of insect-rearing systems information that is available rather sporadically. The Entomological Society of America frequently offers symposia on specialized rearing topics, such as rearing applications in SIT, biological control, or genetic pest management. Attendees of such symposia are exposed to many of the leading experts in the field of insect rearing, as they are at professional meetings that include insect-rearing reports such as posters at various entomological meetings. There are several websites that contain rearing information, such as www.wormspit.com, www.werearinsects.com, and others such as an excellent site that contained information about rearing tobacco hornworms, called the *Manduca* Rearing Site, sponsored by the University of Arizona and funded by the National Science Foundation. Unfortunately, the site was removed from the Internet and is no longer available. Companies that sell insects, especially for educational purposes (e.g., Carolina Biologicals), have web-based information and in house-produced publications such as booklets and pamphlets that cover rearing insects including Drosophila, hornworms, silkworms, and waxworms. A Google search of topics like "how do I rear wax worms?" can turn up dozens of YouTube videos about rearing. Unfortunately, many of these sites are of low quality with very questionable information while others contain excellent and helpful information that can help novices in insect rearing get started on a rearing project.

APPENDIX: STUDENT PROJECTS AND EXAM QUESTIONS FOR INTRODUCTION TO INSECT-REARING COURSES

This appendix provides real-life examples of student work. The students' works represented here were in classes or doing special projects under Allen Carson Cohen's supervision at North Carolina State University (NCSU). Included here are examples of reports on projects completed by students in the insect-rearing courses taught between 2011 and 2019, and examples of tests are also included. The quality of the students' work is a testimony to the abilities of the students and to the possibilities of in-depth learning and well-developed critical thinking capabilities in the domain of insect rearing. The reader is asked to note especially the constant attention to scientific applications to insect rearing. It is also noteworthy that the guidelines for the projects and many of the test questions are aimed at the higher levels of thinking as described in Bloom's Taxonomy (treated in detail in Chapter 12).

Project Posters

In all of the on-site rearing courses (sometimes called, "face-to-face" courses), the students are expected to do a class project as a substantial part of their grade. In the first semester that I taught the insect-rearing course at NCSU, I did not have exams, but instead used three projects as the basis for the students' grades. One project had to be presented as a PowerPoint Presentation; the second had to be presented as a poster, and the third project had to be presented as a written paper in the format of a typical scientific publication (Abstract, Introduction, Materials and Methods, Results, Discussion, and References Cited). The PowerPoint and poster presentations and the projects behind these presentations were done in a Problem-Based Learning context where the students, in teams of twos or threes, conducted the research practicing the group activities of brainstorming, charting ideas for consensus, discussing the best approaches to the problems, and meeting regularly to discuss and interpret results and problems. In all cases, the students were encouraged (urged) to consult me or other resources that were experts in the areas that they worked on. Part of the assignment was to perform a rearing exercise for the insects of their interest or that were conveniently available; and the other part of the assignment was to integrate at least three of the techniques (mainly analytical procedures) that the students had learned in the course. For example, all students were exposed to several biochemical analyses such as total protein determination by the Acid Orange method, free-radical scavengers determined by the 2,2-diphenyl-1-picrylhydrazyl (DPPH) method, and various textural (rheological) analyses that they learned in sessions all students went through in the Department of Food, Bioprocessing, & Nutritional Sciences Rheology Laboratory. Students were also taught use of microscopy techniques for observations of insect feeding and reproductive behaviors. In all class projects, an important part of the report is the students' explanation of an explicitly stated **rationale** and the hypotheses generated by the rationale (Figures A4.1–A4.5).

Project Papers

In the following section of this appendix, two reports are presented (in part) to reflect the type of studies that students completed as research papers on such factors as comparisons of diet components such as different yeast products and their effects on *Heliothis virescens* (by Terri Billeisen) and comparisons of several kinds of bean meal on *H. virescens* (by Diane Silcox). In all cases, students are assigned a project where they: (1) rear an insect and (2) integrate at least two of the principles learned in class to gain experience with rearing to witness the various problems and pitfalls in the rearing process and to integrate their learning about various kinds of analysis, including biochemical analysis of proteins, carbohydrates, free-radical scavengers, lipid analysis, and rheological (textural) analysis. The integration of the analytical procedures demands that students think of their rearing process with the higher levels of thinking emphasized in Blooms taxonomy. The projects also compel students to apply the scientific discipline of stating rationale and hypotheses that help assure that the students are learning the scientific basis of rearing. For example, Ms. (now Dr. Billeisen) asked the question, "does nutritional yeast that was grown on media" have a superior nutritional or palatability value over brewer's yeast in the *H. virescens*' diet because the nutritional yeast was not put through the debittering process? Her hypothesis was that because of the differences in yeast processing, the product that was not debittered would prove to be superior. Ms. Silcox-Reynolds (now Dr. Silcox-Reynolds) based her study on the differences in protein and other nutrient contents of various bean species (her rationale). Her hypotheses were based on prediction regarding these differences.

Influence of Various Yeast Sources on the Larval Development of the Tobacco Budworm (*Heliothis virescens*): Project Draft

Terri Billeisen

Introduction

Function and Types of Yeast

Diet is an important aspect of insect rearing and each individual component functions to fulfill an insect's

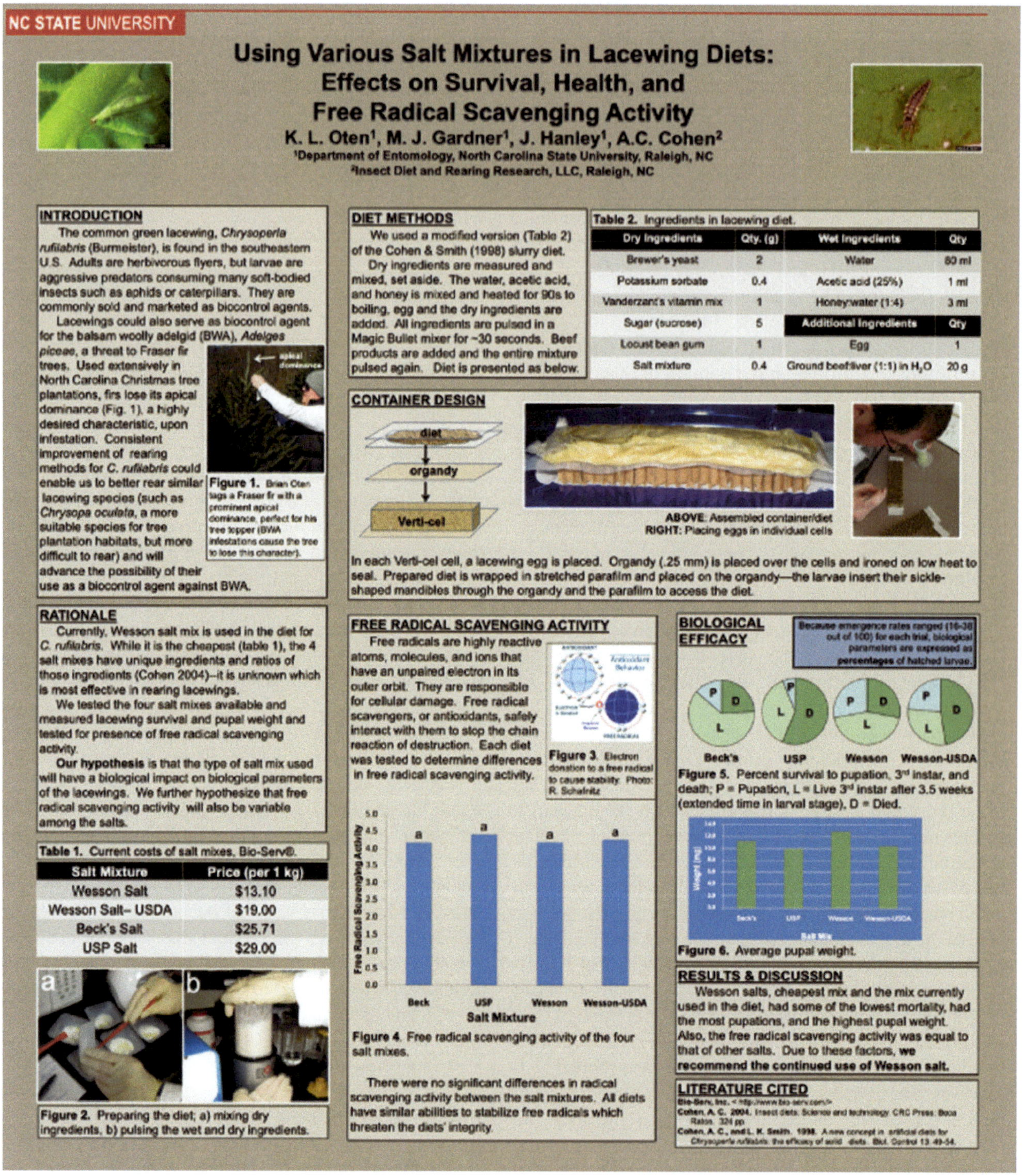

Figure A4.1 Poster presentation of research project on the effects of different salt mixtures in lacewing diets (authors Kelly Oten, Micah Gardner, and John Handley). Note that the students integrated the possible effects of the different salt mixtures on free-radical scavenging activity (with a DPPH test). Note further that the rationale for the experiment's variables is prominent as is the statement of hypotheses.

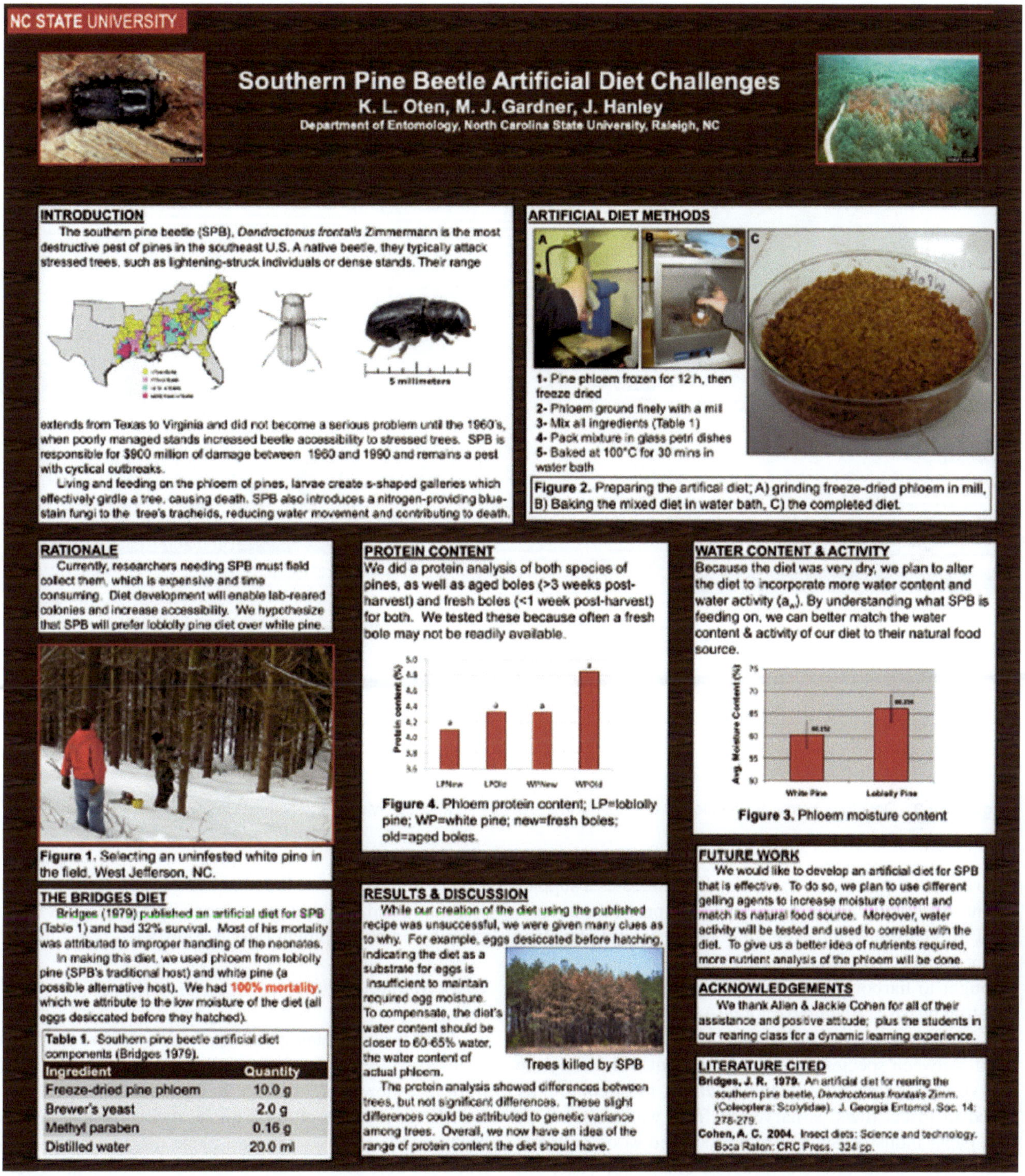

Figure A4.2 Artificial diets for southern pine beetles (K. L. Oten, M. J. Gardner, and J. Hanley) integrated artificial diet trials with analyses of protein content and water activity. Note the statements of rationale and hypotheses to assure that the diet testing was a scientific study, rather than merely a series of anecdotes.

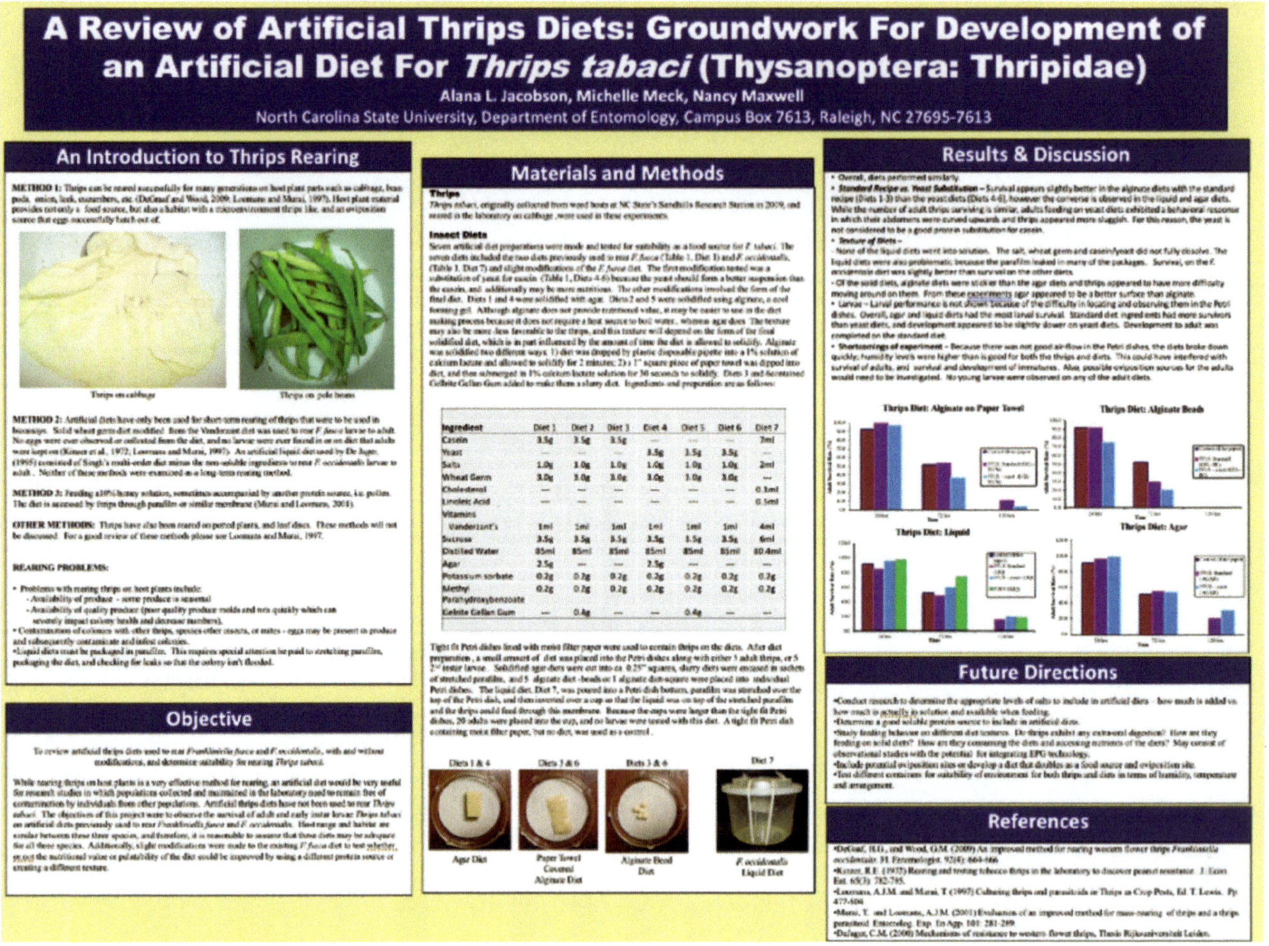

Figure A4.3 Developing an artificial diet for thrips (Alana L. Jacobson, Michelle Meck, and Nancy Maxwell).

nutritional requirements. Yeast is often added as a supplement in artificial insect diets to provide protein, nutrients and B vitamin complexes vital to successful development (Acker 1963; Vanderzant 1974). Yeast has also been thought to harbor and protect these nutrients during the heating phase of diet preparation (Acker 1963).

Yeasts can vary in nutritional quality and composition both between species and substrates from which they are harvested. Torula yeast (*Candida utilis*), most commonly used in insect diets, is harvested from wastewater from the paper industry. Sugars from processed wood pulp and oxygen are metabolized to generate energy for yeast cell growth. Brewer's yeast (*Saccharomyces cerevisiae*), harvested from hops and/or barley, is collected from brewery vats containing beer or ale following the fermentation process. Both torula and Brewer's yeast are generated as a by-product of each manufacturing system. Nutritional yeast, although the same species as Brewer's yeast, is specifically harvested from nutritious growth media (sugarcane, molasses, etc.) with the purpose of providing nutritional benefits. As a result, nutritional yeast is typically sold as a diet supplement for humans rather than for insects or mammals and could have a tremendous impact on insect growth and development (Stone 2006).

Yeast in Insect Diets

Yeast has the potential to supply many of the nutritional inputs that are essential for insect development. Previous work has shown that cabbage loopers can be reared on at least 12 generations on an artificial diet consisting of lima beans, yeast, and ascorbic acid alone (Shorey 1963). Olive fruit flies were successfully reared on diets containing yeast but the authors noted that growth and survivorship could vary depending on the brand of yeast used in the experiment. Variability in yeast influence in olive fruit flies could be attributed to the utilization of different yeast species and yeast substrates utilized throughout the harvesting process, potentially resulting in inconsistent nutritional quality. Brewer's yeast was a common component in early insect diets but has since been replaced with the less-expensive torula yeast as time and research progressed (Vanderzant 1974).

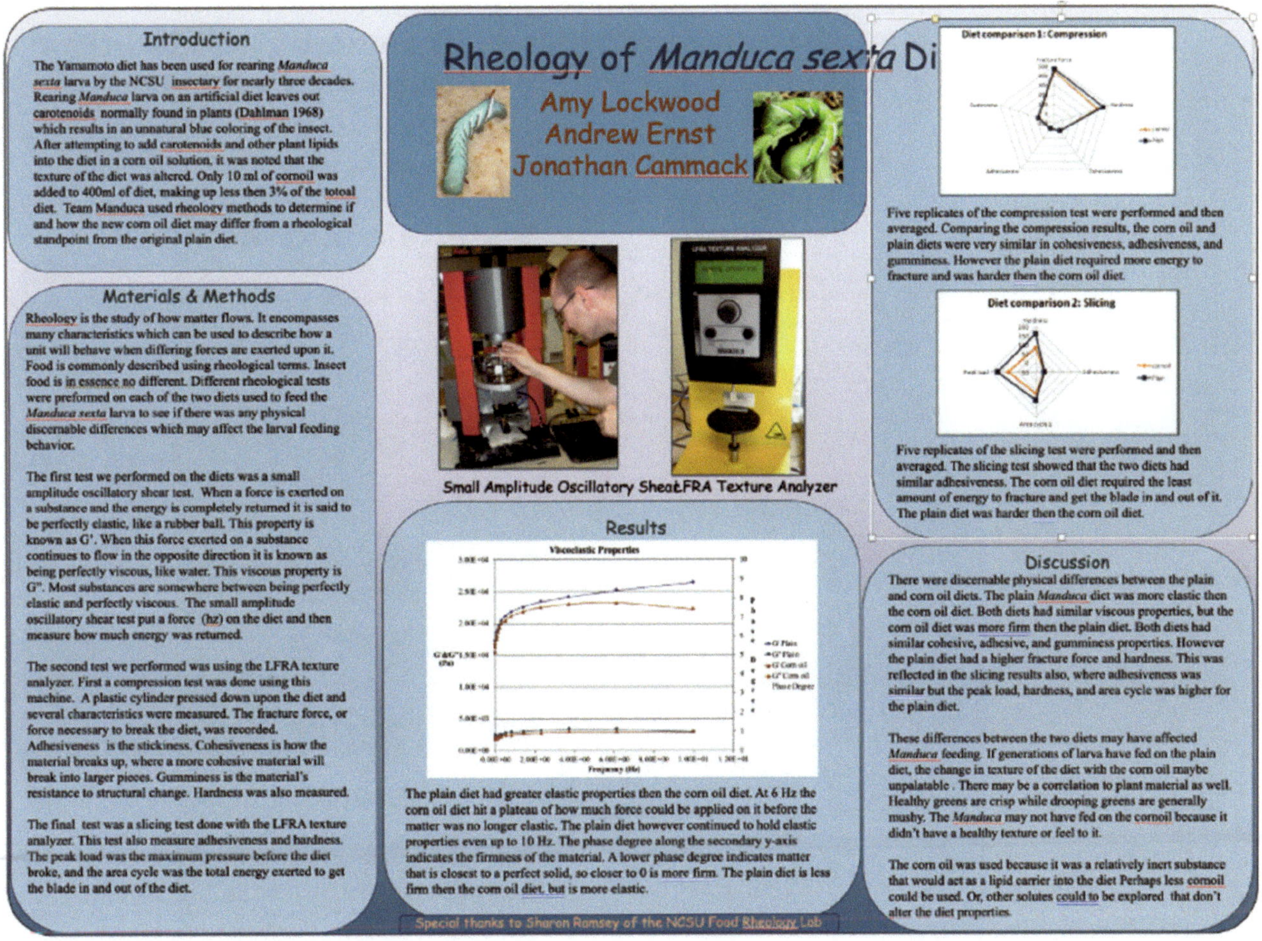

Figure A4.4 Rheology of diets for Manduca sexta (Amy Lockwood, Andrew Ernst, and Jonathan Cammack). Students combined what they learned in the Food Science Rheology Laboratory with diets for tobacco hornworms, testing the textural properties of diets with different components such as corn oil.

Heliothis virescens

Tobacco budworm (*H. virescens*) is a major agricultural crop pest in North Carolina. It is a polyphagous insect that can cause damage to a wide range of crops including tobacco, cotton, tomatoes, corn, and soybean. Although cultural practices have been widely adopted across the United States to minimize *H. virescens* pressure in these areas, control measures still rely significantly on the use of pesticides. As a result, much of the research conducted with this insect seeks to improve control methods and closely examine mechanisms of insecticide-resistance. Successful *H. virescens* rearing for use in laboratory and field studies is necessary in order to better understand the feeding damage potential and behavior of this economically significant pest.

Objectives

With this research, I examine the impact yeast products in tobacco budworm (*H. virescens*) artificial diet using larval development parameters (weight gain, food consumption, and frass production) to assess the nutritional value of the yeasts. If we can improve the nutritional quality of *H. virescens* diet, it may lead to increased success in rearing this insect.

Brewer's yeast and Torula yeast are both produced as by-products of other processes and therefore may not contain beneficial dietary compounds. Additionally, Brewer's yeast has been treated (debittered) to remove harsh flavors and may have lost beneficial nutrients in the debittering process that play a role in insect growth (Stone 2006). Perhaps the substitution of nutritional yeast will supply previously unknown mineral and vitamin complexes, lending additional nutritional value to the development of *H. virescens*. My hypotheses are as follows:

1. Substituting nutritional yeast for torula yeast in the diet of *H. virescens* will increase larval weight gain, larval size, and total frass production due to added nutrient benefits (increased B vitamin concentrations and proteins)

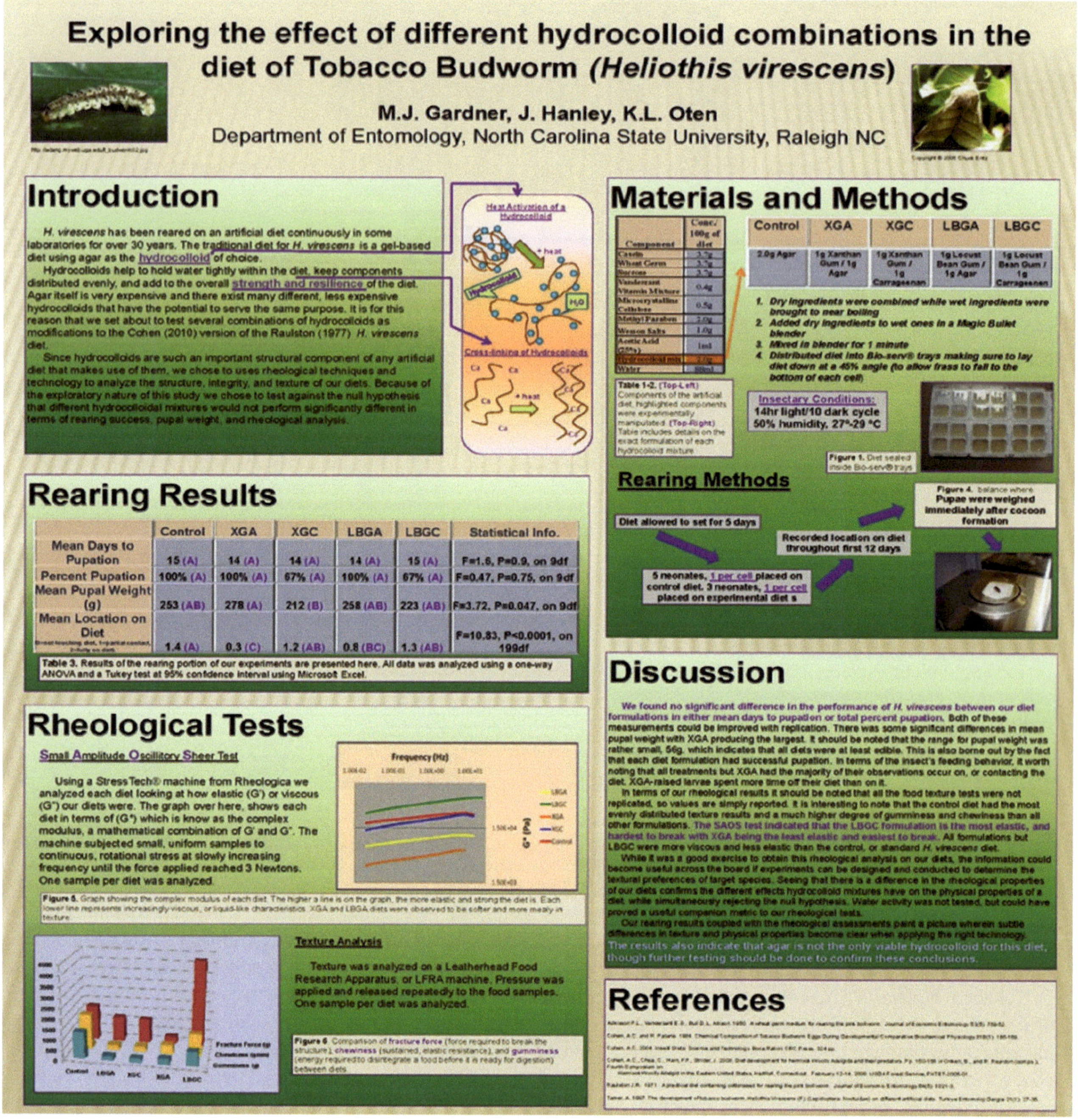

Figure A5.5 Students tested various hydrocolloids (agar, xanthan gum, locust bean gum, and carrageenan) to determine their physical and biological properties. This study allowed students to determine multiple parameter effects of gelling and texturizing agents in responses such as pupal weight, rate of pupation, etc., as well as gel strength and resilience. (Authors: M. J. Gardner, J. Hanley, and K. L. Oten).

2. Substituting Brewer's yeast for Torula yeast in the diet of *H. virescens* will exhibit similar or negative effects on larval weight gain, larval size, and total frass production

I believe that replacing the torula yeast protein source in *H. virescens* artificial diet (Yamamoto 1969) with a nutritional yeast or Brewer's yeast will alter insect development and survivability.

Materials and Methods

The Yamamoto (1969) diet was prepared with modifications by Dr. Allen C Cohen (Table 1). Equal amounts (g)

of nutritional yeast and Brewer's yeast were substituted for torula yeast to yield three different yeast diet treatments. Each liquid diet preparation was poured into a 9.5-oz plastic cylinder container (LockWare™, Glad Company, Oakland, CA) to cool and solidify for two hours. The container was agitated twice to minimize precipitation settling at the bottom of the container and was then placed in a refrigerator at 40°F. Cores of each diet were removed and the weight recorded. One core was placed in each of ten clear, 1-oz plastic cups (Product #100PC, Dart Container Corporation, Mason, Michigan). One newly hatched larva of *H. virescens* was placed on the diet core in each cup. The cups were placed in an incubator at 27°F in five replicates, with two containers of each treatment in one replicate in a randomized complete block design to assign blocks to the potential effect of slight temperature or light gradients. Any variation between blocks will be analyzed in the statistical analysis. Good Larvae will be weighed prior to pupation. Remaining diet cores and total frass weight in each cup will also be recorded.

Growth and Development of Tobacco Budworm *Heliothis virescens* When Reared on Artificial Diet Containing Different Bean Varieties: Project Draft (March 22, 2013)

Diane Silcox

Abstract: Rearing the tobacco budworm, an extensive agriculture pest, is important for developing control programs. Antinutritional qualities of the insect diet can limit the growth and fitness of an insect. This project seeks to identify the optimum bean variety for tobacco budworm development and fitness. Tobacco budworm eggs were placed in a container with a 3-g plug of diet containing either cannelloni flour (control), ground lima beans, ground pinto beans, ground black eyed peas, or lima bean flour. The larvae were allowed to feed for 17 days and then placed in a –20°F freezer to ensure thorough freezing of specimens before they were placed in a freeze drier. The larva/pupa, frass, and diet remaining were weighed to the nearest one-hundredth of a gram. From these measurements approximate digestibility (AD), efficiency of conversion of ingested food (ECI), and efficiency of conversion of digested food (ECD) was calculated. Survival from hatch to pre-pupal stage and energy utilization indexes were analyzed using proc glm and means will be separated using a Student's t-test.

Introduction

The tobacco budworm *H. virescens* is a major pest of tobacco, soybean, and cotton. Rearing this insect is pivotal for developing resistance mechanisms and control measures. Procedures to rear the tobacco budworm have been developed for over 40 years (Raulston and Lingren 1972). These procedures are effective at rearing the tobacco budworm; however, the potential for antinutritional effects from the cannelloni bean component of this diet allows room for improvement. Legumes tend to have lowered nutritional values due to antinutrient effects. These antinutrient factors consist of tannins, which inhibit the digestibility of proteins, and phytic acid, which reduces the bioavailability of some essential minerals (Rehman and Shah 2005).

Project Rationale

This project seeks to determine if there is a bean variety that will be superior to cannelloni beans, allow faster development of tobacco budworms, and reduce the implications of antinutritional effects. Increased development will benefit the researchers and students who work with tobacco budworms. They will be able to run experiments faster which will ultimately produce results for an increased number of publications.

Approach Rationale

I will use the nutritional indices of various bean varieties to understand the protein, vitamin, and mineral sources that promote development in the tobacco budworm. The beans selected will have either similar (lima beans) nutrient content as cannelloni beans or moderate (pinto) to extreme (blackeye peas) differences. These differences will also allow us to extrapolate about antinutritional effects of the cannelloni beans. If there is a better bean variety than cannelloni beans, I will look at the nutritional indices for differences, but if there are not distinct differences then antinutritional factors might have a slight effect on tobacco budworm development. The objective of this research is to determine if varying the bean content of the tobacco budworm diet affects rate of development or utilization of the diet.

Hypotheses

H_0: The tobacco budworms will develop normally feeding on the standard YC5 diet and will develop slower on the YC5 diets with varied bean component.

H_1: The tobacco budworms will develop normally feeding on the standard YC5 diet and will develop faster on the YC5 diets with varied bean component.

H_0: The tobacco budworms will have higher approximate digestibility feeding on the standard YC5 diet than the YC5 diets with varied bean component.

H_1: The tobacco budworms will have higher approximate digestibility feeding on one of the YC% diets with varied bean component than the standard YC5 diet.

H_0: The tobacco budworms will have better growth efficiency feeding on the standard YC5 diet than the YC5 diets with varied bean component.

H_1: The tobacco budworms will have better growth efficiency feeding on one of the YC5 diets with varied bean component than the standard YC5 diet.

H_0: The tobacco budworms will have higher metabolic efficiency feeding on the standard YC5 diet than the YC5 diets with varied bean component.

H_1: The tobacco budworms will have higher metabolic efficiency feeding on one of the YC5 diets with varied bean component than the standard YC5 diet.

Materials and Methods

Insects

This study was conducted using *H. virescens* colony maintained in the insectary on the campus of North Carolina State University. The colony is reared on a corn-soy blend diet at 27°C, 30% relative humidity. Eggs obtained were approximately two days old.

Diet Preparation

Bean varieties were selected based on their nutritional content as compared to the standard cannelloni beans used in the Yamamoto diet. Treatments consisted of cannelloni flour (control), ground lima beans, ground pinto beans, ground black eyed peas, and lima bean flour. Lima bean flour was selected to determine the effects of grinding whole beans vs. using bean flour. Whole cannelloni beans were not available to make this comparison. The whole beans were ground using a hammer mill. To prepare diet treatments 38 g of pre-mixed Yamamoto diet and 6 g of torula yeast were homogenously mixed. Nine grams of bean treatment were added to dry mixture. Water and agar mixture were heated to near boiling and 200 g was placed in a 16-oz plastic container (Ziploc® Twist N Loc containers, S.C. Johnson and Sons, Racine, WI). The dry material was added to the liquid and mixed thoroughly for approximately one minute with an electric whisk. The diet was cooled over-night.

Experiment

Ten, ~3-g plugs were removed from each diet and placed individually in 1-oz cups. A single egg was placed with a fine haired paint brush 1.3 cm from the diet plug. In several instances it was difficult to place a single egg due to clumping. A lid was placed over the cup to prevent larva escape. The cups were placed in an incubator set at 25°C with approximately 21% relative humidity. Rearing conditions were monitored using a HOBO®Pro v2 data logger (Onset Computer Corporation, Bourne, MA). The cups were checked three days after introduction (DAI) and any extra larva were removed, any dead eggs were removed, an extra larva from the same treatment was placed in the respective cup. The containers were subsequently checked 7, 12, and 17 DAI. The cups were held until 17 DAI, and then placed in a freezer at 18°C, survival of the larva will be noted before placement in freezer. The specimens were freeze dried and weight of larva, frass, and remaining diet were measured. From these measurements approximate digestibility (AD), efficiency of conversion of ingested food (ECI), and efficiency of conversion of digested food (ECD) was calculated (Waldbauer 1968). Survival from hatch to pre-pupal stage and energy utilization indexes were analyzed using proc glm and means will be separated using a Student's t-test.

Note: figures were deleted for brevity.

Tests

Insect-Rearing Science (Special Topics in Entomology: Presented as On-site, Face-to-Face Classes): One of the best ways to evaluate what has been taught in a course is to scrutinize tests given in that course. These tests span the ten-year period over which the rearing course was taught at NCSU. The first couple of times the course was taught, the tests were conventional, in-class tests given within the 1.5–2.0 hours of regular class time. In the last three times that the class was taught, I shifted to take-home, open-book tests. The fact that the later tests were extensively essay, I did not worry about students copying answers from one another, and even if there was collaboration, the point of the test—to elicit original, critical thinking and synthesis of complex materials would be accomplished. In many cases, I had extensive interactions with the students who had taken my classes (such as my serving on their graduate committees or having them do part of their thesis work in my laboratory), and I was able to evaluate the extent to which the students grasped the concepts that were covered in the tests. Often, the sophisticated thinking that these students did as reflected by the in-depth interactions was a gratifying part of my teaching rearing courses dedicated to deep, critical thinking about the science of rearing. The following tests reflect some of the learning and critical thinking (Bloom's Taxonomy at its higher levels).

Test 1

Ms. Kristen Hopperstad (now Dr. Hopperstad)

1. List five kinds of programs/organizations that depend upon insect rearing:
    a. Sterile insect technique programs (government)
    b. Private industry (production of biological controls)
    c. Pet industry/museums/aquariums
    d. Product testing research (host-plant resistance programs)
    e. Academic entomological research programs (like my lab)

2. Explain why it is important to know about various aspects of an insect's biology to rear it successfully (please be specific about the aspects to which you are referring):
   a. It is the responsibility of rearing specialists to meet the needs of their insect to rear it successfully. Knowing an insect's biology indicates the suitability of a diet, the necessary environmental conditions, insect density, and rearing habitat, and about frass management, etc. Failure to provide in even one aspect of biology will likely end in poor/no growth, diapause, low fecundity, etc.
3. How do several aspects of an insect's anatomy (or morphology) help us understand the insect's feeding biology/feeding habits?
   a. Mouthpart morphology tells us the types of substances of an insect feed on (piercing/sucking vs. biting/chewing, for example, indicate liquid and solid feeding, respectively). An enlarged salivary gland may indicate extra-oral digestion and that the insect would feed on a slurry over a dilute liquid. A hard, muscular proventriculus indicates a hard solid diet, where a proventriculus that looks like a filter indicates a liquid diet. The PM structure (thick/thin) and gut length also indicate feeding biology. Knowing these and more aspects of insect anatomy are important clues to feeding biology and successful rearing.
4. Pick either the peritrophic matrix (in a lepidopteran larva) OR a diverticulum (in an adult green lacewing) and explain its significance in the feeding (trophic) biology of the insect in question:
   a. The diverticulum in an adult green lacewing is large and highly modified to harbor yeast. Tracheoles to the diverticulum indicate an aerobic yeast, and hairs along the inside aid in retaining yeast. This yeast plays a role in digestion and makes important nutrients bioavailable to lacewings.
5. What are the major groups of nutrients (major classes) that are important to insects?
   a. Nitrogen source (proteins and amino acids)
   b. Lipids (especially sterols)
   c. Vitamins
   d. Minerals
   e. Carbohydrates
6. Explain the differences between essential and non-essential nutrients and please give four examples of nutrients known to be essential to most insects:
   a. Essential nutrients must be obtained from diets while nonessential nutrients can be synthesized inside the insect:
      i. Sterols
      ii. Vitamin C (ascorbic acid)
      iii. B vitamins
      iv. Valine
7. Provide a brief argument supporting the thesis that insect rearing is well served when it is treated as a science rather than an art:
   a. When treated as a science insect rearing is logical and repeatable. Mechanisms can be understood and purposefully manipulated, problems pinpointed and addressed experimentally, and a statistical quality control put into place. "Art" leaves everything in a black box.
8. Drawing from the assigned rearing about formalization of insect rearing, list six ways that the profession of insect rearing and the standing of insect-rearing specialists can be improved:
   a. Rearing specialists add more background and literature cited to add credibility
   b. Rearing Science and Technology section added to ESA journal
   c. Specialists in rearing review publications
   d. Rationale behind experiments explained more thoroughly
   e. Incremental success be publishable—not only major findings
   f. View of rearing change from sub/service science to its own specialized field
9. Prepare five purposes of insect rearing (not organizations or programs, but reasons why insects are reared):
   a. For research (estimated ~60% entomological publications use reared insects)
   b. To reduce natural pest populations (SIT)
   c. As feeder insects (for pet stores, museums with live collections for outreach, etc.)
   d. Product testing for pesticides
   e. For decorative purposes (release butterflies at weddings)
10. List (using specific examples or details) how you could approach understanding what an insect eats? This question is based on the lecture "What does my insect eat?" and it is aimed at getting you to explain five techniques or approaches that you could use to study the specific feeding targets of a target insect:
    a. Direct observation of feeding in field
    b. Addition/subtraction method in diet
    c. Isotope marking of potential food sources
    d. Metabolomics
    e. Dissection of gut
11. Please explain the significance of phytic acid in a framework of the concept of antinutrients:
    a. Phytic acid is able to pull important mineral ions out of solution and precipitate them out of the diet, reducing the nutrition of the diet. Phytic acid can be heated to neutralize its antinutritive effects.
12. What did you learn but I failed to ask? This question invites you to describe something that you learned specifically in this insect-rearing course. I'm asking you to discuss a concept or approach that was new to

you and that you found interesting or possibly useful in your future work:

a. Before the class, I had a rough understanding of "knowing your insect," but always tended to think more about habitat, environmental conditions, and life history. Analyzing internal morphology to help understand the nutritional requirements of insects was a new concept to me and I was previously unaware of the degree of insect gut specialization. I hate to admit I had the impression that one could analyze a natural diet, mix together the components, and give it to an insect. *Knowing* the insect inside and out is vital to developing a successful diet, as are the rheological properties of a diet. I also knew nothing about food science or that springiness, gumminess, and chewiness could be mathematically measured! I'm constantly understanding Mexican fruit fly diet better.

Insect Rearing (ENT 591/791: Special Topics in Entomology)

Test 2

Kristen Hopperstad

1. List the four major groups of microbes that are pathogens to insects:
   a. Bacteria
   b. Fungi
   c. Viruses
   d. Protozoa
2. You hear an insectary worker saying that the chronic mold outbreaks in his/her insectary are only important as potential health hazards to personnel but not to the quality of the insects. Provide a brief argument as to why the insects' quality is likely to be adversely affected:
   a. Even if an insect is tolerant to mold, it is likely still stressed by it. A constant immune response to mold infestation will reallocate energy otherwise put to normal functions like growth or reproduction. Insects constantly fighting off infection are susceptible to opportunistic pathogens as well. Further, the mold may alter the diet and possibly make it less nutritious to the insect.
3. Give an overview of protein analysis: what are two different kinds of protein analysis that may be used, and what are the advantages and liabilities of each type?
   a. Hydrolysis and amino acid analysis
      i. Involves digesting protein completely and then analyzing amino acids
      ii. Most accurate method but is *costly*, involves the use of harsh chemicals, takes time and is a relatively high level of expertise
   b. Acid-orange dye binding method
      i. Sample is ground and thoroughly exposed to dye (protein is precipitated out), then solution is analyzed in spectrophotometer.
      ii. Cheaper, less time-consuming, requires less specific lab equipment (centrifuge and spectrophotometer/plate reader), and reliable. Not useful for collagen and gelatin, however, and may underestimate protein if dye cannot "get" protein to bind.
4. Give a brief overview of the defenses that insects use against invasive microbes:
   a. *Mechanical defenses* physically keep out microbes such as the cuticle, peritrophic matrix, and opening/closing spiracles
   b. *Humoral defenses* such as antibacterial chemical compounds in insect saliva, attack microbes at the molecular level
   c. *Phagocytosis* specialized cells can attack "non-self" particles (including microbes) in an insect
   d. *Encapsulation* non-self particles are wrapped in a matrix to disable and effectively "remove it" from the insect's system
5. What are kinds of insect/microbe relationships that occur in nature and in rearing systems?
   a. Iniquilism
   b. Commensalism
   c. Mutualism
   d. Pathogenic/parasitic
   e. Competition
6. Please choose one of these questions only: (a) discuss the benefits of using enzymatic analysis as part of the development or quality control in rearing systems or (b) explain what we mean by the Pareto principle and give an example of how it is used:
   a. The Pareto principle (80–20 or 90–10 rule) asserts that 10% or 20% of variables in a system may significantly affect the remaining 80–90%. It may be used in an insectary to determine what factors (incorrect diet mixing, microbial contamination, temperature, for example) may impact insect feeding. Controlling for the most important variables will help manage the whole system.
7. Draw a process control chart and label its most important parts:

a.

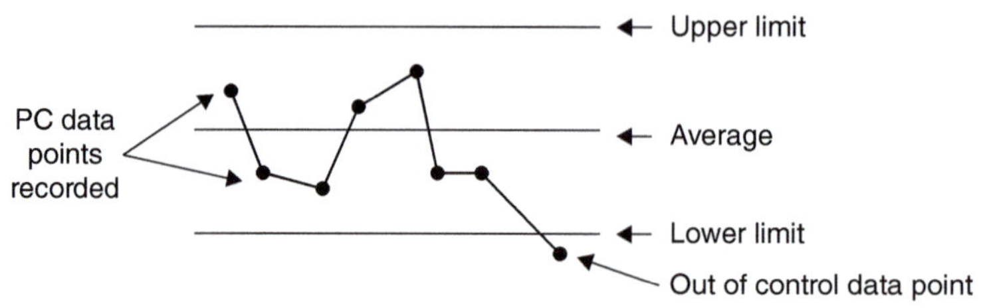

8. Drawing from the assigned reading (Sikorowski and Lawrence 1994), describe how contaminants can enter a rearing system as potential problems:
   a. Contaminants may enter a rearing system through personnel, air flow, or diet ingredients. "Foreign" microbes, especially human-associated bacteria, cause stress among insects that may normally be adapted to different microflora. Because an artificial rearing environment can induce stress by other means (change in diet, temperature, crowding) even normal insect flora can become opportunistically pathogenic. Contaminants in the air, on equipment and transmitted by personnel may result in a decline in quality of insects.
9. Provide five ways that microbial infections (pathogen-based interactions) or symbionts can be diagnosed:
   a. Regularly check colony (air, swab materials, diet) by culturing microbes on various agar/broth and visual identification
   b. Isolate microbe from unhealthy insect and culture on agar
   c. Identification of symptoms associated with a microbe (uncommon symptom associated with one microbe though)
   d. Run 16s RNA on gut of healthy insect
   e. In a controlled experiment inoculate different stages of insect with different concentrations of microbe, then track insects through life cycle and attempt to reisolate microbe at end of experiment
10. List (using specific examples or details) five procedures that help make up analysis in insect-rearing systems:
    a. Add H_2O_2 to a diet ingredient (soy) to check for active enzymes after cooking
    b. Use DPPH to check for antioxidants in diet ingredient
    c. HPLC to separate and analyze sugars in an artificial diet
    d. Acid orange dye bonding test to measure protein quantity in diet ingredient
    e. Soxhlet extraction to measure lipid content of diet ingredient
11. Draw a fishbone diagram that represents components that can cause problems or degrade quality in insect-rearing systems:

a. quality in insect rearing systems (10 points)

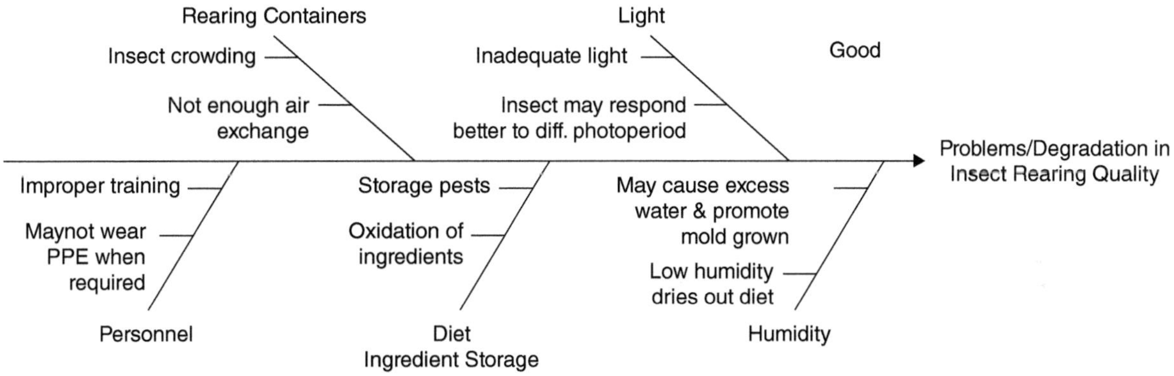

12. What did you learn but I failed to ask? This question invites you to describe something that you learned specifically in this insect-rearing course. I'm asking you to discuss a concept or approach that was new to you and that you found interesting or possibly useful in your future work:
    a. Two concepts stand out most to me. The first is rigorous process control. Though I was familiar with taking measurements throughout the rearing process was important. I was not aware of how data was being used in a meaningful way. Process control diagrams are simple to use and would be highly effective in any rearing system (most labs I've been in put this data in a binder or lab notebook, eyeball the numbers, and then never look at it again!). The second concept is the use of biochemical analyses to assess a rearing system, or really the use of biochemical analyses at all. I have a limited chemistry background and am eager to take biochemistry during my MS or PhD. It is a huge advantage and has an arsenal of useful evaluative tools not only in rearing but in many entomological projects.

Another excellent student's (Diane Silcox, now Dr. Diane Silcox-Reynolds) tests included the following responses to short essay questions:

Question: Explain why it is important to know about the various aspects of an insect's biology to rear it successfully (please be specific about the aspects to which you are referring).

Answer*: Insects are highly specific creatures in that they often have a narrow range with which to live successfully in. You must understand how they select a host in order to prepare the diet in such a manner to reflect this behavior. An example would be if our insect needs a certain token stimulus present that must also be present in the diet. Also,* *if the females need to insert their eggs into a substance it must provide the same force or consistency as the force in nature. Also, if your insect requires a chilling or warming period that must be exposed to have reproduction.*

Question: How do several aspects of an insect's anatomy (or morphology) help us understand the insect's feeding biology/feeding habits.

Answer: *The mouthparts of the insect help us to determine what types of food the insect might feed on. If they have piercing-sucking mouthparts, they are liquid feeders, chewing mouthparts will feed on solid food. Modifications to the proventriculus either for filtering or grinding can indicate types of food ingested. If a large diverticulum is present, the insect might use endosymbionts to help digest food. If a peritrophic membrane is present, this will allow us to determine if they are liquid feeders (no PM unless blood meal) or if it is highly folded PM feed on some form of plant material. Also, the structure of the gut can determine the complexity of the food source. Highly convoluted gut might indicate a fluid feeding and thus the need for counter-current exchange. The composition of the frass can also indicate whether they feed on liquid or solid food.*

Question: Pick either the peritrophic matrix (in a lepidopteran larvae) OR a diverticulum (in an adult green lacewing) and explain its significance in the feeding (trophic) biology of the insect in question.

Answer: *The peritrophic matrix of the lepidopteran larva allows the larvae to feed on complex plant material and be protected from trichomes and sharp edges that the plant material has. The PM is also highly folded to allow maximum absorption of nutrients from the plant material. The PM also provides the larvae protection from infection and various chemicals. There are microvilli near the PM that allow for increased surface area and thus increased efficiency in absorption. Lastly, the PM provides organization to the gut which increases efficiency of at digesting and absorbing the plant materials since digestive enzymes can get right to their action sites.*

Question: What did you learn and find interesting, but I did not ask in this test?

Answer: *One main point that I learned but was not asked was the four components to an insect diet (making it successful). This concept of nutritional aspects, bioavailability, palatability, and stability provides a structure and understanding to insect rearing that I really did not know before. From the nutrition aspect, I learned that it is not enough to just have all the nutrients present, but it was the ratio and antinutritional effects to also account for. I learned that if the diet does not have the correct texture or feeding stimuli, the insect will not feed. Also, if the construction of the nutrients in the diet is not accessible to the insect's digestive system, then the diet's components are not accessible to the insect's digestive system then it does not matter that all the nutrients are present. Also, if there is a nutrient present that blocks active sites for other nutrients this is also issue when rearing the insect. Lastly, if the diet cannot hold up for the duration of the insect's life cycle, this is an obvious but major issue. Also, if it cannot remain stored for some time without changing chemically or physically, this could be an issue you are unaware of that could drastically inhibit your rearing efforts.*

Question: What did you learn and find interesting, but I did not ask in this test?

Answer: *One concept is the idea of the unintended consequences of manipulating a diet in a way you believe will benefit the insect. Some examples would be if you added an antimicrobial agent to your diet to reduce microbe load, but in turn the antimicrobial agent reduced feeding digestibility, metabolism, reduced growth, or fecundity. Or if you decided to heat the diet to reduce microbial load, you could alter the nutrients available to the insect and thus alter their growth and development. Lastly, you could decrease the use of the diet particles to increase digestibility which could increase the surface area for water to bind to which would increase your overall moisture content and could lead to mold formation. This could also alter the texture of the diet leaving it less palatable to the insect.*

Student from the 2017 Insect Rearing Course: Ms. Meredith Favre

Question 1: Make an argument for humane, "homeostatic" conditions in our rearing systems.

Answer*:* *By using humane practices to develop our rearing systems and by focusing on the homeostasis of our insect subjects as our highest priority, we are able to produce high-quality insects that most closely represent wild insect populations and that provide us with accurate data to advance our understanding and research goals. Maintaining conditions within rearing systems that promote the health and well-being of the insects will help to keep their body systems within normal homeostatic conditions which are optimal for survival and development. Insects produced within humane rearing systems are better representations of wild insect populations compared to colonies maintained with sub-optimal conditions. In being attentive to our insects' homeostatic conditions, we can minimize the exposure and impact of stressors that could cause their body systems to be disrupted. Stressors* *within rearing systems originate from various sources, such as exposure to pathogens and adverse stimuli, nutritional deficiencies or food quality issues, inadequate waste management and space allocation, and issues with gas exchange. By recognizing stress factors and by being conscious to maintain healthy, humane conditions which minimize these stressors within rearing systems, this will allow the insects produced to be higher quality subjects in our experiments and to provide results that are more comparable to field populations. Therefore, by centering our rearing around the well-being of our insects through humane treatment, we gather more accurate, higher quality data for our research and promote further expansion of knowledge within the scientific community.*

Question 2: What is the advantage (benefit) of studying the feeding biology of your target insect in relationship to developing and maintaining effective rearing?

Answer*: By understanding the feeding structures and dietary adaptations of our target insects, we can better comprehend and construct effective diets which incorporate the appropriate nutrition and characteristics needed by our insects to develop and thrive within our rearing system. We must "know our insects," to be cognizant of how our insects feed, the intricacies involved in their digestive processes, and how these processes can be influenced by different dietary components to establish and maintain an effective rearing system. We need to study their natural diet, regarding the nutrition, form, texture, chemicals, and phagostimulants involved, to ensure our insects accept our diets and are able to survive and develop on the contents. We must also know the feeding structures and the physiology of our target insects, to understand how their anatomical features and biological processes interact with our diet, to be able to provide adequate nutrition for their long-term survival. For example, in class we discussed that emerald ash borer larvae, Agrilus planipennis, are obligate leaf-edge feeders, being equipped with shearing mouth parts that need access to food with exposed, thin edges in order to feed. By observing and understanding the mouthparts of our target insect and how it uses these structures to consume food, we are able to present diet that is in the appropriate form and texture to facilitate feeding. Presenting a diet that is abrasive to an insect that possesses a peritrophic membrane can disrupt the membrane's structure and interfere with its function to protect the gut lumen. The chemicals we incorporate in a diet must also be evaluated, such as the impacts of chelating agents, binding agents, associated redox potentials, and other features such as pH which can impact absorption of nutrients by microvilli. By studying our target insects' feeding processes and biology, we equip ourselves with the knowledge we need to provide the best diet conditions that complement our insect's feeding strategies. This understanding facilitates the development of an effective and efficient rearing system that can maintain a colony's health in the long term.*

Question 3: How do insectary conditions differ from conditions in nature?

Answer*: Many variables differ between insectaries or laboratory environments and conditions present in nature. These differences involve exposure to environmental factors, effects of temporal cycles, access to food and resources, space availability and population management, waste disposal, microbial exposure, and gene flow. Insectary conditions differ from nature due to human manipulation of the environment and aspects relating to stimuli that the insects are subjected to. The insectary is the insect's universe, and we, the rearing staff, have some level of control in managing the elements that compose its universe.*

Within an insectary, climatic conditions such as temperature, light, and humidity are ideally kept at constant levels, or on a predictable schedule, in order to keep insects healthy and developing at a consistent rate. This occurs because we want to have predictability in our rearing outcomes by holding environmental conditions constant. By keeping photoperiod cycles the same all year, in conjunction with constant regulation of temperature and humidity, we can prevent insects from entering diapause during winter months and prevent this behavior and physiological state of low activity from impacting colony production. Within nature, temperature, humidity, and photoperiod are constantly changing and these conditions allow insects to establish a rhythm corresponding with temporal cycles such as seasons. Another difference in environmental factors present in the lab compared to in nature is the quality and intensity of light. Artificial light provided in insectaries is lower quality compared to sunlight, and insect behavior and well-being could be impacted based on this difference.

Another aspect that differs between the insectary environment and nature is accessibility to food and resources. Food availability varies in nature relating to the seasons, and the nutritional profiles of foods can also change temporally, such as alterations in sugar profiles of plants across developmental stages and different seasons. Compared to the fluctuating availability of food in nature, there is relatively constant access to food and other resources for insects reared in an insectary. The same plants and ingredients used for diets can often be grown or purchased throughout the course of the year. However, even though food is accessible and conveniently provided to insects within an insectary, the type of food provided might differ from their natural diet in the wild. Therefore, it is important to consider our insect's natural diet and try to develop the best representative diet for our rearing system that meets the necessary nutritional requirements for our insect and is affordable.

Space availability and population dynamics are other aspects that differ for insects living in the nature compared those reared in an insectary. In the wild, insects have relatively abundant free space to move in order to find food and shelter, to mate, and to escape from adverse stimuli. In an insectary, insects are limited to their container's useable surface area. Often rearing containers will host populations of insects that are more dense than they would be for the same given area in nature. Therefore, issues associated with crowding can occur, such as the accumulation of waste which promotes the spread of microbes and disease amongst insects. As observed in Cactus moth rearing, the accumulation of frass within the limited space of a rearing container can also cause insects to behave differently from how they would in the wild regarding waste management. Insects in rearing containers also need to have sufficient access to light and shelter, since they are contained and unable to control their exposure to these stimuli as they would in nature. Vibrations from laboratory equipment could also interfere with insects' behavior and their ability to communicate as they would in the wild.

Lastly, the flow of genetic material within wild insect populations differs significantly compared to insect colonies reared in insectaries. In nature, insect populations of different biotypes and from different areas mix and mate with each other, keeping gene pools diverse and promoting selection of characteristics imparting fitness. In a laboratory setting, maintaining genetic diversity is more challenging. It is possible to slow the loss of genetic diversity in a lab-reared colony through occasionally supplementing the colony with wild insects and implementing techniques to minimize inbreeding. However, an insect colony in an insectary will decrease in genetic diversity much faster than compared to wild populations, and issues can arise over time relating to colony health and behavior, such as genetic defects and domestication.

M. Simon Pinilla-Gallego

Final Exam ENT 791: Spring 2019

Please note that the answers are due at 5:00 pm EST on May 2. Each question is worth 50 points. Also, your reports are due at 5:00 pm on May 2. The report is worth 200 possible points.

Question 1: Explain the nature of error (variability, uncertainty) in insect-rearing systems. Give examples of the three major categories of error (random, systematic, and gross error) and explain how to address them.

Answer: *Error refers to the variability or uncertainty that is observed in every system or experiment, including insect-rearing systems. The higher the variability of the system, the higher the error. Many factors can contribute the uncertainty of measurements, including the instruments and techniques used. For example, some scales (balances) are more precise than others, allowing to determine smaller differences between samples, and in that way reducing the level of uncertainty. Another example is, when there is high phenotypic variation of individuals in a population, the variability of, for example, survival or performance, will be high, and the error of the average value of the population will be higher.*

Random error is due to unknown and unpredictable changes in the experiment. Those changes can be above or below the mean, and usually follow the normal distribution. This type of error can be addressed by taking many samples, so the error can be managed with the statistical analysis. Some examples in rearing systems include variation in the compositions of diet components (especially whole ingredients) due to changes in the raw materials used to make the components, changes in providers, or changes in the methods that the provider uses to estimate the nutritional composition of the product. Another example could be error due to the person who is performing the lab work. An employee who is more careful and meticulous will decrease random error, while an employee who is more distracted and less carful will increase the random error in the rearing system.

Systematic error is reproducible and consistent, and it's usually due to problems with the measuring instrument. For example, if the scale is not tared, all measurements made with that scale will be off by a certain amount. Or a pH meter that is not well calibrated will take all measures off. A way to address systematic error is to regularly calibrate the instrument, and to make sure that personal are using the instrument in the correct way, to be sure that measurements are precise and accurate.

Gross error is due to people's mistakes or instrument's failure, and it will appear as an outlier on the data. Usually these outliers can be detected with statistical methods, and deleted if they are having a strong influence on the analysis.

Question 2: Discuss microbial relations in rearing systems.

Answer: *Microbes are common organisms in nature and they can easily grow in insect-rearing facilities when the conditions are met. Dealing with microbial issues can be costly and cause many problems that can be avoided if we follow good practices. Microbes: They can have many different effects on the system, in most cases negative, but in some also positive. For example, many insects need microbial symbionts that help them to digest food or that* *produce vitamins or compounds that the insects need, like in the case of termites that have special protozoa and bacteria in their gut that allow them to digest cellulose. In other cases, the microbes pre-digest the food items in a way that makes them accessible to insects, or the insects may even feed on the microbes, like in the case of Drosophila that feeds on bacteria, mainly yeast growing on the rotting fruit, or the leaf-cutter ants that feed on the fungus growing on the leaves that they collect.*

Microbes can also be pathogens of insects, causing serious mortality in colonies or dramatically reducing performance of the colonies. This includes many species of bacteria, viruses and fungus and protozoans that can either kill or reduce fitness and performance of the insects. In the case of Wolbachia, it can cause male feminization or cytoplasmic incompatibility, and therefore reducing fecundity of the colony. Microbes can also have negative effect on rearing systems when they grow on the diet, causing it to go bad too quickly and promoting chemical changes in the diet composition, making it unpalatable for insects, and in some cases producing compounds that can be toxic. Additionally, changes in conditions like humidity and temperature can trigger the grow of microbes that otherwise wouldn't grow in the lab. Fungi are a particular common problem in some rearing systems.

There are many ways in which we can try to control microbes in rearing systems. The first one, and maybe most obvious, is to try to keep everything sterile, more sanitized, rather than sterile as a more realistic goal, including implements and materials, diet components, and ask the personal to wash their hands and keep clean conditions in the rearing facilities. However, even when these clean and sterile conditions are achieved, microbes can still come into the systems in many ways, for example, in the ventilation system of the facilities, with some raw whole ingredients of the diet (eggs or meat), or with new insects that are brought from the field. To control microbes that can get into the diet, it is useful to heat the diet to try to sterilize it, but too much heating can also degrade some of the diet components and make it less nutritious for the insects. We can also add antifungal and antibacterial compounds to prevent growth of microbes, but we have to be careful because those components can also make the diet less palatable for the insects, or reduce the bioavailability of some nutrients.

Question 3: Discuss genetics and domestication in rearing systems.

Answer: *In nature, insect populations are usually organized in heterogeneous subpopulations, and in many cases those subpopulations can develop genetic population structures and local adaptation. When we sample insects from the wild to start a colony we are taking only a small sample of individuals from one or few of those subpopulations, and therefore we are starting colonies with individuals that represent just a fraction of the genetic diversity of the whole population. This is equivalent to a bottle neck event, where a population is started from just a few individuals, and where many of the alleles of the original population are lost. This loss of heterozygosity can have negative effect on the individuals and the colony that we are starting.*

Some of those negative effects are due to inbreeding depression, caused by the matting of related individuals, that increase the likelihood of an individual to be homozygous for deleterious alleles. This can also make individuals less fit to survive in the field if they are released for biocontrol purposes. On the other hand, this reduction in genetic diversity can also result in selection of breeds of insects with desirable characters that make them more adapted to captivity, or that can increase production of a desired trait, like silk in the case of the silkworm. In those cases, introducing new individuals from the field to try to increase genetic diversity of the colony can actually be bad, because it can dilute the adaptations that the colony has developed over time. In other cases, it is uncertain whether introducing new individuals from the field can actually increase the genetic diversity of the colony, because in some cases captive individuals won't mate with field caught ones, or the number of field individuals can be so small that it won't have an effect on the overall genetic of the colony.

Some of the genetic changes observed in captive insect colonies can also be due to artificial selection imposed by the researcher, sometimes intentionally but in many others unintentionally. For example, the type of containers, diet or environmental conditions can impose selection and end up favoring particular genotypes in the colony. This is particularly important when a gene affects more than one trait (pleiotropy), because selection of one trait can have unanticipated consequences on another trait. For example, selection for individuals with small wings in Drosophila can also reduce the ovary size of females, because there is a common gene that affects both traits. In some cases, after a long period of selection and captivity, captive populations can become unable to reproduce with wild individuals, which indicate that a process of speciation is occurring during insect rearing.

Question 4: Explain (using examples) the basis for understanding the physics of rearing processes and equipment.

Answer: *Insect rearing requires a lot of different equipment and materials, and they can interact in different ways, and the physical and chemical properties of those materials can affect the outcome of different processes. Some of the equipment that can be found in a rearing* *lab and that are common to many other lad settings are balances, pH meters, volumetric equipment, stirring hot plate and incubators. Some of the specialized equipment includes chromatography, spectrophotometers, incubators, heaters, humidifiers, water activity meters, hygrometers, light meters, and data loggers.*

Some processes that can impact the quality of the raw materials include frequent freezing/thawing. Even if you don't take the materials out of storage, if the temperature inside the freezer fluctuates, that will create a freezing/thawing effect. A freezer that can keep a constant temperature and fluctuate less will keep materials in good conditions for a longer period of time.

Another important factor to consider is the effect that the surface/mass ratio can have on different processes that are performed in the lab. For example, a determined volume of diet will cool down faster in a small beaker than in a big beaker, because more area is exposed in the small beaker, and the heat transfer will happen faster. This is important because if you are using big container, the temperature of the diet at the center of the container will be different than at the periphery, and depending on where you measure the temperature, you can end up pouring it when it is too hot or too cold, and that can result in a bad mixing of the diet, making it less homogeneous

The size, shape, and surface/mass ratio of the diet components, like flour and bran, can affect the physical properties and the way they behave and interact in the diet. For example, smaller particle size can increase water retention and viscosity of the diet. Smaller particle size may require more water to achieve the same viscosity, and it is possible that the excess of water could promote fungus growth.

The materials that some of the instruments are made off are also important, as some materials are better conducting head. For example, metals are much better at transmitting heat than glass, and that would result in a faster and cheaper preparation of diets.

Question 5: How can you develop a plan of process control for a rearing system? What tools can you use to build your process control strategy?

Answer: *Having a process control plan is a very important part of having an insect-rearing system, because it allows us to reduce variability and error in the system, reduce loses, creates a more consistent and reliable output, allow to save money and time, and overall give the opportunities for a constant improving of the system. It is important to point out that changes should be implemented in a progressive way, and not by changing the whole system at once. In that way if something goes wrong with the new system, you can correct it without risking the whole operation.*

The first step in developing the process control plan is to identify what are the most important factors that can affect your system. You can also have a list of steps or components of the systems and think about all the things that could go wrong in every step. Although it is impossible to think about every single possibility, this will allow us to be more prepared for likely events. It is important to also establish what are the threshold for each parameter for the system. For example, what are the higher and lower temperatures that are acceptable for a good performance of the system, or the acceptable texture range of diets that the insects would eat.

After knowing what factors should be monitored, and the acceptable range for each one, we have to think about the actual monitoring strategy. It is important to use instruments that are precise and accurate, to get good measurements of the parameters that we need to monitor. It is also important to establish how often each parameter should be monitored to be able to promptly correct any changes in the optimal conditions. Environmental conditions like temperature and relative humidity can be constantly monitored and adjusted, while factors like diet properties can be checked every time a new batch is made. Diet ingredients and components should also be checked every time a new purchase is made, as some properties and the composition can be different even if you are buying from the same provider. If you store materials, they can also change over time, and usually they have a shelf life. Other parameters that can be monitored are the genetic compositions of particular colonies. This information can help to determine whether a colony is getting dangerously close to a genetic bottleneck, which can prevent sudden collapse of the colony.

CHAPTER 13

Future of Insect Rearing

13.1 THE PAST IS PROLOGUE TO THE FUTURE IN REARING-SYSTEM DEVELOPMENT

When we ask where insect rearing is going, we can get help in predictions by studying the past or the historical underpinnings of rearing. From studies of the most successful rearing programs, we see certain features that are common to success. Starting with the oldest, longest standing rearing system, silkworms, we see that there is: (1) a practical purpose for rearing the insect, (2) feasibility of the rearing operation, (3) an insect that lent itself to rearing, and (4) a willing and capable pool of personnel to perform and improve rearing operations. It is arguable that honeybees vie for equal status with silkworms as the record-holders for historical precedent as the longest-domesticated insects. However, I focus on the silkworm model because this is an example of more intensive insect/human interactions in the design of the rearing and domestication systems. This is another way of saying that silkworms and humans have higher order interdependence than humans and honeybees.

13.2 PRACTICAL PURPOSE FOR REARING SUCCESS

The practical (and primary) purpose of rearing silkworms is to obtain the valuable silk fiber from the insects. The rearing of silkworms goes so far back in time (about 5,000 years, starting in ancient China {https://depts.washington.edu/chinaciv/archae/2dwkmain.htm}) that we must rely on legend as much as history and inference as much as direct data to account for the origins. This topic is treated in detail in Chapter 4 (on "Silkworm Rearing Systems"). But the discovery that upon placing a cocoon of *Bombyx mori* into hot water and the silk strands that constitute a cocoon begin to unravel had to be one of the early discoveries that took place most likely by accident. This legend told in Anonymous (http://www.silk-road.com/artl/silkhistory.shtml) is summarized here:

> According to Chinese myth, sericulture and the weaving of silk cloth was invented by Lady Hsi-Ling-Shih, the wife of the mythical Yellow Emperor who is said to have ruled China in about 3,000 BC. Hsi-Ling-Shi is credited with both introducing sericulture and inventing the loom upon which silk is woven. In Chinese texts, she is sometimes referred to as The Goddess of Silk.

Various accounts say that Lady Hsi-Ling-Shih was drinking hot tea under a mulberry tree when a cocoon fell into her cup. Upon seeing the silk begin to unravel in the hot liquid, she began the process of unwinding the silk and then spinning it into thread or yarn. However true or apocryphal that account, we know that once the silk strands begin to unravel, they can be collected by exerting a force (now called reeling) to extend the strands, which can now be twisted into a strong, resilient thread. The context of the evolution of the thread/yarn—> cloth is beyond the scope of this chapter, but there is evidence that woven cloth was of great antiquity in China where plant-derived materials such as hemp were utilized. This anonymous article from the previously cited website explains the reasons why the niche for silk production was so prominent:

> Silk, the most valued of all fabrics, was made in China from Shang times, if not earlier. Silk makes excellent clothing because it is soft, sheer, lightweight, long-lasting, and can be dyed brilliant colors. It can be made light enough to wear on the hottest of days, but also helps protect against the cold because silk floss makes excellent padding. (https://depts.washington.edu/chinaciv/archae/2dwkmain.htm).

Given the demand for silk and the beginning of a technology for extracting it from cocoons, the next challenge was to acquire sufficient cocoons to allow cloth-making to be feasible. The association of the silkworms with the white mulberry tree (*Morus alba*) was evidently fairly obvious, and it must have arisen from the observation that the large worms and their cocoons were present on mulberry trees, which were growing wild in the regions.

13.2.1 Feasibility of the Rearing Operation

The issue of feasibility includes several components that include historical background (has the insect or a similar insect been reared successfully), are the resources such as diets and containers possible to obtain? The insect itself is so central to the feasibility question that it is covered in its own Section 13.4. The question of facilities is prominent. Provisions of oviposition sites and mating sites must be included. Also, accommodations must be made for special behaviors such as nuptial flights or special thermoregulation requirements.

13.2.2 Does the Insect Lend Itself to a Rearing Process?

Over the long duration of my career in rearing, I have speculated and heard others speculate about just what makes an insect a good candidate for rearing? Are generalists easier than specialists in terms of the insects' ecology? Does the insect have features that make rearing overly challenging? For example, dragon flies have stages of their life cycle that may lend themselves to rearing, but adult dragon flies are so large and fast-flying that it is hard to imagine the kind of cages or containers that would lend themselves to efficient rearing. Many social insects such as honeybees, yellow jackets, tarantula hawks (Pompilidae), and certain ants such as *Dorylus* (driver or safari ants) would be challenging to rear in a laboratory setting. It must be emphasized, however, that often insects that are thought of as too challenging to rear can be accommodated in especially creative ways. For example, Banks et al. 1981 and Vogt et al. 2003 demonstrated that imported fire ants and one of their parasitoids could be reared with appropriate accommodations such as special containers for ant confinement and various measures to provide live insects for the ants' diet.

13.2.3 Availability of Competent Rearing Personnel

It is difficult to rank the importance of facilities vs. personnel in importance of successful rearing. No matter how good the rearing system is, how well-vetted and optimized, and no matter how excellent the facilities, if the personnel fail to carefully heed the directions of the standard operating procedures (SOPs), the rearing process will not succeed. This leads to the lessons inherent in Figure 13.1 where personnel are shown to be one of the six key components of insect-rearing infrastructure.

13.3 SOURCES OF FUNDING

In the diagram in Figure 13.1, the relationship between funding and the five other major infrastructural components is shown, indicating that funding is central to the progress of each of the other components. The connection between the components is designed to illustrate that

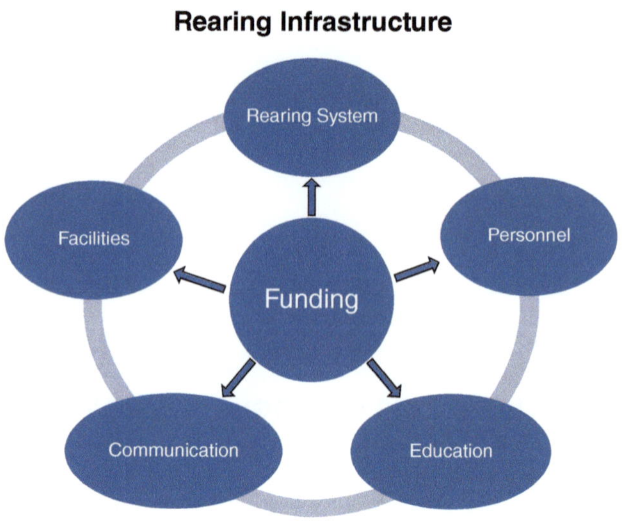

Figure 13.1 Components of insect-rearing infrastructure. (From Cohen 2019.)

Rearing Systems, Personnel, Education, Communication, and Facilities are interconnected with one another. Cohen 2019 discussed the fact that funding or budgeting shapes the rearing program in the nature of its development and in the status quo of the system, once it is developed. Starting with **facilities**, the funding will influence how much space, equipment, and maintenance a rearing program will have. Proceeding to personnel, the amount of funding will strongly shape the number of personnel and the qualifications that can be expected of the personnel. Scant funding will allow too few personnel to be hired as scientists and their support staff. Too often, where a senior scientist is needed to develop a rearing system or to solve rearing problems in existing rearing systems, a post-doctoral research associate or a junior scientist is put in place to save money. Again, too often, because the complexities of rearing systems are underestimated by managers and the scientific community, it is deemed acceptable to bring in a novice to insect rearing. I have seen this repeatedly take place in the USDA, ARS, and other organizations that depend on rearing for their programs. Throughout this book, I have tried to indicate the complexity of rearing systems (treating them as artificial ecological niches, venues for registering various kinds of stress, including genetic drift and epigenetic/transgenerational shifts in the original population), and I have stated as directly as possible that recognizing the nature and consequences of the stress requires a competent rearing professional with eclectic understanding of the biotic and abiotic facets of the rearing system.

This complexity of rearing systems and the employment of rearing personnel to deal with rearing systems relate to funding. If adequate funding were available for salaries, the rearing personnel being selected could be expected to have fuller backgrounds in terms of rearing experience and especially rearing-related education. The funding is also a determinant of the education that is available. If there is no designated fund to hire professors, instructors, teachers, and rearing educational support personnel, then there is no avenue for the establishment of a rearing education system. In my own case, I have conducted nearly all of my educational efforts as a sideline to my main job description. The miniscule exception to my contention about lack of funding for rearing education is in the workshops where personnel are sent by their organizations for a short training period (generally five days). This costs the organization at least $2,500–3,000 per workshop participant (including registration fees, travel, and lodging/food expenses). The total expenditure for rearing education over the past 20 years can be calculated easily by considering the cost per person times about 30 people per year (for the insect-rearing workshops at Mississippi State).

Taking as an example the funding for my program (Insect Rearing Education and Research Program—the IREAR Program at North Carolina State University [NCSU]), my status is research professor, which means that while I am a member of the Department of Entomology and Plant Pathology, I must generate all funding for the program's continuance from grants, contracts, and service fees for online courses and possibly other services that I can offer in insect rearing. I have supported this program with several contracts from Federal funding of projects, some funding from private companies, and portions of course registration fees, which I contribute to the rearing program. It boils down to my getting a contract, for example to develop a diet or rearing system for a government or private organization, and I then transfer the salary/consulting fee portion of the contract to the IREAR program to support education and research. As one would guess from considering this structure, there is not a great deal of leeway for further development of the program, especially the education part. There is little or no opportunity to hire-in incipient rearing specialists, technicians, or any other support personnel.

Because I believe in the need for education and research specifically designated for improvement of insect-rearing systems, especially the education of rearing professionals, I provide my time and most of my waking efforts and energy to furthering the system, but the concept of specific, dedicated IREAR demands a more dependable and broader source of funding. I dwell on this idea further in Section 13.3.2.

13.3.1 Traditional Sources of Funding for Insect Rearing

Cohen 2001 and 2019 explained that one of the major problems in insect-rearing infrastructure is that the nature of rearing has been one of service and subservience to other purposes, programs, and industries. Insect rearing has traditionally been performed for some purpose **other than** rearing per se. Even in the earliest rearing situations—silkworm rearing—the goal was not to rear silkworms; it was to produce silk. When the early researchers in genetics of *Drosophila* made advancements in rearing those flies, the advancements were not purposed to rear flies in a better way; the purpose was to enable the researchers to perform higher quality, more reliable, and more repeatable genetic experiments. When Riley and his associates cultivated Vedalia beetles in tent-covered citrus trees in the late 1880s, it was not to produce the beetles for the sake of rearing progress; it was to use them as a tool to control cottony cushion scale (DeBach 1974). The story goes on with virtually every insect that has been reared that there is a purpose such as research, sterile insect techniques (SITs), biological control, using reared insects as food for other creatures, using them to control infected wounds, for fiber and other products, or for celebrations (butterfly releases), the reason for rearing the insects is to serve some purpose. Because of this service status, and because of several other accidents of rearing history, the rearing process per se is considered secondary in importance compared to the primary purpose of insect production. The structure of this funding system is a sort of food pyramid where the base of many (most?) entomology programs is a rearing system—often a mass-rearing system. The programs such as biological control, pesticide development, SIT, research, etc. are above the base and dependent upon it for their very existence, yet the development of the rearing system that supports the "higher" functions is treated as an afterthought. If the program is based on insect genetics, the best insect geneticists are hired or funded; if it's a toxicology program, the most qualified toxicologists are sought or are screened by funding authorities. And so it goes, except for the personnel who are to develop and maintain the insect-rearing system that exists at the bottom of the pyramid. The rearing has always been treated as merely an adjunct or after-thought to the program's purpose. This would be fine, if the insect-rearing portion of the program were something simple to develop and maintain. However, as has been the point of this entire book, there are so many ways that the rearing program can be derailed.

13.3.2 Special Comments about Funding of Insect Research

Cohen 2019 treated the subject of funding as a special issue in rearing situations and in entomology, in general. I also discuss the important topic in Chapter 11 (on "Publishing Rearing Papers"). However, it is in order here to point out that insect rearing is hampered by its unique position as a service for other disciplines and programs but not as an entity unto itself. By "an entity unto itself," I mean that rearing has been treated as a necessity for other disciplines and programs but not as something worth pursuing as an endpoint. In contrast to rearing, genetics (for example) emerged as a well-respected discipline from an origin in breeding organisms for some practical purpose such as crop improvement or livestock domestication. The practice of breeding organisms that were of practical use or desirable to people had its roots in the domestication of dogs, cats, cattle, maize, rice, wheat, and even lower organisms such as yeasts. In this book, I discuss the domestication of silkworms in considerable detail. In all these and many more examples, people made profound changes in the organisms that they domesticated, such as forming breeds as disparate as Chihuahuas to Great Danes. The breeds were formed by deliberate selection before there was ever a notion of "genetics." The work of Gregor Mendel on sweet peas was also a plant breeding exercise that was a harbinger of genetics (a term coined by William Bateson in 1905); and within the same time frame, genetics became an ever-increasingly rigorous scientific discipline with journals of heredity and related subjects. Probably every other "true scientific discipline" evolved as a "pure science" from an applied or practical ancestor.

Once a discipline is accepted as a respected domain of science, it can be characterized as being worthwhile simply as a means of our understanding the universe, nature, or our world just for the sake of knowledge. It is true that most "pure" science has potential value of helping us solve problems of practical significance. There has been, from the dawn of civilization, an interplay between pure and applied science. Understanding quantum physics and relativity helped develop control over atomic energy. Understanding the nature of microbes helped us develop the medicine dealing with disease. Understanding the basic physiology of muscle contractions helps us deal with heart rhythm disorders.

In rearing, we apply the knowledge of insect nutrition to help us develop diets for insects, and understanding the physics of weather and other aspects of environmental science helps us design better-working rearing rooms that are climate-controlled. Science or knowledge of scientific principles is indeed a useful foundation for developing and improving all aspects of insect-rearing systems. However, the reciprocal is also true and of great importance. **Being able to rear insects under controlled conditions gives us the ability to design experiments that would be difficult or impossible to conduct to discover basic knowledge of most aspects of insect biology. Virtually all the discoveries of genetics that came from studies of *Drosophila melanogaster* were only possible once controlled rearing of these flies was mastered.** This point is made abundantly clear in the works of Delcourt and Guyenot, Baumberger, and especially the many papers from the laboratory (known as "The Fly Room") of Thomas Hunt Morgan (and his most famous assistants, Muller, Sturtevant, and Bridges, known in some circles as "the Fly Boys"). All these pioneering researchers in drosophilia genetics had instances when they had to sort phenotypic outcomes from genotypic causes to find which ones were indeed genetic and which ones were results of some kind of error introduced into their studies by faults in or misunderstandings of the rearing system.

13.3.3 Novel Funding Sources

Figure 13.2 shows a scheme of the future of rearing as proposed by Cohen. The branches from "Rearing Future" show rearing research (right) and rearing education (left).

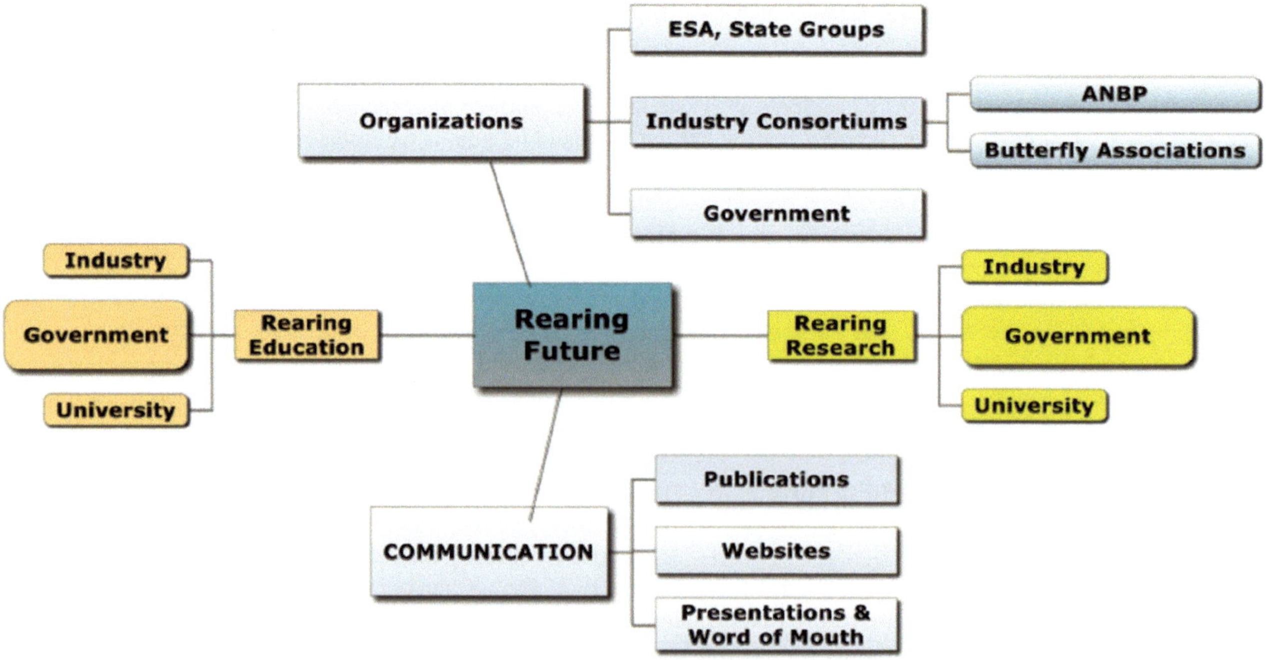

Figure 13.2 Current and potential new funding sources for insect-rearing research and development.

The sources of funding that would benefit and profit from enhanced programs in rearing education and research are government, industry, and universities (which have been the traditional bastion of education and research in various scientific and technological fields). The chart also suggests that organizations (government, non-government organizations, and various "lower level" groups such as entomological societies and even private philanthropic groups—not diagrammed here) are potentially influential in decisions about what kinds of research and education should be funded. The implications of this chart are that where there are no current sources of direct funding for rearing education and research, there should be established efforts, possibly led by grassroots groups, who recognize that while all of these entities profit from and thrive on the outputs of rearing education and research, they pay little, if anything, support for the research and education entities.

The situation is made more difficult because: (1) the tradition has been for rearing research and education to be treated as a minor, low consequence offshoot of "mainstream programs" such as biological control, SIT, pesticide development, research, etc. If there is not educational entity where rearing is being taught, where do well-meaning people and organizations direct the funding? If there is no funding, how do the rearing research and education entities come into being? It's an "no chicken—no egg" phenomenon. Yet the fact remains, the entomology community as a whole would profit greatly from a well-organized rearing education and research infrastructure and such a structure needs support.

13.4 IMPROVEMENTS IN COMMUNICATION OF REARING ADVANCEMENTS

13.4.1 Publications on Rearing Advancements

All of Chapter 11 in this book is dedicated to suggestions for improvements in communication of rearing progress. This topic is discussed, at length, by Cohen 2001 and 2019, and it was treated compellingly by Leppla 1978. However, it bears repeating that the literature on insect rearing has a long history, including many 19th century writings about various aspects of silkworm rearing, and in the early 20th century numerous excellent papers were published on rearing *Drosophila* species, mainly *D. melanogaster*. Many of the early 20th century papers were published in the most prestigious journals of their time (*Science*, *Nature*, *Proceedings of the National Academy of Sciences*, *Journal of Experimental Zoology*, etc.). Often, papers on other scientific discoveries included descriptions of rearing advancements. Yet, despite the tremendous progress made in rearing technology and despite the confluence of rearing information adopted for rearing insects besides *D. melanogaster*, such as use of yeast, agar, and casein, the rearing portion of the discoveries was left behind while the other discoveries and advancements (such as gene linkage, crossing over, and radiation-induced mutations) were heralded and respected. By establishing publication channels where rearing research advancements can be presented prominently (rather than hidden in a small section of materials and methods of some other research), rearing advancements would gain the credibility and respect that they deserve. Along with the prominent and welcoming acceptance of rearing advancements, there must be a clearly expressed set of standards for both incremental and saltatory accomplishments. This is discussed in detail in Chapter 11. This topic is also treated in the related discussion on treatment of personnel's recognition and professional status (Section 13.6).

13.4.2 Communications Other Than Publications

As discussed elsewhere in this chapter, other sources of rearing communication and outlets for rearing advancements include national and international meetings as well as local meetings of The Entomological Society of America (ESA).

13.5 IMPROVEMENTS IN REARING EDUCATION

The infrastructure consists of the workshops offered yearly by Mississippi State University, and occasional workshops offered by NCSU, and regular online and onsite courses offered by NCSU. These entities can serve as models and launching points for further development of insect-rearing education. I have started writing a new book on rearing education and am enhancing my website to provide guidance and direction for developing insect-rearing courses. Part of my personal mission is to try to mentor professionals and would-be professionals who want to start teaching insect-rearing courses.

13.6 IMPROVEMENT IN PERSONNEL ISSUES (RECOGNITION AND PROFESSIONAL STATUS)

The information in Table 13.1, which is taken from USDA, ARS standards for the Research Position Evaluation System (RPES) should be adopted by all organizations that depend on insect rearing. Table 13.1 includes two columns that apply nicely to how insect-rearing specialists should be evaluated: research and development. The thrust of the RPES table is that what ARS calls Category 1 scientists and how these professionals should be evaluated with regard to basic science (research) and applied science (development). Category 1 scientists may

Table 13.1 Research Grade Evaluation Guide (September, 2006) for Evaluating USDA, ARS Scientists

	Research	Development
Purpose	Extending knowledge and understanding	New or improved products, processes, and techniques
Assignments	Problems to be solved: • entail relative freedom to explore promising areas in relation to organizational programs; • may stem from an intent to close gaps in knowledge in a given field, or to develop new theories or explanations of phenomena; and • are difficult to define in terms of expected outcomes and measurable results.	Problems to be solved: • are defined in advance or assigned; • may stem from an intent to exploit an understanding of phenomena and principles; or • have predictable outcomes or measurable results.
Results	Products are: • papers describing new and modified theories and principles; • explanations of phenomena; and • information to improve the understanding of techniques and processes.	Products are: • papers describing application of theories and principles; • design concepts, models, patents, and inventions; and • equipment, techniques, and processes.

have in their job description a primary goal to develop diets or other rearing system components. In accomplishing this goal, the scientist must demonstrate that she/he worked with creativity, originality, and demonstrated accomplishments to solve the problems. ARS scientists, in writing up their RPES cases, must convey that their demonstrated accomplishments (in this instance rearing system accomplishments) had impact on the scientific community along with their originality and depth of scientific advancement character. There is room in this system for the Category 1 scientist to make scientific discoveries that would fit into the "research" category such as discovery of a novel feeding stimulus or an aspect of feeding dynamics that explains how the insect's digestive system works. As with the case of ARS Category 1 scientists, researchers from other government agencies, from universities and from private industry, should have the same kinds of standards and evaluation protocols that are described here. However, although this evaluation system is excellent for its rigor and potential to evoke excellent science, it can only succeed if the scientific community has the communication infrastructure that allows the publications that report the accomplishments to be properly handled in terms of: (1) editorial expertise in rearing, (2) peer reviewers that have fair and appropriate standards and the knowledge to perform a truly objective scrutiny of the reported work, and (3) a journal where papers that meet these high level criteria can be published. A key word here is peer review, but I hasten to add that for peer review to be effective and for it to work properly, it must involve peers—that is fellow scientists who are capable of: (1) understanding the work in question thanks to their familiarity with the field of insect rearing (including a willingness to keep up on the literature), (2) a fairness about reading and commenting objectively on the works of their friends **and** those of their competitors, and (3) a sense of duty that despite the fact that they are not being paid by the journals, they will spend the several hours it takes to perform a fair and thorough review.

Other outlets for the research and development described in Sections 13.4.1 and 13.4.2 can include oral and poster presentations at meetings such as the International Congress of Entomology, The Entomological Society of America, and various societies and organizations where entomological research and development are welcomed. Besides presentations at meetings, rearing work can be presented on websites, and in lectures and courses at universities.

13.7 IMPROVEMENTS IN FACILITIES

Upscaling Insectary Design (sources of new, improved designs such as GLP, OSHA, various industrial standards, hospital standards) is discussed further in Chapter 14 in the context of standards for rearing systems that must follow the standards of the most sanitary, safe, and wholesome conditions. In terms of air quality (e.g., filtration and number of air turnovers per hour), surface cleanliness, and personnel conduct the rearing laboratory/facility should be a wholesome place both for personnel **and** for the insects that are being cultivated.

There is an extensive base of resources on design of facilities to accomplish: (1) safety for personnel (and comfortable, humane working conditions), (2) efficient use of space, (3) engineering and architectural standards for making floor, walls, ceilings, diet preparation areas (i.e., food preparation standards), and environmental conditions in compliance with the concept of "a good place to work." It is discussed in Chapter 14 and other portions of this book how generally there is a congruence between what is good for the insects and what is good for personnel. If the funding improvements are made for insect rearing in general, then it can be expected that some of the financial enhancement of rearing will help improve facilities such as air handling, sanitation, and other workspace issues. Interestingly, during the writing of this topic, the COVID-19 pandemic is afoot, and there are many websites that, though they are intended to address the pandemic, are rich in information on sanitation and safety, in general (such as coronavirus.jhu.edu which is the Johns Hopkins website on COVID-19).

13.8 THE CULTURE OF INSECT REARING IN THE CONTEXT OF ENTOMOLOGY AND SCIENCE AS A WHOLE

I am writing this section in the first person, deliberately, to convey that the ideas that I suggest are my own perspective developed over nearly 50 years in the field. Over the past 20+ years, I have lectured and written about the phenomenon of insect-rearing suffering from perception of the discipline being of lesser status than other scientific disciplines. In one of my classes, a student expressed it aptly when we were discussing this topic, asking, "Is it because insect rearing isn't 'sciency' enough?" That was ten years ago when that student questioned whether rearing was sciency, and since then I have come to decide that in part rearing and especially rearing communication is part of the problem in not being treated internally (within the discipline) with enough scientific **rigor** (see Chapters 11 and 14 for further discussion of rigor). However, the scientific community that houses insect-rearing practices bears some responsibility for failing to recognize the professional nature of rearing, especially when reporting on insect-rearing accomplishments does contain rigor and appropriate scientific content.

My experience with the non-esteemed status of rearing started when I joined the USDA, ARS Biological Control of Insects Lab in Tucson, AZ in 1979. The ARS had just emerged from a review by the National Academy of Sciences whose report rebuked ARS for insufficient publications in peer-reviewed journals. In response to this report, and as a 1976 Congressional mandate, ARS leadership was directed to increase the profile and stature of ARS research by revising the standards of retention and promotion (RPES). The most prominent and heavily weighted component of this system is the Demonstrated Accomplishments, which are expected to be original, high impact research; and the most-weighty aspect of these accomplishments is publication in peer-reviewed journals. To retain their grade, ARS researchers (including rearing specialists or scientists whose predominant assignment is insect rearing) must publish at least one paper per year as senior author. For promotion to the higher grades, more publications than one per year are expected, and "regional" journals do not get as much credit as higher profile journals—ones of national or international stature. Essentially, the 1976 Congressional mandate made ARS into a "publish or perish" system.

13.9 RESEARCH VS. DEVELOPMENT IN REARING SCIENCE AND TECHNOLOGY

What transpired for at least two decades in ARS was: (1) an increase in the attention to quality research and development, which was rewarding to the scientists who were comfortable with publishing in scientific journals, and (2) an increase in tension for scientists who worked in disciplines where publication had not previously been demanded—where only solutions to problems were expected. By the 1970s, a major thrust in insect rearing had become a matter of solving the problem(s) of successfully rearing a given insect was the end-point of research and development expectations. In those days of my early career in ARS, I found that a confusion about what was "publishable" in rearing research was not clear to traditional journal editors and reviewers, let alone to scientists in rearing research positions. The Research Evaluation Guide published by the US Office of Personnel Management (https://www.opm.gov/policy-data-oversight/classification-qualifications/classifying-general-schedule-positions/functional-guides/gsresch.pdf) contains an excellent summary of the similarities and distinctions between research and development and is presented in Table 13.1. The fundamental difference between research and development is that the former is advancement of **basic** science while the latter is **applied** science and technology. As an ARS scientist for 23 years, my position description was that my job was to develop artificial diets for predatory insects (later, for insects in general). My work straddled the two domains (basic and applied science), and whenever I tried to publish basic information (such as explaining how various adaptations of predators involve extra-oral digestion, or how water balance and thermal adaptations were associated with the evolutionary and ecological backgrounds of the insects I studied), publishing came easy. However, when I tried to publish directly rearing accomplishments, I struggled to get my work published in the more respected entomological journals. I learned that all of my colleagues who were working as rearing specialists had the same kind of trouble getting their works published.

13.10 VARIOUS POTENTIALLY NOVEL AND INNOVATIVE LINES OF REARING RESEARCH

13.10.1 Statistically-Based Research and Rearing-System Improvements Using Design and Analysis of Experiments, Statistical Process Control, and Statistical Quality Control

My perspective about the potential for revolutionizing development and optimization of diets and other rearing systems components came from several papers written by

Dr. Steve Lapointe. In my estimation Lapointe's papers (2008, 2010a, 2010b) will have revolutionized approaches to insect rearing by applying the theory and practices of design of experiments (or "design and analysis of experiments" in Montgomery's 2013 terms). I will not treat this subject in detail here as I have devoted part of Chapter 7 and all of Chapter 9 to providing tutorials on using the SAS, JMP (JMP®, version 14.2. SAS Institute, Inc., Cary, NC, 1989–2019) approach to improving rearing systems by using geometric analyses, surface designs, full-factorial designs, Taguchi procedures, and mixture designs. These statistically-based procedures were developed for systems engineering statistics where multiple variables are analyzed at the same time in experiments that test ***responses*** (outcomes) from varying types or amounts of various rearing components (known as ***factors***). Some of the more sophisticated programs such as the JMP Custom Design can handle mixed variables such as different amounts of several diet components, different kinds of components (agar vs. carrageenan gelling agents), and different temperatures (to provide a few of many possible variables or factors). Assemi et al. 2012 used 36 factors in their Taguchi type experiments. I consider this approach revolutionary because it allows the determination of interactions between factors, which can help us understand influences of containers, diet components, temperatures, and several other experimental conditions. For insectary managers and their support personnel, the design of experiments approach will be without parallel for testing which rearing system components are useful, highly important, redundant, and they can go further to use optimization on a component-by-component basis to develop the most highly efficient rearing system possible.

13.10.2 Finding New Diet Components (Biomanufacturing, Other Subjects, and Sources)

Insect-rearing specialists are ever on a quest for the "new wheat germ," that is, the material that will jump start a diet or other rearing factor in a saltatory, rather than incremental way. That is what wheat germ did for insect diets in 1960 and what yeast and agar did for insect diets back in 1910 and 1917, respectively. Once antimicrobial agents (especially antifungals) were added to insect diets, a major avenue for contaminant control was established. In 1996, the application of the twin-screw extruder made a giant leap for insect rearing Figure 13.3.

From these significant advances comes the hope that further exploration will bring us new diet components or new processing techniques. Some of the techniques that are described extensively in Cohen 2015 include biomanufacturing procedures with such production instruments as bioreactors. Bioreactors are illustrated in Figures 13.4 and the schemes for modification of various foods and reaction component inputs are illustrated in Figures 13.4 and 13.5).

Figure 13.3 Using twin-screw extruder to make novel diets by varying process steps in conjunction with varying diet components. The versatility of the extruder shows promise to develop countless new diets, varying the nutritional composition and texture by using varied heating, mixing, and kneading profiles along with varying diet components such as gelling agents and water content. Although Edwards et al. 1996 demonstrated the proof of principle that extruders can solve many intractable diet problems, the potential of extruders and other modern food processing equipment has not been adequately explored and reported.

13.10.2.1 Bioreactor Outputs and Benefits

- Diets with enhanced nutrition
  - Higher, better quality protein content
  - Better lipids (phospholipids, sterols, polyunsaturated fatty acids)
  - Better Vitamins
- Improved bioavailability of minerals (incorporation of organic counterions such as lactates and citrates)
  - Removal of antinutrients
  - Carbohydrate improvement (detoxification of lectins, etc.)
  - Enhanced stability, uniformity, palatability

13.10.3 Analytical Techniques as Potentially Game-Changing Tools to Help Us Understand the Scientific Basis (Mechanisms and Causes and Effects) of Reared Insects' Homeostasis

Every year, new molecular techniques appear for genetic analysis of insect and microbes in rearing systems. And analytical techniques become available (e.g., ultra-high performance liquid chromatography [UHPLC], which provides higher resolution separations and analyses of components in diets and in insect systems): all these allowing fairly easy access of rearing specialists to determine the genetic, molecular, and metabolomic outcomes of rearing process manipulations. The high precision, high throughput analytical procedures coupled with

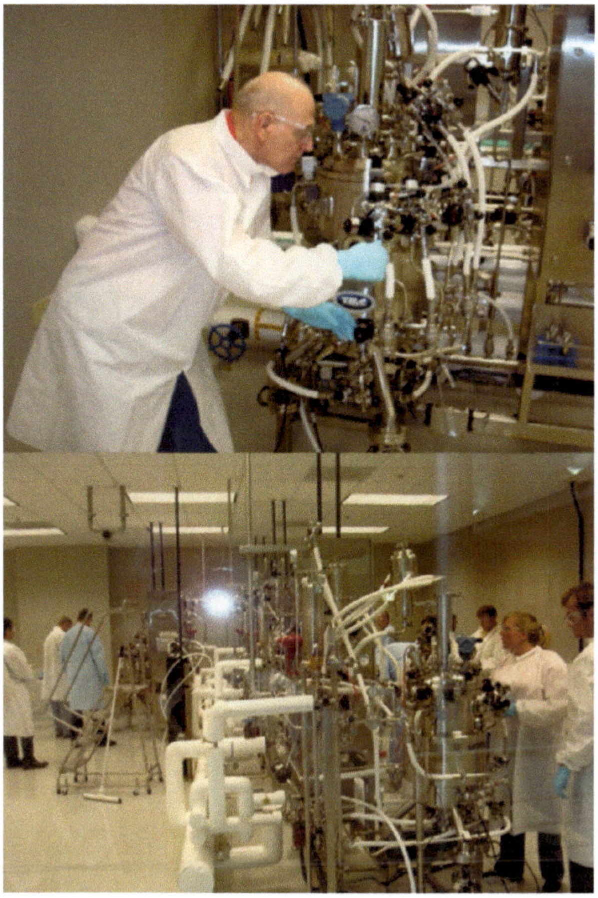

Figure 13.4 Bioreactors for bioprocessing pharmaceuticals and foods. These 30-L bioreactors (at NCSU, Biomanufacturing Training and Education Center [BTEC]). These reactors can be programmed to culture selected microbes to produce various fermentation/microbial processing products. We have been doing preliminary work with several species of yeasts to bio-modify raw materials such as soy flour, rice flour, bean meal, molasses, etc. to improve the nutritional value and palatability of these foods for use in insect diets.

the statistical procedures will allow the discovery-based science-driven rearing applications that I am advocating throughout this book. Also, as discussed by Cohen 2015, novel inquiries such as ascertaining the sensory roles of diet selection and acceptance should prove to be a highly fruitful and informative domain in design and operation of successful rearing systems (Figure 13.6).

13.10.4 Symptomology

Developing a comprehensive base of knowledge-relating stress factors to specific symptoms.

Using molecular genetics for identification of microbial relations and using the analytical power of metabolomics can help serve as a basis for developing a flow chart or algorithm of symptoms and the various disorders that are associated with those symptoms. This approach, following the principles involved in human disease diagnosis patterns such as "metabolic panels" of normal metabolite patterns, can be established, and deviations from these normal patterns can be evaluated through controlled experiments with stress factors. In such a scheme, a "fingerprint" of metabolic components (small molecules such as sugars, amino acids, fatty acids, etc.) can be tied to such stresses as infection or nutritional deficiencies. The molecular analysis mentioned here can also be used to diagnose normal and abnormal microbial associations.

13.10.5 Insects as Food for Humans and Other Animals

In the past several years, there has been a re-awakening of the concept of insects as food for people. This topic is discussed in Chapter 2 of this book and by Cohen 2015 in terms of the practicality and feasibility of using insect-rearing systems as a farming approach with the idea of producing a significant enough biomass of insects to partially satisfy the needs for food security. The Idea of using insects as food has rested on two approaches: (1) harvesting insects that are of natural origin (such as great infestations of locusts, palm beetles, or other species of insects that occur in large number), and (2) using very inexpensive foods for the insects such as waste products from food supply chains or waste products that the insects can convert to higher quality biomass. The black soldier fly is an example of this type of approach as are crickets and mealworms. The research into making insect transduction of low-quality food to high-quality insect biomass is intriguing as a challenge, but it may be questionable whether the tradeoffs in rearing/farming costs for the amount and quality of the biomass allow this concept to become a feasible reality. The question of whether the techniques to make the insects' products (cricket protein or black soldier fly meal, for example) sufficiently inexpensive to provide the food security to a world whose human population will increase be a billion over the next several decades.

13.11 CONCLUDING REMARKS ABOUT THE FUTURE OF INSECT-REARING SYSTEMS AND REARING INFRASTRUCTURE

From the materials presented in this book, I hope that readers will be able to arrive at most of the implications that I have tried to lay out. (1) Insect rearing should be treated as a science. This will serve our purposes of producing abundant, healthy, high quality insects whose production will be enhanced and under much better control than if we treat rearing as a black box. (2) Rearing systems should be approached through an experimental

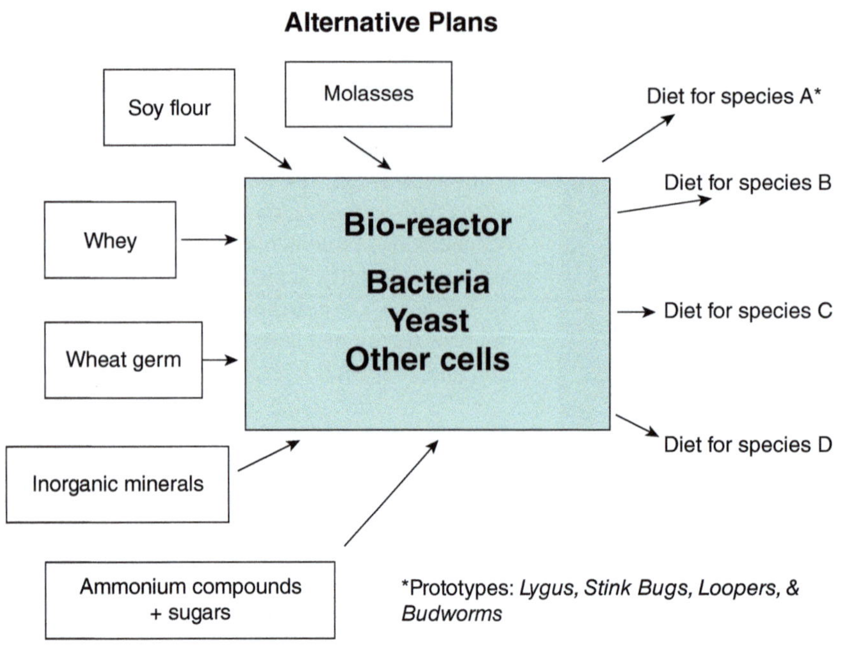

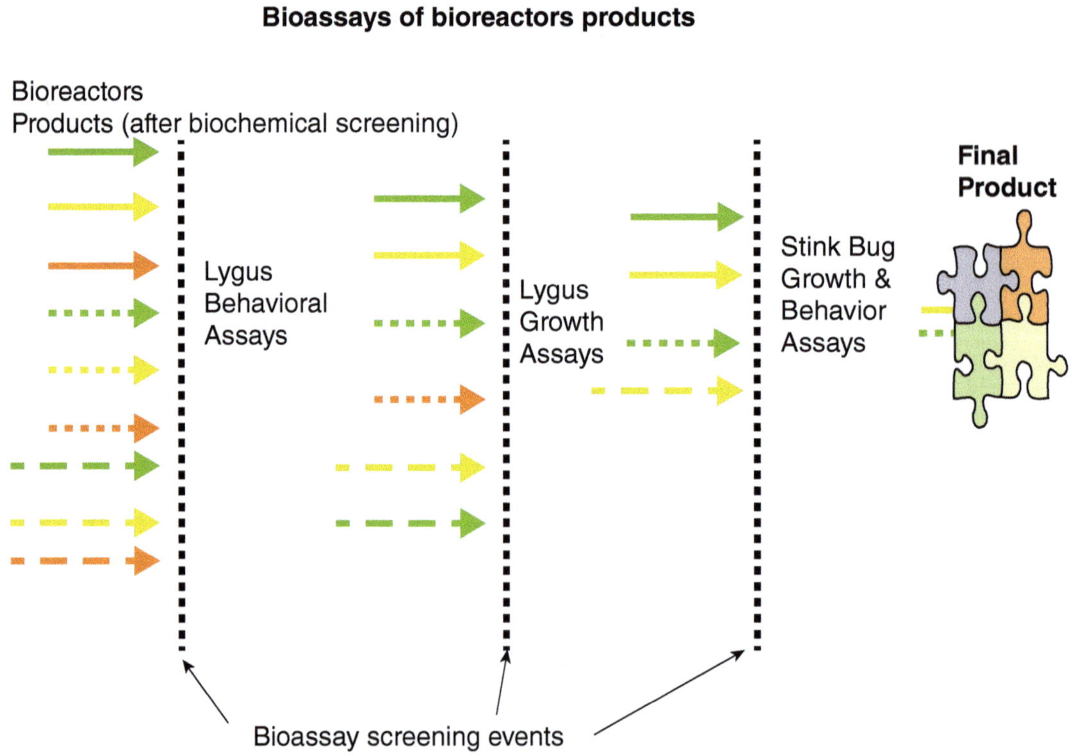

Figure 13.5 Plans for using bioreactors to modify various diet materials to develop totally new products that are improved over the raw materials. The schemes represent the inputs and outputs (top diagram), and the assay techniques for the products are illustrated in the lower diagram.

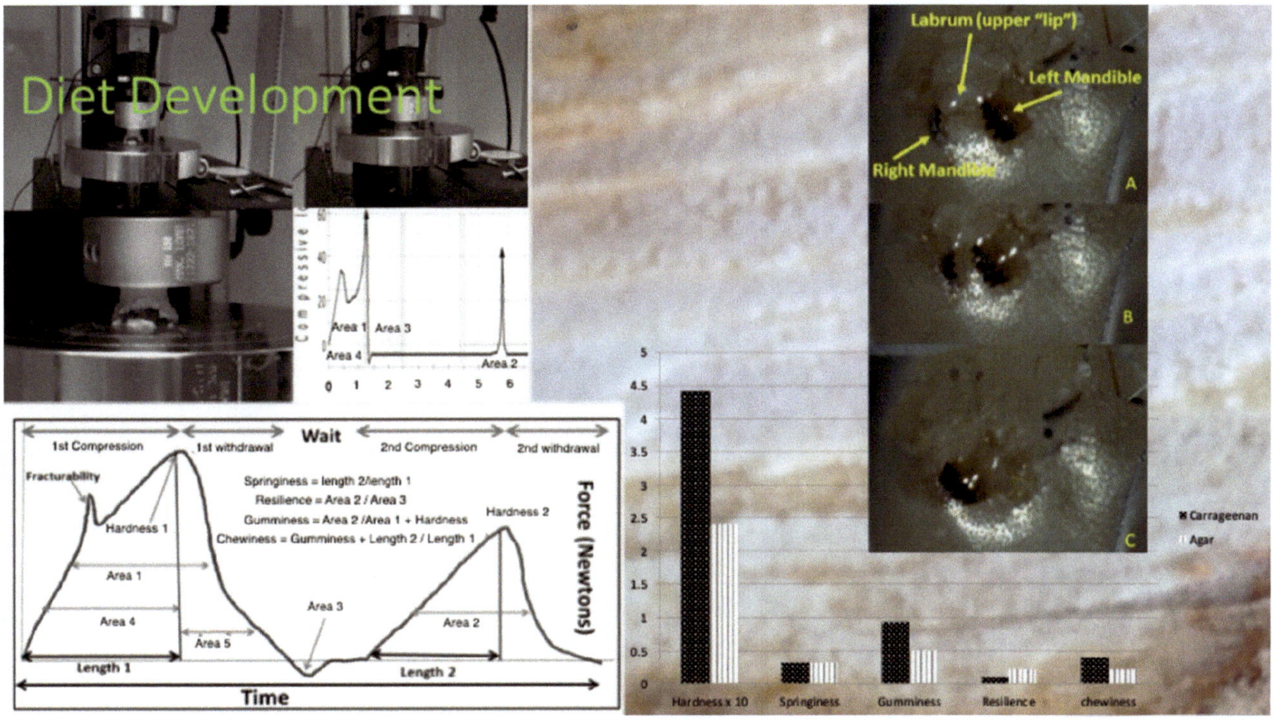

Figure 13.6 Measuring texture with rheological instruments. Among the very promising analytical techniques, the use of food rheology instruments such as the Universal Texture Analyzer (above) opens new horizons in correlating textural sensory properties with diet palatability. Although the chemical sensory properties (e.g., from Dethier's works) have been fairly well explored, the relationships between texture qualities such as fracture force, viscosity, gumminess, stringiness, cohesiveness, and resilience remain open question as to diets' acceptability to insects with various mouth part configurations.

discipline where discovery of underlying biology, chemistry, and physics principles can help us improve the design and operation of the rearing processes. (3) As we continue to develop the scientific understanding of our rearing systems and the complexities of the interplay between the systems' components, we must communicate the discoveries to the scientific community as a whole. In this vein, we must strive to improve the infrastructure of publishing research findings and to teach the emerging generation of rearing specialist and their stakeholders. (4) In being dedicated to the standards of rigor in our studies we can continue to improve the rearing community and those we serve by recognizing the scholars who came before us. (5) We must recognize the complexity of rearing systems and the reality that these systems are artificial ecological niches that demand handling and research approaches that explore the intricate interconnections between our systems' components. (6) Finally, if we recognize the value of our insect subjects as sentient beings that deserve humane treatment, we are more likely to accept our responsibilities to respect them duly. In keeping with the respect to our insect subjects, we will be rewarded with the highest possible quality of these subjects that our diligent attention can render.

CHAPTER **14**

Research Integrity in Insect-Rearing Studies

This chapter is presented as a guide for personnel who perform their work in insect-rearing facilities. It contains information from several background sources that can be consulted online for more complete treatment of the standards that these organizations prescribe. The sites representing these organizations are used here:

Good Laboratory Practices (GLP)
Organisation for Economic Co-operation and Development (OECD)
World Health Organization (WHO)
Environmental Protection Association (EPA) Federal Insecticide, Rodenticide and Fungicide Act and the Toxic Substances Control Act

Please note that using each of these terms in a search engine will allow access to the organizations and their recommendations.

In keeping with growing trends in most domains of scientific disciplines, formal guidelines are needed in the insect-rearing community. It would be appropriate for the rearing community to conduct discussions and contribute to considerations of how the rearing community can insure high levels of integrity. I have suggested in several contexts that integrity, high standards of conduct in our research, and quality of our research and inquiry outcomes will lead to improved standing and respect from our scientific audience and the general public (Figure 14.1).

For starters, I recommend that rearing researchers and their stakeholders become familiar with the discussions and outputs of the Office of Research Integrity (see the image and website above). The ORI provides resources such as PDF documents, lists of publications, and other materials that raise key questions and perspectives on issues of research integrity.

For my perspective, I feel that hallmark concepts that interplay toward integrity are professionalism (or professional conduct of rearing experts), dedication to quality of insects produced by rearing programs, commitment to humane treatment of our research subjects with special attention to the safety and well-being of insectary personnel, and integrity go hand-in-hand. I believe that essential components of integrity include continuing education in rearing-related disciplines, dedication to the concept of continuous improvement of all facets of insect-rearing infrastructure, and recognition of the contributions of the people who established or introduced the systems and system components that we use. Acknowledgement of how we got where we are in our rearing operations is essential to maximization of further rearing accomplishments.

There is a wealth of information and guidelines regarding all aspects of research integrity available at: https://ori.hhs.gov/.

One of the richest sources of in-depth information about research conduct as used by ORI comes from the National Institutes of Health (NIH). To give the reader an idea of the scope of the free information available from NIH, the following cover page and table of contents is presented at oir.nih.gov (https://oir.nih.gov/sites/default/files/uploads/sourcebook/documents/ethical_conduct/guidelines-conduct_research.pdf). Also, a valuable document (PDF) is available at https://www.nih.gov/sites/default/files/about-nih/nih-director/testimonies/nih-policies-procedures-promoting-scientific-integrity-2012.pdf.

Contents

Figure 14.1 Logo for the Office of Research Integrity at this site (https://ori.hhs.gov/).

The NIH document on guidelines for scientific integrity begins its narrative with the following statement:

> Scientific integrity, in this context, refers to maintaining the quality and objectivity of the research activities that the National Institutes of Health (NIH) funds and conducts, such that they are sound and worthy of the public's confidence. NIH's commitment to sound, objective science also strengthens the public's trust in policy decisions informed by scientific data. In fostering scientific integrity, NIH aims to ensure that (1) scientific findings are objective, credible, and readily available to the public and (2) the development of policies based on science is conducted with appropriate transparency.

While the statement is directly written for NIH, the basis of the concept of scientific integrity pertains to all areas of science, including insect rearing. I came upon the materials from ORI and NIH during a course that I was required by my university to take as a principal investigator in Federal grants to my institution. I found the materials in the course, which drew heavily from ORI and NIH, very useful in framing and reinforcing procedures, policies, and attitudes that I consider essential to making insect-rearing science and technology a potentially more respected practice (Figure 14.2).

Guidelines and Policies for the

Conduct of Research

in the Intramural Research Program at NIH

National Institutes of Health
Office of the Director

Figure 14.2 The cover page from the document from the National Institutes of Health, Office of the Director.

The NIH document contains clear and useful discussions of and guidelines for proper data management, authorship, publication of research, responsibilities of research mentors and mentees, collaboration, scientific rigor, and research misconduct. The guidelines advise very stringent expectations for all these facets of research such as an insistence that data be properly recorded with records maintained in ways the poor record-keeping and loss of information is unlikely. The formula for authorship is simple: all authors must have made substantial contribution to the research, including the planning, conduct, and interpretation of the research results. Authorship must not be a gift to a lab director or someone loosely associated with the work, nor should it be awarded to persons who made minor contributions such as reagents, specimens, etc. The treatment of misconduct, unsafe, or unhospitable workplaces (including issues of sexual harassment, bullying, exploitation of workers, etc.) is covered in meticulous detail. A topic covered in the NIH document and that deserves special attention in the discussion of integrity in rearing research laboratories is scientific rigor. Though I have covered this topic extensively in Chapter 11 on "Publishing Rearing Papers," it is important to discuss scientific rigor in the context of integrity in rearing research and in rearing production systems. The NIH document presents the following discussion under **Scientific Rigor and Reproducibility:**

> Scientific Rigor is the robust and unbiased application of the scientific method to well-defined research questions. To facilitate robust results, valid data are obtained by utilizing independent approaches to address research questions and ensuring that experiments are sufficiently controlled and documented to be reproducible. Reproducibility or replication of experiments, observations, and results is an integral component of the scientific method and allows science to be "Self-correcting."

Combined, rigor and reproducibility are essential to the generation of knowledge and the elaboration of new principles. Rigor involves 4 areas of focus:

1. Careful assessment of prevailing knowledge in field of study to identify strengths and weaknesses in prior research and any gaps in knowledge.
2. How applications of the scientific methodology will ensure robust and unbiased experimental design, methodology, analysis, interpretation, and dissemination of information.
3. The careful consideration of biologic variables that can influence experimental design and scientific methodology, i.e., species, sex, age, weight, animal models, and environmental considerations.
4. Validation of reagents, chemicals, biologics, and tests used in the research.

This definition and areas of focus of rigor apply fully to insect-rearing research (and the ensuing insect culture propagation). The characterization of scientific rigor in rearing research starts with unbiased (objective) application of the scientific research. The exhortation for researchers to begin with well-defined research questions is in line with stating rationale and statement of hypotheses that are tested with experiments that can be reproduced or replicated. These characteristics lend themselves to one of the most fundamental tenets that the science is continuously in a process of self-correction. Finally, the rearing research leads to "elaboration of new principles." The NIH discussion of scientific rigor and reproducibility goes on to focus on four areas of rigorous science, beginning with careful and honest evaluation of existing knowledge. Too often in rearing research, there is not an adequate examination of the existing or prevailing state of knowledge and a failure to properly examine and cite the existing knowledge. I have treated this deficiency in several places (Cohen 2001, 2015, 2018a, 2018b, and 2019), but the failing bears further discussion here. Insect rearing, being a practical process, suffers from a perception that rearing procedures are self-evident. Practices that have become commonly exercised are so taken for granted and passed on from laboratory to laboratory and from publication to publication end up with no trace of credit to the discoverer. For example, many insect diets contain agar as a solidifying agent; however, I have not been able to trace a single citation crediting Baumberger 1917b and Baumberger and Glasser 1917 for the idea of using agar in diets for *Drosophila*. Despite the facts that both papers were published in mainstream journals (*American Naturalist* and *Science*) and that "solid media for *Drosophila*-rearing" was in the papers' title, the later researchers who employed this important breakthrough failed to mention the source of the idea. Likewise, many other rearing innovations (such as casein, wheat germ, antimicrobial agents, and many other rearing breakthroughs) fail to be credited by rearing researchers who come along and use the discoveries as though they were free for the taking with no recognition or acknowledgment of the originator. I argue that the failure to acknowledge sources of ideas and techniques is a violation of NIH's first tenet of Scientific Rigor:

1. In the NIH document, the second tenet is that Scientific Rigor "will ensure robust and unbiased experimental design, methodology, analysis, interpretation, and dissemination of information." These elements, as discussed by Cohen 2001 and 2019, are often lacking in publications on rearing and even in standard operating procedures used in rearing settings. All of these factors add up to meticulous application of good scientific practices in the rearing research and rearing production setting.
2. The "careful consideration of biologic variables" in rearing research and production systems is something that receives little or no attention from rearing researchers, who use reared insects as the centerpiece of their research projects. The NIH document includes as biologic variables "sex, age, weight, animal models, and environmental considerations." I take "animal model" to be the organisms that are the subjects of the research, in our case—insects.
While careful researchers who use insect models or targets document the scrutiny that they afford to nearly every aspect of their studies, they often neglect the quality (fitness) of their insects and how the insects were treated in their rearing origins. Many standards apply to experimental conditions such as listing the source of equipment and materials. Implicit in the source citation is the understanding that there was appropriate quality control by the manufacturer to allow the researchers and their audience to accept the assumption that the instruments and materials were capable of delivering the test results upon which conclusions were drawn. If one claims a level of precision, for example, beyond the scope of a gas chromatograph equipped with a flame ionization detector or an HPLC equipped with a refractive index detector, informed reviewers and readers can discern that the claims of a researcher are beyond the scope for that instrument. A simple example of our ability to detect that a procedure was interpreted (or extrapolated) beyond the inherent bounds of the equipment, if a researcher reported using pH paper to measure the pH of a solution and that the precision was two decimal places (e.g., 7.12), then alerting reviewers and readers could reject the claim as being beyond the bounds of the instrument being used. Conversely, if the researcher reports a 7.12 pH having been obtained from an Orion™ Versa Star Pro™ pH meter, the assumption has greater credibility. But the most neglected components of too many studies of various aspects of insects' biology are the insects themselves. The requirements of journals and other sources of communication about research programs and the products of such programs are haphazard and lack a consistent set of guidelines. This topic is covered in greater depth in

Chapters 10 (on "Relationships between Fitness, Quality Control, Health, Homeostasis, and Stress") and 11 (on "Publishing Rearing Papers"). Suffice it to say for now, that the insects derived from insect-rearing systems remain a kind of "black box" where the rearing inputs and the consequences of these inputs remain poorly or entirely unknown.

3. The NIH document prescribes "Validation of reagents, chemicals, biologics, and tests used in the research": indicating the importance of careful attention to the materials and protocols for research, and these components are equally important in rearing research and production systems. In Topic 3 regarding Scientific Rigor and Reproducibility, the "consideration" of the experimental components was discussed, and in Topic 4, the "validation" becomes the operant word. In their wisdom and vast background in experimentation, NIS recognizes that there must be an ever-vigilant scrutiny of the materials, equipment, procedures, and biologics. Implied in the exhortation by NIH to validate is the implication that the researchers are responsible for assuring the scientific community that the experimental components were treated with consideration of degrees of purity, freshness, and authenticity of reagents and chemicals and that the tests were conducted according to documented standards. For the biologics (especially in the entomology community), the standards are (as mentioned in Topic 3) frequently neglected. Often, the mention that chemicals used in a reported experiment were obtained from a supplier whose name is provided provides some assurance that there is some level of quality control. It is further assumed that a reputable supplier will have provided a chemical or reagent of acceptable quality. For procedures, it is assumed (perhaps too permissively) that citation of a reference where the procedure was used previously assures the correctness and quality of the procedure. Here again there is a tacit expectation that the procedure has been properly vetted and that it is being repeated properly in the current study.

It should be noted that the sense of the NIH document's term "biologics" is the state and quality of the biological components of the systems in question. A suitable synonym for the term and the concept behind it is "organisms." In this sense, the statements about validation of reagents, chemicals biologics (organisms), and tests apply to our insect-rearing systems in the context of the insects themselves, the diets, environment, and other aspects of the rearing systems that are to be treated with the rigor and diligence suggested here.

14.1 GOOD LABORATORY PRACTICES

Another set of standards that are well established and reasonable protocols designed for other purposes but are applicable to rearing research and rearing production operations is what has come to be known as Good Laboratory Practices.

Robinson 2003 describes GLP as follows:

> Good laboratory practice (GLP) is a standard by which laboratory studies are designed, implemented, and reported to assure the public that the results are correct and the experiment can be reproduced exactly, at any time in the future. In less technical terms, GLP is the cornerstone of all laboratory-based activities in any organization that prides itself on the quality of the work it performs. And, despite its immediate association with the pharmaceutical sector (Figure 1), GLPs can (and should) be applied to virtually all industries in which laboratory work is performed, including companies involved in drug manufacturing, food and drink production, and engineering.

There are several sources that give extensive backgrounds in GLP, including several websites, PDF documents, and courses specially designed for entry-level laboratory personnel as well as those with extensive experience in laboratory work.

Included in these sources are WHO Handbook 2009, GLP Master Handbook, and Weinberg 2003.

14.2 ETHICAL AND HUMANE TREATMENT OF SUBJECTS

In my 2nd edition of Insect Diets: Science and Technology 2015, I included the following statements regarding humane treatment of organisms, especially insect, which are the centerpiece of my professional activity:

14.2.1 Humane Treatment of Organisms: Insect-Rearing Education and Research Program

The Insect-Rearing Education and Research Program coordinated by Allen C. Cohen subscribes to a policy of humane and ethical treatment of organisms. The values in this policy include these features: (1) Recognizing the diversity and complexity of life systems, we value all living things. (2) In light of this value, we try to design our work to help us understand the nature of life and living systems with a minimum of harm to or discomfort of living organisms. (3) We treat the subjects of our studies with respect, dignity, and with efforts to inflict a minimum of pain, trauma, or damage. This includes using, whenever possible, physical models instead of living systems to conduct our experiments. (4) We first consider ways that can inflict a minimum of potential harm to organisms and we ask the question: is the experiment necessary to give us the information we seek? Is the information of sufficient value to merit such sacrifices? (5) Furthermore, we subscribe to the tenet that the information derived from studies of organisms can be no better than the health, homeostasis, and minimization of stress of the organisms involved. Therefore, we strive to make our rearing systems and every other aspect of our handling of insects compatible with insect well-being.

Translation of these tenets into practice:

1. We use the concept of insect homeostasis or "comfort" in the sense that sensitive and reasonable thought about our insect handling can suggest guidelines about issues like food quality, space, management of waste, access to avoidance of adverse stimuli, sufficient gas exchange, protection from contaminants and pathogens, and all other aspects of insects' well-being as it would pertain to the populations of insects in the wild.
2. We subscribe to the model which teaches that well-treated insects are closer to "normal" insects from the wild. Such insects are of higher quality than ones that have been reared under suboptimal conditions. Therefore, we believe that our science is better when all aspects of our inquiries are conducted with the highest possible quality of components.
3. Therefore, our research into development and improvement of rearing systems keeps at its forefront the homeostasis of the insect subjects in our studies, and we further strive continuously to improve the base of knowledge for the scientific community to improve the conditions of rearing systems and the quality and well-being of the subjects.

In re-reading these paragraphs, I feel somewhat satisfied that I have touched on major points and motivating ideas that encourage thoughtful treatment of our insect subjects; however, I now feel that there are far-reaching implications of the concepts stated above, and these concepts merit further discussion and amplification. I feel that my 50 years as a professional biologist and the many experiences I have had with a great variety of organisms gives me an unusual perspective. I also feel that many of my experiences where I and sometimes my colleagues violated many of the premises of humane treatment further add to my authenticity in suggesting that there is a better way to do our science than what I had done. Even as I write these statements, I continue to have a nagging question of whether I am acting ethically and morally, and I have further questions of what entitles me to decide for another living thing how it is to be treated. This rather self-questioning and confession having been said, I proceed to raise some topics that I feel worth considering in our quest to improve our rearing standards.

Appendix 1: The *Drosophila* Rearing System History

A1.1 METHODS OF REARING *DROSOPHILA MELANOGASTER* (FORMERLY *D. AMPELOPHILA*)

I have selected several sections from older, historical publications to provide the rationale for both saltatory and incremental improvements in the rearing system for *Drosophila*. The role of *Drosophila* rearing in the evolution and history of insect-rearing systems is covered in Chapters 3 and 6 ("Design and Operation of Insect Rearing Systems: Science, Technology, and Infrastructure" and "Rearing Systems as Ecological Niches or N-Dimensional Hypervolumes"), but in this appendix, I have presented the original works in representative segments. I have chosen to present the authors' original wording because it conveys the eloquence and depth of thought about rearing systems that have presaged the features of current rearing systems that are highly successful in producing high-quality insects.

A1.1.1 Part 1: Rationale on Improving the Rearing System for *Drosophila*

In their 1933 paper, Bridges and Darby explained in the most fundamental terms the rationale for tuning rearing systems to ever-finer and more useful processes and components. The points made by Bridges and Darby here and Bridges elsewhere in several of his works are too often ignored by members of the research community who make tacit assumptions that their insects, produced by rearing techniques, are representative of insects in nature. In the most basic terms, the insects produced in rearing systems are taken for granted to provide faithful representation of organisms, wild counterparts that have not been subjected to the stresses and selection that went into the rearing process:

> **Part I. Improvements in Culture Methods: The Role of Yeast**
>
> *Gametic and Observed Ratios*
>
> IN the two decades during which *Drosophila melanogaster* has been bred as material for studies on heredity frequent improvements have been made in the culture media and methods in order to secure unhampered development of the full number of eggs laid by a female during a limited sampling period of ten days or less. This is to obtain the closest possible approximation or parallelism between the directly observed phenotypic series or ratios of zygotes on the one hand and the gametic series or ratios which are the real objects of study. The discrepancy is greatest where relatively weak mutant forms must be raised in competition with wild-type or strong forms. In many classes of experiments the accuracy and importance of the results are directly dependent on absence of distortion in the zygotic ratios, for example, in studies where linkage or non-disjunction is involved. Wherever possible the experiments are planned so as to avoid use of mutant types with high mortality. ***But often no mutant of normal viability is available for some special region of a chromosome, and of course in all experiments to discover the nature and behavior of those mutant types which happen to have low viability the only resource is to employ optimum culture conditions already developed and tested.*** In testing for the optimum methods, these mutants of low viability are good indices, for where "Weak flies do emerge consistently in the closest approximation to the number of wild-type sibs, there conditions are optimum for both kinds."
>
> **{Italic and bold font are Cohen's}**

The bottom line of Bridges and Darby's opening discussion of the paper on diet improvements that took place over the past two decades of *Drosophila* research (from about 1910 through 1932) was that in order to accommodate the flies that were less fit due to their genetic makeup, improvements had to be made in diets, as well as in other rearing system components (such as the temperature regulation system devised by Bridges 1932). The authors point out that the very nature of the phenotypes whose genotypes and other genetic makeup presented included individuals that were so frail, that they would not be noticed in the examination and counting process that went into analysis of such issues as mutations and chromosomal abnormalities. As early researchers from Delcourt and Guyenot 1910 through Bridges and Darby 1933 pointed out, only with the best rearing techniques could the best genetic analysis be performed!

The rationale for high levels of attention to the rearing issues is stated brilliantly by Bridges 1913:

> A certain kind of exception that has come up during the course of my work on *Drosophila ampelophila*, with Dr. T. H. Morgan, bears directly upon the problem of sex determination, and especially upon the view that the chromosomes are the carriers of the differentiators of the hereditary characters. The evidence which I am to present deals with the X-chromosome of Drosophila, concerning which the evidence relating to sex linkage and the linear arrangement of factors is more definite and complete than in any other case. Ordinarily, when a female with white eyes is mated to a wild male with red eyes, the daughters have red, and the sons, white eyes. This is a typical case of the criss-cross inheritance characteristic of sex linkage.

A1.1.2 Part 2: Rearing Methods from Morgan Et Al. 1915

The first publication on rearing *Drosophila* spp. was stated by Delcourt and Guyenot 1910 in "The Possibility of Studying Certain Diptera in a Defined Environment" where the authors explained the difficulties in performing genetic analyses with any precision in the face of rearing chaos that ensued with poorly controlled conditions. This issue was noted by other authors who were racing to discover the secrets of heredity that was the hotbed of scientific inquiry in the first two decades of the 20th century, including Baumberger, Loeb, and the Morgan group at Columbia University. The most coherent and regimented procedures for *Drosophila* rearing were presented in the comprehensive work: *The Mechanism of Mendelian Heredity* by T. H. Morgan, A. H. Sturtevant, H. J. Muller, and C. B. Bridges, all of Columbia University when that book was published in 1915. The team leader, Morgan, and the three students, Sturtevant, Muller, and Bridges using *Drosophila melanogaster* as their model system, would go on to publish several hundreds of papers that helped define the principles of genetics, and two of those researchers would go on to receive Nobel Prizes in Medicine for their work, Morgan in 1933 and Muller in 1946. In fact, as influential as the 1915 work was, the subsequent works done by these researchers in gene linkage, mutations, crossovers, and other genetic and chromosomal mechanisms revolutionized the science of genetics and modern biological thinking. All of this discovery and scientific revolution came from the ability to rear *D. melanogaster* in **operations** that were **designed** to give the researchers **control** over their subjects.

A1.1.2.1 Methods of Breeding Drosophila

Drosophila ampelophila has shown itself to be so generally useful for class work in genetics and is being so widely employed for this purpose that it may not be out of place here to give a few directions concerning apparatus, methods, and material.

A1.1.2.1.1 Culture Bottles

Large-mouthed bottles of about 500-cc capacity should be used. Pint milk bottles can be purchased at reasonable rates from wholesale dealers, and serve admirably as culture bottles. Stoppers of raw cotton are used, which should be tight but not packed in the mouth of the bottle.

A1.1.2.1.2 Temperature

The optimum is about 25°C. Extreme summer heat kills the flies in culture bottles unless special precautions are taken. Cold retards the development of the larvae indefinitely, but the flies themselves can withstand almost freezing temperature. Ordinary room temperature suffices, as a rule, but a controlled temperature of about 25°C is better.

A1.1.2.1.3 Food

Ripe fermented banana is the best food. If raw bananas with intact skin are peeled and put into old juice, contamination is not likely to occur, but for ordinary purposes the bananas should be peeled, broken up into pieces, covered with water, and slowly brought to nearly the boiling point, or else steamed. The pieces are then put into the old juice. This juice is made in the first instance by adding a little yeast to the water that covers the bananas (after they have cooled). The juice can then be used over and over again if it is occasionally greatly diluted with water to keep it from becoming too acid. The food is at its best from one to two days after being in the juice, although it may be used for a week or more. Keep food in glass-stoppered large-mouthed jars. Scrupulously avoid leaving the margin of the jar wet after removing food.

A1.1.2.1.4 Feeding

When a culture is to be made up the most approved way is to put about a teaspoonful of food on the bottom of the bottle and over it a folded piece of absorbent paper. The flies, while still under ether, may be dropped into a cornucopia of paper which is placed in the bottle. Often, however, the flies are brushed into a dry bottle and the cotton plug put in. An hour or more later, when the flies have recovered, the food, wrapped up in paper, is added. Toweling paper is cheap and serves excellently the purpose of wrapping the food, etc.

By putting in at first more food than recommended above, it is possible to carry a culture to the end without further feeding. But more flies and greater accuracy result if a small amount of food is first added, then as much more at the end of six or seven days. When the parent flies are taken out at about the tenth day, food may be added for a third time. New food should not be allowed to cover the old food and thereby drown the pupae.

A1.1.2.1.5 Precautions

In most cases a single female with one or in some cases more males should be put into a bottle, except for stock breeding, when more flies should usually be used. Under favorable conditions 200–300 flies should be obtained within the first ten days after the beginning to avoid overlapping of generations. If desirable they may then be transferred to a second, or even to a third culture bottle. In summer the parents should not remain over eight days in the old bottle; in winter they may be left more than ten days with safety. To obtain virgin females for mating, the bottles should be thoroughly emptied—it may be necessary to remove the old paper, etc., in order to make certain that all of the old flies are removed—and females obtained not later than six hours after the bottle was emptied. Females obtained 12 hours after the bottle was emptied are certainly virgin. If the old males have not been removed, females so young as to have their wings not yet unfolded may in rare cases have already undergone copulation.

A1.1.2.1.6 Examination

The flies go toward the light. Therefore, if the bottle is held with its mouth away from the light, the flies are not likely to escape. The plug may then be removed and another smaller bottle, with a mouth that fits the larger one, placed mouth to mouth against the larger bottle. The combination is then turned around, and the flies fly into the smaller bottle or the smaller bottle may be held firmly underneath the other and the flies shaken into it by jarring. The small bottle can then be plugged, and a bit of cotton with four or five drops of ether put into it. In a minute or two, the flies are under the influence of the ether and may be emptied out on to a piece of white paper, or a white glass plate. Some workers prefer to have the small bottle already saturated with ether before the flies are shaken into it; in this case they become etherized almost immediately. They can then be examined with a hand lens or the binocular microscope. Some of the characters require for study the latter or an ordinary microscope.

With a camel's hair brush the flies are pushed out into a row and then sorted out, from right to left, into an upper and a lower row, each of which may be subdivided. If overetherized, the wings stand out above and at right angles to the body. If insufficiently etherized, so that they recover before they can be examined, they may be etherized again.

The pure stock is kept in pint bottles and new cultures made up each week. Descriptions of the mutants as far as published will be found in the journals listed in Table A1.1.

In the introduction to their paper on culture media for *Drosophila*, Bridges and Darby 1933 express eloquently the importance of highest possible quality of rearing procedures in the attainment of good science. They provide rationale for the improvements in rearing technology over the two decades of genetic research on *D. melanogaster*, pointing out that losses of some flies due to rearing problems were disproportionate in wild-type flies and those with genetic features that resulted in weak (or unfit) flies. The authors point out that the observed phenotypic ratios appear to be skewed and therefore the true genetic relationships (which are the subjects of the genetic inquiries) are hidden by differential loses in certain genotypes or products of other genetic events phenomena such as linkage, chromosomal crossovers, or mutations (Figures A1.1–A1.4).

Table A1.1 The Phenotypic Characteristics Studied by the Morgan Group in Early *Drosophila* Genetics Studies

Bar.—"A New Sex-linked Character in Drosophila." *Biol. Bull.*, XXVI. 1914.
Beaded.—"The Analysis of a Case of Continuous Variation, Etc." *Amer. Nat.*, XLIII. 1914.
Bent.—"A Gene for the Fourth Chromosome of Drosophila." *Jour. Exper. Zoöl.*, XVII. 1914.
Black.—"Heredity of Body Color in Drosophila." *Jour. Exper. Zoöl.*, XIII. 1912.
Cherry.—"A New Eye Color Mutation in Drosophila, Etc." *Biol. Bull.*, XXV. 1913.
Ebony.—"A Third Group of Linked Genes in Drosophila." *Science*, XXXVII. 1913.
Eosin.—"Dilution Effects and Bicolorism, Etc." *Jour. Exper. Zoöl.*, XV. 1913.
Eyeless.—"Another Gene in the Fourth Chromosome of Drosophila." *Amer. Nat.*, XLIX. 1915.
Lethal.—"Two Sex-linked Lethal Factors in Drosophila, Etc." *Jour. Exper. Zoöl.*, XVII. 1914.
Miniature.—"A Modification of the Sex Ratio, Etc." *Zeits. f. ind. Abst.- u. Vererb.-Lehre.*, VII. 1912.
Pink.—"Dilution Effects and Bicolorism, Etc." *Jour. Exper. Zoöl.*, XV. 1913.
Rudimentary.—"A Modification of the Sex Ratio, Etc." *Zeits. f. ind. Abst.- u. Vererb.-Lehre.*, VII. 1912.
Spot.—"Another Case of Multiple Allelomorphs in Drosophila." *Biol. Bull.*, XXVI. 1914.
Vermilion.—"Dilution Effects and Bicolorism, Etc." *Jour. Exper. Zoöl.*, XV, 1913.
Vestigial.—"No Crossing Over in the Male of Drosophila, etc." *Biol. Bull.*, XXVI. 1914.
White.—"Sex Limited Inheritance in Drosophila." *Science*, XXXII. 1910.
Yellow.—"Heredity of Body Color in Drosophila." *Jour. Exper. Zoöl.*, XIII. 1912.

In the early work, prior to about 1916, fermented banana was generally used as food for the larvae and adults. The usual technique has been published in the 1915 edition of "The Mechanism of Mendelian Heredity" (page 229). The essentials of the method are the following: bananas with intact skins were put into closed jars, as a protection against the chance of egg deposition by stray flies, and were kept there until they became thoroughly ripe. They were then peeled, and the pulp broken into pieces and dropped into another covered jar containing water seeded with yeast. Usually the yeasty liquid used in the jar was the fermented juice from the previous lot of bananas. Fermentation converted the pieces of banana into a spongy mass floating on top of the liquid. The pulp was at its best about a day after starting the fermentation and became too acid for use after the second day. This rather dry fermented pulp was put into the culture bottles on the bottom, in amounts of about 5 grams for each half-pint

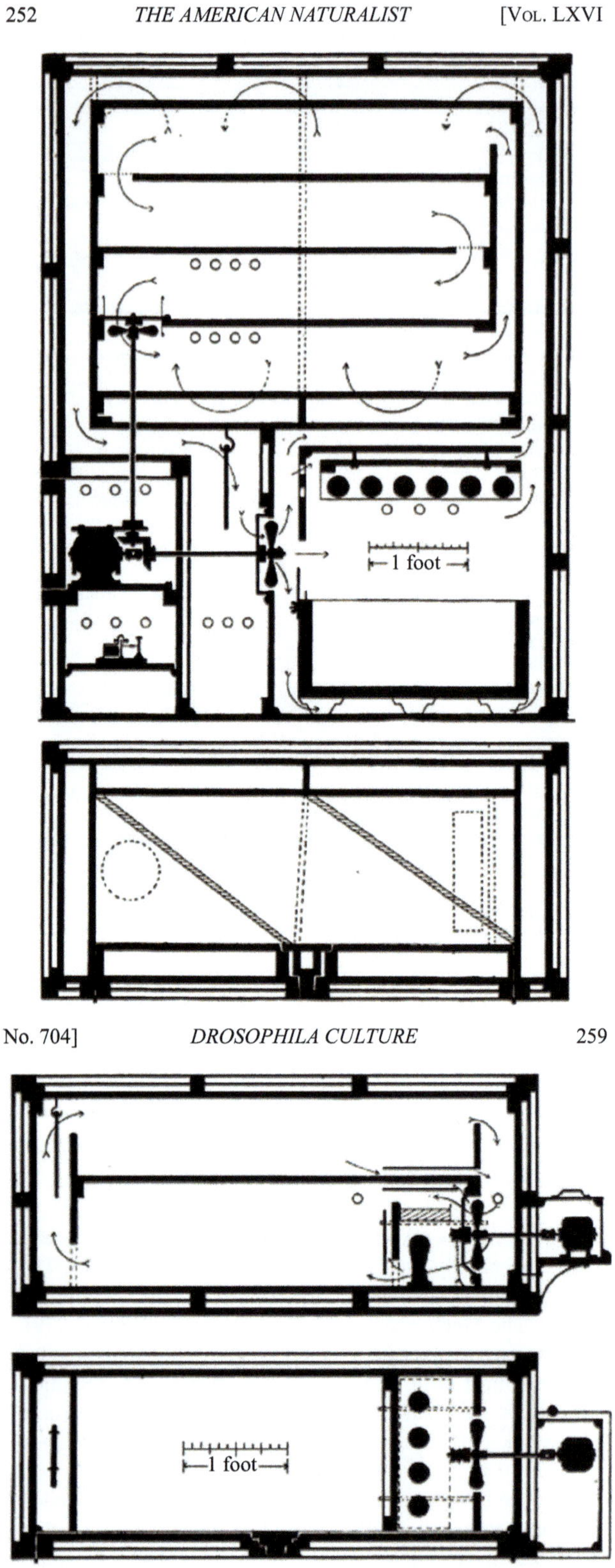

Figure A1.1 (a) and (b) The design for Bridges' incubator. Bridges had devised the incubator to regulate temperatures for the *Drosophila* colonies back in the early days of genetic research at Columbia.

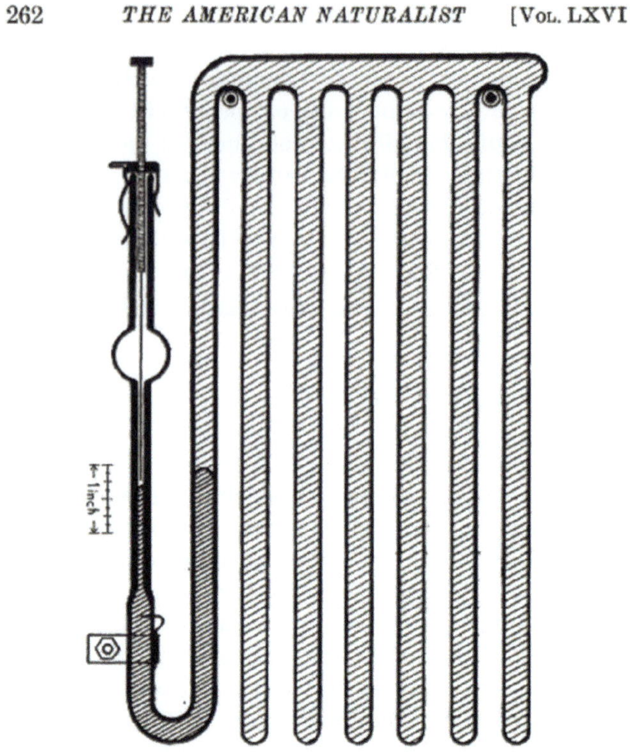

As the mercury column rises with increase in temperature it makes contact with the tip of a three-inch, half-millimeter diameter, nickel-silver wire. This wire is carried by a three-inch 6–32 brass machine screw with fillister head. The screw runs in a nut which is held in place at the top of the trap arm by a strong spring clip to which it is soldered. The arms of the spring clip are grooved on their inner faces to make the alignment positive, and hold firmly by virtue of a three-contact system, two contacts on one arm and the third on the other.

Figure A1.2 A schematic diagram of the thermostat devised by Bridges and described in his 1932 paper on apparatuses and culture methods for *Drosophila*. Of all of the famous scientists in T. H. Morgan's Columbia lab (the "Fly Boys"), Calvin Bridges was the one who gave the most impressive attention to the design and control of the *Drosophila* rearing operation.

bottle and was then covered with a layer or two of paper toweling. The flies were next inserted and the bottle stoppered with cotton. If the flies were inserted while still under ether they were first put into a small paper cornucopia which was stuck to the side of the bottle and in which the flies remained safe while recovering.

The above description was for the medium that had been used for the early years of *D. melanogaster* cultivation (from about 1913 until the early 1920s). However, Calvin Bridges was an ever-vigilant researcher with respect to the diets and environmental factors in the rearing of the flies that provided so much information about genetics:

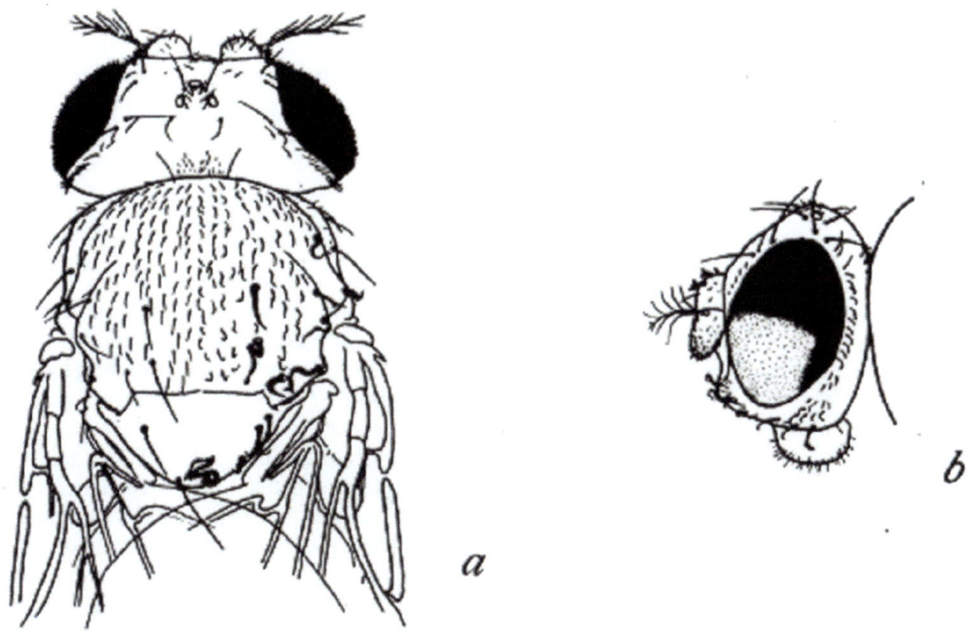

FIGURE 1

"Piebalds." Fig. 1*a* shows a Minute-n female in which the spot (singed but not Minute-n) covers the right dorsal half of the thorax. Fig. 1*b* shows a piebald in which the left eye has a spot of male-type eosin color with the adjacent tissue singed. An independent spot of singed occurs on the antenna.

Figure A1.3 Drawings of some mutant variants from Bridges 1925. (Bridges, C. B. 1925. Elimination of chromosomes due to a mutant (minute-N) in *Drosophila melanogaster.* Proc. Natl. Acad. Sci. 11: 701–706.)

The Alcoholized Banana Medium

In 1915-16 comparative tests were made by Bridges, as the result of which the procedure was improved somewhat (Bridges, 1921). The fermentation was started fresh each time, instead of using the old juice from the last batch of bananas. This delayed the development of acetic acid. The presence of a slight amount of alcohol (optimum 1.5 per cent) at the beginning of the fermentation was found by test to be conducive to higher outputs. The favorable action of the alcohol is probably exercised in two ways; first, alcohol is a definite stimulant to egg-laying; and second, it is a powerful retarder of the development of competing molds and bacteria. Thus, Baumberger (1919) quotes Lafar (1910, II, p. 238) "From the standpoint of the ecological theory of fermentation, the alcohol produced by yeast should be regarded as a weapon capable of hindering the appearance of other fungoid competitors in saccharine nutrient media. However, when accumulated in the medium during the progress of fermentation, it also restricts the further development of the producer." In this case, as with yeast poisons in general, the first result is the diminution and later the cessation (at about 6 per cent) of cell reproduction. A still greater concentration reduces fermentation and brings it to a stop (at 14-21 per cent), while still more alcohol kills the cells.

Recently Richards (1928) has studied the inhibiting effect of alcohol on the growth of yeast and on fermentation and finds it very great. But in the usual bacteria-contaminated yeast cultures used for food for flies the alcohol does not accumulate markedly. Instead, it is changed over by bacterial action into acetic acid. It has long been thought that the increase in acidity of the culture medium is itself a very important obstacle to the culture of flies. The exhaustion of the sugars and yeast foods, from too little or too poor medium, is perhaps the primary factor in limiting the growing of flies.

Readers are directed to the elaborate thinking about the ecological aspects of the relationship between the medium, yeast, bacteria, and drosophilid flies. The balance of alcohol (ethanol) content being produced by the yeast vs. the acetic acid content being produced by competitive bacteria (apparently *Acetobacter* or a similar, acetic acid-producing bacteria) causes the media to turn into acetic acid (= vinegar; hence the name vinegar flies for *D. melanogaster*). These issues are dealt with later in the paper by Bridges and Darby where they explore the pH of the medium to arrive at the optimal pH for the flies, which they determine to be 4.5. The authors go on to elaborate further on the optimization of the alcoholized banana medium:

In the alcohol method the pulp of ripe sound bananas was weighed and put to ferment in a large, shallow, tightly

In 1925, Curt Stern visited our laboratory and was interested in culture-bottle design. Upon his return to

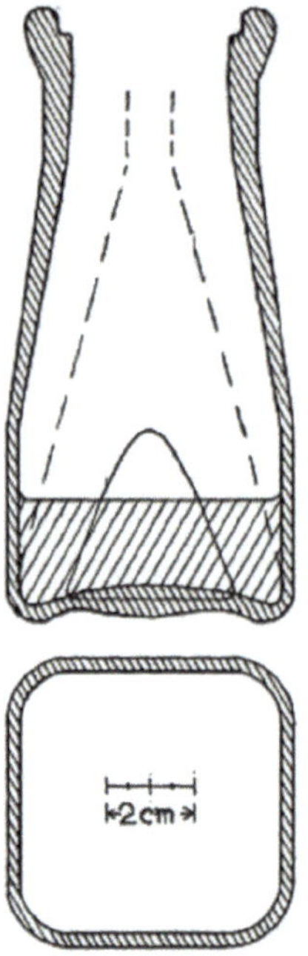

Germany he designed and had made a conical type of bottle, which had extra safeguards against displacement of the food cake by projections from the inner wall. The large size of the circular bottom seems a slight disadvantage of this excellent design.

Figure A1.4 Design of a replacement for milk bottles for rearing *Drosophila melanogaster*, from Bridges 1932. The elaborate design of this container reflects Bridges' meticulous attention to rearing details, including seemingly minor factors such as containers, diet optimization, and regulation of environmental conditions. Based on comments throughout Bridges' publications, it is evident that this excellent scientist with major discoveries of genetics phenomena realized that without the proper rearing conditions, many of the important scientific details could not be derived. Bridges saw that under chaotic and stressful rearing conditions, important phenotypic and genotypic outcomes could be lost or not recorded and interpreted.

covered glass dish with an equal weight of 3 per cent alcohol (optimum percentage). Each banana pulp was crushed slightly, little more than enough to separate it into three longitudinal strips. Yeast was added. The optimum amount of drained pulp, fermented 24 hours, for each half pint bottle was found to be 25 grams-more than twice as much as formerly used. Similarly, the optimum amount of paper was 1.0 gram per bottle, instead of about 0.5 gram as formerly used. This paper was used in the form of strips 5 cm long and 0.7 cm wide, matted down on the food, which was placed directly on the bottom. This method was almost entirely free of trouble from the food mass coming loose. No refeeding was necessary. It was found that pint bottles, with a larger area of bottom, gave larger progenies than half-pint bottles, but the greater convenience in handling the smaller bottles led to their adoption as standard. Another innovation was the use of cheesecloth to cover and reinforce the cotton plugs before insertion. The plugs lasted longer and when removed from the bottle did not, as formerly, leave a fringe of cotton fibers which entangled the flies.

The Banana Agar Medium

In the autumn of 1916, Bridges received from R. W. Glaser directions for making the transparent banana juice and agar culture medium which he had developed in cooperation with Baumberger (Baumberger and Glaser, 1917) for the clear observation of egg-laying and of the growth and behavior of larvae. Tests were made which showed (Bridges, 1921) that the concentration of food in this medium was inadequate for general use, but that the use of agar to hold the medium in place was a decided improvement over matted absorbent paper. Accordingly, the amount and concentration of food was increased by grinding up and using all the banana pulp instead of using only the clear strained juice. The method was modified and developed until in the spring of 1917 it was generally adopted in the laboratory as the standard method. The details of this method have been published (Bridges, 1921; also in the Laboratory Manual by Morgan, Sturtevant, Muller and Bridges; Henry Holt and Company, 1923), and need only brief comment here. The minimum percentage of agar needed to hold the food fairly solid was determined at roughly 0.7 per cent. One per cent was adopted as standard, since with increased stiffness of medium there was less loss of pupae by being worked into the food and smothered or drowned. For this same reason considerable paper was used also (about 0.5 gram) and pupation would then generally occur on the paper rather than in the food cake. With this method the risk of contamination of cultures through the food supply was greatly lowered, since one step in the preparation involved heating to a temperature more than sufficient to kill any eggs or larvae which might be present in the bananas before they were run through the potato masher and mixed with the agar solution. The concentration of food in the medium was kept high by using only as much water as weighed banana pulp. The amount of food per bottle was kept high by using 50 cc of the medium, equivalent to the 25 grams of banana pulp previously determined as optimum. This method remained standard in the Columbia laboratories for over ten years and is still widely used elsewhere.

The above paragraph has sweeping historical value that reaches beyond rearing of *D. melanogaster.* It chronicles the transfer of technology of application of gelling to insect diets. Prior to the works of Baumberger and Glaser 1917, the standard diets for *D. melanogaster* (and the few other species of insects being reared on artificial diets) were made from rotting fruits or purees of rotting fruit (or meat purees, in the case of flesh-eating flies).

Table A1.2 Culture Medium (Medium A) Described by Bridges and Darby 1932 to Replace Previously Described Medium for Rearing *Drosophila melanogaster*

Ingredient (per 100 g Diet)	Amount
Water	68 mL
Agar	1 g
Cornmeal	16 g
Molasses	8 mL
Karo corn syrup	7 mL

pH of diets under various conditions.

The diet is seeded with live yeast. It must be noted that the expression "seeded with yeast" is used frequently in literature on diets for *Drosophila*, but the amounts of yeast used in proportion to the base diets are rarely, if ever, specified. This is apparently because the yeast is assumed to be a growing entity, which will increase its own biomass in the early stages of post-inoculation dynamics.

One of the ubiquitous features of diets for *D. melanogaster* (and other "fruit flies" both tephritid and drosophilid types) is the presence of yeast (Tables A1.2 and A1.3). The topic of yeast/fly relations had been a lively topic since the 1st decade of the 20th century and remains a matter of interest today (Figures A1.5 and A1.6). The summary by Bridges and Darby 1933 expresses the relationship in highly useful detail:

The Role of Yeast

In 1907, E. Guyenot had published an investigation of the nutrition of flies in relation to microorganisms and concluded that for uniformity of results, whether genetic or physiological, the cultures should be entirely free of living microorganisms. Delcourt and Guyenot finally accomplished this aim (1911). They had first used as culture media potatoes or apples, on which many microorganisms flourished. They then transferred flies to fresh culture tubes containing sterilized potato and selected those lines in which the flies had happened to carry over the smallest variety of microorganisms. They thus secured lines from which they had eliminated one by one all microorganisms except living yeast. For a considerable period they carried on their routine cultures of flies satisfactorily on a medium of potato and the pure culture of living yeast that the flies carried over in the transfers. They were of the opinion that the principal food of the flies was the yeast. Finally they used dead sterilized baker's or brewer's yeast held in a mat by absorbent cotton. The stock of nutrient medium consisted of 300 grams of commercial compressed yeast made up to 1,000 cc with water (30 per cent yeast). About 50 cc of this medium and sufficient cotton to absorb all free water were placed in Ehrlenmeyer flasks and sterilized. The dead yeast

Table A1.3 Changes in pH From the Initial Day of Culturing (Day 0) Through the 18th Day of Culturing on Medium A with Flies.

TABLE 3

MEDIUM A, WITH FLIES; pH READINGS ON SUCCESSIVE DAYS

No.	0	1	2	3	4	5	6	8	10	14	16	18
1	5.8	5.8	4.7	3.8	3.7	3.7	3.7	3.8	4.4	4.8	4.9	4.8
2	5.8	5.6	4.7	3.9	3.6	3.6	3.6	3.6	3.8	4.8	5.0	4.8
3	5.8	5.3	4.6	4.0	3.8	3.7	3.6	3.7	3.9	4.8	5.0	4.8
4	5.8	5.6	4.4	3.7	3.5	3.5	3.5	3.5	3.7	4.2	4.4	4.6
5	5.8	5.6	4.6	3.9	3.8	3.8	3.8	3.8	4.6	4.8	4.8	4.8
6	5.8	5.6	4.4	3.7	3.5	3.4	3.4	3.6	3.6	4.1	4.2	5.0
7	5.8	5.7	4.7	3.9	3.8	3.7	3.7	3.7	4.4	4.4	4.6	4.8
8	5.8	5.5	4.6	3.8	3.7	3.7	3.7	3.8	4.0	4.5	4.9	4.8
9	5.8	5.5	4.0	3.6	3.6	3.7	3.7	4.0	4.0	4.1	4.1	4.6
10	5.8	5.8	4.6	3.9	3.6	3.8	3.9	4.0	4.1	4.8	5.0	4.6
11	5.8	5.6	4.5	3.8	3.7	3.6	3.6	3.8	3.8	4.6	4.7	4.9
12	5.8	5.6	4.6	3.8	3.6	3.6	3.6	3.6	3.8	4.2	4.4	4.6
Av. pH	5.8	5.6	4.5	3.8	3.7	3.7	3.7	3.7	4.0	4.5	4.7	4.8

The uniformity of the pH is to be noted, and the drop in pH, starting on days 2 and 3 is also noteworthy. Bridges and Darby point out that the drop in pH in media with flies is faster by 1–2 days the drop determined without flies. Though the authors do not discuss the reasons for these pH drops, it is likely that the production of CO_2 by microbes and by the fly larvae would lead to acidification of the media as CO_2 dissolves in water, forming carbonic acid.

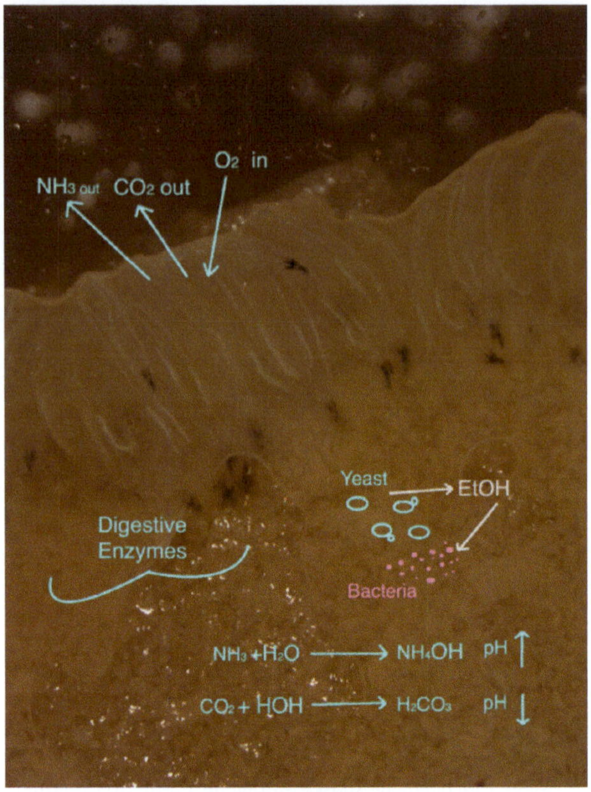

Figure A1.5 *Drosophila melanogaster* larvae lined up on banana agar diet seeded with yeast. Larvae feed collectively on the exposed surfaces of the diet, secreting digestive enzymes that communally modify the food to be more digestible and bioavailable. Larvae also interact with the medium by putting out carbon dioxide and ammonia (as part of their overall nitrogenous waste excretion), and they take in oxygen. The CO_2 helps to drop the pH of the medium from an initial 5.8 to something in the high 3's or low 4's. There is also a continuous interaction between the diet, the microbes, and the fly larvae where changes in pH and addition of digestive enzymes and nitrogenous wastes modify the fitness of the environment for the various competing microbes such as the yeast/bacteria interactions. Bridges and Darby discussed the effects of alcohol (ethanol) interaction with microbes and larvae, pointing out that the higher the alcohol content, the less fit the medium would be for various bacteria.

proved excellent for the flies but unfavorable for the living yeast. By repeated transfers they were able to secure strains which were entirely free from living yeast, as well as other micro-organisms, and which thrived well on the dead yeast medium.

The work of Delcourt and Guyenot was confirmed by Northrop and Loeb (1917) and by Baumberger (1917a). Northrop and Loeb raised flies many generations with success by essentially the same method as that used by Delcourt and Guyenot, viz., upon dead sterilized yeast held in a mat by absorbent cotton. They rendered their flies free from microorganisms by treatment of the eggs with HgCl, rather than by repeated transfers. Baumberger secured sterile races much more easily by washing pupae or eggs for 10 minutes in 85 per cent alcohol. These

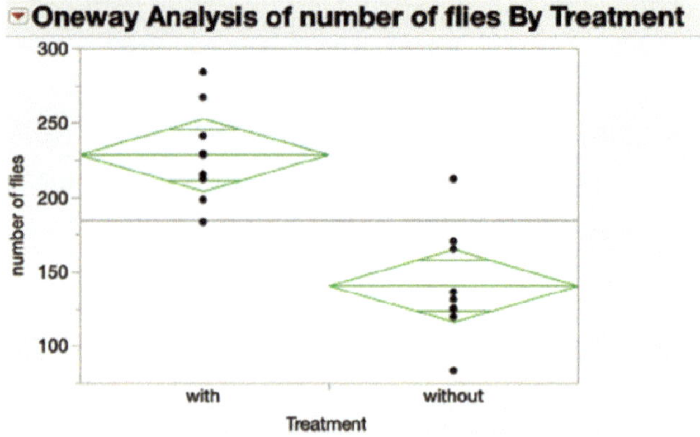

Figure A1.6 ANOVA. The means for with KH_2PO_4 and without it were 220 and 141, respectively, with an F ratio of 29.1 (prob. 0.001).

researches made it clear that normally the yeast growing on the banana or other media, rather than the banana or fruit itself, was the principal food of the larvae, although other microorganisms and the fruit pulp were undoubtedly utilized as well. Baumberger (1917a, 1919) was of the opinion that a loose symbiosis exists in the fact of the universal transference of yeast by eggs, larvae and the feet of adults. The importance of the yeast as food for the larvae lies principally in its high protein count, approximately 11 per cent, whereas the per cent) that larvae restricted to sterile banana pulp protein concentration in the banana itself is so low (1.3 require two to six times as long to reach pupation, and then only in greatly reduced numbers and of small size (Baumberger, 1919).

Baumberger stated (1917b) that yeast and flies thrived well on Pasteur's nutrient agar medium, the formula for which he reprinted.

The common name of Drosophila melanogaster has been "pomace fly," "vinegar fly" "fruit fly," "banana fly," etc., but in the light of these researches it is seen that it should be called the "yeast fly." Another "fruit fly," Anastrepha ludens, the Mexican fruit fly, has been reared by Darby (unpublished) on a nutrient medium with yeast as the food. The question may be asked whether all "fruit flies" may not be supported by a diet of microorganisms. {Cohen's italics and bold fonts}.

According to Baumberger (1919) those substances tend to attract adult insects and also to induce egg deposition that are the normal products of the activity of the particular microorganism on which the larvae of the insect principally feed. Not "fruity" odors, but odors characteristic of the activity of microorganisms growing on the fruit are the attractions. Hence alcohol derives its attractiveness to *Drosophila* from the fact that the principal normal food of *Drosophila* larvae is yeast, the alcohol odor being a sign-post to the presence of yeast.

In the commercial growing of yeast, molasses is used widely, and Baumberger (1917b) reported that for raising flies Pasteur's solution could be approximated by diluting molasses with three parts of water. During 1917 Dr. H. J. Muller used a molasses agar medium at Columbia. His best success came from substitution of Karo corn syrup for about half of the molasses content. One may now guess that the improvement lay in reducing the amount of SO_2 which was present in the bleached molasses. While this method was fairly successful, he abandoned it presently in favor of the banana agar method that had been improved and standardized.

The differences in mean numbers of flies from each culture were clearly significant, and if it was not from the pH effects of the buffer's presence, Bridges and Darby considered other factors as they explain in the following:

The buffering effect of the phosphate on the pH of the medium had been negligible in this pH range and in the amount used. But phosphates have other effects, such as modifying sugar metabolism, and it was thought such effects had been important here. Accordingly, two sets of medium-A cultures were run simultaneously, both with three-day flies from Florida wild stock. One set, cultures 1-9, Table 6, served as control and the other cultures (10-18) were modified by the additions of 0.2 gm KH_2PO_4 per culture.

The authors' comments are noteworthy and exemplary in several respects. First, they had stated a rationale for using a potassium phosphate buffer in the experiment would be expected to benefit the flies by keeping a steady pH in the medium that is otherwise acted upon by the microbes and the flies. The implied hypothesis from this rationale is that the buffer would influence a steady pH and a steady, close-to-optimum pH would benefit the flies in terms of the response of numbers of surviving flies. The measurement of pH was a meticulous step in ascertaining whether the buffer actually stabilized the pH. Evidently, the buffer did not accomplish this, but something about the buffer's presence benefitted the flies' survival. The authors suggest that addition of phosphate may have benefitted the flies by some metabolic effects such as enhancing sugar metabolism. The implications of this finding should encourage further experimentation with diet supplements such as other phosphate compounds or possibly other sources of potassium (Tables A1.4–A1.6).

Table A1.4 The Meticulous Measurements of pH Reported in the Studies of Bridges and Darby 1933

MEDIUM A; CULTURES 1–10; WITHOUT FLIES; CULTURES 11–16, WITHOUT FLIES, WITH KH_2PO_4; CULTURES 16–25, WITH FLIES, WITH KH_2PO_4

No.	1	2	3	4	5	7	9	11	13	16	18	20	Flies
1	5.9	4.2	4.0	3.8	3.6	3.8	4.3	4.4	4.4	4.3	3.8	......	......
2	6.0	4.1	3.9	3.6	3.6	3.8	4.0	4.1	4.3	4.2	4.2	......	......
3	5.9	4.0	4.0	3.9	3.6	4.1	4.1	4.2	4.2	4.4	4.5	......	......
4	6.0	4.0	4.0	3.8	3.9	3.9	4.1	4.3	4.3	4.3	4.3	......	......
5	6.0	4.1	4.0	3.8	3.8	3.8	4.2	4.0	4.2	3.8	4.2	......	......
6	6.0	4.2	4.0	3.6	3.8	3.8	4.1	4.1	4.4	4.1	4.4	......	......
7	5.9	4.1	3.9	3.8	3.8	4.0	4.1	4.0	4.2	4.2	4.3	......	......
8	6.0	4.3	4.2	4.0	3.7	3.8	3.9	4.4	4.4	4.1	4.1	......	......
9	5.9	4.1	3.9	3.7	3.8	3.8	4.1	4.2	4.1	4.0	4.2	......	......
10	5.9	4.0	4.0	3.8	3.7	3.7	4.0	4.2	4.2	4.2	4.1	......	......
Av.	6.0	4.1	4.0	3.8	3.7	3.9	4.1	4.2	4.3	4.2	4.2	......	......
11	5.4	4.0	3.7	3.7	3.8	4.0	4.1	4.1	4.1	4.1	4.1	......	......
12	5.5	4.0	3.8	3.8	3.8	4.4	4.2	4.1	4.1	4.1	4.1	......	......
13	5.4	3.9	3.8	3.8	3.8	4.1	4.4	4.1	4.1	4.1	4.1	......	......
14	5.5	4.0	3.8	3.7	3.7	4.2	4.3	4.2	4.2	4.1	4.1	......	......
15	5.4	4.0	3.8	3.8	3.8	4.2	4.2	4.2	4.1	4.1	4.1	......	......
Av.	5.4	4.0	3.8	3.8	3.8	4.2	4.2	4.1	4.1	4.1	4.1	......	......
16	6.0	4.1	4.0	3.7	4.0	4.0	4.0	3.9	3.9	4.0	4.0	4.0	141
17	6.0	4.0	4.0	3.8	3.8	4.0	4.0	3.9	4.0	4.2	4.1	4.2	165
18	5.9	4.1	4.0	3.7	3.7	4.0	4.0	3.9	4.0	4.0	4.0	4.1	151
19	5.9	4.2	3.9	3.8	3.9	3.9	4.0	3.9	4.0	3.9	4.1	4.1	186
20	6.0	4.1	4.0	3.7	3.7	3.8	3.9	3.8	4.0	3.8	4.0	4.0	223
21	6.0	4.0	3.9	3.8	3.6	4.0	3.8	3.8	3.9	3.9	4.0	4.0	303
22	5.9	4.0	4.0	3.8	3.8	3.9	4.1	3.9	4.0	4.1	4.3	4.3	189
23	6.0	4.1	3.8	3.6	3.8	4.0	4.1	4.0	4.0	4.0	4.1	4.0	151
24	6.0	4.0	4.0	3.7	3.8	4.0	4.1	4.1	4.2	4.1	4.4	4.2	124
25	6.0	4.1	4.0	3.7	3.6	4.1	4.1	4.1	4.2	4.2	4.1	4.0	89
Av.	6.0	4.1	4.0	3.7	3.8	4.0	4.0	3.9	4.0	4.0	4.1	4.1	172

Table A1.5 From Bridges and Darby 1933. The pH of the Media Measured at 1, 3, and 7 Days Post-Inoculation with Flies. The Target pH was 4.5, Determined by the Researchers as Optimal for *D. melanogaster*

TABLE 7
MEDIUM B

No.	Days			Flies
	1	3	7	
1	5.8	4.9	4.0	191
2	5.7	4.9	4.0	292
3	5.8	5.0	4.1	273
4	5.8	4.9	4.0	334
5	5.8	5.0	3.9	320
6	5.7	4.9	4.1	272
7	5.8	5.0	4.0	277
Av.	5.8	4.9	4.0	266

TABLE 8
MODIFIED MEDIUM B

No.	1	3	7	Flies
1	4.9	4.3	4.3	208
2	4.9	4.3	4.4	......
3	5.0	4.4	4.4	33
4	4.9	4.3	4.2	341
5	4.9	4.3	4.4	144
6	5.0	4.3	4.3	......
7	4.9	4.4	4.4	289
8	4.9	4.3	4.3	358
9	5.0	4.3	4.4	464
Av.	4.9	4.3	4.3	204

Table A1.6 Comparison of Cultures of Flies on Medium A without and with KH_2PO_4.

TABLE 6
CULTURES 1–9, MEDIUM A; 10–18, MEDIUM A, WITH KH_2PO_4

No.	1	3	5	9	11	13	15	Flies
1	5.5	4.4	3.6	3.6	3.8	4.1	4.3	125
2	5.6	4.4	3.6	3.6	3.6	3.6	3.8	165
3	5.6	4.4	3.6	3.6	3.6	3.6	3.6	136
4	5.6	4.7	3.6	3.7	3.6	3.6	3.7	212
5	5.6	4.1	3.6	3.6	3.6	3.6	3.6	170
6	5.6	4.7	3.6	3.6	3.6	3.6	3.6	131
7	5.7	4.4	3.7	3.7	3.8	4.0	4.0	83
8	5.6	4.4	3.6	3.8	3.7	4.0	4.0	119
9	5.6	4.5	3.6	3.7	3.7	3.6	3.6	124
Av.	5.6	4.5	3.6	3.7	3.7	3.8	3.9	142
10	5.5	3.7	3.6	3.6	3.8	3.8	3.8	267
11	5.6	3.7	3.7	3.8	3.8	3.8	3.8	241
12	5.5	3.7	3.6	3.6	3.8	4.0	3.9	215
13	5.4	3.6	3.5	3.6	3.7	3.7	3.8	284
14	5.5	3.9	3.7	3.6	3.7	3.8	3.8	228
15	5.5	3.7	3.7	3.9	3.9	4.0	4.0	198
16	5.4	3.7	3.3	3.8	3.8	3.8	3.8	229
17	5.5	3.9	3.7	3.9	3.7	3.8	3.8	183
18	5.4	3.9	.3.7	3.8	4.0	4.0	4.0	212
Av.	5.5	3.7	3.6	3.7	3.8	3.9	3.9	229

The authors noted that the KH_2PO_4.was added to the medium to serve as a buffering agent, but that it evidently had little or no effect on stabilizing the pH as seen for the successive days that the culture was held at 25°C.

THE MECHANISM

OF

MENDELIAN HEREDITY

BY

T. H. MORGAN

PROFESSOR OF EXPERIMENTAL ZOOLOGY
COLUMBIA UNIVERSITY

A. H. STURTEVANT

CUTTING FELLOW, COLUMBIA UNIVERSITY

H. J. MULLER

ASSISTANT IN ZOOLOGY, COLUMBIA UNIVERSITY

C. B. BRIDGES

FELLOW IN ZOOLOGY, COLUMBIA UNIVERSITY

NEW YORK
HENRY HOLT AND COMPANY
1915

Figure A1.7 The cover of the original book on heredity from T. H. Morgan et al. 1915. The book contains complete information about the rearing system used for the early studies of *Drosophila melanogaster*.

The Bridges and Darby 1933 paper should stand as a model for publications on exploration of the science of insect rearing. It contains the authors' in-depth thinking about the complex interrelationships that take place in rearing system components. As I have tried to convey throughout this book, often what are seemingly unrelated phenomena can be discovered and tested to determine mechanisms behind the biological determinants of fitness (Figure A1.7).

Appendix 2: The Culture of the Mulberry Silkworm

U. S. DEPARTMENT OF AGRICULTURE,

DIVISION OF ENTOMOLOGY—BULLETIN NO. 39, NEW SERIES.

L. O. HOWARD, Entomologist.

THE CULTURE

OF

THE MULBERRY SILKWORM.

BY

HENRIETTA AIKEN KELLY,

SPECIAL AGENT IN SILK INVESTIGATIONS.

WASHINGTON:

GOVERNMENT PRINTING OFFICE.

1903.

KFN 1 →LETTER OF TRANSMITTAL

US Department of Agriculture

Division of Entomology

Washington, DC, January 31, 1903

Sir: I have the honor to transmit the manuscript of a bulletin which has been prepared under my direction by Miss Henrietta Aiken Kelly, a special agent of this Division, and which treats the culture of the mulberry silkworm in a plain yet complete way. It is intended to distribute it among people who wish to begin the culture of the mulberry silkworm in different parts of the United States. I recommend that this manuscript be published as Bulletin No. 29, new series, of this Division.

Respectfully, L. O. Howard

Entomologist

Hon. James Wilson,

Secretary of Agriculture

A2.1 PREFACE

KFN 1 → To inaugurate the public instruction in silk culture on a modern scientific basis, this bulletin, the first of a series, is prepared.

The information given was gained at original sources, during a long residence in Italy and France, and by practical experience under skillful sericulturists. Many Italian and French authorities have been cited, and special acknowledgment is due to Dr. A. Hugues for permission to quote from his "Practical instructions," recently published in L'Intermediare des Bombyculteurs et Entomologistes, Ardeche, France.

HENRIETTA AIKEN KELLY

CHARLESTON, S.C., January 15, 1903

A2.1.1 Contents

A2.1.2 Illustrations

A2.2 THE CULTURE OF THE MULBERRY SILKWORM

KFN 2 → The caterpillars of many moths and of a few butterflies produce silk, but certain of those belonging to the family *Bombycidae*, or true silk spinners, particularly *Bombyx: (Sericaria) mori*, or the mulberry silkworm, yield the most and the best silk. The races of *Bombyx mori* today are the result of domestication and artificial rearing, and the wild type is uncertain, though most authorities assign the foot of the Himalaya as the cradle of the mulberry silkworm. It has been industrially cultivated in China from time immemorial and in Europe since the sixth century.

A2.3 THE LIFE OF THE SILKWORM

Like all insects of its class, before arriving at the perfect winged state, it exists: (1) as a caterpillar or larva, (2) in a chrysalid state, and (3) as a moth.

A2.3.1 The Larva or Caterpillar

The larva (Figure A2.1) has a cylindrical body composed of 12 rings; each of the first three rings has a pair of jointed legs, and each of the sixth, seventh, eighth, ninth, and twelfth ring bears a pair of false legs, destined later to disappear.

The black elliptical spots on the side are the orifices for breathing and are called stigmata or spiracles.

KFN 3 → The head is a small mass covered with a hard scale and is provided with jaws that move laterally like the wings of a folding door. The alimentary canal extends throughout the entire length of the body, and on each side of it is placed a silk gland (Figure A2.2). These

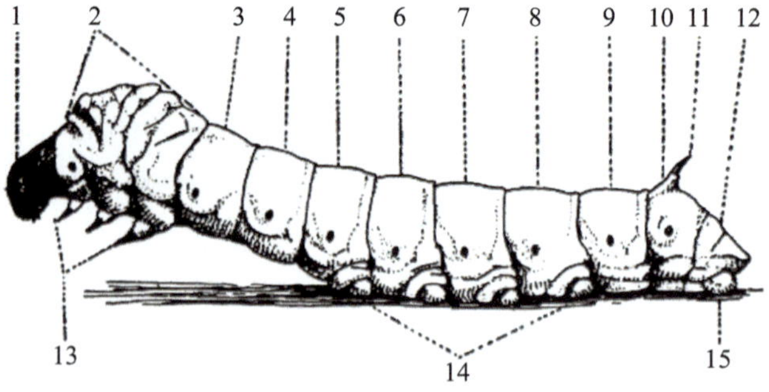

Figure A2.1 Adult silkworm. 1: head; 2, 3, 4, 5, 6, 7, 8, 9, 10, and 12: rings; 11: horn; 13: three pairs of articulated legs; 14: four pairs of abdominal or false legs; and 15: a pair of false legs on the last ring.

consist of two whitish or amber-colored cords, which after innumerable curves unite in the spinneret in the region of the mouth. There are also two glands, whose excretory

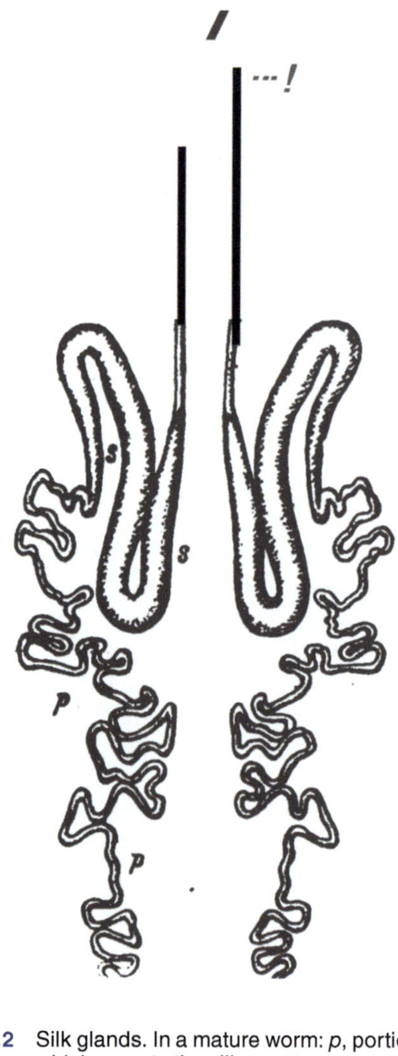

Figure A2.2 Silk glands. In a mature worm: *p*, portion of glands which secrete the silky matter; *a*, reservoir; *c*, conducting canal; *f*, spinneret; *g*, accessory glands. (Redrawn from Verson and Quajat.)

canal opens in the spinneret, and covers the silk as it comes out with an impermeable varnish rendering it insoluble.

In acids and alkalis, this varnish is about a fifth of the weight of the thread.

Hatchings usually occur annually in the spring. Simple contact with the air causes the new-born insect immediately to acquire a volume larger than it had in the egg, and it quickly begins to gnaw the under surface and edges of the mulberry leaf. It eats day and night at all hours, except when asleep, and in about 30 days grows 14,000 times larger than it was at birth.

As the silkworm grows larger, it becomes paler in color, because its dark chestnut brown hairs are scattered over a larger surface, thus showing more of the true color of the skin.

About five days from its birth the vitality of the larva decreases, and it eats scantily or not at all, and becomes thin and whitish in color. Then it moves around unquietly, and finding a convenient place attaches itself to it, holding on by its false feet. It thus remains motionless, with the front part of its body raised up, for a period of time varying according· to temperature, and takes its first so-called "sleep," or molt, during which time the body undergoes extraordinary modifications. The skin is entirely shed, and all the tissues that cannot keep up with the rapid growth of the insect are changed.

The scale which covers the snout is the first part of the case to fall, and a new case appears under the former one. The worm then pushes itself forward through its first ring, sets at liberty the legs of the thorax, and by a wriggling movement comes out of its old sheath.

To aid the insect in this difficult change a liquid is secreted between the old skin and the one forming beneath it, which facilitates the separation.

The life of the larva is usually divided into five ages, varying in length according to temperature, frequency of feeding, race, and the robustness of the worm. The following is about the average: KFN 1:

- First age: From birth to first molt, five to six days
- Second age: From first to second molt, four days
- Third age: From second to third molt, four to five days
- Fourth age: From third to fourth molt, five to seven days
- Fifth age: From fourth molt to maturity, seven to twelve days

Sometime after each molt, perhaps the time needed to regain lost strength and to solidify the newly formed organs, the worm remains in a state of relative torpor. Not much practice is needed to recognize when it has come out of its "sleep." It moves its head and thorax, which are whitish, while the rest of the body is gray, and has completely lost the shining aspect which it had when the worm began to molt. When fairly through its first molt the larva begins again to eat, and its hunger does not cease until it is ready for a new molt.

Four molts having been made, the worm eats a prodigious quantity of leaf until it reaches its maximum growth, when its appetite diminishes and ceases altogether. It then stops moving and remains for some time in repose, evacuating, meanwhile, its digestive canal, thus losing up to 12% of its weight. Its lean body is now white or yellow, according to the race, and semitransparent.

Very soon it begins to move about again, lifts up its head, which is longer and more pointed, and turns in every direction seeking to find a convenient angle, finding which, it throws out a silk thread from its spinneret. First a net is formed to hold the cocoon which is to be spun, then the regular spinning begins and the form of the cocoon is designed. For some time through the veil which very soon is to surround it, the diligent larva, with its hack turned outward, may be seen completing its task. It is calculated that with its head alone the silkworm makes 69 movements every minute, describing arcs of circles, crossed in the form of the digit 8 (Figure A2.3(a)).

Meanwhile the web grows closer and the veil thickens, and in about 72 hours the worm is completely shut up in its cocoon, which serves it as a protective covering.

A2.3.2 The Chrysalis

KFN 4 → In the cocoon, the silkworm goes through the last phase of its larval life. After four or five days the skin breaks, and the insect which issues from this old covering is the chrysalis, whose weight is only half that of the larva at its highest development (Figure A2.3).

The chrysalis seems to have neither head nor feet, is golden-colored at first, and then turns chestnut brown. The skin dries rapidly and forms a hard case, on which the lines of only the posterior rings are seen, the place of the first three rings being covered with the wing cases.

The chrysalid state is in certain respects a sleep and in others a period of great activity, in which the entire being is transformed. Wings, antennae, reproductive organs, and legs are all now developed. This state lasts from 18 to 20 days, according to the temperature. Then the metamorphosis is complete, and the sheath breaks in the region of

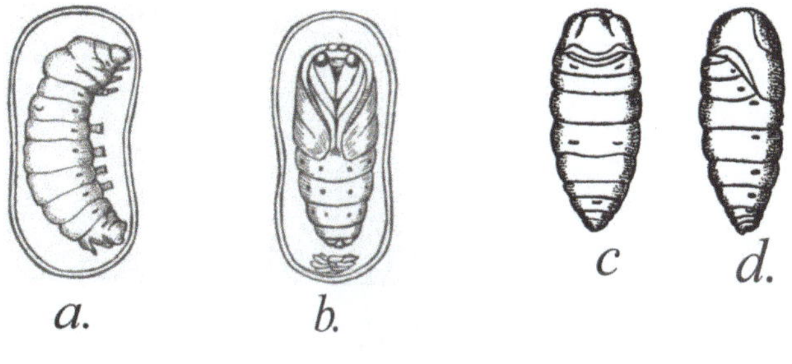

Figure A2.3 The chrysalis. (a) Silkworm completing its cocoon; (b) cocoon and chrysalis-cast-off skin of larva beneath; (c) back view of chrysalis; and (d) side view of chrysalis. (Redrawn from Maillot.)

the head and out of the old case the moth or perfect insect issues.

A2.3.3 The Moth

The larva in spinning the cocoon leaves one end less dense, so that the threads open freely to permit the egress of the moth. By the aid of an alkaline fluid, the moth softens and parts the threads and liberates itself.

KFN 5 → The moth (Figure A2.4) comes out of the cocoon, as the larva out of the egg, in the early morning hours. It has a distinct head, thorax, and abdomen, four wings, two comb-shaped antennae, three pairs of legs, and a pair of compound eyes.

Shortly after emerging, the moth deposits a liquid substance, generally white, sometimes colorless and sometimes reddish, and then the union occurs, lasting several hours, after which the eggs are laid either immediately or in the course of four or five hours.

A gelatinous substance supplied by two glands near the extremity of the oviduct covers the eggs as they come out and causes them to adhere to the substance on which they are laid.

The laying, consisting of 300–700 eggs, is generally completed in three days, 70–80% being deposited the first day, 20–30% the second day, and a few the third day. The mother moth dies six to twelve days later, her death being usually preceded by that of the male. Death occurs more or less speedily, according to the robustness of the moth, the temperature, and the tranquility in which it has been left.

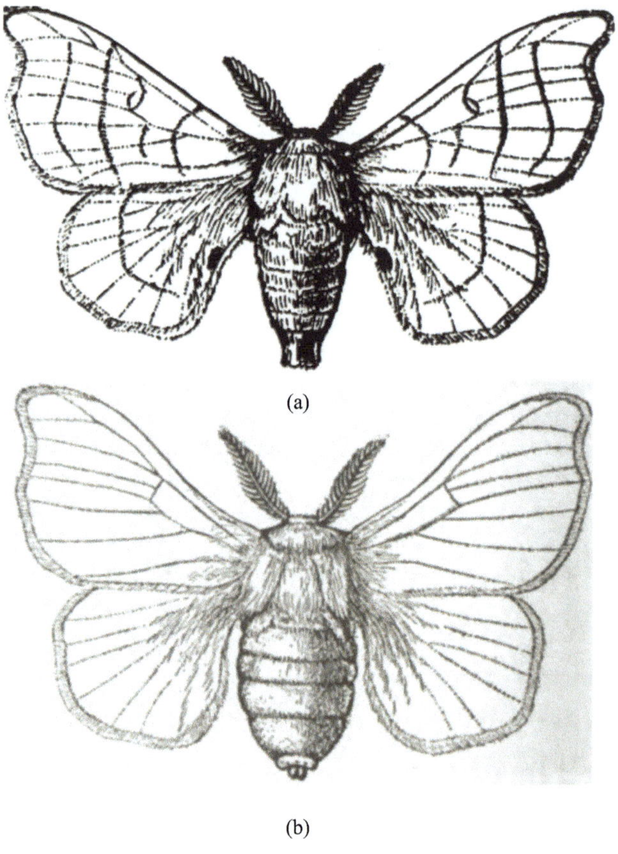

(a)

(b)

Figure A2.4 The moth. (a) The male and (b) the female.

Thus in about 65 days the silkworm has completed its cycle of existence, its three periods being 30–40 days in a larval state, 15–20 days as a chrysalis, and 8–20 days as a moth or perfect insect.

KFN 6 → The rapid development of the silkworm and its marvelous transformations indicate extraordinary power and very active functions. Its respiration is almost equal to that of the frog and of certain large birds, which is equivalent to saying that it must be always surrounded by plenty of pure air.

A2.4 THE FOOD OF THE SILKWORM

KFN 7 → The leaf of the white mulberry is the natural and the best food for the silkworm. There are many varieties of the white mulberry—some much better adapted than others to commercial silk culture, and some better suited to certain localities.

The race, the size of the worms, the variety and age of the mulberry, the nutritive quality of the leaf, the year, the season, and the climate make the requisite quantity of leaf very variable (*see Bulletin 34, Bureau of Plant Industry). The following is given as a basis of calculation, all circumstances being considered, and the leaves not being cleaned: For the larvae hatched from 1 ounce of eggs, during the first age, 11 pounds of leaves; during the second age, 26–33 pounds; during the third age, 88–145 pounds; during the fourth age, 264–352 pounds; during the fifth age, 1,540–1,760 pounds; or about 2,200 pounds in all, of which one-half is consumed in the last six or seven days of the fifth age. ←KFN 7

The age of the leaf should be relative to that of the worm. Young worms fed on old leaf, or old worms fed on young leaf, are apt to become diseased, and even though they may not die, will scarcely mount or will spin indifferently. A change of leaf, too, should, if possible, be avoided, or made gradually. Fresh leaf only should be used, and never when wet with dew or heated, or before it has in a measure acquired the temperature of the room in which the worms are being reared. It should be gathered early in the morning or in the evening and should not be bruised or torn, nor should the baskets or aprons used to carry it be the same which are used to remove litter. To avoid fermentation, the leaves must be spread out in a cellar or cool, darkened room. Cut up only a limited quantity of leaf at a time and cover with a damp cloth to keep fresh, but never submerge the leaf in water, as this is apt to occasion

* For complete information regarding the mulberry, see Bulletin No. 34, Bureau of Plant Industry, U. S. Department of Agriculture.

flacherie, a very destructive disease. From the fourth age on there should always be a day's supply ahead, so that in case of rain the worms will not have to fast.

KFN 8 → In gathering leaf, always strip the branches from base to top, so as not to tear the bark and injure the new buds. It is above all essential to strip the tree entirely of its leaves, for if leaves are left on some branches they will there attract the sap, while the bare branches will be imperfectly nourished. Always strip the young trees first. The sacks for holding the leaves should have a hoop around the opening and a hook to suspend them to the branches.

Restricting the height of the trunk of the mulberry to 5 or 6 feet makes it possible for old women and children to gather leaf, thus diminishing the cost of labor one-half, a most important point in commercial silk culture.

Figure 5 Rearing Shelves.

It is also important to cultivate trees that bear little or no fruit, for the production of fruit not only consumes part of the strength of the tree, but also much labor is involved in being compelled to divest the branches of fruit before they can be used as food for silkworms.

The stump mulberry, or that growing low like a shrub, the hedge mulberry, and that which grows along walls vegetate much earlier than the medium and high trunk trees. Silkworm rearers should always have a ready supply of leaf for the first ages of the worm, and especially is this necessary if early cultures are desired with a view to escaping the heated days of May and June.

A2.5 IMPLEMENTS NECESSARY TO SILKWORM CULTURE

KFN 9 → Commercial silk culture requires a smaller outlay of capital than almost any other industry. The net gain the first year may pay for an outfit which will last for many years. The following articles are indispensable:

1. Some very light movable shelves, permeable to air, for the first age 1:1; and, for the following ages, latticed shelves about 3 feet wide, and stands to support them
2. Unsized ordinary wrapping paper or newspapers to cover the shelves
3. A small ladder, if necessary, to reach high shelves
4. Small trays to remove worms
5. Knives to cut leaves and baskets to distribute them
6. Coarse tulle and nets or perforated paper for changing beds and equalizing the worms
7. A supply of brush, straw, or shavings to construct the spinning place
8. A thermometer

Wire, twine, laths, or canes are suitable for the lattice work of the shelves. Make the space between the shelves about 14 inches. If possible, do not arrange the shelves along the wall and allow a good passage between the tiers of shelves (Figures A2.5 and A2.6).

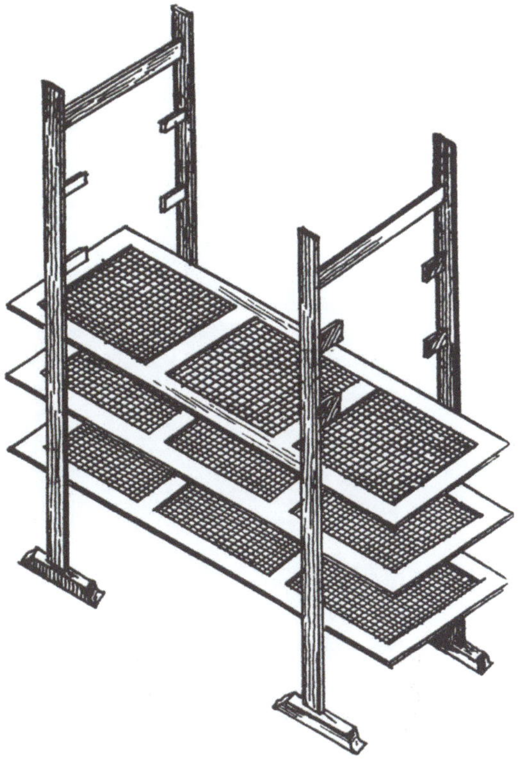

Figure A2.5 Light movable shelves.

A2.6 SILKWORM EGGS: HOW TO WINTER AND HATCH THEM

KFN 10 → There are two kinds of silkworm culture: one for production and one for reproduction. The object in the first case is to get the greatest yield of cocoons, and with a little training, may be carried on by anyone of ordinary intelligence.

The aim in culture for reproduction is to secure eggs free from hereditary taint of disease, and experts only can be depended upon to conduct it. Besides a careful physiological examination throughout the rearing, the body of the mother moth is microscopically tested after death, and her eggs are not retained if signs of disease are discovered. In this way the birth of healthy worms is insured. Pasteur first applied this method of selecting silkworm eggs, and thus checked the plague (pebrine) which was rapidly destroying silkworm culture in Europe.

Formerly, through an indiscriminate use of eggs, disease invaded the rearings to such a degree that from 65 to 90 pounds of cocoons was considered a good yield per ounce of eggs. At the present low price of silk such cultures are no longer remunerative, and industrial silk

Figure A2.6 The rearing room. (After Gobin.)

culture now demands the exclusive use of scientifically tested eggs.

The grainage, or preparation of eggs for market, constitutes a special department of silk culture. In Italy there are over 400 establishments which supply eggs to raise the annual silk crop. The poorest peasant, though well skilled in the art of rearing silkworms, would not risk a rearing with eggs which have not been selected and preserved by experts.

The eggs of crossed races are best for cultures with a view to production of silk, and here, as much as in the examinations which have been referred to, the knowledge of experts is needed.

KFN 11 → The life of the egg, in those races which have but one generation each year, has three phases, the first lasting about three months, full of activity; the second lasting from about October to the middle of February, one of inactivity, in which there are no signs of life; the third from the middle of February to the moment of hatching, in which the vital activity recommences, and the germ begins to organize as soon as the temperature rises a little above 50°F. The measure of the activity of the egg, in this stage, is the measure of its danger, for any sudden change of temperature would injure or destroy the delicate embryo, or cause the larva to be born before its food was ready. To guard against such accidents, eggs must be wintered in a high region or in a refrigerator at a uniform temperature of about 35°F, from December 15, until the mulberry begins to bud or until hatchings are desired. There should always be good ventilation, the air should not be moist, and great care must be taken to keep the eggs out of the reach of mice and insects.

Natural hatchings are almost always irregular, extending over eight or ten days, thus multiplying the divisions and rendering the rearing difficult and costly; hence, the necessity for artificial incubation.

Eighteen days before the time decided on for the hatching, spread out the eggs in thin layers in the incubating room or incubator. The temperature should be 55°F.

From the fourth day on, gradually increase the temperature two degrees in 24 hours until 73°F is reached, when, at this uniform temperature, hatchings will occur in ten days on an average. This period, however, varies from eight to fifteen days, according to race, the cold supported during the winter, the first grade of heat, and the highest during the incubation, and the number of days taken to pass through these two grades of heat and also according to the humidity. To obtain a good and complete hatching, a slight humidity is necessary, especially during the last four or five days. To secure this keep an open vessel of water near occasionally.

KFN 12 → The temperature may be raised during the hatching to 75°F, but sudden changes of heat must be avoided, and, unless the newborn worms can be kept in the same temperature, it is dangerous to have the maximum temperature of the incubator so high.

It is better for the period of incubation to be protracted than suddenly shortened. Holding the eggs at a certain temperature, or slowly lowering the temperature a little, does no harm. When the season is not propitious, the hatching may in this way be retarded.

An incubating room is preferable, because it also serves for the first two stages of the worms, but in small rearings, an incubator is more economical, both with regard to service and to fuel.

For a large quantity of eggs (5–10 ounces), a small incubator, which is very much used in chemical laboratories to dry substances, is recommended. Any ordinary tinsmith can make it. It consists of a double case, cubical in form, with a zinc bottom. The space between the outer and inner walls is filled with water. The front face of the cube is furnished with a glass, so that the temperature within, indicated by the thermometer hanging on the glass, may be seen without the necessity of opening the incubator. There are two openings below on opposite sides to allow the cold air to come in, and an opening in the center of the top to permit the outward flow of the heated air. Having filled the space between the walls with water, the incubator is placed on a support and heated by means of a gas or oil lamp to the desired degree, and, by raising or lowering the flame, a constant or progressive temperature can be maintained (Figure A2.7). A self-regulating incubator, such as is used to hatch chickens, would be more convenient, but would cost more.

The whitening of the eggs denotes the near approach of the hatching. Double pieces of tulle or sheets of perforated paper sprinkled over with finely cut-up white mulberry leaves should then be lightly placed over the eggs to allow the outward passage of the worms as soon as hatched. The object in employing two pieces of tulle or paper is to prevent the unhatched eggs which cling to the sheet from being removed with the newborn worms. This process must be repeated daily during the hatching, the second sheet always being renewed.

The duration of the hatching varies from three to five days. The eggs hatching is as follows: during the first day, 5%; during the second, 33%; during the third, 50%; during the fourth, 5%; and during the fifth, 7%.

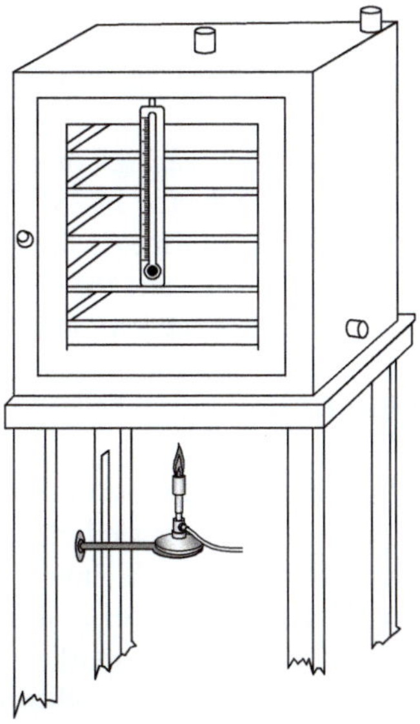

Figure A2.7 Hot water incubator. (Redrawn from Verson and Quaint.)

Twenty-five grams of eggs will give about 17 g of worms. In small rearings, most cultivators raise only the worms that are hatched on the second and third days, to avoid the necessity of forming too many classes. The worms must be classed according to the date of birth, and the insignificant number hatched on the first and last days scarcely compensate for the trouble of rearing them. Different races must also be reared separately.

Where several in one neighborhood are engaged in silk culture, it greatly reduces the cost to have all the eggs hatched in one incubator. The person best acquainted with silk culture can undertake the incubation and distribute the young worms on the second or third days to those who are to rear them. This is the plan adopted among the Italian peasantry, the wife of the supervising farmer hatching the eggs for whole villages.

A2.7 THE REARING OF SILKWORMS

KFN 13 → Before entering into the details of a rearing, some general directions must be given concerning the rearing room—the heating, ventilation, and disinfection.

A2.7.1 General Directions

The place chosen for a rearing should be relatively high, and not exposed to malaria or bad odors, and mulberry trees should grow around it. Any room that can be properly heated and ventilated will answer the purpose. An open fireplace is the best means of heating, but is expensive, as much of the heat is lost. Hot water pipes, such as are used to heat a greenhouse, are good for a building specially built for silkworm rearing. Iron stoves should not be used, unless placed in an adjoining room with communicating pipes. Never employ charcoal as fuel.

A2.7.1.1 Ventilation

The domesticated worm should be surrounded continually by pure air. The amount of carbonic-acid gas given out by the worms and their attendants is very considerable; besides this, a quantity of deleterious gas is generated by the litter of the beds, and the lights and fires consume a great deal of oxygen. Myriads of spores and germs of organic life float in the air of the rearing room, and their influence paralyzes the vital energy of the skin and of the organs of respiration, on whose normal functions the robustness of the worm so much depends. Hence, it is evident that the quantity of vitiated air which should be expelled from the room requires the introduction of a large quantity of fresh air. For this, a double system of ventilation is necessary, which may be obtained by double openings in the windows, to allow the heated bad air to pass out above and the cool fresh air to come in below. To renew the air in every part of the room, and to avoid a single and often violent current, there should be more than one window. An open fireplace is the best means of ventilation. When the difference between the external and internal air is slight, or there is no difference at all, artificial means· must be used to create a current. Light and frequent fires, or a burning lamp in the fireplace, or a revolving fan, may be used to prevent stagnation of the air.

A2.7.1.2 Disinfection

Eight or ten days before introducing the worms into their quarters, all the shelves and implements should be washed in a solution of chloride of lime (11 pounds of chloride of lime to 88 quarts of water, or in a solution of sulfate of copper to 100 by weight).

When everything is in order—tools, perforated paper, material for the worms to spin their cocoons on, etc., each in its own room—close the doors and windows as tightly as possible and fumigate the rooms with sulfur (11 pounds of sulfur to every 100 cubic yards of space). To fumigate properly, powder the sulfur and place it in an earthen or metallic vessel over a slow fire. The sulfur will ·gradually melt and take tire. Place it immediately in the rearing room and leave it there, with the doors and windows completely shut, for 24 hours.

Nets should not be exposed to sulfur fumes, for this would soon rot them, but should be washed in a solution of sulfate of copper, and immediately afterwards in plain water.

Twenty-four hours after the fumigation the floors should be washed with a solution of chloride of lime or sulfate of copper, and the walls should be whitewashed with lime.

When dead worms are seen on the shelves, change the beds and create in the rearing room sulfurous gas by burning a pound of sulfur during six hours, or make a strong wood smoke, which is a good disinfectant and will not harm the worms.

Wise precaution having been taken to destroy germs of disease in the rearing room, the new-born worms may now be safely installed there.

A2.7.1.3 Space Required

KFN 14 → The worms from 1 ounce of eggs should cover at birth 1 square yard. Doubling this space on the fourth day, they would require 2 square yards, and at their change of beds after molting, 4 square yards. By the spacing of the third day of the second age, and the doubling of beds preceding the second molt, they will need for the second age 8 square yards. For the third age 16 square yards will be required; for the fourth age 32 square yards; and for the fifth age 60 square yards. The more space that is accorded to the worms in their first ages, the more robust they will be; and if the space can be tripled instead of doubled during the fourth age, and for the fifth age be 70, 80, or 90 instead of 60 square yards, the harvest of cocoons could be raised from 60 kg to 70, 80, or 90 kg per ounce of eggs, the quality of silk also being superior.

A2.7.1.4 Temperature

KFN 15 → The silkworm is not a tropical insect and attains its best development between the temperatures of 68°F and 77°F. It is safe to adopt the mean between these two temperatures for the general rearing. Each cultivator, however, may suit his convenience, remembering that to fall below or to exceed the mentioned limits of heat is detrimental to the worm, and will affect the quantity and the quality of its spinning. From the second age the temperature should be from 70°F to 72° F and should be kept as uniform as possible to the end of the rearing. The time which elapses between one change and the following one may be much shortened by raising the temperature and feeding oftener. Such hasty rearings may be made from 22 to 24 days.

They are, however, to be condemned, as contrary to the nature of the silkworm. Meals following each other too closely cannot be properly digested and are likely to cause disease. Besides, hasty rearings require more labor, and the service must be kept up night and day. As there is danger in too high a temperature, so there is danger of one that is too low (64–68°F.). A rearing that is too prolonged, lasting over 32 days, is to be avoided to escape the heat of June, under which the beds are more likely to ferment, causing disease; the worms have less appetite and leave more leaf from one meal to another; the changes are slower and less likely to occur instantaneously; and there is more risk of muscardine or calcino, a disease due to a mold.

Both hasty and tardy rearings are, therefore, to be proscribed, and those conducted in 28–32 days alone are recommended. This lapse of time permits the leaves of the mulberry to acquire maturity, and the growth of the worm should be relative to that of the leaf on which it feeds.

A2.7.2 The First Age

KFN 16 → Hatchings usually occur early in the morning. The worms which have crawled up through the holes of the tulle or paper to get food, should not be removed before 10 am to the latticed shelves covered with paper to receive them. Each shelf must be marked with the date of the birth of the worms put upon it, and care must be taken to place on the same shelf only worms born on the same day, as a remunerative rearing demands that such alone be raised together.

Should the hatching occur at 68–70°F, keep this temperature during the first ages, and feed eight times during 24 hours; if the temperature at birth is 75–77°F, slowly diminish the temperature one or two degrees, and feed ten times in 24 hours. The appetite of the worm increases or diminishes with the heat. The second-day, in case of the worms hatched at the maximum, adjust the temperature to the degree proposed to conduct the rearing. In feeding sprinkle finely cut tender leaves frequently over the worms, and toward the fourth day begin to regulate the number of meals so that it will range from four to eight, according to the temperature. Before cutting the leaves, remove the stems. Distribute the leaves uniformly and equally on the shelves, in order to prevent the worms from crowding more on one side than another, and in order that they may be equally nourished and make their changes simultaneously. Cut only a small quantity of fresh leaves at a time and keep the rest in jars or baskets covered with a damp cloth. Never submerge the leaves in water. For the first two or three ages, the white ungrafted mulberry is recommended, it being lighter and more digestible for delicate worms.

It is well during the feeding to open the door and windows to insure a good supply of fresh air. After feeding, close the door and windows, unless the day is warm, when they may be permanently left open, protected by curtains through which the air passes freely. The worms should never be exposed to direct sunlight or to a strong current of air, and during a thunderstorm the windows and doors should be closed.

Worms of the same age and development should be classed together. To obtain this equalization, do not feed newborn worms until all that have been hatched on one day have been removed to shelves, then give a general meal. Ifat the end of two or three days it is noticed that on certain shelves there are smaller worms than on others, in order to allow the less developed worms to catch up with the more advanced ones, place the former nearer the fire or on the highest shelves, where the air is warmer, and give them one or two more meals than the larger worms. For this reason, it is well to have light movable shelves (Figure A2.5).

Many cultivators of silkworms do not change the beds during the first age, and it is not absolutely necessary, if the leaf has been well cleaned of stems and very finely cut up, and, above all, if the air is dry. Change of bed, however, must be made if the litter is damp, and the weather rainy, for the worms are going to molt in two or three days, and this crisis should not occur in unhealthy conditions. It is always more prudent to change beds on the fourth day, and is, therefore, advised. The space occupied by the worms must be doubled when the change of beds is made.

The bed on which the leaf and excrement accumulate is, perhaps, the greatest source of danger to the worms. When there is not a free circulation of air, gases are developed which almost always cause fermentation, paving the way for future disease. Hence the necessity for frequent change of beds. This is made in various ways. The practice of doing this by hand is to be condemned because it consumes too much time and is apt to injure the worms. Threadnets (Figure A2.8) and perforated paper are the best means to employ. They save two-thirds of the hand labor and thus allow beds to be oftener renewed, which is a most important consideration. In the first age tulle or mosquito net may be employed instead of nets or paper.

The manner of proceeding is as follows: Place the last meal at night on the nets and extend them over the worms. By morning the worms will have mounted above the opening in search of fresh leaf. Then lift up the nets, beginning at the top shelf, and place them on clean shelves. Carefully detach from the nets any portion of the old bed, and, if the worms are not molting, gather up the few worms that have remained behind, and tenderly place them with the others. The change of beds is thus rapidly effected with the least labor.

It is very important that the tension of the net be such as to prevent the worms from being crowded together in the middle.

Perforated paper (Figure A2.9) is another means often used to effect change of beds; but it does not allow the worms to mount with the same facility. It is also apt to

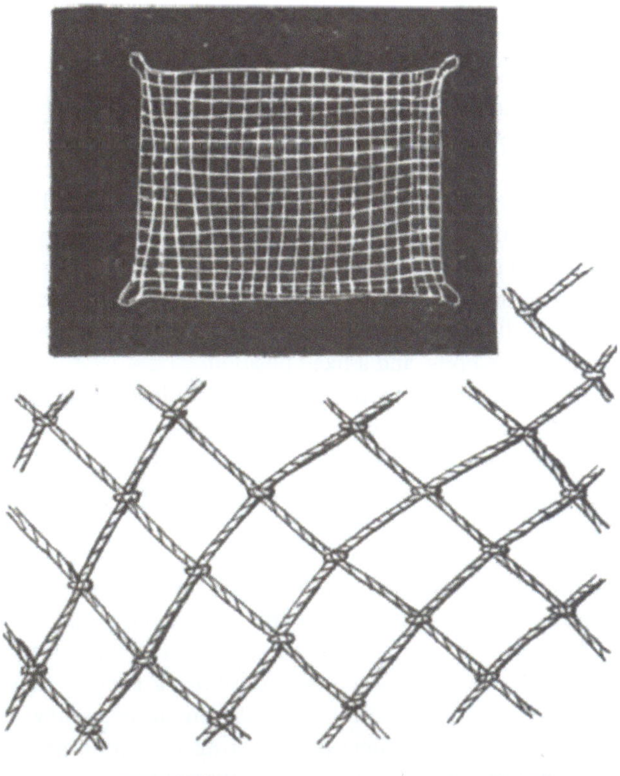

Figure A2.8 Net used in changing beds.

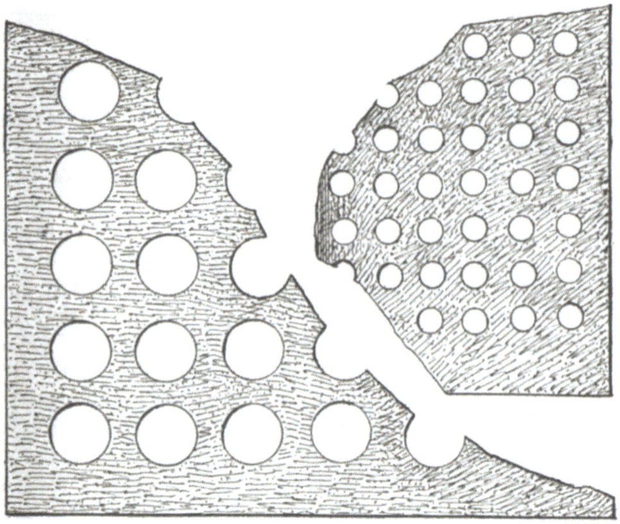

Figure A2.9 Perforated paper used in changing beds in the second and third ages.

break when the worms become heavy, and in many cases it has to be renewed annually, so, in the end, it is no cheaper than nets.

In changing beds, do not feed the worms that are first taken up until all from the old bed have been removed; then give a general meal, for all the worms born on the same day and forming one class should have the same number of meals to preserve their equality of growth, which is necessary for a successful rearing.

Having adopted hours for feeding, these should be adhered to throughout the rearing. When four meals are given, the best hours are 5 to 6 am, 10 to noon, 3 to 6 pm, and 9 to 11 pm.

Toward the sixth day worms begin to eat less. This is a sign that they are going to molt. Then another change of beds and doubling of space are necessary.

The molt or change of the worm is easily recognized by a swelling of the head, whitening of the skin, transparency of the body, and a fixed position.

To change the beds, proceed as before, only leave undisturbed on the old beds the worms that are molting. When all the tardy worms have been taken up and placed on shelves, give them frequent sprinklings of finely cut-up leaf to enable them to catch up with the worms already molting. Diminish the feeding as the backward worms begin to molt, and cease feeding entirely as soon as a single worm comes out of the molt. Then wait for about 24 hours so that the worms may be well over the change before giving a general meal. In this way the equality of development necessary for a methodical and successful rearing is maintained. A fast of 24 hours will not hurt the advanced worms, while the extra feeding given to the backward ones may enable them to become equal to the former. The beds of the worms on the old litter may be changed when the general meal is given.

Many cultivators do not change the beds and double the space until after the first molt, allowing all the worms to change in the same bed. In this case, owing to the distribution of leaf for one or two days after the molting has begun, the molting worms are covered by a more or less thick coat of litter, and exposed to emanations of bad gases in a critical period of life, which is likely to cause disease. Besides, in the first age, the worms are so small that they are likely to be lost in the litter, or to perish from suffocation. Hence, it is healthier to change beds and double the space before the molt.

A2.7.3 The Second Age

KFN 17 → The coming out of the molt is announced by the appearance of a small triangular-shaped livid spot on the worm's head, and the changed skin is grayish in color. The worm takes several hours to recover from a change; then it begins to search for food, which, however, as before stated, must not be given before all the worms have recovered from the molt. Then a slight meal is given by means of perforated paper or nets. The worms crawl up promptly and can be easily gathered up and placed on fresh shelves.

If the two sets of worms recover from their molt at the same time, they may be classed together; if there is a difference of a day, it will be necessary to keep them separate throughout the rest of the rearing, for the equality of age has disappeared, and, if they are put together, the second change will not occur simultaneously for all the worms, but will extend over several days, and occasion the greatest trouble to reestablish the equality of size necessary for the best results.

In case no worms have had change of bed before the molt, do not recommence feeding until the greater part of the worms are awake. Do not fear that they will suffer from hunger. Then form a new division of those still molting. It is frequently better to have two divisions, but if, to simplify the work, one is desired, by putting the backward worms on the highest shelves, and feeding them oftener than the advanced set, an equality may be reestablished.

Three days after the first molt the beds must be renewed, and at the same time more space must be allowed for the worms.

The second age is the shortest, being less by two or three days than any of the others. Toward the fifth day of this age, the worms begin to molt again. Then act as before—that is, by aid of nets or perforated paper, remove the backward worms, in order to place them elsewhere, and try by more heat and abundant food to make them catch up with the forward worms.

A2.7.4 The Third Age

KFN 18 → The worms double their size in their third age; consequently the space allotted to them must be doubled; that is, they must have 16 square yards instead of 8 square yards, as in the second age.

From the second to the third molt the same care is to be given as has been prescribed for the first two ages, except that, if a small incubating room has been used instead of an incubator, the worms must now be transferred to a larger room to complete the rearing. Care must be taken to heat this room the day before to 80°F, and the next day lower the heat to the degree adopted for the rearing. Feed six times daily either with whole leaves or leaves which are coarsely cut up. For this *as* well as all the other ages the best rule to follow in feeding is to give only a light sprinkling of leaf at the beginning and end of each age, gradually increasing the ration up to the middle of each age, and then diminishing to the time of a molt. The appetite of the worm will serve as a guide. Give more or less leaf according as the preceding meal has been more or less eaten. In this way leaf will not be wasted and a large quantity of litter will not accumulate under the worms, to their detriment.

Toward the sixth or seventh day of the third age, according to the temperature, the worms begin to be languid and lose appetite, as before, and are ready to make a third change. This is the most difficult of all and the one in which they seem to suffer the most. It is also the period when diseases due to bad eggs or to a poor incubation are developed. Excepting accidental diseases, a good result may generally be predicted if the third change is safely passed. With these facts in view, from the beginning of the third age, keep the worms sparse on the shelves and see that the beds are dry and changed with scrupulous care, the litter being far removed from the rearing room. Avoid feeding with wet leaf and to favor the molting raise the temperature a degree. It will be noticed that the head and body of the worm are more swollen than in other molts. It is this superabundance of liquid that renders this molt so critical and necessitates a drier atmosphere and a bed which is very dry and not apt to ferment.

The worms increase three times in volume after the third molt and must have space accordingly. They must be separated into three divisions in the following manner: Instead of waiting, as in the first two changes, until half of the worms have begun to molt, let down the nets, or otherwise prepare for the removal of the backward worms when one-third or even less of the worms show signs of molting. About two-thirds will then crawl up on the fresh leaves, and must be placed on a shelf where, after one or two meals, they proceed to molt, being again divided after the first or second distribution of leaves, according to the rapidity of the molting.

If only a few worms mount when the first division is made, the operation was delayed too long, and it is unnecessary to divide those first taken up; but after the change the division of those left on the old bed can be made. To allow the worms to spread out, each division should occupy but one-third of the shelf on which it is placed.

A2.7.5 The Fourth Age

KFN 19 → When the worms are over their third molt they cease to be gray and take the characteristic color of their race. If they are too long in molting it is because the temperature is too low, that is, below 68°F. Increasing the heat a few degrees will excite the worms and enable them to complete their change. After this the temperature adopted for the rearing must gradually be resumed.

When all the worms have molted the third time a change of beds must be made as in preceding ages. Do not be in haste to change. Wait a few hours to permit the worms to recover their strength a little. The recommendation to feed lightly at first applies especially to the beginning of the fourth age.

If the outside temperature is normal, fires need not be kept up, and the doors and windows may be left open, guarded by light curtains. In this age the worms eat enormously and more help will be required to gather and distribute leaf. Small branches of leaf may now be given in place of whole leaves, and the number of meals may be reduced to four, if the temperature is 68–70°F, at which temperature the age will last nine days. If it is desired to reduce this age to seven or eight days, raise the heat to 72°F and give five or six meals daily. Should change of weather retard the growth of the mulberry trees, and temporarily cut off the supply of leaf, adopt the rearing to such a con tingency by lowering the temperature slowly to 66°F or 68°F and giving only three or even two meals daily. Then, when leaf is obtained, gradually raise the temperature to the degree adopted for the rearing. During the fourth age four changes of beds are made, including the one which follows the third molt. A single division of the worms is sufficient, which will be the last, and particular care must be taken to divide the worms into two equal parts. For this, spread the nets or perforated paper over the worms when half are molting and proceed as before.

A2.7.6 The Fifth Age

KFN 20 → Do not be in haste to change beds as soon as the worms have molted. This precaution is necessary to allow the new organs to acquire consistency and to prevent worm from being lost in the litter. Consequently, wait several hours and change beds after the second meal, the first being only a slight sprinkling of leaves. Worms are not strong enough immediately after a change to digest a heavy meal.

The space during this age should not be limited. See that between two worms another could be easily placed. Experience has proved that the harvest of cocoons is often in proportion to the space accorded to the worms during this age.

Feed from four to six times daily, according to the temperature, and spread the branches or leaves out regularly, not to form a mass. For a simultaneous mounting into the brush all the worms should eat an equal quantity of food.

As ripe berries are very indigestible, and also cause beds to ferment, care must be taken to shake the trees well before gathering the leaves, and whatever ripe fruit remains must be taken off before feeding.

If possible, change beds daily, especially if more than five meals are given or if the weather is very warm and damp and there are signs of disease.

The first five days after the change the worms grow enormously, and it is very difficult to satisfy their appetite. At the end of this time the body suddenly diminishes in circumference, the excrement, formerly dry and firm, now becomes moist and soft, and the appetite diminishes and becomes capricious. This state generally lasts three days, then suddenly the worm ceases to eat, and tries to get away from its food. It prolongs its head, and, changing its former lazy habit of scarcely moving except to get food, runs about in every direction, stopping from time to time, and moving its head (now transparent gold or white, according to race) like a blind person seeking the way. These signs indicate that the worm is hunting a convenient place to spin its cocoon. The worm is now mature, and a spinning place should be ready to facilitate its metamorphosis. The humidity, which always exists at this time on account of the mass of litter, is especially dangerous to the worms; and it is increased if the worms do not all mature and mount at the same time, for those that remain below are wet by the liquid dejections of those that are the first to mount. For this reason, do not put worms in the spinning place until they are perfectly mature. They then mount, and crawl around some time seeking a favorable place for their cocoons. Finding this they evacuate their digestive canal and begin to throw around them an irregular net in which the cocoon spun later will be suspended.

A2.8 PREPARATIONS FOR SPINNING

KFN 21 → A considerable loss may occur in the spinning place even when the rearing has been most successful. To avoid such loss, observe the following precautions: (1) prepare the spinning place in time; (2) arrange it so that the worms may regularly mount, and have abundant room; (3) have it well made, yet economical; and (4) regulate the heat and ventilate the room.

Any convenient dry bushy brush, odorless and free from gum, will serve to construct the spinning place; or if such is not available, bundles of straw, or shavings, or finely split up wood may be substituted. The best and most economical arrangement is small bundles of brush or straw placed upright between the feeding shelves, in rows, about 16 inches apart. The bundles are cut a half inch taller than the space between the shelves, and their tops are spread out to form arches, and to allow the worms plenty of room to spin (Figure A2.10).

Branches of elm, oak, birch, etc., are used to place the worms in the spinning place. These branches are spread over the shelves at the end of the fifth age. Very soon they will be covered with mature worms which have ceased to eat and are turning away from the mulberry. In this way it is easy to select the worms that should be transferred. If the worms are equally developed, in 30 or 40 hours they will be shut up in their cocoons. The few that remain behind should be placed elsewhere; fed with fresh leaf on (lean beds they will soon catch up with the others).

The fifth day after the mounting, the worms that have not begun to spin should be placed in bundles of twigs and covered with straw or leaves, or put in a basket of shavings, where they will be forced to spin.

The temperature during the spinning should be 75°F, and the humidity throughout the rearing about 65°F. A good practical test of humidity is a saucer of salt; when the salt is moist, reduce the humidity. Carefully avoid disturbing the worms while spinning, and then, as during all the ages, keep the room as quiet as possible.

The most scrupulous cleanliness should always be observed, both with regard to the quarters and the attendants; to keep from raising dust, wipe the floor with a damp cloth instead of sweeping it.

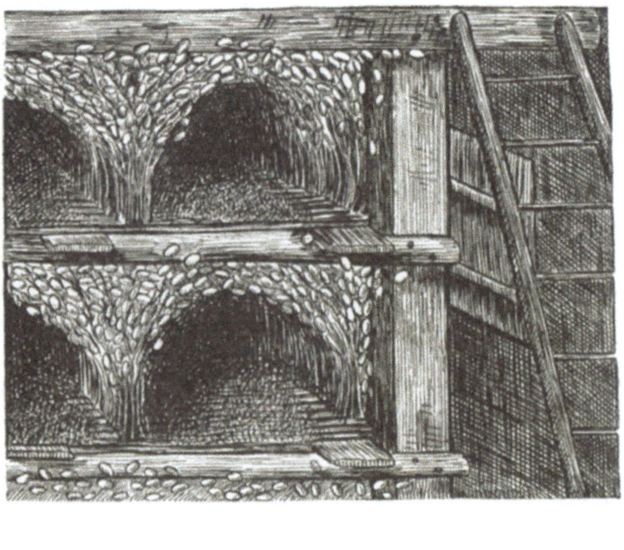

Figure A2.10 Arrangement of spinning places. (Redrawn from Pasteur.)

A2.9 PREPARING COCOONS FOR THE MARKET

KFN 22 → The transformation of the larva into the chrysalis is, according to the temperature, completed in from seven to ten days from the time at which the first worm begins to spin. The cocoons are then said to be mature, and this is the best time to gather them. If gathered earlier, the producer will run the risk of having his cocoons rejected in the market; and if later, he will sustain a very sensible loss in their weight, as they grow lighter from the time of their maturity until the moth comes out. The best authority estimates the minimum loss as 4.1%, and the maximum as 23.3%. To avoid the risk of soiling them, gather the cocoons from the lowest shelf up. They may be freed from the web or floss by a very simple instrument (Figure A2.11) or by hand.

After the removal of the web, the cocoons are sorted into three classes: (1) the perfect; (2) the double; and (3) the defective or spoiled. They are then spread out in 4-inch layers, on clean shelves, in an airy, dry room, not exposed to sunlight, and very carefully guarded from rats, mice, and insects. The defective and discolored ones are put in a separate room.

The thread of a cocoon is continuous with that of the web and diminishes in diameter within. Its length varies from 1,200 yards to 1,600 yards, and its value accordingly. Different races, sexes, and conditions of rearing often produce notable differences in weight of cocoons. Thus the weight may vary from 155 to 320 cocoons to the pound (340–700 to the kilogram).

Often two or more worms are enclosed in the same cocoon. Cocoons formed from such collaboration are larger than single ones, irregular in form, and cottony in texture. They cannot be unreeled and consequently are far less valuable than single ones.

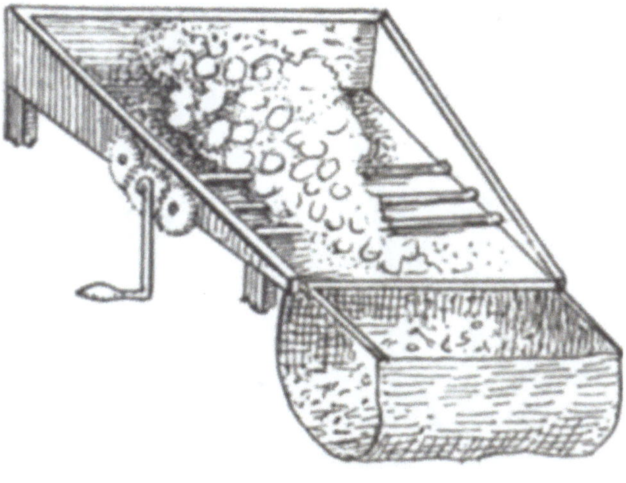

Figure A2.11 Device for removing floss from cocoons. (Redrawn from Nenei.)

The proportion of silk in a cocoon varies according to the race and the regime to which the worm has been subjected. The average normal cocoon at the time it is sold is thus composed:

Water: 68.2%
Silk: 14.3%
Web and veil: 0.7%
Chrysalis: 16.8%

If the cocoons are not sold as soon as gathered, the chrysalides should be killed without delay unless they are to be reserved for reproduction. Otherwise the moths may pierce the cocoons, thus rendering them unfit to be reeled.

The chrysalides are usually killed either by heat or suffocation. The means most commonly employed are: (1) the heat of the sun; (2) hot dry air in a stove; (3) hot humid air in a stove; (4) steam; (5) oil of turpentine; and (6) carbon bisulphid or some other gas. Probably the best of these means are steam at a temperature not above 212°F applied without pressure, or hot damp air at a temperature of 196°F. The killing of the chrysalides is an important operation and one requiring care and judgment. If some are left alive, the moths will issue, thus rendering the cocoons of little value. On the other hand, if the operation is continued too long, the silk may be injured. The best methods are those in which the heat is carefully controlled and excessive dryness is avoided.

The following is a very simple and easy way to destroy the chrysalides by the use of steam.

Place a cauldron of water on a stove. When boiling begins set over the cauldron a white hollow wooden cylinder, about 3 feet high and 2 feet in diameter, with a perforated bottom and open at the top. Arrange in this cylinder round baskets three-fourths filled with cocoons, then cover the cylinder with a perforated lid. In about 30 minutes, the operation will be completed, after which remove the cylinder, take out the baskets, and spread out the cocoons to dry before storing, to prevent them from spoiling.

Mr. T. A. Keleher of this office has adopted the following plan:

> The cocoons are placed in an air-tight box of about 24 cubic feet capacity; about half an ounce of bisulphid of carbon in a small dish is placed in the box and left over night. It is best to open one or two of the cocoons to find if the chrysalides are dead; if not the operation must be repeated. Care should be exercised that no fire of any kind be brought into the vicinity during this operation as the bisulphid of carbon is very inflammable.
>
> In shipping cocoons, care must be taken to pack them in baskets or cases permeable to air, but sufficiently close to keep out rats and mice, which are very destructive to cocoons.

A2.10 DISEASES OF SILKWORMS

KFN 23 → In every successful rearing of an ounce of eggs about 40,000 worms are hatched, and 30,000 succeed in spinning cocoons. The rest either die from casual wounds or from diseases incidental to restricted action. But sometimes whole chambers are destroyed by hereditary and contagious diseases, and it is of supreme importance to cultivators to learn how these scourges may be avoided.

In this limited treatise only a bare mention can be made of the most fatal diseases and of the necessary precautions to be taken to guard against them. The general cause of disease is the domestication of the worm. By using good eggs, however, and following the methods which are actually employed by successful rearers, remunerative results are usually obtained. To obtain good eggs, it is necessary to adopt new methods. These chiefly involve the use of the microscope.

Among the many diseases of silkworms, the principal ones are pebrine, flacherie or flaccidity, gattine or macilenza, calcino or muscardine, and grasserie.

A2.10.1 Pebrine

This disease was first noticed in epidemic form in France in 1845. Since then it has appeared in Italy, Spain, Portugal, Turkey, Turkestan, the Caucasus, Kashmir, China, and Japan; threatening to destroy the silk industry.

Between 1833 and 1865, the annual crop of cocoons in France was reduced by pebrine from 57,200,000 pounds to 8,800,000 pounds. No remedy has been found for the disease, but the Pasteur microscopical selection of eggs, insuring the birth of healthy worms, is a sure preventive. The universal adoption of this method has made pebrine almost a thing of the past; and following Pasteur's line of research, means have now been discovered for avoiding every kind of silkworm disease.

Worms affected with pebrine develop slowly, irregularly, and very unequally. Black spots are the most marked outward characteristics of the disease; the internal signs are oval corpuscles only visible through the microscope.

Worms healthy born may contract pebrine during life, but this may not prevent their spinning as the disease does not reach it; climax before the chrysalid or moth stage, and in its incipiency the worm is strong enough to spin, though the moth will produce diseased eggs. Hence the necessity of repeating the microscopical examination for each generation of worms.

Pebrine is not always visible, and when latent induces other diseases. When only one crop of cocoons is made annually, it is comparatively easy to resist pebrine, as the germ of it, outside of an egg, retains its vitality not longer than seven months. The disease takes 30 days to develop; therefore, if worms from pebrinized eggs can be made to spin within 25 days after hatching, they may yield a fair harvest of cocoons. In any case, however, it is only safe to use pure eggs, as pebrine, even in undeveloped stages, renders the worm more liable to contract all other diseases.

A2.10.2 Flacherie or Flaccidity

This is now the most dreaded disease among European silkworms. In general, worms are struck with it after their fourth molt, when they are mature, or even while spinning (Figure A2.12).

Without any apparent cause, they begin to languish, then remain completely still, and shortly die. They blacken

Figure A2.12 Worms affected with flacherie dying in the brush after Pasteur.

Figure A2.13 Worm which died of flacherie, putrefying after death. (Redrawn from Pasteur.)

after death (Figure A2.13) and give out a disagreeable odor. Often entire chambers perish in a day. Again, the progress of the disease may be slow, the worms even spinning their cocoons, but, dying in the chrysalid state, they putrefy and soil the cocoon, thus greatly diminishing the value of the harvest.

Flacherie is but another name for indigestion. Pasteur and many other scientists assert that flacherie is due to ferments and vibrioni developing in the intestinal canal of the worm; other authorities maintain that the disease may exist independently of these. However, as these microorganisms, in the majority of cases, play a prominent part in the development of flacherie, it is well to guard against them.

The principal causes of flacherie are: (1) eggs being spoiled through careless preservation; (2) hereditary tendency; (3) overfeeding of worms: (4) wet, sweating, dewy, and fermented leaf; (5) leaf submerged in water or full of mud; (6) leaf from a new plantation or from a shaded spot, coarse leaf, or change of leaf; (7) lack of ventilation; (8) excessive heat; (9) dust; (10) keeping worms too thick on trays; (11) accidental deaths of worms from injuries, these putrefying, and the ferments thus created being communicated to other worms; and (12) debility.

If these causes are avoided, flacherie is not likely to invade a rearing. To prevent contagion, eggs should be dipped in a solution of sulfate of copper before being incubated; and in cleaning shelves and nets, wherever a dead worm is seen, powdered sulfate of lime or copper should be applied.

Unlike the corpuscles of pebrine, the microscopic organisms, which are probably the immediate cause of flacherie, remain alive from one year to another, and the dust of a rearing room may contain them in considerable quantities and become the means of infection. Hence, in cases of flacherie, immediately after the rearing, the walls, shelves, and all the implements should be washed in a solution of chloride of lime or some other germicide, and the room should be fumigated with sulfur.

A2.10.3 Gattine

The external signs of gattine are indifference to food, torpor, dysentery, and emaciation. It attacks the worm in the first ages and is especially manifested after a molt. Sometimes it is associated with flacherie, and, in its incipient stage, is confounded with this disease. Later the worm

Figure A2.14 Worm emaciated by Gattine after the fourth molt. (Redrawn from Nenei.)

becomes extraordinarily emaciated and sufficiently transparent to be mistaken for a mature larva. The hooks of the prolegs are lengthened out and strongly attach the worm to whatever it touches. Meanwhile torpor creeps on and soon ends its life (Figure A2.14).

Worms having flacherie or gattine do not always die before mounting into the brush, and if the disease has not entirely invaded the organism they may even arrive at spinning. But instead of mounting with the promptness and rapidity of healthy worms, they stop hesitatingly at the base of the brush, then begin slowly to mount, stopping on the first little twigs and distending themselves as though asleep, sometimes with the bead turned toward the base. Again, especially in case of gattine, the worm wanders restlessly here and there, seeking as it were power to eject the silky matter, but too impotent to do more than throw out a scanty thread to weave a web or veil of a cocoon, in which it generally falls and dies.

Eggs free from disease and capable of resistance to disease are the prime requisite in guarding against flacherie and gattine. The moment some deaths are noticed, proceed as follows: (1) change beds immediately, briskly shaking the worms; (2) place the worms on disinfected shelves; (3) burn the diseased and suspected worms that do not mount on fresh beds; (4) if possible move the whole rearing to another room previously aired and disinfected, and also aired after disinfection; (5) do not feed during the three or four hours in which the change is being made; (6) keep up a little wood smoke in the room; (7) give a few scanty meals of light leaf; and (8) diminish the temperature a little.

A2.10.4 Calcino or Muscardine

This disease, at first, has no visible appearance, but by degrees the vitality of the worm is impaired, and it eats and moves slowly. The body turns rose-colored or red, beginning with the stigmata, and then contracts and loses its elasticity, after which the worm stands still as though paralyzed, and finally dies 20–30 hours from the appearance of the first symptoms. After death the body dries up and is covered with a white efflorescence, causing it to look like a stick of white chalk (Figure A2.15); hence the name of the disease.

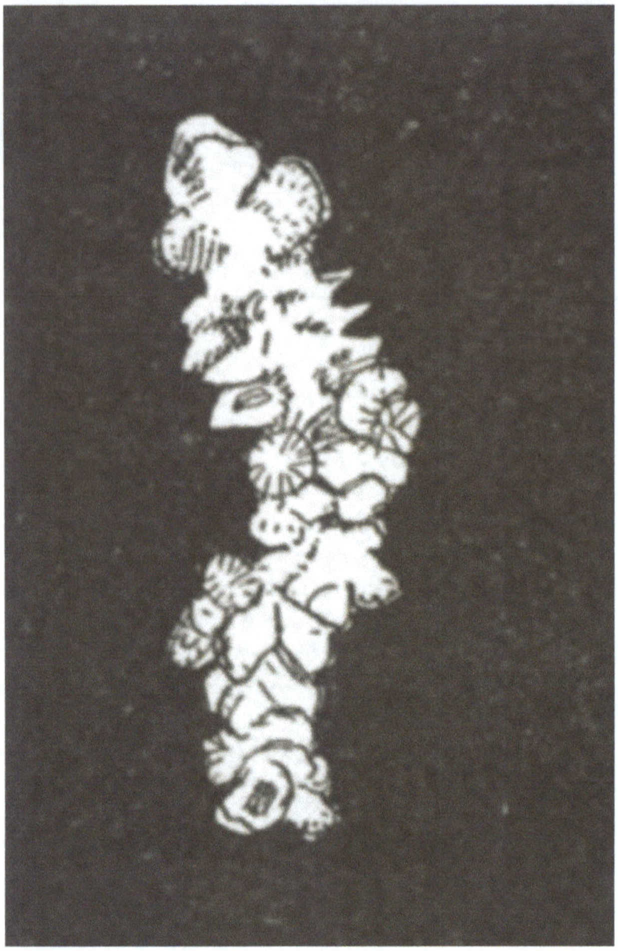

Figure A2.15 Calcinated worm. (Redrawn from Verson and Quajat.)

Calcino is caused by a mold or minute fungus. There are two varieties of this fungus: *Botrytis bassiana* and *B. tenella*. They both attack the worm in the same way. The spores of the mold by chance get on the body of the worm when it is in a molting condition, and there take root, penetrating below the skin. The thread-like mycelium ramifies until it fills the entire body. Later some of the branches fructify on the surface, and the fruit bursting envelopes the worm with innumerable spores resembling a white powder.

Each spore is capable of settling on a molting worm and giving it calcino, hence the necessity of taking steps to avoid contagion. Calcino is more contagious than other silkworm diseases. Darkness, stagnant air, dirt, warmth, and moisture are the five things that favor mold, and calcino is due to a mold.

The chief cause of the disease is neglecting to change the beds and keeping litter in and around the room. When only one or two worms have died from calcino, all the shelves should at once be cleaned and divested of dead worms. The floor should be washed with a solution of sulfate of copper (1–200 by weight), and a pound of sulfur should be burned, or a strong wood smoke created in the room, which should then be shut up five or six hours, after which the worms should be fed. Should any worms die the next day the beds should again be changed and an ounce of sulfur burned. The quantity of sulfur fumes that would kill rats, bats, and lizards and even human beings does no harm to silkworms. No hesitation, therefore, need be felt in fumigating the rearing room with sulfur; but eggs and thread nets must not be subjected to sulfur fumes. Silkworms affected with calcino die before the moth stage; therefore, it is impossible for the disease to be hereditary. But loose spores of the mold creating the disease may get on healthy eggs. These may be washed off by a good bath of fresh water. Some recommend a bath with a solution of sulfate of copper (one-half per cent of copper). In cases of calcino, the room should be disinfected immediately after the cocoons are gathered, and the paper and brush used should be burned.

As calcino is never due to infected eggs no attention need be paid to the presence of spores of the *Botrytis* in the microscopic examination to select eggs.

A2.10.5 Grasserie

Silkworms having this disease become restless, bloated, and yellow. If punctured they exude a purulent matter full of minute, polyhedral granular crystals.

Grasserie is neither hereditary nor contagious. Unlike pebrine, flacherie, and calcino, it is not caused by microbes capable of multiplying and creating plagues. Grasserie does little harm to silkworms in Europe, but in warm countries, as in Bengal, sometimes assumes an epidemic form.

Worms first fed on mature leaf, and afterward on young leaf, are apt to take grasserie. The propagation of large trees is the best means of checking the disease. The main cause of the sporadic appearance of grasserie is mismanagement of the worms at the molting periods. Feeding should not be stopped before all the worms have begun to molt and should not be recommenced until all the worms are well out of the molt; otherwise they are likely to have grasserie. This disease often leads to flacherie, and when it occurs in an exaggerated form indicates latent pebrine.

Appendix 3: System for Mass Rearing Boll Weevil in a Laboratory

Technical Bulletin 95 April 1979

Technical Bulletin 95 **April 1979**

System for Mass Rearing Boll Weevil in a Laboratory

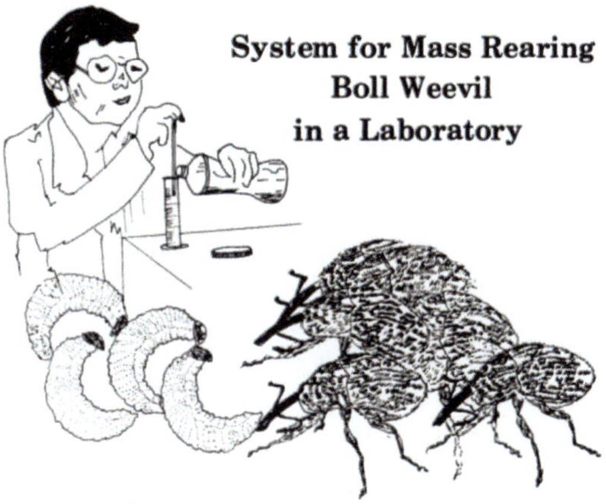

O. H. Lindig, Research Entomologist,

J. G. Griffin
Agricultural Engineer,
USDA, SEA, AR, Boll Weevil Research Laboratory,
Mississippi State University

O. H. Lindig,
Research Entomologist,
USDA, SEA, AR, Boll Weevil Research Laboratory,
Mississippi State University

J. Roberson,
Entomologist
USDA, APHIS, R. T. Gast Rearing Laboratory,
Mississippi State University

P. Sikorowski,
Insect Pathologist,
Department of Entomology
Mississippi State University
USDA, SEA, AR
Delta States Area
In Cooperation with

MAFES MISSISSIPPI AGRICULTURAL & FORESTRY EXPERIMENT STATION
R. RODNEY FOIL, DIRECTOR MISSISSIPPI STATE, MS 39762
Mississippi State University
James D. McComas, President Louis N. Wise, Vice President

A3.1 INTRODUCTION

GFN1 → Large- scale boll weevil (*Anthonomus grandis* Boheman) eradication programs requiring millions of sterile male weevils called for better technology for rearing boll weevils in the laboratory. This bulletin describes the methods and equipment developed at the Robert T. Gast Rearing Laboratory (GRL) and used to produce more than six million weevils per week.

The GRL was built to meet the need for mass-reared boll weevils for the 1972 and 1973 Pilot Boll Weevil Eradication Experiment (PBWEE) in south Mississippi and adjoining states. A component of the program was the release of sterile male weevils.

The boll weevil had been reared in the laboratory on a small scale since the 1950s. Weevils were fed and oviposited on cotton squares in laboratory rearing described by Vanderzant and Davich 1958, Brazzel et al. 1959, Earle et al. 1959, Vanderzant et al. 1959, and Soloman and Hunter 1960. They planted the weevil eggs one at a time on artificial diet in a small vial or tube stoppered with a piece of sterile cotton.

New methods and equipment that made it possible to raise large numbers of weevils in the laboratory were described in the 1960s by Gast 1961, Gast and Vardell 1963, Gast and Davich 1966, and Betz 1966. However, these methods took a lot of labor and time, so the number of weevils that could be raised in the laboratory was limited.

Producing millions of boll weevils for research and control programs means caring for successive generations throughout the life cycle, from egg to adult collection. The operation must be efficient if mass-rearing for research and control programs is to be economically feasible. The important factors are equipment to mechanize the various steps, environment and diet appropriate to growth and reproduction and sanitation to protect weevils from microbial contamination.

The equipment and techniques described here were developed for and used successfully in the boll weevil mass-rearing tests conducted during the summers of 1976 and 1977.

A3.2 DIETS

GFN2 → We use two diets—one for hatching eggs and rearing larvae, the other for adult feeding and oviposition. Both diets contain a diet dry mix (Tables A3.1 and A3.2).

A3.2.1 Equipment

- Two top-loading balances (2 and 5 kg capacity)
- Vibratory separator with a 32mesh TBC screen, 24 inches in diameter (made by SWECO)

Table A3.1 Ingredients and Quantity of Each Used to Make a Batch of the Boll Weevil Diets

	Diet	
Ingredient	Adult (g)	Larval (g)
Cottonseed meal (sifted)	28,500	31,800 g
Agar	8,400	8,280
Sugar	6,850	7,800
Promine D	10,800	12,280
Com cob grits (60-mesh)	6,000	—
Cholesterol	300	360
Wesson® salt	1,125	1,260
Vitamin mix[a]	1,170	1,260
K sorbate	585	630
Tegosept	375	400
Ascorbic acid	270	290
Methenamine	60	—
Cottonseed meats	10,800	10,724

[a] See Table A3.2.

- Hammermill with 3/128 inch (0.6 mm) screen openings, made by Jacobson (Model 160-D)
- Horizontal mixer-blender, paddle type, 6 cubic feet capacity (made by Kelley-Duplex)
- Two vibrating volumetric feeders with 4-inch spout (made by Eriz)

A3.2.2 Procedure

1. Sift cottonseed meal with the separator. Discard what doesn't go through
2. Weigh on the balances the exact amount of each ingredient (see Table A3.1). Proportions may be adjusted for desired batch size
3. Put all ingredients except agar, promine D, and corncob grit through the hammermill
4. Put the milled mixture into the mixer-blender with the portions of agar, promine D, and corncob Grit and blend for 30–40 minutes
5. Sift the blended mixture through the separator, removing large particles and cotton lint
6. Use volumetric feeders to meter and feed materials into hammermill and separator
7. Store mixture in closed dry containers until needed

Table A3.2 Ingredients and Quantity of Each Used in Making a Batch of Vitamin Mix

Ingredient	Quantity(g)
Niacinamide	20
Ca Pantothenate	20
Riboflavin	10
Thiamine HCl	5
Pyridoxine HCl	5
Folic acid	5s
Inositol	40
Sugar	6,160

The ingredients are carefully weighed on balances and ball milled for 1 hour to give a homogenous blend.

A3.2.3 Diet Ingredients

Larvae Diet (100-lb batch)
- 100 lb of water
- 9-mL B12 vitamin
- 9-mL biotin
- 40-mL HCl
- 18 3/4 lb dry diet mix (see Table A3.1)

Adult Diet Pellets (100-lb batch)
- 100-lb water
- 12-mL Bl2 vitamin
- 9-mL biotin
- 40-mL HCl
- 19 lb dry diet mix (see Table A3.1)

A3.2.4 Preparing and Sterilizing Diets

A3.2.4.1 Equipment

Cherry-Burrell No-Bae Unitherm IV® sterilizer unit (Griffin et al. 1974) with one 30-gallon wash and sanitizing tank and two 30-gallon diet mixing tanks (Figure A3.1)
Barber Coleman 530 series automatic, solid-state temperature control and auxiliary modulating valve for controlling heating and cooling temperatures
100-lb capacity scale, 2 ounce divisions
Two 30-gallon plastic tanks and stands
100-mL graduated cylinder
Two 1/3 HP, variable speed mixers
3/8-inch outside diameter (OD) stainless steel tubes and control valves from sterilizer unit to pellet making and larval diet dispensing machines

A3.2.4.2 Procedure

1. Weigh and/or measure all ingredients according to size of batch required
2. Put water into sterilizer diet mixing tank
3. Turn mixer in mixing tank to medium speed, then add other fluid ingredients and dry mix

Figure A3.1 Sterilizer unit.

4. Mix ingredients until homogenous and free of lumps, and reduce mixer to the speed required to keep ingredients in suspension
5. Set the temperature control to bring diet mixture to 141°C for sterilization. Set automatic cooling to 49°C for adult diet or to 42°C for larvae diet, adjusting as necessary so that larval diet will spread in trays
6. Circulate water from wash and sanitizing tank through sterilizer back to tank until these temperatures are reached and pump speed is set
7. Turn off circulating water and turn on valve of diet mixing tank to route diet through the pipes from the tank into the sterilizer and through the heating and cooling coils
8. Adjust valves in 3/8-inch transfer tubes to route sterilized adult diet to pellet machine or larvae diet to egg-planting machine
9. Clean and sanitize when finished

A3.2.5 Preparing Wax for Wax-Coated Adult Diet Pellets

A3.2.5.1 Equipment

- GFN3 → A self-contained, electrically heated steam kettle, 40-gallon capacity
- Two non-self-contained steam kettles, one 60-gallon and one 80-gallon capacity
- Two 6-gallon stock (transfer) pots

A3.2.5.2 Procedure

1. Wash, drain, and boil in 80-gallon kettle the wax recovered from the egg extraction operation to remove the water. Transfer this wax plus more fresh wax, as needed, to 60-gallon kettle. The wax is a mixture of 40% beeswax and 60% paraffin
2. Melt waxes together in the kettle, sterilize by boiling, transfer to self-contained kettle and maintain at 66°C
3. Transfer wax as needed from kettle to wax vat of pellet machine (using stock pots)

A3.2.6 Making Adult Diet into Wax-Coated Pellets

A3.2.6.1 Equipment

GFN4 → Automated equipment described by Griffin and Lindig 1977 includes pellet-forming and cutting machine, wax-coating unit, fan, pellet catch bins, and waste-water tank (Figure A3.2)

A3.2.6.2 Procedure

1. Set sterilizer pumps and control valves to circulate steam through diet transfer lines and tubes to and through pellet-making machine, catching water in waste tank. Be sure cooling water supply valve to

Figure A3.2 Pellet-making and wax-coating equipment.

machine is off while circulating steam through the tubes

2. When lines have been steamed and sterilized for 10–15 minutes, turn directional valve to direct steam flow back to water tank, turn on cooling water valve to sterilizer and let water circulate through sterilizer
3. Move waste tank from under discharge end of forming tubes on machine and replace with waxing unit
4. Open valve to heat-exchanger, cooling water supply. Turn on power switch controlling air valves for cylinders operating discharge valves on manifold
5. Turn on and set speed of the adjustable rotating cutting wheel to cut the rods into 1/2-inch long pellets
6. Turn on motor and adjust speed of wire belt in waxing unit
7. Shut off water tank valve and open the valve on diet tank of sterilizer to route flow of prepared diet through sanitized tubing into the manifold of the pellet-making machine
8. The diet flows into the forming tubes and congeals as cold water (7.5°C) flows around the tube in the heat exchanger
9. The 5/16-inch diameter rod of congealed diet is forced out of the discharge end of tube as more diet enters the manifold end of the tube on each cycle
10. A diversion shield around the cutter wheel directs pellets into the waxing vat of warm (66°C) wax
11. The rotating wire belt elevates pellets out of the wax and drops them into a catch bin as a fan blowing across the falling pellets hardens the wax
12. Store pellets in a clean, dry, and covered container until needed but not more than 24 hours
13. Clean pellet machine and waxing unit immediately

A3.3 EGG PRODUCTION PROCESSING AND PLANTING

A3.3.1 Production

GFN5 → Adult weevils are held for egg production in laying cages (Figure A3.3).* Diet pellets, which are both food and oviposition sites, are placed in these cages with the weevils and serve as the egg collector units.

A3.3.1.1 Equipment

- Oviposition (laying) cages: stainless steel with removable solid tops and 12 × 12 stainless steel mesh wirecloth bottoms. Cages measure 2 × 18 × 38 inches. Built-in feed baskets are stainless steel 1-inch deep with 4 × 4 mesh wirecloth bottom and four 3/8-inch legs
- Cage conveyor (Figure A3.4), a modified laundry bundle conveyor. Shelves are six tiers high, spaced 12 inches apart. Conveyor is controlled by switch near feeding station
- Cage shaker unit (Figure A3.4) at feeding station

A3.3.1.2 Procedure

1. Place 2,250 pellets (measured by volume, 5/6 gallon) in a thin layer in feed basket in cage
2. Dispense over the pellets 9,000 newly emerged and mixed sex boll weevils (measured by weight, 12 mg/weevil or 108 g equals 9,000)
3. Close cage tightly and place on conveyor shelf located in oviposition room

Figure A3.3 Laying cage.

* Detailed specifications are available from the Boll Weevil Research Laboratory, Mississippi State.

Figure A3.4 Cage conveyor and shaker unit.

4. Replace pellets once each day:
   a. Move cages on conveyor to feeding station
   b. Take each cage off conveyor and vibrate on shaker unit for 8–10 seconds before opening. This causes weevils to feign death, falling from pellets in feeding basket to bottom of cage for the feeding operation
   c. Open cage and remove old pellets, saving them for egg extraction
   d. Add 5/6 gallon of new pellets and spread over bottom of feed basket
   e. Close cover tightly and replace cage on conveyor shelf
5. Remove weevils from cage on 14th day, place in escape-proof bag and destroy
6. Clean and sanitize cages and let air dry before re-use
7. Maintain oviposition room temperature at 30 ± 1°C, relative humidity at 50 ± 5%. There is no natural light in this room. Artificial lights are used only during feeding and cleanup

A3.3.2 Extracting and Cleaning

GFN6a and GFN6b → Weevil eggs are extracted from the egg-infested pellets collected from the laying cages, using mechanical equipment (Figure A3.5) described by Griffin and Lindig 1977.

A3.3.2.1 Equipment

- Mechanical equipment unit includes a vibratory feeder open tank with two compartments equipped with discharge valves, a wax cracking unit, two disintegrators, two mixers, a transfer pump, water spray heads and liquid separator with two decks of mesh wirecloth, the top deck 25-mesh and the bottom deck 50-mesh.
- 50-mesh wirecloth egg collecting basket

Figure A3.5 Egg extracting unit.

- 50-mesh sieve
- Two 100-gallon plastic tanks on metal frame with legs
- A 15-gallon plastic tank with bottom tapered to a one-inch drain
- Wax hand skimmer with 14-mesh wirecloth bottom
- Egg hand skimmer with 50-mesh wirecloth
- 1000-mL separatory funnel
- Gallon beaker egg container
- Eight 100-mL graduated cylinders

A3.3.2.2 Procedure

1. Place infested pellets that have been removed from laying cages into the hopper of feeder (Figure A3.5)
2. A water spray cools the pellets as they are flushed through the wax cracker, which partially dislodges wax from diet pellets
3. Direct the flow of wax, diet pellets, and water mixture into the first compartment of the separation tank. Turn on mixer in compartment, where mixer agitation helps to separate wax from diet material and weevil eggs
4. When the first compartment is nearly full, switch flow into second tank, turn on the mixer in the second tank, and stop mixer in first compartment
5. Skim wax from first compartment with wax hand skimmer
6. Agitate the mixture again for a minute or two; hand-skim again
7. Repeat step 6 until all wax is removed from tank
8. Turn on mixer and transfer pump and open discharge valve of first compartment. Mixture goes into the disintegrator units which chop pellets into smaller particles while another water spray flushes out the weevil eggs
9. Discharge egg mixture from disintegrator to top screen deck of liquid separator
10. After the second compartment of tank is filled, skim wax from this compartment, following steps 5, 6, and 7 above and transfer to disintegrators while first compartment is refilling

11. As each batch goes through liquid separator, top deck removes coarse particles and bottom deck catches eggs and fine particles. Coarse particles and water are washed into waste system while egg mixture goes into the wire basket
12. Mix water with salt to make a saturated brine in the large plastic tanks
13. Fill the 15-gallon tank about half full of this brine
14. Add weevil egg mixture from wire basket plus enough brine to bring contents of tank to within 4 or 5 inches of the top
15. Stir brine-egg mixture for one or two minutes. Let settle for about five minutes. Eggs will float
16. Skim eggs from brine with hand skimmer; put eggs in gallon beaker
17. Open tank drain and flush out brine and sediment in the bottom
18. Spray fresh water on the eggs in the beaker. Eggs will settle to the bottom while remaining bits of wax or insect parts float to surface
19. Decant water with trash from eggs
20. Fill separatory funnel two thirds full of brine
21. Add weevil eggs to funnel
22. Rotate until eggs and brine are thoroughly mixed
23. Let funnel stand until eggs float to surface and diet sediment settles to bottom
24. Open petcock on funnel and allow dark layer of diet residue and other trash to drain out. Close petcock before eggs begin to drain
25. Fill funnel with fresh brine and repeat steps 21 through 24
26. Repeat this procedure until visible diet and insect parts are removed
27. Fill funnel with cold tap water, rotate, and let eggs settle. Any remaining trash floats to surface
28. Open petcock and drain eggs into 50-mesh sieve. Close petcock before any trash drains out. Empty funnel
29. Wash eggs back into funnel and repeat step 27 until eggs are clean
30. Drain eggs into graduated cylinders, 1 mL = 4,157 eggs

A3.3.3 Sterilizing Surface Eggs

GFN6b → The surface of the boll weevil eggs is sterilized by the method of Sikorowski et al. 1977 to kill any surface microbes before planting.

A3.3.3.1 Equipment

- Millipore filter unit with 0.45 µ or smaller pore size filter element connected to water faucet
- Four I-gallon jugs
- Six 100-mL graduated cylinders
- Laboratory shaker
- Clean air hood 48 inches wide

Figure A3.6 Shaker used for surface sterilizing eggs.

A3.3.3.2 Procedure

1. Filter 2,000-mL fresh water into a jug
2. Add 500 mL of 37% formaldehyde solution
3. Add 200 mL of cleaned eggs
4. Cap jug, lay it on its side and shake for 20 minutes on laboratory shaker (Figure A3.6)
5. Working under clean air hood, decant liquid from eggs in jug
6. Add 2,500 mL of fresh filtered water
7. Return to shaker for ten minutes
8. Decant liquid
9. Repeat steps 5 through 8
10. Place eggs in graduated cylinders
11. Leave eggs in cylinders under hood until they are mixed with planting solution

A3.3.4 Preparing Planting Solution

GFN 7a → The planting solution is a sugar syrup thickened with starch to the viscosity and specific gravity required to hold eggs in suspension for planting.

A3.3.4.1 Equipment

- Metric scales with at least 10-kg capacity
- 20- to 30-gallon steam kettle
- Eight to ten 3- to 4-quart jars with heat-resistant vacuum seal lids

A3.3.4.2 Procedure

1. Weigh 200-g corn starch, 9-kg water, and 2-kg sugar
2. Mix a little water with starch to make a thin paste
3. Add remaining water to sugar in kettle and heat to boiling
4. Add starch gradually with stirring and bring back to boiling
5. Remove from heat and strain through 16-mesh strainer
6. Pour the solution into heat-safe jars, cover, and autoclave for 20 minutes at 121°C or above
7. Cool jars slowly in autoclave, remove, and allow to cool to room temperature, then refrigerate until needed

A3.3.5 Mixing Eggs with Planting Solution

GFN7b → Mix one part of eggs to 5.5 parts planting solution under clean air hood just before eggs are needed for planting.

A3.3.5.1 Equipment

- Six to eight 3- or 4-quart jars
- Strainer (16-mesh)
- Clean air hood
- Graduated cylinders (100 mL)

A3.3.5.2 Procedure

1. Put sterile jar under hood and add 400-mL eggs that have been surface-sterilized and held under hood
2. Strain 2,200-mL planting solution through sterile strainer into jar
3. Add 8 2/3 mL 37% formaldehyde solution (1 mL for each 300 mL of egg-and-planting solution)
4. Cap jars and shake on a laboratory shaker or by hand until eggs are distributed evenly in a homogeneous mixture through the solution
5. Use the eggs-in-suspension mixture the same day it is prepared

A3.3.6 Sterilizing Sand

GFN7c → Masonry or mortar sand, sterilized and mixed with antimicrobial agents, is used to provide a thin cover for the eggs on the surface of the larval diet trays.

A3.3.6.1 Equipment

- Vibratory sifter with 16-mesh hardware cloth screen
- 6-gallon stock pots
- 3½- or 4-inch outside diameter metal pipes 12 inches long (one for each stock pot)
- 5-gallon heavy duty plastic jugs with screw on lids (one for each stock pot)
- Ball mill type roller with three decks, made to fit jugs
- Autoclave

*A3.3.6.2 Procedure_*

1. Dry sand in sun or in drying oven
2. Screen sand through hardware cloth sifter
3. Put metal pipe upright in the center of stockpot and place sand around it
4. Cover pot with aluminum foil. Fold foil tightly around top of pot
5. Autoclave for at least 6 hours with autoclave at 121°C or above
6. Cool stock pot in clean, dry place to 50°C or below
7. Working under clean air hood, weigh 20-kg sand
8. Put sand in plastic jug and mix with 1.25 g of streptomycin sulfate, 1.25 g of neomycin sulfate, and 6 g of phaltan (75%' Folpet by Ortho, Chevron Chemical Co., P. P. Sikorowski, unpublished data)
9. Cover jugs and put on roller unit for 1.5 hours to get uniform blend
10. Hold jugs unopened until needed

A3.3.7 Planting Eggs

GFN8 → The egg-planting operation has been mechanized successfully by combining several pieces of equipment into an integrated system (Figure A3.7).

A3.3.7.1 Equipment

- Formseal® in-line form-fill-seal machine that forms tray units from roll stock of plastic covers, seals, and trims the trays
- Laminar clean airflow wall module installed four feet from Formseal machine to provide clean air environment during planting
- Roll stock of high impact polystyrene plastic, 6 inches wide and 15–17 mils thick, for making trays
- Roll stock of Tyvek® with heat-sealing adhesive for covering
- Mateer-Burt Model 31—a semiliquid filler unit
- Cooler tunnel with refrigerated air at 0 ± 1°C
- Egg dispenser and planting unit (Figure A3.8):* Filmatic Model FKS filler unit (Spraying System's pneumatic atomizing spray nozzle with air tip No. 134255-5 and liquid tip No. 60100 operated automatically with 6–7 psi air pressure), air filter, air cylinders, spray shield, hose pinch clamp, and air valves
  a. Mateer-Burt Model 31—a granular filler unit
  b. Holding cart with 448 shelves, 896 tray capacity (Figure A3.9)*

A3.3.7.2 Procedure

1. Stock the automated system with materials:
   a. Put plastic and Tyvek roll stock on Formseal machinery
   b. Put jars with suspension of eggs in holder rack
   c. Put sterile sand mixed with chemicals in dispenser unit
2. Close switch to start machine that moves the plastic strip along process line to forming die and makes a continuous ribbon of trays, each with a 4.5 × 10.2 × .75 inches cavity
3. Pump larval diet from sterilizer to Mateer-Burt filler unit
4. Trays fill automatically with 185 mL of warm, sterile diet
5. Cooling tunnel congeals diet
6. Egg planter dispenses 4 mL of egg solution containing about 2,550 eggs over surface of diet

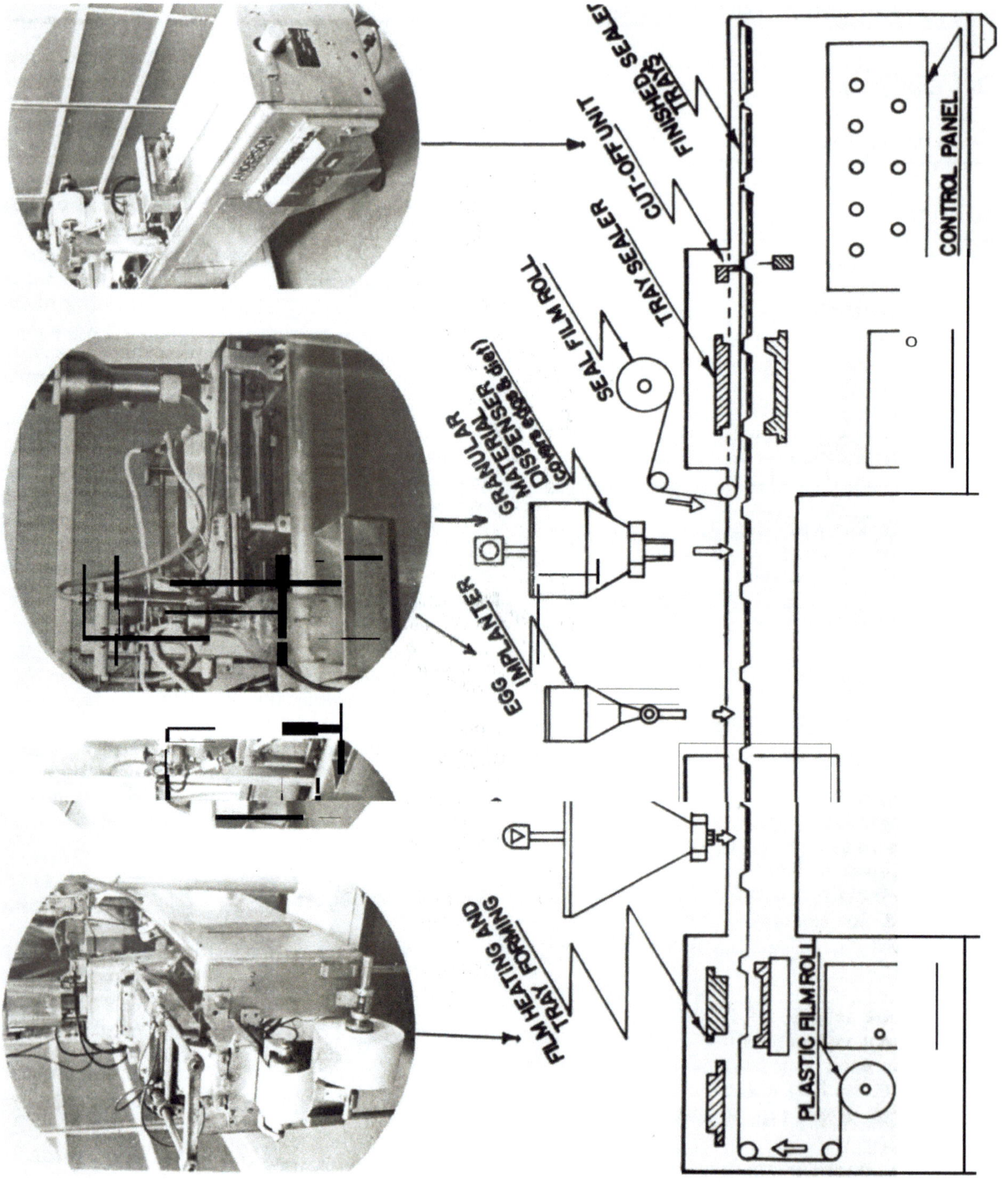

Figure A3.7 Egg extracting and planting device.

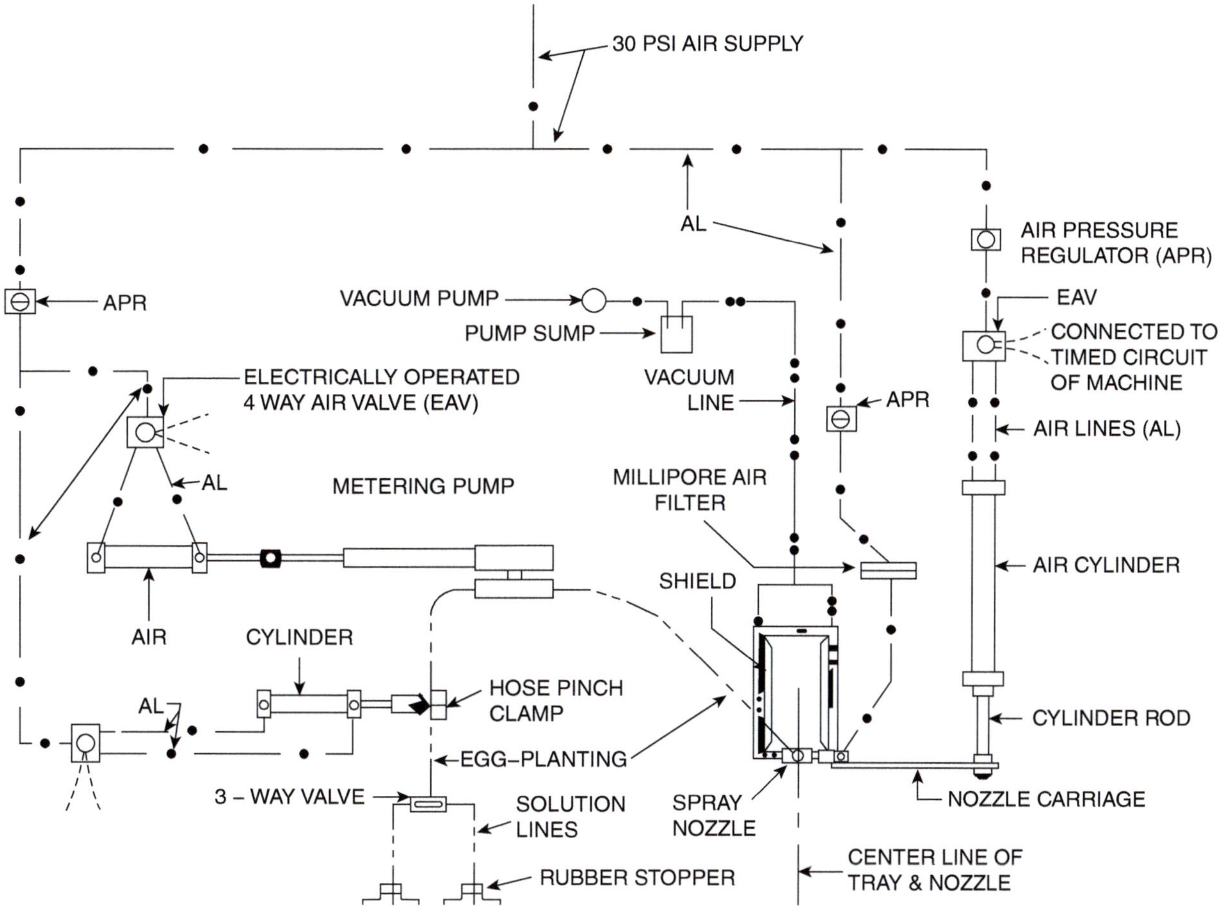

Figure A3.8 Diagram of egg-planting unit.

7. Granular filler dispenses 48 cc of sand containing antimicrobial agents
8. Trays are covered with Tyvek, heat-sealed and cut into double-tray units automatically
9. Take each double-tray unit off the assembly line and give it a rotary shake to distribute sand
10. Stack one double-tray unit per shelf on cart, starting at bottom of each stack column

Figure A3.9 Rearing tray holding cart.

A3.4 WEEVIL DEVELOPMENT AND COLLECTION

A3.4.1 Hatching Eggs and Developing Weevils

GFN9 → Stacked carts with planted trays are held for 13 days in a room with temperature constant at 30 ± 1°C humidity constant at 45–55% and continuous artificial light. Eggs will have hatched larvae developed and 15–20% of adults emerged by the end of 13 days.

A3.4.2 Collecting Adults

Carts are moved from development room to a transfer area on 13th day, or when 15–20% of the adults have developed. Covers are removed from the trays just before they are stacked in emergence boxes.

A3.4.2.1 Equipment

- Emergence box (Figure A3.10),* double-compartment cabinet with light-tight doors; each compartment (27 × 33 × 60 inches) has a four-inch slot cut from top to bottom. Shelf brackets in each compartment hold 23 removable solid metal shelves, each holding 10 double trays (with some overlap).
- Removable collecting cagemetal with 14-mesh screen sides that fit over compartment slots. Light through screen attracts emerging weevils into cage
- Stainless steel table (30 inches × 72 inches) with attached chute
- Emergence room with temperature and humidity control and continuous artificial light
- Gallon beakers

Figure A3.10 Emergence box.

* Detailed specifications are available from the Boll Weevil Research Laboratory, Mississippi State.

A3.4.2.2 Procedure

1. Begin collection when 15–20% of larvae have developed into adult weevils
2. Pull the Tyvek covers off the trays and stack trays in cabinet, IO double trays to a shelf
3. Set temperature control in emergence room at 30–32°C and humidity at 45–50%
4. Collect weevils from cage each morning for six or seven days:
   a. Remove cage from cabinet
   b. Overturn cage on table top and brush weevils from cage onto table
   c. Replace cage on cabinet
   d. Rake weevils into beaker in which a thin layer of petroleum jelly is spread in a 2-inch wide band around the top to keep weevils from crawling out
5. Weigh weevils and calculate number, based on 12 mg per weevil. Put adult weevils for egg production into lots of 9,000 for placement in oviposition cages. Divide remainder into groups for other intended uses
6. When diminishing numbers of collected insects indicate that collection period is complete, remove old trays from cabinet, put trays into heavy-duty paper refuse bags, sew openings closed, and destroy the bags
7. Wash boxes, spray with sanitizing solution, and air dry before reuse

A3.5 SANITATION†

GFN10 → Sanitation is vital to successful procedures to maintain maximum mass-rearing of boll weevils in the egg and weevil production and to laboratory. Following are minimal avoid microbial contamination and mortality.

A3.5.1 Materials (Sikorowski 1975)

1. 10% Chlorox® contains 5.25% sodium hypochlorite solution
2. 0.5% Mikro-Quat® solution (Klenzade product)
3. 1.25% Lysol® solution (National Laboratories, Lehn and Fink Industrial Products)
4. Surgi-Bac® Solution (Economics Laboratory, Inc.)

A3.5.2 Personnel Hygiene

1. Wash and rinse hands in Surgi-Bac Solution before beginning any laboratory procedure

† For more complete contamination and microbial control suggestions, see Sikorowski, P. P. **Microbiological Monitoring in the Boll Weevil Rearing Facility**, MAFES Technical Bulletin 71, April 1975.

2. Only authorized personnel allowed in egg-planting and larval development rooms:
   a. Change to laboratory clothes, shoes, and headgear before entering
   b. Wear mask over nose and mouth during planting operation
3. Wear clean clothes and shoes in all production areas

A3.5.3 Procedure for Sanitizing

Diet Sterilizer

1. Mix Klenzade HC-90®· powder and PL3® solution according to manufacturer's recommendations
2. Wash out lines with HC-90 immediately after each use
3. Reset the discharge valve to recirculate the HC-90 solution through the sterilizer for 30–45 minutes
4. Drain the first solution and recirculate PL3 solution through the sterilizer for 10–15 minutes
5. Drain and rinse with clear water
6. Take the sterilizer apart for drying, following manufacturer's directions

Pellet Machine

1. Immediately after use, remove wax vat and replace with waste collection tank
2. Pump sanitizing solution No. 2 or No. 3 from sterilizer through lines and tubes for 10–15 minutes. Leave lines and tubes filled with the solution between uses
3. Steam out lines and tubes just before using

Waxing Machine

1. Drain wax from wax vat and place in sterilizing kettle after each use
2. Remove wire belt and frame, remove excess wax, wash and spray with sanitizing solution
3. Clean wax vat and sanitize with solution No. 2 or No. 3
4. Replace belt and refill vat with sterile wax from wax warmer kettle

Movable Equipment (Oviposition Cages, Emergence Boxes) and Containers:

1. Wash after each use
2. Dip or spray with sanitizing solutions No. 2 or No. 3
3. Dry before storage or reuse

Table Tops and Stationary Equipment

1. Wash after each use or at the end of the day or more often if needed
2. Spray with sanitizing solution

Form-Fill-Seal Machine

3. Clean and spray surfaces with sanitizing solution after and just before each use

Diet Filler

1. Pump cleaning and sterilizing solution from sterilizer through line to filler for 10–15 minutes after each use
2. Dismantle after cleaning
3. Clean and spray hopper (inside and outside) with sanitizing solution
4. Clean pump parts and hold in sanitizing solution between uses
5. Reassemble and pump steam from sterilizer through lines and into hopper for 10–15 minutes just before each use

Cooling Tunnel

6. Wash and spray inside and outside with sanitizing solution after and before each use

Egg-Planting Unit

1. After use, dismantle parts that come in contact with planting solution
2. Wash parts and leave them in sanitizing solution between use
3. Reassemble and spray with sanitizing solution just before each use

Egg-Planting Room

1. Spray walls with sanitizing solution each day
2. Mop floor and spray with sanitizing solution after and before each operation
3. Filter incoming air through an absolute, high efficiency particulate air (HEPA) filter system
4. Spray and wash carts afterwards, put in separate room and fumigate with formaldehyde; leave carts in fumigation room overnight
5. Observe personnel sanitation

Larval Development Room

1. Mop floor with sanitizing solution each day
2. Filter incoming air through an absolute (HEPA) filter system
3. Observe personnel sanitation

Other Floors

4. Mop with sanitizing solution at the end of every day or more often, as needed

Diet Preparation Room

1. Use a separate building for storage, weighing, and premixing dry ingredients
2. Sweep floor daily
3. Mop floor weekly or more often as needed

A3.6 CONCLUSION

The system described was tested for six weeks in 1976 and for eight weeks in 1977. The production rate of about 5,058,000 weevils per week in 1976 increased to 6,243,000 weevils per week in 1977, after antimicrobial agents were added to the sand. Fewer than 1% of the trays had microbial contamination in the 1977 tests. Our results are conclusive that large numbers of boll weevils can be reared in the laboratory with currently available equipment and technology.

A3.7 REFERENCES

Betz, N. L. 1966. Improved laboratory methods for rearing the boll weevil. J. Econ. Entomol. 59: 374–376.

Brazzel, J. R., T. B. Davich, and Klaus Raven. 1959. Rearing boll weevils on an artificial diet. Texas Agricultural Experimental Station's Miscellaneous Publications 353.

Earle, N. W., R. C. Gaines, and J. S. Roussel. 1959. A larval diet for the boll weevil containing an acetone powder of cotton squares. J. Econ. Entomol. 52: 710–712.

Gast, R. T. 1961. Some shortcuts in laboratory rearing of boll weevils. J. Econ. Entomol. 54: 395–396.

Gast, R. T., and T. B. Davich. 1966. Boll weevils. In D. N. Smith (ed.), *Insect Colonization and Mass Production.* Academic Press, New York and London, pp. 405–419, 618.

Gast, R. T., and H. Vardell. 1963. Mechanical devices to expedite boll weevil rearing in the laboratory. US Agric. Res. Serv., ARS 33-89, 10.

Griffin, J. G., and O. H. Lindig. 1977. System for mechanical harvesting of boll weevil eggs from diet pellets. Trans. ASAE 20: 454–456.

Griffin, J. G., O. H. Lindig, and R. E. McLaughlin. 1974. Flash sterilizers: sterilizing diets for insects. J. Econ. Entomol. 67: 689.

Sikorowski, P. P. 1975. Microbiological monitoring in the Boll Weevil Rearing Facility. Miss. Agric. For. Exp. Stn. Tech. Bull. 71: 21.

Sikorowski, P. P., J. M. Wyatt, and O.H. lindig. 1977. Method for surface-sterilization of boll weevil eggs and the determination of microbial contamination of adults. Southw. Entomol. 2: 32–36.

Soloman, J. D., and R. D. Hunter. 1960. Laboratory rearing of the cotton boll weevil. Arkansas Farm Res. 9 (3): 5.

Vanderzant, E. S., and T. B. Davich. 1958. Laboratory rearing of the boll weevil: a satisfactory larvae diet and oviposition studies. J. Econ. Entomol. 51: 288–291.

Vanderzant, E. S., C. D. Richardson, and T. B. Davich. 1959. Feeding and oviposition by the boll weevil on artificial diets. J. Econ. Entomol. 52: 1138–1143.

A3.8 ADDENDA: UPDATED TO 2020 STANDARDS

A3.8.1 Appendix Gast Boll Weevil Clean Room Standards

Griffin 1984 stated that the standards for the clean areas in boll weevil mass-rearing must comply with Federal Standard 209 a. In concert with improvements in clean room technology, the standards have been revised several times since the Gast Facility was established, and as of 2021 when this book is being completed, the International Standards for Cleanrooms and Associated Controlled Environments (ISO 14644-1) are now the accepted standard for organizations that require highly clean air, such as hospitals and research laboratories.

A3.8.2 Notice of Cancellation FED-STD-209 Notice 1 (November 29, 2001)

A3.8.2.1 Federal Standard

A3.8.2.1.1 Airborne Particulate Cleanliness Classes in Cleanrooms and Clean Zones

Federal Standard 209E dated September 11, 1992 is hereby canceled and superseded by International Organization for Standardization (ISO) Standards. International Standards for Cleanrooms and associated controlled environments, ISO 14644-1 Part 1: Classification of air cleanliness; and ISO 14644-2 Part 2: Specifications for testing and monitoring to prove continued compliance with ISO 14644-1. Application for copies of ISO Standards 14644-1 Part 1, and 14644-2 Part 2; may be addressed to the Institute of Environmental Sciences and Technology (IEST), 940 East Northwest Highway, Mount Prospect, IL 60056-3444. Phone: 847-255-1561, Fax: 847-255-1699, Web site: www.iest.org, E-mail: publicationsales@iest.org.

Preparing Activity: GSA-FSS

FSC 3694

An example of these standards that are available in documents from various private companies such as Terra Universal.com and other companies under the auspices of IEST provide commercial services to help organizations requiring clean areas. The following tables are provided online by Terra Universal (Fullerton, California).

Airborne Particulate Cleanliness Classes (by cubic meter):						
CLASS	**Number of Particles per Cubic Meter by Micrometer Size**					
	0.1 micron	0.2 micron	0.3 micron	0.5 micron	1 micron	5 microns
ISO1	10	2				
ISO2	100	24	10	4		
ISO3	1,000	237	102	35	8	
ISO4	10,000	2,370	1,020	352	83	
ISO5	100,000	23,700	10,200	3,520	832	29
ISO6	1,000,000	237,000	102,000	35,200	8,320	293
ISO7				352,000	83,200	2,930
ISO8				3,520,000	832,000	29,300
ISO9				35,200,000	8,320,000	293,000

In cleanrooms, particulate concentration changes over time — from the construction and installation of equipment to its operational status. ISO delineates three cleanroom classification standards: as-built, at-rest and operational. As instruments and equipment are introduced and particulates rise, an "as-built" cleanroom becomes an "at-rest" cleanroom. When people are added to the matrix, particulate levels rise still further in the "operational" cleanroom.

ISO 14644-2 describes the type and frequency of testing required to conform to certain standards. The following tables indicate mandatory and optional tests.

Required Testing (ISO 14644-2)			
Schedule of Tests to Demonstrate Continuing Compliance			
Test Parameter	Class	Maximum Time Interval	Test Procedure
Particle Count Test	<= ISO 5	6 Months	ISO 14644-1 Annex A
	> ISO 5	12 Months	
Air Pressure Difference	All Classes	12 Months	ISO 14644-1 Annex B5
Airflow	All Classes	12 Months	ISO 14644-1 Annex B4

Optional Testing (ISO 14644-2)			
Schedule of Additional Optional Tests			
Test Parameter	Class	Maximum Time Interval	Test Procedure
Installed Filter Leakage	All Classes	24 Months	ISO 14644-1 Annex B6
Containment Leakage	All Classes	24 Months	ISO 14644-1 Annex B4
Recovery	All Classes	24 Months	ISO 14644-1 Annex B13
Airflow Visualization	All Classes	24 Months	ISO 14644-1 Annex B7

Today, in addition to ISO 14644-1 and ISO 14644-2, eight other cleanroom standards documents are being prepared. Many are in the final voting stage and can be legally used in the trade (see chart).

886 JOURNAL OF ECONOMIC ENTOMOLOGY *Vol. 61, No. 3*

Robert Theodore Gast

1923 - 1967

At 8:00 PM on Sunday, December 10, Bob Gast was severely burned in his laboratory at the Boll Weevil Research Laboratory, State College, Mississippi, in an ether vapor fire. He died at approximately 1:30 AM on Monday, December 18, 1967, in a hospital at Jackson, Mississippi. The attending doctor was optimistic about Bob's chances for survival through Saturday, December 16.

Bob was born April 14, 1923, in Cleveland, Ohio. His interest in entomology stemmed from childhood. While in grade school he repeated one of Fabré's classical experiments on the attraction of male moths to a newly emerged female moth. In grade school, his teachers often had Bob give lectures on insects to his schoolmates. He continued his early interest in entomology with studies at Ohio State University, where he earned B.S. (cum laude) and M.S. degrees and at Cornell University, which granted him the Ph.D. in 1951. He was on the staff of the Entomology Department at North Carolina State University, Raleigh, for 10 years before joining the staff of the Boll Weevil Research Laboratory, ARS, USDA, in 1961.

His major assignment at the laboratory was to develop a procedure for mass rearing the boll weevil. When Bob first came to the laboratory, boll weevil rearing was done manually at a cost of approximately $120.00 per thousand weevils. Bob developed, improved, and semi-automated techniques for the various steps involved in rearing boll weevils and reduced production costs to $5.00 per thousand. He had completed approximately 90% of a fully automated procedure that would have lowered production costs to less than $1.50 per thousand. He designed, built, and installed almost all the intricate equipment needed in automating the rearing procedure.

Bob was the recipient of two Certificates of Merit presented by the U.S. Department of Agriculture in June of 1965 and in May of 1967 in recognition of the outstanding performance of his duties. In addition, he received the Departments' Special Merit Award for outstanding cost reduction achievement in March 1967 from Secretary of Agriculture Orville Freeman in ceremonies in Washington, D.C.

Bob was consulted frequently by research personnel of many laboratories for advice on, or solutions to, various insect-rearing problems. Because of his inventiveness, ingenuity, and keen mind, his colleagues at the Boll Weevil Research Laboratory sought his advice on problems even though they were not related to his own immediate project. He always made good suggestions on approaches they might use in their research program.

Bob was a member of the Entomological Society of America, the American Institute of Biological Sciences, Sigma Xi, and Phi Kappa Phi.

He is survived by his widow, Barbara S., and two young children, Barbara Marie and Robert T. Jr., of State College, Miss.; his parents, Mr. and Mrs. Frederick C. Gast of Cleveland, Ohio; three brothers, Frederick C. Gast, Jr., of Blairsville, Pennsylvania, the Rev. William Gast of Denver, Colorado, and George L. Gast of Euclid, Ohio; and a sister, Miss Marie Gast of Ithaca, New York.

Bob's untimely death came as a shock to all his friends and associates. He was a devoted husband, father, and friend. His driving energy, keen mind, and sense of humor will be sorely missed by those who had the good fortune to know and work with him.

T. B. DAVICH
State College, Mississippi

References

Aceituno-Medina, M., C. Ventura, and E. Hernández. 2016. A novel approach for quality assurance of bulking agents for larval mass-rearing of *Anastrepha ludens*: mycotoxins content. In *9th Meeting of the Tephritid Workers of the Western Hemisphere*, Buenos Aires, Argentina. Abstract Book. p. 119. http://9twwh.senasa.gob.ar/en/meeting-documents.

Acker, L. 1963. Enzymic reactions in foods of low moisture content. In *Advances in Food Research*, E. M. M. C.O. Chichester and G. F. Stewart, Eds.Vol. 11, Academic Press, Inc., New York and London, pp. 263–330.

Adkisson, P. L., D. L. Bull, and W. E. Allison. 1960b. A comparison of certain artificial diets for laboratory cultures of the pink bollworm. *J. Econ. Entomol.* 53: 791–793.

Adkisson, P. L., E. S. Vanderzant, D. L. Bull, and W. E. Allison. 1960a. A wheat germ medium for rearing the pink bollworm. *J. Econ. Entomol.* 53: 759–762.

Alley, D. A., and B. G. Hightower 1966. Mating behavior of the screwworm fly as affected by differences in strain and size. *J. Econ. Entomol.* 59: 1499–1502.

Alverson, J., and A. C. Cohen. 2002. Effect of antifungal agents on biological fitness of *Lygus hesperus* Knight (Heteroptera: Miridae). *J. Econ. Entomol.* 95: 256–260.

Anagnostou, C., M. Dorsch, and M. Rolfs. 2011. Influence of dietary yeasts on *Drosophila melanogaster* life-history traits. *Entomol. Exp. Appl.* 136: 1–11. doi: https://doi.org/10.1111/j.1570-7458.2010.00997.x.

Anderson, P., and S. Anton. 2014. Experience-based modulation of behavioural responses to plant volatiles and other sensory cues in insect herbivores. *Plant Cell Environ.* 37: 1826–1835. doi: 10.1111/pce.12342.

Anderson, G. S., and S. L. VanLaerhoven. 1996. Initial studies on insect succession on carrion in southwestern British Columbia. *J. Forensic Sci.* 41: 617–625.

Anonymous. 2011. History of Sericulture (PDF). Government of Andhra Pradesh (India) - Department of Sericulture. file:///C:/Users/Allen/Desktop/1_History%20of%20Sericulture.pdfArchived from the original (PDF) on 21 July 2011. Retrieved 7 November 2010.

Anonymous. 2013. http://faculty.sites.uci.edu/chem1l/files/2013/11/RDGerroranal.pdf.

Anonymous. 1828. Manual and Report on the growth and manufacture of silk. Document Number 158. US Congress, House of Representatives. https://books.google.com.

Anonymous. 2015. http://yeastgrrl.blogspot.com/2015/02/the-fly-room-new-film-about-thomas-hunt.html.

Anonymous. 1999. Product quality control, irradiation and shipping procedures for mass-reared tephritid fruit flies for sterile insect release programmes. Joint FAO/IAEA Programme. Nuclear Techniques in Food and Agriculture. pp. 1–51. (Available at http://www.iaea.org/programmes/nafa/d4/public/d4 pbl 5 1.html).

Anonymous. 1922. Pebrine in silkworms. *Nature.* 109: 253–254 (reporting on a work by C. M. Hutchinson. 1920. Memoirs Dept. Agric. in India, Bacteriological Series. Vol. 1, No. 8, November, 1920, pp. 75, 26 plates).

Anonymous. 2019a. International Sericultural Commission. United Nations Reg. No. 10418. https://inserco.org/en/organizational-structure.

Anonymous. 2019b. A concept paper on the development of sericulture and silk industry in Latin American regions. In: International Sericultural Commission. United Nations Reg. No. 10418. https://inserco.org/en/organizational-structure.

Anonymous 2019c. http://www2.ece.rochester.edu/courses/ECE111/error_uncertainty.pdf.

Armstrong, P. https://cft.vanderbilt.edu/guideothers-sub-pages/blooms-taxonomy/. From Vanderbilt University's Center for Teaching.

Arunkumar, K. P., M. Muralidhar, and J. Nagaraju. 2006. Molecular phylogeny of silkmoths reveals the origin of domesticated silkmoth, *Bombyx mori* from Chinese *Bombyx mandarina* and paternal inheritance of *Antheraea proylei* mitochondrial DNA. *Mol. Phylogenet. Evol.* 40 (2): 419–427.

Assemi, H., M. Rezapanah, R. Vafaei-Shoushtari, and A. Mehrvar. 2012. Modified artificial diet for rearing of tobacco budworm, *Helicoverpa armigera*, using the Taguchi method and Derringer's desirability function. *J. Insect Sci.* 12 (100): 1–18.

Avosani, S., T. E. Sullivan, M. Ciolli, V. Mazzoni, and D. M. Suckling. 2020. Can vibrational playbacks disrupt mating or influence other relevant behaviours in bactericera cockerelli (Triozidae: Hemiptera)? *Insects.* 11: 299.

Babilonia, E. S., and D. L. Maki. (Website posted by FAO). Mass production of sterile New World screwworm flies in southern Mexico. Food and Agriculture Organization. http://www.fao.org/docrep/u4220t/u4220T0g.htm#facilities%20for%20mass%20production

Baird, D. C. 1962. *Experimentation: An Introduction to Measurement Theory and Experiment Design Prentice-Hall.* Upper Saddle River, NJ.

Bakri, A., K. Mehta and D. R. Lance. 2005. Sterilizing insects with ionizing radiation. In *Sterile Insect Technique. Principles and Practice in Area-wide Integrated Pest Management*, V. A. Dyck, J. Hendrichs, and A. S. Robinson, Eds. Springer, Dordrecht, The Netherlands, pp. 233–264. Chapter 3.3.

Balavenkatasubbaiah, M., and A. R. Narasimha Nayaka. 2008. *Technical Bulletin 38. How to Manage Silkworm Diseases during Rearing. Central Sericultural Research & Training Institute (ISO 9001: 2008 Certified).* Central Silk Board, Min. of Textiles. Govt. of India, Srirampura.

Banks, W. A., C. S. Lofgren, D. P. Jouvenaz, C. E. Stringer, P. M. Bishop, D. F. Williams, D. P. Wojcik, and B. M. Glancey. 1981. Techniques for collecting, rearing, and handling imported fire ants. USDA Technical Bulletin AATS-S-21, National Agricultural Library, USDA, ARS. 9p.

Barber, E. J. W. 1992. *Prehistoric Textiles: The Development of Cloth in the Neolithic and Bronze Ages with Special Reference to the Aegean* (reprint, illustrated Ed.). Princeton University Press, Princeton, NJ, p. 31. ISBN 978-0-691-00224-8.

Bartlett, A. 1994. Maintaining genetic diversity in laboratory colonies of parasites and predators. In *Applications of Genetics to Arthropods of Biological Control Significance*, S. Karl Narang, A. C. Bartlett, and R. M. Faust, Eds. CRC Press, Boca Raton, FL, pp. 133–146.

Bartlett, A. C. 1984. Genetic changes during insect domestication. In *Advances and Challenges in Insect Rearing*, E. G. King, N. C. Leppla, E. G. King, and N. C. Leppla, Eds. USDA, ARS, New Orleans, LA, pp. 2–8.

Bartlett, A. C. 1994. Maintaining genetic diversity in laboratory colonies of parasites and predators. In *Applications of Genetics to Arthropods of Biological Control Significance*, S. Karl Narang, A. C. Bartlett, and R. M. Faust, Eds. CRC Press, Boca Raton, FL, pp. 134–145.

Bartlett, A. C. 1985. Guidelines for genetic diversity in laboratory colony establishment and maintenance. In *Handbook of Insect Rearing*, P. Singh and R. F. Moore, Eds. Vol. I. Elsevier Science Publishers BV, Amsterdam, The Netherlands, pp. 7–18.

Baruah, B., and M. Gogoi. 2013. Economics of silk industry. *Inter. Res. J. Agric. Econ. Stat.* 4 (2): 196–199.

Baumberger, J. P. 1917a. The food of *Drosophila melanogaster* Meigen. *Proc Natl Acad. Sci. USA*. 3: 122–126.

Baumberger, J. P.1917b. Solid media for rearing drosophila. *Am. Nat.* 51: 447–448.

Baumberger, J. P.1919. A nutritional study of insects, with special reference to microorganisms and their substrata. *J. Exp. Zool.* 28 (1-2): 1–81.

Baumberger, J. P., and R. W. Glaser. 1917. The rearing of drosophila Loew on solid media. *Science*. 45: 21–22.

Baumhover, A. H. 1965. Sexual aggressiveness of male screwworm flies measured by effect on female mortality. *J. Econ. Entomol.* 58: 544–548.

Bebas, P., and B. Cymborowski. 1999. Effect of constant light on male sterility in the cotton leafworm *Spodoptera littoralis*. *Physiol. Entomol.* 24: 165–170.

Bebitha, B., P. Mohanraj, S. Manimegalai, and C. A. Mahalingam. 2016. Silkworm disease diagnosis through molecular approach and their management. *In.t J. Plant Protect.* 9 (1): 343–352.

Beck, S. D. 1970. Neural and hormonal control of pupation in *Galleria mellonella* (Lepidoptera: Galleriidae). *Ann. Entomol. Soc. Am.* 63 (1): 143–149.

Beck, S. D. 1960. Growth and development of the greater wax moth, *Galleria mellonella* (L.). *Trans. Wis. Acad. Sci. Arts Lett.* 49: 137–48.

Beckman, H. F., S. M. Bruckart, and R. Reiser.1953. Laboratory culture of the pink bollworm on chemically defined media. *J. Econ. Entomol.* 46 (4): 627–630.

Bell, R. A., and F. G. Joachim. 1976. Techniques for rearing laboratory colonies of tobacco hornworms and pink bollworms (Lepidoptera-Sphingidae-Gelechiidae. *Ann. Entomol. Soc. Am.* 69 (2): 365–373.

Bell, R. A., D. C. Owens, M. Shapiro, and J. R. Tardif. 1984. Development of mass rearing technology. In *The Gypsy Moth: Research toward Integrated Pest Management*, C. C. Doane and M. L. McManus, Eds. U.S. Dep. Agric. Tech. Bull., pp. 599–633.

Benbow, M. E., A. J. Lewis, J. K. Tomberlin, and J. L. Pechal. 2013. Seasonal necrophagous insect community assembly during vertebrate carrion decomposition. *J. Med. Entomol.* 50: 440–450.

Benedict, M. Q., B. G. J. Knols, H. C. Bossin, C. Herve, P. I. Howell, E. Mialhe, C. Caceres, and A. S. Robinson. 2009. Colonisation and mass rearing: learning from others. *Malar. J.* 8 (Suppl 2):S4 doi: 10.1186/1475-2875-8-S2-S4.

Bennett, R. R., and R. H. White. 1989. Influence of carotenoid deficiency on visual sensitivity, visual pigment and P-face particles of photoreceptor membrane in the moth *Manduca sexta*. *J. Comp. Physiol. A*. 164: 321–331.

Bexfield, A., A. E. Bond, C. Morgan, J. Wagstaff, R. P. Newton, N. A. Ratcliffe, E. Dudley, and Y. Nigam. 2010. Amino acid derivatives from *Lucilia sericata* excretions/secretions may contribute to the beneficial effects of maggot therapy via increased angiogenesis. *Br. J. Dermatol.* 162: 554–562.

Bianconi, A., C. J. Von Zuben, A. B. S. Serapião, and J. S. Govone 2010. Artificial neural networks: a novel approach to analysing the nutritional ecology of a blowfly species, *Chrysomya megacephala*. *J. Insect Sci.* 10:58. insectscience.org/10.58

Biedermann, P. H. W., and M. Rohlfs. 2017. Evolutionary feedbacks between insect sociality and microbial management. *Curr. Opin. Insect Sci.* 22: 92–100.

Bloem, K. A., and S. Bloem. 2000. SIT for codling moth eradication in British Columbia, Canada. In *Proceedings of Area-Wide Control of Fruit Flies and Other Insect Pests. International Conference on Area-Wide Control of Insect Pests, and the 5th International Symposium on Fruit Flies of Economic Importance*, K. H. Tan, Ed. Penang, Malaysia, 28 May–5 June 1998, pp. 207–214. (Cited by Dyck 2010)

Bloem, S., K. A. Bloem, J. E. Carpenter, and C. O. Calkins.1999a. Inherited sterility in codling moth (Lepidoptera: Tortricidae): effect of substerilizing doses of radiation on insect fecundity, fertility and control. *Ann. Entomol. Soc. Am.* 92: 222–229.

Bloem, S., K. A. Bloem, J. E. Carpenter, and C. O. Calkins. 1999b. Inherited sterility in codling moth (Lepidoptera: Tortricidae): effect of substerilizing doses of radiation on field competitiveness. *Environ. Entomol.* 28: 669–674.

Bloem, S., K. A. Bloem, J. E. Carpenter, and C. O. Calkins. 2001. Season-long releases of partially sterile males for control of codling moth, *Cydia pomonella* (Lepidoptera: Tortricidae), in Washington apples. *Environ. Entomol.* 30: 763–769.

Bloom, D. 2019. Henrietta Aiken Kelly (1844-1916): the correspondence of a special field agent. In *Southern Women of the Progressive Era*, A. Reader, G. Roberts, and M. Walker, Eds. University of South Carolina Press, Columbia, SC. ISBN-978-1-6117-925-5, pp. 141–179.

Bloom, B. S., M. D. Engelhart, E. J. Furst, W. H. Hill, and D. R. Krathwohl. 1956. Taxonomy of educational objectives: the classification of educational goals. *Handbook I: Cognitive Domain*. David McKay Company, New York, NY.

Bodenheimer, F. 1951. *Insects as Human Food.* Junk, The Hague.

Bogdanow, E. A.1906. About the breeding of larvae in usual flesh flies (*Calliphora vomitoria*) in sterilized nutrients. *Arch. Eur. J. Physiol.* 113: 97–105.

Bogdanow, E. A. 1908. About the dependence of the growth of larvae of bacteria and enzymes and about variability and inheritance in the flesh flies. *Arch. Anat. Physiol.* 1908 (Suppl.): 173–200.

Boggs, C. 1987. Ecology of nectar and pollen feeding in Lepidoptera. In *Nutritional Ecology of Insects, Mites, Spiders and Related Invertebrates*, F. Slansky, and J. G. Rodriguez, Eds. Wiley, New York, NY, pp. 369–391.

Boggs, C. L., and L. E. Gilbert. 1979. Male contribution to egg production in butterflies: evidence for transfer of nutrients at mating. *Science.* 206: 83–84.

Boggs, C. L. 1986. Reproductive strategies of female butterflies-variation in and constraints on fecundity. *Ecol. Entomol.* 11: 7–15.

Boggs, C. L. 1997. Dynamics of reproductive allocation from juvenile and adult feeding: radiotracer studies. *Ecology* 78: 192–202.

Boggs, C. L. 2003. Environmental variation, life histories, and allocation. In *Butterflies: Ecology and Evolution Taking Flight*, C. L. Boggs, W. B. Watt, and P. R. Ehrlich, Eds. University of Chicago Press, Chicago, IL, pp. 185–206.

Boggs, C. L. 2009. Understanding insect life histories and senescence through a resource allocation lens. *Funct. Ecol.* 23: 27–37.

Boller, E. F., and C. O. Calkins. 1984. Measuring, monitoring and improving the quality of mass-reared Mediterranean fruit flies, Ceratitis capitata (Weid.). *Improvement of quality by selection. Zeitschrift für angewandte Entomologie* 98: 1–15.

Boller, E. F., and D. L. Chambers. 1977. Quality control: an idea book for fruit fly workers. *IOBC/WPRS Bull.* 1977 (5): 162.

Boller, E. F., B. I. Katsoyannos, U. Remund, and D. L. Chambers. 1981. Measuring, monitoring, and improving the quality of mass-reared Mediterranean fruit flies, Ceratitis capitata Wied. 1. The RAPID quality control system for early warning. *Z. Angew. Entomol* 92: 67–83.

Brazzel, J. R., and L. D. Newsom. 1959. Diapause in anthonomus grandis boh. *J. Econ. Entomol.* 52 (4): 603–611. doi:10.1093/jee/52.4.603. ISSN 1938-291X.

Brazzel, J. R., C. O. Calkins, D. L. Chambers, and D. B. Gates. 1986. Required quality control tests, quality specifications, and shipping procedures for laboratory produced Mediterranean fruit flies for sterile insect control programs, APHIS 81–51, USDA-APHIS, Hyattsville, MD.

Bridges, C. B., and H. H. Darby. 1933. Culture media for Drosophila and the pH of media. *Am. Nat.* 67 (712): 437–472.

Bridges, C. B. 1932. Apparatus and methods for Drosophila culture. *Am. Nat.* 66 (704): 250–273.

Bridges, C. B. 1913. Non-disjunction of the sex chromosomes of Drosophila. *J. Exp. Zool.* 15 (4): 587–606.

Bridges, C. B. 1925. Elimination of chromosomes due to a mutant (minute-N) in *Drosophila melanogaster. Proc. Natl. Acad. Sci.* 11: 701–706.

Brinton, F. E., M. D. Proverbs, and B. E. Carty. 1969. Artificial diet for mass production of the codling moth, *Carpocapsa pomonella* (Lepidoptera: Olethreutidae). *Can. Entomol.* 101: 577–584.

Brooks, M. A. 1984. Academic training for insect rearing. In *Advances and Challenges in Insect Rearing*, E. G. King, and N. C. Leppla, Eds. USDA, ARS, New Orleans, LA, pp. 261–263.

Brown, H. E. 1984. Mass production of screwworm flies, *Cochliomyia hominivorax.* In *Advances and Challenges in Insect Rearing*, E. G. King, and N. C. Leppla, Eds. USDA, ARS, New Orleans, LA, USA, pp. 193–199.

Bruch, M. K., and F. W. Smith. 1968. Improved method for pouring Rodac plates. *Appl. Microbiol.* 16 (9): 1427–1428.

Bruell, J. H. 1967. Behavioral heterosis. In *Behavior-Genetic Analysis*, J. Hirsch, Ed. McGraw-Hill, New York, NY, pp. 27–86, 522.

Burgess, A. F. 1908. Description of new devices for rearing insects. *J. Econ. Entomol.* 1: 267–269. doi: https://doi.org/10.1093/jee/1.4.267.

Bush, G. L., and R. W. Neck. 1976. Ecological genetics of screwworm fly. *Cochliomyia hominivorax* (Diptera-Calliphoridae) and its bearing on quality-control of mass-reared insects. *Environ. Entomol.* 5 (5): 821–826.

Bush, G. L. 1975. Genetic variation in natural insect populations and its bearing on mass-rearing programmes. In *Controlling Fruit Flies by the Sterile Insect Technique*, IAEA Panel Proc. Ser., Vienna, pp. 9–17, 175.

Bush, G. L., R. W. Neck, and G. B. Kitto. 1976. Screwworm eradication: inadvertent selection for noncompetitive ecotypes during mass rearing. *Science.* 193 (4252): 491–493.

Bushland, R. C., A. W. Lindquist, and E. F. Knipling. 1955. Eradication of screw-worms through release of sterilized males. *Science.* 122 (3163): 287–288.

Butler, G. D., T. J. Henneberry, and T. Clayton. 1983. Bemisia-tabaci (Homoptera, Aleyrodidae) - development, oviposition, and longevity in relation to temperature. *Ann. Entomol. Soc. Am.* 76 (2): 310–313.

Butler, G. D. 1966. Development of several Predaceous Hemiptera in relation to temperature. *J. Econ. Entomol.* 59: 1306–1310. doi: https://doi.org/10.1093/jee/59.5.1306.

Büyükgüzel, K., E. Büyükgüzel, E. Chudzińska, A. Lewandowska-Wosik, R. Gaj, and Z. Adamski. 2020. *Drosophila melanogaster* response to feeding by neomycin-based medium expressed in fluctuating asymmetry. *Insects. 11*(6), 1-17; https://doi.org/10.3390/insects11060378

Byrd, J. H., and J. L. Castner. 2009. Insects of forensic importance. In *Forensic Entomology: The Utility of Arthropods in Legal Investigations*, J. H. Byrd and J. L. Castner, Eds. 2nd Edition. CRC Press, Boca Raton, FL, p. 681.

Cáceres, C., P. Rendón, and J. Andrew. 2012. The FAO/IAEA Spreadsheet for Designing and Operating Insect Mass-rearing Facilities. FAO Plant Production and Protection Paper 205.

Cáceres, C., and P. Rendón. 2008. The FAO/IAEA interactive spreadsheet for design and operating insect mass-rearing facilities. In *Proceedings of the 7th International Symposium on Fruit Flies of Economic Importance*, R. Sugayama, R. Zucchi, S. Ovruski, and J. Sivinski, Eds. Bahia, Brazil, 2006, pp. 307–312.

Caceres, C. P. R., and A. Jessup. 2012. The FAO/IAEA Spreadsheet for Designing and Operating Insect Mass-rearing Facilities: Procedures Manual. Sponsored by the Joint FAO/IAEA Programme of Nuclear Techniques in Food and Agriculture. ISBN 978-92-5-107315-5 ISSN 0259-2517.

Caceres, C., J. P. Cayol, W. Enkerlin, G. Franz, J. Hendrichs, A. S. Robinson. 2004. Comparison of Mediterranean fruit fly (*Ceratitis capitata*: Tephritidae) bisexual and genetic sexing strains: development, evaluation and economics. In *Proceedings, Symposium: 6th International Symposium on Fruit Flies of Economic Importance*, B. N. Barnes, Ed. 6–10 May 2002, Stellenbosch, South Africa. Isteg Scientific Publications, Irene, South Africa, pp. 367–381.

Carroll, L. E., A. L. Norrbom, F. Christian Thompson, and N. L. Evenhuis. 2019. Fruit fly bibliography. (contains 5,368 references on Tephritidae).

Calkins C. O., W. Klassen, and P. Liedo, Eds. 1994. *Fruit Flies and the Sterile Insect Technique*. CRC Press, Boca Raton, FL.

Calkins, C. O., and A. G. Parker. 2005. Sterile insect quality. In *Sterile Insect Technique Principles and Practice in Area-Wide Integrated Pest Management*, V. A. Dyck, J. Hendrichs, and A. S. Robinson, Eds. AA Dordrecht, The Netherlands, pp. 269–296.

Calkins, C. O., A. L. Knight, G. Richardson, K. A. Bloem. 2000. Area-wide population suppression of codling moth. In *Proceedings: Area-Wide Control of Fruit Flies and Other Insect Pests. International Conference on Area-Wide Control of Insect Pests, and the 5th International Symposium on Fruit Flies of Economic Importance*, K. H. Tan, Ed. Penang, Malaysia, 28 May–5 June 1998, pp. 215–219.

Calkins, C. O., K. Bloem, S. Bloem, and D. L. Chambers. 1994. Advances in measuring quality and assuring good field performance in mass reared fruit flies. In *CDC Guidelines for Sanitation: Excellent Resource for Lab Cleaning*, pp. 85–96.

Callier, V., and H. F. Nijhout. 2011. Control of body size by oxygen supply reveals size-dependent and size-independent mechanisms of molting and metamorphosis. *Proc. Natl. Acad. Sci. USA*. 108: 14664–14669.

Cammack, J. A., A. C. Cohen, K. L. Kreitlow, R. M. Roe, and D. W. Watson. 2016. Decomposition of concealed and exposed porcine remains in the North Carolina piedmont. *J. Med. Entomol.* 53 (1): 67–75.

Cappellozza, L., S. Cappellozza, A. Saviane, and G. Sbrenna. 2005. Artificial diet rearing system for the silkworm *Bombyx mori* (Lepidoptera: Bombycidae): effect of vitamin C deprivation on larval growth and cocoon production. *Appl. Entomol. Zool.* 40 (3): 405–412.

Carrington, D. 2011. Insects could be the key to meeting food needs of growing global population, The Guardian 1 August 2010. Retrieved 27 February 2011.

Cazander, G., K. E. B. Van Veen, A. T. Bernards, and G. N. Jukema. 2009. Do maggots have an influence on bacterial growth? A study on the susceptibility of strains of six different bacterial species to maggots of Lucilia sericata and their excretions/secretions. *J. Tissue Viability*. 18 (3): 80–87. doi:10.1016/j.jtv.2009.02.005. PMID 19362001.

Cazander, G., M. W. J. Schreurs, L. Renwarin, C. Dorresteijn, Dörte Hamann, and G. N. Jukema. 2012. Maggot excretions affect the human complement system. *Wound Repair Regen.* 20 (6): 879–86. doi:10.1111/j.1524-475X.2012.00850.x.

CDC 2021. Guidance for cleaning and disinfecting public spaces, workplaces, businesses, schools, and homes guidelines for sanitation. https://www.cdc.gov/coronavirus/2019-ncov/community/reopen-guidance.htmlexcellent resource for lab cleaning..

CDC. (2021) Handwashing: Clean hands save lives. https://www.cdc.gov/handwashing/index.html.

Chabora, P. C. 1969. Mutant genes and the emigration behavior of *Phaenicia sericata*. *Evolution*. 23: 65–71.

Chambers, D. 1977. Quality control in mass rearing. *Ann. Rev. Entomol.* 22: 289–308.

Chambers, D. L. 1975. Quality in mass-produced insects: definition and evaluation. In *Controlling Fruit Flies by the Sterile-Insect Technique - Panel Proceedings Series*. International Atomic Energy Agency, Vienna, Austria, STI/Pub/392, pp.19–32

Chang, C. L. 2009. Fruit fly liquid larval diet technology transfer and update. *J. Appl. Entomol.* 133: 164–173.

Chang, C. L., C. Cáceres, and E. B. Jang. 2004. A novel liquid larval diet and its rearing system for melon fly, *Bactrocera cucurbitae* (Diptera: Tephritidae). *Ann. Entomol. Soc. Am.* 97: 524–528.

Chang, C. L., R. I. Vargas, C. Cáceres, E. Jang, and I. L. Cho. 2006. Development and assessment of a liquid larval diet for *Bactrocera dorsalis* (Diptera: Tephritidae). *Ann. Entomol. Soc. Am.* 99: 1191–1198.

Chang, H. T. 1946. Studies on the use of fluorescent dyes for marking anopheles quadrimaculatus. *Mosquito News*. 6: 122–125.

Chaudhury, M. F., and L. A. Alvarez. 1999. A new starch-grafted gelling agent for screwworm (Diptera: Calliphoridae) larval diet. *J. Econ. Entomol.* 92: 1138–1141.

Chaudhury, M. F., L. A. Alvarez, and J. B. Welch. 1998. An alternative source of blood protein for screwworm (Diptera: Calliphoridae) larval diet. *J. Econ. Entomol.* 91: 1397–1400.

Chaudhury, M. F., L. A. Alvarez, and L. L. Velazquez. 2000 A new meatless diet for adult screwworm (Diptera: Calliphoridae). *J. Econ. Entomol.* 93: 1398–1401.

Chaudhury, M. F., S. R. Skoda, and A. Sagel. 2011. Solidifying agent and processing of blood used for the larval diet affect screwworm (Diptera: Calliphoridae) life-history parameters. *J. Econ. Entomol.* 104 (3): 1103–1107.

Chen, M., N. Zhang, H. Jiang, X. Meng, K. Qiang, and J. Wang. 2019. Transcriptional regulation of *heat shock protein 70* genes by class I histone deacetylases in the red flour beetle, *Tribolium castaneum*. *Insect Mol. Biol.* https://doi.org/10.1111/imb.12627.

Chen, W., and J. F. Hillyer. 2013. FlyNap (triethylamine) increases the heart rate of mosquitoes and eliminates the cardioacceleratory effect of the neuropeptide CCAP. *PLoS One*. 8 (7): e70414. https://doi.org/10.1371/journal.pone.0070414.

Claret, J., and N. Volkoff. 1992. Vitamin-A is essential for 2 processes involved in the photoperiodic reaction in *Pieris brassicae*. *J. Insect Physiol.* 38 (8): 569–574.

Cohen, A. C. 1981. An artificial diet for *Geocoris punctipes* (Say). *Southwest Entomol.* 6: 109–113.

Cohen, A. C. 1982. Water and temperature relations of two hemipteran members of a predator-prey complex. *Environ. Entomol.* 11: 715–719.

Cohen, A. C. 1983a. Improved method of encapsulating artificial diet for rearing predators of harmful insects. *J. Econ. Entomol.* 76: 957–959.

Cohen, A. C. 1983b. A simple, rapid and highly sensitive method of separation and quantification of uric acid, hypoxanthine and xanthine by HPLC. *Experientia.* 39: 435–436.

Cohen, A. C. 1984. Food consumption, food utilization and metabolic rates of *Geocoris punctipes* fed Heliothis virescens eggs. *Entomophaga.* 29: 361–367.

Cohen, A. C. 1985a. Simple method for rearing the insect predator *Geocoris punctipes* (Heteroptera: Lygaeidae) on a meat diet. *J. Econ. Entomol.* 78: 1173–1175.

Cohen, A. C. 1985b. Metabolic rates of two hemipteran members of a predator-prey complex. *Comp. Biochem. Physiol.* 81A: 833–836.

Cohen, A. C. 1989. Ingestion efficiency and protein consumption by a heteropteran predator. *Ann. Entomol. Soc. Am.* 82: 495–499.

Cohen, A. C. 1990. Feeding adaptations of some predaceous Hemiptera. *Ann. Entomol. Soc. Am.* 83: 1215–1223.

Cohen, A. C. 1992. Using a systematic approach to develop artificial diets for predators. In *Advances in Insect Rearing for Research and Pest Management*, T. E. Anderson, and N. C. Leppla, Eds. Westview Press, Boulder, Co, pp. 77–91.

Cohen, A. C. 1993. Organization of proteinases and preliminary characterization of salivary trypsin-like enzymes in the predaceous heteropteran. *J. Insect Physiol.* 39: 823–829.

Cohen, A. C. 1995. Extra-oral digestion in predatory arthropods. *Ann. Rev. Entomol.* 40: 85–103.

Cohen, A. C. 1998a. Artificial media for rearing entomophages comprising cooked, whole egg. US Patent # 5,834,177. November 10, 1998.

Cohen, A. C. 1998b. Solid-to-liquid feeding: the inside(s) story of extra-oral digestion in predaceous Arthropoda. *Am. Entomol.* 44: 103–116.

Cohen, A. C. 1999. Robert Frost's arthropods. *Am. Entomol.* 45 (2): 70–72.

Cohen, A. C. 2000a. Feeding fitness and quality of domesticated and feral predators: effects of long-term rearing on artificial diet. *Biol. Control.* 17: 50–54.

Cohen, A. C.2000b. A review of feeding studies of Lygus spp. with emphasis on artificial diets. *Southwestern Entomol.* Suppl. No 23: 111–119.

Cohen, A. C. 2000c. New oligidic production diet for *Lygus hesperus* Knight and L. lineolaris (Palisot de Beauvois). *J. Entomol. Sci.* 35: 301–310.

Cohen, A. C. 2001. Formalizing insect rearing and artificial diet technology. *Am. Entomol.* 47: 198–206.

Cohen, A. C. 2003a. Artificial diet for arthropods. U.S. Patent 6: 506,597. Jan. 14, 2003.

Cohen, A. C. 2003b. *Insect Diets: Science and Technology.* CRC Press, Boca Raton, FL.

Cohen, A. C. 2015. *Insect Diets: Science and Technology.* 2nd Edition. CRC Press. Boca Raton, FL.

Cohen, A. C. 2018a. Ecology of insect rearing systems: a mini-review of insect rearing papers from 1906-2017. *Adv. Entomol.* 6: 86–115. doi: https://doi.org/10.4236/ae.2018.62008.

Cohen, A. C. 2018b. Standards for effective insect rearing science and technology papers. *Adv. Entomol.* 6: 256–284.

Cohen, A. C. 2019. The nature of unnatural insects: infrastructure of insect rearing. *Am. Entomol.* 65 (2): 122–135.

Cohen, A. C., and C. G. Jackson. 1986. Ontogenetic changes in waste purines from eggs of two species of insects. *Comp. Biochem. Physiol.* 85B: 269–271.

Cohen, A. C., and J. D. Pinto. 1977. An evaluation of xeric adaptiveness of several species of blister beetles (Meloidae). *Ann. Ent. Soc. Am.* 70: 741–749.

Cohen, A. C., and J. L. Cohen. 1976. Nest structure and microclimate of the desert cockroach *Arenivaga apacha. Bull. Southern California Acad. Sci.* 75: 272–277.

Cohen, A. C., and J. W. Debolt. 1983. Rearing *Geocoris punctipes* on insect eggs. *Southwest Entomol.* 8: 61–64.

Cohen, A. C., and R. T. Staten. 1994. Long term culturing and quality assessment of predatory big-eyed bugs, *Geocoris punctipes. In Applications of Genetics to Arthropods of Biological Control Significance*, S. Karl Narang, A. C. Bartlett, and R. M. Faust, Eds. CRC Press, Boca Raton, FL, pp. 122–132.

Cohen, A. C., and L. Smith. 1998. A novel concept in artificial diets for *Chrysoperla rufilabris*: the efficacy of solid diet. *Biol. Control.* 13: 49–54.

Cohen, A. C., and P. Crittenden. 2004. Deliberately added and "Cryptic" antioxidants in three artificial diets for insects. *J. Econ. Entomol.* 97 (2). 265–272.

Cohen, A. C., and R. Patana. 1982. Ontogenetic and stress-related changes in the hemolymph chemistry of beet armyworms. *Comp. Biochem. Physiol.* 71A: 193–198.

Cohen, A. C., and R. Patana. 1984. Efficiency of food utilization by Heliothis zea (Lepidoptera: Noctuidae) fed artificial diets or green beans. *Can. Entomol.* 116: 139–146.

Cohen, A. C., and R. Patana. 1985. Chemical composition of tobacco budworm eggs during development. *Comp. Biochem. Physiol.* 81B: 165–169.

Cohen, A. C. C. C., K. Kidd, and T. Hodgson. 2011. Defining PC/QC standards for mass-rearing HWA predators. In *Implementation and Status of Biological Control of the Hemlock Woolly Adelgid*, Brad Onken and Richard Reardon, Eds. U.S. Forest Service Publication, pp. 139–147.

Cohen, A. C., R. B. March, and J. D. Pinto. 1981. Water relations of the desert blister beetle *Cysteodemus armatus. Physiol. Zool.* 54: 179–187.

Cohen, A. C., R. B. March, and J. D. Pinto. 1986. Effects of water stress and rehydration on hemolymph volume and amino acid content in the blister beetle, *Cysteodemus armatus. Comp. Biochem. Physiol.* 85A: 743–746.

Cohen, A. C., and C. G. Jackson. 1989. Using rubidium to mark a predator (Hemiptera: Lygaeidae). *J. Entomol. Sci.* 24: 57–61.

Cohen, A. C., and J. D. Pinto. 1977. An evaluation of the xeric adaptiveness of several species of blister beetles (Meloidae). *Ann. Entomol. Soc. Am.* 70: 741–749.

Cohen, A. C., and N. M. Urias. 1986. Meat-based artificial diets for *Geocoris punctipes. Southwest. Entomol.* 11: 171–176.

Cohen, A. C., and N. M. Urias. 1988. Nutritional indices of a predator *Geocoris punctipes* (Heteroptera: Lygaeidae) fed artificial diets with rutin. *J. Entomol. Sci.* 23: 174–179.

Cohen, A. C., and R. T. Staten.1994. Long-term culturing and quality assessment of predatory big-eyed bugs, *Geocoris punctipes.* In *Applications of Genetics to Arthropods of Biological Control Significance*, S. Karl Narang, A. C. Bartlett, and R. M. Faust, Eds. CRC Press, Boca Raton, FL, pp. 122–132.

Cohen, A. C., R. A. Smith, and D. K. Harsh. 2004. A device for stretching Parafilm for presenting artificial diet to insects. United States Patent 6550419. Application Number: 09/944343, Publication Date: 04/22/2003

Cohet, Y., and J. Bouletre. 1974. Epigenetic influences upon reproduction of an insect-variation of copulation reactivity of females Drosophila in relation to their growth temperature. *Comptes Rendus Hebdomadaires des seances de la Academie des sciences Serie D.* 278 (25): 325–328.

Collins, S. R., C. W. Weldon, and P. W. Taylor. 2008. Effects of field cage colour and supplementary shade on environmental conditions and mating behaviour of Queensland fruit flies, *Bactrocera tryoni. Entomol. Exp. Appl.* 129 (2): 142–147.

Conchou, L. P. L., C. Meslin, M. Proffit, M. Staudt, and M. Renou. 2019. Insect odorscapes: from plant volatiles to natural olfactory scenes. *Front. Physiol.* 10: 972. doi:. 10.3389/fphys.2019.00972.

Connor, E. F. 2006. Effects of the light environment on oviposition preference and survival of a leaf-mining moth, *Cameraria hamadryadella* (Lepidoptera: Gracillariidae), on *Quercus alba* L. *Ecol. Entomol.* 31 (2): 179–184.

Cook, M. 2019. www.wormspit.com.

Cothran, V. R., and G. G. Gyrisco. 1966. A container for rearing phytophagous insects with potential application to controlled humidity experiments. *J. Econ. Entomol.* 59 (2): 481.

Cotter, S. C., S. J. Simpson, D. Raubenheimer, and K. Wilson. 2011. Macronutrient balance mediates trade-offs between immune function and life history traits. *Funct. Ecol.* 25: 186–198.

Crane, E. 1999. *The World History of Beekeeping and Honey Hunting.* Routledge, New York, NY.

Crombie, A. C. 1945. On competition between species of graminivorous insects. *Proc. R. Soc. Ser. B-Biol. Sci.* 132 (569): 362–395.

Cui, S. F., L. Wang, J. P. Qiu, X. Q. Geng, and Z. C. Liu. 2019. Effects of hypoxida/hypercapnia on the metabolism of *Callosobruchus chinensis* (L.) larvae. *J. Stored Prod. Res.* 83: 322–330.

Dahlman, D. L. 1969. Cuticular pigments of tobacco hornworm (Manduca sexta) larvae: effects of diet and genetic differences. *J. Insect Physiol.* 15: 807–814.

Damodaran, S. 2008. Amino acids, peptides, and proteins. In *Fennema's Food Chemistry,* S. Damodaran, K. L. Parkin, and O. R. Fennema, Eds. 4th Edition. CRC Press, Boca Raton, FL.

Darwin, C. 1859. On The Origin of Species by Means of Natural Selection (or the Preservation of Favoured Races in the Struggle for Life). London. John Murray, Albemarle Street.

Dattel, E. R. 2019. Mississippi Historical Society. Mississippi History Now, a publication of the Mississippi Society http://mshistorynow.mdah.state.ms.us/articles/291/cotton-and-the-civilwar

Davich, T. B. 1968. Robert Theodore Gast 1923–1967. *J. Econ. Entomol.* 61 (3): 886. https://doi.org/10.1093/jee/61.3.886.

Davich, T. B. 1984. Introduction. In *Boll Weevil Mass Rearing Technology*, P. P. Sikorowski, J. G. Griffin, J. R. Roberson, and O. H. Lindig, Eds. University Press of Mississippi, Jackson, MS.

Lide, D. R. Ed. *CRC Handbook of Chemistry and Physics*, Internet Version 2005, <http://www.hbcpnetbase.com>, CRC Press, Boca Raton, FL, 2005. If a specific table is cited, use the format: Physical Constants of Organic Compounds.

DeBach, P. 1974. *Biological Control by Natural Enemies.* Cambridge University Press, New York, NY. 323.

Debolt, J. 1982. Meridic diet for rearing successive generations of *Lygus hesperus. Ann. Entomol. Soc. Am.* 75: 119–122.

DeFoliart, G. R. 1999. Insects as food. *Annu. Rev. Entomol.* 44: 21–50.

Defoliart, G. R. 1975. Insects as a source of protein. *Bull. Entomol. Soc. Am.* 21: 161–163.

Delcourt, A., and E. Guyenot. 1910. The possibility of studying certain Diptera in a defined environment. *Comptes rendus hebdomadaires des séances de l'Académie des sciences (0001-4036).* 151: 255–257.

Deming, W. E. 1982. Out of the Crisis: Quality, Productivity and Competitive Position. ISBN 0-521-30553-5.

Deming, W. E. 1964 [1943]. Statistical Adjustment of Data. Dover. ISBN 0-486-64685-8. LCCN 64-24416.

Deming, W. E. 1966 [1950]. Some Theory of Sampling. Dover. ISBN 0-486-64684-X. LCCN 66-30538.

Denlinger, D. L. 2002. Regulation of diapause. *Annu. Rev. Entomol.* 47: 93–122.

Denlinger, D. L., and R. E. Lee. 1998. Physiology of cold sensitivity. In *Temperature Sensitivity in Insects and Application in Integrated Pest Management*, G. J. Hallman and D. L. Denlinger, Eds. Westview Press, Inc., Boulder, CO.

Denlinger, D. L., R. E. Lee, G. D. Yocum, and O. Kukal. 1992. Role of chilling in the acquisition of cold tolerance and the capacitation to express stress proteins in diapausing pharate larvae of the gypsy moth, *Lymantria dispar. Arch. Insect Biochem. Physiol.* 21: 271–280.

Deshpande, S. A., R. Yamada, C. M. Mak, B. Hunter, A. S. Obando, S. Hoxha, and W. W. Ja. 2015. Acidic food pH increases palatability and consumption and extends drosophila lifespan. *J. Nutr.* 145, 2789–2796. doi: https://doi.org/10.3945/jn.115.222380.

Dethier, V. G. 1941. Chemical factors determining the choice of food plants by Papilio Larvae. *Am. Nat.* 75 (756): 61–73.

Dethier, V. G. 1962. *To Know a Fly.* Holden-Day, San Francisco, CA.

Dethier, V. G. 1976. *The Hungry Fly.* Harvard University Press, Cambridge, Mass London.

Dethier, V. G., and D. Bodenstein1958. Hunger in the blowfly. *Z. Tierpsychol.* 15: 129–140.

Dethier, V. G., and M. T. Yost. 1979. Oligophagy and absence of food-aversion learning in tobacco hornworms, *Manduca sexta. Physiol. Entomol.* 4: 125–130.

Dethier, V. G., and E. Bowdan. 1984. Relations between differential threshold and sugar receptor mechanisms in the blowfly. *Behav. Neurosci.* 98: 79–803.

Dethier, V. G., and L. E. Chadwick. 1947. Rejection thresholds of the blowfly for a series of aliphatic alcohols. *J. Gen. Physiol.* 30: 247–253.

Dethier, V. G., D. R. Evans, and M. V. Rhoades. 1956. Some factors controlling the ingestion of carbohydrates by the blowfly. *Biol. Bull.* 111: 204–222.

Diallo, S., M. T. Seck, J. B. Rayaisse, A. G. Fall, M. D. Bassene, B. Sall, A. Sanon, M. J. B. Vreysen, P. Takacl, A. G. Parker, G. Gimonneauet, and J. Bouyer. 2019. Chilling, irradiation and transport of male Glossina palpalis gambiensis pupae: effect on the emergence, flight ability and survival. *PLoS One.* 14 (5): e0216802. https://doi.org/10.1371/journal.pone.0216802.

Dijksterhuis, J., J. Houbraken, and R. A. Samson. 2013. F. Kempken. In *Agricultural Applications*, F. Kempken, Ed. Springer, Berlin/Heidelberg, Germany, pp. 35–56. [Google Scholar]

Dillon, R. J., and V. M. Dillon. 2004. The gut bacteria of insects: nonpathogenic interactions. *Annu. Rev. Entomol.* 49: 71–92.

Dobzhansky, T., and B. Spassky. 1947. Evolutionary changes in laboratory cultures of Drosophila pseudoobscura. *Evolution* 1: 191–216.

Douglas, A. E., and W. A. Prosser. 1992. Synthesis of the essential amino-acid tryptophan in the pea aphid (*Acyrthosiphon pisum*) symbiosis. *J. Insect Physiol.* 38 (8): 565–568.

Douglas, A. E. 1998. Nutritional interactions in insect-microbial symbioses: aphids and their symbiotic bacteria Buchnera. *Annu. Rev. Entomol.* 43: 17–37.

Dowell, R. V., J. Worley, and P. J. Gomes. 2005. Sterile insect supply, emergence, and release. In *Sterile Insect Technique Principles and Practice in Area-Wide Integrated Pest Management*, V. A, Dyck, J. Hendrichs, and A. S. Robinson, Eds. Springer, AA Dordrecht, The Netherlands, pp. 298–319.

Dubovskiy, I. M., N. A. Kryukova, V. V. Glupov, and N. A. Ratcliffe. 2016. Encapsulation and nodulation in insects. *Invertebr. Surviv. J.* 13: 229–246. ISSN 1824-307X Review.

Dyck, V. A. 2010. Rearing codling moth for the sterile insect technique. Food and Agriculture Organization of the United Nations. FAO Plant Production and Protection Paper 199. Rome, Italy. FAO and IAEA. Rome, Italy. Now available as a PDF document at http://www.fao.org/3/a-i1537e.pdf.

Dyck, V. A., J. Hendrichs, and A. S. Robinson. 2005. *Sterile Insect Technique Principles and Practice in Area-Wide Integrated Pest Management.* Springer, AA Dordrecht, The Netherlands.

Economopoulos, A. P., N. D. Bruzzone, and S. Judt. 1987. Effect of continuous light and oviposition hole size on the egg-production of Mediterranean fruit-flies mass-reared for sterile insect technique programs. In conference: 4th International Symposium on Regulation of Insect Reproduction, M. Tonner, T. Soldan, and B. Bennetova, Eds. Zinkovy, Czechoslovakia. Regulation of insect reproduction IV, pp. 193–195.

Economopoulos, A. P. 1992. Adaptation of the Mediterranean fruit fly (Diptera: Tephritidae) to artificial rearing. *J. Econ. Entomol.* 85: 753–758.

Edney, E. B. 1977. *Water Balance in Land Arthropods.* Springer-Verlag, Berlin.

Edwards, R. H., E. Miller, R. Becker, A. P. Mossman, and D. W. Irving. 1996. Twin screw extrusion processing of diet for mass rearing the pink bollworm. *Trans. Am. Soc. Agric. Eng.* 39 (5): 1789–1797.

Ekesi, S., and S. A. Mohamed. 2011. Mass Rearing and Quality Control Parameters for Tephritid Fruit Flies of Economic Importance in Africa, Wide Spectra of Quality Control, Isin Akyar, IntechOpen, doi: 10.5772/21330. Available from: https://www.intechopen.com/books/wide-spectra-of-quality-control/mass-rearing-and-quality-control-parameters-for-tephritid-fruit-flies-of-economic-importance-in-afri

Enkerlin, W. R. 2005. Impact of fruit fly control programmes using the sterile insect technique. In *Sterile Insect Technique: Principles and Practice in Area-Wide Integrated Pest Management*, V. A. Dyck, J. Hendrichs, and A. S. Robinson, Eds. Joint FAO/IAEA Programme, Vienna, Austria. Springer, Dordrecht, The Netherlands, pp. 651676.

Erickson, E., A. C. Cohen, D. Brummett, W. Lusby, and B. Cameron. 1997. Tyrosine nodules in the gasters of adult honeybees. *J. Invertebr. Pathol.* 70: 27–32.

Esquivel, J. F. 2007. Evaluation of puncture types as indicators of boll weevil (Coleoptera: Curculionidae) oviposition in cotton squares. *Environ. Entomol.* 36 (1): 183–186.

Everett, T. R., and J. O. Ray. 1964. Observations of puncturing and oviposition behavior of boll weevils. *J. Econ. Entomol.* 57 (1): 121–123.

Everett, T. R., and N. W. Earle. 1964. Boll weevil oviposition responses in cotton squares and various other substances. *J. Econ. Entomol.* 57: 651–656.

FAO/IAEA/USDA. 2003. Manual for Product Quality Control and Shipping Procedures for Sterile Mass-Reared Tephritid Fruit Flies. Version 5.0, May 2003.

FAO/IAEA/USDA. 2014. Product Quality Control for Sterile Mass-Reared and Released Tephritid Fruit Flies, Version 6.0. International Atomic Energy Agency, Vienna, Austria. 164 pp.

Fay, H. A. C., and V. Wornoayporn. 2002. Inert reusable substrates as potential replacements for wheat bran in larval diets for Mediterranean fruit fly, *Ceratitis capitata* (Wied.) (Dipt., Tephritidae). *J. Appl. Entomol.* 126: 92–96.

Fay, R. W., and G. B. Craig. 1969. Genetically marked *Aedes aegypti* in studies of field populations. *Mosquito News.* 29: 121–127.

Fellows, P. J. 2009. *Food Processing Technology: Principles and Practice.* 3rd Edition. Woodhead Publishing; Oxford, UK: Boca Raton, FL, USA. [Google Scholar]

Finke, M. D. 2002. Complete nutrient composition of commercially raised invertebrates.

Finke, M. D. 2005. Nutrient composition of bee brood and its potential as human food.

Finke, M. D. 2007. Estimate of chitin in raw whole insects. *Zoo. Biol.* 26 (2):105–115. doi: 10.1002/zoo.20123.

Finke, M. D. 2013. Complete nutrient content of four species of feeder insects. *Zoo. Biol.* 32 (1): 27–36. doi: 10.1002/z00.21012.

Finney, G. L., and T. W. Fisher.1964. Culture of entomophagous insects and their hosts. In *Biological Control of Insect Pests and Weeds*, P. DeBach, Ed. Reinhold, New York, pp. 328–355. 844 pp.

Flagg, R. O. 2021 Item 45-2620. *Carolina Drosophila Manual.* Carolina. Burlington, NC.

Frazier, M. R., H. A. Woods, and J. F. Harrison. 2001. Interactive effects of rearing temperature and oxygen on the development of *Drosophila melanogaster. Physiol. Biochem. Zool.* 74 (5): 641–650.

Funke, B. R. 1983. Mold control for insect-rearing media. *Bull. Entomol. Soc. Am.* 29: 41–44.

Gambichler, T., S. Boms, and M. M. Freitag. 2004. Contact dermatitis and other skin conditions in instrumental musicians. *BMC Dermatol.* 4: 3. doi:10.1186/1471-5945-4-3.

Garbutt, J. S., and T. J. Little. 2017. Bigger is better: changes in body size explain a maternal effect of food on offspring disease resistance. *Ecol. Evol.* 7: 1403–1409. doi: https://doi.org/10.1002/ece3.2709.

Gast, R. T. 1965. Mass producing artificial diet pellets for adult boll weevils. *J. Econ. Entomol.* 58: 1024–1025.

Gast, R. T. 1966. A spray technique for implanting boll weevil eggs on artificial diet. *J. Econ. Entomol.* 59: 239–240.

Gast, R. T., and H. Vardell. 1963. Mechanical devices to expedite boll weevil rearing in the laboratory. *ARS (Ser.)* 10: 33–89.

Gast, R. T., and T. B. Davich. 1966. Boll weevils. In *Insect Colonization and Mass Production*, C. N. Smith, Ed. Academic Press, Inc., New York and London, pp. 405–408, 618.

Gause, G. F. 1932. Experimental studies on the struggle for existence: 1. Mixed population of two species of yeast. *J. Exp. Biol.* 9: 389–402.

Gause, G. F., O. K. Nastukova, and W. W. Alpatov. 1934. The influence of biologically conditioned media on the growth of a mixed population of *Paramecium caudatum* and *P. aurelia. J. Anim. Ecol.* 3 (2): 222–230.

Gerroranal, R. D. 2020. Analysis of Error in https://ucirvine.instructure.com/courses/1095/files/362956RDGerroranal.pdf

Gersabeck, E. F., S. D. McCombs, T. E. McGovern, J. Reyes-Flores, and M. de los Santos Ramos. 2010. Fruit Fly Emergence and Release Facilities Review. Animal and Plant Health Inspection Service United States and Mexico Lower Rio Grande Valley Mexican Fruit Fly Eradication Program Review: Final Report.

Gingrich, R. E. 1972. Nutritional studies: their bearing on the development of practical oligidic diets for mass rearing larvae of the screwworm, Cochliomyia hominivorax. In *Insect and Mite Nutrition*, J. G. Rodriguez, Ed. North-Holland Publishing Company, Amsterdam, pp. 257–268.

Gingrich, R. E., A. J. Graham, and B. G. Hightower. 1971. Media containing liquefied nutrients for mass-rearing larvae of the screwworm. *J. Econ. Entomol.* 64: 673–683.

Glasi, D. A., and J. Drouillard. 2002. Composition and Feeding Value of Cottonseed Feed Products for Beef Cattle. Contribution No. 02-426-E from the Kansas Agricultural Experiment Station.

GLP Master Handbook. U S. Food and Drug Administration FDA (Federal Drug Administration) GLP. US FDA Title 21 CFR Parts. GMP Publications, Inc., Online Book Store.

Goettel, M. S., and G. D. Inglis. 2006. Methods for assessment of contaminants of invertebrate biological control agents and associated risks. In *Environmental Impact of Invertebrates for Biological Control of Arthropods: Methods and Risk Assessment*, F. Bigler, D. Babendreier, and U. Kuhlmann, Eds., pp. 145–165.

Goos, J. M., C. J. Swain, S. B. Munch, and M. R. Walsh. 2018. Maternal diet and age alter direct and indirect relationships between life-history traits across multiple generations. *Funct. Ecol.* 33: 491–502.

Gould, F. 2008. Broadening the application of evolutionarily based genetic pest management. *Evolution.* 62: 500–510.

Graham, H. M., and G. L. Mangum. 1971. Larval diets containing dyes for tagging pink bollworm moths internally. *J. Econ. Entomol.* 64: 377–379.

Griffin, J. G. 1982. Effect of oviposition cage size on egg production and hatch when mass rearing boll weevils. *J. Econ. Entomol.* 75: 351–352.

Griffin, J. G. 1982. Influence of relative humidity and temperature on egg production and Hatch from boll weevils (Coleoptera: Curculionidae). *Environ. Entomol.* 2: 1029–1031.

Griffin, J. G. 1983. Water-cooling system for pelleting machine in a boll weevil (Coleoptera: Curculionidae) rearing operation. *J. Econ. Entomol.* 76 (3): 671–673.

Griffin, J. G. 1984. Facility and production equipment. In *Boll Weevil Mass Rearing Technology*, P. P. Sikorowski, J. G. Griffin, J. R. Roberson, and O. H. Lindig, Eds. University Press of Mississippi, Jackson, MS.

Griffin, J. G., and O. H. Lindig. 1977. System for mechanical harvesting of boll weevil eggs from diet pellets. *Trans. ASAE* 20: 454–456.

Griffin, J. G., and O. H. Lindig. 1975. Granular materials used in implanting boll-weevil (Coleoptera: Curculionidae) eggs on artificial diet. *J. Econ. Entomol.* 68 (4): 433–434.

Griffin, J. G., O. H. Lindig, J. Roberson, and P. P. Sikorowski. 1979. System for rearing boll weevils in a laboratory. Mississippi agricultural and experimental station. *Tech. Bull.* 95: 15.

Griffin, J. G., O. H. Lindig, and R. E. McLaughlin. 1974. Flash sterilizers: sterilizing artificial diets for insects. *J. Econ. Entomol.* 67 (5): 689.

Gu, H., and W. W. Danthanarayana. 1992. Influence of larval rearing conditions on the body size and flight capacity of Epiphyas postvittana moths. *Aust. J. Zool.* 40 (6): 573–581.

Guyenot, E. 1913a. A biological study of a *Drosophila ampelophila* Low fly I – the possibility of an aseptic life for an individual and the line. *Comptes Rendus des Séances de da Société de Biologie Et de ses Filiales (0037-9026).* 74: 97–99.

Guyenot, E. 1913b. Biological study in a fly, *Drosophila ampeltophila* low II – the role of yeast in food. *Comptes Rendus des Seances de la Societe de Biologie et de Ses Filiales.* 74. 178–180.

Guyenot, E. 1913c. Biological studies on a fly, *Drosophila ampelophila* low V – nutrition of adults and fertility. *Comptes Rendus des Séances de la Société de Biologie et de Ses Filiales (0037-9026).* 74: 332–334.

Guyenot, E. 1914a. Biological studies on a fly *Drosophila ampelophila* low. Need to create a definite feeding environment. *Comptes rendus des séances de la Société de biologie et de ses filiales (0037-9026).* 76: 483–485.

Guyenot, E. 1914b. Primary trials of determination of an artificial nutritive environment for the raising of a fly, *Drosophila ampelophila* low. *Comptes rendus des séances de la Société de biologie et de ses filiales (0037-9026).* 76: 548–550.

Hagen, K. S., and R. L. Tassan. 1965. A method of providing artificial diets to Chrysopa larvae. *J. Econ. Entomol.* 58: 999–1000.

Hagler, J. R., A. C. Cohen, D. Bradley-Dunlop, and F. J. Enriquez. 1992. New approach to mark insects for feeding and dispersal studies. *Environ. Entomol.* 21: 20–25.

Hagler, J. R., and A. C. Cohen. 1990. Effects of time and temperature on digestion of purified antigen by *Geocoris punctipes* (Hemiptera: Lygaeidae) reared on artificial diet. *Ann. Entomol. Soc. Am.* 83: 1177–1180.

Hagler, J. R., and A. C. Cohen. 1991. Prey selection by in vitro- and field-reared *Geocoris punctipes. Entomol. Exp. Appl.* 59: 201–205.

Hagler, J. R., A. C. Cohen, F. J. Enriquez, and D. Bradley-Dunlop. 1991. An egg specific monoclonal antibody to *Lygus hesperus. Biol. Control.* 1: 75–80.

Hainsworth, F. R., and T. Hamill. 1993. Foraging rules for nectar: food choices by painted ladies. *Am. Nat.* 142: 857–867.

Hamamura, Y. 2001. Feeding mechanisms and artificial diet of silkworms. In *Silkworm Rearing on Artificial Diet (Japanese),* Y. Hamamura, Ed. Science Publishers, Inc., Enfield, NY, pp. 3–32.

Hamamura, Y. 2001. Silkworm Rearing on Artificial Diet. Science Publishers (translated from the 1973 Japanese version). #138963 | ISBN: 1578081742.

Hamamura, Y. 1959. Food selection by silkworm larvae. *Nature.* 183: 1746–1747.

Hamilton, D. W., and D. O. Hathaway. 1966. Codling moths (Ch. 22). In *Insect Colonization and Mass Production,* C. N. Smith, Ed. Academic Press, New York, pp. 339–354.

Harak, M, A, Kuusik, and K., Hiiesaar. 1998. Calorimetric investigations on physiological stress in Tenebrio molitor (Coleoptera, Tenebrionidae) pupae. Thermochim. Acta. 309 (1–2): 57–61. Conference: 10th Conference of the International Society for Biological Calorimetry, Ascona, Switzerland, April 27–30, 1997.

Harrell, E. A., and J. G. Griffin. 1981. *Facility for Mass-Rearing Boll-Weevils—Engineering Aspects.* USDA. Science and Education Administration Publications, Vol AAT, NS-19, pp. 1–77.

Harris, E. J., W. C. Mitchell, A. H. Baumhover, and L. F. Steiner 1968. Mutilation and survival of sterile oriental fruit flies and melon flies emerging in drop boxes. *Econ. Entomol.* 61: 493–496.

Harris, F. A., E. P. Lloyd, and H. C., Lane1969. Influence of light on diapause in boll weevil. 2. Dependence of diapause response on narrow bands of visible radiation and a broad band of infrared radiation used to extend photoperiod. *J. Econ. Entomol.* 62 (4): 854.

Harrison, J. F., A. Kaiser, and J. M. VandenBrooks. 2009. *Mysteries of oxygen and insect size.* In S. Morris, and A. Vosloo Bologna; S. Morris, and A. Vosloo Bologna, Eds. Medimond Publishing, Italy, pp. 293–302.

Harrison, J. F., A. Kaiser, and J. M. VandenBrooks 2010. Atmospheric oxygen level and the evolution of insect body size. *Proc. R. Soc. B: Biol. Sci.* 277: 1937–1946.

Harrison, J. F., and G. G. Haddad, 2011. Effects of oxygen on growth and size: synthesis of molecular, organismal and evolutionary studies with *Drosophila melanogaster. Annu. Rev. Physiol.* 73: 95–113.

Hartley, G. G., E. G. King, F. D. Brewer, and C. W. Gantt. 1982. Rearing of the Heliothis sterile hybrid with a multicellular larval rearing container and pupal harvesting. *J. Econ. Entomol* 75: 7–10.

Hayes, D. K., N. O. Morgan, and R. E. Webb. 1991. Space travel shortens diapause in gypsy-moth eggs. *Chronobiol. Int.* 8 (1): 75–83.

Haykin, S. 1999. *Neural Networks: A Comprehensive Foundation.* 2nd Edition. Prentice Hall. Upper Saddle River, NJ.

Heed, W. B., W. T. Starmer, M. Miranda, M. W. Miller, and H. J. Phaff. 1976. An analysis of yeast flora associated with cactiphilic Drosophila and their host plants in the Sonoran Desert and its relation to temperate and tropical associations. *Ecology.* 57: 151–160.

Hendrichs, J., and A. Robinson. 2009. Sterile insect technique. In *Encyclopedia of Insects,* Vincent H. Resh and Ring T. Card, Eds. 2nd Edition. London, Oxford, Boston, New York and San Diego: Academic Press, Elsevier Science Publisher,. pp. 953–957.

Henneberry, T. J., A. N. Kishaba, M. Z. Iqbal, and B. B. Klingler. 1968. Reproduction, longevity, and flight of cabbage looper moths treated topically with tepa. *J. Econ. Entomol.* 61: 1536–1540.

Henry, C. S., M. L. M. Wells, and C. M. Simon. 1999b. Convergent evolution of courtship songs among cryptic species of the carnea-group of green lacewings (Neuroptera: Chrysopidae: Chrysoperla). *Evolution.* 53 (4): 1165–1179.

Henry, C. S., S. J. Brooks, P. Duelli, and J. B. Johnson. 1999a. Revised concept of *Chrysoperla mediterranea* (Hölzel), a green lacewing associated with conifers: courtship songs across 2800 kilometers of Europe (Neuroptera: Chrysopidae). *Syst. Entomol.* 24 (4): 335–350.

Herman, W. S., and S. H. Dallmann. 1981. Endocrine biology of the painted lady butterfly *Vanessa cardui. J. Insect Physiol.* 27 (3): 163–168.

Hightower, B. G., J. J. O'Grady Jr., and J. J. Garcia. 1972. Ovipositional behavior of wild-type and laboratory-adapted strains of screw-worm flies. *Environ. Entmol.* 1: 227–229.

Hildebrandt, J. P., A. A. Wiesenthal, and C. Müller. 2018. Phenotypic plasticity in animals exposed to osmotic stress—is it always adaptive. *BioEssays.* 40: 1–9.

Hobischak, N. R., S. L. VanLaerhoven, and G. S. Anderson. 2006. Successional patterns of diversity in the insect fauna on carrion in sun and shade in the Boreal Forest region of Canada, near Edmonton, Alberta. *Can. Entomol.* 138: 376–383.

Hodkinson, I. D. 2005. Terrestrial insects along elevation gradients: species and com- munity response to altitude. *Biol. Rev.* 80 (3): 489–513.

Hori, M., K. Shibuya, M. Sato, and Y. Saito. 2014. Lethal effects of short-wavelength visible light on insects. *Sci. Rep.* 4: 1–7.

Howell, J. F. 1971. Problems involved in rearing codling moth. Lepidoptera: Olethreutidae on diet in trays. *J. Econ. Entomol.* 64: 631.

Howell, J. F. 1972. Modifications of artificial diet for codling moths, Lepidoptera: Olethreutidae to improve larval acceptance and production moths. *J. Econ. Entomol.* 65: 57-.

Howell, J. F. 1970. Rearing codling moth on an artificial diet. *J. Econ. Entomol.* 63: 1148.

http://faculty.sites.uci.edu/chem1l/files/2013/11/RDGerroranal.pdf

http://www.digipac.ca/chemical/sigfigs/experimental_errors.htm

http://www.upscale.utoronto.ca/PVB/Harrison/ErrorAnalysis/

Huaijian, L., L. Shi, and W. Liu. 2017. Effects of light intensity on the flight behaviour of adult *Tirumala limniace* (Cramer) (Lepidoptera: Nymphalidae: Danainae). *J. Insect Behav.* 30 (2): 139–154.

Hussa, E., and H. Goodrich-Blair. 2012. Rearing and injection of *Manduca sexta* larvae to assess bacterial virulence. *J. Vis. Exp.* (70): e4295. doi: 10.3791/4295.

Hutchinson, G. E. 1965. *The Ecological Theater and the Evolutionary Play.* Yale University Press, New Haven, CT.

Hutchinson, G. E. 1957. Concluding remarks. *Cold Spring Harbor Symposia Quant. Biol.* 22: 415–427.

Ignoffo, C. M., and O. P. Boening. 1970. Compartmented disposable plastic trays for rearing insects. *J. Econ. Entomol.* 63 (5): 1696.

Imai, M. 1986. *Kaizen: The Key to Japan's Competitive Success.* McGraw-Hill/Irwin, New York, NY. ISBN 0-07-554332-X.

Inglis, G. D., and A. C. Cohen. 2004. Influence of antimicrobial agents on the spoilage of a meat-based entomophage diet. *J. Econ. Entomol.* 97 (2): 235–250.

Inglis, G. D., and Sikorowski. 2009a. Microbial contamination and insect rearing. In *Principles and Procedures for Rearing High Quality Insects*, J.C. Schneider, Ed. Mississippi State University, Mississippi State, MS, pp. 150–222.

Inglis, G. D., and P. P. Sikorowski. 2009b. Entomopathogens and insect rearing. In *Principles and Procedures for Rearing High Quality Insects*, J. C. Schneider, Ed. Mississippi State University, MS State, MS. 352, pp 223–288.

Inoda, T, M, Hasegawa, S, Kamimura and S., Hori. 2009. Dietary program for rearing the larvae of a diving beetle, *Dytiscus sharpi* (Wehncke), in the laboratory (Coleoptera: Dytiscidae). *Coleopterist Bull.* 63(3):340350.

Ito, K., H. Sugiyama, and N. Salleh. 1993. Effects of lunar phase on light trap catches of the Malayan black rice bug, *Scotinophara coarctata* (Heteroptera, Pentatomidae). *Bull. Entomol. Res.* 83 (1): 59–66.

Ito, T., and Y. Horie. 1962. Nutrition of the silkworm, *Bombyx mori* VII. An aseptic culture of larvae on semi-synthetic diets. *J. Insect Physiol.* 8: 569–578.

Ito, T., Y. Horie, K. Watanabe, K. Takamika, M. Furuyama, M. Miyabaya, K. Yamamoto, and M. Nagashim. 1974. Rearing of larvae of silkworm, *Bombyx mori* entirely on semisynthetic diets. *J. Agric. Chem. Soc. Jpn.* 48: 403–407.

Ito, T., Y. Horie, and S. Nakasone. 1975. Deterrent effect of soybean meal on feeding of the silkworm *Bombyx mori. J. Insect Physiol.* 21: 995–1006.

Ja, W. W., G. B. Carvalho, B. M. Zid, E. M. Mak, T. Brummel, and S. Benzer. 2009. Water and nutrient-dependent effects of dietary restriction on Drosophila lifespan. *Proc. Natl. Acad. Sci. USA.* 106: 18633–18637.

Jang, J. S. R. 1993. ANFIS: adaptive-network- based fuzzy inference systems. *IEEE Trans. Syst. Man Cybern.* 23: 665–685.

Jaumanna, S., and E. C. Snell-Rood. 2019. Adult nutritional stress decreases oviposition choosiness and fecundity in female butterflies. 30 (3): 852–863. doi:10.1093/beheco/arz022.

Jones, R. W. 2001. Evolution of the host plant associations of the *Anthonomus grandis* species group Coleoptera: Curculionidae): phylogenetic tests of various hypotheses. *Ann. Entomol. Soc. Am.* 94 (1): 51–58.

Joseph, V. R., and C. F. J. Wu. 2002. Performance measure in dynamic parameter design. *J. Jpn. Quality Eng.* 10: 82–86.

Kajiwara, H., and R. Murakami. 2019. Rapid detection of silkworm microsporidia by matrix-assisted laser desorption/ionization time-of-flight mass spectrometry. *J. Invert. Pathol.* 166: 1–4.

Keena, M. 1993. The abnormal performance syndrome case. In *Proceedings U.S. Department of Agriculture*

Interagency Gypsy Moth Research Forum 1993, Annapolis, Maryland, p. 1321.

Keena, M. A., T. M. ODell, and J. A. Tanner. 1998. Environmentally based maternal effects are the primary factor in determining the developmental response of Gypsy moth (Lepidoptera: Lymantriidae) to dietary iron deficiency. *Ann. Entomol. Soc. Am.* 91: 710718.

Keena, M. H. Nadel, and J. Gould. 2015. Survival and phenology of *Agrilus planipennis* (Coleoptera: Buprestidae) reared on a newly developed artificial diet free of host material. *Great Lakes Entomol.* 98 (1-2): 931.

Keiper, J. B., and A. Foote. 1996. A simple rearing chamber for lotic insect larvae. *Hydrobiologia.* 339 (1–3): 137–139.

Keiser, 1, and E. L., Schneider. 1969. Need for immediate sugar and ability to with stand thirst by newly emerged oriental fruit flies, melon flies and Mediterranean fruit flies untreated or sexually sterilized with gamma irradiation. *Econ. Entomol.* 62:53940.

Keller, J. C., F. G. Maxwell, and J. N. Jenkins. 1962. Cotton extracts as arrestallts and feeding stimulants for the boll weevil. *J. Econ. Entomol.* 55: 800801.

Kelly, H. A. 1903. The culture of the mulberry silkworm. Bulletin No. 39 1-45. New Series. US Department of Agriculture. L. O. Howard, Entomologist. US Government Printing Office. Washington, DC.

Kelly, L., and D. M. Debinski. 1999. Effects of larval food-limitation on *Vanessa cardui* Linnaeus (Lepidoptera: Nymphalidae). *Am. Midland Nat.* 141 (2): 315–322.

Kim, K.-N., U.-C. Sin, Y.-C. Jo, Z.-J. Huang, A. Hassan, Q.-Y. Huang, and C.-L. Le. 2019. Influence of green light at night on Juvenile hormone in the oriental armyworm *Mythimna separate* (Lepidoptera: Noctuidae). *Physiol Entomol.* doi: 10.1111/phen.12300.

Klepzig, K. D., and D. L. Six. 2004. Bark beetle-fungal symbiosis: context dependency in complex associations. *Symbiosis* 37: 189–205.

Klok, C. J., A. J. Kaiser, and J. J. Socha. 2016. Multigenerational effects of rearing atmospheric oxygen level on the tracheal dimensions and diffusing capacities of Pupal and adult *Drosophila melanogaster.* In *Hypoxia: Translation in Progress, Book Series: Advances in Experimental Medicine and Biology*, R. C. Roach, P. D. Wagner, and P. H. Hackett, Eds. Vol. 903, pp 285–300.

Klunder, H. C., J. Wolkers-Rooijackers, J. M. Korpela, and M. J. R. Nout. 2012. Microbiological aspects of processing and storage of edible insects. *Food Control.* 26 (2): 628.

Knipling, E. F. 1955. Possibilities of insect control or eradication through use of sexually sterile males. *J. Econ. Entomol.* 48: 459–462.

Knipling, E. F. 1959. Sterile-male method of population control. *Science.* 130 (# 3380): 902–904.

Knipling, E. F. 1960. Use of insects for their own destruction. *J. Econ. Entomol.* 53 (3): 415–420.

Knipling, E. F. 1979. *The Basic Principles of Insect Population Suppression and Management.* USDA Agric. Handb. 512.

Kobayashi, Y., Y. Nojima, T. Sakamoto, K. Iwabuchi, T. Nakazato, H. Bono, A. Toyoda, A. Fujiyama, M. R. Kanost, and H. Tabunoki. 2019. Comparative analysis of seven types of superoxide dismutases for the ability to respond to oxidative stress in *Bombyx mori. Sci. Rep. Nat.* 9: 2170.

Kollisch, G. V., P. D. Verhaert, and K. H. Hoffmann. 2003. *Vanessa cardui* adipokinetic hormone (Vanca-AKH) in butterflies and a moth. *Comp. Biochem. Physiol. Part A.* 135: 303–308.

Kollisch, G. V., P. D. Verhaert, M. W. Lorenz, R. Kellner, G. Gade, and K. H. Hoffmann. 1999. Structure elucidation of Mas-AKH as the major adipokinetic hormone in the butterfly *Vanessa cardui* (Lepidoptera: Nymphalidae). *Eur. J. Entomol.* 96: 309–315.

Krathwohl, D. R. 2002. A revision of Bloom's taxonomy: an overview. *Theory Pract.* 41 (4): 212–218. doi: 10.1207/s15430421tip4104_2. ISSN 0040-58410040-5841.

Krause, B., and A. Farina. 2016. Using ecoacoustic methods to survey the impacts of climate change on biodiversity, Bernie Krause, Almo Farina. *Biol. Conserv.* 195: 245–254.

Kwauk, K. J., D. K. Hasegawa, M. S. Lehnert, C. E. Beard, P. D. Gerard, K. G. Korney, and P. H. Adler. 2014. Drinking with an unsealed tube: fluid uptake along the butterfly proboscis. *Ann. Entomol. Soc. Am.* 107 (4): 886–892. http://dx.doi.org/10.1603/AN14027.

Kyei-Poku, G. K., K. D. Floate, B. Benkel, and M. S. Goettel 2003. Elimination of Wolbachia from *Urolepis rufipes* (Hymenoptera: Pteromalidae) with heat and antibiotic treatments: implications for host reproduction. *Biocontrol Sci. Technol.* 13 (3): 341–354.

Lapointe, S. L., R. P. Niedz, and T. J. Evens. 2010a. An artificial diet for Diaprepes abbreviates (Coleoptera: Curculionidae) optimized for larval survival. *Florida Entomol.* 93 (1): 56–62.

Lapointe, S. L., T. J. Evens, and R. P. Niedz. 2008. Insect diets as mixtures: optimization for a polyphagous weevil. *J. Insect Physiol.* 54: 1157–1167.

Lapointe, S. L., T. J. Evens, R. P. Niedz, and D. G. Hall. 2010b. Artificial diet optimized to produce normative adults of diaprepes abbreviatus (Coleoptera: Curculionidae). *Environ. Entomol.* 39 (2): 670–677.

Lee, G., and J. H. Park. 2004. Hemolymph sugar homeostasis and starvation induced hyperactivity affected by genetic manipulations of the adipokinetic hormone-encoding gene in *Drosophila melanogaster. Genetics.* 167: 311–323.

Lee, K. P., S. J. Simpson, and K. Wilson. 2008. Dietary protein-quality influences melanization and immune function in an insect. *Funct. Ecol.* 2008 (22): 1052–1061.

Lee, R. E., C. P. Chen, and D. L. Denlinger. 1987. A rapid cold-hardening process in insects. *Science.* 238: 1415–1417.

Lee, W. J., & J. A., Lucey. 2004. Structure and physical properties of yogurt gels: effect of inoculation rate and incubation temperature. *J. Dairy Sci.* 87 (10): 3153.

Lee, W. J., and J. A. Lucey. 2010. Formation and physical properties of yogurt. *Asian-Australas. J. Anim. Sci.* 23 (9): 1127–1136.

Lees, A. D. 1953. The significance of the light and dark phases in the photoperiodic control of diapause in Metatetranychus ulmi Koch. *Ann. Appl. Biol.* 40 (3): 487–497.

Lemaitre, B., and J. Hoffmann. 2007. The host defense of *Drosophila melanogaster. Annu. Rev. Immunol.* 25: 697–743.

Lemaitre, B., J. M. Reichhart, and J. A. Hoffmann. 1997. *Drosophila* host defense: differential induction of antimicrobial peptide genes after infection by various classes of microorganisms. *Proc. Natl. Acad. Sci. USA.* 94: 14614–14619.

Leopold, R. A.1998. Cold storage of insects for integrated pest management. In *Temperature sensitivity in insects and application in integrated pest management*, G. J. Hallman and D. L. Denlinger, Eds. Westview Press, Inc., Boulder, CO, pp. 235–267.

Leopold, R. A. 2000. Insect cold storage: using cryopreservation and dormancy as aids to mass rearing. In *Proceedings: Area-Wide Control of Fruit Flies and other Insect Pests, and the 5th International Symposium on Fruit Flies of Economic Importance*, K. H. Tan, Ed. 28 May–5 June 1998, Penang, Malaysia. Penerbit Universiti Sains Malaysia, Pulau Pinang, Malaysia, pp. 315–324.

Leppla, N. C., Eds. Proc. 7th Workshop of the IOBC Global Working Group "Quality Control of Mass Reared Arthropods", 13–16 September 1993, Rimini, Italy. (Available at http://users.ugent.be/~padclerc/AMRQC/proceedings.html).

Leppla, N. C. 1978. Publication of entomological techniques. *Bull. Entomol. Soc. Am.* 24: 21–22.

Leppla, N. C., M. D. Huettel, D. L. Chambers, T. R. Ashley, D. H. Miyashita, T. T. Y. Wong, and E. J. Harris. 1983. Strategies for colonization and maintenance of the Mediterranean fruit fly. *Entomol. Exp. Appl.* 33: 89–96.

Leyva Salas, M., J. Mounier, F. Valence, M. Coton, A. Thierry, and E. Coton. 2017. Antifungal microbial agents for food biopreservation-a review. *Microorganisms.* 5 (3): 37. doi:10.3390/microorganisms5030037.

Li, T., G. Q. Pan, C. R. Vossbrinck, J. S. Xu, C. F. Li, J. Chen, and Z. Y. Zhou. 2017. SilkPathDB: a comprehensive resource for the study of silkworm pathogens. *Database.* 2017 (1): bax001. doi:10.1093/database/bax001.

Lindig, O. H. 1979. A replacement for cottonseed meal and meats in boll weevil diets. *J. Econ. Entomol.* 72: 291–292.

Lindig, O. H. 1984. Development of boll weevil diets. In *Boll Weevil Mass Rearing Technology*, Sikorowski, P. P, J. G. Griffin, J. R. Roberson, and O. H. Lindig, Eds. University Press of Mississippi, Jackson.

Lindroth, R. L., P. B. Reich, M. G. Tjoelker, J. C. Volin, and J. Oleksyn. 1993. Light environment alters response to ozone stress in seedlings of Acer saccharum Marsh. and hybrid Populus L. *New Phytol.* 124: 647–651.

Lindsey, A. R. I., and R. Stouthamer. 2017. Penetrance of symbiont-mediated parthenogenesis is driven by reproductive rate in a parasitoid wasp. *Peer J.* doi:10.7717/peerj.3505.

Lloyd, E. P., F. C. Tingle, and R. T. Gast. 1967. Environmental stimuli inducing diapause in the Boll Weevil. *J. Econ. Entomol.* 60 (1): 99–102.

Loeb, J. 1915. The simplest constituents required for growth and the completion of the life cycle in an insect (Drosophila). *Science* 41 (1048): 169–170.

Loeb, J., and J. H. Northrop. 1917. On the influence of food and temperature on the duration of life. *J. Biol. Chem.* 32: 103–121.

Longvah, T. K. Mangthya, and P. Ramulu. 2011. Nutrient composition and protein quality evaluation of eri silkworm (*Samia ricinii*) prepupae. *Food Chem.* 128 (2): 400403.

Lorraine, H., and N. D. Bruzzone. 1992. Process control: an untapped resource for managers of sterile insect programs. *Technol. Forecasting Social Change.* 41: 57–69.

Ludemann, L. R., B. R. Funke, and C. E. Goodpasture 1979. Mold control in insect rearing media - survey of agricultural fungicides and evaluation of the use of humectants. *J. Econ. Entomol.* 72 (4): 579–582.

Ludvik, D. F. 1972. A device for producing artificial squares for boll weevil oviposition and feeding. *J. Econ. Entomol.* 65 (4): 1200–1201.

Lundy, M. E., and M. P. Parrella. 2015. Crickets are not a free lunch: protein capture from scalable organic side-streams via high-density populations of *Acheta domesticus. PLoS One.* 10 (4): e0118785. https://doi.org/10.1371/journal.pone.0118785.

Lysyk, T. J. 1992. Effect of larval rearing temperature and maternal photoperiod on diapause in the horn fly (Diptera, Muscidae). *Environ. Entomol.* 21 (5): 1134–1138.

Mackauer, M. 1972. Genetic aspects of insect production. *Entomophaga* 17: 27–48.

Malainey, S. L., and G. S. Anderson. 2020. Impact of confinement in vehicle trunks on decomposition and entomological colonization of carcasses. *PLoS One.* 15 (4): e0231207. https://doi.org/10.1371/journal.pone.0231207.

Manel, S., and D. Debouzie. 1997. Modeling insect development time of Two or more larval stages in the Field under variable temperatures. *Environmental Entomology* 26 (2): 163–169.

Mao, R., J. Tang, and B. G. Swanson. 2000. Texture properties of high and low acyl mixed gellan gels. *Carbohydr. Polym.* 41 (4): 331–338.

Martin, D. F. 1966. Pink bollworm. In *Insect Colonization and Mass Production*, C. N., Smith. Ed. Academic Press, New York, NY, pp. 355–366.

Mascarenhas, R. O., L. F. Prezotto, A. L. P. Perondini, C. L. Mario, and D. Selivon. 2016. Wolbachia in guilds of Anastrepha fruit flies (Tephritidae) and parasitoid wasps (Braconidae). *Genet. Mol. Biol.* 39 (4): 600–610.

Matsuda, N., T. Kanbe, S. Akimoto, and H. Numata. 2017. Transgenerational seasonal timer for suppression of sexual morph production in the pea aphid, *Acyrthosiphon pisum. J. Insect Physiol.* 101: 1–6.

Mayr, E. 1942. *Systematics and the Origin of Species.* Columbia University Press. New York, p. 334.

Mbande, A., M. Tedder, and F. Chidawanyika. 2020. Offspring diet supersedes the transgenerational effects of parental diet in a specialist herbivore Neolema abbreviata under manipulated foliar nitrogen variability. *Insect Sci.* 27: 361–374. doi: 10.1111/1744.12644.

McCloud, E. S., and M. R. Berenbaum. 1994 Stratospheric ozone depletion and plant-insect interactions-effects of UVB radiation on foliage quality of *Citrus jambhiri* for Trichoplusia ni. *J. Chem. Ecol.* 20 (3): 525–539.

McGowan, J. E. Jr. 1985. Role of the microbiology laboratory in prevention and control of nosocomial infections. In *Manual of Clinical Microbiology*, E.H. Lennette, A. Balows, W.J. Hausler, Jr., and H.J. Shadomy, Eds. 4th Edition. American Society for Microbiology, Washington, DC, pp. 110–122.

McGrew, R. 1985. *Encyclopedia of Medical History*. McGraw Hill, New York, NY.

McIndoo, N. E. 1926. Senses of the cotton boll weevil-an attempt to explain how plants attract insects by smell. *J. Agric. RCS*. 33: 1095–1141.

Melis, R., A. Braca, R. Sanna, S. Spada, G. Mulas, M. L. Fadda, M. M. Sassu, G. Serra, and R. Anedda. 2019. Metabolic response of yellow mealworm larvae to two alternative rearing substrates. *Metabolomics*. 15: 113. https://doi-org.prox.lib.ncsu.edu/10.1007/s11306-019-1578-2.

Melvin, R., and R. C. Bushland. 1936. A method of rearing Chochliomyia Americana C. & P. on artificial media. U.S. Bureau of Entomology and Plant Quarantine, Report et-88, Washington DC.

Melvin, R., and R. C. Bushland. 1940. The nutritional requirements of screwworm larvae. *J. Econ. Entomol.* 33: 850–852.

Mende, S., M. Peter, K. Bartels, H. Rohm, and D. Jaros. 2013. Addition of purified exopolysaccharide isolates from *S. thermophilus* to milk and their impact on the rheology of acid gels. *Food Hydrocolloids*. 32 (1): 178–185.

Merritt, R. W., K. W. Cummins, and M. B. Berg, Eds. 2008. *An Introduction to the Aquatic Insects of North America*. Kendall/Hunt Publishing Co, Dubuque, Iowa.

Meshrif, W. S., M. Rohlfs, and T. Roeder. 2016. The effect of nutritive yeasts on the fitness of the fruit fly *Drosophila melanogaster* (Diptera: Drosophilidae). *Afr. Entomol.* 24 (1): 90–99.

Michelbacher, A. E., W. M. Hoskins, and W. B. Herms. 1932. The nutrition of flesh fly larvae, Lucilia sericata (Meig.). *J. Exp. Zool.* 64. 109–131.

Miller, E., F. Stewart, A. Lowe, and J. Bomberg. 1996. New method of processing diet for mass rearing pink Bollworm, *Pectinophora gossypiella* Saunders (Lepidoptera: Gelichiidae). *J. Agric. Entomol.* 13 (2): 129.

Mishra, N., N. C. Hazarika, K. Narain, and J. Mahanta. 2003. Nutritive value of non-mulberry and mulberry silkworm pupae and consumption pattern in Assam, India. *Nut. Res.* 23: 1303–1311.

Mitlin, N., D. H. Vickers, and R. T. Gast. 1964. Estimation of nitrogenous compounds in the feces of boll weevils, *Anthonomus grandis*, fed different diets. *Ann. Entomol. Soc. Am.* 57: 757–759.

Montgomery, D. C. 1991. *Introduction to Statistical Quality Control*. 4th Edition. John Wiley and Sons, New York, NY.

Montgomery, D. C. 2009. *Introduction to Statistical Quality Control*. 6th Edition. John Wiley and Sons, New York, NY.

Montgomery, D. C. 2013. *Introduction to Statistical Quality Control*. 7th Edition. John Wiley and Sons Inc., New York, NY. ISBN-13: 9781118146811I

Moore, R. F. Jr., F. F. Whisnant, and H. M. Taft. 1967. A laboratory diet containing egg albumin for larval and adult boll weevils. *J. Econ. Entomol.* 60: 237–241.

More, D. 1976. *The Bee Book: The History and Natural History of the Honeybee*. Universe Books, New York, NY.

Moreno, F. J., and A. Clemente. 2008. 2S albumin storage proteins: what makes them food allergens? *Open Biochem. J.* 2: 16–28. doi:10.2174/1874091X00802010016.PMC 2570561.

Morgan, T. H., A. N. H. Sturtevant, H. J. Muller, and C. B. Bridges. 1915. *The Mechanism of Mendelian Heredity*. Henry Holt and Company, New York, NY.

Morgan, T. H. 1912. Heredity of body color in Drosophila. *J. Exp. Biol.* 13 (1): 27–45.

Morgan, T. H., A. H. Sturtevant, H. J. Muller, and C. B. Bridges. 1913. *The Mechanism of Mendelian Heredity*. Henry Holt and Company, New York, NY.

Morrison, R. K., V. S. House, and R. L. Ridgway. 1975. Improved rearing unit for larvae of a common green lacewing. *J. Econ. Entomol.* 68 (6): 821–822.

Muir, G., and P. Vantomme. 2013. Edible insects: future prospects for food and feed security. FAO Forestry Paper 171. Wageningen UR.

Mukherjee, S. 2016. *The Gene: An Intimate History*. Scribner, New York, NY.

Needham, J. G., Ed. 1937 *Culture Methods for Invertebrate Animals*. Dover, New York, NY, p. 590.

Nguyen, K.-O., S. Al-Rashid, M. Clarke Miller, J. Tom Diggs, and E. C. Lampert. 2019. Trichoplusia ni (Lepidoptera: Noctuidae) qualitative and quantitative sequestration of host plant carotenoids. *Environ. Entomol.* 48 (3): 540–545. doi: 10.1093/ee/nvz029.

NIH. 2019. *Guidelines and Policies for the Conduct of Research in the Intramural Research Program at NIH*. 6th Edition. National Institutes of Health: Office of the Director. oir.nih.gov

Nishide, Y., T. Naoko, T. Onodera, M. Tanahashi, M. Moriyama, T. Fukatsu, and R. Koga. 2017. Aseptic rearing procedure for the stinkbug *Plautia stali* (Hemiptera: Pentatomidae) by sterilizing food-derived bacterial contaminants. *Appl. Entomol. Zool.* 52: 407–415. doi:10.1007/s13355-017-0495-y.

Nojima, Y., H. Bono, T. Yokoyama, K. Iwabuchi, R. Sato, K. Arai, and H. Tabunoki. 2019. Superoxide dismutase down-regulation and the oxidative stress is required to initiate pupation in *Bombyx mori*. *Sci. Rep. Nat.* 9: 14693. https://doi.org/10.1038/s41598-019-51163-3.

Nunney, L. 2002. The population genetics of mass-rearing. In *Proc. Quality Control for Mass-Reared Arthropods, 8th and 9th Workshops of the IOBC Working Group on Quality Control of Mass-Reared Arthropods*, N.C. Leppla, K.A. Bloem and R.F. Luck, Eds., pp. 43–49. (Available at http://biocontrol.ifas.ufl.edu/ amrqc/IOBCproceedings/amrqcbook.htm).

O Dell, T. M., and W. D. Rollinson. 1966. A technique for rearing the gypsy moth, Porthetria dispar (L.). *J. Econ. Entomol* 59: 741–742.

O Dell, T. M., R. A. Bell, V. C. Mastro, J. A. Tanner, and L. F. Kennedy. 1984. Production of the gypsy moth,

Lymantria dispar. In *Advances and Challenges in Insect Rearing*, E. G. King, and N. C. Leppla, Eds. USDA, ARS, New Orleans, LA, pp. 156–166.

ODell, T. M. 1992. Straggling in gypsy moth production strains: a problem analysis for developing research priorities. In *Advances in Insect Rearing for Research and Pest Management*, T. A. Anderson, and N. C. Leppla, Eds. Westview, Boulder, CO, pp. 325–350.

ODell, T. M., M. A. Keena, and R. B. Willis. 1997. Dietary influence of iron formulation on the development of gypsy moth (Lepidoptera: Lymantriidae) in laboratory colonies. *Ann. Entomol. Soc. Am.* 90: 149–154.

ODell, T. M., M. A. Keena, and R. B. Willis. 1997. Dietary influence of iron formulation on the development of gypsy moth (Lepidoptera: Lymantriidae) in laboratory colonies. *Ann. Entomol. Soc. Am.* 90: 149–154.

ODell, T. M., R. A. Bell, V. C. Mastro, J. A. Tanner, and J. F. Kennedy. 1984. Production of gypsy moth, *Lymantria dispar*, for research and biological control. In *Advances and Challenges in Insect Rearing*, E. G. King and N. C. Leppla, Eds. U.S. Department of Agriculture, Agricultural Research Service, New Orleans, LA, pp. 155–166.

Ormerod, K. G., O. K. LePine, P. S. Abbineni, J. M. Bridgeman, J. R. Coorssen, A. J. Mercier, and G. J. Tattersall. 2017. *Drosophila* development, physiology, behavior, and lifespan are influenced by altered dietary composition. *Fly.* 11:3: 153–170. doi: 10.1080/19336934.2017.1304331.

Orozco, D. O., A. G. Schwarz, and A. Perez Romero. 1983. Manual de procedimientos de control de calidad. Dirección Generál de Sanidad Vegetal, Secretaría de Agricultura y Recursos Hidraulicos. Talleres Graficos de Nacion. Mexico, D.F., 137 pp.

Ouye, M. T. 1962. Effects of antimicrobial agents on microorganisms and pink bollworm development. *J. Econ. Entomol* 55: 854–857.

Owsley, F. L. Jr. 1959. King Cotton Diplomacy: Foreign Relations of the Confederate States of America. Chicago: The University of Chicago Press, 1959. In Mississippi History Now: An Outline of the Mississippi Historical Society in an article by Frank L. Owsley Jr. "Cotton and the Civil War."

Owsley, F. L. Jr. http://mshistorynow.mdah.state.ms.us/articles/291/cotton-and-the-civilwar.

Park, T. 1954. Experimental studies of interspecies competition. II. Temperature, humidity and competition in two species of Tribolium. *Physiol. Zool.* 27: 177–238.

Park, T. 1957. Experimental studies of interspecies competition. III. Relation of initial species proportion to the competitive outcome in populations of Tribolium. *Physiol. Zool.* 30: 22–40.

Park, T. 1948. Experimental studies of interspecies competition. I. Competition between populations of the flour beetles *Tribolium confusum* Duval and *Tribolium castaneum* Herbst. *Ecol. Monogr.* 18: 265–308.

Parker, A. G. 2005. Mass-rearing for sterile insect release. In *Sterile Insect Technique: Principles and Practice in Area-Wide Integrated Pest Management*, V.A. Dyck, J Hendrichs, A.S. Robinson, Eds. Springer, Dordrecht, The Netherlands, pp. 209–232. ISBN 1-4020-4050-4

Parsell, D. A., and S. Lindquist. 1993. The function of heat-shock proteins in stress tolerance: degradation and reactivation of damaged proteins. *Annu. Rev. Genet.* 27: 437–496.

Pascacio-Villafán, C., L. Guillén, and M. Aluja. 2020. Agar and carrageenan as cost-effective gelling agents in yeast-reduced artificial diets for mass-rearing fruit flies and their parasitoids. *Insects.* 2020 (11): 131. doi: 10.3390/insects11020131.

Pascacio-Villafán, C., L. Guillén, T. Williams, and M. Aluja. 2018. Effects of larval density and support substrate in liquid diet on productivity and quality of artificially reared *Anastrepha ludens* (Diptera: Tephritidae). *J. Econ. Entomol.* 111 (5): 2281–2287.

Pascacio-Villafán, C., A. Birke, T. Williams, and M. Aluja. 2017. Modeling the cost-effectiveness of insect rearing on artificial diets: a test with a tephritid fly used in the sterile insect technique. *PLoS One.* 12: e0173205.

Pascacio-Villafán, C., T. Williams, A. Birke, and M. Aluja. 2016. Nutritional and non-nutritional food components modulate phenotypic variation but not physiological trade-offs in an insect. *Sci. Rep.* 6: 29413.

Passioura, J. B. 1980. The meaning of matric potential. *J. Exp. Bot* 31: 1161–1169.

Patana, R. 1969. Rearing cotton insects in the laboratory. *U.S. Dep. Agric. Prod. Res. Rep.* 108.

Patana, R. 1977. Rearing Selected Western Cotton Insects in the Laboratory. Volume 51 of ARS-W, United States Agricultural Research Service. United States. Agricultural Research Service. Western Region. Department of Agriculture, Agricultural Research Service, Western Region, 8 pages.

Patana, R. 1982. Disposable diet packet for feeding and oviposition of *Lygus hesperus* (Hemiptera, Miridae). *J. Econ. Entomol.* 75 (4): 668–669.

Patana, R., and J. W. Debolt. 1985. Rearing *Lygus hesperus* in the laboratory. In *Handbook of Insect Rearing*, Vol. I and II. Pritam Singh and R. F. Moore, Eds. Vol. I, viii + 488 pp., ISBN 0444-424665-2; Vol. II, viii+ 514 pp., ISBN 0444-424665-0. Elsevier Science Publishers B.V., Amsterdam.

Patana, R., and J. W. Debolt. 1986. Rearing Lygus hesperus in the Laboratory. USDA-ARS ARS-45.

Patana, R., and W. C. McAda. 1973. Tobacco budworms: use of dry diet flakes in rearing. *J. Econ. Entomol.* 66: 817–818.

Payne, J. A. 1965. A summer carrion study of the baby pig Sus scrofa Linnaeus. *Ecology.* 46: 592–602.

Pearson, T. C., R. H. Edwards, A. P. Mossman, D. F. Wood, P. C. Yu, and E. L. Miller. 2002. Insect egg counting on mass rearing oviposition pads by image analysis. *Trans. Am. Soc. Agric. Eng.* 18: 128–135.

Pedigo, L. P. 1994. Painted lady. In *Handbook of Soybean Insect Pests*, L.G. Higley and D. J. Boethel, Eds. Entomological Society of America, p. 77.

Perry, J. J. P., D. S. Shin, E. D. Getzoff, and J. A. Tainer. 2010. The structural biochemistry of the superoxide dismutases. *Biochim. Biophys. Acta.* 1804 (2): 245–262.

Phillips, A. M., and M. C. Swingle. 1940. Rearing of mosquito larvae and effect of diet on their resistance to rotenone and nicotine. *J. Econ. Entomol.* 33: 172–176.

Pianka, E. R. 1974. Niche overlap and diffuse competition. *Proc. Natl. Acad. Sci. USA.* 71 (5): 2141–2145.

Pijanowski, B. C., L. J. Villanueva-Rivera, S. L. Dumyahn, A. Farina, B. L. Krause, Brian M. Napoletano, S. H. Gage, and N. Pieretti. 2011. Soundscape ecology: the science of sound in the landscape. *BioScience.* 61 (3): 203–216. www.biosciencemag.org,

Ponton, F., K. Wilson, S. C. Cotter, D. Raubenheimer, and S. J. Simpson. 2011. Nutritional immunology: a multi-dimensional approach. *PLoS Pathog.* 7 (12): e1002223. doi:10.1371/journal.ppat.1002223.

Pruitt, J. N., D. I. Bolnick, A. Sih, N. DiRienzo, and N. Pinter-Wollman. 2016. Behavioural hypervolumes of spider communities predict community performance and disbandment. *Proc. R. Soc.* B283. http://dx.doi.org/10.1098/rspb.2016.1409.

Pugh, E. M., and G. H. Winslow. 1966. *The Analysis of Physical Measurements.* Addison-Wesley, Boston, MA.

Raulston, J. R., and P. D. Lingren. 1972. Methods for large-scale rearing of the tobacco budworm. *USDA Prod. Res. Rep.* 145: 10.

Rehman, Z., and W. H. Shah. 2005. Thermal heat processing effects on antinutrients, protein and starch digestibility of food legumes. *Food Chem.* 91: 327–331.

Raman, A. 2014. Discovery of *Kerria lacca* (Insecta: Hemiptera: Coccoidea), the lac insect, in India in the late 18th century. *Curr. Sci.* 106 (6): 886.

Ramsey, S. D., R. Ochoa, G. Bauchan, C. Gulbronson, J. D. Mowery, A. Cohen, D. Lima, J. Joklik, J. M. Cicero, J. D. Ellis, D. Hawthorne, and D. van Engelsdorp. 2019. *Varroa destructor* feeds primarily on honey bee fat body tissue and not hemolymph. *Proc. Natl. Acad. Sci. USA.* 116 (5): 1792–1801.

Rani, A., A. Sharma, R. Rajagopal, T. Adak, and R. K. Bhatnagar. 2009. Bacterial diversity analysis of larvae and adult midgut microflora using culture-dependent and culture-independent methods in lab-reared and field-collected *Anopheles stephensian* Asian malarial vector. *BMC Microbiol.* 9: 96.

Rao, C. G. P., S. V. Seshgiri, C. Ramesh, B. K. Ibrahim, H. Nagaraju, and Chandrashekaraiah. 2006. Evaluation of genetic potential of the polyvoltine silkworm (*Bombyx mori* L.) germplasm and identification of parents for breeding programme. *J. Zhejiang Univ. Sci. B.* 215–220.

Redfern, F. E. 1964. Concentrate medium for rearing the codling moth. *J. Econ. Entomol.* 57: 607–608.

Redfern, R. E. 1963. Concentrate media for rearing red-banded leaf roller. *J. Econ. Entomol.* 56: 240–241.

Reed, H. B. 1958. A study of dog carcass communities in Tennessee, with special reference to the insects. *Am. Midl. Nat.* 59: 213–245.

Reinecke, J. P. 1991. Increased field trap capture of the mass-reared boll weevil (Coleoptera: Curculionidae) fed diet supplemented with beta-carotene. *J. Econ. Entomol.* 84 (2): 633–637.

Richmond, C. A., and D. F. Martin. 1966. Technique for mass rearing the pink bollworm. *J. Econ. Entomol.* 59: 762–763.

Ridgway, R. L., R. K. Morrison, and M. Badgley. 1970. Mass rearing a Green Lacewing. *J. Econ. Entomol.* 63 (3): 834–836.

Rinehart, J. P., G. D. Yocum, and D. L. Denlinger. 2000. Thermotolerance and rapid cold hardening ameliorate the negative effects of brief exposures to high or low temperatures on fecundity in the flesh fly, *Sarcophaga crassipalpis. Physiol. Entomol.* 25: 330–336.

Ritchie, G. L., C. W. Bednarz, P. H. Jost, and S. M. Brown. 2007. Cotton growth and development. Bulletin 1252. University of Georgia Cooperative Extension. 17 pp.

Ritossa, F. 1962. A new puffing pattern induced by temperature shock and DNP in drosophila. *Cell. Mol. Life Sci.* 18 (12): 571–573. doi:10.1007/BF02172188.

Ritossa, F. 1996. Discovery of the heat shock response. *Cell Stress Chaperones.* 1 (2): 97–98. PMC 248460. PMID 9222594.

Roberson, J. R. 1984. Laboratory rearing. In *Boll Weevil Mass Rearing Technology*, P. P. Sikorowski, J. G. Griffin, J. R. Roberson, and O. H. Lindig, Eds. University Press of Mississippi, Jackson, MS, pp. 85–113.

Roberts, R. H. 1927. Relation of composition to growth and fruitfulness of young apple trees as affected by girdling, shading, and photoperiod. *Plant Physiol.* 2 (3): 273–286.

Roberts, W. S. 1945. Silk culture in America. http://www.worm-spit.com/library.htm

Robinson, K. 2003. *BioPharm International*-08-01-2003. Volume 16, Issue 8: 40-45. https://www.biopharminternational.com/view/glps-and-importance-standard-operating-procedures

Rock, G. C. 1967. Aseptic rearing of codling moth on synthetic diets—ascorbic acid and fatty acid requirements. *J. Econ. Entomol.* 60 (4): 1002–1005.

Rogers, D. J., and C. J. Winks. 1993. Quality Control in Laboratory-Reared Codling Moth at Mt Albert Research Centre, Auckland, New Zealand. In Proceedings of the 7th Workshop of the Global IOBC Working Group "Quality Control of Mass Reared Arthropods", G. Nicoli, M. Benuzzi, and N. C. Leppla, Eds. Rimini, Italy, 13–16 September, pp. 13–23. Available at http://users.ugent.be/~padclerc/AMRQC/proceedings.html. Accessed March 13, 2021.

Roubaud, E. 1930. Evolutionary suspension and obligatory larval hibernation provoked by the heat, in the common mosquito, *Culex pipiens* L. the true diapause and the pseudo-diapause in insects. *Comptes Rendus Hebdomadaires des Seances de l Academie des Sciences* 190: 324–326.

Ruberson, J. R., L. Bush, and T. J. Kring. 1991. Photoperiodic effect on diapause induction and development in the predator *Orius insidiosus* (Heteroptera, Anthocoridae). *Environ. Entomol.* 20 (3): 786–789.

Russell, L. M. 1978. Leland Ossian Howard: a historical review. *Ann. Rev. Entomol.* 1978 (23): 1–15.

Saito, H. 1998. Purification and properties of two blue biliproteins from larval hemolymph and integument of *Rhodinia fugax* (Lepidoptera: Saturniidae). *Insect Biochem. Mol. Biol.* 28: 995–1005.

Sall, J. A. Lehman, M. Stephens, and Sheila Loring. 2017. A guide to statistics and data analysis using JMP. *JMP® Start Statistics: A Guide to Statistics and Data Analysis using JMP®*. 6th Edition. SAS Institute Inc, Cary, NC.

Sandstrom, J. P., J. A. Russell, J. P. White, and N. A. Moran. 2001. Independent origins and horizontal transfer of bacterial symbionts of aphids. *Mol. Ecol.* 10 (1): 217–228.

SAS Institute Inc. 2018. JMP® 14 Design of Experiments Guide. Cary, NC: SAS Institute Inc. ISBN 978-1-63526-504-0 (Web PDF)

SAS Institute. JMP 2007. *Design of Experiments Guide. Release 7.* SAS Institute, Cary, NC.

Schilman, P. E., J. S. Walters, and J. F. Harrison. 2011. Effects of temperature on responses to anoxia and oxygen reperfusion in *Drosophila melanogaster. J. Exp. Biol.* 214 (8): 1271–1275.

Schlesinger, M. J. 1990.Heat shock proteins. *J. Biol. Chem.* 265 (21): 12111–12114. PMID 2197269.

Schmidt, F. S., and A. Skerra. 1994. The bilin-binding protein of *Pieris brassicae* cDNA sequence and regulation of expression reveal distinct features of this insect pigment protein. *Eur. J. Biochem.* 219: 855–863.

Schmidt, P. J. 1917. The diapause in earthworms. *Comptes Rendus des Seances de la Societe de Biologie et des ses Fialiales.* 80: 366–368.

Schneider, J. 2013. *Principles and Procedures for Rearing High Quality Insects.* John C. Schneider, Ed. Mississippi State University Press, Mississippi State, MS.

Schneider, J., N. Leppla, M. Chaudhury, L. Castrillo, S. Ng, W. Fisher, P. Ebling, M. Caprio, and T. Riddell. 2018. Educating the next generation of insect rearing professionals. *Am. Entomol.* 64: 102–111.

Scholl, C., Y. Wang, and M. Krischke et al. 2015. Light exposure leads to reorganization of microglomeruli in the mushroom bodies and influences juvenile hormone levels in the honeybee. *Dev. Neurobiol.* 74: 1141–1153.

Schrodinger, E. 1944. *What Is Life? The Physical Aspect of the Living Cell.* Cambridge University Press, Cambridge.

Schultz, A., R. Wieland, and G. Lutze 2000. Neural networks in agroecological modelling – stylish application or helpful tool. *Comput. Electron. Agric.* 29: 73–97.

Schultz, A., and R. Wieland 1997. The use of neural networks in agroecological modelling. *Comput. Electron. Agric.* 18: 73–90.

Schwarz, A. J., A. Zambada, D. H. S. Orozco, J. L. Zavala, and C. O. Calkins. 1985. Mass production of the Mediterranean fruit fly at Metapa, Mexico. *Fla. Entomol.* 68: 467–477.

Scruggs, C. G. 1978. The origin of the screwworm control program. In *The Screwworm Problem—Evolution of Resistance to Biological Control*, R. H. Richardson, Ed. University of Texas Press. Austin, TX, pp. 11–18.

Sharp, J. E. F. B., and D. L. Chambers. 1983. Selection for flight propensity of laboratory and wild strains of Anastrepha suspensa and *Ceratitis capitata* (Diptera: Tephridtidae). *J. Econ. Entomol.* 76: 302–305

Shaver, T. N., and J. R. Raulston. 1971. A soybean-wheat germ diet for rearing the tobacco budworm. *Ann. Ent. Soc. Am.* 64: 1077–1079.

Sherman, R. A., M. J. R. Hall, and S. Thomas. 2000. Medicinal maggots: an ancient remedy for some contemporary afflictions. *Annu. Rev. Entomol.* 45: 55–81. doi:10.1146/annurev.ento.45.1.55. PMID 10761570.

Shewhart, W. A. 1931. *Economic Control of Quality of Manufactured Product.* Bell Telephone Laboratories, Murray Hill, NJ. ISBN 0-87389-076-0.

Shewhart, W. A. 1939. *Statistical Method from the Viewpoint of Quality Control.* Dover Publications, Mineola, NY. ISBN 0-486-65232-7

Shimizu, I., Y. Yamakawa, T. Minamoto, and K. Sakamoto. 1998. Cloning of genes encoding the visual pigments in the silkworm, *Bombyx mori. Appl. Entomol. Zool.* 33 (1): 199–204.

Shorey, H. H., and R. L. Hale. 1965. Mass rearing of the larvae of nine noctuid species on a simple artificial medium. *J. Econ. Entomol.* 58: 522–524.

Shorey, H. H. 1963. A simple artificial rearing medium for the cabbage looper. *J. Econ. Entomol.* 56: 536–537. doi: https://doi.org/10.1093/jee/56.4.536a.

Sikorowski, P. P., and A. M. Lawrence. 1994. Microbial contamination and insect rearing. *Am Entomol* 40: 240–253.

Sikorowski, P. 1984. Microbial contamination in insectaries: occurrence, prevention, and control. In *Advances and Challenges in Insect Rearing*, E. G. King, and N. C. Leppla, Eds. USDA, ARS, New Orleans, LA, pp. 143–153.

Sikorowski, P., and A. Lawrence. 1994. Microbial contamination and insect rearing. *Am. Entomol.* 40: 240–253.

Sikorowski, P. P., J. G. Griffin, J. R. Roberson, and O. H. Lindig. 1984. *Boll Weevil Mass Rearing Technology.* University Press of Mississippi, Jackson, MS.

Silva, M., and F. Lidon. 2016. Food preservatives—an overview on applications and side effects. *Emir. J. Food Agric.* 28 (6). doi: 10.9755/ejfa.2016-04-351.

Simmons, G. S., A. R. McKemey, N. I. Morrison, S. O'Connell, B. E. Tabashnik, J. Claus, G. Fu, G. Tang, M. Sledge, A. S. Walker, C. E. Phillips, E. D. Miller, R. I. Rose, R. T. Staten, C. A. Donnelly, and L. Alphey. 2011. Field performance of a genetically engineered strain of pink bollworm. *PLoS One.* 6 (9): e24110. doi:10.1371/journal.pone.0024110.

Singh, P., and H. L. House. 1970. Antimicrobials: safe levels in a synthetic diet of an insect, *Agria affinis. J. Insect Physiol.* 16 (9): 1769.

Singh, P., and M. R. Surrey. 1980. A plastic container for rearing insects on artificial diets, *N. Z. J. Zool.* 7 (3) 441–442. doi: 10.1080/03014223.1980.10423799.

Singh, P. 1977. *Artificial Diets for Insects, Mites and Spiders.* Plenum Press, New York, NY.

Singh, P. 1984. Insect diets. Historical developments, recent advances and future prospects. In *Advances and Challenges in Insect Rearing*, E. G. King, and N. C. Leppla, Eds. New Orleans, LA, pp. 32–44.

Singh, P. 1985. Multiple species rearing diets. In *Handbook of Insect Rearing*, P. Singh and R. F. Moore, Eds. Vol. I. Elsevier Science Publishers BV, Amsterdam, The Netherlands, pp. 19–44.

Singh, P., and H. L. House. 1970a. Antimicrobials: "safe" levels in a synthetic diet of an insect, *Agria affinis. J. Insect Physiol.* 16: 1769–1782.

Singh, P., and H. L. House. 1970b. Antimicrobial agents: their detrimental effects on size of an insect, *Agria affinis*. *Can. Entomol* 102: 1340–1344.

Singh, P., and R. F. Moore. 1985. *Handbook of Insect Rearing*, Pritam Singh and R. F. Moore. Eds. Vol. I and II. Vol. I, viii + 488 pp., ISBN 0444-424665-2; Vol. II, viii+ 514 pp., ISBN 0444-424665-0. Elsevier Science Publishers B.V., Amsterdam, The Netherlands.

Sinkins, S. P., H. R. Braig, and S. L. O'Neill. 1995. Wolbachia superinfections and the expression of cytoplasmic incompatibility. *Proc. Biol. Sci.* 261 (1362): 325–330.

Six, D. L. 2012. Ecological and evolutionary determinants of bark beetle fungus symbioses. *Insects*. 3: 339–366.

Skinner, T. 2018. How Connecticut's Homegrown Silk Industry Unraveled: The problems began with a new variety of mulberry and ended with lumpy thread. (Available at https://www.atlasobscura.com/articles/connecticut-silk-industry).

Smith, C. N., Ed. 1966. *Insect Colonization and Mass Production*. Academic Press, Inc., New York and London, p. 618.

Smith, J. W., and W. D. Swink. 2003. Boll weevil eradication: a model for sea lamprey control? *J. Great Lakes Res.* 29 (Supplement 1): 445–455.Contribution # 1218 (OP).

Smith, R. C. 1931. A study of temperature and humidity conditions in common types of insect rearing cages. *J. Agric. Res.* 43: 547–557.

Smith, R. A. 1999. Mechanization of mass rearing operations. In *Proceedings: ASAE/CSAE/SCGR Annual International Meeting*, 18–21 July 1999, Toronto, Ontario, Canada. The American Society of Agricultural Engineers, St. Joseph, MI.

Smith, R. A., and D. A. Nordlund. 1999. Automation in insect rearing a key to the development of competitive augmentative biological control. *Nat. Enemies Insects*. 21: 70–81.

Sørensen, J. G., M. F. Addison, and J. S. Terblanche. 2012. Mass-rearing of insects for pest management: challenges, synergies and advances from evolutionary physiology. *Crop Prot.* 38: 87–94. DOI: 10.1016/j.cropro.2012.03.023

Sprugeon, D. W., and C. S. Brent. 2016. Development, survival, and hatching periodicity of *Lygus hesperus* (Hemiptera: Miridae) eggs under constant and variable temperatures. *J. Entomol. Sci.* 51 (4): 292–304.

Stadler, F. 2020. The maggot therapy supply chain: a review of the literature and practice. *Med. Vet. Entomol.* 34 (1): 1–9. doi: 10.1111/mve.12397.

Stalter, R., G. Nadal, and D. Kincaid. 2000. Toxicity of *Bacillus thuringiensis* var. Kurstaki to the Painted Lady Butterfly, *Vanessa cardui*. *Am. Biol. Teacher.* 62 (3): 207–210.

Stark, J., and H. S. Tan. 2003. Natamycin. In *Food Preservatives*, N. J. Russell and G. W. Could, Eds. Springer, New York, NY, pp. 179–195.

Starmer, W. T., J. S. F. Barker, H. J. Phaff, and J. C. Fogleman. 1986. Adaptations of Drosophila and yeasts: their interactions with the volatile 2-propanol in the cactus-microorganism-Drosophila model system. *Aust. J. Biol. Sci.* 39: 69–77.

Starmer, W. T., and J. C. Fogleman. 1986. Coadaptation of Drosophila and Yeasts in their natural habitat. *J. Chem. Ecol.* 12, 1037–1055. doi: https://doi.org/10.1007/BF01638995.

Stewart, F. D. 1984. Mass rearing the pink bollworm, *Pectinophora gossypiella*. In *Advances and Challenges in Insect Rearing*, E. G. King, and N. C. Leppla, Eds. USDA, ARS, New Orleans, LA, pp. 176–187.

Stewart, F. D., R. B. Bell, A. J. Martinez, J. J. Roberson, and A. M. Lowe. 1976. The surface sterilization of pink bollworm eggs and spread of cytoplasmic polyhedrosis virus in rearing containers. *U. S. Anim. Plant Health Insp. Serv.* [Rep.] 81-27: 7.

Stinner, R. E., A. P. Gutierrez, and G. D. Butler. 1974. Algorithm for temperature-dependent growth-rate simulation. *Can. Entomol.* 106 (4): 519–524.

Stockard, J. E. 2002. On women's work. Silk reeling: gender, labor, and technology in the historical silk industries of Connecticut and South China. Textile Society of America Symposium Proceedings. 549. https://digitalcommons.unl.edu/tsaconf/549.

Stone, C. W. 2006. Yeast products in the feed industry: practical guide for feed professionals. https://en.engormix.com/feed-machinery/articles/yeast-products-in-feed-industry-t33489.html.

Strong, F. E., and E. Kruitwagen. 1970. Gustatory discrimination between meridic diets by the bug, *Lygus hesperus*. *J. Insect Physiol.* 16: 521–530.

Strong, F. E., and E. Kruitwagen. 1969. Feeding and nutrition of *Lygus hesperus*. III. Limited growth and development on a meridic diet. *Ann. Entomol. Soc. Am.* 62: 148–155.

Sturtevant, A. H. 1913. The linear arrangement of six sex-linked factors in Drosophila, as shown by their mode of association. *J. Exp. Zool.* 14 (1): 43–59.

Sturtevant, A. H. 1965. The fly room. *Am. Sci.* 53 (326A): 303–307.

Subramanyam, B., and D. W. Hagstrum. 1991. Quantitative analysis of temperature, relative humidity, and diet influencing development of the larger grain borer, *Prostephanus truncatus* (Horn) (Coleoptera: Bostrichidae). *Trop Pest Manag.* 37 (3): 195–202. doi: 10.1080/09670879109371581.

Sun, J., F. Ren, Y. Chang, P. Wang, Y. Li, H. Zhang, and J. Luo. 2018. Formation and structural properties of acid-induced casein-agar double networks: role of gelation sequence. *Food Hydrocolloids.* 85: 291–298.

Swanson, H. K., M. Lysy, M. Power, A. D. Statso, J. D. Johnson, and J. D. Reist. 2015. A new probabilistic method for quantifying n-dimensional ecological niches and niche overlap. *Ecology* 96 (2): 316–324.

Tabor, K. L., C. C. Brewster, and R. D. Fell. 2004. Analysis of the successional patterns of insects on carrion in Southwest Virginia. *J. Med. Entomol.* 41: 785–795.

Tabor, K. L., R. D. Fell, and C. C. Brewster. 2005. Insect fauna visiting carrion in Southwest Virginia. *Forensic Sci. Int.* 150: 73–80.

Tanada, Y., and H. K. Kaya 1993. *Insect Pathology*. Academic Press Inc, San Diego, CA.

Tang, J., M. A. Tung, and Y. Zeng. 1996. Compression strength and deformation of gellan gels formed with mono- and divalent cations. *Carbohydr. Polym.* 29 (1): 11–16.

Tao, H., X. Li, and J.-F. Qiu. 2017. The light cycle controls the hatching rhythm in *Bombyx mori* via negative feedback loop of the circadian oscillator. *Arch. Insect Biochem. Physiol.* 96 (2): Article Number 321408.

Tayal, M. K., and T. P. S. Chauhan. 2017. Silkworm diseases and pests. In *Industrial Entomology*, Omkar, Ed. Springer, Singapore. https://doi.org/10.1007/978-981-10-3304-9_9.

Taylor, J. R. 1982. *An Introduction to Error Analysis*. University Science Books, Mill Valley, CA.

Taylor, R. 1973. *Butterflies in My Stomach or Insects in Human Nutrition*. Woodbridge, Santa Barbara, CA.

Terblanche, J. S., C. J. Klok, and S. L. Chown. 2004. Metabolic rate variation in *Glossina pallidipes* (Diptera: Glossinidae): gender, ageing and repeatability. *J. Insect Physiol.* 50 (5): 419–428.

Thomas, D. B. 1993. Fecundity and oviposition in laboratory colonies of the screwworm fly (Diptera, Calliphoridae). *J. Econ. Entomol.* 86 (5): 1464–1472.

Tilden, J. W. 1962. General characteristics of the movements of *Vanessa cardui*. *J. Res. Lepidoptera*. 1: 43–49.

Tillman, P. G., G. McKibben, S. Malone, and D. Harsh. 1997. Form-fill-seal machine for mass rearing noctuid species. In *Office of Agricultural Communications, Division of Agriculture, Forestry, and Veterinary Medicine*, K. H. Remy, Ed. Mississippi State University, Mississippi State, MS, MS. Technical Bulletin 213, pp. 1–4.

Tomberlin, J. K., and P. H. Adler. 1998. Seasonal colonization and decomposition of rat carrion in water and on land in an open field in South Carolina. *J. Med. Entomol.* 35: 704–709.

Tomita, M., and T. Nishioka. 2005. *Metabolomics: The Frontier of Systems Biology*. Springer. ISBN 4-431-25121-9.

Tong, L., and H. Liu. 2011. Rearing of silkworm larva in a closed gas revitalizing plant system. *Ecol. Eng.* 37 (9): 1259–1266.

Triggs, A. M., and R. J. Knell. 2012. Parental diet has strong transgenerational effects on offspring immunity. *Funct. Ecol.* 26: 1409–1417.

Tyree, M. T., and P. G. Jarvis 1982. Water in tissues and cells. In *Encyclopedia of Plant Physiology N.S. Physiological Plant Ecology*, O. Lange, P.S. Nobel, and C.B. Osmond, Eds. Vol. 12B. Springer-Verlag: Berlin, pp. 35–77.

Tyree, M. T., and A. J. Karamanos. 1980. Water stress as an ecological factor. In *Plants and Their Atmospheric Environment*, J. Grace, E. D. Ford, and P. G. Jarvis, Eds. Oxford, pp. 237–261.

Vacek, D. C. W. T. S., and W. Heed. 1979. Relevance of the ecology of citrus yeasts to the diet of Drosophila. *Microb. Ecol.* 5: 43–49.

Vail, P. V., T. J. Henneberry, A. N. Kishaba, and K. W. Arakawa. 1968. Sodium hypochlorite and formalin as antiviral agents against nuclear-polyhedrosis virus in larvae of the cabbage looper. *J. Invertebr. Pathol* 10: 84–93.

Vainker, S. 2004. *Chinese Silk: A Cultural History*. Rutgers University Press, New Brunswick, NJ, p. 20. ISBN 0813534461.

van Huis, A., J. Van Itterbeeck, H. Klunder, E. Mertens, A. Halloran, G. Muir, and P. Vantomme. 2013. Edible insects future prospects for food and feed security. *FAO Forestry Paper*. 171. Wageningen, UR. 1–189.

VandenBrooks, J. M., G. Gstrein, J. Harmon, J. Friedman, M. Olsen, A. Ward, and G. Parker. 2018. Supply and demand: how does variation in atmospheric oxygen during development affect insect tracheal and mitochondrial networks? *J. Insect Physiol.* 106: 217–233.

Vanderzant, E. S. 1957. Growth and reproduction of the pink bollworm on an amino acid medium. *J. Econ. Entomol* 50 (2): 219–221.

Vanderzant, E. S. 1958. The amino acid requirements of the pink bollworm. *J. Econ. Entomol* 51 (3): 309–311.

Vanderzant, E. S. 1967. Wheat-germ diets for insects: rearing the Boll Weevill and the Salt-Marsh caterpillar. *Ann. Entmol. Soc. Am.* 60: 1062–1066.

Vanderzant, E. S. 1967. Rearing Lygus bugs on artificial diets. *J. Econ. Entomol.* 60 (3): 813–816.

Vanderzant, E. S. 1969. An artificial diet for larvae and adults of Chrysopa carnea, an insect predator of crop pests. *J. Econ. Entomol.* 62 (1): 256–257.

Vanderzant, E. S. 1973. Improvements in the rearing diet for *Chrysopa carnea* and the amino acid requirements for growth. *J. Econ. Entomol.* 66 (2): 336–338.

Vanderzant, E. S. 1974. Development, significance, and application of artificial diets for insects. *Annual Rev. Entomol.* 19: 139–160 (Volume publication date January 1974). https://doi.org/10.1146/annurev.en.19.010174.001035.

Vanderzant, E. S., and R. Reiser. 1956a. Aseptic rearing of the pink bollworm on synthetic media. *J. Econ. Entomol.* 49 (1): 7–10.

Vanderzant, E. S., and R. Reiser. 1956b. Studies of the nutrition of the pink bollworm using purified casein media. *J. Econ. Entomol.* 49 (4): 4540458.

Vanderzant, E. S., and T. Davich. 1958. Laboratory rearing of the boll weevil: a satisfactory larval diet and oviposition studies. *J. Econ. Entomol.* 51 (3): 288–291.

Vanderzant, E. S., and T. B. Davich. 1961. Artificial diets for the adult boll weevil and techniques for obtaining eggs. *J. Econ. Entomol.* 54: 923–928.

Vanderzant, E. S., C. D. Richardson, and T. B. Davich. 1959. Feeding and oviposition by the boll weevil on artificial diets. *J. Econ. Entomol.* 52: 1138–1143.

Vanderzant, E. S., D. Kerur, and R. Reiser. 1957. The role of dietary fatty acids in the development of the pink bollworm. *J. Econ. Entomol.* 50 (5): 606–608.

Vanderzant, E. S., R. Reiser, and E. E. Ivy. 1956. Methods for the mass rearing of the pink bollworm. *J. Econ. Entomol.* 49 (4): 559–560.

Vargas, R. I., and S. Mitchell. 1987. Two artificial larval diets for rearing *Dacus latifrons* (Diptera; Tephritidae). *J. Econ. Entomol.* 80: 1337–1339.

Vargas, R. I. 1989. Mass production of tephritid fruit flies. In *Fruit Flies, their Biology, Natural Enemies, and Control. World Crop Pests*, G. Hooper and W. Robinson, Eds. Elsevier, Amsterdam, pp. 141–152.

Vargas, R. I., H. Chang, and D. L. Williamson. 1983. Evaluation of a sugarcane bagasse larval diet for mass production of the Mediterranean fruit fly (Diptera; Tephritidae) in Hawaii. *J. Econ. Entomol.* 76: 1360–1362.

Villalba, J. J., K. A. Beauchemin, P. Gregorini, and J. W. MacAdam. 2019. Pasture chemoscapes and their ecological services. *Transl. Anim. Sci.* 3 (2): 829–841, https://doi.org/10.1093/tas/txz003.

Visscher, S. N., R. Lund, and W. Whitmore. 1979. Host plant-growth temperatures and insect rearing temperatures influence reproduction and longevity in the grasshopper, Aulocara Elliotti (Orthoptera, Acrididae). *Environ. Entomol.* 8 (2): 253–258.

Vodovar, Nicolas, Marisa Vinals, Peter Liehl, Alan Basset, Jeril Degrouard, Paul Spellman, Frédéric Boccard, and Bruno Lemaitre. 2005. Drosophila host defense after oral infection by an entomopathogenic Pseudomonas species. *Proc. Natl. Acad. Sci. USA.* 102 (32): 1141411419.

Vogelweith, F., J. Moreau, D. Thiery, and Y. Moret. 2015. Food-mediated modulation of immunity in a phytophagous insect: an effect of nutrition rather than parasitic contamination. *J. Insect Physiol.* 77: 55–61.

Vogt, J. T., S. D. Porter, D. A. Nordlund, and R. Smith. 2003. A modified rearing system for production of *Pseudacteon curvatus* (Diptera: Phoridae), a parasitoid of imported fire ants. *Biol. Control.* 28: 346–353.

Waldbauer, G. P. 1968. The consumption and utilization of food by insects. *Adv. Insect Physiol.* 5: 229–288.

Walker, J. K. 1967. Studies in the fall and winter of oviposition prior to diapause in the boll weevil with observations on the reversions of diapause to reproduction. *J. Econ. Entomol.* 60 (3): 799–802.

Wasserman, P. D. 1993. *Advanced Methods in Neural Computing.* Prentice Hall, Upper Saddle River, NJ.

Watson, E. J., and C. E. Carlton. 2003. Spring succession of necrophilous insects on wildlife carcasses in Louisiana. *J. Med. Entomol.* 40: 338–347.

Weber, E. 2012. Apis mellifera: the domestication and spread of European honey bees for agriculture in North America. *Univ. Michigan Undergraduate Res. J.* 9: 20–23.

Weckwerth, W. 2006. *Metabolomics: Methods and Protocols (Methods in Molecular Biology).* Humana Press. ISBN 1-588-29561-3. OCLC 493824826

Weinberg, S. 2003. *Good Laboratory Practices.* 2nd Edition. Marcel Dekker, Inc., New York, Basel.

Weldon, C. W., S. Mnguni, F. Dé mares, E. E. du Rand, K. Malod, A. Manrakhan, and S. W. Nicolson. 2019. Adult diet does not compensate for impact of a poor larval diet on stress resistance in a tephritid fruit fly. *J. Exp. Biol.* 222 (1-11): jeb192534. doi:10.1242/jeb.192534.

Werren, J. H. 1997. Biology of Wolbachia. *Annu. Rev. Entomol.* 42: 587–609.

Werren, J. H., L. Baldo, and M. E. Clark. 2008. *Wolbachia*: master manipulators of invertebrate biology. *Nat. Rev. Microbiol.* 6: 741–751. doi: 10.1038/nrmicro1969.

WHO. *Handbook: Good Laboratory Practice (GLP): Quality Practices for Regulated Non-clinical Research and Development*, 2nd ed. Copyright © World Health Organization on behalf of the Special Programme for Research and Training in Tropical Diseases 2009. ISBN 978 92 4 154755 0.

Wolfram. 2021. https://reference.wolfram.com/applications/eda/ExperimentalErrorsAndErrorAnalysis.html.

Williams, C. M. 1953. Control of pupal diapause by direct action of light on insect brain. *Science.* 140 (356): 386.

Williams, D. D., and S. S. Williams. 2017. Aquatic insects and their potential to contribute to the diet of the globally expanding human population insects. *Insects.* 8: 72. doi: 10.3390/insects8030072 www.mdpi.com/journal/insects.

Xia, Z., S. Wu, S. Pan, and J. M. Kim. 2012. Nutritional evaluation of protein from *Clanis bilineata* (Lepidotpera) an edible insect. *J. Sci. Food, Agric.* 92: 1479–1482.

Xiang, H., X. Liu, M. Li, Ya'nan Zhu, L. Wang, Y. Cui, L. Liu, G. Fang, H. Qian, A. Xu, W. Wang, and S. Zhan. 2018. The evolutionary road from wild moth to domestic silkworm. *Nat. Ecol. Evol.* 2: 1268–1279.

Yamamoto, R. T. 1968. Mass rearing of tobacco hornworm: 1 egg production. *J. Econ. Entomol.* 61 (1): 170–174.

Yamamoto, R. T. 1969. Mass rearing of tobacco hornworm: 2 larval rearing and pupation. *J. Econ. Entomol.* 62: 1427–1431.

Yoshinori Hatakeyama and Shoji Hayasaka. 2003. A new method of pebrine inspection of silkworm egg using multiprimer PCR. *J. Invert. Pathol.* 82: 148–151.

Yoshitake, N. 1968. Phylogenetic aspects on the origin of Japanese race of the silkworm, *Bombyx mori* L. *J. Sericol. Sci. Jpn.* 37: 83–87.

Yukuhiro, K.; Sezutsu, H.; Itoh, M.; Shimizu, K. & Banno, Y. 2002. Significant levels of sequence divergence and Gene rearrangements have occurred between the mitochondrial genomes of the wild mulberry silkmoth, *Bombyx mandarina*, and its close relative, the domesticated silkmoth, *Bombyx mori. Mol. Biol. Evol.* 19 (8): 1385–1389.

Zamboni Costa, K., M. L. Zamboni Costa, V. W. Botteon, K. M. Faggioni, N. Costa, and T. Mastrangelo. 2019. Quality control and characterization of the testicles and ovaries of irradiated *Anastrepha obliqua* from Brazil. *Sci. Agricola Sci. Agric.* 77 (1): e20170253.

Zha, C. and A. C. Cohen. 2014. Effects of anti-fungal compounds on feeding behavior and nutritional ecology of tobacco budworm and painted lady butterfly larvae. *Entomol. Ornithol. Herpetol.* 3: 1–9.

Zhang Z. J., Zhang S. S., Niu B. L., Ji D. F., Liu X. J., et al. 2019. A determining factor for insect feeding preference in the silkworm, *Bombyx mori. PLoS Biol.* 17 (2): e3000162. https://doi.org/10.1371/journal.pbio.3000162.

Zhang, M. 2018. *Effects of Different Light Conditions on Biological Characteristics of Five Insect Species.* Master's Thesis, Huazhong Agricultural University, Wuhan, China. (In Chinese with English abstract).

Zhangshuai He, Qi Ni, Yue Song, Rui Wang, Yunlin Tang, Yujiao Wu, Lulu Liu, Jialing Bao, Jie Chen, Mengxian Long, Junhong Wei, Chunfeng Li, Tian Li, Zeyang Zhou, Guoqing Pan. 2019. Development of a nucleic acid lateral flow strip for rapid, visual detection of *Nosema bombycis* in silkworm eggs. *J. Invert. Pathol.* 164: 59–65.

Zhou, J. and D. Han. 2006. Proximate, amino acid and mineral composition of pupae of the silkworm *Antharaea pernyi* in China. *J. Food Comp. Anal.* 19: 850–853.

Zizzari, Z. V., N. M. Van Straalen, and J. Ellers. 2016. *J. Evol. Biol.* 29: 1317–1327.

Index

Note: Locators in *italics* represent figures and **bold** indicate tables in the text.

E

F

G

H

I

J

K

L

M

N

O

P

W

Y